Multivariable Calculus
with Vectors

Multivariable Calculus with Vectors

HARTLEY ROGERS, Jr.

Massachusetts Institute of Technology

PRENTICE HALL / Upper Saddle River, New Jersey 07458

Library of Congress Cataloging-in-Publication Data

Rogers, H. (Hartley)
    Multivariable calculus with vectors / Hartley Rogers, Jr.
         p.   cm.
    Includes index.
    ISBN 0-13-605643-1
    1. Calculus.   I. Title.
QA303.R576   1999   98-3635
515--dc 21   CIP

Editorial director, Tim Bozik
Editor-in-chief, Jerome Grant
Acquisition editor, George Lobell
Executive managing editor, Kathleen Schiaparelli
Managing editor, Linda Mihatov Behrens
Editorial assistants, Gale Epps, Nancy Gross
Assistant VP production/manufacturing, David W. Riccardi
Manufacturing manager, Trudy Pisciotti
Manufacturing buyer, Alan Fischer
Creative director, Paula Maylahn
Art Director, Jane Conte
Composition, Marjorie Potvin
Illustrations, Martin Stock
Cover Photo, Stephen Wilkes
Marketing manager, Melody Marcus

© 1999 by Prentice-Hall, Inc.
Simon & Shuster / A Viacom Company
Upper Saddle River, New Jersey 07458

Printed in the United States of America

10 9 8 7 6 5 4 3 2 1

ISBN: 0-13-605643-1

Prentice Hall International (UK) Limited, London
Prentice Hall of Australia Pty. Limited, Sydney
Prentice Hall Canada, Inc., Toronto
Prentice Hall Hispanoamericana, S.A., Mexico
Prentice Hall of India Private Limited, New Delhi
Prentice Hall of Japan, Inc., Tokyo
Simon & Schuster Asia Pte. Ltd., Singapore
Editora Prentice Hall do Brasil, Ltda., Rio de Janeiro

For Adrianne, Hartley, Campbell, and Caroline

Contents

Preface

I

Multivariable calculus is traditionally known as "the calculus of functions of several real variables." As the word *calculus* suggests, multivariable calculus provides algorithms of power and versatility. At the same time it supplies definitions, concepts, and intuitions for much of physical theory.

This text has been developed for three purposes:

To present, for multivariable calculus, a conceptual underpinning which is natural and intuitively simple.

To give readers an awareness of the logical structure of multivariable calculus and to help them recognize that mathematics provides an independently existing conceptual reality which can be used to build mathematical models for observed phenomena.

To combine, in a single text, material which can be used satisfactorily at any desired level of formal rigor and conceptual sophistication (from a simple qualitative introduction, to a course for future mathematics majors), with material which may give the book value as a reference work.

In the effort to achieve these purposes, the following features of the text are prominent:

The synthetic, coordinate-free geometry of the two- and three-dimensional *Euclidean spaces* ($\mathbf{E}^2$ and $\mathbf{E}^3$) have a central role. These spaces, together with their assigned distance functions ("metrics"), are assumed as given and known structures. Basic definitions and properties are described, but a self-contained logical treatment is not provided. (For example, the Jordan curve theorem for piecewise-smooth curves in $\mathbf{E}^2$ is stated and assumed, but not proved.) In accord with this geometrical emphasis, *scalar fields* (mappings from a region of $\mathbf{E}^2$ or $\mathbf{E}^3$ to the real numbers) are given an equal place with functions of two and three variables.

Wherever possible, *coordinate-free* definitions are used. For example, the definitions of the differential operators *divergence* and *curl*, and of the integral operators *double integral* and *triple integral*, make no mention of a coordinate system. Furthermore, the physicist's $\{\hat{i},\hat{j},\hat{k}\}$ notation is used for vectors in a Cartesian coordinate system (instead of canonical-coordinate triples) to emphasize the subordinate status of any chosen Cartesian system. Where coordinates appear in the text, they are used primarily for purposes of computation or proof.

New versions of certain calculus concepts are introduced, to obtain a better fusion of mathematical argument with geometrical intuition. In particular, the central concept of *limit* is defined in terms of *funnel functions*. (In the terminology of mathematical logic, a *funnel function* is the inverse of a monotone Skolem function for a true statement of the form "for every $\varepsilon > 0$, there exists a $\delta > 0...$".) Funnel functions are especially useful in multivariable calculus where, for example, they give a simple and readily visualizable definition of *uniform continuity*. Similarly, the central concept of *definite integral* is defined in terms of *proper sequences* of Riemann sums (a form of direct limit). For many students, this definition appears to be simpler and less artificial than the more familiar definition in terms of upper and lower sums. Similarly, *finitely additive measures* and their derivatives are used to provide a more natural framework for the great theorems of vector integral calculus. The amount of emphasis given to these technical concepts will depend on the level of rigor to which a student or teacher aspires. For some students, these technical concepts will be suggestive but peripheral features of their study.

Some proofs are initially presented as plausibility arguments. Full proofs are then indicated in later portions of the text, or in problem material.

The transition to Euclidean spaces of more than three dimensions occurs near the end of the text. The treatment of linear algebra leads, in Chapter 34, to the emergence of the synthetic geometries of higher-dimensional Euclidean spaces, to which the geometric intuitions developed in three dimensions can be largely transferred. This has seemed preferable to the loss of these geometric intuitions, which can accompany a premature introduction of canonical coordinates for dimensions two and three. (It is intended that a future edition of the text will include several additional chapters: a chapter on multivariable calculus in n dimensions, including a geometry-based development of vector integral calculus; a chapter on complex-valued fields and functions on Euclidean spaces; a chapter on series; and a chapter on futher physical applications of linear algebra and n-dimensional calculus.)

In summary, this book is intended to serve as a text in which the role of geometric intuition is emphasized, in which the role of rigorous argument can be emphasized or de-emphasized as the reader or teacher may choose, and in which rigorous arguments, when used, serve to support and illuminate intuition.

II

The study of multivariable calculus is required of all undergraduates at the Massachusetts Institute of Technology. The present text has evolved, over a period of 16 years, as lecture notes for a one-term course offered in the fall term to entering students who have advanced placement in elementary (one-variable) calculus. Enrollment has varied between 350 and 500 students per year. There are three hours of large lecture each week and, for each student, two additional hours of small recitation class. Successive lecture topics are drawn from the successive chapters of the text, except that there is no time for material from Chapters 33 and 34. On average, each lecture is limited to the first two-thirds of the corresponding text chapter. Students' responsibility is limited to material covered in lecture or in recitation class. A few chapters (3 and 29, for example) are covered in their entirety. In Chapter 23, the last three sections are covered, respectively, in the subsequent lectures on Chapters 24, 25, and 26. Lectures on Chapters 27 and 28 are limited to a few chosen examples. A total of five or six lectures is usually allowed for Chapters 14, 15, 16, and 17. Proofs in the text are either omitted, or presented in qualitative summary, and students are not responsible for them. Students have weekly problems sets of six to ten problems from the text, four closed-book hour tests (usually after Chapters 5, 13, 20, and 28) and a three-hour, closed-book final examination covering the whole term.

It will be evident to the reader that, if time is available, a fuller and more rewarding use of this text will be achieved over two quarters or two semesters.

III

During the past 16 years, more than 7000 undergraduates have taken this course, and more than 200 faculty members or instructors have served as its recitation teachers. Many students and teachers have provided valuable ideas, suggestions, and criticisms. I am deeply grateful for their help, and regret that neither memory nor space allow me to acknowledge their individual contributions.

I owe a special debt of gratitude to my faculty colleagues at MIT, to the Department of Mathematics at MIT, and to its Head, Professor David J. Benney. The Department and faculty have provided me with ideal circumstances of freedom and encouragement. In particular, I wish to thank

Professors Arthur P. Mattuck, Michael Artin, and Daniel J. Kleitman. Professor Mattuck has shown me, by long-standing example, that successful undergraduate teaching, both in content and in method, requires intellectual commitment and seriousness of a high order. Professor Artin suggested the possibility of using measures and Radon-Nikodym derivatives to formulate the principal theorems of vector integral calculus. Professor Kleitman has provided lively encouragement as well as continuing volunteered service as a recitation instructor. About one-fifth of the recitation classes have been taught by graduate teaching assistants. MIT's graduate students in mathematics are an extraordinary intellectual resource, and their help and advice has been invaluable. I also wish to thank the Department's Undergraduate Mathematics Office and its Administrator, Joanne E. Jonsson, for their faithful and patient help over the past 16 years with this course and with the distribution of its evolving text.

At Prentice Hall, I am grateful to David Ostrow and George Lobell for their early, patient, and persistent encouragement, and, more recently, to my production editor, Nicholas Romanelli, for his authoritative and gracious editorial help. I also thank the following reviewers for their suggestions and corrections: James T. Vance of Wright State University, Robert Boyer of Drexel University, Ernest Stitzinger of North Carolina State University, and Stephen Agard of the University of Minnesota, and Jianguo Cao of the University of Notre Dame. I am especially indebted to Donald Kreider of Dartmouth College for his encouraging comments after reading an early version of this book.

In the final preparation of the typescript, I am grateful to Mr. Constantin Chiscanu of the MIT Class of 2000, who volunteered the eye of an IMO Gold Medalist for a careful and thoughtful reading of the entire text, and to my friend and neighbor Martin Stock (mstock@mit.edu) for his skillful preparation of the figures. Above all, I am grateful to Marjorie Rockwood Potvin who has, for the past 10 years, been solely responsible for the typing and retyping of successive versions of the text and who, in 1998 and at a distance of 1000 miles, undertook the heroic responsibility of preparing, in her spare time, the final copy of the typescript. Her help has been indispensable.

Finally, and most deeply, I thank my wife, Adrianne, for her steadfast patience and support, and my children, Hartley, Campbell, and Caroline, and other family and friends, for their affectionate respect and enthusiasm.

Hartley Rogers, Jr.
Winchester, Massachusetts

Chapter 1

Euclidean Geometry in Three Dimensions

§1.1   Introduction; Euclidean Geometry in Two Dimensions

Introduction. In order to understand and use multivariable calculus, a student must develop an ability to visualize geometrical relationships in three dimensions. Geometric visualization is based on an imaginary universe of precisely defined mathematical objects called *points*, *lines*, *planes*, *circles*, *spheres*, etc. This mathematical universe was first identified and described by Euclid, a mathematician of ancient Greece.

Our study of multivariable calculus will require the following background in geometry:

(1)   An understanding of what the *basic objects* in three-dimensional geometry are, and of how we think and reason about them.

(2)   Knowledge of the basic *definitions* and *terminology* of this geometry.

(3)   A good command of the elementary *facts* of this geometry.

(4)   Familiarity with the use of *deductive argument* to prove new facts (of whose truth one may be as yet unsure) from old facts (of whose truth one is already certain).

1

Most of the present chapter is concerned with presenting basic definitions and elementary facts. The student's primary goal should not be to memorize this material but rather to achieve a level of experience and familiarity with it where he or she can understand geometric statements and can instinctively recognize the truth or falsity of statements which are simple and elementary. The student's secondary goal should be to develop some feeling for the use and power of deductive argument in geometry.

Euclid presented a single, overarching, deductive structure for geometry, where certain limited and simple facts were assumed as *axioms* and *postulates*, and where further geometrical facts were then rigorously deduced and called *theorems*.

We do not present such a structure. Instead we make available a general collection of facts which the student can apply directly in later work or from which the student can deduce the truth of further facts. All of the facts given here can be deduced, in the manner of Euclid, from a small set of elementary assumptions. Although we do not present geometry in this form, we will give a few illustrative deductions as we proceed. In §1.7, we comment further on the underlying assumptions of Euclidean geometry and on the development of Euclidean geometry as a deductive system.

Euclidean geometry in two dimensions. In secondary school, the study of geometry is chiefly concerned with the *Euclidean plane*, also known as *Euclidean two-dimensional space*. We henceforth refer to this plane as $\mathbf{E}^2$. $\mathbf{E}^2$ may be viewed as a set of mathematical objects called *points*. Certain subsets of $\mathbf{E}^2$ are called *straight lines* or, more briefly, *lines*. (We visualize straight lines as extending infinitely far in both directions.) Points and lines are fundamental in the study of $\mathbf{E}^2$. Certain other sets of points are also fundamental. These include *line-segments*, *rays* (or *half-lines*), and *half-planes*. In the study of $\mathbf{E}^2$, the relationship of two lines being *parallel* is defined, and so is the relationship of two lines being *perpendicular*. We shall assume that the reader is familiar with these basic concepts for $\mathbf{E}^2$ and with elementary facts about them.

$\mathbf{E}^2$ includes a measure of *distance* between points; that is to say, we assume that for every pair of points P and Q, we are given a non-negative real number, which we write d(P,Q), and which we call the *distance from P to Q*. This distance measure satisfies the law that for any two given points P and Q in $\mathbf{E}^2$,

(1.1) $$d(P,Q) = 0 \Leftrightarrow P = Q$$

and the law that for any three given points P, Q, and R in $\mathbf{E}^2$,

(1.2) $$d(P,R) \leq d(P,Q) + d(Q,R).$$

(1.2) is called the *triangle inequality* for $\mathbf{E}^2$.

Subsets of $\mathbf{E}^2$ are usually referred to as *figures*. A function f (sometimes called a *mapping* or *correspondence*) from all points in $\mathbf{E}^2$ onto all points in $\mathbf{E}^2$ is said to be an *isometry* if, for every pair of points P and Q in $\mathbf{E}^2$, d(f(P),f(Q)) = d(P,Q). Two figures in $\mathbf{E}^2$ are said to be *congruent*, if there is an isometry which carries one figure onto the other. We express this informally by saying that two figures are congruent if we can take a rigid copy of one figure, without changing any of its measurements, and fit it exactly over the other figure, where we are allowed to turn the copy over, if necessary, to do so.

An *angle* in $\mathbf{E}^2$ is a figure formed by two rays with a common endpoint (or *vertex*). By using the concept of congruence, it is possible to define a measure of *size* for angles. This measure is usually given in *degrees* (where a right angle has 90 degrees) or in *radians* (where a right angle has $\pi/2$ radians). It is customary, for brevity, to refer to the size of a given angle simply as "the angle"; for example, we say "the angle A is $\pi/4$" instead of "the size of the angle A is $\pi/4$."

Other figures are also prominent in the study of $\mathbf{E}^2$. These include *triangles, quadrilaterals, polygons, parallelograms, rectangles, squares*, and *circles*. In the study of $\mathbf{E}^2$, we use the distance measure, together with the idea of limit, to define *area* for these and certain other figures in $\mathbf{E}^2$.

Sometimes, in studying $\mathbf{E}^2$, we begin with basic assumptions, called *axioms* or *postulates*, and then deduce further facts about $\mathbf{E}^2$ from these assumptions. These assumptions arise, in part, from our direct experience of the physical world. *Points* in $\mathbf{E}^2$ can be thought of as locations on a flat surface (such as a piece of paper on a desk) in physical space. *Lines* can be thought of as paths on the surface given by a straight-edge, where a straight-edge, in turn, is physically determined and checked by using the path of a light ray.

It is important, however, to keep in mind that $\mathbf{E}^2$ itself is an idealized, abstract, and imaginary mathematical concept; it does not have a fully precise physical counterpart. There is no way for us to locate and identify infinitely small *points*, infinitely long and thin *lines*, and perfectly flat surfaces in the physical universe. Nevertheless, we shall use $\mathbf{E}^2$ to help construct precise *mathematical models* of structures and processes in the physical universe. The theory of $\mathbf{E}^2$ is known as *Euclidean plane geometry*.

The geometric theory of $\mathbf{E}^2$ can be extended to a geometric theory for an enlarged mathematical space known as *Euclidean three-dimensional space*. We henceforth refer to this space as $\mathbf{E}^3$. In the first half of the twentieth century, the study of $\mathbf{E}^3$ was a standard feature of secondary-school mathematics and was known as *Euclidean solid geometry*. In the remainder of this chapter, we describe the basic concepts in the geometry of $\mathbf{E}^3$ and survey the elementary facts of this geometry. To do multivariable calculus, the reader must learn to use the concepts and terminology of this geometry with precision and understanding.

§1.2   Points, Lines, and Planes

$\mathbf{E}^3$, like $\mathbf{E}^2$, is a collection of *points*.   Certain subsets of $\mathbf{E}^3$ are called *lines*, which we visualize as straight lines extending to infinity in both directions; and certain subsets are called *planes*, which we visualize as flat surfaces extending to infinity in all directions.   By the *intersection* of two given subsets of $\mathbf{E}^3$, we mean the set of all points common to both subsets.   If a set C of points contains at least one point, we say that the set C is *non-empty*.   If C contains no points, we say that C is *empty*.   If the intersection of given sets A and B is non-empty, we say that "A and B intersect."   If C is their non-empty intersection, we also say that "A and B intersect at C" or that "A and B intersect in C."   If the intersection of A and B is empty, we say that "A and B do not intersect."

We list some elementary facts about interrelationships among points, lines, and planes in $\mathbf{E}^3$. These facts are known as *incidence statements* (from the Latin word "incidere," meaning *to fall upon* or *encounter*).

(2.1)   *Given two distinct points* P *and* Q, *there is a unique* (one and only one) *line that contains both* P *and* Q.

Definition. We say that a set S of points in $\mathbf{E}^3$ is *collinear* if there is a single line which contains all of S.

(2.2)   *Given three non-collinear points* P, Q, *and* R, *there is a unique plane that contains* P, Q, *and* R.

(2.3)   *Given two distinct points* P *and* Q *on a plane* M, *then* M *contains the unique line determined by* P *and* Q *(by (2.1)).*

(2.4)   *If two distinct planes intersect, then they intersect in a line.*

(2.5)   *If* L_1 *and* L_2 *are two distinct lines, then either* L_1 *and* L_2 *intersect at a single point, or else* L_1 *and* L_2 *do not intersect.*

(2.6)   *Given a line* L *and a point* P *not on* L, *there is a unique plane that contains both* L *and* P.

(2.7)   *If a plane* M *and a line* L *intersect, then either they intersect at exactly one point, or else* L *is contained in* M.

(2.8)   *If two distinct lines intersect, then there is a unique plane that contains both those lines.*

Statements (2.1), (2.2),…,(2.8) are not logically independent of each other. In particular, (2.5), (2.6), (2.7), and (2.8) can be deduced from (2.1), (2.2), and (2.3). These deductions also rely on certain assumptions of Euclidean geometry which are not usually stated explicitly: for example, that a line has infinitely many points; that for every line, there is a plane containing that line; that for any line, a plane has infinitely many points not on that line; and that for any plane, E^3 has infinitely many points which are not on that plane.

As an illustration of a deductive argument, we prove (2.8) from (2.1), (2.2), and (2.3).

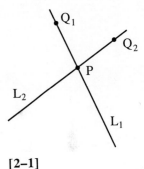

[2–1]

Proof. Let L_1 and L_2 be two distinct intersecting lines. As is usual in a proof of uniqueness, there are two separate proofs to carry out.

First, we prove that there is *at least one* plane containing L_1 and L_2. By (2.1), L_1 and L_2 intersect at a single point. Call this point P. Let Q_1 be a point on L_1 different from P, and let Q_2 be a point on L_2 different from P. See [2–1]. The points P, Q_1, and Q_2 are not collinear (otherwise, by (2.1), L_1 and L_2 would coincide). By (2.2), there is a plane M which contains P, Q_1, and Q_2. By (2.3), M contains L_1 and L_2. This completes the first proof.

Second, we prove that there is *at most one* plane containing L_1 and L_2. Let P, Q_1, and Q_2 be as in the first proof. If there were more than one plane containing L_1 and L_2, then there would be more than one plane containing the non-collinear points P, Q_1, and Q_2. This would contradict (2.2). This completes the second proof.

Like E^2, E^3 has a measure of distance $d(P,Q)$ which satisfies a triangle inequality:

(2.9)   *For any three given points* P, Q, *and* R *in* E^3,

$$d(P,R) \le d(P,Q) + d(Q,R).$$

Definitions. If we have a line L and a point P on L, then a half-line of L with end-point P is called a *ray* with *vertex* P. Similarly, if we have a plane M and a line L in M, we can have a *half-plane* of M with *edge* L. This consists of all points which lie on L or to a chosen side of L in M. Finally, if we are given a plane M in E^3, we can take a *half-space* of E^3 with *face* M. This consists of all points which lie on M or to a chosen side of M in E^3. (For the meaning of "side," see problem 4 for §1.2.)

A large body of additional facts is implied by the following:

(2.10)   *Given any plane in E^3, then all the facts of Euclidean plane geometry hold for the points, lines, and figures of this plane.   (In other words, every plane in E^3 may be viewed as a copy of E^2.)*

Without going into details, we note that isometries for E^3 can be defined as for E^2: an *isometry* is a function f from all points of E^3 onto all points of E^3 such that for every pair of points P and Q in E^3, $d(f(P),f(Q)) = d(P,Q)$. Two figures in E^3 are said to be *congruent* if there is an isometry which carries one figure onto the other. In particular, we have, as a fundamental fact:

(2.11)   *Given six points P, Q, R, P′, Q′, and R′ in E^3 such that $d(P,Q) = d(P′,Q′)$, $d(Q,R) = d(Q′,R′)$, and $d(R,P) = d(R′,P′)$, then there is an isometry f of E^3 such that $f(P) = P′$, $f(Q) = Q′$, and $f(R) = R′$.*

(2.11) allows us to use, in E^3, the familiar congruent triangle theorems of E^2.

§1.3   Parallelism

Definitions. We say that two lines are *parallel* if (i) they lie in a common plane, and (ii) they do not intersect.
        If two lines do not lie in a common plane, they are said to be *skew*. (By (2.8), skew lines cannot intersect.)
        If a line and a plane do not intersect, they are said to be *parallel.*
        If two planes do not intersect, they are said to be *parallel.*

The following facts about parallelism are fundamental.

(3.1)   *Given a line L and a point P not on L, there is a unique line through P parallel to L, [3–1].*

(3.2)   *Given a plane M and a point P not on M, there is a unique plane through P parallel to M, [3–2].*

Three further facts are:

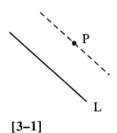

[3–1]

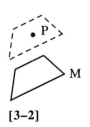

[3–2]

(3.3)   *Let* M_1 *and* M_2 *be parallel planes. Let the plane* M *intersect* M_1 *in* L_1 *and intersect* M_2 *in* L_2*. Then* L_1 *and* L_2 *must be parallel lines.*

(3.4)   *If* M_1 *and* M_2 *are distinct planes, and each is parallel to the plane* M_3*, then* M_1 *must be parallel to* M_2*.*

(3.5)   *If* L_1 *and* L_2 *are distinct lines and each is parallel to the line* L_3*, then* L_1 *must be parallel to* L_2*.*

We note that (3.3) and (3.4) are easily deduced from (3.1) and (3.2). No simple and direct deduction of (3.5) from (3.1) and (3.2) is known.

Definitions. We say that two line-segments are *parallel* if they lie on parallel lines.

We say that a set S of points in $\mathbf{E}^3$ is *coplanar* (or, simply, *planar*) if there is a single plane which contains all of S.

A set of points in $\mathbf{E}^3$ is usually referred to as a *figure* in $\mathbf{E}^3$.

The following definitions and facts will be important in our later work with vectors.

Definition. A figure S in $\mathbf{E}^3$ is called a *quadrilateral* if there are four points P_1, P_2, P_3, P_4, no three of which are collinear, such that S consists of the four line-segments P_1P_2, P_2P_3, P_3P_4, and P_4P_1 where these segments intersect only at common endpoints, [3–3]. P_1, P_2, P_3, and P_4 are called the *vertices* of S, and S is referred to as the *quadrilateral* $P_1P_2P_3P_4$. A quadrilateral need not be planar, since the vertices need not be coplanar.

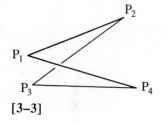

[3–3]

Definition. We say that a quadrilateral $P_1P_2P_3P_4$ is a *parallelogram* if P_1P_2 is parallel to P_3P_4 and P_2P_3 is parallel to P_4P_1. By the definition of parallel lines and by (2.3), the points of a parallelogram must lie in a single plane.

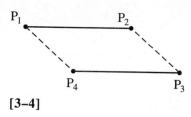

[3–4]

From the geometry of $\mathbf{E}^2$, we have:

(3.6)   *Let* S *be the quadrilateral* $P_1P_2P_3P_4$*. Then:* S *is a parallelogram* $\Leftrightarrow$ [P_1P_2 *is parallel to* P_3P_4 *and* $d(P_1,P_2) = d(P_3,P_4)$]. See [3–4].

(3.7)   *For any three non-collinear points* P_1,P_2,P_3*, there is a unique point* Q *such that the quadrilateral* $P_1P_2P_3Q$ *is a parallelogram,* [3–5].

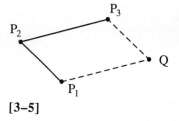

[3–5]

§1.4 Angles and Perpendicularity

A figure in $\mathbf{E}^3$ which consists of two rays with a common vertex is called an *angle*. Given an angle, there must always be, by (2.8), some plane is which this angle lies. (If the rays are not collinear, then this plane is unique.) Hence we can use the system of angle measurement developed for measuring angles in $\mathbf{E}^2$ to measure the size of an angle given in $\mathbf{E}^3$, and we can show that two angles are equal in size $\Leftrightarrow$ they are congruent.

Two rays are said to be *parallel* if they lie on parallel lines. Evidently, two parallel rays can have either the same or opposite directions. We make this precise with the following definition: we say that two parallel rays have the *same direction* if there is a parallelogram which has the endpoints of the rays as adjacent vertices and has segments of the rays as two of its sides, [4–1]. Otherwise we say that the two parallel rays have *opposite directions*. Similarly, for two collinear rays, we say that they have the same direction if one of the rays is contained in the other; otherwise we say that the collinear rays have *opposite directions*. It follows from (3.1) that:

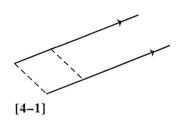

[4–1]

(4.1)   *If* R *is a ray and* P *is a point, then there is a unique ray which has vertex* P *and the same direction as* R.

The following fact about angles in $\mathbf{E}^3$ is fundamental.

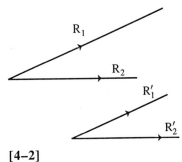

[4–2]

(4.2)   *Let* R_1, R_2, R_1' *and* R_2' *be given rays such that*: R_1 *and* R_2 *have a common vertex*; R_1' *and* R_2' *have a common vertex*; R_1 *and* R_1' *have the same direction; and* R_2 *and* R_2' *have the same direction. Then the angle* R_1R_2 *is equal (in size) to the angle* $R_1'R_2'$, [4–2].

As a further illustration of deductive argument, we give a proof of (4.2) from the facts in §1.2 and §1.3.

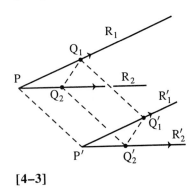

[4–3]

Proof of (4.2). Let P be the vertex of angle R_1R_2 and let P′ be the vertex of angle $R_1'R_2'$ [4–3]. Take points Q_1 on R_1 and Q_1' on R_1' so that $d(P,Q_1) = d(P',Q_1')$. (We usually write this: $\overline{PQ_1} = \overline{P'Q_1'}$.) Similarly, take Q_2 on R_2 and Q_2' on R_2' so that $\overline{PQ_2} = \overline{P'Q_2'}$. By (3.6) $PQ_1Q_1'P'$ and $PQ_2Q_2'P'$ are parallelograms. Hence PP' is parallel to Q_1Q_1' and to Q_2Q_2'. Thus, by (3.5), Q_1Q_1' is parallel to Q_2Q_2'. Furthermore, by (3.6), $\overline{PP'} = \overline{Q_1Q_1'}$ and $\overline{PP'} = \overline{Q_2Q_2'}$. Hence $\overline{Q_1Q'} = \overline{Q_2Q_2'}$. By (3.6), we see that $Q_1Q_1'Q_2'Q_2$ is a parallelogram. By (3.6), this implies that $\overline{Q_1Q_2} = \overline{Q_1'Q_2'}$. Hence by the "side-side-side" congruent triangle theorem, the triangles PQ_1Q_2 and $P'Q_1'Q_2'$ are congruent. It follows that angle R_1R_2 = angle $R_1'R_2'$ as desired.

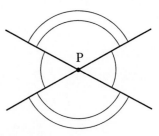

Definitions.   Two lines are *perpendicular* if they intersect at a point P and form right angles at P.

   If two distinct lines intersect at P, then (by the geometry of E^2) these lines form two pairs of opposite and equal angles, called *vertical angles*, at P [4–4]. If the lines are not perpendicular, we refer to the size of the smaller angles at P as the *acute angle* between these lines. If two lines are parallel, we say that the *acute angle* between them is zero.

   A line L and a plane M are defined to be *perpendicular* if L and M intersect at a point P and if L is perpendicular to every line in M that goes through P. If a line L and a plane M are perpendicular, we sometimes say that the line L is *normal* to the plane M.

[4–4]

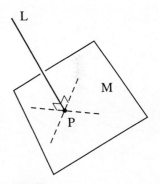

   We have the following further facts about perpendicularity.

(4.3)   *If a line L intersects a plane M at a point P, and if L is perpendicular to two distinct lines each of which lies in M and goes through P, then L is normal to M,* [4–5].

[4–5]

(4.3) follows from:

(4.4)   *Let line L intersect plane M at point P. Let L be perpendicular to two distinct lines through P which lie in M. Then for any third line L' through P:*

$$L' \text{ is perpendicular to } L \Leftrightarrow L' \text{ lies in } M.$$

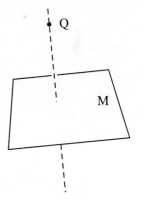

(4.5)   *Given a plane M and a point Q (which may or may not lie on M), then there is a unique line through Q normal to M,* [4–6].

[4–6]

(4.6)   *Given a line L and a point Q (which may or may not lie on L), then there is a unique plane through Q perpendicular to L.*

(4.7)   *Given a line L and a point Q which does not lie on L, then there is a unique line through Q perpendicular to L,* [4–7].

(4.8)   *Given a line L perpendicular to a plane M, then for any other line L':*

[4–7]

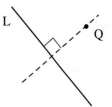

L' is perpendicular to M $\Leftrightarrow$ *L' is parallel to* L.

It follows that if two distinct lines are normal to the same plane, the lines must be parallel.

(4.9)   *Given a line* L *perpendicular to a plane* M, *then for any other plane* M':

M' *is perpendicular to* L $\Leftrightarrow$ M' *is parallel to* M.

It follows that if two distinct planes have a common normal line, the planes must be parallel.

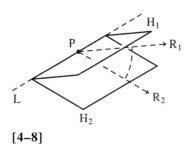

A figure which consists of two half-planes H_1 and H_2 with a common edge L is called a *dihedral angle*. We can measure the size of a dihedral angle as follows, [4–8]. Let H_1, H_2, and L be as above. Choose a point P on L. Let R_1 be a ray in H_1 perpendicular to L with vertex P, and let R_2 be a ray in H_2 perpendicular to L with vertex P. We define the size of the given dihedral angle to be the size of the angle $R_1 R_2$. By (4.2), the size of this angle does not depend on our choice of the point P.

[4–8]

Definitions. We say that two planes are *perpendicular* if they intersect in a line and if the dihedral angles formed at this line are right angles.

We note that for a given point Q and a given plane M, there are infinitely many planes through Q perpendicular to M.

We have:

(4.10)   *Given planes* M_1 *and* M_2 *which intersect in line* L, *then for any plane* M:

L *is normal to* M $\Leftrightarrow$ M *is perpendicular to both* M_1 *and* M_2.

(4.11)   *Given a plane* M *and a line* L *which is not normal to* M, *then there exists a unique plane* M' *such that* (i) M' *contains* L *and* (ii) M' *is perpendicular to* M.

If two distinct planes intersect at L, then these planes form two pairs of opposite and equal *vertical dihedral angles* at L. If the planes are not perpendicular, we refer to the size of the smaller dihedral angles at P as the *acute angle* between these planes. If two planes are parallel, we say that the *acute angle* between them is zero.

Assume that a line L intersects a plane M at a point P. If L is not perpendicular to M, we can define the *acute angle* between L and M as follows.

Take another point Q on L. Let Q′ be the intersection of M with the normal to M through Q. Then the acute angle between L and M is the size of the angle formed by PQ and PQ′.

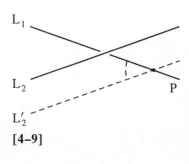

[4–9]

It follows from (4.2) that there is a simple and natural way to define the size of the angle between two given rays in $\mathbf{E}^3$ which do *not* have a common vertex. Let R_1 and R_2 be the given rays, where R_1 has vertex P. Let R_2' be a ray with vertex P and with the same direction as R_2. *The size of the angle between* R_1 *and* R_2 *is defined to be the size of the angle* R_1R_2'.

This construction suggests the following. Let L_1 and L_2 be skew lines, [4–9]. Let P be a point on L_1. Let L_2' be the line through P parallel to L_2. We define L_1 to be *orthogonal* to L_2 if L_2' is perpendicular to L_1. If L_1 is not orthogonal to L_2, then the size of the *acute angle* between L_1 and L_2 is defined to be the size of the acute angle between L_2' and L_1. (It is easy to show that this size is independent of our choice of P.)

§1.5 Projections

Let P be a point in $\mathbf{E}^3$, and let M be a plane in $\mathbf{E}^3$ such that M does not contain P, [5–1]. Let L be the unique line through P normal to M. Let P′ be the unique point of intersection of L with M. Then P′ is called the *projection of* P *on* M. The following fact about P, P′, and M is often useful:

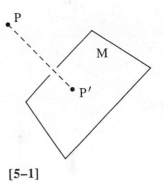

[5–1]

(5.1)   *For any other point* Q *in* M,

$$d(P,Q) > d(P,P').$$

Let S be a figure in $\mathbf{E}^3$, and let M be a plane in $\mathbf{E}^3$, [5–2]. Then the set of all projections on M of points in S is called the *projection of* S *on* M. It follows from (4.11) that:

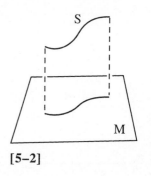

[5–2]

(5.2)   *The projection of a line onto a plane must be either a line or a single point.*
We also have:

(5.3)   *Let* S *be the projection of two parallel lines onto a plane.   Then* S *is either a single line, two parallel lines, or two points.*

From (5.3), it follows that:

(5.4)   *If* S *is a parallelogram and* M *is a plane not perpendicular to the plane of* S, *then the projection of* S *on* M *must be a parallelogram.*

We also have the following useful fact.

(5.5)   *Let* M *and* M' *be two distinct and non-perpendicular planes such that the acute angle between* M *and* M' *is* θ. *Let* S *be a figure in the plane* M *whose area in* M *is* A. *Let* S' *be the projection of* S *on* M'. *Let* A' *be the area of* S' *in* M'. *Then* A' = A cos θ.

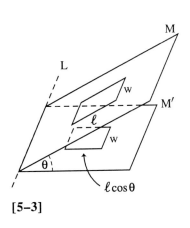

[5–3]

Proof of (5.5). If M and M' intersect, let L be their line of intersection. We first prove the result for the special case where S is a rectangle in M with a side of length w parallel to L and a side of length ℓ perpendicular to L, [5–3]. By elementary trigonometry, S' must be a rectangle in M' with a side of length w parallel to L and a side of length ℓ cos θ perpendicular to L. Hence A = wℓ and A' = wℓ cos θ, as asserted by the theorem.

We prove the result for an arbitrary figure in S by using a collection of small, adjacent, non-overlapping rectangles, with sides parallel and perpendicular to L, as an approximation to S. Call these rectangles $r_1, r_2, \ldots, r_n$. Let their areas be $a_1, a_2, \ldots, a_n$, and let $r'_1, r'_2, \ldots, r'_n$ be their projections on M. Then, by the special case just proved, the areas of $r'_1, r'_2, \ldots, r'_n$ must be $a_1 \cos \theta, a_2 \cos \theta, \ldots, a_n \cos \theta$. Hence we have

$$A' \approx a_1 \cos \theta + a_2 \cos \theta + \cdots + a_n \cos \theta = (a_1 + a_2 + \cdots + a_n) \cos \theta \approx A \cos \theta,$$

where "$\approx$" means "approximately equal." We can make these approximations as accurate as we like by taking sufficiently many rectangles of sufficiently small size. Hence we have A' = A cos θ.

If the planes M and M' do not intersect, then M and M' are parallel. It follows that S and S' are congruent and that A = A'. Since θ is defined to be zero in this case (see §1.4), we again have A' = A cos θ as asserted.

Projections onto a given line can be defined in the same way as projections onto a given plane. Let P be a point in E^3, and let L be a line in E^3, [5–4]. Let M be the unique plane through P perpendicular to L. Let P' be the unique point of intersection of M with L. Then P' is called the *projection* of P on L. As in (5.1), we have:

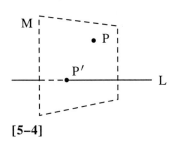

[5–4]

(5.6)   *If* P *is not on* L, *then for any other point* Q *on* L,

$$d(P,Q) > d(P,P').$$

Let S be a figure in $\mathbf{E}^3$, and let L be a line in $\mathbf{E}^3$. The set of all projections on L of points in S is called the *projection* of S on L.

§1.6 Figures in $\mathbf{E}^3$

We give several definitions and facts relating to figures in $\mathbf{E}^3$.

Definitions. A figure S in $\mathbf{E}^2$ or $\mathbf{E}^3$ is said to be *bounded* if there is some point P and some real number k such that for every point Q in S, $d(P,Q) \le k$.

Let $P_1, P_2, \ldots, P_n$ be a finite sequence of distinct points in $\mathbf{E}^3$. Consider the line-segments $P_1P_2, P_2P_3, \ldots, P_{n-1}P_n, P_nP_1$. The bounded figure S consisting of these n segments is called a *polygon* or n-*gon* in $\mathbf{E}^3$. The points $P_1, P_2, \ldots, P_n$ are called the *vertices* of S, and the segments $P_1P_2, \ldots, P_nP_1$ are called the *edges* (or "sides") of S. We say that a polygon S is *simple* if the following two conditions hold: (i) no two adjacent edges are collinear, and (ii) except for the vertices of S, no point in S lies on more than one edge. In what follows, we limit our attention to simple polygons.

If the vertices of a given simple polygon S are coplanar, we say that S is a *planar polygon*.

We give some examples and comments about polygons.
(1)   The following statements are always true:

S is a 3-gon $\Leftrightarrow$ S is a triangle.
S is a 3-gon $\Rightarrow$ S is planar.

(2)   S is a 4-gon $\Leftrightarrow$ S is a *quadrilateral* as defined in §1.3.

Quadrilaterals in $\mathbf{E}^3$ need not be planar, but parallelograms must be planar. (See §1.3).

(3) Various examples and properties of planar polygons will be familiar to the reader from plane geometry. Examples include *squares* and *rectangles*. A planar polygon S divides its plane into two parts: (i) the region of all points on S or enclosed by S; and (ii) the remaining points, which lie outside S. We refer to the first part as the *inside* of S or the *face* formed by S.

(4) A planar polygon is said to be *convex* if, for every pair P and Q of points inside S, the segment PQ lies entirely inside S.

(5) A planar polygon is said to be *regular* if it satisfies three conditions: (i) the polygon is convex, (ii) all edges have the same length, and (iii) the size of the angle formed by the two adjacent edges at a vertex is the same for all vertices.

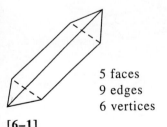

5 faces
9 edges
6 vertices

[6–1]

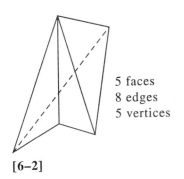

5 faces
8 edges
5 vertices

[6–2]

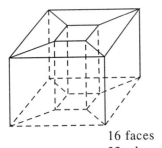

16 faces
32 edges
16 vertices

[6–3]

4-hedron

pyramid

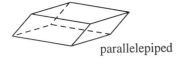

parallelepiped

prism

[6–4]

A *polyhedron* is a bounded figure in $\mathbf{E}^3$ analogous to a planar polygon in $\mathbf{E}^2$. It can be informally defined as consisting of a finite number of planar-polygon faces joined along their edges so that they form a surface which fully encloses some region of $\mathbf{E}^3$. Three examples are given in [6–1], [6–2], and [6–3].

The above informal definition of polyhedron will suffice for our present purposes. [For the reader's interest, we include the following more precise definition:

A *polyhedron* is a figure in $\mathbf{E}^3$ consisting of a finite number of planar-polygon faces joined along their edges in such a way that: (i) every edge is shared by exactly two faces; (ii) every vertex is shared by three or more faces; (iii) no two faces can intersect except at a joining edge or common vertex; (iv) for every pair of faces F_1 and F_2, there is a chain of successively adjacent faces which joins F_1 to F_2, where we define two faces to be *adjacent* if they have a common edge; and (v) for every pair of faces F_1 and F_2 which share a common vertex, there is a chain of successively adjacent faces which joins F_1 to F_2 and whose members all share that vertex.]

A polyhedron with n-faces is called an n-*hedron*. It follows from the definition that there is no 3-hedron, and that all the faces of a 4-hedron must be triangles. A 4-hedron is also called a *tetrahedron*. Two 5-hedrons are shown in [6–1] and [6–2].

A polyhedron S divides $\mathbf{E}^3$ into two parts: (i) the region of all points on S or enclosed by S; and (ii) the remaining points, which lie outside S. We refer to the first part as the *inside* of S. A polyhedron is *convex* if, for every pair of points P and Q inside S, the segment PQ lies entirely inside S. Some examples of convex polyhedrons are given in the figure [6–4]. (The polyhedron in [6–1] is convex, while the polyhedrons in [6–2] and [6–3] are not convex.)

A 6-hedron whose faces lie in three pairs of parallel planes is called a *parallelepiped*. Parallelepipeds will have special importance in our future work. It is easy to show that:

(6.1)   *Each face of a parallelepiped is a parallelogram and is congruent to its opposite face.*

Other general kinds of polyhedrons include *pyramids* and *prisms*. A *pyramid* is obtained by taking a planar-polygon face S and a point P not in the plane of S. If we join P to each of the vertices of S by means of line segments, we get triangular faces which, together with S itself, form the faces of the pyramid. P is called the *vertex* of the pyramid and S is called the *base*.

A *prism* is formed by taking two congruent planar polygons S_1 and S_2 which lie in parallel planes so that the line-segments joining corresponding vertices of S_1 and S_2 are parallel. If we picture S_1 and S_2 as the top and bottom faces, then the side faces are parallelograms. S_1 and S_2 are called the *bases* of the prism. In particular, a parallelepiped is a prism whose bases are parallelograms.

Definition. A polyhedron is said to be a *regular* polyhedron if: (i) the polyhedron is convex; (ii) for each face, the edges of that face form a regular polygon; (iii) each vertex is shared by the same number of faces; and (iv) for every pair of faces F_1 and F_2, F_1 is congruent to F_2.

Regular tetrahedron

[6–5]

It can be shown that a regular polyhedron must have one of the five following forms:

(6.2)   4 *triangular faces and* 4 *vertices*, [6–5].
         6 *square faces and* 8 *vertices*, [6–6].
         8 *triangular faces and* 6 *vertices*, [6–7].
         12 *pentagonal faces and* 20 *vertices*, [6–8].
         20 *triangular faces and* 12 *vertices*, [6–9].

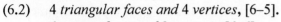

These five forms of regular polyhedron are known as *regular tetrahedrons, cubes, regular octahedrons, regular dodecahedrons,* and *regular icosahedrons.* It can be shown that if two regular polyhedrons have the same number of faces and the same edge-length, then they are congruent.

    In addition to polyhedrons, certain bounded figures in $\mathbf{E}^3$ with curved surfaces will be important for our later work. These include *cylinders, cones,* and *spheres.*

Regular hexahedron
(or cube)

[6–6]

Definitions. A *bounded circular cylinder* (often referred to as a "closed cylinder" or simply a "cylinder") is formed by taking two circular disks which have the same radius and are perpendicular to the line through their centers. The cylinder is then completed by taking all line segments which are normal to both disks and which go from the edge of one disk to the edge of the other.

    A *bounded circular cone* (often referred to as a "closed cone" or simply a "cone") is formed by taking a circular disk C in a plane M, and a point P not in M, where the projection of P on M is the center of C. The cone is then completed by taking all line segments from P to the circular edge of C.

    A *sphere of radius* r *with center at* P is defined as the set of all points Q in $\mathbf{E}^3$ for which $d(P,Q) = r$. We note the following definitions and facts.

(6.3)   *A sphere and a line can intersect at no point, at one point, or at two points.*

Regular octahedron

[6–7]

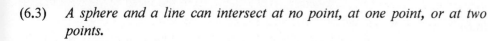

Regular dodecahedron

[6–8]

Regular icosahedron

[6–9]

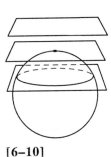

[6–10]

If a line L intersects a sphere S at exactly one point Q, we say that L is *tangent* to S at Q. In this case, it is easy to show that L is perpendicular to QP, where P is the center of S.

(6.4)   *A sphere and a plane can intersect at no point, at one point, or in a circle,* [6–10].

If a plane M intersects a sphere S at a single point Q, we say that M is *tangent* to S at Q. In this case, it is easy to show that M is perpendicular to QP, where P is the center of S.

A cylinder, cone, or sphere divides E^3 into two parts, the *inside* of the figure and the *outside*. *Volume* can be defined for the inside of a polyhedron, cylinder, cone, or sphere in the same way that *area* was defined in plane geometry for regions bounded by polygons or circles. More specifically, in E^3, we approximate the inside of a given figure by a collection of adjacent and similarly oriented rectangular blocks. The approximate volume of the figure is then given by the sum of the volumes of the blocks, where the volume of each block is the product of its three rectangular dimensions. The exact volume of the given figure is then defined as the limit of these approximate values as we use more and more blocks of smaller and smaller size. Using elementary calculus to find these limits, we get the following:

(6.5)   Volume of a sphere of radius a $= \frac{4}{3}\pi a^3$.

Volume of a cylinder of radius a and height b $= \pi a^2 b$.

Volume of a cone of radius a and height b $= \frac{1}{3}\pi a^2 b$.

Volume of a pyramid $= \frac{1}{3}Ah$,

    where A is the area of the base and h (the *altitude*) is the perpendicular distance from the vertex to the plane of the base.

Volume of a prism $= Ah$,

    where A is the area of a base and h (the *altitude*) is the perpendicular distance between the planes of the bases.

The problem of defining and calculating areas of curved surfaces in E^3 is considered in Chapter 21. For reference, we list the following results:

(6.6)   Area of a sphere $= 4\pi a^2$.

Area of a closed cylinder of radius a and height b $= 2\pi ab + 2\pi a^2$

                                                                   $= 2\pi a(a + b)$.

Area of a closed cone of radius a and height b
$$= \pi a^2 + \pi a \sqrt{a^2 + b^2}$$
$$= \pi a\left(a + \sqrt{a^2 + b^2}\right).$$

§1.7 Synthetic Geometry

The geometry of points, lines, and planes in $\mathbf{E}^2$ and $\mathbf{E}^3$, as presented by Euclid, is known as *synthetic geometry*, because the basic objects of study are "synthesized" collections of points (such as lines and planes) rather than individual points and their coordinates as in *analytic geometry*.

In this chapter, we have not presented the synthetic geometry of $\mathbf{E}^3$ as an integrated system of axiomatic assumptions and demonstrated theorems. We note, however, that all the facts about $\mathbf{E}^3$ given in this chapter can be deduced from the geometry of $\mathbf{E}^2$ together with the four additional assertions (2.1), (2.2), (2.3), and (3.1). The chief methods and proofs for $\mathbf{E}^3$ can be found in Book 11 of Euclid's *Elements of Geometry* (300 BC).

As was mentioned in §1.2, there are several logically necessary assumptions which are not made explicit in the usual deductive expositions of Euclidean geometry. The first fully explicit and technically complete system of axioms and postulates was provided by David Hilbert in 1899. It is natural to ask the following question. If the formulations of Euclidean geometry from 300 BC to 1898 were not complete, on what basis could mathematicians, prior to 1899, believe that Euclidean geometry was a precise and well-defined mathematical subject? The answer is that all of Euclid's axioms and postulates were natural abstractions from our experience in the physical world and that the additional implicit assumptions, except for one, were directly rooted in our physical experience. The exceptional assumption was mathematically natural and was suggested, but not directly confirmed, by physical experience. Thus Euclidean geometry and its deductions were, in fact, well-defined and precise. Hilbert's formulation served primarily to validate what had already been accomplished over a period of two thousand years. The exceptional implicit assumption has come to be called *Euclid's axiom* and can be restated as (3.1). It considers the non-intersection of straight lines in a common plane when these lines have been extended, in Euclid's words, "ever so far."

In the present text, our approach to multivariable calculus is based primarily on the synthetic geometry of $\mathbf{E}^3$. Coordinate geometry will be used for computation and technical analysis, but our central definitions and results will be "coordinate-free" (that is to say, they will not depend on a particular choice of coordinate system).

We conclude with two more illustrations of deductive argument. We use the concepts and facts of this chapter to give two different proofs of the same fact:

(7.1)   *If* L_1 *and* L_2 *are skew lines, then there exists a line* L *which intersects, and is perpendicular to, both* L_1 *and* L_2.

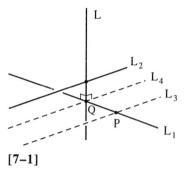

[7–1]

First proof. Let P be a point on L_1, [7–1]. Let L_3 be the line through P parallel to L_2 (by (3.1)). Let M be the plane of L_1 and L_3 (by (2.8)). Let L_4 be the projection of L_2 on M. Then L_2 is parallel to L_4 (see problem 8 for §1.3). Let Q be the intersection of L_1 and L_4. Let L be the normal line to M at Q. Then L is perpendicular to L_1 and L_4. Since L_2 is parallel to L_4, L must also be perpendicular to L_2. (Recall from the geometry of $\mathbf{E}^2$, that if L is perpendicular to one of two parallel lines, it must be perpendicular to the other.) Hence L is the desired line.

Second proof. Let d_m be the minimum distance between points on L_1 and points on L_2. Let P_1 on L_1 and P_2 on L_2 be points such that $d(P_1, P_2) = d_m$. Let L be the line of P_1 and P_2. Then, by (5.5), L must be perpendicular to both L_1 and L_2.

Like many good proofs, these deductions have the merit of taking a statement whose truth is initially far from clear and making that statement seem almost obvious.

Remark. The reader should keep in mind that $\mathbf{E}^3$, with its points, lines, and planes, is an idealized mathematical universe. Depending on the circumstances, $\mathbf{E}^3$ may or may not be useful for making a mathematical model of a physical reality. For example, basic features of $\mathbf{E}^3$ may not carry over to astronomical distances or to submicroscopic distances. At astronomical distances, it may not be true that the sum of the angles of a triangle, as determined by light rays, is 180°; and at submicroscopic distances, the concept of straight line may not have direct physical meaning. Other forms of geometry may prove to be better suited to these situations. Such alternative geometries, different from the geometries of $\mathbf{E}^2$ and $\mathbf{E}^3$, are called *non-Euclidean geometries*. Studies of non-Euclidean geometry often begin by altering or dropping (3.1). Indeed, if we interpret *straight line* as *path of a light ray in empty space*, (3.1) is not true as a statement about the physical universe.

For a simple example of a non-Euclidean geometry in two-dimensions, take as "points" the points on the surface of a given sphere in $\mathbf{E}^3$ (we think of this sphere as being very large), and take as "lines" the great circles on this sphere (a *great circle* on a sphere S is the intersection of S with a plane through the center of S). In the immediate vicinity of any given "point" on S, the geometrical relationships of "points," "lines," and "angles" will be approximately the same as for $\mathbf{E}^2$. On the other hand, the total length of each "line" is finite, and any two given lines must eventually intersect.

Using an appropriate definition of four-dimensional Euclidean space (which we refer to as $\mathbf{E}^4$, see Chapter 34), we can also define a "sphere" in $\mathbf{E}^4$. This "surface" has a three-dimensional geometry of "points," "lines," "planes," and "angles" whose geometrical relationships in the vicinity of any chosen "point" are approximately the same as for $\mathbf{E}^3$, but where each "line" has finite total length, each "plane" has finite total area, the three-dimensional "surface" has finite total volume, and any two "lines" lying in a common "plane" must eventually intersect. This three-dimensional non-Euclidean geometry provides one plausible version of the geometry of a finite physical universe as considered by cosmologists. From this point of view, the idea of a finite physical universe is no more startling than the idea of a spherical earth.

§1.8 Problems

§1.2

1. Show that (2.5) follows from (2.1), (2.2), and (2.3).

2. Show that (2.6) follows from (2.1), (2.2), and (2.3).

3. Show that (2.7) follows from (2.1), (2.2), and (2.3).

4. (a) For a line L, a point P on L, and two other points Q_1 and Q_2 on L, give a geometrical definition for "Q_1 and Q_2 lie on the same side of P." (*Hint.* Consider the line-segment Q_1Q_2.)
    (b) For a plane M, a line L in M, and two points Q_1 and Q_2 in M but not on L, give a geometrical definition for "Q_1 and Q_2 lie on the same side of L."
    (c) For a plane M in $\mathbf{E}^3$ and two points Q_1 and Q_2 not on M, give a geometrical definition for "Q_1 and Q_2 lie on the same side of M."

5. Deduce the following from facts in §1.2. *For every line* L, *there are two planes whose intersection is* L.

§1.3

1. Which of the following are true, and which are false?
    (a) Two distinct lines parallel to a third line must be parallel to each other.
    (b) Two distinct lines parallel to the same plane must be parallel.
    (c) Two distinct planes parallel to the same line must be parallel.
    (d) Two distinct planes parallel to a third plane must be parallel to each other.
    (e) If two parallel planes are intersected by a third plane, the two lines of intersection must be parallel to each other.

2. Deduce each true statement in problem 1 from facts in the text.

3. For each false statement in problem 1, show that it is false.

4. Deduce (3.3) from (3.1).

5. Deduce (3.4) from (3.2).

6. Let L_1 and L_2 be skew lines. Let P_1 and Q_1 be any two distinct points on L_1, and let P_2 and Q_2 be any two distinct points on L_2. Show that the segments P_1P_2 and Q_1Q_2 must lie on skew lines.

7. Let L_1, L_2, P_1, Q_1, P_2, and Q_2 be as in problem 6. Let M be a plane parallel to both L_1 and L_2. (Problem 6 of §1.4 shows that such a plane exists.) Show that the intersections of M with the four lines determined by the line-segments P_1P_2, P_1Q_2, Q_1P_2, and Q_1Q_2 form the vertices of a parallelogram.

8. Line L lies in plane M. Let L' be a line parallel to L. Use facts in the text to show that L' is either contained in M or is parallel to M.

§1.4

1. Consider the statement about $\mathbf{E}^3$, "For every point P and line L, there is a unique line L' which contains P and is perpendicular to L." If this statement is true, deduce it from facts in the text. If it is false, give an example showing that it is false.

2. Which of the following are true and which are false?
   (a) Two distinct lines perpendicular to a third line must be parallel.
   (b) Two distinct lines perpendicular to the same plane must be parallel.
   (c) Two distinct planes perpendicular to the same line must be parallel.
   (d) Two distinct planes perpendicular to the same plane must be parallel.

3. (a) Let R be a fixed ray with vertex P. Let PQ_0 and PQ be two segments of length 1, each making the same fixed angle with R. PQ_0 is held fixed. PQ is free to rotate about R while maintaining a constant fixed angle with R. Show that the angle between the two segments is greatest when PQ and PQ_0 lie in a common plane and on opposite sides of the line of R. (*Hint.* Use the law of cosines.)
   (b) In $\mathbf{E}^2$, the sum of the vertex angles of a quadrilateral must be 2π. Consider a quadrilateral in $\mathbf{E}^3$. What can you conclude about the sum of its vertex angles when it is not planar?

4. Show that (4.1) follows from preceding facts.

5. A *counterexample* to a given statement is an example which shows that the statement is false. In each of following cases, outline a proof if the statement is true, and give a counterexample if the statement is false.
   (a) Given two parallel lines in $\mathbf{E}^3$ and a point P not on either line, there exists at least one plane which contains P and is parallel to both lines.
   (b) Given two parallel lines in $\mathbf{E}^3$ and a point P not on either line, there exists a unique plane which contains P and is parallel to both lines.

6. In each case, give a proof.
   (a) Let L_1 and L_2 be skew lines in $\mathbf{E}^3$. Then there exists a unique plane M_1 which contains L_1 and is parallel to L_2 and a unique plane M_2 which contains L_2 and is parallel to L_1. Furthermore, M_1 and M_2 are parallel. (*Hint.* See the last paragraph of §1.4.)
   (b) Let two skew lines be given, then for any given point P there is a unique plane M such that (i) M contains P, and (ii) either M is parallel to both lines, or M contains one line and is parallel to the other.

§1.5

1. A line L lies in a plane M. Let P be a point not on M. Let P_1 be the projection of P on L. Let P_2 be the projection of P on M, and let P_3 be the projection of P_2 on L. Must P_1 and P_3 always be the same point? Justify your answer.

2. A rectangle of length 2 and width 1 lies in a plane M. Its projection on plane M' is a square of side c. What is the value of c and what is the acute angle between M and M'. Justify your answer.

§1.6

1. Consider conditions (i), (ii), and (iii) in the definition of *regular polygon*. Show by examples that if any one of these conditions is dropped, there is a planar polygon which satisfies the remaining two conditions but is not regular.

2. Show that the midpoints of the six edges of a cube which do not meet a particular interior diagonal are coplanar and are the vertices of a regular hexagon. (*Hint.* Consider successive rotations of the cube, with the given diagonal as axis of rotation and with each rotation having size $\pi/3$.)

3. Show, from the definition of *polyhedron*, that:
   (a) there is no 3-hedron, and
   (b) all the faces of a 4-hedron must be triangles.
   (The proofs are short.)

4. Show that if two edges of a 4-hedron do not have a common vertex, then the lines of these edges must be skew.

5. (a) Show from the definition of *polyhedron* that every face of a 5-hedron is either a triangle or a quadrilateral.
   (b) Show by examples that it is possible for a convex 5-hedron to have exactly one quadrilateral face and that it is possible for a convex 5-hedron to have exactly three quadrilateral faces.

6. Every convex polyhedron satisfies the equation

$$F + V - E = 2,$$

where F = the number of faces, V = the number of vertices, and E = the number of edges. This relationship was discovered and proved by the mathematician Euler and is known as *Euler's equation*. Its proof is outlined in problem 9 below.
   (a) Show from Euler's equation that a convex 5-hedron must have at least one quadrilateral face.
   (b) Show from Euler's equation that it is not possible for a convex 5-hedron to have exactly two or exactly four quadrilateral faces.
   (c) Show from Euler's equation that it is not possible for a convex 5-hedron to have all its faces be quadrilaterals.

7. Consider conditions (i), (ii), (iii), and (iv) in the definition of *regular polyhedron*. Show by examples that if any one of these conditions is dropped, there is a polyhedron which satisfies the remaining three conditions but is not regular.

8. Show from the definition of *parallelepiped* that opposite faces of a parallelepiped must be congruent parallelograms.

9. (a) Let $\mathcal{P}$ be any given convex polyhedron. Let O be a point inside $\mathcal{P}$ but not on a face of $\mathcal{P}$, and let S be a sphere with center at O. Show that $\mathcal{P}$ can be "projected" onto S, where each point Q on $\mathcal{P}$ "projects" to the point Q' on S such that Q and Q' lie on a common ray from O. Show that distinct points of $\mathcal{P}$ are "projected" to distinct points of S and that each edge of $\mathcal{P}$ is "projected" to an arc of a great circle on S.
   (b) *Euler's equation* for a convex polyhedron $\mathcal{P}$ was stated in problem 6 above. Prove Euler's equation for a given $\mathcal{P}$ by proving it for the "projection" of $\mathcal{P}$ as described in (a). Do so by removing successive "edges" from the projection until you are left with a single "edge," the two "vertices" of that "edge," and a single "face" consisting of the entire sphere. (As an "edge" is removed, any currently shared "vertex" of that edge is left behind, and if that edge was shared by two distinct "faces" then those faces are merged into a single "face." At intermediate stages of the process, newly defined "faces" need not have "convex" boundaries, and a remaining "edge" need not be shared by two distinct "faces.") Devise a rule for choosing successive edges to be removed so that the quantity F + V − E must remain unchanged as each successive edge is removed.

10. Use Euler's equation, stated in problem 6 above, together with the definition of *regular polyhedron*, to show that the only regular polyhedrons are those listed in (6.3).

11. The acute angle between planes M_1 and M_2 is θ. A circle C of diameter d lies in M_1. Let E be the projection of C on M_2. Assume that this projection is an ellipse. (This will be proved in the problems for Chapter 4.)
    (a) In terms of d and θ, what is the maximum diameter of E? What is the minimum diameter through the center of E? What is the area enclosed by E in M_2?
    (b) Let a = one-half of this maximum diameter and let b = one-half of this minimum diameter. Give a formula for this area in terms of a and b.

12. S_1 is a sphere of radius 1 and S_2 is a sphere of radius 2, [8–1]. The distance between their centers is 4.

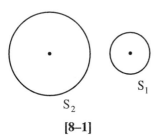

[8–1]

The figure shows a plane through their centers. For each of the following values of n, identify the points in this plane through which there are exactly n lines, such that each line is tangent to both spheres. (These lines need not lie in the plane of the figure.) n = 0, 1, 2, 3, 4, ∞. Make six copies of the figure to show your answers.

13. Do the same as in problem 12 for the case where the distance between centers is 3.

14. Do the same as in problem 12 for the case where the distance between centers is 2.

15. Explain why the polyhedrons in [6–2] and [6–3] are not convex.

§1.7

1. Show that the line L in (7.1) is unique. (*Hint.* Use the result of problem 2b for §1.4.)

Chapter 2

Geometric Vectors and Vector Algebra

§2.1   Vectors

In applications of physical science, scientists and engineers define and measure *physical quantities* and then use *mathematical models* and *natural laws* to describe and predict relationships among these physical quantities. For example, in the case of a moving particle, we define physical quantities called *time*, *position* of the particle, *velocity* of the particle, *acceleration* of the particle, *mass* of the particle, and *force* acting on the particle. We then use calculus and Newton's second law of motion to describe and predict the motion of the particle.

If the result of measuring a given physical quantity is a single real number and nothing more, we say that the given quantity is a *scalar quantity*, because the entire result of measuring that quantity can be pictured as the position of a pointer on a numerical "scale." In our example of the moving particle, *time* and *mass* are scalar quantities. Other examples of scalar quantities include *temperature* at a given point and moment of time in a given substance, *pressure* at a given point and moment, and *density* at a given point and moment. Since individual real numbers are the measured values that we get for a scalar quantity, the real numbers themselves are often referred to by scientists and engineers as *scalars*.

There are other physical quantities for which the result of a measurement combines both (i) a single *non-negative real number* and (ii) a single *geometrical direction* in space. We say that such a quantity is a *vector quantity*. In our example of the moving particle, the *velocity* of the particle at a given moment, the *acceleration* of the particle at a given moment, and the *force* acting on the particle at a given moment are vector quantities. (The definitions of *velocity, accelera-tion,* and *force* are considered further in Chapter 5.) The *position* of the particle at a given moment can also be viewed as a vector quantity if we measure position as *distance and direction* from a given fixed reference-point in space.

We shall treat the result of measuring a vector quantity as a single mathe-matical entity which combines: (i) a non-negative real number and (ii) a direction in $\mathbf{E}^3$ or $\mathbf{E}^2$. A mathematical entity of this kind is called a *geometrical vector*, or simply a *vector*. If the direction is given in $\mathbf{E}^3$, we say that we have a *vector in* $\mathbf{E}^3$. If the direction is in $\mathbf{E}^2$, we have a *vector in* $\mathbf{E}^2$. A *direction* in $\mathbf{E}^3$ or $\mathbf{E}^2$ can be thought of as a family of commonly directed parallel rays. We can use any single ray from the family to represent that direction, just as we can draw a single north-pointing ray on a nautical chart to represent the direction *North*. We say that two rays with distinct vertices have the *same direction* if the angle between them, as defined in §1.4, is zero.

A vector is a *single entity* in the same sense that a real number is a single entity. A vector is the "observed value" that we get when we measure a vector quantity.

Notation. We use

$$\vec{A},\vec{B},...,\vec{a},\vec{b},...$$

to denote vectors.

The non-negative real-number part of a vector is called the *magnitude* of that vector. The geometrical direction part of a vector is called, simply, the *direc-tion* of that vector.

Notation. $\left|\vec{A}\right|$ denotes the magnitude of the vector $\vec{A}$. Hence $\left|\vec{A}\right| \geq 0$ for every vector $\vec{A}$.

It is convenient and useful to represent a given vector $\vec{A}$ as an *arrow*, where the direction of the arrow indicates the direction of $\vec{A}$, and the length of the arrow gives $\left|\vec{A}\right|$, the magnitude of $\vec{A}$. We make the following formal definitions: (1) an *arrow* is a line segment with one endpoint chosen as its *initial point* (or "tail") and the other endpoint chosen as its *final point* (or "head"). We use the notation $\overrightarrow{PQ}$ to denote the arrow that has initial point P and final point Q, [1–1]. (2) We say that an arrow $\overrightarrow{PQ}$ *represents* a vector $\vec{A}$ if the length of $\overrightarrow{PQ}$ is equal to $\left|\vec{A}\right|$ and if the direction of $\overrightarrow{PQ}$ is the direction of $\vec{A}$. (3) Different arrows can be used to represent the same vector, provided that those arrows have the same magnitudes and same directions. This can be expressed geometrically in the following simple way.

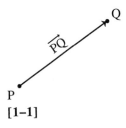

P

[1–1]

(1.1)   *Two arrows* $\overrightarrow{P_1Q_1}$ *and* $\overrightarrow{P_2Q_2}$ *not on the same line represent the same vector* $\Leftrightarrow$ *the quadrilateral* $P_1Q_1Q_2P_2$ *is a parallelogram*, [1–2]. (This requires, for arrows in E^3, that there be a single plane which contains the four points P_1, Q_1, Q_2, and P_2.)

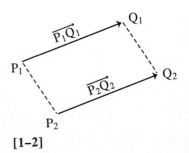

[1–2]

For example, we say that two moving particles have the *same velocity* if their velocities have the same magnitudes and directions, even though the particles may be at different positions in space, and even though we may use two differently located arrows to represent those velocities. It is customary among scientists and engineers to refer informally and briefly to an individual given arrow as a "vector" rather than to speak of it as an "arrow representing a vector." We shall often follow this informal usage. *It is customary to use the sign = between arrows to mean "represents the same vector as."* Thus we can restate (1.1) above as: *if* $\overrightarrow{P_1Q_1}$ *and* $\overrightarrow{P_2Q_2}$ *are not on a common line, then*

$$\overrightarrow{P_1Q_1} = \overrightarrow{P_2Q_2} \Leftrightarrow P_1Q_1Q_2P_2 \text{ is a parallelogram.}$$

We also write $\overrightarrow{PQ} = \vec{A}$ or $\vec{A} = \overrightarrow{PQ}$ to mean that the arrow $\overrightarrow{PQ}$ represents the vector $\vec{A}$.

Sometimes, when we measure a vector quantity, we find that the magnitude is zero and that no specific direction in space is given. This is the case, for example, when we measure the velocity of a particle which happens to be at rest. To represent such a measurement, we introduce a special vector, called the *zero vector*, and we write this vector as $\vec{0}$. $\vec{0}$ has magnitude 0 ($|\vec{0}| = 0$) and no specific direction. It is sometimes useful to think of $\vec{0}$ as represented by an arrow $\overrightarrow{PP}$ whose initial and final points are the same. In a figure, this arrow would appear as a single point P.

It follows from the geometry of E^2 and E^3 that:

(1.2)   *if we are given a vector* $\vec{A}$ *and a point* P, *then there must be a unique point* Q *such that* $\vec{A} = \overrightarrow{PQ}$.

§2.2 Applications of Vectors

(1) In that part of mechanics known as *kinematics*, vectors are used with such vector quantities as *velocity* of a moving point, *acceleration* of a moving point, and *angular velocity* of a rotating body. For angular velocity, the magnitude of the vector is the angular rate of rotation (in radians/sec), and the direction of the vector is given by the axis of rotation and a *right-hand rule*, which states

that if we place a right hand with the fingers curling around with the rotation, then the thumb indicates the vector's direction along the axis of rotation.

(2) A vector can also be used to describe the *displacement* (change in position) of a particle which has moved from a point P to a point Q. We simply define the *displacement vector* for the given change of position from P to Q to be the vector represented by the arrow $\overrightarrow{PQ}$.

(3) Vectors can also be used to describe various kinds of *flow*. For example, the flow of heat at a given point P, at a given moment in a given substance, can be described by a *heat-flow vector* $\vec{H}$, where the direction of $\vec{H}$ is the direction of the flow at P at the given moment, and the magnitude of $\vec{H}$ is

$$\left|\vec{H}\right| = \lim_{a \to 0} \frac{h(a)}{\pi a^2},$$

where h(a), at that moment, is the rate, per unit time, at which heat energy passes through an imaginary disk of radius a (and area πa^2) having its center at P and lying in a plane perpendicular to the direction of flow at P.

Similarly in a fluid flow, the flow of mass at a given point P at a given moment in a given fluid can be described by a *mass-flow vector* $\vec{m}$, where the direction of $\vec{m}$ is the direction of the fluid flow at P at the given moment, and the magnitude of $\vec{m}$ is

$$\left|\vec{m}\right| = \lim_{a \to 0} \frac{f(a)}{\pi a^2},$$

where f(a), at that moment, is the rate, per unit time, at which mass passes through an imaginary disk of radius a (and area πa^2) having its center at P and lying in a plane perpendicular to the direction of flow at P. The mass-flow vector is sometimes called the "mass-transfer vector."

(4) Vectors allow us to make brief but informative statements of certain laws of nature. For example, *Newton's second law of motion* can be expressed as

$$\vec{F} = m\vec{a},$$

where $\vec{F}$ is a vector measuring the force acting on a moving particle at a given moment, and $m\vec{a}$ is a vector whose direction is the direction of the measured acceleration $\vec{a}$ of the particle and whose magnitude is the product of $\left|\vec{a}\right|$ with the mass m of the particle. Newton's law asserts that these two vectors are the same.

(5) Vectors are also useful in *mathematics*. We shall give some geometrical uses of vectors in §2.7 below. As already mentioned in §2.1, we can use a vector to describe the geometrical *position* of a point P: We first choose a fixed reference-point (or *origin*) O. Then the position of any given point P can be ex-

pressed by the *position arrow* $\overrightarrow{OP}$, [2–1]. The vector represented by the arrow $\overrightarrow{OP}$ is called the *position vector of the point* P *relative to* O. We shall often use the symbol $\vec{R}$ for a *position vector*, since a position vector is represented by a "radial" arrow from the origin O.

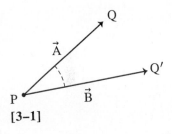

[2–1]

§2.3 Some Geometrical Definitions

The following definitions are used in geometric applications of vectors. Let $\vec{A}$ and $\vec{B}$ be given non-zero vectors. Choose a point P and then, by (1.2), choose points Q and Q′ so that $\overrightarrow{PQ} = \vec{A}$ and $\overrightarrow{PQ'} = \vec{B}$, [3–1]. We define the *size of the angle between* $\vec{A}$ *and* $\vec{B}$ (which we shall later usually refer to as simply as "the angle between $\vec{A}$ and $\vec{B}$") to be the size of the angle QPQ′. It is immediate from (4.1) in §1.4 that the size of this angle does not depend on our choice of the point P. If $\vec{A}$ and $\vec{B}$ are initially given by arrows $\overrightarrow{P_1Q_1}$ and $\overrightarrow{P_2Q_2}$, we can find the size of the angle between $\vec{A}$ and $\vec{B}$ by taking a new arrow $\overrightarrow{P_1Q'}$ so that $\overrightarrow{P_1Q'} = \overrightarrow{P_2Q_2}$ and by then using the angle Q_1P_1Q'. See [3–2].

[3–1]

We say that two non-zero vectors have *opposite directions* if the angle between them is π (= 180°). We say that two non-zero vectors are *perpendicular* if the angle between them is $\pi/2$ (= 90°). We say that two non-zero vectors are *collinear* if the angle between them is either 0 or π. Thus two given vectors are collinear if and only if it is possible to represent them by two arrows which lie on a common line. (Note that if two particles are moving along parallel lines, then their velocities must be collinear vectors, even though the particles themselves are not moving on a common line.)

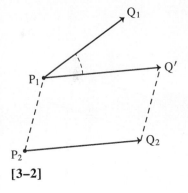

Similarly, we say that three non-zero vectors are *coplanar* if it is possible to represent them by three arrows which lie in a common plane. (Note that if three particles are moving in parallel planes, then their velocities must be coplanar vectors, even though the particles themselves are not moving in a common plane.)

[3–2]

§2.4 The Four Operations of Vector Algebra

We now define the four basic operations of *vector algebra*. These operations are: (I) *multiplication of a vector by a scalar*; (II) *addition of two vectors*; (III) *dot-product of two vectors*; and (IV) *cross-product of two vectors*. We shall see that operation II is analogous to the operation of addition in ordinary arithmetic and that each of the operations I, III, and IV is, to some extent, analogous to the operation of multiplication in ordinary arithmetic. The four operations will allow us to write algebraic equations about vectors and to carry out algebraic computations. Vector algebra is a powerful tool in applications of mathematics and is used for theoretical derivations as well as for practical calculations.

After each definition (of an operation), we list the algebraic laws which that operation satisfies. In §2.5, we shall indicate the proofs of these laws.

(I) *Multiplication of a vector by a scalar.* Let c be a given scalar (that is to say, a given real number), and let $\vec{A}$ be a given vector. The result of this operation, *multiplying the vector $\vec{A}$ by the scalar* c, will be denoted as $c\vec{A}$ (or sometimes as $\vec{A}c$). The product $c\vec{A}$ is a *vector* and is defined from c and $\vec{A}$ as follows:

(4.1)   *If $\vec{A} \neq \vec{0}$ and c > 0, then $c\vec{A}$ is the vector whose direction is the same as the direction of $\vec{A}$ and whose magnitude is $c|\vec{A}|$. (Here $c|\vec{A}|$ is the product, in ordinary arithmetic, of the scalars c and $|\vec{A}|$.)*

If $\vec{A} \neq \vec{0}$ and c < 0, then $c\vec{A}$ is the vector whose direction is opposite to the direction of $\vec{A}$ and whose magnitude is $-c|\vec{A}|$.

If c = 0, then for all $\vec{A}$, $c\vec{A} = \vec{0}$.

If $\vec{A} = \vec{0}$, then for all c, $c\vec{A} = \vec{0}$.

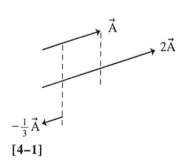

[4–1]

Thus, for example, $2\vec{A}$ has the same direction as $\vec{A}$ but twice the magnitude, and $(-1/3)\vec{A}$ has the opposite direction to $\vec{A}$ and one-third the magnitude, [4–1]. (The operation of multiplication by a scalar appears in example 4 of §2.2.)

We have the following *algebraic laws.*

(Ia) $|c\vec{A}| = |c||\vec{A}|$. Here $|c|$ is the absolute value of c, while $c|\vec{A}|$ and $|\vec{A}|$ are magnitudes of vectors.

(Ib) $c(d\vec{A}) = (cd)\vec{A}$. Here cd is the ordinary arithmetical product of the scalars c and d.

(Ic) $0\vec{A} = \vec{0}$.

(Id) $c\vec{0} = \vec{0}$.

The operation of multiplication by a scalar leads us to the following definitions and notations.

Definition. If $\left|\vec{A}\right| = 1$, we say that $\vec{A}$ is a *unit vector*.

If $\vec{A} \neq \vec{0}$, then, using operation I, we define

$$\hat{A} = \left(\frac{1}{\left|\vec{A}\right|}\right)\vec{A}.$$

By (Ia), $\left|\hat{A}\right| = \left|\frac{1}{\left|\vec{A}\right|}\vec{A}\right| = \left|\frac{1}{\left|\vec{A}\right|}\right|\left|\vec{A}\right| = \frac{1}{\left|\vec{A}\right|}\left|\vec{A}\right| = 1$. Thus $\hat{A}$ is the unit vector whose

direction is the same as the direction of $\vec{A}$. From now on, when we know that a given vector is a unit vector, we will usually write the sign $\wedge$ in place of $\rightarrow$ to indicate that fact.

Definition. For any given $\vec{A}$, we define

$$-\vec{A} = (-1)\vec{A}.$$

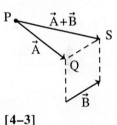

[4–2]

Thus $-\vec{A}$ has the same length as $\vec{A}$, but opposite direction, [4–2].

(II) *Addition of vectors.* Let $\vec{A}$ and $\vec{B}$ be given vectors. The result of this operation, *adding the vector $\vec{B}$ to the vector $\vec{A}$*, will be called the *sum* of the vectors $\vec{A}$ and $\vec{B}$ and will be denoted as $\vec{A}+\vec{B}$. The sum $\vec{A}+\vec{B}$ is a *vector* and is defined from $\vec{A}$ and $\vec{B}$ as follows:

(4.2) *Choose some arrow $\overrightarrow{PQ_1}$ which represents $\vec{A}$. Then, by (1.2), find the point S such that $\overrightarrow{QS}$ represents $\vec{B}$, [4–3]. Then $\overrightarrow{PS}$ represents the vector $\vec{A}+\vec{B}$.*

[4–3]

The addition operation can also be defined in a different but equivalent way.

(4.3) *Take arrows $\overrightarrow{PQ_1} = \vec{A}$ and $\overrightarrow{PQ_2} = \vec{B}$ from a common initial point P, [4–4]. Find the point S such that PQ_1SQ_2 is a parallelogram (see (6.1b) in §1.6). Then the arrow $\overrightarrow{PS}$, given by the diagonal from P in this parallelogram, represents $\vec{A}+\vec{B}$.*

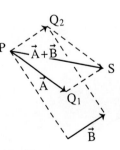

[4–4]

The equivalence of the two definitions is evident from the figures. (4.3) implies the following.

(4.4) *Let PQ_1SQ_2 be a given quadrilateral, then*

$$PQ_1SQ_2 \text{ is a parallelogram} \Leftrightarrow \vec{PQ}_1 + \vec{PQ}_2 = \vec{PS}.$$

The following algebraic laws can be proved.

(IIa) $\vec{A} + \vec{B} = \vec{B} + \vec{A}$ (IId) $\vec{A} + (-\vec{A}) = \vec{0}$

(IIb) $(\vec{A}+\vec{B}) + \vec{C} = \vec{A} + (\vec{B}+\vec{C})$ (IIe) $(c+d)\vec{A} = c\vec{A} + d\vec{A}$

(IIc) $\vec{A} + \vec{0} = \vec{A}$ (IIf) $c(\vec{A}+\vec{B}) = c\vec{A} + c\vec{B}$

It follows from (IIf) that $-(\vec{A}+\vec{B}) = (-\vec{A})+(-\vec{B})$. (IIb) allows us to omit parentheses when we write vector sums and thus to write both $(\vec{A}+\vec{B})+\vec{C}$ and $\vec{A}+(\vec{B}+\vec{C})$ as $\vec{A}+\vec{B}+\vec{C}$. Moreover, (IIa) allows us to change the order of the terms in a finite vector sum without altering the value of this sum.

The operations of addition and of multiplication by a scalar enable us to define an operation of *subtraction*:

Definition. We define $\vec{A}-\vec{B} = \vec{A}+(-\vec{B})$. Note, in [4–5], how an arrow for $\vec{A}-\vec{B}$ is given by the third side of the triangle formed by arrows for $\vec{A}$ and $\vec{B}$ from a common initial point P. The reader should become familiar with this triangular construction for $\vec{A}-\vec{B}$. Thus: *If a parallelogram is formed by arrows from a common point P for $\vec{A}$ and $\vec{B}$, then the diagonal from P represents $\vec{A}+\vec{B}$ and the other diagonal, from $\vec{B}$ to $\vec{A}$, represents $\vec{A}-\vec{B}$.*

Note that the symbol $+$ is used in vector algebra in two different ways. In (IIe), it appears on the left (in $c+d$) as the familiar addition of real numbers and on the right (in $c\vec{A}+d\vec{B}$) as the newly defined addition of vectors. For a given occurrence of $+$, it will always be clear, from the quantities being added, which use of $+$ is intended.

(III) ***Dot-product of vectors.*** Let $\vec{A}$ and $\vec{B}$ be given vectors. The result of this operation, the *dot-product* of $\vec{A}$ and $\vec{B}$, will be denoted as $\vec{A}\cdot\vec{B}$. We shall see from the definition below that $\vec{A}\cdot\vec{B}$ is a *scalar*. For this reason, $\vec{A}\cdot\vec{B}$ is sometimes called the *scalar product* of $\vec{A}$ and $\vec{B}$. $\vec{A}\cdot\vec{B}$ is defined as follows:

(4.5) For $\vec{A} \neq \vec{0}$ and $\vec{B} \neq \vec{0}$, let θ be the angle between $\vec{A}$ and $\vec{B}$, [4–6]. Then $\vec{A}\cdot\vec{B} = |\vec{A}|\,|\vec{B}|\cos\theta$. For $\vec{A} = \vec{0}$ or $\vec{B} = \vec{0}$ (or both), $\vec{A}\cdot\vec{B} = 0$.

The following facts for non-zero $\vec{A}$ and $\vec{B}$ are immediate:

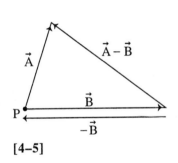

[4–5]

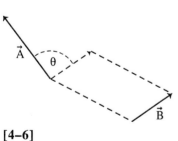

[4–6]

$$\vec{A} \cdot \vec{B} = |\vec{A}| \, |\vec{B}| \iff \theta = 0,$$

$$\vec{A} \cdot \vec{B} > 0 \iff \theta < \frac{\pi}{2},$$

$$\vec{A} \cdot \vec{B} = 0 \iff \theta = \frac{\pi}{2},$$

$$\vec{A} \cdot \vec{B} < 0 \iff \theta > \frac{\pi}{2},$$

$$\vec{A} \cdot \vec{B} = -|\vec{A}| \, |\vec{B}| \iff \theta = \pi.$$

Moreover, for any $\vec{A}$ and $\vec{B}$, we have

(4.6) $\vec{A} \cdot \vec{B} = 0 \iff [\textit{either } |\vec{A}| = 0 \textit{ or } |\vec{B}| = 0 \textit{ or } \theta = \frac{\pi}{2} \,].$

If we define the *scalar projection of* $\vec{A}$ *on* $\vec{B}$ (for non-zero $\vec{A}$ and $\vec{B}$) to be $|\vec{A}| \cos \theta$, then

(4.7) $\vec{A} \cdot \vec{B} = |\vec{A}| \; (\textit{scalar projection of } \vec{B} \textit{ on } \vec{A})$

$\qquad\qquad\qquad = |\vec{B}| \; (\textit{scalar projection of } \vec{A} \textit{ on } \vec{B}).$

Note that a scalar projection is > 0 when $\theta < \frac{\pi}{2}$, is $= 0$ when $\theta = \frac{\pi}{2}$, and is < 0, when $\theta > \frac{\pi}{2}$.

From (4.5), $\vec{A} \cdot \vec{A} = |\vec{A}|^2$. Hence, for any vector $\vec{A}$, $|\vec{A}| = \sqrt{\vec{A} \cdot \vec{A}}$. In vector algebra, the dot product $\vec{A} \cdot \vec{A}$ is often abbreviated as $\vec{A}^2$. Thus we can write $|\vec{A}| = \sqrt{\vec{A}^2}$.

The following algebraic laws can be proved.

(IIIa) $\vec{A} \cdot \vec{B} = \vec{B} \cdot \vec{A}$

(IIIb) $\vec{A} \cdot (\vec{B} + \vec{C}) = \vec{A} \cdot \vec{B} + \vec{A} \cdot \vec{C}$

(IIIc) $\vec{A} \cdot \vec{0} = 0$

(IIId) $c(\vec{A} \cdot \vec{B}) = (c\vec{A}) \cdot \vec{B} = \vec{A} \cdot (c\vec{B}).$

These laws permit us to carry out calculations for dot products that are similar in algebraic form to calculations carried out in the elementary algebra of real numbers. For example,

$$
\begin{aligned}
(\vec{A} - \vec{B}) \cdot (\vec{A} + \vec{B}) &= (\vec{A} - \vec{B}) \cdot \vec{A} + (\vec{A} - \vec{B}) \cdot \vec{B} && \text{(by IIIb)} \\
&= \vec{A} \cdot (\vec{A} - \vec{B}) + \vec{B} \cdot (\vec{A} - \vec{B}) && \text{(by IIIa)} \\
&= \vec{A} \cdot \vec{A} - \vec{A} \cdot \vec{B} + \vec{B} \cdot \vec{A} - \vec{B} \cdot \vec{B} && \text{(by IIIb and IIId)} \\
&= \vec{A} \cdot \vec{A} - \vec{A} \cdot \vec{B} + \vec{A} \cdot \vec{B} - \vec{B} \cdot \vec{B} && \text{(by IIIa)} \\
&= \vec{A} \cdot \vec{A} - \vec{B} \cdot \vec{B} \\
&= \vec{A}^2 - \vec{B}^2 .
\end{aligned}
$$

Definition. We say $\vec{A}$ and $\vec{B}$ are *orthogonal* if $\vec{A} \cdot \vec{B} = 0$.

Saying that two vectors $\vec{A}$ and $\vec{B}$ are orthogonal is the same as saying that either (i) $\vec{A} = \vec{0}$, or (ii) $\vec{B} = \vec{0}$, or (iii) the arrows for $\vec{A}$ and $\vec{B}$ from a common point are perpendicular.

Observe that the notation $(\vec{A} \cdot \vec{B}) \cdot \vec{C}$ has no meaning, since $\vec{A} \cdot \vec{B}$ is a scalar and not a vector. On the other hand, $(\vec{A} \cdot \vec{B})\vec{C}$ does have meaning; it gives the result of multiplying the vector $\vec{C}$ by the scalar $(\vec{A} \cdot \vec{B})$.

Definition. For non-zero $\vec{A}$ and $\vec{B}$, take arrows from a common point P for $\vec{A}$ and $\vec{B}$. Let Q be the projection of the head of the arrow for $\vec{A}$ onto the line of the arrow for $\vec{B}$. Then the vector represented by the arrow $\overrightarrow{PQ}$ is called the *vector projection of $\vec{A}$ on $\vec{B}$*. This vector projection is $(|\vec{A}| \cos\theta)\hat{B}$ and is given by a simple formula of vector algebra as follows:

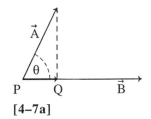

[4–7a]

(4.8)
$$
(|\vec{A}| \cos\theta)\hat{B} = |\vec{A}| \cos\theta \left(\frac{|\vec{B}|}{|\vec{B}|}\right)\left(\frac{1}{|\vec{B}|}\right)\vec{B} = \left(\frac{\vec{A} \cdot \vec{B}}{\vec{B} \cdot \vec{B}}\right)\vec{B},
$$

see [4–7 a,b].

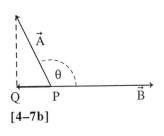

[4–7b]

(IV) *Cross-product of vectors.* Let $\vec{A}$ and $\vec{B}$ be given vectors. The result of this operation, the *cross-product of $\vec{A}$ with $\vec{B}$*, will be denoted as $\vec{A} \times \vec{B}$. We shall see from the definition below that $\vec{A} \times \vec{B}$ is a *vector*. For this reason, $\vec{A} \times \vec{B}$ is sometimes called the *vector product* of $\vec{A}$ and $\vec{B}$. $\vec{A} \times \vec{B}$ is defined as follows:

(4.9) *We specify the magnitude and direction of the vector $\vec{A} \times \vec{B}$.*

Magnitude: *For* $\vec{A} \neq \vec{0}$ *and* $\vec{B} \neq \vec{0}$, *let* θ *be the angle between* $\vec{A}$ *and* $\vec{B}$.
Then $\left| \vec{A} \times \vec{B} \right| = \left| \vec{A} \right| \left| \vec{B} \right| \sin \theta$.

For $\vec{A} = \vec{0}$ *or* $\vec{B} = \vec{0}$ *(or both)*, $\left| \vec{A} \times \vec{B} \right| = 0$.

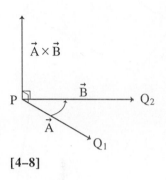

Direction: *If* $\left| \vec{A} \times \vec{B} \right| \neq 0$, *then* $\vec{A} \neq \vec{0}$, $\vec{B} \neq \vec{0}$, *and* $0 < \theta < \pi$. *Hence, if*

we represent $\vec{A}$ *and* $\vec{B}$ *by arrows* $\overrightarrow{PQ_1}$ *and* $\overrightarrow{PQ_2}$, *the points* P, Q_1, *and* Q_2 *are non-collinear and determine a unique plane* M, [4–8]. *We take the direction of* $\vec{A} \times \vec{B}$ *to be normal to* M *according to the following right-hand rule: if a right hand is placed at* P *with fingers curling from* $\overrightarrow{PQ_1}$ *toward* $\overrightarrow{PQ_2}$ *(through the angle* $\theta < \pi$*), then the thumb indicates the direction of* $\vec{A} \times \vec{B}$ *normal to* M.

[4–8]

It is immediate from the figures that the magnitude of the cross-product has the following geometrical significance.

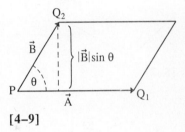

(4.10)　*Let* $\overrightarrow{PQ_1}$ *and* $\overrightarrow{PQ_2}$ *represent* $\vec{A}$ *and* $\vec{B}$ *with* $0 < \theta < \pi$, [4–9]. *Then* $\left| \vec{A} \times \vec{B} \right|$ *is the area of the parallelogram that has* $\overrightarrow{PQ_1}$ *and* $\overrightarrow{PQ_2}$ *as two adjacent sides.*

[4–9]

It follows from (4.9) that *if arrows for* $\vec{A}$ *and* $\vec{B}$ *form two sides of a triangle, then the area of this triangle is* $\frac{1}{2} \left| \vec{A} \times \vec{B} \right|$.

We also have:

(4.11)　$\vec{A} \times \vec{B} = \vec{0} \Leftrightarrow$ *either* $\vec{A} = \vec{0}$ *or* $\vec{B} = \vec{0}$ *or* $\vec{A}$ *and* $\vec{B}$ *are* collinear.

The following algebraic laws can be proved.

(IVa)　$\vec{A} \times \vec{B} = -(\vec{B} \times \vec{A})$

(IVb)　$\vec{A} \times \vec{A} = \vec{0}$

(IVc)　$\vec{A} \times (\vec{B} + \vec{C}) = (\vec{A} \times \vec{B}) + (\vec{A} \times \vec{C})$

(IVd)　$(\vec{B} + \vec{C}) \times \vec{A} = (\vec{B} \times \vec{A}) + (\vec{C} \times \vec{A})$

(IVe)　$\vec{A} \times \vec{0} = \vec{0}$

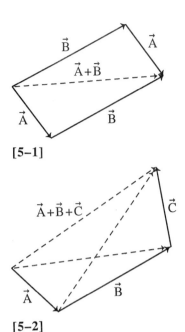

[5–1]

[5–2]

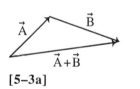

[5–3a]

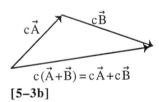

$$c(\vec{A}+\vec{B})=c\vec{A}+c\vec{B}$$

[5–3b]

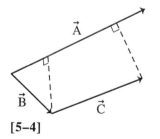

[5–4]

(IVf)　　$c(\vec{A}\times\vec{B}) = c\vec{A}\times\vec{B} = \vec{A}\times c\vec{B}.$

Further laws and identities concerning the cross-product will be given in Chapter 3.

The cross-product does *not* always satisfy the equation $\vec{A}\times\vec{B} = \vec{B}\times\vec{A}$ (commutative law) or the equation $(\vec{A}\times\vec{B})\times\vec{C} = \vec{A}\times(\vec{B}\times\vec{C})$ (associative law). For example, let $\hat{u}$, $\hat{v}$, and $\hat{w}$ be three mutually orthogonal unit vectors such that $\hat{u}\times\hat{v}=\hat{w}$. Then $\hat{v}\times\hat{u}=-\hat{w}$, and $(\hat{u}\times\hat{v})\times\hat{v}=-\hat{u}$ while $\hat{u}\times(\hat{v}\times\hat{v})=\hat{u}\times\vec{0}=\vec{0}$. The fact that the cross-product does not satisfy these familiar laws, which hold for the multiplication of scalars, suggests that we must treat the cross-product with special care in algebraic calculations. In particular, it suggests that we pay close attention to the order of factors and to the position of parentheses in any formula that uses a cross-product. For more information on the failure of the cross-product to satisfy the commutative and associative laws, see problem 17 for §2.4 and problem 20 for §3.6.

Remark. Our vector operations have been defined for vectors in $\mathbf{E}^3$. Operations I, II, and III can be defined in the same way for vectors in $\mathbf{E}^2$, and the same algebraic laws will hold. Operation IV, however, requires vectors in $\mathbf{E}^3$. It is occasionally useful, in working with vectors in $\mathbf{E}^2$, to consider $\mathbf{E}^2$ as a plane contained in $\mathbf{E}^3$. This permits us to make temporary use of operation IV.

§2.5　Proving the Laws of Vector Algebra

In §2.4, the operations of vector algebra are defined geometrically. It follows that the algebraic laws given in §2.4 may themselves be viewed as geometric statements. All but two of these statements (IVc and IVd are the exceptions) can be proved as direct consequences of the definitions or can be deduced as geometric theorems from elementary facts of Euclidean geometry. For example, Ia and Ic are immediate from the definition of multiplication by a scalar, while Ib is equivalent to the assertion that $|k|(|\ell|\|\vec{A}\|)=(|k||\ell|)\|\vec{A}\|$, and this last assertion follows from the associative law for multiplication of real numbers. Similarly the proof of IIa follows from [5–1] and the proof of IIb follows from [5–2]. (Note that the four points in [5–2] need not lie in a common plane.) The proof of IIf is suggested in figure [5–3 a,b], which implies, by similar triangles, that $c\vec{A}+c\vec{B}$ must equal $c(\vec{A}+\vec{B})$.

The proof of IIIb is indicated in [5–4], where we see that the sum of the scalar projections of $\vec{B}$ on $\vec{A}$ and of $\vec{C}$ on $\vec{A}$ must equal the scalar projection of $\vec{B}+\vec{C}$ on $\vec{A}$. (Note that the perpendiculars constructed in the figure need not lie in a common plane.) The proof of IVc, the distributive law for cross-products, is more complex (see the problems). IVd follows from IVc by IVa, IIa, and IIf.

§2.6   Applications of Vector Operations

Vector operations are used in the formulation of natural laws such as Newton's laws of motion and Maxwell's equations for an electromagnetic field. They are also used throughout natural science and engineering for various other theoretical and computational purposes.

Perhaps the single most important application of vectors in elementary physics is the *law of composition of forces*:

(6.1)   *If a point (in a physical object) is acted on by* n *forces, where these forces are given by the vectors* $\vec{F}_1,...,\vec{F}_n$, *then the combined simultaneous effect of these forces is the same as the effect would be of a single force at that point given by the vector* $\vec{F}_1 + \vec{F}_2 + \cdots + \vec{F}_n$.

Much of the usefulness of vector algebra in elementary physics follows from this law. The law is not a mathematical fact; it is a physical fact (that is to say, a law of nature).

The composition-of-forces law has the following useful consequence.

(6.2)   *Let* $\vec{F}_1, \vec{F}_2,...,\vec{F}_n$ *be forces acting at a point* P *in* $\mathbf{E}^3$. *Then the effect of these forces can be exactly balanced (or "canceled") by a single additional force* $\vec{G}$ *acting at* P, *provided that* $\vec{G} = -(\vec{F}_1 + \vec{F}_2 + \cdots + \vec{F}_n)$.

Used in conjunction with Newton's third law of motion, (6.2) is a fundamental principle of engineering mechanics.

A further useful consequence is the *principle of resolution of forces*. For simplicity, we state it for $\mathbf{E}^2$.

(6.3)   *Let* $\hat{u}$ *and* $\hat{v}$ *be non-collinear unit vectors in* $\mathbf{E}^2$. *Let* $\vec{G}$ *be a force acting at a point* P *in* $\mathbf{E}^2$. *Then we can always find two forces* $\vec{F}_1$ *and* $\vec{F}_2$ *such that* $\vec{F}_1$ *is collinear with* $\hat{u}$, $\vec{F}_2$ *is collinear with* $\hat{v}$, *and* $\vec{F}_1 + \vec{F}_2 = \vec{G}$. *The forces* $\vec{F}_1$ *and* $\vec{F}_2$ *are uniquely determined by* $\vec{G}$, $\hat{u}$, *and* $\hat{v}$. *By the law of composition of forces, the combined physical effect of* $\vec{F}_1$ *and* $\vec{F}_2$ *acting at* P *in the absence of* $\vec{G}$ *must be the same as the physical effect of* $\vec{G}$ *itself.*

When engineers calculate the forces $\vec{F}_1$ and $\vec{F}_2$ for given $\vec{G}$, $\hat{u}$, and $\hat{v}$, they say that they have *resolved* the force $\vec{G}$ into a force collinear with $\hat{u}$ and a force collinear with $\hat{v}$.

The proof of the principle of resolution of forces is apparent from [6–1], where we use arrows from P to represent $\hat{u}$ and $\hat{v}$ and choose Q so that $\overrightarrow{PQ} = \vec{G}$. We find $\vec{F}_1$ and $\vec{F}_2$ by constructing lines through Q parallel to $\hat{u}$ and $\hat{v}$. [6–1] is referred to by engineers as the *parallelogram of forces* for the given force $\vec{G}$ and the directions $\hat{u}$ and $\hat{v}$.

The following mathematical version of the resolution principle is immediate from the same figure.

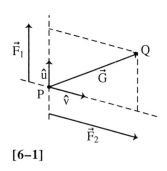

[6–1]

(6.4) *Let $\vec{G}$ be a given vector in $\mathbf{E}^2$, and let $\hat{u}$ and $\hat{v}$ be non-collinear unit vectors in $\mathbf{E}^2$. Then there exist unique scalars a and b such that $\vec{G} = a\hat{u} + b\hat{v}$.*

This algebraic fact is known as the *basis theorem for vectors in $\mathbf{E}^2$*. Three-dimensional versions of the resolution principle and of the basis theorem will be stated and proved in §3.4.

§2.7   Vector Proofs in Geometry

In the following two examples, we show how vector algebra can be used to derive facts of trigonometry and of plane and solid geometry. Further examples appear in the problems.

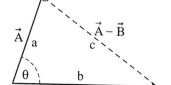

[7–1]

(7.1) *The law of cosines* (in plane trigonometry) *asserts: if three sides of a triangle have lengths a, b, and c, and if θ is the angle between the sides a and b, then it is always true that*

$$a^2 + b^2 - 2ab \cos\theta = c^2.$$

We use vector algebra to prove this theorem. We draw the three sides of the given triangle as arrows (see [7–1]). Let $\vec{A}$, $\vec{B}$, and $\vec{A} - \vec{B}$ be the vectors represented by these arrows. Then $|\vec{A}| = a$, $|\vec{B}| = b$, $|\vec{A} - \vec{B}| = c$. We have $c^2 = |\vec{A} - \vec{B}|^2 = (\vec{A} - \vec{B}) \cdot (\vec{A} - \vec{B}) = (\vec{A} - \vec{B}) \cdot \vec{A} - (\vec{A} - \vec{B}) \cdot \vec{B} = \vec{A}^2 + \vec{B}^2 - 2\vec{A} \cdot \vec{B} = a^2 + b^2 - 2ab\cos\theta$. This completes the proof.

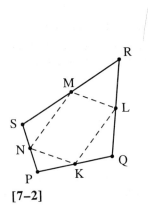

[7–2]

(7.2) **Theorem.** *Let PQRS be an arbitrary quadrilateral in $\mathbf{E}^3$. Let K, L, M, N be, respectively, the midpoints of the segments PQ, QR, RS, and SP*, [7–2].

Then the points K, L, M, N *are coplanar and form the vertices of a parallelogram.* (This is a theorem of three-dimensional geometry, since P, Q, R, and S need not be coplanar.)

Proof.

$$\vec{RM} = \tfrac{1}{2}\vec{RS},$$

$$\vec{RL} = \tfrac{1}{2}\vec{RQ},$$

$$\vec{LM} = \vec{RM} - \vec{RL} = \tfrac{1}{2}\vec{RS} - \tfrac{1}{2}\vec{RQ} = \tfrac{1}{2}(\vec{RS} - \vec{RQ}) = \tfrac{1}{2}\vec{QS}.$$

Similarly, $\vec{KN} = \tfrac{1}{2}\vec{QS}$. Therefore $\vec{LM} = \vec{KN}$, and, by (6.1a) in Chapter 1, KLMN must be a parallelogram.

We give further examples of vector proofs in the problems. In particular, we obtain the *law of sines* of plane trigonometry. In the problems for Chapter 3, we shall obtain, in a similar way, the *law of cosines* and *law of sines* for spherical trigonometry. These last two laws are used in navigational and astronomical calculations.

§2.8 Problems

§2.4

1. $\vec{A}$ and $\vec{B}$ are given vectors. Show that the points whose position vectors are $\vec{A}$, $\vec{B}$, and $4\vec{A} - 3\vec{B}$ must be collinear points.

2. Find non-zero scalars a, b, and c such that for all vectors $\vec{A}$ and $\vec{B}$, $a\vec{A} + b(\vec{A} - \vec{B}) + c(\vec{A} + \vec{B}) = \vec{0}$.

3. $\vec{A}, \vec{B}, \vec{C}, \vec{D}, \vec{E}, \vec{F}$, and $\vec{G}$ are represented by the arrows in [8–1].

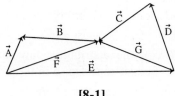

[8-1]

(a) Find $\vec{G}$ in terms of $\vec{A}, \vec{B},$ and $\vec{E}$.

(b) Find $\vec{D}$ in terms of $\vec{A}, \vec{B}, \vec{C}$ and $\vec{E}$.

(c) Find $\vec{X}$ such that $\vec{X} + \vec{B} = \vec{E} - \vec{G}$.

4. Let S be a regular hexagon whose sides are given as a chain of arrows pointing clockwise around S. Two successive adjacent sides represent the vectors $\vec{A}$ and $\vec{B}$. In terms of $\vec{A}$ and $\vec{B}$, find expressions for the vectors represented by the remaining four arrows.

5. R_1 and R_2 are rays with a common vertex. $\vec{A}_1$ and $\vec{A}_2$ are vectors in the directions of R_1 and R_2. In terms of $\vec{A}_1$ and $\vec{A}_2$, find a vector in the same direction as the ray that bisects the angle between R_1 and R_2.

6. Explain why (4.3) implies (4.4). (*Suggestion.* For the '⇔' which appears in (4.4), consider the cases '⇒' and '⇐' separately.)

7. Let a and b be the lengths of two sides of a triangle, and let α and β be the opposite vertex angles, respectively. Use vectors to prove that

$$a = b \Leftrightarrow \alpha = \beta.$$

(*Hint.* Represent sides a and b as vectors $\vec{A}$ and $\vec{B}$, and the third side as $\vec{A} - \vec{B}$. Use the definition and laws for the dot product.)

8. Use vectors to show that for any parallelogram, the sum of the squares of (the lengths of) the four sides is equal to the sum of the squares of (the lengths of) the two diagonals.

9. (a) Prove that for any vectors $\vec{A}$ and $\vec{B}$,

$$|\vec{A} + \vec{B}| \le |\vec{A}| + |\vec{B}|.$$

    (*Hint.* Square both sides and use the definition of the dot product.)

    (b) Similarly, prove that for any $\vec{A}$ and $\vec{B}$,

$$|\vec{A} - \vec{B}| \ge ||\vec{A}| - |\vec{B}||.$$

10. $\vec{A}$ and $\vec{B}$ are given non-zero vectors. We wish to resolve $\vec{A}$ into a vector collinear with $\vec{B}$ and a vector orthogonal to $\vec{B}$. That is to say, we seek vectors $\vec{C}$ and $\vec{D}$ such that $\vec{B} \times \vec{C} = \vec{0}$, $\vec{C} \cdot \vec{D} = 0$, and $\vec{A} = \vec{C} + \vec{D}$. Find formulas of vector algebra for $\vec{C}$ and $\vec{D}$ in terms of $\vec{A}$ and $\vec{B}$. (*Hint.* Note that $\vec{C}$ must be a scalar multiple of $\vec{B}$.)

11. Let $\vec{A}$ and $\vec{B}$ be given non-zero vectors in $\mathbf{E}^3$, and let O be a fixed reference point. Describe the set of points determined by the equation $(\vec{R} - \vec{A}) \cdot (\vec{R} - \vec{B}) = 0$, where $\vec{R}$ is a variable position vector for a point.

12. Let two sides of a triangle be arrows representing vectors $\vec{A}$ and $\vec{B}$. Justify the vector expression in terms of $\vec{A}$ and $\vec{B}$ for the area of this triangle. (*Hint.* See (4.9).)

13. For $\vec{A} \ne 0$, show whether or not it is always true that $[\vec{A} \cdot \vec{B} = \vec{A} \cdot \vec{C}$ and $(\vec{A} \times \vec{B}) = (\vec{A} \times \vec{C})] \Rightarrow \vec{B} = \vec{C}$.

14. Let c be a non-zero scalar; let $\vec{A}$ be a non-zero vector; and let $\hat{u}$ be a unit vector such that $\vec{A} \cdot \hat{u} = 0$. Let O be a fixed reference point.

(a) Give a geometrical description of the set of all points whose position vectors satisfy the equation $\vec{R} \cdot \vec{A} = c$.

(b) Give a geometrical description of the set of all points whose position vectors satisfy the equation $\vec{R} \times \hat{u} = \vec{A}$.

15. Given an arbitrary 4-hedron, we associate with each of the four triangular faces a vector outwardly normal to that face with magnitude equal to the area of that face. Show that the sum of these four vectors is $\vec{0}$. (*Hint:* Choose a vertex and use vectors for the three edges from that vertex. Then use the vector formula for the area of a triangle; see (4.9).)

16. Let S be a regular octahedron. With each triangular face, associate an outward normal vector as in problem 15. Show that the sum of these eight vectors is $\vec{0}$. (*Hint.* Divide S and its interior into four adjacent but non-overlapping 4-hedrons. Add the vectors for the sixteen faces of these 4-hedrons, and apply problem 15. Note that each interface inside the original 8-hedron contributes two vectors of equal magnitudes but opposite directions.)

17. Prove: $\vec{A} \times \vec{B} = \vec{B} \times \vec{A} \Leftrightarrow \vec{A} \times \vec{B} = \vec{0}$.

§2.5

1. Give a full proof for each of the following laws from §2.4. For each law and each scalar coefficient, be sure to consider all three possibilities: that the coefficient be positive, be zero, or be negative.

    (a) Ia   (b) Ib
    (c) IIId.

2. Prove (IVc), the distributive law for the cross-product. Suggestion: Do so in three steps.

    (1) Prove for the case where $\vec{A}$, $\vec{B}$, or $\vec{C}$ is zero.

    (2) Prove for the case where $|\vec{A}| = 1$ and $\vec{B}$ and $\vec{C}$ are not zero.

(3) Prove for the general case where $\vec{A}$, $\vec{B}$, and $\vec{C}$ are any non-zero vectors.

(*Hint.* (1) is immediate. (3) follows from (2) if we multiply both sides of the law for $\hat{A}$ by $|\vec{A}|$. To prove (2), fill in details of the following. Show that an arrow for $\hat{A} \times \vec{B}$ can be obtained by projecting an arrow for $\vec{B}$ onto a plane perpendicular to $\hat{A}$ and then giving this projection a positive rotation in this plane (with respect to $\hat{A}$) of size $\pi/2$. By (4.3), arrows for $\vec{B}$ and $\vec{C}$ can either lie on a line or form a parallelogram with an arrow for $\vec{B}+\vec{C}$ as a diagonal. By (5.4) of Chapter 1, this line or parallelogram projects to a line or parallelogram in the plane perpendicular to $\hat{A}$, and this projected figure remains a line or parallelogram after the rotation. By (4.4), this gives (2).)

§2.7

1. Use vector algebra to prove that an angle inscribed in a semicircle must be a right angle. (*Hint*: See [8-2].)

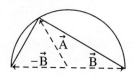

[8-2]

2. (a) Prove that the diagonals of a parallelogram bisect each other.
   (b) Prove that if the diagonals of a quadrilateral (in $\mathbf{E}^3$) bisect each other, then the quadrilateral is a parallelogram.

3. (a) Prove that the two lines drawn between midpoints of opposite sides of a parallelogram bisect each other.
   (b) Consider the statement: "If the two lines drawn between midpoints of opposite sides of a quadrilateral bisect each other, then the quadrilateral is a parallelogram." If this statement is always true, prove it. If not, give an example where it fails.

4. The *law of sines* of plane trigonometry asserts that, in any triangle,
$$\frac{\sin \alpha}{a} = \frac{\sin \beta}{b} = \frac{\sin \gamma}{c},$$
where a, b, and c are the lengths of the sides, and α, β, and γ are the sizes of the respectively opposite vertex angles. Give a vector proof of this law. (A vector proof of the law of cosines is given in §2.7.) (*Hint.* It is enough to prove the equation for a,b,α,β. Let P,Q,R be vertex points for angles α,β,γ. Let $\vec{A} = \overrightarrow{RQ}$, $\vec{B} = \overrightarrow{RP}$. Then $a = |\vec{A}|$, $b = |\vec{B}|$, $c = |\vec{A} - \vec{B}|$. Use appropriate cross products to get expressions for $\sin \alpha$ and $\sin \beta$.)

5. Prove that the three interior diagonals of a parallelepiped meet at a common point and bisect each other.

6. Show that the three line segments which join the midpoints of opposite edges of a 4-hedron meet at a common point and bisect each other.

7. Let θ be the angle between two given unit vectors $\hat{u}$ and $\hat{v}$.

   (a) Show that $\frac{1}{2}|\hat{u} - \hat{v}| = \sin \frac{\theta}{2}$.

   (b) By calculating $(\hat{u} - \hat{v})^2$, deduce the half-angle formula of elementary trigonometry:
$$\sin \frac{\theta}{2} = \left(\frac{1 - \cos \theta}{2} \right)^{1/2}.$$

8. Use vectors to prove that the segment joining the midpoints of two sides of a triangle must be parallel to the third side and have half the length of the third side.

Chapter 3

Vector Algebra with Cartesian Coordinates

§3.1   Cartesian Coordinates in $\mathbf{E}^3$

The reader may be familiar with the use of *Cartesian* (also called "rectangular") *coordinates* in $\mathbf{E}^2$. The use of Cartesian coordinates in $\mathbf{E}^3$ is closely analogous to their use in $\mathbf{E}^2$. First, a point is selected, denoted as O, and called the *origin*. Then three mutually perpendicular rays, R_x, R_y, and R_z are chosen with vertex O, [1–1]. These rays are called the *positive* x *axis*, *positive* y *axis*, and *positive* z *axis*. (It would be more accurate to speak of them as *non-negative* axes.) Let L_x be the line containing R_x. Each point P on L_x can be associated with a distinct real number r(P) by the rule:

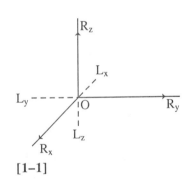

[1–1]

For P *in* R_x, r(P) = d(O,P); *and for* P *in* L_x *but not in* R_x, r(P) = –d(O,P).

(Here d is the measure of distance in $\mathbf{E}^3$.) Thus every point on L_x can be identified with a real number, and every real number becomes associated with some point on L_x. L_x is therefore called a *number-line* or *number-axis*. We refer to L_x as the x axis. The y *axis*, L_y, and the z *axis*, L_z, are defined from R_y and R_z in the same way.

We adopt a standard order in which we refer to the three axes: *first* x, *then* y, *then* z. We assume that the figure formed by the three rays R_x, R_y, and R_z is *right-handed*; that is to say, a right-hand placed at O, with fingers curling from R_x to R_y, will have its thumb pointing in the direction of R_z. R_x, R_y, and R_z are most commonly sketched as in [1–1].

Let P be any point in space. Take the projections of P on our chosen axes, [1–2]. The projection-points on the axes provide three real numbers x, y, and z, which we call the *coordinates* of P. It is clear from geometry that each point P in $\mathbf{E}^3$ now has a unique set of coordinates (x,y,z) and that each ordered triple of real numbers (x,y,z) now serves as the coordinates of a unique point P. (*Of course, a different choice of axes would provide a different assignment of coordinates to points in* $\mathbf{E}^3$.)

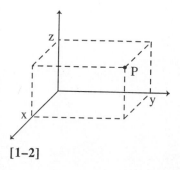

[1–2]

If a set of axes is chosen as described above, it is called a *Cartesian* (or *rectangular*) *coordinate system*. Cartesian systems were invented by the French mathematician and philosopher Descartes and are named after him. The x, y, and z axes are sometimes called *coordinate axes*. The plane determined by R_x and R_y is called the xy *plane*. Similarly, we have the yz *plane* and the xz *plane*. These three planes are called the *coordinate planes*.

Cartesian coordinate systems allow us to use the familiar algebra of real numbers in the study of geometric problems. This approach to geometry is known as *analytic geometry*. We consider analytic geometry in Chapter 4. In §3.2 below, we shall see how, in a chosen Cartesian system, we can assign coordinates to *vectors*. Then, in §3.3, we shall find *coordinate formulas* for the four basic vector operations. These formulas enable us to use the algebra of real numbers for vector calculations.

§3.2 Vectors in Cartesian Coordinates

Assume that we have chosen a Cartesian coordinate system in $\mathbf{E}^3$.

The three unit vectors in the directions of the positive x, y, and z axes are customarily expressed as $\hat{i}$, $\hat{j}$, and $\hat{k}$. (Possible arrows for $\hat{i}$, $\hat{j}$, and $\hat{k}$ are shown in the figure [2–1].) Let $\vec{A}$ be a given vector in $\mathbf{E}^3$. Find a point P such that $\overrightarrow{OP} = \vec{A}$, that is to say, a point P such that $\vec{A}$ serves as the position vector for P with respect to O. Let (a_1, a_2, a_3) be the coordinates of P. The coordinates of P are then also called the *coordinates of the vector* $\vec{A}$ in the chosen Cartesian system.

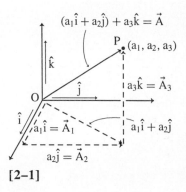

[2–1]

Now let $\vec{A}_1$, $\vec{A}_2$, and $\vec{A}_3$ be the vectors $\vec{A}_1 = a_1\hat{i}$, $\vec{A}_2 = a_2\hat{j}$ and $\vec{A}_3 = a_3\hat{k}$. $\vec{A}_1$, $\vec{A}_2$, and $\vec{A}_3$ are represented by the indicated arrows in [2–1]. By the definition of vector addition, we have $\vec{A} = \vec{A}_1 + \vec{A}_2 + \vec{A}_3$. Hence we have $\vec{A} = a_1\hat{i} + a_2\hat{j} + a_3\hat{k}$. *Thus, given a Cartesian system with unit vectors* $\hat{i}, \hat{j}, \hat{k}$ *and coordinates* a_1, a_2, a_3 *for a given vector* $\vec{A}$, *we can write*

(2.1) $$\vec{A} = a_1\hat{i} + a_2\hat{j} + a_3\hat{k}$$

as a vector-algebra expression for the vector $\vec{A}$.

This vector-algebra expression for $\vec{A}$ is called the $\hat{i}, \hat{j}, \hat{k}$ -*representation* of $\vec{A}$ in the chosen Cartesian system. ([2–1] is drawn for the case where a_1, a_2, and a_3 are all positive; the reader can easily verify that (2.1) holds for all vectors $\vec{A}$). If a_1, a_2, or a_3 is 0, we usually omit the corresponding term from (2.1). Thus if $\vec{A}$ has coordinates (2,0,–4), we write $\vec{A} = 2\hat{i} - 4\hat{k}$. In the special case $\vec{A} = \vec{0}$, we write the $\hat{i}, \hat{j}, \hat{k}$ -representation of $\vec{A}$ simply as $\vec{A} = \vec{0}$. The coordinates a_1, a_2, a_3 of a vector $\vec{A}$ are also called the *scalar components* of $\vec{A}$, and the vectors $a_1\hat{i}, a_2\hat{j}, a_3\hat{k}$ are called the *vector components* of $\vec{A}$, in the chosen Cartesian system.

Note from [2–1] that $a_1 = \vec{A} \cdot \hat{i}, a_2 = \vec{A} \cdot \hat{j},$ and $a_3 = \vec{A} \cdot \hat{k}$ by the definition of dot-product. Hence the $\hat{i}, \hat{j}, \hat{k}$ -representation can also be expressed as

$$(2.2) \qquad \vec{A} = (\vec{A} \cdot \hat{i})\hat{i} + (\vec{A} \cdot \hat{j})\hat{j} + (\vec{A} \cdot \hat{k})\hat{k} .$$

((2.1) can also be thought of as the resolution of $\vec{A}$ in the directions of $\hat{i}, \hat{j},$ and $\hat{k}$. See §2.6, and §3.4.)

§3.3 Coordinate Formulas for the Vector Operations

We use $\hat{i}, \hat{j}, \hat{k}$ -representations and the laws of vector algebra to obtain coordinate formulas for the four basic operations.

(I) *Multiplication by a scalar.* Let c be a given scalar and let $\vec{A} = a_1\hat{i} + a_2\hat{j} + a_3\hat{k}$. Then

$$c\vec{A} = c(a_1\hat{i} + a_2\hat{j} + a_3\hat{k}) .$$

Therefore, by laws Ib and IIf in §2.4,

$$(3.1) \qquad c\vec{A} = ca_1\hat{i} + ca_2\hat{j} + ca_3\hat{k} .$$

This is our desired coordinate formula. *To multiply a vector $\vec{A}$ by a scalar c, we need only multiply each scalar component by* c. For example, if $\vec{A} = 2\hat{i} + 0\hat{j} - 3\hat{k}$ (which we would usually write: $\vec{A} = 2\hat{i} - 3\hat{k}$), we can compute $-4\vec{A} = -8\hat{i} + 12\hat{k}$.

(II)   *Addition of vectors.*   Let $\vec{A} = a_1\hat{i} + a_2\hat{j} + a_3\hat{k}$ and $\vec{B} = b_1\hat{i} + b_2\hat{j} + b_3\hat{k}$. Then

$$\vec{A} + \vec{B} = (a_1\hat{i} + a_2\hat{j} + a_3\hat{k}) + (b_1\hat{i} + b_2\hat{j} + b_3\hat{k}).$$

By laws IIa and IIb,

$$\vec{A} + \vec{B} = a_1\hat{i} + b_1\hat{i} + a_2\hat{j} + b_2\hat{j} + a_3\hat{k} + b_3\hat{k}.$$

Therefore, by law IIe, we have

(3.2) $\vec{A} + \vec{B} = (a_1 + b_1)\hat{i} + (a_2 + b_2)\hat{j} + (a_3 + b_3)\hat{k}.$

This is our desired coordinate formula. *To add two vectors, we need only add corresponding components.* Thus if $\vec{A} = 2\hat{i} - 3\hat{k}$ and $\vec{B} = \hat{i} - 2\hat{j} + \hat{k}$ we get

$$\vec{A} + \vec{B} = (2 + 1)\hat{i} + (0 - 2)\hat{j} + (-3 + 1)\hat{k} = 3\hat{i} - 2\hat{j} - 2\hat{k},$$
and $\vec{A} - \vec{B} = (2 - 1)\hat{i} + (0 + 2)\hat{j} + (-3 - 1)\hat{k} = \hat{i} + 2\hat{j} - 4\hat{k}.$

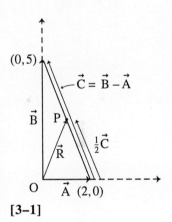

[3–1]

We give a simple geometric application. In a given Cartesian sysem, we wish to find the coordinates of the midpoint P of the line segment in E^2 whose endpoints are $(2,0)$ and $(0,5)$. We use arrows to define vectors as in [3–1]. The components of the position vector $\vec{R} = \overrightarrow{OP}$ will be the desired coordinates. From our geometric definition of vector addition, we see that $\vec{R} = \vec{A} + \frac{1}{2}\vec{C}$ and that $\vec{C} = \vec{B} - \vec{A}$. In our given Cartesian system, we have $\vec{A} = 2\hat{i}, \vec{B} = 5\hat{j}$, and $\vec{C} = -2\hat{i} + 5\hat{j}$. Hence $\frac{1}{2}\vec{C} = -\hat{i} + \frac{5}{2}\hat{j}$, and

$$\vec{R} = \vec{A} + \frac{1}{2}\vec{C} = 2\hat{i} - \hat{i} + \frac{5}{2}\hat{j} = \hat{i} + \frac{5}{2}\hat{j}.$$

Thus the coordinates of P are $(1, \frac{5}{2})$.

(III) *Dot-product of vectors.*   Let $\vec{A} = a_1\hat{i} + a_2\hat{j} + a_3\hat{k}$ and $\vec{B} = b_1\hat{i} + b_2\hat{j} + b_3\hat{k}$. Using the definition of dot-product, the reader may verify that

(3.3)
$$\begin{cases} \hat{i} \cdot \hat{i} = \hat{j} \cdot \hat{j} = \hat{k} \cdot \hat{k} = 1, \text{ and} \\ \hat{i} \cdot \hat{j} = \hat{j} \cdot \hat{i} = \hat{j} \cdot \hat{k} = \hat{k} \cdot \hat{j} = \hat{i} \cdot \hat{k} = \hat{k} \cdot \hat{i} = 0. \end{cases}$$

We have $\vec{A} \cdot \vec{B} = (a_1\hat{i} + a_2\hat{j} + a_3\hat{k}) \cdot (b_1\hat{i} + b_2\hat{j} + b_3\hat{k})$. By laws IIIb, IIId, and Ib, we have

$$\vec{A} \cdot \vec{B} = a_1\hat{i} \cdot (b_1\hat{i} + b_2\hat{j} + b_3\hat{k}) + a_2\hat{j} \cdot (b_1\hat{i} + b_2\hat{j} + b_3\hat{k})$$
$$+ a_3\hat{k} \cdot (b_1\hat{i} + b_2\hat{j} + b_3\hat{k})$$
$$= a_1b_1(\hat{i} \cdot \hat{i}) + a_1b_2(\hat{i} \cdot \hat{j}) + a_1b_3(\hat{i} \cdot \hat{k}) + a_2b_1(\hat{j} \cdot \hat{i}) + a_2b_2(\hat{j} \cdot \hat{j})$$
$$+ a_2b_3(\hat{j} \cdot \hat{k}) + a_3b_1(\hat{k} \cdot \hat{i}) + a_3b_2(\hat{k} \cdot \hat{j}) + a_3b_3(\hat{k} \cdot \hat{k}).$$

By (3.3), this becomes

(3.4)
$$\vec{A} \cdot \vec{B} = a_1b_1 + a_2b_2 + a_3b_3.$$

This is our desired coordinate formula. *To take the dot-product of two vectors, we need only multiply corresponding scalar components and then take the sum of these products.* For example, if $\vec{A} = 2\hat{i} - 3\hat{k}$ and $\vec{B} = \hat{i} - 2\hat{j} + \hat{k}$, we get

$$\vec{A} \cdot \vec{B} = (2 \cdot 1) + (0 \cdot -2) + (-3 \cdot 1) = 2 - 3 = -1.$$

Here are a few simple applications. For each of these examples, as for almost all later examples in this book, the reader should try to find his or her own solution to the proposed problem before reading the solution given in the example.

Example 1. Let $\vec{B} = \hat{i} - 2\hat{j} + \hat{k}$. *Find* $|\vec{B}|$, *and then find* $\hat{B}$, *the unit vector in the direction of* $\vec{B}$. We get $|\vec{B}| = \sqrt{\vec{B} \cdot \vec{B}} = \sqrt{1 + 4 + 1} = \sqrt{6}$, and

$$\hat{B} = \frac{1}{|\vec{B}|}\vec{B} = \frac{1}{\sqrt{6}}(\hat{i} - 2\hat{j} + \hat{k}) = \frac{\sqrt{6}}{6}\hat{i} - \frac{\sqrt{6}}{3}\hat{j} + \frac{\sqrt{6}}{6}\hat{k}.$$

Example 2. Let P_1 and P_2 be points with coordinates (x_1, y_1, z_1) and (x_2, y_2, z_2) in a given Cartesian system. *Find the* $\hat{i}, \hat{j}, \hat{k}$-*representation for the vector given by* $\overrightarrow{P_1P_2}$. From [3-2], $\overrightarrow{P_1P_2} = \overrightarrow{OP_2} - \overrightarrow{OP_1} = (x_2\hat{i} + y_2\hat{j} + z_2\hat{k}) - (x_1\hat{i} + y_1\hat{j} + z_1\hat{k})$. Hence

(3.5)
$$\overrightarrow{P_1P_2} = (x_2 - x_1)\hat{i} + (y_2 - y_1)\hat{j} + (z_2 - z_1)\hat{k}.$$

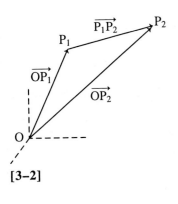

[3-2]

Example 3. *For* P_1 *and* P_2 *as in Example 2, find a formula for* $d(P_1,P_2) =$ *distance from* P_1 *to* P_2. We get $d(P_1,P_2) = \left|\overrightarrow{P_1P_2}\right| = \sqrt{(\overrightarrow{P_1P_2})^2}$. Thus

(3.6) $\qquad d(P_1,P_2) = \sqrt{(x_2 - x_1)^2 + (y_2 - y_1)^2 + (z_2 - z_1)^2}$.

Example 4. *Find the angle between the vectors* $\vec{A} = 2\hat{i} - 3\hat{k}$ *and* $\vec{B} = \hat{i} - 2\hat{j} + \hat{k}$.

We know that $\vec{A} \cdot \vec{B} = |\vec{A}|\,\|\vec{B}|\cos\theta$.

We have $|\vec{A}| = \sqrt{\vec{A}^2} = \sqrt{4+9} = \sqrt{13}$ and $|\vec{B}| = \sqrt{\vec{B}^2} = \sqrt{1+4+1} = \sqrt{6}$.

By (3.4), we have $\vec{A} \cdot \vec{B} = 2 - 3 = -1$.

Thus $-1 = \sqrt{13}\sqrt{6}\cos\theta, \cos\theta = -\dfrac{1}{\sqrt{78}} = -0.113$, and $\theta = \cos^{-1}(-0.113) = 96.5°$.

(IV) **Cross-product of vectors.** Let $\vec{A} = a_1\hat{i} + a_2\hat{j} + a_3\hat{k}$ and $\vec{B} = b_1\hat{i} + b_2\hat{j} + b_3\hat{k}$. From the definition of cross-product, the reader may verify that

(3.7) $\qquad \begin{cases} \hat{i}\times\hat{i} = \hat{j}\times\hat{j} = \hat{k}\times\hat{k} = \vec{0}, \text{ and} \\ \hat{i}\times\hat{j} = -\hat{j}\times\hat{i} = \hat{k},\ \hat{j}\times\hat{k} = -\hat{k}\times\hat{j} = \hat{i},\ \hat{k}\times\hat{i} = -\hat{i}\times\hat{k} = \hat{j}. \end{cases}$

We have $\vec{A} \times \vec{B} = (a_1\hat{i} + a_2\hat{j} + a_3\hat{k})\times(b_1\hat{i} + b_2\hat{j} + b_3\hat{k})$. The calculation is similar to that for the dot product. Using laws IVc, IVd, and IVf, and then applying (3.7) we get

(3.8a) $\qquad \vec{A} \times \vec{B} = (a_2b_3 - a_3b_2)\hat{i} + (a_3b_1 - a_1b_3)\hat{j} + (a_1b_2 - a_2b_1)\hat{k}$.

Determinant notation provides us with a helpful memory rule for this coordinate formula. We have

(3.8b) $\qquad \vec{A} \times \vec{B} = \begin{vmatrix} \hat{i} & \hat{j} & \hat{k} \\ a_1 & a_2 & a_3 \\ b_1 & b_2 & b_3 \end{vmatrix}$.

(We assume that the reader has some knowledge of 2×2 and 3×3 determinants. Their definitions, basic properties, and evaluation procedures are reviewed in §3.7 below.) Thus if $\vec{A} = 2\hat{i} - 3\hat{k}$ and $\vec{B} = \hat{i} - 2\hat{j} + \hat{k}$, we get

$$\vec{A} \times \vec{B} = \begin{vmatrix} \hat{i} & \hat{j} & \hat{k} \\ 2 & 0 & -3 \\ 1 & -2 & 1 \end{vmatrix} = (0 \cdot 1 - (-2 \cdot -3))\hat{i} + (-3 \cdot 1 - 2 \cdot 1) \cdot \hat{j} + (2 \cdot -2 - 0 \cdot 1)\hat{k}$$

$$= -6\hat{i} - 5\hat{j} - 4\hat{k}.$$

(Recall, from the definition of cross-product, that $\vec{A} \times \vec{B}$ must be orthogonal to $\vec{A}$ and also to $\vec{B}$. We can use this fact as a check on our calculation of $\vec{A} \times \vec{B}$:

we have $(\vec{A} \times \vec{B}) \cdot \vec{A} = (-6\hat{i} - 5\hat{j} - 4\hat{k}) \cdot (2\hat{i} - 3\hat{k}) = -12 + 12 = 0,$

and $(\vec{A} \times \vec{B}) \cdot \vec{B} = (-6\hat{i} - 5\hat{j} - 4\hat{k}) \cdot (\hat{i} - 2\hat{j} + \hat{k}) = -6 + 10 - 4 = 0,$

as expected.)

§3.4 Frames and the Frame Identity

Definition. A set of three vectors $\{\vec{A}, \vec{B}, \vec{C}\}$ in $\mathbf{E}^3$ is said to be a *frame* if: (i) each vector is a unit vector, and (ii) the three vectors are mutually orthogonal, [4–1]. In other words, $\{\vec{A}, \vec{B}, \vec{C}\}$ is a frame if (i) $\vec{A}^2 = \vec{B}^2 = \vec{C}^2 = 1$ and (ii) $\vec{A} \cdot \vec{B} = \vec{B} \cdot \vec{C} = \vec{C} \cdot \vec{A} = 0$.

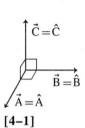

[4–1]

For example, in a given Cartesian system, the set $\{\hat{i}, \hat{j}, \hat{k}\}$ is a frame. So is the set $\left\{ \dfrac{\hat{i} + \hat{j}}{\sqrt{2}}, \dfrac{\hat{i} - \hat{j} + \hat{k}}{\sqrt{3}}, \dfrac{\hat{i} - \hat{j} - 2\hat{k}}{\sqrt{6}} \right\}$, as the reader may verify.

The following fact about frames is fundamental: *Let* $\{\hat{u}, \hat{v}, \hat{w}\}$ *be a frame, and let* $\vec{G}$ *be a given vector. Then* (i) *there exist unique constants* a, b, *and* c *such that* $\vec{G} = a\hat{u} + b\hat{v} + c\hat{w}$, *and* (ii) *these constants are given by* $a = \vec{G} \cdot \hat{u}$, $b = \vec{G} \cdot \hat{v}$, $c = \vec{G} \cdot \hat{w}$. In §3.2, we saw that this is true for the frame $\{\hat{i}, \hat{j}, \hat{k}\}$. It holds for any frame $\{\hat{u}, \hat{v}, \hat{w}\}$ by the same argument. We summarize this fact as follows.

(4.1)   ***The frame identity.*** *For any vector* $\vec{G}$ *and any frame* $\{\hat{u}, \hat{v}, \hat{w}\}$,

$$\vec{G} = (\vec{G} \cdot \hat{u})\hat{u} + (\vec{G} \cdot \hat{v})\hat{v} + (\vec{G} \cdot \hat{w})\hat{w}.$$

For example, if we wish to express the vector $\vec{G} = 2\hat{i}$ in terms of the frame $\hat{u} = \frac{1}{\sqrt{2}}(\hat{i} + \hat{j})$, $\hat{v} = \frac{1}{\sqrt{3}}(\hat{i} - \hat{j} + \hat{k})$, $\hat{w} = \frac{1}{\sqrt{6}}(\hat{i} - \hat{j} - 2\hat{k})$, we immediately have $\vec{G} \cdot \hat{u} = \sqrt{2}$, $\vec{G} \cdot \hat{v} = \frac{2\sqrt{3}}{3}$, $\vec{G} \cdot \hat{w} = \frac{\sqrt{6}}{3}$. Hence we have $2\hat{i} =$

$$\sqrt{2}\left(\frac{1}{\sqrt{2}}(\hat{i} + \hat{j})\right) + \frac{2\sqrt{3}}{3}\left(\frac{1}{\sqrt{3}}(\hat{i} - \hat{j} + \hat{k})\right) + \frac{\sqrt{6}}{3}\left(\frac{1}{\sqrt{6}}(\hat{i} - \hat{j} - 2\hat{k})\right).$$

What if we have a set of three vectors $\{\vec{A}, \vec{B}, \vec{C}\}$ which is not from a frame? Can we still express an arbitrarily given vector in terms of $\{\vec{A}, \vec{B}, \vec{C}\}$? The following *basis theorem for* $\mathbf{E}^3$ answers this question.

(4.2)   *Let $\{\vec{A}, \vec{B}, \vec{C}\}$ be a given non-coplanar set of non-zero vectors, and let $\vec{G}$ be any vector. Then there exist unique scalars* a, b, *and* c *such that*

$$\vec{G} = a\vec{A} + b\vec{B} + c\vec{C}.$$

This theorem is proved by the simple geometric construction indicated in [4–2]. We represent $\{\vec{A}, \vec{B}, \vec{C}\}$ and $\vec{G}$ as arrows from a common initial point P. Let Q be the final point of the arrow $\overrightarrow{PQ}$ which represents $\vec{G}$. Let L_A be the line of the arrow for $\vec{A}$. Let M_A be the unique plane through Q parallel to the plane of the arrows for $\vec{B}$ and $\vec{C}$. Let S_A be the point where M_A intersects L_A. Then $\overrightarrow{PS}_A$ is collinear with $\vec{A}$, and hence there is a scalar a such that $\overrightarrow{PS}_A = a\vec{A}$. Similarly, we use a plane through Q parallel to $\vec{A}$ and $\vec{C}$, and a plane through Q parallel to $\vec{A}$ and $\vec{B}$, to find S_B and S_C such that $\overrightarrow{PS}_B = b\vec{B}$ and $\overrightarrow{PS}_C = c\vec{C}$ for scalars b and c. We see from the parallelepiped in the figure that $\vec{G} = \overrightarrow{PQ} = \overrightarrow{PS}_A + \overrightarrow{PS}_B + \overrightarrow{PS}_C = a\vec{A} + b\vec{B} + c\vec{C}$.

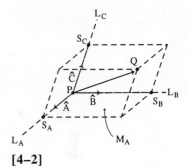

[4–2]

Can we find formulas of vector algebra which give a, b, and c (in (4.2)) in terms of $\vec{G}$, $\vec{A}$, $\vec{B}$ and $\vec{C}$? (The frame identity (4.1) no longer holds, since $\vec{A}$, $\vec{B}$, and $\vec{C}$ are not necessarily orthogonal or unit vectors.) In problem 14 for §3.6, we find these formulas.

In §2.6, we considered the *resolution* of a given force, acting at point P in $\mathbf{E}^2$, into two forces acting along two arbitrarily given distinct lines through P. The basis theorem for $\mathbf{E}^3$ (4.2), together with the principle of composition of forces, implies a similar *principle of resolution for* $\mathbf{E}^3$: *Given any three non-coplanar lines through a point* P, *then the physical effect of any* *given force $\bar{G}$ acting at P is the same as the combined effect would be, in the absence of $\bar{G}$, of three appropriately chosen forces acting along those lines at P.* We say that the given force is *resolved* into the chosen forces.

§3.5 Applications

Here are three elementary examples of the use of vector algebra for geometric and physical problems.

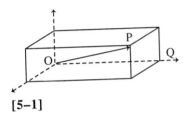

[5–1]

Example 1.   Trigonometry. A rectangular block is four feet long, one foot wide and one foot deep.   Find the angle between a 4-foot edge and an adjacent interior diagonal.

We begin by introducing Cartesian axes as in [5–1]. We seek the angle between $\overrightarrow{OP} = \hat{i} + 4\hat{j} + \hat{k}$ and $\overrightarrow{OQ} = 4\hat{j}$.   We have $\overrightarrow{OP} \cdot \overrightarrow{OQ} = |\overrightarrow{OP}|\,|\overrightarrow{OQ}|\cos\theta$, where θ is the desired angle. Thus

$$\cos\theta = \frac{\overrightarrow{OP} \cdot \overrightarrow{OQ}}{|\overrightarrow{OP}|\,|\overrightarrow{OQ}|} = \frac{16}{4\sqrt{18}} = \frac{2}{3}\sqrt{2}$$

and

$$\theta = \cos^{-1}\left(\frac{2\sqrt{2}}{3}\right) = 19.5°.$$

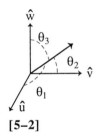

[5–2]

Example 2. Geometry. Let $\bar{G}$ be a given vector and let $\{\hat{u}, \hat{v}, \hat{w}\}$ be a given frame. Let θ_1, θ_2, and θ_3 be the angles that $\bar{G}$ makes, respectively, with $\{\hat{u}, \hat{v}, \hat{w}\}$, [5–2]. The quantities $\cos\theta_1$, $\cos\theta_2$, and $\cos\theta_3$ can be used to specify the direction of $\bar{G}$ with respect to the frame $\{\hat{u}, \hat{v}, \hat{w}\}$.   They are called the *direction cosines* of $\bar{G}$ for the frame.   They can be calculated as $\cos\theta_1 = \hat{G} \cdot \hat{u}$, $\cos\theta_2 = \hat{G} \cdot \hat{v}$, and $\cos\theta_3 = \hat{G} \cdot \hat{w}$, where $\hat{G}$ is the unit vector for $\bar{G}$. In other words, the direction cosines of $\bar{G}$ in a given frame are simply the scalar components of the unit vector $\hat{G}$ in that frame. Thus, for example, the direction cosines of $\bar{G} = \hat{i} - 2\hat{j} + \hat{k}$ in the frame $\{\hat{i}, \hat{j}, \hat{k}\}$ must be $\frac{1}{\sqrt{6}} = \frac{\sqrt{6}}{6}$,

$-\frac{2}{\sqrt{6}} = -\frac{\sqrt{6}}{3}$, and $\frac{1}{\sqrt{6}} = \frac{\sqrt{6}}{6}$.

Example 3. Physics. A solid, rigid object is rotating with an angular velocity of ω radians/sec ($\omega > 0$) about an axis of rotation L through a point O.

Let the unit vector û give a direction along L *such that the rotation is right-handed with respect to û . As previously described in example 1 in §2.2, the vector* $\vec{A} = \omega\hat{u}$ *is called the* **angular velocity** *of the rotating object. At a given moment, a certain particle* P *in the solid object has position vector* $\vec{R}$ *with respect to* O. *What is the velocity of* P *at that moment?*

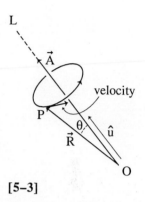

[5–3]

From [5–3], we see that P travels around a circle of radius $|\vec{R}|\sin\theta$ lying in the plane through P perpendicular to L. Hence the velocity of P must have its magnitude = $\omega|\vec{R}|\sin\theta$ and its direction tangent to this circle. It follows from the definition of cross-product that this velocity is $\vec{A} \times \vec{R}$. For example, if, at a given moment, $\omega = 6$, $\hat{u} = \hat{k}$ and $\vec{R} = 2\hat{i} + \hat{j} + 2\hat{k}$, we find that the velocity of P is

$$6\hat{k} \times (2\hat{i} + \hat{j} + 2\hat{k}) = 12\hat{j} - 6\hat{i} = -6\hat{i} + 12\hat{j}.$$

§3.6 Triple Products

Operations of dot-product and cross-product may be repeated or combined in a single formula. Two such combinations are especially useful and important. They are the *scalar triple product* and the *vector triple product*.

Scalar triple product. This arises when we take a cross-product of two given vectors and then combine the resulting vector with a third given vector by a dot-product; for example, $(\vec{A} \times \vec{B}) \cdot \vec{C}$. The final result of this combined operation must be a *scalar*; hence the name "scalar triple product."
Calculation with scalar triple products is simplified by theorems (6.1), (6.2), and (6.3).

(6.1)   *Theorem.* $(\vec{A} \times \vec{B}) \cdot \vec{C} = \vec{A} \cdot (\vec{B} \times \vec{C})$.

(6.1) tells us that the *cross* and the *dot* in a scalar triple product can be interchanged without changing the value of the product. For this reason, the scalar triple product $(\vec{A} \times \vec{B}) \cdot \vec{C}$ is sometimes written, simply, as $[\vec{A}, \vec{B}, \vec{C}]$ without specifying particular positions for the cross and the dot.

(6.2)   *Theorem.* $[\vec{A}, \vec{B}, \vec{C}] = [\vec{B}, \vec{C}, \vec{A}] = [\vec{C}, \vec{A}, \vec{B}] = -[\vec{A}, \vec{C}, \vec{B}] = -[\vec{C}, \vec{B}, \vec{A}]$
$$= -[\vec{B}, \vec{A}, \vec{C}].$$

(6.3)   **Theorem.** *Let* $\vec{A}=a_1\hat{i}+a_2\hat{j}+a_3\hat{k}, \vec{B}=b_1\hat{i}+b_2\hat{j}+b_3\hat{k},$ *and* $\vec{C}=c_1\hat{i}+c_2\hat{j}+c_3\hat{k}.$

Then
$$[\vec{A},\vec{B},\vec{C}] = \begin{vmatrix} a_1 & a_2 & a_3 \\ b_1 & b_2 & b_3 \\ c_1 & c_2 & c_3 \end{vmatrix}.$$

Proofs. (6.1) and (6.3) can be proved by calculating the values of $(\vec{A}\times\vec{B})\cdot\vec{C}$, of $\vec{A}\cdot(\vec{B}\times\vec{C})$, and of the determinant in (6.3). (6.2) then follows by property (7.5a) of determinants in §3.7 below.

The scalar triple product has the following geometrical significance:

(6.4)   *Let given vectors $\vec{A}$, $\vec{B}$ and $\vec{C}$ be represented by arrows $\overrightarrow{PQ}_1$, $\overrightarrow{PQ}_2$, and $\overrightarrow{PQ}_3$ from a common initial point. Then the volume of the parallelepiped with adjacent edges $\overrightarrow{PQ}_1$, $\overrightarrow{PQ}_2$, and $\overrightarrow{PQ}_3$ is given by $\left|[\vec{A},\vec{B},\vec{C}]\right|$, (that is to say, by the absolute value of $[\vec{A},\vec{B},\vec{C}]$.)*

To prove (6.4), we note that $\left|\vec{A}\times\vec{B}\right|$ is the area of the face determined by $\overrightarrow{PQ}_1$ and $\overrightarrow{PQ}_2$, and that $\left|\left|\vec{C}\right|\cos\theta\right|$ is the distance from this face to its opposite parallel face, where θ is the angle between $\vec{C}$ and the perpendicular line L given by $\vec{A}\times\vec{B}$, [6–1]. By §1.6, the volume is

$$\text{(area of base)(altitude)} = \left|\vec{A}\times\vec{B}\right|\left|\left|\vec{C}\right|\cos\theta\right| = \left|(\vec{A}\times\vec{B})\cdot\vec{C}\right| = \left|[\vec{A},\vec{B},\vec{C}]\right|.$$

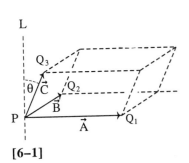

[6–1]

(6.4) gives, as a corollary, a useful test for deciding whether three given vectors are coplanar:

(6.5)   *Let $\vec{A}$, $\vec{B}$, and $\vec{C}$ be three given non-zero vectors. Then $\vec{A}$, $\vec{B}$, and $\vec{C}$ are coplanar $\Leftrightarrow$ $[\vec{A},\vec{B},\vec{C}] = 0$.*

Vector triple product. This arises when we take a cross-product of two given vectors and then take the cross-product of the resulting vector with a third given vector; for example, $(\vec{A}\times\vec{B})\times\vec{C}$. The final result of this combined operation must be a *vector*; hence the name "vector triple product."

Since the cross-product operation is not associative, the products $(\vec{A}\times\vec{B})\times\vec{C}$ and $\vec{A}\times(\vec{B}\times\vec{C})$ usually result in different vectors. (For more information, see problem 20 for §3.6.) Note that $(\vec{A}\times\vec{B})\times\vec{C}$ must be coplanar with

$\vec{A}$ and $\vec{B}$, since it is perpendicular to $\vec{A} \times \vec{B}$, and $\vec{A} \times \vec{B}$, in turn, is perpendicular to both $\vec{A}$ and $\vec{B}$. Similarly, $\vec{A} \times (\vec{B} \times \vec{C})$ must be coplanar with $\vec{B}$ and $\vec{C}$. The following theorem gives vector-algebra formulas for evaluating these products.

(6.6) **Theorem.** $(\vec{A} \times \vec{B}) \times \vec{C} = (\vec{C} \cdot \vec{A})\vec{B} - (\vec{C} \cdot \vec{B})\vec{A}$,

$\vec{A} \times (\vec{B} \times \vec{C}) = (\vec{A} \cdot \vec{C})\vec{B} - (\vec{A} \cdot \vec{B})\vec{C}$.

These formulas may be verified by direct calculation. See problem 10 for §3.6. The formulas in (6.6) should be memorized for future use. A convenient memory-rule for recalling these formulas is

$$(\overset{\rightarrow}{\text{separate}} \cdot \overset{\rightarrow}{\text{further}}) \overset{\rightarrow}{\text{nearer}} - (\overset{\rightarrow}{\text{separate}} \cdot \overset{\rightarrow}{\text{nearer}}) \overset{\rightarrow}{\text{further}} .$$

The vector triple product appears in physical problems of rotational motion and fluid flow. It also often occurs in general vector algebra calculations.

Quadruple products. The expression $(\vec{A} \times \vec{B}) \cdot (\vec{C} \times \vec{D})$ is called a *quadruple scalar product*, and the expression $(\vec{A} \times \vec{B}) \times (\vec{C} \times \vec{D})$ is called a *quadruple vector product*. The reader need not memorize formulas for evaluating these products or for evaluating more complex "higher-order" products. Such expressions can always be simplified, by the use of (6.1), (6.2), and (6.6), to expressions in which scalar triple products and ordinary dot- and cross-products are combined under the operations of vector addition, of multiplication by a scalar, and of elementary algebra.

For example, in the case of the quadruple scalar product, we have

$$(\vec{A} \times \vec{B}) \cdot (\vec{C} \times \vec{D}) = ((\vec{A} \times \vec{B}) \times \vec{C}) \cdot \vec{D} \qquad \text{(by 6.1)}$$

$$= ((\vec{C} \cdot \vec{A})\vec{B} - (\vec{C} \cdot \vec{B})\vec{A}) \cdot \vec{D} \qquad \text{(by 6.6)}$$

$$= (\vec{C} \cdot \vec{A})(\vec{B} \cdot \vec{D}) - (\vec{C} \cdot \vec{B})(\vec{A} \cdot \vec{D}).$$

Similarly, in the case of the quadruple vector product, we have

$$(\vec{A} \times \vec{B}) \times (\vec{C} \times \vec{D}) = ((\vec{C} \times \vec{D}) \cdot \vec{A})\vec{B}) - ((\vec{C} \times \vec{D}) \cdot \vec{B})\vec{A} \qquad \text{(by 6.6)}$$

$$= [\vec{C}, \vec{D}, \vec{A}]\vec{B} - [\vec{C}, \vec{D}, \vec{B}]\vec{A} .$$

§3.7 Review of 2×2 and 3×3 Determinants; Cramer's Rule

For this chapter, we have assumed that the reader is familiar with 2×2 and 3×3 determinants. We review the basic facts about 2×2 and 3×3 determinants in the present section. (In Chapter 31, we consider determinants of size n×n, for arbitrary $n > 0$.)

The expression $\begin{vmatrix} a & b \\ c & d \end{vmatrix}$, where a,b,c,d are real numbers, is *called a determinant with 2 rows and 2 columns* or, more briefly, a 2×2 ("2 by 2") or *second-order* determinant. a,b,c,d are called *elements* of the determinant. The determinant has a numerical value, namely,

(7.1)
$$\begin{vmatrix} a & b \\ c & d \end{vmatrix} = ad - bc.$$

It is easy to remember this fact by the pattern $\begin{vmatrix} a & b \\ c & d \end{vmatrix}$, where the product along the diagonal \ is counted positively, and the product along the diagonal / is counted negatively.

Example 1. $\begin{vmatrix} 1 & -4 \\ -3 & 2 \end{vmatrix} = 1 \cdot 2 - (-3)(-4) = -10.$

The 2×2 determinant has the following geometrical significance in $\mathbf{E}^2$-with-coordinates.

(7.2)   *The absolute value of the determinant* $\begin{vmatrix} a & b \\ c & d \end{vmatrix}$ *is the area of the parallelogram that has the position-vector arrows for* $a\hat{i} + b\hat{j}$ *and* $c\hat{i} + d\hat{j}$ *as adjacent sides.*

To prove this, introduce a z axis, recall the geometric significance of the magnitude of a cross product (§2.4), and calculate $(a\hat{i} + b\hat{j}) \times (c\hat{i} + d\hat{j})$.

The expression $\begin{vmatrix} a & b & c \\ d & e & f \\ g & h & i \end{vmatrix}$ is called a 3×3 (or third-order) determinant. It has the numerical value given by the formula:

(7.3)
$$\begin{vmatrix} a & b & c \\ d & e & f \\ g & h & i \end{vmatrix} = aei + bfg + cdh - ceg - afh - bdi .$$

We can remember this fact by the pattern

where the first and second columns have repeated and where the product along each of the three diagonals \ is counted positively and along each of the three diagonals / is counted negatively.

Example 2.
$$\begin{vmatrix} 1 & 2 & -1 \\ 0 & -1 & 3 \\ 2 & 1 & 1 \end{vmatrix} = \begin{vmatrix} 1 & 2 & -1 \\ 0 & -1 & 3 \\ 2 & 1 & 1 \end{vmatrix} \begin{matrix} 1 & 2 \\ 0 & -1 \\ 2 & 1 \end{matrix} = -1 + 12 - 2 - 3 = 6.$$

(7.4)   *The determinant* $\begin{vmatrix} a & b & c \\ d & e & f \\ g & h & i \end{vmatrix}$ *has* (by (6.4)) *the following geometrical*

significance in $\mathbf{E}^3$-*with-coordinates: The determinant's absolute value is the volume of the parallelepiped that has the position-vector arrows:* $a\hat{i} + b\hat{j} + c\hat{k}$, $d\hat{i} + e\hat{j} + f\hat{k}$ *and* $g\hat{i} + h\hat{j} + i\hat{k}$ *as adjacent edges.*

Any 3×3 determinant can be expressed in terms of 2×2 determinants as follows, as the reader may verify.

$$\begin{vmatrix} a & b & c \\ d & e & f \\ g & h & i \end{vmatrix} = a\begin{vmatrix} e & f \\ h & i \end{vmatrix} - b\begin{vmatrix} d & f \\ g & i \end{vmatrix} + c\begin{vmatrix} d & e \\ g & h \end{vmatrix}.$$

The coefficients are the elements in the first row. The 2×2 determinant that appears with a given first row element is called the *minor* of that element and is found by deleting the row and column of that element from the original 3×3 determinant. Finally, we see that + and − alternate. The expression above, in terms of elements and their minors, is called a *Laplace expansion* of the determinant.

In example 2 above, we get

$$1\begin{vmatrix} -1 & 3 \\ 1 & 1 \end{vmatrix} - 2\begin{vmatrix} 0 & 3 \\ 2 & 1 \end{vmatrix} + (-1)\begin{vmatrix} 0 & -1 \\ 2 & 1 \end{vmatrix} = -4 + 12 - 2 = 6.$$

A similar expression using the first column also holds. In fact, any column or row can be used to make a Laplace expansion, with the provision that when the *second row* or *second column* is used, the alternating signs (in the general expansion formula) must begin with −. Thus, using the second row in example 2, we get:

$$-0\begin{vmatrix} 2 & -1 \\ 1 & 1 \end{vmatrix} + (-1)\begin{vmatrix} 1 & -1 \\ 2 & 1 \end{vmatrix} - 3\begin{vmatrix} 1 & 2 \\ 2 & 1 \end{vmatrix} = 0 - 3 + 9 = 6.$$

(The assertions above are special cases (for 3×3 determinants) of the *fundamental theorem on Laplace expansions* given in Chapter 31.)

The determinant concept can be viewed as a special *procedure* (or *function*) which, given a square array of numbers as *input*, produces a single number as *output*. Both as procedure and as notation, the determinant concept is useful in many areas of mathematics and physics.

Important properties of determinants include the following:

(7.5a) If we alter a determinant by interchanging two rows, this multiplies the determinant's value by −1.

(7.5b) If two rows are identical, a determinant's value is 0.

(7.5c) If we alter a determinant by multiplying each element in some chosen row by the same constant, this multiplies the value of the determinant by that constant.

(7.5d) If we alter one row in a determinant by adding to it some multiple of another row, this leaves the determinant's value unchanged.

(7.5e) A determinant's value = 0 ⇔ there is some row that can be expressed as a linear combination of the other rows. (We say that *row* $[a_1,a_2,a_3]$ *can be expressed as a linear combination of rows* $[b_1,b_2,b_3]$ *and* $[c_1,c_2,c_3]$ *if* there are constants α and β such that

$$a_1 = \alpha b_1 + \beta c_1,$$
$$a_2 = \alpha b_2 + \beta c_2,$$
$$a_3 = b_3 + \beta c_3.)$$

(7.5f) Statements (a)–(e) all hold with "column" in place of "row."

Statements (a)-(d), in both *row* and *column* form, are immediate from the defining formulas for 2×2 and 3×3 determinants. In (e), the result $\Leftarrow$ follows from (d), but the result $\Rightarrow$ is less obvious. It is proved in Chapter 31 (§31.7), where we also see that the six properties above hold for n×n determinants, for any $n > 0$.

Cramer's rule. If a set of two simultaneous linear equations in two unknowns has a unique solution, then 2×2 determinants can be used to give standard formulas for the solution values in terms of the coefficients appearing in the given equations. In particular:

(7.6)   *If the given system is*

$$\begin{pmatrix} ax + by = e \\ cx + dy = f \end{pmatrix},$$

then the system has a unique solution if and only if

$$\begin{vmatrix} a & b \\ c & d \end{vmatrix} \neq 0,$$

and this unique solution is given by the formulas

$$x = \frac{1}{D}\begin{vmatrix} e & b \\ f & d \end{vmatrix} \quad and \quad y = \frac{1}{D}\begin{vmatrix} a & e \\ c & f \end{vmatrix}, \ where \ \ D = \begin{vmatrix} a & b \\ c & d \end{vmatrix}.$$

This fact and these formulas are known as *Cramer's rule* for a system of two linear equations in two unknowns. (Note that D is formed from the coefficients for x and y; that in the numerator of the formula for x, the column $\begin{pmatrix} a \\ c \end{pmatrix}$ of

coefficients for x (in D) is replaced by the column $\begin{pmatrix} e \\ f \end{pmatrix}$ of constant terms; and

that in the numerator of the formula for y, the column of coefficients for y (in D) is replaced by the column of constant terms.)

Example 3. *Given the system* $\begin{pmatrix} 2x - 4y = 3 \\ x + 3y = -1 \end{pmatrix}$, we get $D = \begin{vmatrix} 2 & -4 \\ 1 & 3 \end{vmatrix} =$

10, $x = \frac{1}{10}\begin{vmatrix} 3 & -4 \\ -1 & 3 \end{vmatrix} = \frac{5}{10} = \frac{1}{2}$, and $y = \frac{1}{10}\begin{vmatrix} 2 & 3 \\ 1 & -1 \end{vmatrix} = \frac{-5}{10} = -\frac{1}{2}$.

Cramer's rule for a system of three equations in three unknowns uses 3×3 determinants and is exactly analogous. Given the system $\begin{pmatrix} ax + by + cz = j \\ dx + ey + fz = k \\ gx + hy + iz = \ell \end{pmatrix}$, we

take $D = \begin{vmatrix} a & b & c \\ d & e & f \\ g & h & i \end{vmatrix}$. Then the system has a unique solution if and only if $D \neq 0$, and, for example, we find

$$y = \frac{1}{D} \begin{vmatrix} a & j & c \\ d & k & f \\ g & \ell & i \end{vmatrix}.$$

In chapter 31, Cramer's rule is formulated and proved for systems of n simultaneous linear equations in n unknowns, for any n > 0. The general formulation of Cramer's rule is analogous to the formulations above.

§3.8 Problems

§3.3

1. Let $\vec{A} = 3\hat{i} - 4\hat{k}$, $\vec{B} = \hat{i} + \hat{j} - \hat{k}$, and $\vec{C} = \hat{j} + 2\hat{k}$.
   Find:
   (a) $\vec{A} \cdot \vec{B}$ (c) $\vec{A} \times \vec{B}$
   (b) $|\vec{A}|$ (d) $2\vec{A} - \vec{B} + 3\vec{C}$

2. Let $\vec{A} = \hat{i} + 2\hat{j} + 2\hat{k}$, $\vec{B} = 2\hat{i} + \hat{j} - 2\hat{k}$, and $\vec{C} = -2\hat{i} + 2\hat{j} - \hat{k}$. Find each of the following:
   (a) $|\vec{A}|$ (d) $\vec{A} \cdot \vec{B}$
   (b) $\vec{A} + \vec{B}$ (e) $\vec{A} \times \vec{B}$
   (c) $\vec{A} - \vec{B}$

3. Let $\vec{A} = \hat{i}$, $\vec{B} = \hat{i} + \hat{j}$, and $\vec{C} = \hat{i} + \hat{j} + \hat{k}$. Let θ be the angle between $\vec{B}$ and $\vec{C}$. Find each of the following:
   (a) $|\vec{C}|$ (d) $\vec{B} \times \vec{C}$
   (b) $\hat{C}$ (e) $\cos \theta$
   (c) $\vec{B} \cdot \vec{C}$

4. Let $\vec{A} = \hat{i}$, $\vec{B} = \hat{i} + 2\hat{j}$, and $\vec{C} = \hat{i} + \hat{j} + \hat{k}$. Let θ be the angle between vectors $\vec{B}$ and $\vec{C}$. Evaluate each of the following:
   (a) $|\vec{C}|$ (d) $\vec{A} \cdot (\vec{B} \times \vec{C})$
   (b) $\vec{B} \times \vec{C}$ (e) $(\vec{A} \times \vec{B}) \times \vec{C}$
   (c) $\cos \theta$

5. You are given the following information about vectors $\vec{A}$, $\vec{B}$, and $\vec{C}$: $\vec{A} \cdot \vec{C} = 5$, $\vec{B} \cdot \vec{C} = 1$, $\vec{B} \cdot \vec{B} = 1$, $\vec{C} \cdot \vec{C} = 4$, $(\vec{A} \cdot \vec{B})^2 = |\vec{A}|^2 |\vec{B}|^2$. Find the following values:
   (a) $\vec{A} \cdot \vec{B}$ (b) $\vec{A} \cdot \vec{A}$
   (*Hint.* Visualize. Use "coordinate-free" definitions from §2.4. Avoid long calculations. §3.6 is not needed.)

6. You are given that $\vec{A} \cdot \vec{A} = 4$, $\vec{B} \cdot \vec{B} = 4$, $\vec{A} \cdot \vec{B} = 0$, $(\vec{A} \times \vec{B}) \times \vec{C} = \vec{0}$, and $(\vec{A} \times \vec{B}) \cdot \vec{C} = 8$. Find each of the following:
   (a) $\vec{C} \cdot \vec{A}$ (c) $\vec{C} \times \vec{B}$
   (b) $|\vec{C}|$

(*Hint.* Visualize. Use basic definitions from §2.4. Avoid long calculations. §3.6 is not needed.)

7. You are given that $\vec{A} \cdot \vec{A} = 9$, $\vec{B} \cdot \vec{B} = 1$, $\vec{A} \cdot \vec{B} = 0$, $(\vec{A} \times \vec{B}) \times \vec{C} = 0$, and $(\vec{A} \times \vec{B}) \cdot \vec{C} = 6$. Find each of the following:

   (a) $\vec{C} \times \vec{A}$

   (b) $|\vec{C}|$

   (c) The scalar k such that $\vec{C} \times \vec{B} = k\vec{A}$.

8. In a system of Cartesian coordinates, a triangle has vertices at (1,3,2), (1,0,4), (0,3,4). Find its area.

9. (a) Use vector algebra to show that if $\vec{A} + \vec{B} + \vec{C} = 0$, then $\vec{A} \times \vec{B} = \vec{B} \times \vec{C} = \vec{C} \times \vec{A}$.

   (b) Give a geometric interpretation of this result.

§3.4

1. Let $\vec{A} = \hat{i} + \hat{j} - 2\hat{k}, \vec{B} = \hat{i} - \hat{j}, \vec{C} = \hat{i} + \hat{j} + \hat{k}$.

   (a) Show that $\hat{A}$, $\hat{B}$, and $\hat{C}$ form a frame.

   (b) Find the components of $\vec{V} = \hat{i} + 2\hat{j} + 3\hat{k}$ in this frame.

2. Three of the following four vectors form a set of mutually orthogonal vectors: $2\hat{i} - \hat{j} + \hat{k}$, $\hat{i} + 2\hat{j}$, $\hat{j} + \hat{k}$, $\hat{i} + \hat{j} - \hat{k}$.

   (a) Identify these three vectors.

   (b) Find the scalar components of the remaining vector in the frame determined by these three vectors.

3. Let $\vec{A}$, $\vec{B}$, and $\vec{C}$ be as in problem 2 of §3.3.

   (a) Show that the set $\{\hat{A}, \hat{B}, \hat{C}\}$ is a frame.

   (b) Let $\vec{V} = 3\hat{i} - 3\hat{j} + 3\hat{k}$. Find scalar values a, b, and c such that $\vec{V} = a\hat{A} + b\hat{B} + c\hat{C}$.

4. In the following list of statements about $\vec{A}$, $\vec{B}$, and $\vec{C}$, cross out each statement that can be deduced from one or more of the preceding statements in the list.

$$\vec{C} = \vec{A} \times \vec{B}$$
$$\vec{A} \cdot \vec{B} = 0$$
$$\vec{A}^2 = 1$$
$$\vec{B}^2 = 1$$
$$\vec{B} \cdot \vec{C} = 0$$
$$\vec{C}^2 = 1$$
$$\vec{A} = \vec{B} \times \vec{C}$$
$$\vec{C} \cdot \vec{A} = 0$$
$$\vec{B} = \vec{C} \times \vec{A}.$$

(*Hint.* At what point in the list do you know that $\vec{A}, \vec{B}, \vec{C}$ form a frame?)

5. Given $\vec{A} = \hat{i} + \hat{j}$ and $\vec{B} = \hat{j} + \hat{k}$, find a frame $\{\hat{d}, \hat{e}, \hat{f}\}$, where $\hat{d}$ has the same direction as $\vec{A}$, and $\hat{e}$ is coplanar with $\vec{A}$ and $\vec{B}$. (See problem 10 for §2.4.)

§3.5

1. (a) Find the area of the triangle whose vertices have the Cartesian coordinates (0,0,0), (3,0,0), and (1,1,1).

   (b) Do the same for the triangle with vertices at (0,0,0), (1,2,0), (0,2,1).

2. Let L_1 and L_2 be two interior diagonals of a cube. Find the acute angle at which L_1 and L_2 intersect.

3. A rectangular block has length 3, width 2, and height 1. Find the possible values that can occur for the acute angle formed when two interior diagonals intersect.

4. A regular tetrahedron is placed on Cartesian axes so that it has vertex A at (0,0,0), vertex B at (1,0,0), vertex C in the first quadrant of the xy plane, and vertex D with positive z coordinate. Find the coordinates of C and D. (*Hint.* Use vector methods.)

5. Use the result of problem 4 and vector methods to find the dihedral angle between two faces of a regular tetrahedron.

6. Given a regular tetrahedron, find the angle between an edge and a face which does not contain that edge. (Use the result of problem 4.)

7. Consider two faces of a regular octahedron which have a common edge. Use vector methods to find the dihedral angle between these faces.

8. A pyramid has a square base of area 1. Its vertex is at distance 1 from the base, and the projection of the vertex on the base is the center-point of the base. Find the size of the dihedral angle between two adjacent triangular faces. (*Hint.* Introduce Cartesian coordinates with origin at one corner of the base and x and y axes along two sides of the base.)

9. Let $\vec{A} = \hat{i} + \hat{j}$ and $\vec{B} = \hat{i} + \hat{j} + \hat{k}$. Find a scalar b and vectors $\vec{C}$ and $\vec{D}$ such that $\vec{C} = b\vec{B}$, $\vec{C} + \vec{D} = \vec{A}$, and $\vec{C} \cdot \vec{D} = 0$. (This is a special case of problem 10 in §2.4.)

§3.6

1. Evaluate:

(a) $\hat{i} \times (\hat{i} \times \hat{k})$

(b) $[\vec{A}, \vec{B}, \vec{C}]$, given that $\vec{A} = \hat{i} - \hat{j} + \hat{k}$, $\vec{B} = 2\hat{i} + \hat{k}$, $\vec{C} = 2\hat{i} - \hat{j}$.

2. The $\vec{A}$, $\vec{B}$, and $\vec{C}$ in problem 2 of §3.3 are mutually orthogonal. Find:

(a) $\vec{A} \cdot (\vec{B} \times \vec{C})$ (c) $(\vec{A} \times \vec{B}) \times \vec{C}$

(b) $\vec{A} \times (\vec{B} \times \vec{C})$

3. Let $\vec{A} = \hat{i}$, $\vec{B} = \hat{i} + \hat{j}$, $\vec{C} = \hat{i} + \hat{j} + \hat{k}$. Find

(a) $\vec{A} \cdot (\vec{B} \times \vec{C})$ (c) $(\vec{A} \times \vec{B}) \times \vec{C}$

(b) $\vec{A} \times (\vec{B} \times \vec{C})$

4. Let $\vec{A} = 3\hat{i} - 4\hat{k}$, $\vec{B} = \hat{i} + \hat{j} - \hat{k}$, $\vec{C} = \hat{j} + 2\hat{k}$. Find

(a) $[\vec{A}, \vec{B}, \vec{C}]$ (c) $\vec{B} \times (\vec{C} \times \vec{A})$

(b) $\vec{A} \cdot (\vec{B} \times \vec{C})$

5. Find the volume of the parallelepiped whose edges at one vertex are given by arrows representing $3\hat{i} + 4\hat{j}$, $2\hat{i} + 3\hat{j} + 4\hat{k}$, and $5\hat{k}$.

6. Find the volume of the tetrahedron with vertices $(0,0,0)$, $(1,-1,1)$, $(2,1,-2)$, and $(-1,2,-1)$.

7. Simplify:

(a) $[\vec{A} \times (\vec{A} \times \vec{B})] \times \vec{C} \cdot (\vec{A} + \vec{B})$

(b) $(\vec{A} + (\vec{B} \times \vec{C}) - (\vec{C} \times \vec{A})) \cdot (\vec{B} - \vec{A})$

8. Simplify:

(a) $(\vec{A} \times \vec{B}) \cdot ((\vec{A} \times \vec{C}) \times (\vec{B} \times \vec{C}))$

(b) $\vec{A} \times ((\vec{A} \times \vec{B}) \times (\vec{A} \times (\vec{B} \times \vec{C})))$

(c) $\vec{A} \cdot ((\vec{A} \times \vec{B}) \times (\vec{A} \times (\vec{B} \times \vec{C})))$

9. (a) Prove (6.1) and (6.3) by calculation.

(b) Deduce (6.2) from (6.3).

10. Verify (6.6) by calculation.

(*Hint:* The calculations can be simplified by introducing an appropriate Cartesian system. For example, choose coordinate axes so that $\vec{A} = a\hat{i}$, $\vec{B} = b_1\hat{i} + b_2\hat{j}$, $\vec{C} = c_1\hat{i} + c_2\hat{j} + c_3\hat{k}$. This simplifies the calculation for the identity for $\vec{A} \times (\vec{B} \times \vec{C})$. The identity for $(\vec{A} \times \vec{B}) \times \vec{C}$ then follows by (IVa) in §2.4.)

Note. In each of the next three problems, be sure to check that the solutions which you deduce satisfy the given conditions of the problem.

11. $\vec{A}$, $\vec{B}$ and $\vec{C}$ are given vectors such that $\vec{A} \cdot \vec{B} \neq 0$ and $\vec{B} \cdot \vec{C} = 0$. k is a given scalar. Find a vector-algebra expression in terms of $\vec{A}, \vec{B}, \vec{C}$ and k, for a vector $\vec{X}$ such that $\vec{X} \cdot \vec{A} = k$ and $\vec{X} \times \vec{B} = \vec{C}$. See the note above. (*Hint.* Take the vector product of $\vec{A}$ with the last equation.)

12. $\vec{A}$ and $\vec{B}$ are given non-zero vectors such that $\vec{A} \cdot \vec{B} = 0$. d is a given scalar. $\vec{X}$ and $\vec{Y}$ are unknown vectors such that:

$$2\vec{X} + \vec{Y} = \vec{A}, \quad \vec{X} \times \vec{Y} = \vec{B}, \quad \vec{X} \cdot \vec{A} = d.$$

Use vector algebra to find $\vec{X}$ and $\vec{Y}$ in terms of $\vec{A}$, $\vec{B}$, and d. See the note above.

13. $\vec{A}$ and $\vec{B}$ are given vectors such that $\vec{A} \times \vec{B} \neq \vec{0}$. Find vector-algebra expressions for vectors $\vec{X}$ and $\vec{Y}$ which satisfy the following conditions: $\vec{X} + \vec{Y} = \vec{A} + \vec{B}$, $\vec{X} \cdot \vec{A} = 0$, $\vec{Y} \cdot \vec{B} = 0$, $\vec{X} \cdot (\vec{A} \times \vec{B}) = 0$. See the note above.

14. Prove theorem (4.2) by finding vector-algebra formulas for a, b, and c in terms of $\vec{A}$, $\vec{B}$, $\vec{C}$, and $\vec{G}$. (*Hint.* To find an expression for a, apply $\times \vec{B}$ and then $\cdot \vec{C}$ to the desired equation. Once you have the expressions for a, b, and c, be sure to show that the desired equation does, in fact, hold and to explain why a, b, and c are unique.)

15. *Heron's formula for the area* A(T) *of a plane triangle* T *in terms of the lengths of its sides.* The formula asserts that $A(T) = \sqrt{s(s-a)(s-b)(s-c)}$, where a,b,c are the lengths of the sides and $s = \frac{1}{2}(a+b+c)$ is the *semi-perimeter.* Use vector algebra to prove Heron's formula. (*Hint.* Let $\vec{A}$ and $\vec{B}$ be vectors for sides a and b. Note that $A(T) = \frac{1}{2}|\vec{A} \times \vec{B}|$ and hence that

$$A(T)^2 = \frac{1}{4}(\vec{A} \times \vec{B}) \cdot (\vec{A} \times \vec{B}).$$

Expanding this quadruple product gives an expression in $\cos^2\gamma$ where γ is the angle between $\vec{A}$ and $\vec{B}$. Use the law of cosines to eliminate $\cos^2\gamma$ from this expression for $A(T)^2$ and show that the resulting algebraic expression in a,b,c gives $A(T)^2 = s(s-a)(s-b)(s-c).$)

Note. The following four problems concern spherical triangles. A *spherical triangle* is a "triangle" drawn on the curved surface of a sphere, where the *sides* of the triangle are arcs of great circles. (Recall that a great circle on a sphere is the intersection of the sphere with a plane through its center.) The size of a *vertex angle* of a spheri-

cal triangle is defined to be the size of the dihedral angle formed by the planes for the great circles for the sides which meet at that vertex. We assume that our sphere has radius 1. The length of a side is then equal to the angle, in radians, of its arc as part of a great circle. We limit our attention to triangles where each vertex angle is $< \pi$. It then follows that each side must have length $< \pi$ and that there is a hemisphere which contains the entire triangle. We call the vertex points A, B, and C. Using the center of the sphere as reference point, α, β, γ we let $\hat{A}, \hat{B}, \hat{C}$ be the position vectors for A,B,C. α, β, γ are the respective vertex angles, and a,b,c are the lengths of the sides opposite to A,B,C.

Computation with spherical triangles is the mathematical basis of celestial navigation. In the following problems, we see that vector algebra can be used to deduce the principal mathematical laws used.

16. Given a spherical triangle as above, find vector-algebra expressions in $\hat{A}, \hat{B}, \hat{C}$ for the following:
     (a) sin a, sin b, sin c
     (b) cos a, cos b, cos c
     (c) sin α, sin β, sin γ
     (d) cos α, cos β, cos γ
     (*Hint.* For example, sin a = $|\vec{B} \times \vec{C}|$, and cos a = $\hat{B} \cdot \hat{C}$. sin α and cos α use quadruple products.)

17. Use the results of problem 16 to verify that

$$\frac{\sin\alpha}{\sin a} = \frac{\sin\beta}{\sin b}.$$

This is the *law of sines* of spherical trigonometry.

18. Use the results of problem 16 to verify that:
     (a) cos a = cos b cos c + sin b sin c cos α
     (b) cos α = $-$cos β cos γ + sin β sin γ cos a
     These are the two *laws of cosines* of spherical trigonometry.

19. Show that the area A(T) of a spherical triangle T is given by

$$A(T) = \alpha + \beta + \gamma - \pi.$$

(*Hint.* Prove this result in the following steps:

(i) Let H_a, H_b, H_c be the three hemispheres which are bounded by the great circles for a,b,c and which contain T. Let $H_a \cup H_b \cup H_c$ be the set of those points which lie in at least one of the hemispheres. Show that:

$$A(H_a \cup H_b \cup H_c) =$$

$$A(H_a) + A(H_b) + A(H_c) - A(H_a \cap H_b) -$$

$$A(H_a \cap H_c) - A(H_b \cap H_c) + A(T).$$

Where, for example, $H_a \cap H_b$ is the set of points common to H_a and H_b.

(ii) Show that the points on the sphere which do *not* lie in either H_a, H_b, or H_c form the interior of a triangle T′ which is congruent to T. (T′ is the *antipodal* triangle to T.) Hence A(T′) = A(T), and

$$4\pi = A(H_a \cup H_b \cup H_c) + A(T')$$

$$= \text{area of whole sphere}$$

$$= A(H_a \cup H_b \cup H_c) + A(T).$$

Thus, from (i), $4\pi = A(H_a) + A(H_b) + A(H_c) -$ $A(H_a \cap H_b) - A(H_a \cap H_c) + A(H_b \cap H_c) +$ 2A(T).

(iii) Note that $A(H_a) = A(H_b) = A(H_c) = 2\pi$. Show that $A(H_b \cap H_c) = 2\alpha$, $A(H_a \cap H_c) = 2\beta$, and $A(H_a \cap H_b) = 2\gamma$.

(iv) Conclude that $4\pi = 6\pi - 2(\alpha+\beta+\gamma) +$ 2A(T), and hence that $A(T) = \alpha + \beta + \gamma - \pi$.)

20. Prove: $(\vec{A} \times \vec{B}) \times \vec{C} = \vec{A} \times (\vec{B} \times \vec{C})$

$$\Leftrightarrow \text{either } \vec{A} \times \vec{C} = \vec{0} \text{ or } \vec{B} \cdot \vec{A} = \vec{B} \cdot \vec{C} = 0.$$

21. *A three-dimensional analogue to Heron's formula.* (See problem 15 above.) Let A, B, C, and D be the vertices of a tetrahedron. Let $b = |\overrightarrow{AB}|$, $c = |\overrightarrow{AC}|$, $d = |\overrightarrow{AD}|$, $m = |\overrightarrow{BC}|$, $n = |\overrightarrow{CD}|$, and $p = |\overrightarrow{DB}|$. Use vector algebra to find the volume of the tetrahedron in terms of b, c, d, m, n, and p. (*Hint.* Use (7.4).)

§3.7

1. Verify (7.2).

2. Use properties (7.5) to find values for the following determinants. Do not use direction evaluation.

(a) $\begin{vmatrix} 2 & 1 & 2 \\ 2 & 3 & 2 \\ 4 & 2 & 4 \end{vmatrix}$.

(c) $\begin{vmatrix} 2 & 1 & -1 \\ 2 & 3 & 5 \\ 4 & 2 & -2 \end{vmatrix}$.

(b) $\begin{vmatrix} 2 & 1 & 3 \\ 0 & 0 & 0 \\ 2 & -1 & 4 \end{vmatrix}$.

3. Use Cramer's rule to solve the system:

$$\begin{pmatrix} x + 2y = 4 \\ x - y = 1 \end{pmatrix}.$$

4. Use Cramer's rule with 3×3 determinants to solve the system:

$$\begin{pmatrix} x + y = 2 \\ y + z = -3 \\ x + z = 1 \end{pmatrix}.$$

Chapter 4

Analytic Geometry in Three Dimensions

§4.1 Review of Analytic Geometry in $\mathbf{E}^2$; Conic Sections

Cartesian coordinate systems enable us: (1) to represent certain geometrical figures in algebraic form or (2) to visualize certain algebraic equations in geometrical form. We can then either: (1) use algebraic ideas and methods to help solve geometric problems, or (2) use geometric ideas and methods to help solve algebraic problems. *Analytic geometry* is the study of this interconnection between Euclidean geometry and algebra. In secondary-school mathematics, students learn basic facts of *analytic geometry for* $\mathbf{E}^2$. We review these facts in the remainder of this section.

In what follows, $\mathcal{F}[x,y]$ stands for an algebraic formula in the variables x and y (for example, x^2+y^2-4), and $\mathcal{F}[x,y] = 0$ stands for an equation whose left-hand side is such a formula and whose right-hand side is 0. Note that any equation in x and y of the form

$$\mathcal{F}_1[x,y] = \mathcal{F}_2[x,y]$$

is equivalent to an equation of the form

$$\mathcal{F}[x,y] = 0 ,$$

since we can always transpose the right side to the left and then view the resulting left side, $\mathcal{F}_1[x,y] - \mathcal{F}_2[x,y]$, as a single formula in x and y.

Definitions. Let a Cartesian coordinate system be given in $\mathbf{E}^2$. Let $\mathcal{F}[x,y]$ be a formula in x and y. If a and b are real numbers, and if it is true that $\mathcal{F}[a,b] = 0$, we say that the ordered pair (a,b) *satisfies* the equation $\mathcal{F}[x,y] = 0$. (Here $\mathcal{F}[a,b]$ is the formula obtained from $\mathcal{F}[x,y]$ by replacing variables x and y by the real numbers a and b.) For example, if $\mathcal{F}[x,y]$ is the formula x^2+y^2-4, then $(\sqrt{3},1)$ satisfies the equation $\mathcal{F}[x,y] = 0$, since $\mathcal{F}[\sqrt{3},1] = 0$.

Let S be the set of all points in $\mathbf{E}^2$ whose coordinates satisfy the equation $\mathcal{F}[x,y] = 0$. We then say that the equation $\mathcal{F}[x,y] = 0$ *determines* the figure S. For example, if $\mathcal{F}[x,y]$ is the formula x^2+y^2-4, then the figure S determined by $\mathcal{F}[x,y] = 0$ is the circle of radius 2 with center at O.

Definition. An equation $\mathcal{F}[x,y] = 0$ is said to be *linear* if $\mathcal{F}[x,y]$ has the form $Ax+By+C$, where the coefficients A, B, C are real numbers, and $A^2+B^2 > 0$. (To say that $A^2+B^2 > 0$ is the same as saying that at least one of the coefficients A and B is not zero.) We observe that: $\mathcal{F}[x,y] = 0$ *is linear* $\Leftrightarrow$ $\mathcal{F}[x,y]$ *is a polynomial of first degree in* x *and* y. For this reason, linear equations are also known as *first-degree equations*. We have the following basic theorem about linear equations.

(1.1)   *In* $\mathbf{E}^2$ *with Cartesian coordinates:*
    (a)   *every linear equation determines a unique straight line; and*
    (b)   *for every straight line, there is some linear equation which determines that line.*

(It is possible for several distinct linear equations to determine the same line. For example, $x+y-1 = 0$ and $2x+2y-2 = 0$ determine the same line.)

We can use a given linear equation to get geometric information about the line that it determines. In particular, the coefficients of the equation tell us how the line is situated on the coordinate axes. We have:

(1.2)   *In* $\mathbf{E}^2$ *with Cartesian coordinates:*
    (a)   *The vector* $A\hat{i} + B\hat{j}$ *is normal to the line determined by the linear equation* $Ax+By+C = 0$. *See* [1–1].
    (b)   *The intercepts (intersections with the axes) for the line determined by* $Ax+By+C = 0$ *are:* $(-\frac{C}{A},0)$ *for the* x *axis (if* $A \neq 0$*) and* $(0,-\frac{C}{B})$ *for the* y *axis (if* $B \neq 0$*).* (Note that a line must always have at least one intercept.)

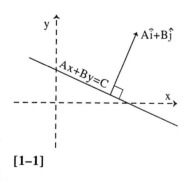

[1–1]

(c)   *The line with intercepts* (a,0) *and* (0,b), a *and* b ≠ 0, *is determined by the equation* $\frac{x}{a}+\frac{y}{b}=1$.

In secondary-school mathematics, result (a) is usually stated in the equivalent form: for B ≠ 0, the *slope* of the line determined by Ax+By+C = 0 is $-\frac{A}{B}$, while for B = 0, the line is parallel to the y axis. (1.1) and (1.2) are proved in secondary-school mathematics. We do not prove these results here.

Definition. An equation F [x,y] = 0 is said to be a *second-degree equation* if $\mathcal{F}$[x,y] has the form $Ax^2+By^2+Cxy+Dx+Ey+F$, where the coefficients A, B, C, D, E, and F are real-numbers, and $A^2+B^2+C^2 > 0$. We observe that: F [x,y] = 0 *is a second-degree equation* ⇔ $\mathcal{F}$[x,y] *is a polynomial of second degree in* x *and* y.

In previous work on analytic geometry in $\mathbf{E}^2$, the reader may have studied certain special cases of second-degree equations and may have learned the following:

(1.3)   (a)   *For* $\mathcal{F}$[x,y] *in the form* Ax^2+By^2+F *(with C = D = E = 0):*

(i)   [A > 0, B = A, *and* F < 0]) ⇒ *the equation* F [x,y] = 0 *determines a circle of radius* $\sqrt{\frac{-F}{A}}$ *with center at* **O**.

(ii)   [A > 0, B > 0, A ≠ B, *and* F < 0] ⇒ *the equation* $\mathcal{F}$[x,y] = 0 *determines an* ellipse *with center at* **O**, *axes on the coordinate axes, semi-major axis =* $max\left[\sqrt{\frac{-F}{A}},\sqrt{\frac{-F}{B}}\right]$, *and semi-minor axis =* $min\left[\sqrt{\frac{-F}{A}},\sqrt{\frac{-F}{B}}\right]$.

(iii)   [AB < 0 *and* F ≠ 0] ⇒ *the equation* F [x,y] = 0 *determines a* hyperbola *(of two branches) with center at the origin, axes on the coordinate axes, and asymptotes given by the equations* Ax+By = 0 *and* Ax − By = 0. (AB < 0 tells us that either [A > 0 and B < 0] or [B > 0 and A < 0].)

(b)   *For* $\mathcal{F}$[x,y] *in the form* $Ax^2+Ey+F = 0$ *(with B = C = D = 0), the equation* $\mathcal{F}$[x,y] = 0 *determines a* parabola *whose axis is the* y *axis, and for* $\mathcal{F}$[x,y] *in the form* By^2+Dx+F *(with A = C = E = 0), the equation* $\mathcal{F}$[x,y] = 0 *determines a* parabola *whose axis is the* x *axis.*

For the benefit of readers who have not made a full study of the analytic geometry of second-degree equations in $\mathbf{E}^2$, we summarize several further results here (see (1.4) below) and in the problems. To prove these results, we must know how to alter the equation for a given figure by changing from one Cartesian coordinate system to another. We do this by introducing a new and different set of axes. We begin with two examples.

Example 1. *Consider the equation* $x^2+y^2+2x+2y = 0$. This equation determines a certain figure S in $\mathbf{E}^2$. In §4.7, we show that we can replace our given Cartesian coordinate system by a new and different Cartesian system (with different origin and different axes) such that the same figure S is determined by the equation $x^2+y^2-2 = 0$ in the new coordinate system. This shows us, by (1.3), that the figure S is a circle of radius $\sqrt{2}$. The fact that S is a circle is a geometric fact independent of any particular coordinate system. The new Cartesian system serves to reveal to us that S is a circle. (In this example, the new system is located with respect to the old system as follows: *new origin* at the point whose old coordinates are $(-1, -1)$; *new axes* parallel to the old axes.)

Example 2. *Consider the equation* $xy = 1$. Let S be the figure in $\mathbf{E}^2$ determined by this equation. See [1–2]. We shall show in §4.7 that we can find a new Cartesian system in which S is determined by the equation $x^2-y^2-2 = 0$. By (1.3), this latter equation reveals to us that S is a hyperbola with perpendicular asymptotes. (In this example, the new system is located as follows: origin unchanged; new axes making an angle of $\frac{\pi}{4}$ with the old axes.)

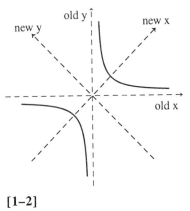

[1–2]

The following general result is proved in §4.7.

(1.4)   *Let* $\mathcal{F}[x,y] = 0$ *be a second-degree equation, and let* S *be the figure in* $\mathbf{E}^2$ *determined by this equation. Then either:*
        (I) *we can find a new Cartesian system in which* S *is determined by an equation of the form*

$$Ax^2+By^2+F = 0, \text{ with } A^2+B^2 > 0 \text{ and } F = -1 \text{ or } 0; \text{ or}$$

        (II) *we can find a new Cartesian system in which* S *is determined by an equation of the form*

$$Ax^2+y = 0, \text{ with } A \neq 0.$$

The new coordinate system in (1.4) depends upon the given formula $\mathcal{F}[x,y]$. §4.7 gives a method for finding, for any given second-degree polynomial $\mathcal{F}[x,y]$, a corresponding new coordinate system and new equation as described in (1.4).

In general, in (I) and (II) of (1.4), the new coefficients A, B, and F (for the given figure S in the new coordinate system) will differ from the old coefficients in the original second-degree equation. When we find the new equation for S, the coefficients in this new equation provide us with direct geometric information about S. The following tables summarize this information.

(I) *Table for* $Ax^2 + By^2 + F = 0$ *with* $A^2 + B^2 > 0$ *and* F = −1 *or* 0. The left-hand column lists possible ways in which A and B, in one order or the other, can be positive, negative, or zero. Thus, for example, '- 0' is the case where one of A and B is negative and the other is 0.

			F = -1	F = 0
+	+		ellipse or circle	single point
−	−		empty figure	single point
+	−		hyperbola	two intersecting lines
+	0		two parallel lines	single line
−	0		empty figure	single line

(1.5a)

("Empty figure" means that there is no point in $\mathbf{E}^2$ which satisfies the equation.)

(II) *Table for* $Ax^2 + y = 0$ *with* $A \neq 0$. The left-hand column lists possible ways for A to be positive or negative.

(1.5b)

+	parabola
−	parabola

By (1.4), the tables above give a complete list of the figures in $\mathbf{E}^2$ which are determined by second-degree equations in Cartesian coordinates. The cases of *ellipse, circle, hyperbola,* and *parabola* are called *principal cases*. The cases of *empty figure, point, line,* or *two lines* are called *special cases* (or "borderline" or "degenerate" cases). It can be shown that the principal cases, together with all the special cases except the cases of *empty figure* and *two parallel lines*, are the same as the possible intersections which can occur between the plane and a two-sheeted cone in $\mathbf{E}^3$. These possible intersections between a plane and a cone have been known since the time of Euclid as "conic sections." Today, we define *conic sections* to cover all the cases in the tables above, including *empty figure* and *two parallel lines*. The results in (1.4) and (1.5) can now be summarized: *every second-degree equation in* x *and* y *determines a conic section in* $\mathbf{E}^2$; *and for every conic section (in* $\mathbf{E}^2$), *there is a second-degree equation in* x *and* y *which determines that conic section.*

(*Note.* A *two-sheeted cone* S in $\mathbf{E}^3$ can be defined as follows. Let P_0 be a fixed point, and let L_0 be a fixed line through P_0. Let θ_0 be a fixed value, $0 < \theta_0 < \frac{\pi}{2}$. Then let S consist of the points on all lines through P_0 which make an angle of size θ_0 with L_0.)

§4.2 Planes in $\mathbf{E}^3$

We now begin the study of analytic geometry in $\mathbf{E}^3$. Let a Cartesian coordinate system be given in $\mathbf{E}^3$. Our basic definitions are analogous to the definitions in §4.1. We use $\mathscr{F}[x,y,z]$ to stand for an algebraic formula in x, y, and z. We say that an ordered triple of real numbers (a,b,c) *satisfies* an equation $\mathscr{F}[x,y,z] = 0$ if $\mathscr{F}[a,b,c] = 0$, and we say that an equation $\mathscr{F}[x,y,z] = 0$ *determines* a figure S in $\mathbf{E}^3$ if S consists of exactly those points whose coordinates satisfy $\mathscr{F}[x,y,z] = 0$.

Definition. An equation $\mathscr{F}[x,y,z] = 0$ is said to be *linear* if $\mathscr{F}[x,y,z]$ has the form

$$Ax + By + Cz + D = 0,$$

where the coefficients A,B,C,D are real numbers, and $A^2 + B^2 + C^2 > 0$.

As before: $\mathscr{F}[x,y,z] = 0$ *is linear* $\Leftrightarrow$ $\mathscr{F}[x,y,z]$ *is a polynomial of first degree in* x, y, *and* z. Hence linear equations are also known as *first-degree equations.*
The chief results of this section are the following analogues to (1.1) and (1.2). (In §4.6, we present analogues to (1.3), (1.4), and (1.5).)

In $\mathbf{E}^3$ *with Cartesian coordinates:*

(2.1a) *every linear equation determines a unique plane, and*

(2.1b) *every plane is determined by some linear equation.*

In $\mathbf{E}^3$ *with Cartesian coordinates:*

(2.2a) *The vector* $\vec{N} = A\hat{i} + B\hat{j} + C\hat{k}$ *is normal to the plane M determined by the linear equation* Ax+By+Cz+D = 0. *(We say that* $\vec{N}$ *is a* normal vector *to* M.)

(2.2b) *The intercepts (intersections with the axes) for the plane determined by* Ax+By+Cz+D = 0 *are* $(-\frac{D}{A}, 0, 0)$ *for the x-axis (if* A ≠ 0*),* $(0, -\frac{D}{B}, 0)$ *for the y axis (if* B ≠ 0*), and* $(0, 0, -\frac{D}{C})$ *for the z axis (if* C ≠ 0*).* (Note that a plane must always have at least one intercept.)

(2.2c) *The plane with intercepts* (a,0,0), (0,b,0), *and* (0,0,c), a, b, *and* c ≠ 0, *is determined by the equation* $\frac{x}{a} + \frac{y}{b} + \frac{z}{c} = 1$.

To prove (2.1a), let Ax+By+Cz = –D be a given linear equation, and let $P_0 = (a, b, c)$ 7 be a fixed point whose coordinates satisfy that equation. (We can always find such a point by setting two coordinates equal to 0 and solving the equation for the third coordinate. For example 2x+z = 4 is satisfied by (2,0,0).) Let $\vec{N}$ be the vector $A\hat{i} + B\hat{j} + C\hat{k}$. Since $P_0 = (a, b, c)$ satisfies the given equation, we have

$$Aa + Bb + Cc = -D.$$

Let P = (x,y,z) be a variable point. Let $\vec{R}_0$ be the fixed position vector for P_0, and let $\vec{R}$ be a variable position vector for the point P. Let Q_0 be chosen so that $\overrightarrow{OQ_0} = \vec{N}$, and let L be the line through O and Q_0. See [2–1].

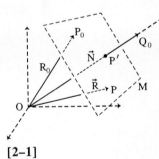

[2–1]

Then P = (x,y,z) satisfies the given equation ⟺

Ax+By+Cz = –D ⟺

Ax+By+Cz = Aa+Bb+Cc ⟺

$\vec{R} \cdot \vec{N} = \vec{R}_0 \cdot \vec{N}$ (by the coordinate formula for dot-product) ⟺

$\frac{\vec{R} \cdot \vec{N}}{\vec{N} \cdot \vec{N}} \vec{N} = \frac{\vec{R}_0 \cdot \vec{N}}{\vec{N} \cdot \vec{N}} \vec{N}$ ⟺

$\vec{R}$ and $\vec{R}_0$ have the same vector projection on $\vec{N}$ (see (4.8) in §3.4) ⟺

P_0 and P project to the same point P′ on L ⟺

P lies in the unique plane M which contains P_0 and is perpendicular to L.

Thus our given equation determines the plane M. This proves (2.1a). Since M is perpendicular to L, the vector $\vec{N}$ is normal to M. This proves (2.2a).

Conversely, to prove (2.1b), reverse the argument used to prove (2.1a). Let M be a given plane. Let $\vec{N} = A\hat{i} + B\hat{j} + C\hat{k}$ be a vector normal to M, and let $P_0 = (a,b,c)$ be some point in M. Let $\vec{R}_0$ be the fixed position vector for P_0, and let $\vec{R}$ be the variable position vector for P. Let L be the line through O normal to M. Then

$$P = (x,y,z) \text{ is in } M \Leftrightarrow$$

$$P \text{ and } P_0 \text{ project to the same point on } L \Leftrightarrow$$

$$\vec{R} \text{ and } \vec{R}_0 \text{ have the same vector projection on } \vec{N} \Leftrightarrow$$

$$\frac{\vec{R} \cdot \vec{N}}{\vec{N} \cdot \vec{N}} = \frac{\vec{R}_0 \cdot \vec{N}}{\vec{N} \cdot \vec{N}} \Leftrightarrow \vec{R} \cdot \vec{N} = \vec{R}_0 \cdot \vec{N} \Leftrightarrow$$

$$Ax + By + Cz = Aa + Bb + Cc \Leftrightarrow$$

$$Ax + By + Cz + D = 0, \text{ where } D = -(Aa + Bb + Cc).$$

This proves (2.1b). The proofs of (2.2b) and (2.2c) are easy and are left to the reader.

Example 1. *Find the equation of the plane* M *which contains the point* $P_0 = (2,1,-3)$ *and has* $\vec{N} = \hat{i} - \hat{j} + 2\hat{k}$ *as a normal vector.* From (2.1) and (2.2), we know that M has an equation of the form

$$x - y + 2z + D = 0.$$

This equation must be satisfied by (2,1,-3). This gives

$$2 - 1 - 6 + D = 0,$$
$$D = 5.$$

Hence $x - y + 2z + 5 = 0$ is the desired equation.

Example 2. *What is the distance from the origin to the plane* M *in example 1?* From the proof and figure above, we see that this distance is $\left| \vec{R}_0 \cdot \hat{N} \right|$, where $\hat{N} = \dfrac{\vec{N}}{|\vec{N}|}$ is the unit vector of $\vec{N}$, and $\vec{R}_0$ is the position vector of some chosen point in M. In example 1, we have $\vec{R}_0 \cdot \vec{N} = -5$ and $|\vec{N}| = \sqrt{6}$. Hence

$$\vec{R}_0 \cdot \hat{N} = \frac{\vec{R}_0 \cdot \vec{N}}{|\vec{N}|} = -\frac{5}{\sqrt{6}}, \text{ and } |\vec{R}_0 \cdot \hat{N}| = \frac{5}{\sqrt{6}} = \frac{5\sqrt{6}}{6}.$$

Thus $\frac{5\sqrt{6}}{6}$ is the desired distance from the origin to the plane M.

From example 2, we have the following general result:

(2.3) *Let a plane* M *be determined by the equation* $Ax+By+Cz+D = 0$. *Then*

(a) *the distance from the origin to* M *is* $\dfrac{|D|}{|\vec{N}|}$ *where* $\vec{N} = A\hat{i} + B\hat{j} + C\hat{k}$;

hence, in particular,

(b) M *contains the origin* $\Leftrightarrow D = 0$.

The following result is also immediate from the figure above and the proof of (2.1).

(2.4) *Let* $\vec{R}_0$ *and* $\vec{N}$ *be given vectors with* $\vec{N} \neq \vec{0}$. *Let* $\vec{R}$ *be a variable position vector for a point* P. *Consider the vector-algebra equation* $\vec{R} \cdot \vec{N} = \vec{R}_0 \cdot \vec{N}$, *which can also be written*

$$(\vec{R} - \vec{R}_0) \cdot \vec{N} = 0.$$

Then: (a) *the points* P *whose position vectors* $\vec{R}$ *satisfy this vector-algebra equation form a plane, and*

(b) *every plane can be represented by a vector-algebra equation of this form.*

The vector-algebra equation in (2.4) does not use Cartesian coordinates. It assumes only that we have a fixed reference point for position vectors.

§4.3 Applications

We illustrate the results of §4.2 with several examples. We assume a given Cartesian coordinate system in $\mathbf{E}^3$.

Example 1. Let M_0 be the plane with equation $x+2y-3z = 4$. *Find an equation for the plane* M *which is parallel to* M_0 *and contains the point* $(1,0,1)$. By (2.2), $\vec{N} = \hat{i} + 2\hat{j} - 3\hat{k}$ is a normal vector to M_0. By geometry (§1.4), $\vec{N}$ must also be a normal vector to M. Hence M must have the equation $x+2y-3z = d$ for some d. Since $(1,0,1)$ is in M, we must have $1+2(0)-3(1) = d$; hence $d = -2$. Therefore, $x+2y-3z = -2$ is an equation for M, as desired.

Example 2.   What are the intercepts of the plane M_0 in example 1? Setting $y = z = 0$, we get $x = 4$. Similarly $x = z = 0 \Rightarrow y = 2$, and $x = y = 0 \Rightarrow z = -\frac{4}{3}$. Thus the intercepts are $(4,0,0)$, $(0,2,0)$ and $(0,0,-\frac{4}{3})$.

Example 3.   The plane M_1 is given by the equation $x+y+z = 4$, *and the plane* M_2 *by* $x-y-z = 2$. *Are these planes perpendicular to each other? If not, find the acute angle between them.*   By (2.2), we have normal vectors $\vec{N}_1 = \hat{i} + \hat{j} + \hat{k}$ for M_1 and $\vec{N}_2 = \hat{i} - \hat{j} - \hat{k}$ for M_2. By geometry, M_1 and M_2 are perpendicular $\Leftrightarrow$ $\vec{N}_1$ and $\vec{N}_2$ are orthogonal $\Leftrightarrow$ $\vec{N}_1 \cdot \vec{N}_2 = 0$. Calculating, we find $\vec{N}_1 \cdot \vec{N}_2 = 1-1-1 = -1$. Hence M_1 and M_2 are not perpendicular. The angle between $\vec{N}_1$ and $\vec{N}_2$ is $\cos^{-1}\left(\dfrac{\vec{N}_1 \cdot \vec{N}_2}{|\vec{N}_1|\,|\vec{N}_2|}\right) = \cos^{-1}\left(\dfrac{-1}{\sqrt{3}\sqrt{3}}\right) = \cos^{-1}\left(-\dfrac{1}{3}\right) = $ 109.5°. By geometry, this must be one of the angles between the planes M_1 and M_2. Since 109.5° is not acute, the acute angle between M_1 and M_2 must be 180° - 109.5° = 70.5°.

Example 4.   Is there a unique plane which contains the three points $P_1 = (1,0,0)$, $P_2 = (1,2,0)$, *and* $P_3 = (2,1,3)$? *If so, find an equation for this plane.*

By geometry, there must be at least one plane containing P_1, P_2, and P_3. Hence the arrows $\overrightarrow{P_1 P_2}$ and $\overrightarrow{P_1 P_3}$ must lie in such a plane. Now $\overrightarrow{P_1 P_2} = \overrightarrow{OP_2} - \overrightarrow{OP_1} = 2\hat{j}$, and $\overrightarrow{P_1 P_3} = \overrightarrow{OP_3} - \overrightarrow{OP_1} = \hat{i} + \hat{j} + 3\hat{k}$. The lines determined by $\overrightarrow{P_1 P_2}$ and $\overrightarrow{P_1 P_3}$ are not the same, since $\overrightarrow{P_1 P_3}$ is not a scalar multiple of $\overrightarrow{P_1 P_2}$. Hence, by geometry, there must be a unique plane M containing P_1, P_2, and P_3. To find an equation for M, let

$$\vec{N} = \overrightarrow{P_1P_2} \times \overrightarrow{P_1P_3} = \begin{vmatrix} \hat{i} & \hat{j} & \hat{k} \\ 0 & 2 & 0 \\ 1 & 1 & 3 \end{vmatrix} = 6\hat{i} - 2\hat{k}.$$

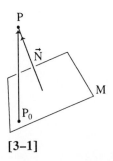

[3–1]

$\vec{N}$ must be normal to M, since $\vec{N}$ is normal to $\overrightarrow{P_1P_2}$ and $\overrightarrow{P_1P_3}$. Hence M has the equation 6x–2z = d for some d. Using P_1, we find d = 6. Thus 6x–2z = 6 (or, equivalently, 3x–z = 3) is an equation for M.

Example 5. *We wish to find the distance from the point* P = (2,1,–1) *to the plane* M *given by* x+y+z = 1. [3–1] shows us one way to proceed. $\vec{N} = \hat{i} + \hat{j} + \hat{k}$ is a normal vector to M. Let P_0 be some chosen point in M. For example, we can take P_0 to be the intercept (1,0,0). Then $\overrightarrow{P_0P} = \overrightarrow{OP} - \overrightarrow{OP_0} = \hat{i} + \hat{j} - \hat{k}$. The desired distance must be the absolute value of the scalar projection

of $\overrightarrow{P_0P}$ on $\vec{N}$. This gives us: distance from P to M = $\left| \dfrac{\overrightarrow{P_0P} \cdot \vec{N}}{|\vec{N}|} \right| = \left| \dfrac{1}{\sqrt{3}} \right| = \dfrac{\sqrt{3}}{3}$.

(If we recall (2.3), we can also get the same result directly.)

§4.4 Lines in $\mathbf{E}^3$

For $\mathbf{E}^3$, there is no simple and readily identifiable kind of scalar equation of the form F [x,y,z] = 0 such that every equation of this kind determines a straight line and every straight line is determined by an equation of this kind. Nevertheless, there are two useful ways to represent lines algebraically: (1) by a *system of linear scalar parametric equations or by a single linear vector-algebra parametric equation*, and (2) by a *system of linear scalar non-parametric equations or by a single linear vector-algebra non-parametric equation*.

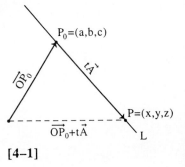

[4–1]

Parametric equations. This is the most useful and common way to represent a line in $\mathbf{E}^3$. We assume a Cartesian coordinate system. Let L be a given fixed line. Let P_0 be a fixed point on L given by (a,b,c), and let $\vec{A} = A_1\hat{i} + A_2\hat{j} + A_3\hat{k}$ be a non-zero vector parallel to L. See [4–1]. Then, for a variable point P given by (x,y,z):

P is on L $\Leftrightarrow$ $\overrightarrow{P_0P}$ is a scalar multiple of $\vec{A}$

$\Leftrightarrow$ $\overrightarrow{P_0P} = t\vec{A}$ for some scalar t

$$\Leftrightarrow (x-a)\hat{i} + (y-b)\hat{j} + (z-c)\hat{k} = tA_1\hat{i} + tA_2\hat{j} + tA_3\hat{k} \quad \text{for some } t$$
$$\Leftrightarrow \text{for some } t, \ x = a + A_1t, \ y = b + A_2t, \text{ and } z = c + A_3t.$$

The scalar quantity t is called a *parameter*. Intuitively, we can think of t as a measure of time and of P as the position of a moving point at time t. As t increases through all the real numbers, P traces out the line L in the direction of $\vec{A}$. When $t = 0$, P is at P_0.

Definition. The system of equations

$$(4.1) \qquad \left. \begin{array}{l} x = a + tA_1 \\ y = b + tA_2 \\ z = c + tA_3 \end{array} \right\} \quad t \text{ in } (-\infty, \infty),$$

where a, b, c, A_1, A_2, and A_3 are given real numbers such that $A_1^2 + A_2^2 + A_3^2 > 0$, is called a *system of linear scalar parametric equations on* $(-\infty,\infty)$ *in* $\mathbf{E}^3$. The notation $(-\infty,\infty)$ is another name for the set of all real numbers. (It expresses the set of all real numbers as an infinite interval from $-\infty$ to ∞.) (4.1) gives the coordinates x, y, and z (of P) as functions of t. We have the following general result:

(4.2)   (a)   *Every system of linear scalar parametric equations on* $(-\infty,\infty)$ *in* $\mathbf{E}^3$ *determines a unique straight line in* $\mathbf{E}^3$.

      (b)   *Every straight line in* $\mathbf{E}^3$ *is determined by some system of linear scalar parametric equations on* $(-\infty,\infty)$ *in* $\mathbf{E}^3$. (Note that we can get a different linear parametric system for the same line L by using a different point P_0 on L or by using a different scalar multiple of $\vec{A}$ in place of $\vec{A}$.)

If we let $\vec{R}_0$ be a fixed position vector for P_0 and let $\vec{R} = x\hat{i} + y\hat{j} + z\hat{k}$ be a position vector for P, then $\vec{R}$ may be thought of as a vector-valued function $\vec{R}(t)$, and the system (4.1) can be expressed by the single *linear vector-algebra parametric equation*

$$(4.3) \qquad\qquad \vec{R}(t) = \vec{R}_0 + t\vec{A}, \quad t \text{ in } (-\infty, \infty).$$

(We consider vector-valued functions in Chapter 5. Equation (4.3) does not use Cartesian coordinates and assumes only that we have a fixed reference-point for position vectors.)

Example 1. Find parametric equations for the line L *through the point* (1,2,–2) *and parallel to* $\hat{i}+\hat{j}$. Here we have $\vec{R}_0 = \hat{i}+2\hat{j}-2\hat{k}$ and $\vec{A}=\hat{i}+\hat{j}$. Our parametric vector equation for L is given by

$$x\hat{i}+y\hat{j}+z\hat{k} = \vec{R}(t) = \vec{R}_0 + t\vec{A} = (\hat{i}+2\hat{j}-2\hat{k}) + t(\hat{i}+\hat{j}) = (1+t)\hat{i} + (2+t)\hat{j} - 2\hat{k}.$$

This gives the system of scalar parametric equations

$$\left. \begin{array}{l} x = 1+t \\ y = 2+t \\ z = -2 \end{array} \right\} \quad t \text{ in } (-\infty, \infty).$$

Non-parametric equations. Consider a system of two equations, $\mathcal{F}[x,y,z] = 0$ and $\mathcal{G}[x,y,z] = 0$.

Definition. We say that an ordered triple (a,b,c) *satisfies* this system if (a,b,c) satisfies both equations in the system. We say that the system *determines* the figure S in $\mathbf{E}^3$ if S consists of all points whose coordinates satisfy the system.

The intersection of two non-parallel planes is a straight line L. If we have an equation for each plane, those two equations form a system which determines L. We then say that the equations of this system are *non-parametric equations for* L. Moreover, by geometry, for every line L, there must exist two non-parallel planes which intersect in L. Hence we have:

(4.4) (a) *The linear equations for two non-parallel planes form a system which determines a unique straight line in* $\mathbf{E}^3$.

  (b) *Every straight line in* $\mathbf{E}^3$ *is determined by some system of two linear equations for non-parallel planes.*

Let $A_1x + B_1y + C_1z + D_1 = 0$ and $A_2x + B_2y + C_2z + D_2 = 0$ be equations for the planes M_1 and M_2. It is easy to tell whether or not M_2 and M_2 are non-parallel planes. We simply check whether $\vec{N}_1 = A_1\hat{i} + B_1\hat{j} + C_1\hat{k}$ is a scalar multiple of $\vec{N}_2 = A_2\hat{i} + B_2\hat{j} + C_2\hat{k}$. We thus have: M_1 *and* M_2 *are non-parallel* $\Leftrightarrow \vec{N}_1$ *is not a scalar multiple of* $\vec{N}_2$.

Example 2. Consider the system x+y–1 = 0 *and* y–z–2 = 0. Here $\vec{N}_1 = \hat{i}+\hat{j}$ and $\vec{N}_2 = \hat{j}-\hat{k}$. Since $\vec{N}_1$ is not a scalar multiple of $\vec{N}_2$, this system gives non-parametric linear equations for a line L.

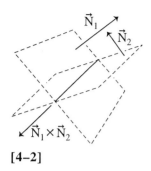

[4–2]

If we have a system of linear scalar *non-parametric* equations for a line L, we can use these equations to find a system of linear scalar *parametric* equations for L. Let $\vec{N}_1$ and $\vec{N}_2$ be normal vectors to the planes. Then $\vec{A} = \vec{N}_1 \times \vec{N}_2$ is a non-zero vector perpendicular to both $\vec{N}_1$ and $\vec{N}_2$, [4–2]. Hence, by geometry, $\vec{A}$ must be parallel to L. Setting one variable in the system = 0 and solving the system for the remaining two variables, we can find a point P_0 on L. P_0 and $\vec{A}$ then give parametric equations as in (4.1).

Example 3. *Find parametric equations for the line* L *in example 2 above.* For $\vec{A}$, we have $\vec{A} = \vec{N}_1 \times \vec{N}_2 = -\hat{i} + \hat{j} + \hat{k}$. Setting $z = 0$ in the system, we have

$$\left. \begin{array}{r} x + y = 1 \\ y = 2 \end{array} \right\} .$$

This gives $y = 2$ and $x = -1$. Hence the point $P_0 = (-1, 2, 0)$ lies on L. We therefore have the vector equation

$$\vec{R}(t) = (-\hat{i} + 2\hat{j}) + t(-\hat{i} + \hat{j} + \hat{k}),$$

and this gives the system of scalar equations:

$$(4.5) \qquad \left. \begin{array}{l} x = -1 - t \\ y = 2 + t \\ z = t \end{array} \right\} \quad t \text{ in } (-\infty, \infty).$$

Example 4. It is also easy to go from linear scalar *parametric* equations for a line L to linear scalar *non-parametric* equations for L, as we now show.

Case 1. If the parameter t appears in all three of the given parametric equations, we find two non-parametric equations by taking the three equations in two pairs and eliminating t from each pair. For example, if we begin with the parametric system in (4.5) and eliminate t between the first and second equations, we get $t = -x-1 = y-2 \Rightarrow x+y = 1$. If we also eliminate t between the first and third equations, we get $t = -x-1 = z \Rightarrow x+z = -1$. Then $\{x+y-1 = 0, x+z+1 = 0\}$ is a non-parametric system for the same line.

Case 2. If t appears in two of the given parametric equations but not in the third, we eliminate t from that pair. This result, together with the third equation, gives us our two non-parametric equations.

Case 3. If t appears in only one of the given parametric equations, we use the remaining two equations as our non-parametric equations.

A straight line can also be represented by a single *linear vector-algebra non-parametric equation.* Let $\vec{R}_0$ be the position vector of some fixed point on a line L, and let $\vec{A}$ be a vector parallel to L. Let $\vec{R}$ be a variable position vector for a point P. Then

(4.6) $$P \text{ is on } L \Leftrightarrow \vec{R} \times \vec{A} = \vec{R}_0 \times \vec{A}.$$

For a proof of (4.6), see the problems.

Remark. Let $\mathscr{F}[x,y,z] = 0$ and $\mathscr{G}[x,y,z] = 0$ form a system of two non-parametric linear equations for a straight line L. Then

$$(\mathscr{F}[x,y,z])^2 + (\mathscr{G}[x,y,z])^2 = 0$$

is a single *second-degree* equation which determines the line L. In this way, a line L can, in fact, be represented by a single scalar algebraic equation. Unfortunately, individual second-degree equations for lines are difficult to recognize and inconvenient to use. (See §4.6 below.)

§4.5 More Applications

We give several illustrative examples. For each example, we assume a given Cartesian system in $\mathbf{E}^3$.

Example 1. *Find* (i) *scalar parametric equations and* (ii) *scalar non-parametric equations for the line* L *through* $P_1 = (1,2,3)$ *and* $P_2 = (3, -2,1)$.

(i)   For *parametric* equations, we use $\vec{R}_0 = \overrightarrow{OP_1} = \hat{i} + 2\hat{j} + 3\hat{k}$ and $\vec{A} = \overrightarrow{OP_2} - \overrightarrow{OP_1} = 2\hat{i} - 4\hat{j} - 2\hat{k}$. This gives

$$\vec{R}(t) = \vec{R}_0 + t\vec{A} = (1 + 2t)\hat{i} + (2 - 4t)\hat{j} + (3 - 2t)\hat{k},$$

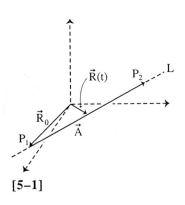

and we have the parametric system

$$\left.\begin{array}{l} x = 1 + 2t \\ y = 2 - 4t \\ z = 3 - 2t \end{array}\right\} \quad t \text{ in } (-\infty, \infty) \text{ ; see [5–1] .}$$

(ii) For *non-parametric* equations for L, we eliminate t from the first two parametric equations to get $2x+y = 4$ and from the last two parametric equations to get $y-2z = -4$. Thus we have

$$\left.\begin{array}{l} 2x + y = 4 \\ y - 2z = -4 \end{array}\right\}$$

as a non-parametric system.

Example 2. *Let* L *be the line in example 1. Does* L *intersect the plane* M *given by* x–y+z = 10? *If so, describe and find this intersection.*

We proceed as follows. $\vec{N} = \hat{i} - \hat{j} + \hat{k}$ is a normal vector to M. Take $\vec{A}$ is in example 1. Since $\vec{A} \cdot \vec{N} = 2 + 4 - 2 = 4 \neq 0$, we see, by geometry, that L cannot lie in M or be parallel to M. Hence L must intersect M at a single point. To find this point, we use the parametric equations for L and find the value of the parameter t for which x, y, and z satisfy the equation of the plane. That is to say, we substitute the parametric expressions for x, y, and z in the equation for M and then solve for t. We get

$$1 + 2t - 2 + 4t + 3 - 2t = 10$$
$$4t = 8$$
$$t = 2 .$$

Setting t = 2 in the parametric equations, we get the coordinates of our desired point: x = 5, y = –6, z = –1. Thus (5,–6, –1) is the point where L intersects M.

Example 3. *Let* M_1 *be the plane given by* x+y–z = 1, *and let* M_2 *be the plane given by* 2x–y+z = 2. *Does* M_1 *intersect* M_2? *If so, describe and find the intersection of* M_1 *and* M_2.

We proceed as follows: $\vec{N}_1 = \hat{i} + \hat{j} - \hat{k}$ and $\vec{N}_2 = 2\hat{i} - \hat{j} + \hat{k}$ are normal vectors to M_1 and M_2, respectively. Since $\vec{N}_1$ is not a scalar multiple of $\vec{N}_2$, we know that M_1 and M_2 must be distinct and non-parallel. Hence, by geometry, M_1 and M_2 intersect in a line L. The equations given for M_1 and M_2 are *non-parametric equations* for L. We can also find *parametric equations* for L from these non-parametric equations by the method of example 3 in §4.4. We get $\vec{A} = \vec{N}_1 \times \vec{N}_2 = -3\hat{j} - 3\hat{k}$. We get P_0 by setting $z = 0$ in the non-parametric equations. We get $x+y = 1$ and $2x-y = 2$. Solving these, we have $x = 1$ and $y = 0$. Hence $P_0 = (1,0,0)$. Thus we find

$$\vec{R}(t) = \vec{R}_0 + t\vec{A} = \hat{i} + t(-3\hat{j} - 3\hat{k}) = \hat{i} - 3t\hat{j} - 3t\hat{k}.$$

Our parametric system is:

$$\left.\begin{array}{l} x = 1 \\ y = -3t \\ z = -3t \end{array}\right\} \quad t \text{ in } (-\infty, \infty).$$

(Note that these parametric equations lead us, in turn, to an even simpler non-parametric system:

$$\left.\begin{array}{l} x = 1 \\ y = z \end{array}\right\} \quad .)$$

Example 4. *Let L_1 be the line given by $\vec{R}_1(t) = (2\hat{i} + 2\hat{j}) + t(\hat{i} + \hat{j} + \hat{k})$ and let L_2 be the line given by $\vec{R}_2(t) = 3\hat{i} + t(2\hat{i} - \hat{j} + \hat{k})$. Do L_1 and L_2 intersect? If so, describe and find their intersection.*

We proceed as follows. Since $\vec{A}_1 = \hat{i} + \hat{j} + \hat{k}$ is not a scalar multiple of $\vec{A}_2 = 2\hat{i} - \hat{j} + \hat{k}$, L_1 and L_2 cannot coincide. It follows that if they intersect, they must intersect at a single point. The parametric equations for L_1 and L_2 do not help us directly to test for, or find, a point P of intersection, since the parameter value from $\vec{R}_1(t)$ for such a P might be different from the parameter value from $\vec{R}_2(t)$. We therefore use non-parametric equations for L_1 and L_2. We find these from the parametric systems by the method of example 4 in §4.4. For L_1, we get $x-y = 0$ and $y-z = 2$. For L_2, we get $x+2y = 3$ and $y+z = 0$. (In each case, we have used parametric equations 1 and 2, and then 2 and 3, as our pairs.) We view these four equations as a system of four equations in three unknowns. Then L_1 and L_2 have a point of intersection $\Leftrightarrow$ this system has a solution. Solving the first three equations (in this case, direct substitution is easier than Cramer's rule), we get x = 1, y = 1, z = −1. We see that this is also a solution for the fourth equation. Hence L_1 and L_2 do intersect, and P = (1,1, −1) is their point of intersection.

Remark. In part (ii) of example 1, we calculated

and

$$\left.\begin{array}{l} x = 1 + 2t \\ y = 2 - 4t \end{array}\right\} \Rightarrow \frac{x-1}{2} = t = \frac{y-2}{-4} \Rightarrow 2x + y = 4 ,$$

$$\left.\begin{array}{l} y = 2 - 4t \\ z = 3 - 2t \end{array}\right\} \Rightarrow \frac{y-2}{-4} = t = \frac{z-3}{-2} \Rightarrow y - 2z = - 4 .$$

Here, the intermediate equations can be written in the combined form

$$\frac{x-1}{2} = \frac{y-2}{-4} = \frac{z-3}{-2} .$$

An expression of this combined form is sometimes called a "standard equation" for a line. Note that the denominators give the components of a vector $\vec{A}$ parallel to L, and that negatives of the coordinates of a point P_0 lying on L appear in the numerators. (This is the point for t = 0 in the given parametric equations.) In order for a line L to have a "standard equation," the parameter t must appear in all three parametric equations for L.

§4.6   Quadric Surfaces

Assume a given Cartesian system in $\mathbf{E}^3$.   Consider a single equation $\mathcal{F}[x,y,z] = 0$, where $\mathcal{F}[x,y,z]$ is a polynomial of second degree in x, y, and z with real-number coefficients. $\mathcal{F}[x,y,z]$ then has the form

$$Ax^2 + By^2 + Cz^2 + Dxy + Eyz + Fzx + Gx + Hy + Iz + J ,$$

with $A^2+B^2+C^2 > 0$. We call the equation $\mathcal{F}[x,y,z] = 0$ a *second-degree equation*.

In §4.1 we saw, for $\mathbf{E}^2$, that a figure S, determined by a second-degree equation $\mathcal{F}[x,y] = 0$, can always be represented, after an appropriate change of Cartesian coordinate system, by a second-degree equation in one of two standard, simplified forms. (See (1.4).) The same is true in $\mathbf{E}^3$ for a figure determined by a second-degree equation $\mathcal{F}[x,y,z] = 0$. The proof for $\mathbf{E}^2$ is given below in §4.7. The proof for $\mathbf{E}^3$ is indicated in §34.5. The result for $\mathbf{E}^3$ is as follows.

(6.1)   *For a given Cartesian coordinate system in* $\mathbf{E}^3$, *let* $\mathcal{F}[x,y,z] = 0$ *be a second-degree equation, and let* S *be the figure in* $\mathbf{E}^3$ *determined by this equation. Then either:*

(I)   *we can find a new Cartesian system in which* S *is determined by an equation of the form*

$$Ax^2 + By^2 + Cz^2 + J = 0, \text{ with } A^2 + B^2 + C^2 > 0 \text{ and } J = -1 \text{ or } 0; \text{ or}$$

(II)   *we can find a new Cartesian system in which* S *is determined by an equation of the form*

$$Ax^2 + By^2 + z = 0, \text{ with } A^2 + B^2 > 0.$$

In general, in (I) and (II) of (6.1), the new coefficients A,B,C, and J (for the given figure S in the new coordinate system) will differ from the old coefficients in the original second-degree equation.

When we have found a new equation for S of form (I) or form (II), the coefficients in this equation provide us with direct geometric information about S. The catalogue of possibilities is given in the following tables.

(I)   *Table for* $Ax^2 + By^2 + Cz^2 + J = 0$ *with* $A^2 + B^2 + C^2 > 0$ *and* $J = -1$ *or* 0.

The left-hand column of the table lists possible ways in which A, B, and C, in some order, can be positive, negative, or 0. Thus, for example, $+ - 0$ is the case where one of A,B,C is positive, one of A,B,C is negative, and one of A,B,C is 0.

(6.2a)

			$J = -1$	$J = 0$
+	+	+	*ellipsoid or sphere*	*single point*
−	−	−	*empty figure*	*single point*
+	+	−	*hyperboloid of one sheet*	*cone of two sheets*
+	−	−	*hyperboloid of two sheets*	*cone of two sheets*
+	+	0	*elliptical or circular cylinder*	*single line*
−	−	0	*empty figure*	*single line*
+	−	0	*hyperbolic cylinder*	*two intersecting planes*
+	0	0	*two parallel planes*	*single plane*
−	0	0	*empty figure*	*single plane*

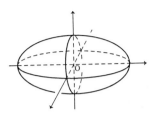

Ellipsoid

[6–1]

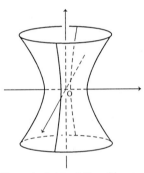

Hyperboloid of One Sheet

[6–2]

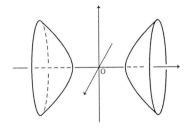

Hyperboloid of Two Sheets

[6–3]

(II)   *Table for* $Ax^2 + By^2 + z = 0$ *with* $A^2 + B^2 > 0$. The left-hand column lists possible ways for A and B, in either order, to be positive, negative, or 0.

(6.2b)

+	+	*elliptic paraboloid*
–	–	*elliptic paraboloid*
+	–	*hyperbolic paraboloid*
+	0	*parabolic cylinder*
–	0	*parabolic cylinder*

The tables above give a complete list of figures in $\mathbf{E}^3$ determined by second-degree equations in Cartesian coordinates. The cases of *ellipsoid*, *sphere*, *hyperboloids*, *paraboloids*, *cylinders*, and *cone* are called *principle cases*. The cases of *empty figure*, *point*, *line*, *plane*, or *two planes* are called *special cases* (or "borderline" or "degenerate" cases). The figures occurring as principal cases are known as *quadric surfaces*. It is easy to show, by elementary algebra and a change of Cartesian coordinates, that

(6.3)   *the intersection of a plane and a quadric surface is always a conic section.*

The reader need not memorize the detailed information in (6.2a) and (6.2b). The reader should, however, learn the names and general forms of the quadric surfaces, should remain aware of the connection between geometric form and coefficients in (I) and (II), and should know where to find the information given in (6.2) The general forms of principal-case quadric surfaces, except for cylinders, are indicated in [6–1], [6–2], [6–3], [6–4], [6–5], and [6–6].

§4.7 Changing from One Cartesian System to Another in $\mathbf{E}^2$

In what follows, we require a *Cartesian system in* $\mathbf{E}^2$ to be "right-handed" in the sense that a counterclockwise rotation by angle of $\frac{\pi}{2}$ carries the positive x axis to the positive y axis.

If we have a given Cartesian system in $\mathbf{E}^2$, there are two especially simple ways to introduce a new Cartesian system.

(1) We can keep the same origin and rotate the old axes by a given angle θ to obtain the new axes. This change of coordinates is called a *rotation* of the old coordinates.

(2) We can choose a new origin but keep the new x and y axes parallel, respectively, to the old x and y axes. This change of coordinates is called a *translation* of the old coordinates.

The following result is immediate.

(7.1) *We can reach any desired Cartesian system, from an initially given system, by proceeding in two steps:* (1) *We use a rotation to get axes in the desired axis directions. This gives us an intermediate system.* (2) *We then use a translation of this intermediate system to get our final desired system.*

The following theorem gives us (1.4).

Let S *be a figure in* $\mathbf{E}^2$ *determined, in a given Cartesian coordinate system, by the second-degree equation*

$$Ax^2 + By^2 + Cxy + Dx + Ey + F = 0, \text{ where } A^2 + B^2 > 0.$$

(7.2a) *By an appropriate rotation, we can get a new Cartesian system in* $\widetilde{x}$ *and* $\widetilde{y}$, *where the same figure* S *is determined by an equation of the form*

$$\widetilde{A}\widetilde{x}^2 + \widetilde{B}\widetilde{y}^2 + \widetilde{D}\widetilde{x} + \widetilde{E}\widetilde{y} + F = 0, \text{ with } \widetilde{A} \neq 0 \text{ and } F \text{ unchanged.}$$

(7.2b) *Case* (i). *If* $\widetilde{B} \neq 0$ *or* $\widetilde{E} = 0$ *in the result of (7.2a), we can then, by an appropriate translation, obtain a new Cartesian coordinate system in* $\overline{\widetilde{x}}$ *and* $\overline{\widetilde{y}}$, *in which the same figure* S *is determined by an equation of the form*

$$\widetilde{A}\overline{\widetilde{x}}^2 + \widetilde{B}\overline{\widetilde{y}}^2 + \overline{F} = 0, \text{ where } \widetilde{A} \text{ and } \widetilde{B} \text{ are the same as in (7.2a) and } \overline{F} \text{ is new.}$$

Case (ii). *If* $\widetilde{B} = 0$ *and* $\widetilde{E} \neq 0$ *in the result of (7.2a), we can then, by an appropriate translation, find a new Cartesian coordinate system in* $\overline{\widetilde{x}}$ *and* $\overline{\widetilde{y}}$ *in which the same figure* S *is determined by an equation of the form*

$$\widetilde{A}\overline{\widetilde{x}}^2 + \widetilde{E}\overline{\widetilde{y}} = 0, \text{ with } \widetilde{A} \text{ and } \widetilde{E} \text{ the same as in (7.2a).}$$

(7.2c) *Finally, in the result of case* (i) *of (7.2b), if* $\overline{F} \neq 0$, *we can divide by* $-\overline{F}$ *to get the equation*

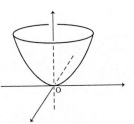

Elliptic Paraboloid

[6–4]

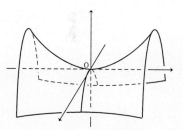

Hyperbolic Paraboloid

[6–5]

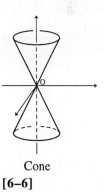

Cone

[6–6]

$$\widetilde{\widetilde{A}}\widetilde{x}^2 + \widetilde{\widetilde{B}}\widetilde{y}^2 - 1 = 0$$

for S, *where* $\widetilde{\widetilde{A}} = -\widetilde{A}/\widetilde{F}$ *and* $\widetilde{\widetilde{B}} = -\widetilde{B}/\widetilde{F}$; *and in the result of case* (ii) *of* (7.2b), *we can divide by* $\widetilde{E}$ *to get the equation*

$$\widetilde{\widetilde{A}}\widetilde{x}^2 + \widetilde{y} = 0 \ \ \textit{for } S, \textit{where} \ \ \widetilde{\widetilde{A}} = \widetilde{A}/\widetilde{E}.$$

Proof of (7.2a). For any given rotation, let x and y be coordinates in our original Cartesian system with $\hat{i}$ and $\hat{j}$ as coordinate vectors. Let $\widetilde{x}$ and $\widetilde{y}$ be coordinates in the new rotated system with $\hat{i}'$ and $\hat{j}'$ as coordinate vectors. Let P be a point with old coordinates (x,y) and new coordinates $(\widetilde{x},\widetilde{y})$. Let $\vec{R}$ be the position vector for P with respect to the origin (this origin is the same for both systems). Then

$$\vec{R} = x\hat{i} + y\hat{j} = \widetilde{x}\hat{i}' + \widetilde{y}\hat{j}'.$$

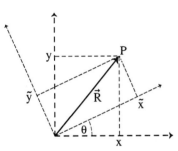

[7–1a]

It follows that

$$x = \vec{R} \cdot \hat{i} = \widetilde{x}\hat{i}' \cdot \hat{i} + \widetilde{y}\hat{j}' \cdot \hat{i} \ , \text{ and}$$
$$y = \vec{R} \cdot \hat{j} = \widetilde{x}\hat{i}' \cdot \hat{j} + \widetilde{y}\hat{j}' \cdot \hat{j}.$$

If we have rotated by the angle θ, we see from [7–1a,b] that $\hat{i}' \cdot \hat{i} = \cos\theta$, $\hat{j}' \cdot \hat{i} = -\sin\theta$, $\hat{i}' \cdot \hat{j} = \sin\theta$, and $\hat{j}' \cdot \hat{j} = \cos\theta$. Hence we have the defining equations for our change of coordinates:

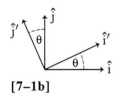

[7–1b]

(7.3)
$$\left. \begin{array}{l} x = \cos\theta\widetilde{x} - \sin\theta\widetilde{y} \\ y = \sin\theta\widetilde{x} + \cos\theta\widetilde{y} \end{array} \right\}.$$

Substituting (9.3) in the original second-degree equation, we get

(7.4)
$$A(\cos\theta\widetilde{x} - \sin\theta\widetilde{y})^2 + B(\sin\theta\widetilde{x} + \cos\theta\widetilde{y})^2 +$$
$$C(\cos\theta\widetilde{x} - \sin\theta\widetilde{y})(\sin\theta\widetilde{x} + \cos\theta\widetilde{y}) +$$
$$D(\cos\theta\widetilde{x} - \sin\theta\widetilde{y}) + E(\sin\theta\widetilde{x} + \cos\theta\widetilde{y}) + F = 0.$$

Multiplying out and collecting terms in (9.6), we find that our new equation is:

(7.5)
$$\tilde{A}\tilde{x}^2 + \tilde{B}\tilde{y}^2 + \tilde{C}\tilde{x}\tilde{y} + \tilde{D}\tilde{x} + \tilde{E}\tilde{y} + \tilde{F} = 0.$$

where

(7.6)
$$\tilde{A} = A\cos^2\theta + B\sin^2\theta + C\cos\theta\sin\theta$$
$$\tilde{B} = A\sin^2\theta + B\cos^2\theta - C\cos\theta\sin\theta$$
$$\tilde{C} = C(\cos^2\theta - \sin^2\theta) + 2(B-A)\sin\theta\cos\theta$$
$$\tilde{D} = D\cos\theta + E\sin\theta$$
$$\tilde{E} = -D\sin\theta + E\cos\theta$$
$$\tilde{F} = F.$$

By trigonometric identities, the expression for $\tilde{C}$ is

$$\tilde{C} = C\cos 2\theta + (B-A)\sin 2\theta.$$

Hence we can make $\tilde{C} = 0$ by choosing θ so that

$$(A-B)\sin 2\theta = C\cos 2\theta \Leftrightarrow \frac{A-B}{C} = \cot 2\theta.$$

This equation is satisfied by $\theta = \frac{1}{2}\cot^{-1}\left(\frac{A-B}{C}\right)$ and by $\theta = \frac{1}{2}\cot^{-1}\left(\frac{A-B}{C}\right) + \frac{\pi}{2}$.
Both choices of θ give $\tilde{C} = 0$. At least one choice gives $\tilde{A} \neq 0$, as we show in the problems. In this way we find a rotation that carries out (7.2a).

Example 1. Let $xy-1 = 0$ be our original equation. (Thus $A = B = D = E = 0$, $C = 1$, and $F = -1$.) Then $\theta = \frac{1}{2}\cot^{-1}(0) = \frac{\pi}{4}$, or $\theta = \frac{1}{2}\cot^{-1}(0) + \frac{\pi}{2} = \frac{3\pi}{4}$. For $\theta = \frac{\pi}{4}$, we have $\cos^2\theta = \sin^2\theta = \sin\theta\cos\theta = \frac{1}{2}$. By (7.6), this gives $\tilde{A} = \frac{1}{2}, \tilde{B} = -\frac{1}{2}, \tilde{C} = \tilde{D} = \tilde{E} = 0$, and $F = -1$. Our new equation is

$$\frac{1}{2}\tilde{x}^2 - \frac{1}{2}\tilde{y}^2 - 1 = 0.$$

Similarly, $\theta = \frac{3\pi}{4}$ gives the new equation

$$-\frac{1}{2}\tilde{x}^2 + \frac{1}{2}\tilde{y}^2 - 1 = 0.$$

***Proof of* (7.2b).** For case (i), we complete the squares $\tilde{x}$ and $\tilde{y}$ as indicated in the following example: Assume that after (7.2a) we have

$$2\tilde{x}^2 + \tilde{y}^2 + 4\tilde{x} - 2\tilde{y} - 7 = 0.$$

Completing the squares, we have

$$2\tilde{x}^2 + 4\tilde{x} + 2 + \tilde{y}^2 - 2\tilde{y} + 1 - 7 - 3 = 0$$

(7.7) $$2(\tilde{x}+1)^2 + (\tilde{y}-1)^2 - 10 = 0.$$

We then translate to a new coordinate system in $\tilde{\tilde{x}}$ and $\tilde{\tilde{y}}$, where the new origin is at $\tilde{x} = -1$, $\tilde{y} = 1$. The $\tilde{\tilde{x}}\,\tilde{\tilde{y}}$ system is related to the $\tilde{x}\,\tilde{y}$ system by the defining equations

$$\left.\begin{array}{l}\tilde{x} = \tilde{\tilde{x}} - 1\\\tilde{y} = \tilde{\tilde{y}} + 1\end{array}\right\}.$$

Substituting in (7.7), we have

(7.7a) $$2\tilde{\tilde{x}}^2 + \tilde{\tilde{y}}^2 - 10 = 0,$$

which is in the desired form.

For case (ii), we complete the square in $\tilde{x}$ as indicated in the following example: Assume that after (7.2a) we have

$$2\tilde{x}^2 + 4\tilde{x} - 2\tilde{y} - 7 = 0.$$

Completing the square, we have

$$2\tilde{x}^2 + 4\tilde{x} + 2 - 2\tilde{y} - 7 - 2 = 0$$

(7.8) $$2(\tilde{x}+1)^2 - 2(\tilde{y}-\frac{9}{2}) = 0.$$

We translate to a new coordinate system in $\tilde{\tilde{x}}$ and $\tilde{\tilde{y}}$ given by

$$\left.\begin{array}{l} \tilde{x} = \tilde{\tilde{x}} - 1 \\ \tilde{y} = \tilde{\tilde{y}} + \dfrac{9}{2} \end{array}\right\}.$$

Substituting in (7.8), we have

(7.8a) $\qquad\qquad\qquad\qquad 2\,\tilde{\tilde{x}}^2 - 2\,\tilde{\tilde{y}} = 0,$

which is in the desired form.

      Proof of (7.2c). This result is immediate. For (7.7a), it gives

$$\tfrac{2}{10}\,\tilde{\tilde{x}}^2 + \tfrac{1}{10}\,\tilde{\tilde{y}}^2 - 1 = 0,$$

and for (7.8a), it gives

$$-\tilde{\tilde{x}}^2 + \tilde{\tilde{y}} = 0.$$

      This concludes the proof of (7.2).

      Is there some way to get information about the result of (7.2a) without the computations required by the proof above? The following theorem gives an affirmative answer to this question.

(7.9)   *Let* A,B,C,D,E,F *and* $\tilde{A},\tilde{B},\tilde{C},\tilde{D},\tilde{E},\tilde{F}$ *be coefficients of second-degree equations for a figure* S *before and after an arbitrary rotation of axes. Then the following equations always hold:*

    (i) $\tilde{A} + \tilde{B} = A + B,$

    (ii) $4\tilde{A}\tilde{B} - \tilde{C}^2 = 4AB - C^2.$

In particular, for a rotation which yields $\tilde{C} = 0,$ *we have*

    (iii) $\tilde{A}\tilde{B} = AB - \tfrac{1}{4}C^2,$ *and*

    (iv) $\tilde{A}$ *and* $\tilde{B}$ *are the roots of the quadratic equation* $\lambda^2 - (A{+}B)\lambda +$ $(AB - \tfrac{1}{4}C^2) = 0.$

(i) and (ii) follow from (7.6) by straightforward algebraic calculation. (iii) is immediate from (ii). For (iv), observe that, by (i) and (ii),

$$(\lambda - \tilde{A})(\lambda - \tilde{B}) = \lambda^2 - (\tilde{A} + \tilde{B})\lambda + \tilde{A}\tilde{B} = \lambda^2 - (A + B)\lambda + (AB - \tfrac{1}{4}C^2).$$

(iv) enables us to calculate values for $\tilde{A}$ and $\tilde{B}$ (with $\tilde{C} = 0$) directly from A, B, and C. Knowledge of $\tilde{A}$ and $\tilde{B}$ then tells us which line in tables (1.5a) and (1.5b) applies. Thus, if our figure S turns out to be a principal case, we know ahead of time, without further calculation, what that principal case must be.

Example 2. For $xy-1 = 0$, we have $A+B = 0$ and $AB - \tfrac{1}{4}C^2 = -\tfrac{1}{4}$. Hence by (iv) of (9.9), $\lambda^2 + (-\tfrac{1}{4}) = 0 \Rightarrow \lambda = \pm\tfrac{1}{2} \Rightarrow [\tilde{A} = \tfrac{1}{2}$ and $\tilde{B} = -\tfrac{1}{2}]$ or $[\tilde{A} = -\tfrac{1}{2}$ and $\tilde{B} = \tfrac{1}{2}]$. (Which of these last alternatives is correct depends upon whether we rotate by $\tfrac{\pi}{4}$ or by $\tfrac{\pi}{4} + \tfrac{\pi}{2} = \tfrac{3\pi}{4}$, as we saw in example 1.) From this information about $\tilde{A}$ and $\tilde{B}$, we immediately conclude, without further calculation, that our figure S must be either a hyperbola or a pair of intersecting lines.

Analogous results hold for second-degree equations in $\mathbf{E}^3$. In particular, an analogous version of (7.9) is described in §34.5. The proofs for $\mathbf{E}^3$ use concepts and methods of *linear algebra*.

§4.8   Problems

§4.1

1. Show that: (1.2a) $\Leftrightarrow$ the line determined by $Ax+By+C = 0$ has slope $-\dfrac{A}{B}$. (*Note.* The symbol $\Rightarrow$ between two statements $\mathcal{A}$ and $\mathcal{B}$ asserts that whenever $\mathcal{A}$ is true, then $\mathcal{B}$ is true. When $\mathcal{A} \Rightarrow \mathcal{B}$, we say: "$\mathcal{A}$ *implies* $\mathcal{B}$" or "*if $\mathcal{A}$ then $\mathcal{B}$*" or "$\mathcal{A}$ *only if $\mathcal{B}$*" or "$\mathcal{A}$ *is a sufficient condition for $\mathcal{B}$*" or "$\mathcal{B}$ *if $\mathcal{A}$*" or "$\mathcal{B}$ *is a necessary condition for $\mathcal{A}$.*" The symbol $\Leftrightarrow$ between two statements asserts that (1) whenever $\mathcal{A}$ is true, then $\mathcal{B}$ is true, and also that (2) whenever $\mathcal{B}$ is true, then $\mathcal{A}$ is true. When $\mathcal{A} \Leftrightarrow \mathcal{B}$, we say: "$\mathcal{A}$ *is equivalent to $\mathcal{B}$*" or "$\mathcal{A}$ *if and only if $\mathcal{B}$*" or "$\mathcal{A}$ *is a necessary and sufficient condition for $\mathcal{B}$.*" In order to prove a statement of the form $\mathcal{A} \Leftrightarrow \mathcal{B}$, it is usually simplest and easiest to give two separate and independent proofs: first a proof that $\mathcal{A} \Rightarrow \mathcal{B}$, and then, second, a proof that $\mathcal{B} \Rightarrow \mathcal{A}$.)

2. (a) Deduce (1.5b) and the first column (F = −1) of (1.5a) from (1.3).
   (b) Prove the second column F = 0) of (1.5a).

§4.2

1. Find an equation for the plane through $(1,1,0)$ parallel to $\hat{j} - \hat{k}$ and $\hat{i} + 2\hat{j}$.

2. Prove (2.2b).

3. Prove (2.2c).

§4.3

1. In Cartesian coordinates, find the equation of the plane through the points $(0,0,0)$, $(1,1,2)$, $(2,1,-1)$.

2. In Cartesian coordinates, find the equation of the plane through the point $(2,1,-1)$ and parallel to the vectors $2\hat{i} + 3\hat{j} - \hat{k}$ and $\hat{i} + 2\hat{k}$.

3. Find an equation for the plane through the point $(1,-1,1)$ which makes right angles with the plane $2x+y-z = 4$ and with the plane $x+2y-z = 1$.

4. How far is the point P from the plane M, where P has the Cartesian coordinates $(1,-2,0)$ and M is given by $2x-2y+z = 0$?

5. A plane contains the three points: $(0,2,1)$, $(2,1,2)$, $(1,1,3)$.
   (a) Find the equation of the plane.
   (b) Find the distance from the origin to the plane. (See example 2 in §4.2.)
   (c) Find direction cosines, in the frame $\{\hat{i}, \hat{j}, \hat{k}\}$, of a "downward" normal vector from the plane. (See example 2 in §3.5.)

6. Find the distance between the parallel planes given by $x+2y+3z = 5$ and $x+2y+3z = 19$.

7. In Cartesian coordinates, let (a,b,c) be the coordinates of a point P and let $Ax+By+Cz = D$ be an equation for a plane M. Give a formula, in terms of a, b, c, A, B, C, and D for the distance of P from M. (*Hint.* Use [3–1] in the text and coordinates (a,b,c) for the point P_0.)

8. Consider the triangle with vertices $(0,0,0)$, $(1,1,1)$, and $(0,1,0)$, and the triangle with vertices $(0,0,0)$, $(1,1,1)$, and $(1,0,0)$. Find the acute dihedral angle between the planes of these triangles.

9. The planes M_1 and M_2 are given by the equations $x+2y+2z = 1$ and $3x-2y+6z = 3$. M_1 and M_2 intersect in a line L. Find an equation for the plane which contains L and bisects the acute dihedral angle between M_1 and M_2.

§4.4

1. A triangle has vertices $(1,0,0)$, $(2,1,0)$, and $(0,-1,1)$. Find parametric equations for the line which bisects the vertex angle at $(1,0,0)$.

2. A line is parallel to the vector $\hat{i} - 2\hat{j} + \hat{k}$ and contains the point $(2,0,-1)$.
   (a) Give parametric equations for L.
   (b) Give non-parametric equations for L.

3. A straight line is given by $\vec{R}(t) = (1-t)\hat{i}+3t\,\hat{j}+(1+t)\hat{k}$. For what values of c and d will this line lie in the plane $x+y+cz = d$?

4. Prove (4.6).

§4.5

1. In Cartesian coordinates, let L be the line through $(1,1,-2)$ parallel to the vector $(\hat{i} - \hat{j} - 2\hat{k})$. Where does L intersect the plane $2x-3y+z = 0$?

2. Find the coordinates of the point of intersection between the xy plane and the line through $(2,3,1)$ and $(-1,1,2)$.

3. Straight line L includes the two points with Cartesian coordinates $(-1,2,13)$ and $(0,3,15)$. At what point or points does L intersect the surface determined by the equation $z = x^2+y^2$?

4. Find the projection of the point $(1,1,1)$ on the plane $2x+5y-z = 11$.

5. Find the projection of the point $(3,-1,5)$ on the plane $2x-3y+z = 0$.

6. Find the angle between the plane x+y+z = 0 and the line given by the non-parametric equations x = y, y = 2z.

7. Show that for a system of Cartesian coordinates and a given line L:

    L has a *standard equation*

    $\Leftrightarrow$ L is not parallel to any of the three coordinate planes

    $\Leftrightarrow$ L is not orthogonal to any of the three coordinate axes.

    (See the end of §1.4 for a definition of *orthogonality* for skew lines.)

Note on doing analytic geometry without coordinates. Using vector algebra, we can develop general formulas for solving analytic-geometry problems, where these formulas do not use, or depend on, a coordinate system. The content of such formulas can be understood and visualized without reference to coordinates. At the same time, such formulas can be adapted and used in a coordinate system for the purpose of a particular numerical computation.

We begin by choosing a fixed reference point in E^3 (or E^2). We then represent each *point* P by its position vector $\vec{R}_P$. A *line* L can be represented by a formula of the form $\vec{A}t + \vec{B}$, where $\vec{A}$ is some fixed vector parallel to L, and $\vec{B}$ is the position vector of some fixed point on L. Here t is a parameter, and the vector equation $\vec{R}(t) = \vec{A}t + \vec{B}$, $(-\infty < t < \infty)$, gives a parametric expression for L. A plane M can be represented by an equation of the form $\vec{R} \cdot \vec{N} = d$, where $\vec{R}$ is a variable position vector, $\vec{N}$ is a fixed non-zero vector normal to M, and d is a fixed scalar such that $d = \vec{R}_Q \cdot \vec{N}$ for every point Q in M. (See the proofs of (2.1) and (2.2).)

For a very simple illustration, let $\vec{R} \cdot \vec{N} = d$ be a given plane M, and let P be a given point. We seek the plane which is parallel to M and contains P. We immediately find that this plane is $\vec{R} \cdot \vec{N} = \vec{R}_P \cdot \vec{N}$. (See example 1 in §4.3.)

Problems 8–14 below are of this kind.

8. Let a plane M be given by $\vec{R} \cdot \vec{N} = d$ for a given vector $\vec{N}$ and scalar d. Let P be a given point with position vector $\vec{R}_P$. Find a coordinate-free formula, in terms of $\vec{N}$, $\vec{R}_P$, and d, for the distance from P to M. (See example 5 in §4.3.)

9. A line L is represented by $\vec{R} = \vec{A}t + \vec{B}$, and a plane M is represented by $\vec{R} \cdot \vec{N} = d$, where $\vec{A}$, $\vec{B}$, $\vec{N}$, and d are given quantities.

    (a) Find an algebraic statement about $\vec{A}$, $\vec{N}$, and d which will be true $\Leftrightarrow$ L intersects M at a single point.

    (b) If L intersects M at a single point, find the point P of intersection (that is to say, find its position vector) in terms of $\vec{A}$, $\vec{B}$, $\vec{N}$, and d. (*Hint.* See example 2 in §4.5.)

10. A point P has position vector $\vec{R}_P$, and a plane has the equation $\vec{R} \cdot \vec{N} = d$. Find (the position vector for) the projection of P on M, in terms of $\vec{R}_P$, $\vec{N}$, and d.

11. A line L is given by $\vec{R} = \vec{A}t + \vec{B}$, and a plane M is given by $\vec{R} \cdot \vec{N} = d$. Find the projection of L on M in terms of $\vec{A}$, $\vec{B}$, $\vec{N}$, and d. (If the projection is a point, give its position vector. If the projection is a line express this line in the form $\vec{R} = \vec{A}'t + \vec{B}'$.)

12. Find the projection of a point P with position vector $\vec{R}_P$ on a line L given by $\vec{R} = \vec{A}t + \vec{B}$, in terms of $\vec{R}_P$, $\vec{A}$, and $\vec{B}$.

13. A line L_1 is given by $\vec{R} = \vec{A}_1 t + \vec{B}_1$ and a line L_2 by $\vec{R} = \vec{A}_2 t + \vec{B}_2$.

    (a) Find an algebraic statement about $\vec{A}_1, \vec{B}_1, \vec{A}_2,$ and $\vec{B}_2$ which will be true $\Leftrightarrow$ L_1 intersects L_2 at a single point.

(b) If L_1 intersects L_2 at a single point, find the position vector of this point in terms of $\vec{A}_1$, $\vec{B}_1$, $\vec{A}_2$, and $\vec{B}_2$.

14. A line L_1 is given by $\vec{R} = \vec{A}_1 t + \vec{B}_1$ and a line L_2 by $\vec{R} = \vec{A}_2 t + \vec{B}_2$.

   (a) Find an algebraic statement about $\vec{A}_1, \vec{B}_1$, $\vec{A}_2$, and $\vec{B}_2$ which will be true $\Leftrightarrow$ L_1 and L_2 are skew lines.

   (b) If L_1 and L_2 are skew lines, find $\vec{A}'$ and $\vec{B}'$ such that $\vec{R} = \vec{A}'t + \vec{B}'$ is the unique line which is perpendicular to both L_1 and L_2. (Obtain $\vec{A}'$ and $\vec{B}'$ in terms of $\vec{A}_1, \vec{B}_1, \vec{A}_2$, and $\vec{B}_2$.)

   (c) The length of the segment between L_1 and L_2 of the line found in (b) is called the *distance* between the skew lines L_1 and L_2. Find a scalar-valued formula of vector algebra which gives this distance in terms of $\vec{A}_1, \vec{B}_1, \vec{A}_2$, and $\vec{B}_2$.

§4.6

1. Let $\vec{A}$ be a constant vector; let b be a constant scalar; and let $\vec{R}$ be a variable position-vector. What figure is determined by the equation

$$\vec{R}^2 - 2\vec{A} \cdot \vec{R} - b = 0 \ ?$$

(*Hint.* Let $\vec{R} = x\hat{i} + y\hat{j} + z\hat{k}$. [This problem can also be done without coordinates: Try, by "completing the square," to express the left-hand side as the "difference of two squares" and then factorize. Finally, use problem 11 in §2.4.])

2. In a given Cartesian system in $\mathbf{E}^3$, a figure S is determined by a second-degree equation. Show that in any other Cartesian system, S is also determined by a second-degree equation.
   (*Hint.* Show that the defining equations for a change of Cartesian coordinates in $\mathbf{E}^3$ must be linear. See the derivation of (7.5) in §4.7.)

3. Prove (6.3). That is to say, show that the intersection of a plane M and a quadric surface S must be a conic section in M. Deduce that the projection of a circle onto a plane must be either an ellipse or a line segment.
   (*Hint.* Choose a new Cartesian system for which M is the xy plane. Then use the result of problem 2.)

§4.7

1. Prove (9.1) by giving further details.

2. The derivation following (7.6) gives two choices for the value of θ. Show that at least one choice gives $\tilde{A} \neq 0$.

3. Verify that the quadratic equation in (7.9 (iv)) always has real roots.
   (*Hint.* Use the quadratic formula.)

4. Using (7.9) to analyze a certain second-degree equation in x and y, you find that the equation determines a hyperbola. How can you also find, from your analysis, the angle between the asymptotic rays to one branch of the hyperbola?

5. Find the geometric nature of the figures determined by each of the following equations:
   (a) $x^2 + 2xy + y^2 + 1 = 0$
   (b) $x^2 + 2xy + y^2 - 1 = 0$
   (c) $3x^2 + 2\sqrt{2}xy + 4y^2 - 1 = 0$
   (d) $2x^2 - 4xy - y^2 + 1 = 0$

Chapter 5

Calculus of One-Variable Vector Functions

§5.1   One-variable Vector Functions

A *one-variable vector function* is a function which takes real numbers as inputs and yields vectors as outputs. For example, the position vector $\vec{R}(t)$ of a moving particle at time t (with respect to some fixed reference point in space) is a one-variable vector function in $\mathbf{E}^3$. This function has t as its independent variable. It takes values of t as inputs and gives values of the position vector $\vec{R}$ as outputs. The velocity and acceleration vectors of a moving particle are also one-variable vector functions of time. (*Velocity* and *acceleration* are mathematically defined in §5.3 below.) The *force* acting on a particle, considered as a function of time t, is also a one-variable vector function. The set of real numbers which a vector function $\vec{A}(t)$ takes as inputs is called the *domain* of $\vec{A}(t)$. We sometimes say that a function $\vec{A}(t)$ is "defined" on its domain.

If we choose a system of Cartesian coordinates in $\mathbf{E}^3$, then a given vector function $\vec{A}(t)$ can be expressed in this coordinate system as

$$\vec{A}(t) = a_1(t)\hat{i} + a_2(t)\hat{j} + a_3(t)\hat{k},$$

where each of the scalar components $a_1(t)$, $a_2(t)$, and $a_3(t)$ is a one-variable scalar function of the same independent variable t.

We can also *define* vector functions in Cartesian coordinates by specifying their components. For example, we might define a vector function by writing "Let $\vec{A}(t) = 2\cos t\,\hat{i} + 2\sin t\,\hat{j} + 3t\hat{k}$ for t in $[0,4\pi]$." (If we view this $\vec{A}(t)$ as the position-vector of a moving particle, $\vec{R}(t) = \vec{A}(t)$, then $\vec{R}(t)$ traces out the path of the particle: two turns of a helix about the z axis, [1–1].)

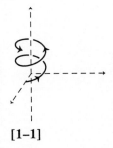

[1–1]

§5.2 Limits and Continuity

Let $\vec{A}(t)$ be a vector function on an interval [a,b] (as domain). Let $\bar{L}$ be a given fixed vector, and let c be a given fixed value in [a,b]. What does it mean to say that

$$\lim_{t \to c} \vec{A}(t) = \bar{L} \ ?$$

Our definition of limit will be based on the following intuitive idea: for $\vec{A}(t)$ to have $\bar{L}$ as a limit, the direction of $\vec{A}(t)$ should have the direction of $\bar{L}$ "as a limit," and the magnitude of $\vec{A}(t)$ should have the magnitude of $\bar{L}$ "as a limit." This suggests that we want

$$\lim_{t \to c} \vec{A}(t) = \bar{L}$$

to mean

$$\lim_{t \to c} (\bar{L} - \vec{A}(t)) = \vec{0} \ ,$$

and that this, in turn, should mean

$$\lim_{t \to c} \left|\bar{L} - \vec{A}(t)\right| = 0 \ .$$

In this last expression, $\left|\bar{L} - \vec{A}(t)\right|$ is a scalar function of t so that $\lim_{t \to c}\left|\bar{L} - \vec{A}(t)\right|$ is a scalar limit as defined in elementary calculus. Summarizing, we have the following.

(2.1)   *Definition.* If $\lim\limits_{t \to c} |\vec{L} - \vec{A}(t)| = 0$, we say that $\lim\limits_{t \to c} \vec{A}(t) = \vec{L}$.

(We review the elementary calculus definition of scalar limit at the end of this section.)

Taking components in a Cartesian system, we have $\vec{A}(t) = a_1(t)\hat{i} + a_2(t)\hat{j} + a_3(t)\hat{k}$ and $\vec{L} = \ell_1\hat{i} + \ell_2\hat{j} + \ell_3\hat{k}$. Then

$$\lim_{t \to c} |\vec{L} - \vec{A}(t)| = 0 \Leftrightarrow \lim_{t \to c} \sqrt{(\ell_1 - a_1(t))^2 + (\ell_2 - a_2(t))^2 + (\ell_3 - a_3(t))^2} = 0$$

$$\Leftrightarrow \lim_{t \to c} ((\ell_1 - a_1(t))^2 + (\ell_2 - a_2(t))^2 + (\ell_3 - a_3(t))^2) = 0$$

$$\Leftrightarrow [\lim_{t \to c} a_1(t) = \ell_1, \ and \ \lim_{t \to c} a_2(t) = \ell_2, \ and \ \lim_{t \to c} a_3(t) = \ell_3].$$

Thus $\lim\limits_{t \to c} \vec{A}(t) = \vec{L} \Leftrightarrow$ *each component of $\vec{A}$ has the corresponding component*

of $\vec{L}$ as its limit (where these component limits are scalar limits as defined in elementary calculus).

Definitions of *continuity* for vector functions have the same form as definitions of continuity for scalar functions as given in elementary calculus:

(2.2)   *Definition.* We say that $\vec{A}(t)$ is *continuous at* c if $\lim\limits_{t \to c} \vec{A}(t) = \vec{A}(c)$.

By our discussion above, this is the same as saying that each scalar component of $\vec{A}(t)$ is continuous at c.

(2.3)   *Definition.* We say that $\vec{A}(t)$ is *continuous on* [a,b] if $\vec{A}(t)$ is continuous at c for every c in [a,b].

This is the same as saying that each scalar component of $\vec{A}(t)$ is continuous on [a,b]. For example, $\vec{A}(t) = 2\cos t\hat{i} + 2\sin t\hat{j} + 3t\hat{k}$ is a continuous vector function, since, from elementary calculus, each of its components is a continuous scalar function.

The following theorems on continuity can be proved by expressing the vector functions in terms of their scalar components in a Cartesian system and then using corresponding theorems about scalar functions as proved in elementary calculus:

(2.4)   If $\bar{A}(t)$ *is a continuous vector function and* $c(t)$ *is a continuous scalar function, then* $c(t)\bar{A}(t)$ *is a continuous vector function.*

(2.5)   If $\bar{A}(t)$ *and* $\bar{B}(t)$ *are continuous vector functions, then* $\bar{A}(t) \cdot \bar{B}(t)$ *is a continuous scalar function, and* $\bar{A}(t) + \bar{B}(t)$ *and* $\bar{A}(t) \times \bar{B}(t)$ *are continuous vector functions.*

Review of the elementary calculus definition of scalar limit. Let f(x) be a scalar function, and let ℓ be a real number. We wish to define

$$\lim_{x \to c} f(x) = \ell \; .$$

Attempts to express the correct mathematical idea of limit in everyday language are almost always inadequate. We can say that f(x) "approaches ℓ as x approaches c" or that f(x) "gets closer and closer to ℓ as x gets closer and closer to c," but neither of these statements is quite right, because, for example, we need to include cases where, as x approaches c, the value of f(x) oscillates around ℓ with the size of these oscillations eventually "approaching zero." The correct and precise mathematical definition of limit is expressed graphically in [2–1a] and [2–1b].

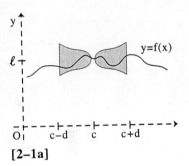

[2–1a]

For $\lim_{x \to c} f(x) = \ell$ to be true, there must be a symmetrical shaded region on some interval [c–d,c+d] with d > 0, where (i) the upper boundary of the region decreases to ℓ , and the lower boundary of the region increases to ℓ , as x approaches c, and where (ii) for x ≠ c, the graph of f(x) on [c–d,c+d] lies entirely in the shaded region. This region is called a *funnel region*. The function on [0,d] which defines the height of the upper right-hand boundary (and thereby defines, by symmetry, the shape of the entire funnel region) is called a *funnel function* for $\lim_{x \to c} f(x) = \ell$. We express this idea in the following definitions.

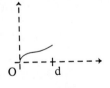

funnel function

[2–1b]

Definitions. A scalar function $\varepsilon(t)$ is called a *funnel function* on [0,d] if (i) $0 \le t_1 < t_2 \le d \Rightarrow 0 \le \varepsilon(t_1) < \varepsilon(t_2)$, and (ii) for every a > 0, there is a value t, 0 < t ≤ d, such that $0 < \varepsilon(t) < a$. Let f be a scalar function, and let c be a real number lying inside the domain of f or on the boundary of the domain of f. We then define

$$\lim_{x \to c} f(x) = \ell$$

to mean that for some $d > 0$ there is a funnel function $\varepsilon(t)$ on $[0,d]$ such that for every x in the domain of f,

$$0 < |x - c| \le d \implies |f(x) - \ell| \le \varepsilon(|x - c|).$$

The funnel function ε is not unique. For the same f, ℓ, and c, there can be many different choices of funnel function which show that $\lim_{x \to c} f(x) = \ell$.

The definition of $\lim_{x \to c} f(x) = \ell$ is similar. There must be an *infinite fun-*

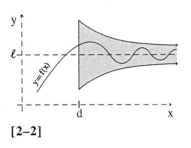

[2–2]

nel region on $[d,\infty)$ (for some d) whose upper boundary decreases asymptotically to ℓ and whose lower boundary increases asymptotically to ℓ, as in [2–2]. An infinite funnel region is defined by an appropriate *infinite* funnel function $\varepsilon(t)$ on $[d,\infty]$.

The funnel-function definition of *limit* allows us to make precise statements about limits.

> **Example.** *We wish to show, for any one variable functions* f *and* g, *that if*
>
> $$\lim_{x \to c} f(x) = k \quad and \quad \lim_{x \to c} g(x) = \ell,$$
>
> *then*
>
> $$\lim_{x \to c} (f(x) + g(x)) = k + \ell.$$

In this case, our hypothesis tells us that there is a funnel function $\varepsilon_1(t)$ on $[0,d_1]$ for $\lim_{x \to c} f(x) = k$, and there is a funnel function $\varepsilon_2(t)$ on $[0,d_2]$ for $\lim_{x \to c} g(x) = \ell$. We now define $\varepsilon(t) = \varepsilon_1(t) + \varepsilon_2(t)$ on $[0,d]$, where $d = \min[d_1, d_2]$. It is easy to show: (i) that $\varepsilon(t)$ on $[0,d]$ is a funnel function, and (ii) that

$$0 < |x - c| \le d \implies |f(x) + g(x) - k - \ell| \le \varepsilon(|x - c|).$$

(See the problems.) Thus the funnel function $\varepsilon(t)$ shows us that

$$\lim_{x \to c} (f(x) + g(x)) = k + \ell.$$

As we shall see later, the precise definition of limit is essential for understanding several subtle but central features of multivariable calculus.

§5.3 Derivatives

The derivative of a one-variable vector function is defined in exactly the same way as the derivative of a one-variable scalar function, except that the defining formulas use vector algebra rather than scalar algebra. Let $\vec{A}(t)$ be a vector function and let c be a given fixed value of t. Let Δt be a non-zero real number. We define the vector

$$\Delta \vec{A} = \vec{A}(c + \Delta t) - \vec{A}(c).$$

Thus $\Delta \vec{A}$ is a vector function of Δt. We next define the vector

$$\frac{\Delta \vec{A}}{\Delta t} = \left(\frac{1}{\Delta t}\right) \Delta \vec{A}$$

(where vector $\Delta \vec{A}$ is multiplied by the scalar $1/\Delta t$.) Then $\frac{\Delta \vec{A}}{\Delta t}$ is also a vector function of Δt.

(3.1) Definition. If $\lim_{t \to c} \frac{\Delta \vec{A}}{\Delta t} = \vec{D}$, for some vector $\vec{D}$, we say that $\vec{D}$ is the *derivative of the vector function* $\vec{A}(t)$ *at* c, and we say that the function $\vec{A}(t)$ is *differentiable at* c. We denote $\vec{D}$ as $\frac{d\vec{A}}{dt}\Big|_c$.

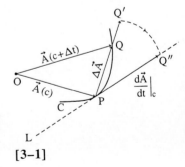

[3–1]

The geometric nature of this definition is indicated in [3–1], where we picture $\vec{A}(t)$ as a position-arrow tracing out a curve C. P is the position of the moving point at $t = c$, and Q is the position at $t = c + \Delta t$. Hence $\overrightarrow{PQ} = \Delta \vec{A}$. $\overrightarrow{PQ'}$ represents $\frac{\Delta \vec{A}}{\Delta t}$. (As $|\Delta t|$ becomes small, $|\overrightarrow{PQ'}|$ need not become small, since $\overrightarrow{PQ'}$ is the result of multiplying the small vector $\Delta \vec{A}$ by the large scalar $\frac{1}{\Delta t}$.)

Finally, $\overrightarrow{PQ''}$ represents $\left.\dfrac{d\vec{A}}{dt}\right|_c$, the limiting vector approached by $\overrightarrow{PQ'}$ as Δt approaches zero. (Note that the arrows $\overrightarrow{PQ}$, $\overrightarrow{PQ'}$, and $\overrightarrow{PQ''}$ in this figure are not drawn as position-arrows from O, but are drawn, instead, from the common initial point P.) If $\left.\dfrac{d\vec{A}}{dt}\right|_c \neq \vec{0}$, let L be the line of the arrow $\overrightarrow{PQ''}$. Since $\overrightarrow{PQ''} = \left.\dfrac{d\vec{A}}{dt}\right|_c$, L is the limiting position of the line through P and Q as Δt approaches zero. It is natural to call the line L the *tangent line at* P *to the curve* C.

(3.2)   *If* $\vec{A}(t)$ *is differentiable at* c, *then* $\vec{A}(t)$ *must be continuous at* c.

This is true because $\lim\limits_{\Delta t \to 0} \dfrac{\Delta\vec{A}}{\Delta t} = \left.\dfrac{d\vec{A}}{dt}\right|_c \Rightarrow \lim\limits_{\Delta t \to 0}\left|\Delta\vec{A}\right| = 0 \Rightarrow \lim\limits_{\Delta t \to 0} \Delta\vec{A} = \vec{0} \Rightarrow$ $\lim\limits_{\Delta t \to 0} \vec{A}(c + \Delta t) = \vec{A}(c) \Rightarrow \vec{A}(t)$ is continuous at c, by the definition of continuity.

Definitions. If $\vec{A}(t)$ is differentiable at c for every c in [a,b], we say that $\vec{A}(t)$ is *differentiable* on [a,b]. If $\vec{A}(t)$ is differentiable on [a,b], we can define a vector function $\vec{D}(t)$ on [a,b] by letting $\vec{D}(c) = \left.\dfrac{d\vec{A}}{dt}\right|_c$ for each c in [a,b]. We denote this vector function as $\dfrac{d\vec{A}}{dt}$ and call it the *derivative of* $\vec{A}(t)$ *on* [a,b].

Let $\vec{A}(t) = a_1(t)\hat{i} + a_2(t)\hat{j} + a_3(t)\hat{k}$. Then, for a given fixed c, we have $\Delta\vec{A} = \Delta a_1\hat{i} + \Delta a_2\hat{j} + \Delta a_3\hat{k}\ \Delta\vec{A}$, where $\Delta a_1 = a_1(c + \Delta t) - a_1(c)$, and where Δa_2 and Δa_3 are defined similarly. Hence $\dfrac{\Delta\vec{A}}{\Delta t} = \dfrac{\Delta a_1}{\Delta t}\hat{i} + \dfrac{\Delta a_2}{\Delta t}\hat{j} + \dfrac{\Delta a_3}{\Delta t}\hat{k}$. It follows from (2.1) that

$$\frac{d\vec{A}}{dt} = \frac{da_1}{dt}\hat{i} + \frac{da_2}{dt}\hat{j} + \frac{da_3}{dt}\hat{k} \ .$$

(3.3)   *Thus* (i) *the components of the derivative of* $\vec{A}(t)$ *are given by the derivatives of the components of* $\vec{A}(t)$, *and*

(ii) *a vector function* $\vec{A}(t)$ *is differentiable at* c $\Leftrightarrow$ *each of the scalar component functions is differentiable at* c.

Example.   If $\vec{A}(t) = 2\cos t\hat{i} + 2\sin\hat{j} + 3t\hat{k}$,   then $\dfrac{d\vec{A}}{dt} = -2\sin t\hat{i} +$ $2\cos\hat{j} + 3\hat{k}$.

If $\dfrac{d\vec{A}}{dt}$ is itself a differentiable function, then we write the derivative of $\dfrac{d\vec{A}}{dt}$ as $\dfrac{d^2\vec{A}}{dt^2}$. Using components, we have

$$\frac{d^2\vec{A}}{dt^2} = \frac{d^2a_1}{dt^2}\hat{i} + \frac{d^2a_2}{dt^2}\hat{j} + \frac{d^2a_3}{dt^2}\hat{k} \ .$$

For the example above, this gives

$$\frac{d^2\vec{A}}{dt^2} = -2\cos t\,\hat{i} - 2\sin t\,\hat{j}.$$

Higher derivatives, $\dfrac{d^3\vec{A}}{dt^3}, \dfrac{d^4\vec{A}}{dt^4}, \ldots,$ may be found in the same way, where each new derivative is a vector function.

In order to indicate the geometric significance of $\left.\dfrac{d^2\vec{A}}{dt^2}\right|_c$, we need a different figure from the figure used above to describe $\left.\dfrac{d\vec{A}}{dt}\right|_c$. For $\left.\dfrac{d^2\vec{A}}{dt^2}\right|_c$, we use a figure in which the vector $\dfrac{d\vec{A}}{dt}$ appears as a position vector tracing out a curve. This curve, traced out by the position vector $\vec{R}(t) = \dfrac{d\vec{A}}{dt}$, is called the

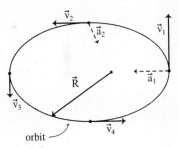

[3–2a]

hodograph of the original vector function $\vec{A}(t)$. [3–2b] shows the hodograph of the elliptical motion [3–2a] of a planet around the sun. It can be shown from Newton's law of gravity and Newton's second law of motion that the hodograph of a planet's elliptical motion around the sun must be a circle. (See §7.7.)

If t is time and $\vec{R}(t)$ is the position vector of a moving particle, then the vector function $\dfrac{d\vec{R}}{dt}$ (if $\vec{R}(t)$ is differentiable) is called the *velocity* of the particle, and the vector function $\dfrac{d^2\vec{R}}{dt^2}$ (if $\dfrac{d\vec{R}}{dt}$ is differentiable) is called the *acceleration* of

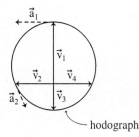

[3–2b]

the particle. (This is the vector quantity which appears as "acceleration" in Newton's second law of motion. See example 4 in §2.2.) $\dfrac{d\vec{R}}{dt}$ is sometimes abbreviated as $\vec{v}$ or $\vec{v}(t)$, and $\dfrac{d^2\vec{R}}{dt^2}$ is sometimes abbreviated as $\vec{a}$ or $\vec{a}(t)$. The terms "velocity" and "acceleration" are also sometimes used to refer to the first and second derivatives of an arbitrary vector function $\vec{A}(t)$, even when $\vec{A}$ is not viewed as a position vector, and t does not represent time.

Dot notation. In order to have more readable formulas, we occasionally use the following abbreviations for derivatives of vector and scalar functions. By convention, this notation is only used when the independent variable is t. For a vector function $\vec{A}(t)$,

$$\frac{d\vec{A}}{dt} \text{ may be abbreviated } \dot{\vec{A}}(t),$$

and

$$\frac{d^2\vec{A}}{dt^2} \text{ may be abbreviated } \ddot{\vec{A}}(t).$$

For a scalar function a(t),

$$\frac{da}{dt} \text{ may be abbreviated } \dot{a}(t),$$

and

$$\frac{d^2a}{dt^2} \text{ may be abbreviated } \ddot{a}(t).$$

§5.4   Differentiation Rules

The following rules can be used to simplify derivative calculations with vector and scalar functions of one variable. Let $\vec{A}(t)$ and $\vec{B}(t)$ be differentiable vector functions, and let a(t) be a differentiable scalar function. Then

(4.1)
$$\frac{d}{dt}(\vec{A} + \vec{B}) = \frac{d\vec{A}}{dt} + \frac{d\vec{B}}{dt},$$

(4.2)
$$\frac{d}{dt}(a\vec{A}) = \frac{da}{dt}\vec{A} + a\frac{d\vec{A}}{dt},$$

(4.3)
$$\frac{d}{dt}(\vec{A} \cdot \vec{B}) = \frac{d\vec{A}}{dt} \cdot \vec{B} + \vec{A} \cdot \frac{d\vec{B}}{dt},$$

(4.4) $$\frac{d}{dt}(\vec{A} \times \vec{B}) = \frac{d\vec{A}}{dt} \times \vec{B} + \vec{A} \times \frac{d\vec{B}}{dt}.$$

The proofs for these rules use the definitions of derivative and limit, and are similar in form to proofs of the differentiation rules for *sum* and *product* in elementary calculus. Like the elementary calculus proofs, the vector proofs begin with appropriate *limit theorems*. These theorems assert:

(4.5) $$\lim_{t \to c} (\vec{A}(t) + \vec{B}(t)) = \lim_{t \to c} \vec{A}(t) + \lim_{t \to c} \vec{B}(t),$$

(4.6) $$\lim_{t \to c} (a(t)\vec{A}(t)) = (\lim_{t \to c} a(t))(\lim_{t \to c} (\vec{A}(t)),$$

(4.7) $$\lim_{t \to c} (\vec{A}(t) \cdot \vec{B}(t)) = (\lim_{t \to c} \vec{A}(t)) \cdot (\lim_{t \to c} \vec{B}(t)), \text{ and}$$

(4.8) $$\lim_{t \to c} (\vec{A}(t) \times \vec{B}(t)) = (\lim_{t \to c} \vec{A}(t)) \times (\lim_{t \to c} \vec{B}(t)).$$

These limit theorems may be proved by using Cartesian components and applying the limit theorems of elementary calculus. The limit theorems of elementary calculus are proved by constructing appropriate new funnel functions from given funnel functions, as illustrated at the end of §5.2.

(4.9) **Chain rule.** The *chain rule for one-variable vector functions* asserts the following. *Let* $\vec{A}(t)$ *be a differentiable vector function of* t, *and let* t(u) *be a differentiable scalar function of* u. *Let* b *be a given value of* u, *and let* c = t(b). *Then* $\vec{A}(t(u))$ *is a differentiable vector function of* u, *and*

$$\left.\frac{d\vec{A}}{du}\right|_b = \left.\frac{d\vec{A}}{dt}\right|_c \left.\frac{dt}{du}\right|_b$$

where $\frac{d\vec{A}}{du} = \frac{d}{du}\vec{A}(t(u))$ *and* $\frac{dt}{du} = \frac{d}{du}t(u)$.

We usually abbreviate this chain rule as $\frac{d\vec{A}}{du} = \frac{d\vec{A}}{dt}\frac{dt}{du}$.

The chain rule can be proved by expressing $\vec{A}(t)$ in terms of its components and applying the chain rule of elementary calculus. (It can also be proved directly as follows. For $\Delta u \neq 0$, let $\Delta t = t(b+\Delta u)-t(b)$ and $\Delta \vec{A} = \vec{A}(c + \Delta t) - \vec{A}(c)$. Define $\vec{\varepsilon}$ as a vector function of Δt by:

$$\vec{\varepsilon} = 0 \quad \text{if} \quad \Delta t = 0,$$

$$\vec{\varepsilon} = \frac{\Delta \vec{A}}{\Delta t} - \frac{d\vec{A}}{dt}\bigg|_c \quad \text{if} \quad \Delta t \neq 0.$$

Then: (i) $\Delta \vec{A} = \dfrac{d\vec{A}}{dt}\bigg|_c \Delta t + \vec{\varepsilon}\Delta t \;$ for all Δt, and

(ii) $\lim\limits_{t \to c} \vec{\varepsilon} = \vec{0}$

Hence $\dfrac{\Delta \vec{A}}{\Delta u} = \dfrac{d\vec{A}}{dt}\bigg|_c \dfrac{\Delta t}{\Delta u} + \vec{\varepsilon}\dfrac{\Delta t}{\Delta u}.$

Taking limits as $\Delta u \to 0$, noting that $\lim\limits_{\Delta t \to 0} \Delta t = 0$ (because $t(u)$ differentiable $\Rightarrow$ $t(u)$ continuous), and applying (4.6), we get

$$\frac{d\vec{A}}{du}\bigg|_b = \frac{d\vec{A}}{dt}\bigg|_c \frac{dt}{du}\bigg|_b \;.)$$

§5.5 Derivatives of Unit Vectors

Let $\vec{A}(t)$ be a one-variable vector function with the special property that for all inputs t, $\left|\vec{A}(t)\right| = 1$. Thus $\vec{A}(t)$ is a vector whose direction may change with time t, but whose length remains constant and equal to 1. We say that $\vec{A}(t)$ is a *unit-vector function*, and we sometimes write it as $\hat{A}(t)$. The following theorem gives two facts about unit-vector functions. These facts will be used in Chapter 6.

Theorem. *Let $\hat{A}(t)$ be a differentiable unit-vector function.*

(5.1)   (i) *For all inputs t, $\dfrac{d\hat{A}}{dt} \cdot \hat{A} = 0$. Thus $\dfrac{d\hat{A}}{dt}$ is always orthogonal to $\hat{A}$.*

(5.2)   (ii) *Let $\Delta \alpha$ be the angle (in radians) between $\hat{A}(c+\Delta t)$ and $\hat{A}(c)$.*

Then $\lim\limits_{\Delta t \to 0} \dfrac{\Delta \alpha}{\Delta t} = \left|\dfrac{d\vec{A}}{dt}\bigg|_c\right|.$ *Thus* $\left|\dfrac{d\hat{A}}{dt}\right|,$ *the magnitude of* $\dfrac{d\hat{A}}{dt},$ *gives the instantaneous rate of angular change of the direction of $\hat{A}$.*

Proof of (i). Since $\hat{A}$ is a unit vector, we have $\hat{A}\cdot\hat{A} = 1$. Differentiating both sides of this equation and using (4.3), we have

$$\frac{d}{dt}(\hat{A}\cdot\hat{A}) = \dot{\hat{A}}\cdot\hat{A} + \hat{A}\cdot\dot{\hat{A}} = 0 \ .$$

Hence $2\dot{\hat{A}}\cdot\hat{A} = 0$, and $\dot{\hat{A}}\cdot\hat{A} = 0$.

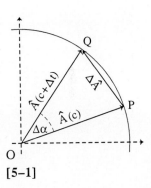

[5–1]

Proof of (ii). In [5–1], $\overrightarrow{OP} = \hat{A}(c)$, $\overrightarrow{OQ} = \hat{A}(c+\Delta t)$, and $\Delta t > 0$. Thus $\left|\overrightarrow{OP}\right| = \left|\overrightarrow{OQ}\right| = 1$, $\left|\Delta\hat{A}\right| = \left|\overrightarrow{PQ}\right|$, and $\Delta\alpha$ is the angle between $\overrightarrow{OP}$ and $\overrightarrow{OQ}$. By the law of cosines, we have $\left|\Delta\hat{A}\right|^2 = 2 - 2\cos\Delta\alpha$, and hence $\dfrac{\left|\Delta\hat{A}\right|^2}{\Delta\alpha^2} = \dfrac{2 - 2\cos\Delta\alpha}{\Delta\alpha^2}$.

By l'Hopital's rule from elementary calculus, we get $\lim\limits_{\Delta\alpha\to 0}\dfrac{\left|\Delta\hat{A}\right|}{\Delta\alpha} = 1$. (See the problems.) From this it follows that

$$\left|\frac{d\vec{A}}{dt}\right|_c = \left|\lim_{\Delta t\to 0}\frac{\Delta\hat{A}}{\Delta t}\right| = \lim_{\Delta t\to 0}\left|\frac{\Delta\hat{A}}{\Delta t}\right| = \lim_{\Delta t\to 0}\frac{\left|\Delta\hat{A}\right|}{\Delta t}$$

$$= \lim_{\Delta t\to 0}\left(\frac{\left|\Delta\hat{A}\right|}{\Delta\alpha}\frac{\Delta\alpha}{\Delta t}\right) = \left(\lim_{\Delta t\to 0}\frac{\left|\Delta\hat{A}\right|}{\Delta\alpha}\right)\left(\lim_{\Delta t\to 0}\frac{\Delta\alpha}{\Delta t}\right) = \lim_{\Delta t\to 0}\frac{\Delta\alpha}{\Delta t} \ ,$$

by the limit theorem of elementary calculus that $[\lim f(x) = k$ and $\lim g(x) = \ell\,] \Rightarrow \lim(f(x)g(x)) = k\ell$.

To summarize (5.1) and (5.2): if $\hat{A}(t)$ is a differentiable unit-vector function, then the *direction* of $\dfrac{d\hat{A}}{dt}$ must be orthogonal to $\hat{A}$, and the *magnitude* of $\dfrac{d\hat{A}}{dt}$ must give the angular rate at which the direction of $\hat{A}$ is changing.

§5.6 An Application: Motion of a Center of Mass

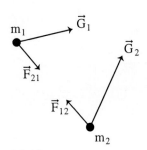

[6–1]

Let P_1 be a particle of mass m_1, and let P_2 be a particle of mass m_2, [6–1]. We treat these particles as a combined *physical system* moving through space. As the particles move, P_1 exerts a force $\vec{F}_{12}$ on P_2, and P_2 exerts a force $\vec{F}_{21}$ on P_1. (These forces could be gravitational or electromagnetic, for example, and they may vary with time and with the positions of P_1 and P_2.) We call them *internal* forces of the system. *External* forces, from sources outside the system, also act on the particles. Let $\vec{G}_1$ be the external force on P_1, and let $\vec{G}_2$ be the external force on P_2. $\vec{G}_1$ and $\vec{G}_2$ may also vary with time and with the positions of P_1 and P_2. Let $\vec{R}_1$ and $\vec{R}_2$ be position vectors for P_1 and P_2. Then $\vec{R}_1$, $\vec{R}_2$, $\vec{F}_{12}$, $\vec{F}_{21}$, $\vec{G}_1$, and $\vec{G}_2$ are all vector functions of time t.

We define a new position vector,

$$(6.1) \qquad \vec{R}(t) = \frac{m_1\vec{R}_1(t) + m_2\vec{R}_2(t)}{m_1 + m_2}.$$

Thus, at any moment t, $\vec{R}(t)$ is a weighted average (by mass) of the position-vectors for P_1 and P_2. This new point is called the *center of mass* of the system at time t. Taking the second derivative of each side of (6.1), we get

$$\ddot{\vec{R}}(t) = \frac{m_1\ddot{\vec{R}}_1(t) + m_2\ddot{\vec{R}}_2(t)}{m_1 + m_2}, \text{ and hence}$$

$$(m_1 + m_2)\ddot{\vec{R}}(t) = m_1\ddot{\vec{R}}_1(t) + m_2\ddot{\vec{R}}_2(t).$$

By the law of composition of forces and Newton's second-law of motion,

$$m_1\ddot{\vec{R}}_1(t) = \vec{F}_{21} + \vec{G}_1 \quad \text{and} \quad m_2\ddot{\vec{R}}_2(t) = \vec{F}_{12} + \vec{G}_2.$$

Thus $(m_1 + m_2)\ddot{\vec{R}} = \vec{F}_{21} + \vec{G}_1 + \vec{F}_{12} + \vec{G}_2$.

By Newton's third law of motion, $\vec{F}_{21} = -\vec{F}_{12}$. This gives us:

$$(6.2) \qquad (m_1 + m_2)\ddot{\vec{R}} = \vec{G}_1 + \vec{G}_2.$$

(6.2) tells us that the acceleration, and hence the motion, of the center of mass must be identical with the acceleration, and hence the motion, that would occur for a single particle of mass $m_1 + m_2$ acted on by the sum of the external forces $\vec{G}_1 + \vec{G}_2$.

The definition of *center of mass* and the result (6.2) are easily generalized to a system with any number of particles (for example, a galaxy of stars or a cloud of water droplets). We thus obtain a fundamental principle of mechanics: *if we are given a system subject to various internal and external forces which act at various places in the system, then the path of the center of mass of this system is the same as the path would be if the entire mass of the system were concentrated in a single particle at the center-of-mass point and if all the external forces were acting on this particle.*

§5.7   Indefinite Integrals

In §5.3 and §5.4, we saw that we can differentiate a vector function by differentiating its scalar components in a Cartesian coordinate system. We now show that we can integrate a vector function by integrating its scalar components in a Cartesian coordinate system. We consider *indefinite integrals in §5.7*, and then *definite integrals in §5.8*.

The following definitions and results about vector functions are closely parallel to definitions and results for scalar functions in elementary calculus.

Definition. Let $\vec{A}(t)$ and $\vec{F}(t)$ be vector functions. We say that $\vec{F}(t)$ is an *indefinite integral* (or *antiderivative*) of $\vec{A}(t)$ if $\vec{F}(t)$ is differentiable and $\dfrac{d\vec{F}}{dt} = \vec{A}(t)$.

We observe that a given vector function $\vec{A}(t)$ can have more than one indefinite integral. For example, $\vec{F}_1(t) = t^2\hat{i} + \hat{j}$ and $\vec{F}_2(t) = t^2\hat{i} - 2\hat{k}$ are both indefinite integrals of $\vec{A}(t) = 2t\hat{i}$.

The notation $\int \vec{A}(t)dt$ is sometimes used to denote a chosen indefinite integral of $\vec{A}(t)$. Thus, for $\vec{A}(t) = 2t\hat{i}$, we could write $\int \vec{A}(t)dt = t^2\hat{i} + \hat{j}$, or $\int \vec{A}(t)dt = t^2\hat{i} - 2\hat{k}$, or $\int \vec{A}(t)dt = t^2\hat{i} + \vec{C}$ for any chosen constant vector $\vec{C}$.

(7.1)   *Let $\vec{A}(t)$ be a given vector function. Let $\vec{F}_1(t)$ and $\vec{F}_2(t)$ be two indefinite integrals of $\vec{A}(t)$. Then $\vec{F}_1(t) - \vec{F}_2(t)$ must be a constant vector function.*

We prove (7.1) by taking scalar components (in Cartesian coordinates) of $\vec{A}(t)$, $\vec{F}_1(t)$, and $\vec{F}_2(t)$. By §5.3, the scalar components of $\vec{A}(t)$ are derivatives of the corresponding scalar components of $\vec{F}_1(t)$ and also of the corresponding scalar components of $\vec{F}_2(t)$. Hence, by elementary calculus, each scalar component of $\vec{F}_1(t)$ must differ from the corresponding scalar component of $\vec{F}_2(t)$ by a scalar constant. Let $\vec{C}$ be the constant vector whose scalar components are these scalar constants. Then $\vec{F}_1(t) - \vec{F}_2(t) = \vec{C}$.

It is immediate from (7.1) that if $\vec{F}(t)$ is a given indefinite integral of $\vec{A}(t)$, then any other indefinite integral $\vec{F}^*(t)$ of $\vec{A}(t)$ must satisfy

$$\vec{F}^*(t) = \vec{F}(t) + \vec{C}$$

for some appropriately chosen constant vector $\vec{C}$. We sometimes use the equation

(7.2) $$\int \vec{A}(t)dt = \vec{F}(t) + \vec{C}$$

(where $\vec{F}(t)$ is some chosen indefinite integral of $\vec{A}(t)$ and $\vec{C}$ is called an *arbitrary constant vector*) to express the collection of all indefinite integrals of $\vec{A}(t)$.

Calculation of indefinite integrals is usually done by using Cartesian components. If we know the components of $\vec{A}(t)$, we can use elementary calculus to find indefinite integrals of these components. These indefinite integrals then give us the components of an indefinite integral of $\vec{A}(t)$. For example, if

$$\vec{A}(t) = 2\cos t\, \hat{i} + 2\sin t\, \hat{j} + 3t\hat{k} \ ,$$

we get

$$\int \vec{A}(t)dt = 2\sin t\, \hat{i} - 2\cos t\, \hat{j} + \tfrac{3}{2}t^2\hat{k} + \vec{C}$$

as a general expression for the collection of all indefinite integrals of $\vec{A}(t)$.

Applications of indefinite integrals.

Example 1. Let a portion of E^3, *with Cartesian coordinates, represent a region of physical space at and immediately above the earth's surface. We assume that the region is sufficiently small that the xy plane serves as a good approximation to the earth's surface. At time* t = 0, *a projectile of mass* m *is fired from a gun located at the origin. The initial velocity of the projectile, as it leaves the gun, is* $\vec{v}_0$. *We neglect air resistance and assume that the only force on the projectile, after it is fired, is the force of gravity.* See [7–1]. *We wish to find* $\vec{R}(t)$, *the path of the projectile.* Let $\vec{F}$ be the gravitational force on the projectile. By Newton's second law and the properties of gravitational force near the earth's surface, we have

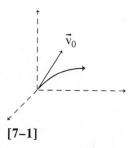

[7–1]

$$\vec{F} = m\ddot{\vec{R}}(t) = -mg\hat{k},$$

where g is the scalar constant known as the *acceleration of gravity at the earth's surface.* Dividing by m, taking the indefinite integral of both sides, and using (7.1), we get

$$\dot{\vec{R}}(t) = -gt\hat{k} + \vec{C},$$

where $\vec{C}$ is an unknown constant vector. We now use the fact that $\dot{\vec{R}} = \vec{v}_0$ when t = 0 to evaluate $\vec{C}$. This gives $\vec{C} = \vec{v}_0$, and we have

$$\dot{\vec{R}}(t) = -gt\hat{k} + \vec{v}_0.$$

Taking the indefinite integral for a second time, we get

$$\vec{R}(t) = \vec{v}_0 t - \frac{1}{2}gt^2\hat{k} + \vec{D},$$

where $\vec{D}$ is an unknown constant vector. We now use the fact that $\vec{R}(t) = \vec{0}$ when t = 0 to evaluate $\vec{D}$. This gives $\vec{D} = \vec{0}$. We thus have our final result:

(7.3) $$\vec{R}(t) = \vec{v}_0 t - \frac{1}{2}gt^2\hat{k}.$$

(Note how this vector-algebra expression separates and clarifies the effects of initial velocity and of gravitational force.)

   Example 2. For a second example from mechanics, let a particle P *move with velocity* $\vec{v}(t)$, *where* $\vec{v}(t)$ *is a known vector function. Let* $\vec{R}(t)$ *be the unknown path of* P. *As in §2.2, we define the* displacement *of* P *between* t =

a *and* $t = b$ *to be* $\vec{R}(b) - \vec{R}(a)$. *We wish to use our known velocity function* $\vec{v}(t)$ *to calculate this* displacement. Let $\vec{F}(t)$ be some specific indefinite integral of $\vec{v}(t)$ which we have found. Since the unknown $\vec{R}(t)$ is also an indefinite integral of $\vec{v}(t)$, we know by (7.1) that for all t,

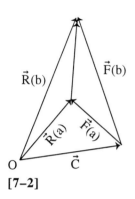

$$\vec{R}(t) = \vec{F}(t) + \vec{C} \text{ for some constant } \vec{C}.$$

Hence
$$\vec{R}(b) = \vec{F}(b) + \vec{C},$$

and
$$\vec{R}(a) = \vec{F}(a) + \vec{C}. \text{ See [7-2].}$$

[7-2]

Subtracting, we get

$$\vec{R}(b) - \vec{R}(a) = \vec{F}(b) - \vec{F}(a).$$

Hence $\vec{F}(b) - \vec{F}(a)$ gives the desired displacement. (Thus $\vec{v}(t)$ enables us to find this displacement even though $\vec{v}(t)$ does not, by itself, give us enough information to find $\vec{R}(t)$.)

§5.8 Definite Integrals

Definition. Let $\vec{A}(t)$ be a given vector function. Let [a,b] be an interval contained in the domain of $\vec{A}(t)$. Then the *definite integral of* $\vec{A}(t)$ *on* [a,b] is denoted

$$\int_a^b \vec{A}(t)dt$$

and is defined as follows. (The definition is closely analogous to the definition of *definite integral* in elementary calculus. For future reference, we give this vector definition in full detail.) Let $t_0, t_1, ..., t_n$ be a chosen set of values in [a,b] such that

$$a = t_0 < t_1 < \cdots < t_n = b.$$

Then $t_0, t_1, ..., t_n$ divide [a,b] into n *subintervals*. For each i, $0 < i \leq n$, let $\Delta t_1 = t_1 - t_{i-1}$ be the length of i-th subinterval, and let t_i^* be a value of t chosen in the i-th subinterval. The values t_i^* are called *sample points*. We call the chosen values $\{t_0, t_1, ..., t_n\}$, together with the chosen sample points $\{t_1^*, t_2^*, ..., t_n^*\}$, *an* n-*decomposition of* [a,b].

For this n-decomposition, we define

$$\vec{S}_n = \sum_{i=1}^{n} \vec{A}(t_i^*)\Delta t_i = \vec{A}(t_1^*)\Delta t_1 + \vec{A}(t_2^*)\Delta t_2 + \cdots + \vec{A}(t_n^*)\Delta t_n .$$

$\vec{S}_n$ is called a *Riemann sum* for $\vec{A}(t)$ on [a,b]. Since $\vec{S}_n$ is a sum of vectors, it must itself be a vector. Let ℓ_n be the maximum value of Δt_i (the maximum length of a subinterval appearing in our chosen n-decomposition). Assume that for each n, an n-decomposition for [a,b] has been chosen (one decomposition for each value of n) and that these choices have been made in such a way that $\lim_{n \to \infty} \ell_n = 0$. We then say that this chosen set of decompositions is a *proper sequence of decompositions*. For any given proper sequence, there is, for each n, a value for $\vec{S}_n$. We note that for such a proper sequence, $\lim_{n \to \infty} \vec{S}_n$ may or may not exist. If, for every possible proper sequence, this limit does exist and always has the same value $\vec{L}$, we say the *definite integral of* $\vec{A}(t)$ *on* [a,b] *exists* and that

$$\int_a^b \vec{A}(t)dt = \vec{L} .$$

This is our definition of *definite integral*. We note that the value of this vector definite integral is itself a vector.

We customarily abbreviate the definition of $\int_a^b \vec{A}(t)dt$ to a single line, using a special notation for the limit:

(8.1)
$$\int_a^b \vec{A}(t)dt = \lim_{\left\{ \begin{smallmatrix} n \to \infty \\ \max \Delta t_i \to 0 \end{smallmatrix} \right\}} \sum_{i=1}^{n} \vec{A}(t_i^*)\Delta t_i .$$

In later chapters, we define various other forms of definite integral, including *multiple integrals*, *line integrals*, and *surface integrals*. In each case we shall use an abbreviated single-line definition like (8.1) rather than give a more detailed definition. With the idea of *proper sequence*, and with the above example as a model, the reader will be able to fill in missing details for these later definitions.

It is easy to express $\int_a^b \vec{A}(t)dt$ in terms of Cartesian components. Let

$$\vec{A}(t) = a_1(t)\hat{i} + a_2(t)\hat{j} + a_3(t)\hat{k} \ .$$

It follows directly from our definition (8.1) that

(8.2)   (i) $\int_a^b \vec{A}(t)dt$ *exists* $\Leftrightarrow$ *each of the scalar integrals* $\int_a^b a_1(t)dt$,

$\int_a^b a_2(t)dt$, *and* $\int_a^b a_3(t)dt$ *exists; and*

     (ii) *if* $\int_a^b \vec{A}(t)dt$ *exists, then*

$$\int_a^b \vec{A}(t)dt = \left(\int_a^b a_1(t)dt \right)\hat{i} + \left(\int_a^b a_2(t)dt \right)\hat{j} + \left(\int_a^b a_3(t)dt \right)\hat{k} \ .$$

Using (8.2), we can apply the methods of elementary calculus to evaluate a vector definite integral. For example, if

$$\vec{A}(t) = 2\cos t\hat{i} + 2\sin t\hat{j} + 3t\hat{k} \ ,$$

then

$$\int_0^{4\pi} \vec{A}(t)dt = \left(\int_0^{4\pi} 2\cos t\, dt \right)\hat{i} + \left(\int_0^{4\pi} 2\sin t\, dt \right)\hat{j} + \left(\int_0^{4\pi} 3t\, dt \right)\hat{k}$$

$$= \frac{3}{2}t^2 \Big]_0^{4\pi} \hat{k} = 24\pi^2\hat{k} \ .$$

Under what mathematical circumstances can we be sure that the definite integral $\int_a^b \vec{A}(t)dt$ exists? A full answer to this question is complex and uses more advanced mathematics than we cover here. The following partial answer is enough for most applications.

(8.3)   *If $\vec{A}(t)$ is continuous on [a,b], then $\int_a^b \vec{A}(t)dt$ exists.*

We can prove this from the analogous result for scalar functions by taking Cartesian components.

The *fundamental theorem of integral calculus (FTIC)* can be formulated for vector definite integrals. As in elementary calculus, this useful fact can be stated in two forms.

(8.4)   (1)   **First form of FTIC.** *Let $\vec{A}(t)$ be a continuous vector function. For each u in [a,b], define*

$$\vec{I}(u) = \int_a^u \vec{A}(t)dt \ .$$

Then $\vec{I}(u)$ is differentiable on [a,b] and

$$\frac{d\vec{I}}{du} = \frac{d}{du}\int_a^u \vec{A}(t)dt = \vec{A}(u) \ ;$$

that is to say, $\vec{I}(u)$ must be an indefinite integral of $\vec{A}(u)$.

Form (1) of the *FTIC* is proved by taking scalar components of $\vec{I}$ and $\vec{A}$ and applying the fact, from elementary calculus, that for a continuous scalar function $f(t)$:

$$\frac{d}{du}\int_0^u f(t)dt = f(u) \ .$$

(8.5)   (2)   **Second form of FTIC.** *Let $\vec{F}(t)$ be an indefinite integral of a continuous vector function $\vec{A}(t)$. Then*

$$\int_a^b \vec{A}(t)dt = \vec{F}(b) - \vec{F}(a) \ .$$

Form (2) of the *FTIC* is proved by noting, from (8.4) and (7.1), that $\vec{I}(t) - \vec{F}(t) = \vec{C}$, for some constant vector $\vec{C}$. This implies that

$$\bar{I}(b) = \int_a^b \vec{A}(t)dt = \vec{F}(b) + \vec{C},$$

and

$$\bar{I}(a) = \int_a^a \vec{A}(t)dt = \vec{F}(a) + \vec{C}.$$

But

$$\int_a^a \vec{A}(t)dt = \vec{0} \qquad \text{by (8.1)}.$$

This gives $\vec{C} = -\vec{F}(a)$. Hence the equation for $\bar{I}(b)$ gives

$$\int_a^b \vec{A}(t)dt = \vec{F}(b) - \vec{F}(a).$$

Remark. If we believe that a given definite integral exists, how can we go about evaluating it? Normally, we will try to use the *FTIC*, either directly or with scalar components. As with scalar integrals, however, other methods, including geometric methods and approximation methods, may be preferable or necessary.

It sometimes occurs that, despite all efforts, we are unable to evaluate a given definite integral. Perhaps calculus methods don't work; perhaps the numerical calculations are too long; perhaps we don't have enough specific information about the integrand. Nonetheless, even when we cannot evaluate such an integral, the integral has a precise meaning. Whether or not we can evaluate them, definite integrals are part of the fundamental vocabulary of science and engineering. As we shall see later, they enable us to define such physical concepts as mass, area, volume, work, and heat, and they enable us to make deductions and to formulate general laws and theorems about these physical concepts.

§5.9 Problems

§5.1

1. A particle's motion is given by the vector function $\vec{R}(t)$, for $t \geq 0$. In each case, describe and sketch the curve traced out by the particle.

   (a) In $\mathbf{E}^2$, $\vec{R}(t) = 3\cos t\,\hat{i} + 3\sin t\,\hat{j}$

   (b) In $\mathbf{E}^2$, $\vec{R}(t) = 3\cos t\,\hat{i} - 3\sin t\,\hat{j}$

   (c) In $\mathbf{E}^2$, $\vec{R}(t) = 2\cos t\,\hat{i} + \sin t\,\hat{j}$

   (d) In $\mathbf{E}^2$, $\vec{R}(t) = t\,\hat{i} + t^2\,\hat{j}$

   (e) In $\mathbf{E}^3$, $\vec{R}(t) = 3\cos t\,\hat{i} + 3\sin t\,\hat{j} + 2\hat{k}$

   (f) In $\mathbf{E}^3$, $\vec{R}(t) = (3+t)\hat{i} + (2-t)\hat{j} + t\hat{k}$

   (g) In $\mathbf{E}^3$, $\vec{R}(t) = 3t\hat{i} + 3\sin t\,\hat{j} - 3\cos t\,\hat{k}$

§5.2

1. The basic *limit theorems* of elementary calculus assert the following.

Let $\lim_{x \to c} f(x) = k$ and $\lim_{x \to c} g(x) = \ell$. Then

(a) $\lim_{x \to c}[f(x) + g(x)] = k + \ell$

(b) $\lim_{x \to c}[f(x)g(x)] = k\ell$

(c) $\ell \neq 0 \Rightarrow \lim_{x \to c}\left[\dfrac{f(x)}{g(x)}\right] = \dfrac{k}{\ell}$

(d) $\lim_{x \to \ell} h(x) = m \Rightarrow \lim_{x \to c}[h(g(x))] = m$

In each case, use funnel functions for the given limits to construct new funnel functions for the desired limits, and thereby prove (a), (b), (c), and (d). In each case, show that the constructed function is a funnel function and that it gives the desired limit. The construction for (a) is indicated at the end of §5.2. (Cases (b) and (c) are less simple than (a) and (d).)

2. Give an appropriate definition of *funnel function* and of *limit* for the case of $\lim_{x \to \infty} f(x) = \ell$.

§5.3

1. A particle moves according to the equation $\vec{R}(t) = t\hat{i} + t^2\hat{j} - t^2\hat{k}$, $(0 \leq t \leq 2)$, where t is time. Find

(a) the velocity when $t = 1$;
(b) the velocity when $t = 2$;
(c) the acceleration when $t = 1$;
(d) the acceleration when $t = 2$;
(e) the speed when $t = 1$;
(f) the speed when $t = 2$.

2. The *normal plane* to a curve C at a point P on C is defined to be the plane normal to the tangent line to C at P. A curve is given by the path $\vec{R}(t) = 2\cos t\hat{i} + \sin t\hat{j} - t^2\hat{k}$. Find equations for the tangent line and the normal plane at the point where:

(a) $t = \pi$ (c) $t = \dfrac{\pi}{4}$ (b) $t = \dfrac{\pi}{2}$

3. Assume that two different paths for the same curve C have derivatives $\neq \vec{0}$ at the same point P on C. Show that the derivatives of the two paths at P must yield the same tangent line at P.
(*Hint.* Try a proof by contradiction.)

§5.4

1. Show that $\dfrac{d}{dt}\left(\vec{R} \times \dfrac{d\vec{R}}{dt}\right) = \vec{R} \times \dfrac{d^2\vec{R}}{dt^2}$.

2. Find $\dfrac{d}{dt}([\vec{R}, \dot{\vec{R}}, \ddot{\vec{R}}])$.

3. Find $\dfrac{d}{dt}((\vec{R} \times \dot{\vec{R}}) \times \dot{\vec{R}})$.

4. Find $\dfrac{d}{dt}((\vec{R} \times \dot{\vec{R}}) \times \ddot{\vec{R}})$.

5. Let $f = f(t)$ and $\vec{B} = \vec{B}(t)$. Find $\dfrac{d^2}{dt^2}(f\vec{B})$.

6. (a) Use (2.1) together with the results of problem 1 in §5.2 to prove (4.5)–(4.8).
(b) Use (4.5)–(4.8) to prove (2.4) and (2.5).
(c) Use (4.5) to prove (3.3).

§5.5

1. *l'Hopital's rule* (in elementary calculus) asserts that if $\lim_{x \to c} f(x) = 0$, $\lim_{x \to c} g(x) = 0$, and $\lim_{x \to c} \dfrac{f'(x)}{g'(x)} = \ell$, then $\lim_{x \to c} \dfrac{f(x)}{g(x)} = \ell$. Use this rule to show that

$$\lim_{x \to 0} \frac{2 - 2\cos x}{x^2} = 1.$$

2. Let $\hat{A} = \hat{A}(t)$. Find $\dfrac{d}{dt}((\hat{A} \times \dot{\hat{A}}) \times \dot{\hat{A}})$.

3. The differentiable unit vectors $\hat{a}(t)$, $\hat{b}(t)$, and $\hat{c}(t)$ form a frame whose orientation in space is changing with time. Show that at each moment t, the vectors $\dot{\hat{a}}(t)$, $\dot{\hat{b}}(t)$, and $\dot{\hat{c}}(t)$ are coplanar.

(*Hint.* Observe that $\hat{a}, \hat{b}, \hat{c}$ form a right-handed frame $\Leftrightarrow \hat{a} = \hat{b} \times \hat{c}$, $\hat{b}^2 = 1$, $\hat{a}^2 = 1$, and $\hat{b} \cdot \hat{a} = 0$. Differentiate these equations and use the results.)

§5.7

1. At time $t = 0$, a particle leaves the point with position vector $\vec{R}(0)$ with initial velocity $\vec{v}(0)$. For all $t \geq 0$, its acceleration is $\vec{a}(t)$. Find expressions for the position $\vec{R}(t)$ and the velocity $\vec{v}(t)$ at any time $t \geq 0$, when:

   (a) $\vec{R}(0) = \vec{0}$, $\vec{v}(0) = -\hat{i} + 2\hat{j} - 2\hat{k}$,
   $\vec{a}(t) = 3t\hat{j} - 2\hat{k}$.

   (b) $\vec{R}(0) = \vec{0}$, $\vec{v}(0) = 2\hat{i} + \hat{k}$,
   $\vec{a}(t) = -2\cos t\hat{i} - 2\sin t\hat{j}$.

   (c) $\vec{R}(0) = a\hat{i} + b\hat{j} + c\hat{k}$, $\vec{v}(0) = 2\hat{i} + \hat{k}$,
   $\vec{a}(t) = -2\cos t\hat{i} - 2\sin t\hat{j}$.

2. A particle moves with velocity $\vec{v}(t) = t\hat{i} - 2t^2\hat{j} - 3\hat{k}$, where t is time. When $t = 1$, the particle's position is given by $\vec{R}(1) = \hat{i} - 3\hat{k}$. Find the position $\vec{R}(t)$ and the acceleration $\vec{a}(t)$ at any later time t.

§5.8

1. The velocity of a moving particle is given by $\vec{v}(t)$. Find the displacement of the particle from $t = 0$ to $t = \frac{\pi}{2}$ when:

   (a) $\vec{v}(t) = \sin t\hat{i} + \cos t\hat{j} + \frac{1}{2}\hat{k}$.

   (b) $\vec{v}(t) = -2\sin t\hat{i} + 2\cos t\hat{j} + 3\hat{k}$

   (c) $\vec{v}(t) = e^t\hat{i} + e^{-t}\hat{j}$.

Chapter 6

Curves

§6.1 Curves and Paths

Certain figures in $\mathbf{E}^3$ and $\mathbf{E}^2$ are called *finite curves*. A finite curve is like a line segment, except that it may have bends, twists, and corners. In this section we state a mathematical definition for *finite curve*. Before we give it, we describe two different intuitive ways of thinking about a finite curve. *First way*: Assume that we have a straight segment of wire, where the material of the wire is pliable and can be stretched, compressed, twisted, and bent but cannot be cut into separate pieces. If a figure in $\mathbf{E}^3$ or $\mathbf{E}^2$ can be modeled by deforming the wire in this way, we say that the figure is a *finite curve*. (We do allow the wire to intersect itself; that is to say, we allow distinct points on the original wire to be merged into a single point on the deformed wire.) *Second way*: Assume that we have a moving particle in $\mathbf{E}^3$ or $\mathbf{E}^2$, which begins its motion at a certain moment and ends its motion at a certain later moment. Then the set of points in $\mathbf{E}^3$ or $\mathbf{E}^2$ which the particle traces out during its motion (the "trajectory" of the particle) is called a *finite curve*.

We give two mathematical definitions for *finite curve*. Our first definition uses Cartesian coordinates in $\mathbf{E}^3$.

Definition. If f(t), g(t), and h(t) are *continuous* functions on [a,b], then the system of equations

$$(1.1) \qquad \left. \begin{array}{l} x = f(t) \\ y = g(t) \\ z = h(t) \end{array} \right\} \quad t \text{ in } [a,b]$$

is called a *system of continuous parametric equations on* [a,b]. t is called the *parameter* of the system. We say that a figure S is *determined* by this system of continuous parametric equations on [a,b] if, for every point P,

> P *is in* S $\Leftrightarrow$ *the coordinates of* P *are given by* (f(t),g(t),h(t)) *for some* t *in* [a,b].

We say that a figure S in $\mathbf{E}^3$ is a *finite curve* if S is determined by some system of continuous parametric equations on some interval [a,b]. The definition for a *finite curve* in $\mathbf{E}^2$ is similar; it uses two parametric equations instead of three.

Given a system of continuous parametric equations on [a,b] (as in (1.1)), the vector-valued function

$$(1.2) \qquad \vec{R}(t) = f(t)\hat{i} + g(t)\hat{j} + h(t)\hat{k} \quad \text{on } [a,b]$$

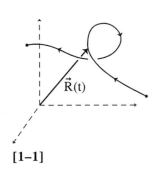

[1–1]

is called a *path* for the curve determined by that system. (See [1–1].) Here $\vec{R}(t)$ is the position vector of the point on the curve given by the parameter value t. We note that a *curve* is a certain *set of points* (a figure) in $\mathbf{E}^3$ or $\mathbf{E}^2$, while a *path* is a certain *function* whose inputs are values of a parameter t and whose outputs are points in $\mathbf{E}^3$ or $\mathbf{E}^2$.

A second and equivalent definition of *finite curve* is as follows. (This definition does not refer to a specific coordinate system and uses only a fixed origin for position arrows.)

Definition. Let $\vec{R}(t)$ be a continuous, one-variable, vector function on [a,b]. The set of points C traced out by a position arrow for $\vec{R}(t)$ is called a *finite curve*, and the vector function $\vec{R}(t)$ is called a *path for* C. The path $\vec{R}(t)$ is sometimes also referred to as a *parameterization of* C.

The following example shows that a given finite curve can have more than one path.

Example 1. *Consider the parametric system*

$$\left.\begin{array}{l} x = t \\ y = t^2 \\ z = 0 \end{array}\right\} \quad t \text{ in } [0,1].$$

The curve determined by this system is the portion of the parabola $y = x^2$ in the xy plane which lies between (0,0,0) and (1,1,0). The path given by this parametric system is

$$\vec{R}_1(t) = t\hat{i} + t^2\hat{j} \quad \text{on } [0,1].$$

This same curve is also given by the path

$$\vec{R}_2(t) = t^2\hat{i} + t^4\hat{j} \quad \text{on } [0,1].$$

The two paths are different, since, for example, $\vec{R}_1(\tfrac{1}{2}) = \tfrac{1}{2}\hat{i} + \tfrac{1}{4}\hat{j} \neq \vec{R}_2(\tfrac{1}{2}) = \tfrac{1}{4}\hat{i} + \tfrac{1}{16}\hat{j}$.

   The mathematical definition of *finite curve* embodies each of the two different intuitive ways of thinking about curves: (1) We can think of the interval [a,b] of values of t as our initial piece of wire and of $\vec{R}(t)$, for each t, as the final position (after the wire is deformed) of the point t in [a,b]. (2) We can think of t as time, of a and b as the initial and final moments of a particle's motion, and of $\vec{R}(t)$, for t in [a,b], as the position of the particle at time t.

   We use the following definitions in connection with paths and curves.

Definitions. Let $\vec{R}(t)$ on [a,b] be a given path. For c in [a,b], we say that parameter value c *maps to* the point P, if $P = \vec{R}(c)$.

   We say that the path is *closed* if $\vec{R}(a) = \vec{R}(b)$. See [1–2].

   We say that a curve C is *closed* if there is a closed path for C. (Thus the deformed wire gives a closed curve if the endpoints of the wire are joined, and the moving particle gives a closed curve as its trajectory if the particle starts and ends at the same point.)

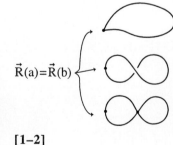

$\vec{R}(a) = \vec{R}(b)$

[1–2]

   We say that the path $\vec{R}(t)$ is *simple* if every pair of distinct values t_1 and t_2 in [a,b], except possibly the pair a and b, maps to distinct points $\vec{R}(t_1)$ and $\vec{R}(t_2)$.

   Thus a *simple path* must be one of two kinds:

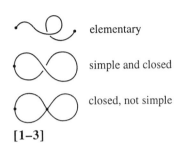

elementary

simple and closed

closed, not simple

[1–3]

(a) $\vec{R}(t)$ is simple, and $\vec{R}(a) \neq \vec{R}(b)$. In this case we say that $\vec{R}(t)$ is an *elementary path*.

(b) $\vec{R}(t)$ is simple, and $\vec{R}(a) = \vec{R}(b)$. In this case, we say that $\vec{R}(t)$ is a *simple, closed path*.

We say that a curve C is *simple* if there is a simple path for C; that a curve C is *elementary* if there is an elementary path for C; and that a curve C is *simple* and *closed* if there is a simple, closed path for C. See [1–3].

Examples 2. *In* $\mathbf{E}^2$, *the curve given by* $\vec{R}(t) = t\hat{i} + t^2\hat{j}$ *on* [0,1] *is elementary.*

In $\mathbf{E}^2$, *the circle given by* $\vec{R}(t) = \cos t\,\hat{i} + \sin t\,\hat{j}$ *on* [0,2π] *is simple and closed.*

In $\mathbf{E}^3$, *the helix (screw-thread) curve given by* $\vec{R}(t) = 2\cos t\,\hat{i} + 2\sin t\,\hat{j} + 3t\hat{k}$ *on* [0,4π] *is elementary.*

In $\mathbf{E}^2$, *the path* $\vec{R}(t) = \cos t\,\hat{i} + \sin t\,\hat{j}$ *on* [0,4π] *is closed but not simple.*

Its curve C, however, *is* simple and closed, since the path $\vec{R}(t) = \cos t\,\hat{i} + \sin t\,\hat{j}$ on [0,2π] is simple and closed and gives the same curve C.

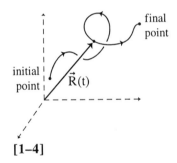

final point

initial point

$\vec{R}(t)$

[1–4]

Definition. We say that a given simple curve C is *directed* if we have specified a direction along C. If C is elementary, we can specify such a *curve-direction* by selecting one endpoint of C as the *initial point* of C. The other endpoint then becomes the *final point* of C, [1–4]. If C is closed, we can specify a direction for C by assigning a direction to an elementary piece of C. When we sketch a directed curve C, we indicate the curve-direction of C by placing one or more arrowheads on C. The curve-direction that we assign to a curve will usually be determined by an intended application. For example, we may wish to indicate the direction of motion of a moving particle in space or the direction of flow of electric current along a thin wire.

If we choose a simple path for a given directed simple curve C, it will often, but not always, be true that the direction of the path (that is to say, the direction of increasing parameter values) is the same as the curve-direction already assigned to C. There will occasionally be situations where it is convenient mathematically to use a path $\vec{R}(t)$ for a given directed curve C, where the direction of the path $\vec{R}(t)$ is opposite to the curve-direction already assigned to C. In problems and examples, however, we shall usually assume that the direction of a path chosen for a directed curve agrees with the curve-direction of that curve. When this assumption is not true, we shall say so.

Remark. We obtain a general definition of *curve* (finite or infinite) by extending (1.2) to include infinite intervals of the form $(-\infty, b]$, $[a, \infty)$, or $(-\infty, \infty)$. We see from the parametric equations considered in §4.4 that straight lines are curves, since $\vec{R}(t) = \vec{R}_0 + t\vec{A}$ is a continuous function on $(-\infty, \infty)$.

§6.2 Finding a Path for a Curve

An algebraically defined *path* for a curve (that is to say, a path where we have algebraic expressions for the continuous functions f(t), g(t), and h(t)) provides us with expressions to which we can apply the analytic techniques of calculus. For this reason, when we begin with a geometric or non-parametric definition of a curve, we may also want to find an algebraic path for that curve. The following examples show how vectors may help us to find such a path.

Example 1. In E^2 *with Cartesian coordinates, let C be a circle in the* xy *plane with radius* a *and center at* O. C *is geometrically defined as the set of points in the* xy *plane which lie at the fixed distance* a *from* O. *To find a path for C, we use Cartesian coordinates, choose a natural seeming parameter, and express the position vector* $\vec{R}$ *of a point P on C in terms of this parameter.* One choice of parameter is the angle α in [2–1], where α is measured in radians and increases as P moves in a counterclockwise direction. Let $\vec{A}$ and $\vec{B}$ be as shown in [2–1]. Then $\vec{A} = a\cos\alpha\,\hat{i}$, $\vec{B} = a\sin\alpha\,\hat{j}$, and

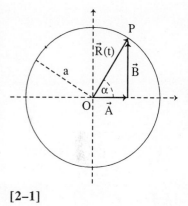

[2–1]

$$\vec{R}(\alpha) = \overrightarrow{OP} = a\cos\alpha\,\hat{i} + a\sin\alpha\,\hat{j} \quad \text{on } [0, 2\pi].$$

This is our desired path with the parameter α.

Other choices of parameter are also possible for example, the angle β in [2–2], which increases as P moves in a clockwise direction. This gives

$$\vec{R}(\beta) = -a\sin\beta\,\hat{i} - a\cos\beta\,\hat{j} \quad \text{on } [0, 2\pi].$$

This is a path for the circle C with the parameter β.

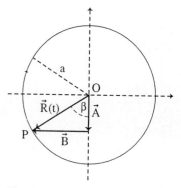

[2–2]

If we cannot find a single algebraic path for a curve C, we may be able to consider C as a chain of finitely many pieces, each of which has its own algebraic path.

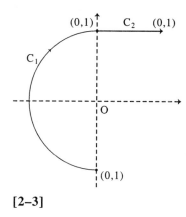

[2–3]

Example 2. In E^2, let C consist of a semi-circle C_1 of radius 1 with center at O as shown in [2–3], together with the line segment C_2 from (0,1) to (1,1). For C_1, we have

$$\vec{R}_1(t_1) = -\sin t_1 \hat{i} - \cos t_1 \hat{j}, \quad 0 \le t_1 \le \pi,$$

and for C_2,

$$\vec{R}_2(t_2) = t_2 \hat{i} + \hat{j}, \quad 0 \le t_2 \le 1.$$

For most purposes, when working with this curve, we can use the separate parameterizations $\vec{R}_1$ and $\vec{R}_2$. If necessary for theoretical purposes, we can combine $\vec{R}_1$ and $\vec{R}_2$ into a single parameterization by

$$\vec{R}(t) = \begin{cases} -\sin t \hat{i} - \cos t \hat{j} & \text{for } 0 \le t \le \pi \\ (t-\pi)\hat{i} + \hat{j} & \text{for } \pi \le t \le \pi+1. \end{cases}$$

Example 3. In E^2, let C be the curve generated by a point P fixed to a moving circle of radius **a**, where this circle rolls, without slipping, along the positive x axis. The curve begins with P at O and ends when P again touches the x axis.

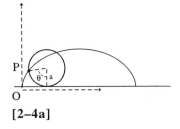

[2–4a]

This curve is called a *cycloid.* As our parameter, we choose θ, the angle through which the moving circle has rotated, [2–4a]. Let $\vec{A}, \vec{B},$ and $\vec{C}$ be as shown in [2–4b]. Then

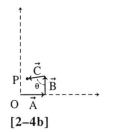

[2–4b]

$$\vec{A} = a\theta\hat{i} \quad \text{(because there is no slipping)},$$
$$\vec{B} = a\hat{j}, \quad \text{and}$$
$$\vec{C} = -a\sin\theta\hat{i} - a\cos\theta\hat{j}, \quad \text{(see [2–2] above).}$$

Hence $\quad \vec{R}(\theta) = \overrightarrow{OP} = \vec{A} + \vec{B} + \vec{C} = a\theta\hat{i} + a\hat{j} - a\sin\theta\hat{i} - a\cos\theta\hat{j}$
$$= a(\theta - \sin\theta)\hat{i} + a(1 - \cos\theta)\hat{j} \quad \text{on } [0,2\pi].$$

This path gives the parametric equations:

$$\left. \begin{array}{l} x = a(\theta - \sin\theta) \\ y = a(1 - \cos\theta) \end{array} \right\} \quad \theta \text{ in } [0,2\pi].$$

Further examples of finding algebraic paths for geometric curves appear in the problems.

Sometimes we need to find a parameterization for a curve C which is initially given by one or more non-parametric algebraic expressions. We illustrate with three examples.

Example 4. *In* E^2, *C is the graph of a function* f(x) *for* $a \le x \le b$. Here we immediately have

$$\vec{R}(t) = t\hat{i} + f(t)\hat{j}, \quad a \le t \le b.$$

Example 5. *In* E^2, *C is the ellipse determined by the equation* $\frac{x^2}{a^2} + \frac{y^2}{b^2} = 1$. (See [2–5].) Here the algebraic form of the equation suggests the parameterization

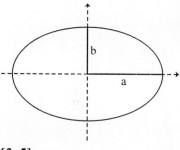

[2–5]

$$\left. \begin{array}{l} x = a \cos t \\ y = b \sin t \end{array} \right\} 0 \le t \le 2\pi.$$

Example 6. *In* E^3 *with Cartesian coordinates, we are given a curve C determined by a pair of equations* f(x,y,z) = 0 *and* g(x,y,z) = 0. In this case, we may be able get a parameterization by first finding the projection of C on one of the coordinate planes and then using vector addition to achieve the final parameterization.

To illustrate, let C be determined by the two equations $x^2 + y^2 = 4$ and x+y+z = 1. (Thus C is the intersection of a plane with a cylinder, [2–6].) The projection of C on the xy plane is immediately given by $x^2 + y^2 = 4$ and has the path $\vec{R}(t) = 2 \cos t\,\hat{i} + 2 \sin t\,\hat{j}$, $0 \le t \le 2\pi$. We achieve our final path for C by adding the vector $(1 - x - y)\hat{k}$ at each point of the projection. We get

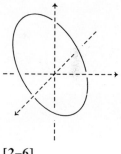

[2–6]

$$\vec{R}(t) = 2 \cos t\,\hat{i} + 2 \sin t\,\hat{j} + (1 - 2 \cos t - 2 \sin t)\hat{k}, \quad 0 \le t \le 2\pi.$$

§6.3 Unit Tangent Vectors; Smoothness

Unit tangent vectors. Let C be a directed, simple curve. Let P be a point on C. Let Q be a variable point on C, [3–1]. We define the unit vector $\hat{u}_{PQ}$ by

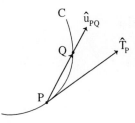

[3–1]

$$\hat{u}_{PQ} = \begin{cases} \dfrac{\overrightarrow{PQ}}{|\overrightarrow{PQ}|} & \text{if Q comes after P in the direction of C;} \\[3em] -\dfrac{\overrightarrow{PQ}}{|\overrightarrow{PQ}|} & \text{if Q comes before P in the direction of C.} \end{cases}$$

Definition. Consider $\lim\limits_{Q \to P} \hat{u}_{PQ}$. If this limit exists, it must be a unit vector; we write it as $\hat{T}_P$ and call it the *unit tangent vector* (for C) *at* P. We immediately have, from the definition of derivative ((3.1) in §5.3), that if $\vec{R}(t)$ is a simple path for C, if $P = \vec{R}(c)$, and if $\left.\dfrac{d\vec{R}}{dt}\right|_c$ exists, then

(3.1) $$\hat{T}_P = \pm \left.\frac{d\vec{R}}{dt}\right|_c \bigg/ \left|\left.\frac{d\vec{R}}{dt}\right|_c\right|,$$

where $+$ holds if the direction of C is the same as the direction of t and $-$ holds if these directions are opposed. For the remainder of this chapter, we assume that the direction of a parameter t for a directed curve is the same as the direction of the curve.

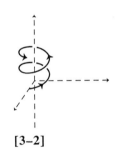

[3–2]

Example 1. Let $\vec{R}(t) = 2\cos t\,\hat{i} + 2\sin t\,\hat{j} + 3t\hat{k}$ *on* $[0, 4\pi]$. (See [3–2].) Then

$$\frac{d\vec{R}}{dt} = -2\sin t\,\hat{i} + 2\cos t\,\hat{j} + 3\hat{k}$$

and $$\left|\frac{d\vec{R}}{dt}\right| = \sqrt{4\sin^2 t + 4\cos^2 t + 9} = \sqrt{13}.$$

Thus, by (3.1), $\hat{T} = \dfrac{-2\sqrt{13}}{13}\sin t\,\hat{i} + \dfrac{2\sqrt{13}}{13}\cos t\,\hat{j} + \dfrac{3\sqrt{13}}{13}\hat{k}$. This expression gives $\hat{T}_P$ at each point P on the curve in terms of the parameter value t at P.

Example 2. Let $\vec{R}(t) = t\hat{i} + t^2\hat{j}$ *on* $[0, 1]$. (This path has, as its curve, a parabolic segment in the xy plane, [3–3].) Then

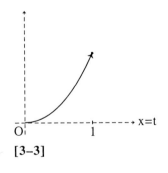

[3–3]

$$\frac{d\vec{R}}{dt} = \hat{i} + 2t\hat{j},$$

and
$$\left|\frac{d\vec{R}}{dt}\right| = \sqrt{1 + 4t^2} \, .$$

Thus
$$\hat{T} = \frac{1}{\sqrt{1+4t^2}} \, \hat{i} + \frac{2t}{\sqrt{1+4t^2}} \, \hat{j}.$$

Smoothness. We wish to define "smoothness" for a curve. Our definition will be geometrical.

Definition. Let C be a directed, simple curve. We say that C is *smooth* if (a) at every point P on C, C has a unit tangent vector $\hat{T}_P$, and (b) $\hat{T}_P$ is a continuous vector function of position on C; that is to say, for every point P on C, $\lim_{Q \to P} \hat{T}_Q = \hat{T}_P$, where Q is a variable point on C. We say that a non-directed simple curve C is *smooth* if, when we assign a direction to C, C satisfies (a) and (b).

We now ask: what conditions on a path $\vec{R}(t)$ on [a,b] will guarantee that the curve C given by this path will be smooth? The following definition, theorem, and corollary supply an answer to this question.

Definition. We say that a path $\vec{R}(t)$ on [a,b] is *differentiable* if $\vec{R}(t)$ is a differentiable vector function on [a,b].

(3.2)   **Theorem.** *Let* C *be a curve with path* $\vec{R}(t)$ *on* [a,b]. *If the following four conditions are all true, then* C *must be smooth.*

> (1) $\vec{R}(t)$ *is elementary.*
> (2) $\vec{R}(t)$ *is differentiable.*
> (3) $\dfrac{d\vec{R}}{dt}$ *is a continuous vector function on* [a,b].
> (4) $\dfrac{d\vec{R}}{dt} \neq \vec{0}$ *on* [a,b].

The proof of (3.2) is immediate from (3.1) and the definition of derivative. A stronger version of (3.2) can be proved as a corollary.

(3.3)   **Corollary.** (3.2) *continues to be true if we replace condition* (4) *with*

> (4') $\dfrac{d\vec{R}}{dt} \neq \vec{0}$ *for* $a < c < b$, *and the limits* $\lim_{t \to a} \hat{T}_{\vec{R}(t)}$ *and* $\lim_{t \to b} \hat{T}_{\vec{R}(t)}$

both exist.

(3.3) shows us, for example, that the one-arched cycloid in example 3 of §6.2 is a smooth curve. See the problems.

Throughout this chapter, we shall assume, unless we state otherwise, that the curves under consideration are smooth. We shall also occasionally assume, for a given path $\vec{R}(t)$, that $\dfrac{d^2\vec{R}}{dt^2}$ (or $\dfrac{d^3\vec{R}}{dt^3}$) exists and is continuous at every point of C.

Definition. A curve C is said to be *piecewise smooth* if it can be viewed as a chain of finitely many smooth elementary curves, where each successive smooth curve is joined to the preceding smooth curve at a common endpoint. Note that a piecewise smooth curve may intersect itself, since the definition of piecewise smooth curve permits two different smooth pieces to intersect each other.

For theoretical purposes, as illustrated in example 2 of §6.2, a single parameterization can always be constructed for a piecewise smooth curve, and the properties *simple, closed*, and *elementary* have the same meanings as before.

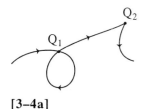

[3–4a]

Example 3. [3–4a] *suggests a curve* C *where smoothness fails for two different reasons:* (a) C intersects itself at the point Q_1, and (b) for every path $\vec{R}(t)$ for C, the sharp change in spatial direction at Q_2 requires either that $\dfrac{d\vec{R}}{dt} = \vec{0}$ at Q_2 or that $\vec{R}(t)$ is not differentiable at Q_2. On the other hand, [3–4b] suggests that the curve C could be a piecewise smooth curve, since it could be viewed as a chain of three smooth pieces: A to B, B to C, and C to D.

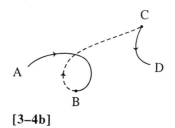

[3–4b]

Example 4. Consider the path in $\mathbf{E}^2$,

$$\vec{R}(\theta) = a(\theta - \sin\theta)\,\hat{i} + a(1 - \cos\theta)\,\hat{j}, \quad \theta \text{ in } [0, 4\pi].$$

The resulting curve C is two adjacent arches of a cycloid and has the form shown in [3–5]. This curve has a sharp change at Q, and we would not want to call it "smooth." In fact $\vec{R}(\theta)$ does not satisfy (3.2), since $\dfrac{d\vec{R}}{d\theta} = \vec{0}$ at $\theta = 2\pi$, and $0 < 2\pi < 4\pi$. On the other hand, C is piecewise smooth.

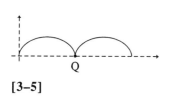

[3–5]

Remark. (3.2) can be modified to cover simple closed curves by changing "elementary" to "simple closed" in (1) and adding "and $\dfrac{d\vec{R}}{dt}\Big|_a$ and $\dfrac{d\vec{R}}{dt}\Big|_b$ have the same direction" to (4).

§6.4 Arc Length; Geometric Properties

A basic geometric feature of a smooth finite curve C is its *arc length*. Informally, the arc length of a finite curve C can be thought of as the value we would get if we put an inelastic thread into the exact shape of C, trimmed off surplus thread at the ends, and then pulled the thread out straight and measured its length. Mathematically, we can define the arc length of a curve C as follows.

(4.1)   *Definition.* Let $\vec{R}(t)$ on [a,b] be a path for C. Choose $n + 1$ points $t_0, t_1, ..., t_n$ in [a,b] such that

$$a = t_0 < t_1 < \cdots < t_n = b.$$

Let $\Delta t_1 = t_1 - t_{i-1}$ for $1 \le i \le n$. Define

$$\Delta \vec{R}_n = \vec{R}(t_i) - \vec{R}(t_{i-1}).$$

Then $\sum_{i=1}^{n} |\Delta \vec{R}_i| = |\Delta \vec{R}_1| + |\Delta \vec{R}_2| + \cdots + |\Delta \vec{R}_n|$ is the total length of a chain of arrows for $\Delta \vec{R}_1, \Delta \vec{R}_2, ..., \Delta \vec{R}_n$ as in [4–1].

Consider $\lim\limits_{\substack{n \to \infty \\ \max \Delta t_i \to 0}} \sum_{i=1}^{n} |\Delta \vec{R}_i|$. (Recall the meaning of this notation from

[4–1]

§5.8.) If this limit exists as a unique finite limit, and if L is the value of this limit, we say that L is the *arc length* of the curve C.

It is easy to give examples of finite curves which do not have arc length (that is to say, for which the finite limit above does not exist). (See the problems for §6.5 below.) If a curve has *arc length*, we say that it is *rectifiable*. Note that if C is rectifiable and C′ occurs as a piece of C, then C′ must be rectifiable.

By more detailed technical arguments, it is possible to prove the following results. (We do not give the proofs.)

(4.2)   *The numerical value of the arc length of C does not depend on the particular path $\vec{R}(t)$ used in the definition above. Hence the arc length of C depends only on C itself (as a set of points in $\mathbf{E}^2$ or $\mathbf{E}^3$).*

(4.3)   *If a finite curve C is smooth, then C is rectifiable.*

(4.3) implies

(4.4)   *If* C *is piecewise smooth, then* C *is rectifiable.*

Definition. Let C be a directed, smooth, finite curve of arc length L, and let $\vec{R}(t)$ on [a,b] be a smooth path for C. For each t in [a,b], let P_t be the point on C given by $\vec{R}(t)$. We define the function s(t) as follows: s(t) is the arc length from P_a to P_t. In particular, s(a) = 0 and s(b) = L. s(t) is called the *distance function* for the smooth path $\vec{R}(t)$.

As $\vec{R}(t)$ is simple, distinct values of t give distinct values of s. It follows that s(t) has an inverse function t(s) which gives the value of t from the value of s. This implies that s itself can be used as a parameter for C. We thus get the parameterization $\vec{R}_g(s) = \vec{R}(t(s))$ on [0,L]. It can be shown that $\vec{R}_g(s)$ must be a smooth path for the given directed curve C. s is called the *geometric parameter* for C, and $\vec{R}_g(s)$ is called the *geometric path* for C. It is usually difficult to find an algebraic formula for $\vec{R}_g(s)$ from an algebraic formula for $\vec{R}(t)$. Nevertheless, as we shall see, s(t) and $\vec{R}_g(s)$ are useful theoretical concepts.
Theoretical calculation with s depends on the following fact.

(4.5)   *Let* C *be a smooth curve. Let* P *be a fixed point on* C, *and let* Q *be a nearby point. Let* Δs = *the arc length from* P *to* Q.

$$Then \quad \lim_{\Delta s \to 0} \frac{|\overrightarrow{PQ}|}{\Delta s} = 1.$$

For a proof of (4.5), see the problems. ((4.5) is not true as a statement about all rectifiable curves or about all differentiable curves.)
(4.5) implies the following:

(4.6)   *If* $\vec{R}(t)$ *is a smooth path for a curve* C, *then* s(t) *is differentiable, and*

$$\frac{ds}{dt} = \left| \frac{d\vec{R}}{dt} \right|.$$

Proof. See [4–2]. (4.6) follows from the identity:

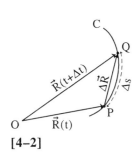

[4–2]

(4.7)
$$\frac{|\Delta\vec{R}|}{\Delta t} = \frac{|\Delta\vec{R}|}{\Delta s} \cdot \frac{\Delta s}{\Delta t}.$$

Taking limits as $\Delta t \to 0$, we have on the left-hand side of (4.7)

$$\lim_{\Delta t \to 0} \frac{\left|\Delta \vec{R}\right|}{\Delta t} = \left|\lim_{\Delta t \to 0} \frac{\Delta \vec{R}}{\Delta t}\right| = \left|\frac{d\vec{R}}{dt}\right| \; ;$$

and on the right-hand side of (4.7)

$$\lim_{\Delta t \to 0} \left(\frac{\left|\Delta \vec{R}\right|}{\Delta t} \cdot \frac{\Delta s}{\Delta t}\right) = \lim_{\Delta t \to 0} \frac{\left|\Delta \vec{R}\right|}{\Delta s} \cdot \lim_{\Delta t \to 0} \frac{\Delta s}{\Delta t} = \lim_{\Delta t \to 0} \frac{\Delta s}{\Delta t} \ \ (\text{by (4.5)}) = \frac{ds}{dt}.$$

This proves (4.6).

(4.6) shows us how to calculate $\dfrac{ds}{dt}$ from $\vec{R}(t)$. This in turn enables us to express the function s(t) as an integral in t:

(4.8)   *Let $\vec{R}(t)$ on [a,b] be a smooth path for* C. *Then*

    (a) $\displaystyle\int_a^b \left|\frac{d\vec{R}}{dt}\right| dt = \int_a^b \frac{ds}{dt} dt = s(b) - s(a) = \textit{arc length of } C.$

    (b)   *For any* t_0 *in* [a,b],

$$s(t_0) = \int_a^{t_0} \left|\frac{d\vec{R}}{dt}\right| dt.$$

(4.8) is immediate from (4.6) by elementary calculus.

Example 1. *For the helix* $\vec{R}(t) = 2\cos t\, \hat{i} + 2\sin t\, \hat{j} + 3t\hat{k}$ *on* $[0, 4\pi]$, *we found* $\left|\dfrac{d\vec{R}}{dt}\right| = \sqrt{13}$ in §6.3. Hence, by (4.8), the arc length of this curve is $\displaystyle\int_0^{4\pi} \left|\frac{d\vec{R}}{dt}\right| dt = \int_0^{4\pi} \sqrt{13}\, dt = 4\sqrt{13}\pi$. Moreover, the distance function for the path $\vec{R}(t)$ is given by $s(t_0) = \displaystyle\int_0^{4\pi} \left|\frac{d\vec{R}}{dt}\right| dt = \sqrt{13}\, t_0$. Thus $s(t) = \sqrt{13}\, t$.

From this, it follows that $t(s) = \dfrac{\sqrt{13}}{13} s$ and that the geometric path for this helix is

$$\vec{R}_g(s) = 2\cos\left(\frac{\sqrt{13}}{13}s\right)\hat{i} + 2\sin\left(\frac{\sqrt{13}}{13}s\right)\hat{j} + \frac{3\sqrt{13}}{13}s\hat{k}.$$

(Example 1 is a case where a simple formula for s(t) is easy to find.)

Example 2. *For the parabolic segment* $\vec{R}(t) = t\hat{i} + t^2\hat{j}$ *on* [0,1], *we found* $\left|\dfrac{d\vec{R}}{dt}\right| = \sqrt{1+4t^2}$. Hence, using a table of integrals, we find that the arc length of this curve is

$$\int_0^1 \sqrt{1+4t^2}\,dt = \frac{1}{2}\left[\sqrt{5} + \frac{1}{2}\ln(2+\sqrt{5})\right] = 1.48.$$

If we write $\vec{R}_g(s)$ simply as $\vec{R}(s)$, we see, by the chain rule for vector functions in §5.4, that

$$\frac{d\vec{R}}{dt} = \frac{d\vec{R}}{ds}\cdot\frac{ds}{dt}.$$

Hence $\dfrac{d\vec{R}}{ds} = \dfrac{d\vec{R}}{dt}\Big/\dfrac{ds}{dt} = \dfrac{d\vec{R}}{dt}\Big/\left|\dfrac{d\vec{R}}{dt}\right| = \hat{T}$, by (3.1) and (4.6), and we have:

(4.9)   **Theorem.** *For a smooth* $\vec{R}(t)$, $\hat{T} = \dfrac{d\vec{R}}{ds}$.

Thus $\hat{T}$ can be viewed as the instantaneous rate of change of $\vec{R}$, per unit of arc length, as we move along C. (4.9) is primarily used for theoretical purposes.

In the case of a moving particle, the scalar quantity $\dfrac{ds}{dt} = \left|\dfrac{d\vec{R}}{dt}\right|$ is the magnitude of the velocity and is often referred to as the *speed* of the particle.

Geometric properties of curves. A property or feature of a curve C is said to be *geometric* if it depends only on the set of points in E^2 or E^3 that forms C. For example, the *arc length* of a smooth curve C is a geometric property, since it can be defined in terms of the geometric figure (or "shape") of C. Similarly, in §6.3, the *unit tangent vector* $\hat{T}$ was defined as a geometric feature of C. Less obviously, the *smoothness* of C is also a geometric property. Recall that C is smooth if C has at least one smooth path. Whether or not C has such a path depends solely on the curve C itself — that is to say, on the set of points that constitutes C.

On the other hand, the vector $\frac{d\vec{R}}{dt}$ at a given point P on C, for a given path $\vec{R}(t)$, is not a geometric feature of C. The vector $\frac{d\vec{R}}{dt}$ at P can have different magnitudes for different paths $\vec{R}$ for C.

Other geometric properties of curves include simpleness, closedness, and differentiability.

Examples. Let C be the parabolic segment given by the path $\vec{R}(t)=$ $t\hat{i}+t^2\hat{j}$ on [0,1]. The fact that this curve starts at the origin is not a geometric property of C, since this fact depends on the Cartesian system we use. The fact that $\frac{d\vec{R}}{dt} \neq \vec{0}$ at the initial point is not a geometric property of C, since $\vec{R}_1(t) = t^2\hat{i}+t^4\hat{j}$ on [0,1] is a different path for C, with $\frac{d\vec{R}_1}{dt} = \vec{0}$ at the initial point.

Remark. The study of how calculus can be used to uncover and analyze geometrical properties of curves is called the *differential geometry of curves*. The formula in (4.8) above, for the arc length of a curve, is a typical result. In Chapter 21 we consider the *differential geometry of surfaces*.

§6.5   Jordan Curves

We consider curves in E^2. A simple, closed curve in E^2 is called a *Jordan curve*. Thus a Jordan curve is simply a "loop" of some kind in a plane.

Definition. We say that a region R in E^2 or E^3 is *bounded* if there are a point P_0 and a real number m such that for every point P in R, $d(P_0,P) \leq m$.

The *Jordan curve theorem* says the following:

(5.1)   *Let C be a Jordan curve in E^2. Then there are two non-empty regions R_i and R_e in E^2 such that, for every point P in E^2, exactly one of the following three statements is true:*

$$P \text{ is in } R_i,$$
$$P \text{ is on } C,$$
$$P \text{ is in } R_e.$$

Furthermore: (i) R_i *is bounded;* (ii) R_e *is not bounded;* (iii) *every pair of points in* R_i *can be connected by a curve lying entirely in* R_i; (iv) *every pair of points in* R_e *can be connected by a curve lying entirely in* R_e; *and* (v) *if* P *is a point in* R_i *and* Q *is a point in* R_e, *then every curve* C *which connects* P *and* Q *must intersect* C. R_i *is called the* interior *of* C, *and* R_e *is called the* exterior *of* C. See [5–1].

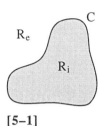

[5–1]

Most students, after they understand (5.1), feel that (5.1) is obviously true. Indeed, for a smooth or piecewise smooth Jordan curve, this feeling is justified, although the proof involves some interesting technical difficulties and details. (We do not give the proof in this text.)

For simple, closed curves in general (which may not be smooth or even piecewise smooth) the theorem is also true, but the proof of this more general fact is surprisingly difficult. The theorem was formulated by Camille Jordan in 1893, but a satisfactory proof was not obtained for another twelve years. Prior to a final proof, some mathematicians came to believe that (5.1) might well be false. Their belief was based, in large part, on experience with a variety of anomalous closed curves whose properties were unexpected if not bizarre. We mention two such properties here. (See the problems for more details.)

(5.2)   *There exists a Jordan curve which is not rectifiable.*

(5.3)   *There exists a region* S *in* E^2, *where* S *consists of a triangle together with all points contained inside the triangle, and where there exists a closed, finite curve* C *such that every point on* C *lies in* S *and every point in* S *lies on* C. (C *is called a* "space-filling" *curve.*)

The curve C in (5.3) cannot be simple. If it were simple (and hence a Jordan curve), the region R_i would be empty, and hence violate the Jordan curve theorem.

§6.6 Curvature; Unit Normal Vectors

Le C be a smooth curve for which the derivative $\dfrac{d\hat{T}}{ds}$ exists at every point. We define the *curvature* of C at each point. By the theorem on unit vectors in §5.5, we know that:

(6.1)   (a) $\dfrac{d\hat{T}}{ds}\cdot\hat{T}=0$ *at every point, and*

(b) $\left|\dfrac{d\hat{T}}{ds}\right|$ *at a given point is the rate of angular change of the*

direction of $\hat{T}$ *per unit of arc length along* C.

(6.2)   *Definition.* For any given point P on C:

(i) The scalar quantity $\left|\dfrac{d\hat{T}}{ds}\right|$ at P is called the *curvature of* C *at* P and is

written as κ_P or simply as κ;

(ii) if $\kappa_P \neq 0$, the unit vector $\dfrac{d\hat{T}}{ds}\Big/\left|\dfrac{d\hat{T}}{ds}\right|$ at P is called the *unit normal*

vector to C *at* P and is written as $\hat{N}_P$ or simply as $\hat{N}$. By (6.1a), $\hat{N}_P$ is
orthogonal to $\hat{T}_P$. See [6–1].

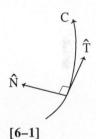

The geometrical significance of κ is expressed by (6.1b).   [6–1]

In general, the scalar κ_P and the vector $\hat{N}_P$ will vary from point to point
along the curve C. We can summarize the definitions and interrelationship in (6.2)
in a single formula:

(6.3) $$\kappa\vec{N}=\dfrac{d\hat{T}}{ds}=\dfrac{d^2\vec{R}}{ds^2}\,.$$

Like the formula (4.9) (that $\dfrac{d\vec{R}}{ds}=\hat{T}$), (6.3) uses the geometric parameter
s and emphasizes the geometric nature of κ and $\hat{N}$. As we have noted above, the
vector $\hat{N}$ is defined for every point P such that $\kappa_P \neq 0$.

Definition. At each point where $\kappa \neq 0$, the scalar quantity $\rho = 1/\kappa$ is called the
radius of curvature of C at that point. A circle of radius ρ has curvature $1/\rho$.
See the problems.

(6.4)   In principle, we could calculate κ and $\hat{N}$ as follows:

$$\hat{T}=\dfrac{d\vec{R}}{ds}=\dfrac{d\vec{R}}{dt}\Big/\dfrac{ds}{dt}\,.$$

By the chain rule (§5.4), $\dfrac{d\hat{T}}{ds}=\dfrac{d\hat{T}}{dt}\Big/\dfrac{ds}{dt}\,.$

Then
$$\kappa = \left| \frac{d\hat{T}}{ds} \right| = \left| \frac{d\hat{T}}{dt} \right| \bigg/ \frac{ds}{dt} \text{, and}$$

$$\hat{N} = \frac{1}{\kappa} \frac{d\hat{T}}{ds}.$$

Although the formulas in (6.3) and (6.4) are useful for theoretical purposes, they can lead to lengthy and complex computations in specific examples. §6.7 gives simpler methods for finding κ and $\hat{N}$.

Remark. *Question:* Suppose that a curve C is given as a piecewise smooth curve with a separate parameterization for each of finitely many smooth pieces. Under what geometrical circumstances can we say that C is, in fact, a smooth curve?

Answer: If at every point P where a smooth piece C_1 joins a smooth piece C_2, C can be directed so that C_1 and C_2 have the same $\hat{T}_P$.

§6.7 Moving Frame; Formulas for κ and $\hat{N}$

Let C be a smooth curve in $\mathbf{E}^3$ for which $\frac{d\hat{T}}{ds}$ exists at every point. Let P be a point on C where $\kappa_P \neq 0$. We have seen that a unit tangent vector $\hat{T}_P$ and a unit normal vector $\hat{N}_P$ are defined at P and that $\hat{T}_P$ and $\hat{N}_P$ are orthogonal. We now define a third unit vector at P:

(7.1)
$$\hat{B}_P = \hat{T}_P \times \hat{N}_P.$$

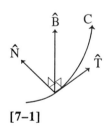

[7–1]

$\hat{B}_P$ is called the *unit binormal vector* to C at P, [7–1]. The three unit vectors $\hat{T}_P$, $\hat{N}_P$ and $\hat{B}_P$ form a frame at P. This frame is called the *moving frame at P* (sometimes called the "moving trihedral at P"). The moving frame is defined at each point P for which $\kappa_P \neq 0$. The orientation in space of the moving frame varies as we move from point to point along the curve. We usually write $\hat{T}_P$, $\hat{N}_P$, $\hat{B}_P$ simply as $\hat{T}$, $\hat{N}$, $\hat{B}$.

Any other vector of interest at a point P on C can be expressed in terms of the moving frame at P. In particular, if we are given a path $\vec{R}(t)$ for C, then the velocity vector $\vec{v}$ $(= \frac{d\vec{R}}{dt})$ at P and the acceleration vector $\vec{a}$ $(= \frac{d^2\vec{R}}{dt^2})$ at P can be expressed in the moving frame at P as follows, where we abbreviate the scalar derivative $\frac{ds}{dt}$ as $\dot{s}$ and the scalar derivative $\frac{d^2s}{dt^2}$ as $\ddot{s}$. By (4.9) and the elementary chain rule ((4.9) in §5.4), we have

$$\vec{v} = \frac{d\vec{R}}{dt} = \frac{d\vec{R}}{ds}\frac{ds}{dt} = \dot{s}\hat{T}\,.$$

Then, using a product-rule for derivatives ((4.2) in §5.4), we get

$$\vec{a} = \frac{d^2\vec{R}}{dt^2} = \frac{d\vec{v}}{dt} = \ddot{s}\hat{T} + \dot{s}\dot{\hat{T}}\,.$$

Since $\dot{\hat{T}} = \frac{d\hat{T}}{ds}\frac{ds}{dt} = \kappa\hat{N}\dot{s}$ (by (6.3)), we have

$$\vec{a} = \ddot{s}\hat{T} + \kappa\dot{s}^2\hat{N}\,. \quad \text{(If } \kappa = 0\text{, this formula reduces to } \vec{a} = \ddot{s}\hat{T}.)$$

We thus have the following expressions for $\vec{v}$ and $\vec{a}$ in the moving frame:

(7.2)
$$\begin{cases} \vec{v} = \dot{s}\hat{T} \\ \vec{a} = \ddot{s}\hat{T} + \kappa\dot{s}^2\hat{N}\,. \end{cases}$$

Furthermore, we see that if $\kappa_P \neq 0$, *then* $\vec{a}, \hat{T}_P,$ *and* $\hat{N}_P$ *are coplanar vectors.*

In physics, these expressions are used for studying the mechanics of a moving particle whose path is given by $\vec{R}(t)$ (where t is time). If we let ρ be the radius of curvature, the equation for $\vec{a}$ can also be written as

(7.3)
$$\vec{a} = \ddot{s}\hat{T} + \frac{\dot{s}^2}{\rho}\hat{N}\,.$$

Using the moving frame, we can find simple and useful formulas for the curvature κ and the unit normal vector $\hat{N}$. By (7.2), we have

(7.4)
$$\vec{v} \times \vec{a} = \kappa\dot{s}^3(\hat{T} \times \hat{N}) = \kappa\dot{s}^3\hat{B}, \quad \text{(since } \hat{T} \times \hat{T} = \vec{0}),$$

and

$$|\vec{v} \times \vec{a}| = \kappa\dot{s}^3 = \kappa|\vec{v}|^3\,.$$

Therefore, we have

(7.5)   *Formula for curvature.* $\kappa = \dfrac{|\vec{v} \times \vec{a}|}{|\vec{v}|^3}$.

From (3.1) and (7.4), we have (7.6) and (7.7).

(7.6)   *Formula for unit tangent vector.* $\hat{T} = \dfrac{\vec{v}}{|\vec{v}|}$.

(7.7)   *Formula for unit binormal vector.* $\hat{B} = \dfrac{\vec{v} \times \vec{a}}{|\vec{v} \times \vec{a}|}$.

Since $\hat{N} = \hat{B} \times \hat{T}$, we have

(7.8)   *Formula for unit normal vector.* $\hat{N} = \dfrac{(\vec{v} \times \vec{a}) \times \vec{v}}{|\vec{v} \times \vec{a}||\vec{v}|}$.

Example 1. In the helix example, we have $\vec{R}(t) = 2\cos t\,\hat{i} + 2\sin t\,\hat{j} + 3t\hat{k}$.

Then $\vec{v} = \dfrac{d\vec{R}}{dt} = -2\sin t\,\hat{i} + 2\cos t\,\hat{j} + 3\hat{k}$, with $|\vec{v}| = \sqrt{13}$.

Similarly, $\vec{a} = \dfrac{d^2\vec{R}}{dt^2} = -2\cos t\,\hat{i} - 2\sin t\,\hat{j}$.

Then $\vec{v} \times \vec{a} = 6\sin t\,\hat{i} - 6\cos t\,\hat{j} + 4\hat{k}$,

and $|\vec{v} \times \vec{a}| = \sqrt{36\sin^2 t + 36\cos^2 t + 16} = \sqrt{52}$.

By formulas (7.5), (7.7), and (7.8), we get:

$$\kappa = \frac{|\vec{v} \times \vec{a}|}{|\vec{v}|^3} = \frac{\sqrt{52}}{13\sqrt{13}} = \frac{2}{13},$$

$$\hat{B} = \frac{\vec{v} \times \vec{a}}{|\vec{v} \times \vec{a}|} = \frac{1}{\sqrt{13}}(3\sin t\,\hat{i} - 3\cos t\,\hat{j} + 2\hat{k}),\text{ and}$$

$$\hat{N} = \frac{(\vec{v} \times \vec{a}) \times \vec{v}}{|\vec{v} \times \vec{a}||\vec{v}|} = \frac{(\vec{v}^2)\vec{a} - (\vec{a} \cdot \vec{v})\vec{v}}{\sqrt{52}\sqrt{13}} = \frac{13\vec{a}}{26} = \frac{-26 \cos t\,\hat{i} - 26 \sin t\,\hat{j}}{26} = -\cos t\,\hat{i} - \sin t\,\hat{j}.$$

Example 2. *In the case of the parabolic segment* $\vec{R}(t) = t\hat{i} + t^2\hat{j}$ *on* [0,1], *we have*

$$\vec{v} = \hat{i} + 2t\hat{j}, \quad |\vec{v}| = \sqrt{1 + 4t^2},$$
$$\vec{a} = 2\hat{j}.$$

Then $\qquad\qquad (\vec{v} \times \vec{a}) = 2\hat{k}$, and $|\vec{v} \times \vec{a}| = 2$.

This gives $\qquad\qquad \kappa = \dfrac{2}{(1 + 4t^2)^{3/2}},$

and $\qquad \hat{N} = \dfrac{(\vec{v} \times \vec{a}) \times \vec{v}}{|\vec{v} \times \vec{a}||\vec{v}|} = \dfrac{-4t\hat{i} + 2\hat{j}}{2(1 + 4t^2)^{1/2}} = \dfrac{-2t\hat{i} + \hat{j}}{\sqrt{1 + 4t^2}}.$ (See example 2 in §6.3.)

Example 3. *Let* f(x) *be a function of one variable. Then the graph of* f *in the* xy *plane is given by the parametric equations*

$$x = t,$$
$$y = f(t),$$
$$z = 0.$$

Thus $\qquad\qquad \begin{aligned} \vec{R}(t) &= t\hat{i} + f(t)\hat{j}, \\ \vec{v}(t) &= \hat{i} + f'(t)\hat{j}, \\ \vec{a}(t) &= f''(t)\hat{j}. \end{aligned}$

Hence, for any value x of t, we have

(7.9) $\qquad\qquad \kappa(x) = \dfrac{|\vec{v} \times \vec{a}|}{|\vec{v}|^3} = \dfrac{|f''(x)\hat{k}|}{(1 + f'(x)^2)^{3/2}} = \dfrac{|f''(x)|}{(1 + f'(x)^2)^{3/2}}.$

This is the familiar formula from elementary calculus for the curvature of a graph in $\mathbf{E}^2$, except that in elementary calculus, the numerator is $f''(x)$ instead of $|f''(x)|$. In elementary calculus, the sign of $f''(x)$ tells us whether the curve is turning to the left or to the right as we move along it in the direction of positive x. For a curve in $\mathbf{E}^3$, however, "left" and "right" have no natural meaning, and (7.5) always gives $\kappa \geq 0$. Other formulas for calculating curvature are considered in the problems.

§6.8 Torsion

We have seen that curvature κ gives the angular rate, per unit of arc length, at which $\hat{T}$ is changing its direction as we move along the curve. We will have a more complete picture of the motion if we also know the angular rates at which $\hat{N}$ and $\hat{B}$ are changing their directions. This information will show how the entire moving frame is changing its orientation in space as we move along the curve. The information is given by the following theorem.

(8.1)   *Frenet's theorem. Let* C *be a smooth curve for which* $\dfrac{d\hat{T}}{ds}$ *and* $\dfrac{d\hat{N}}{ds}$ *exist at every point, and let* P *be a point on the curve where* $\kappa \neq 0$. *Then there exists a scalar* τ_P *(also written as* τ*) called the* torsion *of* C *at* P *(where* τ *may be positive, zero, or negative), such that, at the point* P,

(i) $\dfrac{d\hat{T}}{ds} = \kappa \hat{N}$,

(ii) $\dfrac{d\hat{N}}{ds} = -\kappa \hat{T} + \tau \hat{B}$,

(iii) $\dfrac{d\hat{B}}{ds} = -\tau \hat{N}$.

(Formulas (i), (ii), and (iii) are known as the *Frenet formulas* for a curve.)

For a proof of (8.1), see the problems. The torsion at P gives a measure of the extent to which the curve has a non-planar "cork-screw" shape at P. $\tau > 0$ indicates a right-handed screw as s increases; $\tau < 0$ indicates a left-handed screw. For a given path $\vec{R}(t)$, we can use formula (iii) above to find τ as a function $\tau(t)$. We illustrate with our helix example. We have

$$\hat{B} = \hat{T} \times \hat{N} = \frac{3\sqrt{13}}{13} \sin t\hat{i} - \frac{3\sqrt{13}}{13} \cos t\hat{j} + \frac{2\sqrt{13}}{13}\hat{k} .$$

Then $\dfrac{d\hat{B}}{dt} = \dfrac{3\sqrt{13}}{13} \cos t\hat{i} + \dfrac{3\sqrt{13}}{13} \sin t\hat{j} .$

By (iii), $\qquad -\tau\hat{N} = \dfrac{d\hat{B}}{ds} = \dfrac{d\hat{B}}{dt} \Big/ \dfrac{ds}{dt} = \dfrac{3}{13}\cos t\,\hat{i} + \dfrac{3}{13}\sin t\,\hat{j}.$

But we know from example 1 in §6.5 that $\vec{N} = -\cos t\,\hat{i} - \sin t\,\hat{j}.$ Hence we have

$$\tau = \frac{3}{13}.$$

(In this example, $\tau(t)$ is a constant. In most cases, τ will vary from point to point.)

(7.1) gave us a simple formula for the curvature κ in terms of $\vec{v}$ and $\vec{a}$. In the problems, we deduce an analogous simple formula for the torsion τ in terms of $\vec{v}, \vec{a},$ and $\vec{d}$, where $\vec{d} = \dfrac{d\vec{a}}{dt} = \dfrac{d^3\vec{R}}{dt^3}$. This formula is

(8.2) $\quad$ *Formula for torsion.* $\qquad \tau = \dfrac{[\vec{v}, \vec{a}, \vec{d}]}{(\vec{v} \times \vec{a})^2}.$

Let C be a given smooth, directed curve. Let the functions $\kappa(s)$ and $\tau(s)$ give the curvature and torsion of C as functions of arc length s. Because s, $\hat{T}$, $\hat{N}$, and $\hat{B}$ are geometric features of the curve, and because κ and τ can be expressed in terms of $\dfrac{d\hat{T}}{ds}$ and $\dfrac{d\hat{B}}{ds}$, we see that the functions $\kappa(s)$ and $\tau(s)$ are geometric features of C. Using elementary methods from the theory of differential equations, the following theorem can be proved.

(8.3) $\quad$ *Let* $\kappa(s)$ *and* $\tau(s)$ *be given smooth functions defined on* [0,L], *for some* $L > 0$, *where* $\kappa(s) > 0$ *for all* s *in* [0,L]. *Let an initial point* P *be given, and let initial perpendicular directions for* $\hat{T}$ *and* $\hat{N}$ *at* P *be specified. Then there is one and only one smooth curve in* $\mathbf{E}^3$ *which satisfies the initial conditions at* P, *has arc length* L, *and has its curvature and torsion given by* $\kappa(s)$ *and* $\tau(s)$.

Since the initial conditions, together with the curvature and torsion functions $\kappa(s)$ and $\tau(s)$ on [0,L], uniquely determine a curve C as a set of points in $\mathbf{E}^3$, all other geometric properties of C can be deduced, in principle, from the functions $\kappa(s)$ and $\tau(s)$, provided that $\kappa(s) > 0$ on [0,L].

§6.9 Problems

§6.2

1. A point C moves around the circle in the xy plane of radius 2 with center at the origin. Its angular speed is 1 radian/second, counter-clockwise. A circular disk of radius 1 lies in the xy plane and has its center at C. It rotates with angular speed (with respect to the x axis) of 1 radian/second, clockwise. Point P is fixed on the circumference of the disk. Let t be time. When t = 0, C is at (2,0) and P is at (1,0).
   (a) Find parametric equations for the path of P in terms of the parameter t.
   (b) Sketch the path of P.

2. If a string wound around a fixed circle is unwound counter-clockwise while held taut in the plane of the circle, its endpoint P traces a curve known as an *involute* of the circle. Let the fixed circle be located in the xy plane with its center at the origin O and have radius a. Let the initial position of the tracing point P be the point Q at (a,0). At any later moment, let TP be the unwound portion of the string. Assume that TP is always tangent to the given circle at T. Find parametric equations for the involute, using the angle QOT as the parameter.

3. A turntable of radius a has its surface in the xy plane and its center at the origin. It rotates counter-clockwise at 1 radian per second. A chord is drawn on the table across an angle of $\frac{\pi}{2}$ with endpoints Q_1 and Q_2 on the edge of the table. When t = 0, Q_1 is at (a,0) and Q_2 is at (0,a), and a small ant begins crawling along the chord from Q_1 to Q_2 at a speed of b units per second with respect to the chord. Find parametric equations for the path of the ant in the xy plane for $0 \le t \le \frac{a\sqrt{2}}{b}$.

4. In the xy plane, a circle of radius b rolls (without slipping) around the outside of a fixed circle of radius a with center at origin O. P is a fixed point on the moving circle. Initially, the center C of the moving circle is at (a+b,0), and P is at (a,0). C moves counter-clockwise and φ is the angle through which $\overrightarrow{OC}$ has moved. Find parametric equations, in parameter φ, for the path of P. (This curve is called an *epicycloid*.)

5. In the xy plane, a circle of radius b rolls (without slipping) around the inside of a fixed circle of radius a with center at origin O. P is a fixed point on the moving circle. Initially, the center C of the moving circle is at (a–b,0) and P is at (a,0). C moves counter-clockwise and φ is the angle through which $\overrightarrow{OC}$ has moved. Find parametric equations, in parameter φ, for the path of P. (This curve is called a *hypocycloid*.)

6. A circle has radius a and center C at (a,0) in the xy plane. Let A be a point on the circle other than the origin O. Let B be the intersection of a ray from O through A with the line x = 2a. Let P be the point on $\overline{OB}$ such that d(O,P) = d(A,B). Let D be the point (2a,0), and let θ be the size of the angle ACD.
   (a) Find parametric equations for the path of P with θ as parameter $(-\pi < \theta < \pi)$.
   (b) Sketch the curve. (This curve is known as the *cissoid of Diocles*.)

§6.3

1. (a) Find a simple, closed path for the finite curve in the xy plane determined by the equation

   $$\frac{x^2}{4} + \frac{y^2}{9} = 1.$$

   (b) Is this path smooth? Explain.

2. Given the path $\vec{R}(t) = t\cos t\,\hat{i} + t\sin t\,\hat{j} + t\hat{k}$, $(0 \le t)$, find the unit tangent vector $\hat{T}$ as a function of t.

3. Let $\vec{R}(\theta)$ be given as in example 3 of §6.2.

(a) Show that $\dfrac{d\vec{R}}{d\theta} = \vec{0}$ for $\theta = 0$ and $\theta = 2\pi$.

(b) Use l'Hopital's rule (problem 1 of §5.5) to calculate $\lim\limits_{\theta \to 0} \hat{T}_{\vec{R}(\theta)}$ and $\lim\limits_{\theta \to 2\pi} \hat{T}_{\vec{R}(\theta)}$.

(c) Conclude from (3.3) that the one-arched cycloid in [2–4a] must be a smooth curve.

4. Prove (3.2).

5. Prove (3.3).

§6.4

1. $\vec{R}(t) = e^t\,\hat{i} + te^t\,\hat{j}$ describes the motion of a point. What is $\dfrac{ds}{dt}$ when $t = 0$?

2. A point moves to the right on the curve $y = x^2$ with constant speed $\dfrac{ds}{dt} = 1$. Find $\vec{v} = \dfrac{d\vec{R}}{dt}$ and $\vec{a} = \dfrac{d^2\vec{R}}{dt^2}$ as functions of x.

3. A point moves on the curve $y = e^x$ with constant speed $\dfrac{ds}{dt} = 1$. Find $\vec{v} = \dfrac{d\vec{R}}{dt}$ and $\vec{a} = \dfrac{d^2\vec{R}}{dt^2}$ as functions of x.

4. A curve C in the xy plane is determined by the equation $\dfrac{x^2}{4} + \dfrac{y^2}{9} = 1$.

(a) Find a set of parametric equations for C. (See problem 1 of §6.3.)

(b) Express the length of C as a definite integral.

5. Find the length of the curve in the xy plane defined by equation $y = x^{3/2}$, $0 \le x \le 4$.

6. A curve is given by the path $\vec{R}(t) = t\cos t\,\hat{i} + t\sin t\,\hat{j} + t\hat{k}$, $0 \le t \le \pi$.

(a) Express the arc length of this curve as a definite integral.

(b) Use a table of integrals to evaluate this integral.

7. Find the arc length of the cycloid generated by one full turn of the rolling circle of radius a. (See §6.2.)

8. Prove (4.5).

9. Let C be given by a smooth path $\vec{R}(t)$ on [a,b]. Show that (4.1) is equivalent to the following definition of arc length: Choose n+1 points, $t_0, t_1, ..., t_n$ in [a,b], as in (4.1). For each i, $0 < i \le n$, choose a sample value t_i^*, $t_{i-1} \le t_i^* \le t_i$. Let $P_i = \vec{R}(t_i)$, $0 \le i \le n$. Let $P_i^* = \vec{R}(t_i^*)$, $0 \le t_i^* \le n$. Let L_i be the tangent line to C at P_i^*. Let Δp_i = length of the projection of $\overrightarrow{P_{i-1}P_i}$ onto L_i. Then the arc length of

$$C = \lim_{\substack{n \to \infty \\ \max \Delta t_i \to 0}} \sum_{i=1}^{n} \Delta p_i.$$ (*Hint:* See (4.5) in §6.4 and (7.1) in §17.7.)

§6.5

1. Prove (5.2).

(*Hint.* Define a continuous function $f(x)$ on [0,2] such that (i) $f(x) \ge 0$; (ii) $f(0) = f(2) = 0$; (iii) for every integer $n \ge 1$, $f(\frac{1}{n}) = \frac{1}{n}$; (iv) for every integer $n \ge 1$, there is a c, $\frac{1}{n+1} < c < \frac{1}{n}$, such that $f(c) = 0$. Use the fact that $\lim\limits_{n \to \infty} \sum\limits_{i=1}^{n} \frac{1}{i} = \infty$.)

2. Prove (5.3).

(*Hint.* Begin with a non-isosceles, right triangle S and the interval I = [0,1]. Consider a process where we divide I into smaller and smaller subintervals by repeated bisection and divide S into smaller and smaller similar triangles by repeated use of altitudes to the hypotenuse. Set up an appropriate correspondence between infinite nested sequences of subintervals and infinite nested sequences of subtriangles, so that this correspondence defines a path $\vec{R}(t)$ from I onto S.)

§6.7

1. A path is given by the parametric equations

$$\left.\begin{array}{l} x = 2t \\ y = -t^2 \\ z = \dfrac{1}{t} \end{array}\right\} \ (0 \le t), \text{ where t is time. At } t = 1, \text{ find:}$$

(a) the velocity $\vec{v}$

(b) the speed $\dfrac{ds}{dt}$

(c) the unit tangent vector $\hat{T}$

(d) the acceleration $\vec{a}$

(e) the curvature κ, by using (7.5)

(f) the unit normal vector $\hat{N}$ by using (7.8)

2. A path is given by

$$\vec{R}(t) = t\hat{i} + 2t^2\hat{j} + 2e^{t-1}\hat{k}, \ (0 \le t).$$

At $t = 1$, find:

(a) the velocity $\vec{v}$

(b) the speed $\dfrac{ds}{dt}$

(c) the unit tangent vector $\hat{T}$

(d) the acceleration $\vec{a}$

(e) the curvature κ, by using (7.5)

(f) the unit normal vector $\hat{N}$ by using (7.8)

3. Consider the parabolic path given by $\vec{R}(t) =$ $t\hat{i} + t^2\hat{j}$, $(0 \le t \le 1)$ (example 2 of §6.7), where t is time.   The point whose position-vector is $\vec{R}(t) + \rho(t)\hat{N}(t)$ is called the *center of curvature* at time t.

(a) Find the path (in the parameter t) of the center of curvature.

(b) Sketch the curve given by this path.

4. Let P be a point on a circle C of radius ρ. Using a parameterization in an appropriate Cartesian system, calculate the curvature of C at P.

(*Note*. On the basis of this result, for any point P on any curve C, if we take a circle of radius ρ in the osculating plane whose center is at the center of curvature for P (given by $\vec{R} + \rho\hat{N}$ as in problem 3), then this circle goes through P and has the same curvature at P as C. This circle is called the *circle of curvature* or "osculating circle" at P. It can be shown that this circle is also the limiting circle obtained by taking points P' and P" nearby to P on C, taking the circle through those three points, and letting P' and P" approach P along C.)

5. In the helix example, find the unit binormal vector $\vec{B}$ as a function of t.

6. For the path given in problem 1, find the unit binormal vector $\vec{B}$ at $t = 1$.

7. For the path given in problem 2, find the unit binormal vector $\vec{B}$ at $t = 1$.

8. A point moves so that at time $t = 1$, $\vec{v} = \dot{\vec{R}} = \hat{i} + \hat{j} + \hat{k}$ and $\vec{a} = \ddot{\vec{R}} = \hat{k}$.   Find the values of the following at time $t = 1$.

(a) speed $\dot{s}$

(b) unit tangent vector $\hat{T}$

(c) curvature κ

(d) $\ddot{s} = \dfrac{d^2s}{dt^2}$

(e) unit normal vector $\hat{N}$

9. In a given Cartesian system, assume that the motion of a given particle obeys the law $\vec{a} = C(\vec{v} \times \vec{k})$ where C is a scalar constant. (This is the case, for example, when an electrically charged particle moves in a constant magnetic field which has the direction of the z axis, and in the absence of an electric field.)

(a) Show that the following statements must be true:

(i) $\vec{a}$ is orthogonal to $\vec{v}$

(ii) $|\vec{v}|$ is constant

(iii) $|\vec{a}|$ is constant

(iv) The curvature κ is constant

(b) Describe the path followed by the particle. (*Hint.* For (a), use the moving frame.)

10. Given the path $\vec{R}(t) = e^t \cos t\,\hat{i} + e^t \sin t\,\hat{j} + e^t\hat{k}$, $(0 \le t)$, find the following:

(a) expressions for $s(t)$ and $t(s)$;

(b) curvature as a function of t;

(c) the components of $\vec{v}$ and $\vec{a}$ (as functions of t) in the moving frame;

(d) expressions for $\hat{T}, \hat{N},$ and $\hat{B}$ as functions of t in the frame $\{\hat{i}, \hat{j}, \hat{k}\}$.

11. Find the radius of curvature, as a function of x, of the curve $y = \ln \cos x$, $(-\frac{\pi}{2} < x < \frac{\pi}{2})$, in the xy plane.

12. A curve in the xy plane (in $\mathbf{E}^3$) is given by the path $\vec{R}(t) = x(t)\hat{i} + y(t)\hat{j}$, $(0 \le t)$. Find an expression in terms of $\dot{x}, \dot{y}, \ddot{x},$ and $\ddot{y}$ for curvature κ as a function of t.

13. A curve in $\mathbf{E}^3$ is given by the path $\vec{R}(t) = t\hat{i} + y(t)\hat{j} + z(t)\hat{k}$, $(0 \le t)$. Find an expression in terms of $\dot{y}, \dot{z}, \ddot{y},$ and $\ddot{z}$ for curvature κ as a function of t.

14. A curve is given by the parametric equations

$$\left.\begin{array}{l} x = t \\ y = e^t \\ z = 0 \end{array}\right\} \quad (0 \le t < \infty).$$

(a) In the moving frame, find the tangential and normal components of the velocity $\vec{v}$ and the acceleration $\vec{a}$.

(b) Find $\hat{T}, \hat{N}, \kappa,$ and ρ as functions of t.

15. A path is given as

$$\vec{R}(t) = x\hat{i} + f(x)\hat{j}, \quad (0 \le x).$$

Find expressions for the following as functions of x:

(a) $\dfrac{d\vec{R}}{dx}$

(b) $\left|\dfrac{d\vec{R}}{dx}\right|$

(c) the unit tangent vector $\hat{T}$

(d) $\dfrac{d^2\vec{R}}{dx^2}$

(e) κ

(f) the unit normal vector $\hat{N}$

(g) ρ

16. Calculate the maximum and minimum radius of curvature for the ellipse $\vec{R}(t) = a\cos t\,\hat{i} + b\sin t\,\hat{j}$, $(0 \le t \le 2\pi)$. (*Hint.* Find curvature κ as a function of t. The desired values will be immediate from this formula.)

17. A cycloid is given by the path $\vec{R}(t) = (t - \sin t)\hat{i} + (1 - \cos t)\hat{j}$, $(0 \le t \le 2\pi)$.

(a) Find $\hat{T}, \hat{N},$ and κ as functions of t.

(b) Find the path $\vec{R}(t) + \rho(t)\hat{N}(t)$ of the center of curvature, and show that it gives a cycloid congruent to the original one. (*Hint.* The identity $1 - \cos t = 2\sin^2 \frac{t}{2}$ is useful.)

(c) Make a sketch showing the relation of the two cycloids.

§6.8

1. The path of a particle is given by $\vec{R}(t) = t^2\hat{i} + \frac{2}{3}(t^3 - 1)\hat{j} + (t - 4)\hat{k}$, $(0 \le t)$, where t is time.

(a) Find the curvature when $t = 1$.

(b) Find the tangential and normal components of acceleration when $t = 1$.

(c) Find the torsion when $t = 1$.

2. Prove (8.1). (*Hint.* (i) is the same as (6.3) and already proved. Differentiate the equation

$\hat{B} = \hat{T} \times \hat{N}$. Use §5.5 to conclude that $\hat{T} \times \frac{d\hat{N}}{ds}$ is a scalar multiple μ of $\hat{N}$. Define $\tau = -\mu$. This gives (iii). Differentiate the equation $\hat{N} = \hat{B} \times \hat{T}$. (ii) follows.)

(*Note.* From §5.5, we see that $|\tau|$ is the angular rate of change, per unit distance, of the direction of $\hat{B}$, that $\tau > 0 \Rightarrow \frac{d\hat{B}}{ds}$ has the direction of $-\hat{N}$, and that $\tau < 0 \Rightarrow \frac{d\hat{B}}{ds}$ has the direction of $\hat{N}$.

3. Prove (8.2). (*Hint.* Use Frenet formula (ii) to express $\dot{\hat{N}}$ in the moving frame. Use this to show that $\bar{d} = \frac{d^3\bar{R}}{dt^3}$ in the moving frame is $\bar{d} = (\dddot{s} - \kappa^2 \dot{s}^3)\hat{T} + (3\kappa \dot{s}\ddot{s} + \dot{\kappa}\dot{s}^2)\hat{N} + \kappa\tau\dot{s}^3\bar{B}$. Then show that

$$(\bar{v} \times \bar{d}) \times (\bar{v} \times \bar{a}) = -\kappa^2 \tau \dot{s}^7 \hat{T} = -(\bar{v} \times \bar{a})^2 \tau \dot{s} \hat{T} .)$$

Chapter 7

Cylindrical Coordinates

§7.1   Cylindrical Coordinates

We can often simplify the analysis of a given mathematical or physical problem by using a coordinate system which is appropriate to that particular problem. In some cases, this appropriate coordinate system may be a Cartesian system whose origin and axes have been chosen to fit naturally with the given problem. In other cases, the best coordinate system may be a non-Cartesian system which possesses some of the geometrical symmetries that occur in the given problem. In this chapter, we describe the non-Cartesian coordinate system known as *cylindrical coordinates*. This system is widely used for problems in which some form of cylindrical symmetry appears.

In cylindrical coordinates, each point has three coordinates, r, θ, and z. See [1–1]. In order to define the cylindrical coordinates r, θ, and z, we start with a given Cartesian system in $\mathbf{E}^3$. Let P be a given point in $\mathbf{E}^3$. We use this Cartesian system to find (r,θ,z) for P:

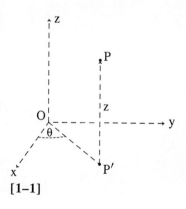

[1–1]

(1)   To find r, we let P′ be the projection of P on the xy plane. We take $r = \left| \overrightarrow{OP'} \right|$.

(2)   To find θ, we consider two cases:
     (a) $r > 0$. In this case we take $\theta =$ an angle (measured in radians) through which a movable arrow, with length r and initial point at O, can be rotated in the xy plane in order to carry that arrow from the positive x axis to

141

$\overrightarrow{OP'}$; where θ is positive for counterclockwise rotation (as viewed from the positive z axis) and negative for clockwise rotation. We note that for $r > 0$, our given point P has infinitely many possible values of θ and that any two values must differ by a multiple of 2π.

(b) $r = 0$. In this case, we take θ to be any real number. With $r = 0$, P again has infinitely many possible values of θ. The point O with $r = 0$ and $z = 0$ is called the *origin* of the cylindrical system.

(3) To find the cylindrical coordinate z at P, we simply give it the value of the Cartesian coordinate z at P.

For example, if P has Cartesian coordinates $(1,1,1)$, then the possible choices for (r,θ,z) are

$$(\sqrt{2},\tfrac{\pi}{4},1), (\sqrt{2},\tfrac{9\pi}{4},1), (\sqrt{2},\tfrac{17\pi}{4},1), ...$$

and $$(\sqrt{2},-\tfrac{7\pi}{4},1), (\sqrt{2},-\tfrac{15\pi}{4},1), (\sqrt{2},-\tfrac{23\pi}{4},1), ...;$$

and if P has Cartesian coordinates $(0,0,7)$, then the choices are $(0,\theta,7)$ for all real numbers θ .

In a cylindrical coordinate system, the possible values of θ and z range over all real numbers, but the possible values of r are limited to the non-negative real numbers.

Coordinate	Range of Possible Values
x	all reals
y	all reals
z	all reals
r	all non-negative reals
θ	all reals

Given a Cartesian system in E^3, the cylindrical system that we define from it, as described above, is called the *associated cylindrical system* for the given Cartesian system. Conversely, this Cartesian system is called the *associated Cartesian system* for the cylindrical system. In working with a cylindrical system, it is often helpful, both for theoretical and for computational purposes, to use the associated Cartesian system at the same time. We shall see examples of this below. There will also, however, be situations where we wish to use a cylindrical system *without* reference to a Cartesian system. We can do this by choosing a plane M to serve as the plane $z = 0$, by choosing a ray in M to serve as the ray for $\theta = 0$ and $z = 0$, and by choosing a ray normal to M, with the same endpoint, to serve as the ray for $r = 0$ and $z \geq 0$.

Given a Cartesian system together with its associated cylindrical system, points in the xy plane can be located by their (x,y) coordinates or by their (r,θ)

coordinates. The (r,θ) coordinates provide a non-Cartesian system of coordinates for $\mathbf{E}^2$ known as *polar coordinates*. This system is used in elementary calculus.

§7.2 Defining Equations

In §7.1, we begin with a given system of Cartesian coordinates in $\mathbf{E}^3$ and use this system to define a new system of cylindrical coordinates in $\mathbf{E}^3$. The following equations give algebraic expressions for the Cartesian coordinates (x,y,z) in terms of the new cylindrical coordinates (r,θ,z):

(2.1)
$$\begin{cases} x = r\cos\theta \\ y = r\sin\theta \\ z = z. \end{cases}$$

These equations are called the *defining equations* of the new system. The defining equations are useful not only because they allow us to calculate x, y, and z from r, θ, and z, but also because we can use the defining equations to convert a relationship in Cartesian coordinates to a relationship in cylindrical coordinates. We accomplish this by using the defining equations to substitute for x, y, and z in terms of r, θ, and z.

For example, in Cartesian coordinates, $x+y+z = 0$ defines a plane through the origin. Using the defining equations, we see that this same plane has the equation $r(\cos\theta+\sin\theta) + z = 0$ in the associated cylindrical coordinates. Similarly, the sphere $x^2 + y^2 + z^2 = 1$ gives the equation $r^2 + z^2 = 1$; the cylindrical surface $x^2 + y^2 = 4$ gives the equation $r = 2$; the cylindrical surface $y^2 + z^2 = 4$ gives the equation $r^2\sin^2\theta + z^2 = 4$; the one-sheeted cone $z = \sqrt{x^2 + y^2}$ gives the equation $z = r$; and the two-sheeted cone $z^2 = x^2 + y^2$ gives the equation $z^2 = r^2$.

We also have *inverse* algebraic formulas for r, θ, and z in terms of x, y, and z. The expressions for r and z are simple:

$$r = \sqrt{x^2 + y^2}$$

and
$$z = z.$$

An expression for θ, however, requires several alternative formulas, depending upon the quadrant or axis-position of (x,y) in the xy plane. Thus for $x > 0$, the formula is

$$\theta = \tan^{-1}\left(\frac{y}{x}\right) + 2n\pi, \qquad\qquad n = 0, \pm 1, \pm 2, \ldots;$$

for $y > 0$, the formula is

$$\theta = \cot^{-1}\left(\frac{x}{y}\right) + 2n\pi, \qquad\qquad n = 0, \pm 1, \pm 2, \ldots;$$

for $x < 0$, the formula is

$$\theta = \tan^{-1}\left(\frac{y}{x}\right) + (2n+1)\pi, \qquad n = 0, \pm 1, \pm 2, \ldots;$$

and for $y < 0$, the formula is

$$\theta = \cot^{-1}\left(\frac{x}{y}\right) + (2n+1)\pi, \qquad n = 0, \pm 1, \pm 2, \ldots.$$

The defining equations (2.1) can be expressed as a single vector equation. Let $\vec{R}$ be the position vector of a point $P = (r,\theta,z)$ in cylindrical coordinates. Then, in the associated Cartesian system, we have

$$\vec{R} = x\hat{i} + y\hat{j} + z\hat{k},$$

and hence, as our *vector defining equation*, we have:

(2.2) $$\vec{R} = r\cos\theta\,\hat{i} + r\sin\theta\,\hat{j} + z\hat{k}.$$

§7.3   Coordinate Surfaces and Coordinate Curves

We use cylindrical coordinates to introduce and illustrate some concepts which will later be used with other non-Cartesian coordinate systems in $\mathbf{E}^3$.

Coordinate surfaces. A *coordinate surface* is obtained by choosing a fixed value for one of the three coordinates and by letting the other two coordinates vary. This usually gives a flat or curved surface in $\mathbf{E}^2$. In special cases, it may give a line, curve, or single point. With cylindrical coordinates, we get

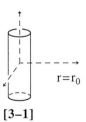

(i) θz-*surfaces.* A θz-surface, [3–1], is obtained by setting $r = r_0$ for some fixed $r_0 \geq 0$ and considering all points (r_0,θ,z) as θ and z vary. For $r_0 > 0$, this surface is an infinitely long cylinder of radius r_0 around the z axis. For $r_0 = 0$,

[3–1]

we get the z axis itself. Thus the points on a θz-surface are obtained by letting θ and z vary in the vector expression

$$\vec{R}(\theta, z) = r_0 \cos \theta \hat{i} + r_0 \sin \theta \hat{j} + z \hat{k}.$$

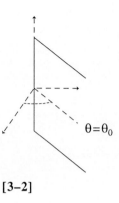

(ii) rz-*surfaces*. For an rz-surface, [3–2], we choose a fixed angle θ_0 and consider all points (r, θ_0, z) as r and z vary. This surface is a half-plane with the z axis as its edge. Thus, an rz-surface is obtained by letting r and z vary in the vector expression

$\theta = \theta_0$

[3–2]

$$\vec{R}(r, z) = r \cos \theta_0 \hat{i} + r \sin \theta_0 \hat{j} + z \hat{k}.$$

(iii) rθ-*surfaces*. For an rθ surface, [3–3], we choose a fixed z_0 and consider all points (r, θ, z_0) as r and θ vary. This surface is a plane perpendicular to the z axis. Thus, an rθ-surface is obtained by letting r and θ vary in the vector expression

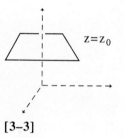

$z = z_0$

$$\vec{R}(r, \theta) = r \cos \theta \hat{i} + r \sin \theta \hat{j} + z_0 \hat{k}.$$

[3–3]

Coordinate curves. A *coordinate curve* is a curve or point in E^3 obtained by choosing fixed values for two of the three coordinates and by allowing the remaining coordinate to vary. With cylindrical coordinates, we get

(i) r-*curves*. For an r-curve, [3–4], we choose values θ_0 and z_0 and consider all points (r, θ_0, z_0) as r varies. These points form a ray in the plane $z = z_0$ with vertex on the z-axis. If we let $t = r$ be the parameter, the defining equations give a path for this curve in the associated Cartesian system:

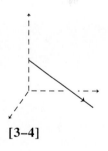

(3.1) $\vec{R}(t) = x\hat{i} + y\hat{j} + z\hat{k}$

$= t \cos \theta_0 \hat{i} + t \sin \theta_0 \hat{j} + z_0 k \quad (0 \le t < \infty).$

[3–4]

(ii) θ-*curves*. For a θ-curve, [3–5], we choose fixed values r_0 and z_0, and consider all points (r_0, θ, z_0) as θ varies. For $r_0 > 0$, these points form a circle of radius r_0 in the plane $z = z_0$ with center on the z axis. Using $t = \theta$ as parameter, we get a path for this curve in the associated Cartesian system:

(3.2) $\vec{R}(t) = r_0 \cos t \hat{i} + r_0 \sin t \hat{j} + z_0 \hat{k} \quad (0 \le t \le 2\pi).$

[3–5]

For $r_0 = 0$, the θ-curve consists of the single point $(0, 0, z_0)$ on the z axis.

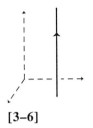

[3–6]

(iii) *z-curves.* For a z-curve, [3–6], we choose fixed values for r_0 and θ_0, and consider all points (r_0, θ_0, z) as z varies. For $r_0 > 0$, these points form a line parallel to the z axis through the point $x = r_0 \cos \theta_0$, $y = r_0 \sin \theta_0$, $z = 0$. Using $t = z$ as parameter, we get a path for this curve in the associated Cartesian system:

$$(3.3) \qquad \vec{R}(t) = r_0 \cos \theta_0 \hat{i} + r_0 \sin \theta_0 \hat{j} + t\hat{k} \quad (-\infty < t < \infty).$$

Thus, for $r_0 = 0$, the z-curve is the z axis.

§7.4 Coordinate Vectors and Coordinate Frames

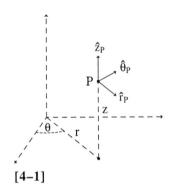

[4–1]

Let P be a fixed point with $r > 0$. Let C_r, C_θ, and C_z be, respectively, the r-curve through P, the θ-curve through P, and the z-curve through P. The *unit tangent vectors* at P to the curves C_r, C_θ, and C_z are denoted, respectively, as $\hat{r}_P$, $\hat{\theta}_P$, and $\hat{z}_P$, [4–1]. $\hat{r}_P$, $\hat{\theta}_P$, and $\hat{z}_P$ are called the *coordinate vectors at* P for our cylindrical coordinate system. Note that the directions of the coordinate vectors $\hat{r}$ and $\hat{\theta}$ may change if we move from coordinate vectors at P to coordinate vectors at another point P'.

If we use (3.1) of §6.3 to find unit tangent vectors to the paths for the coordinate curves, we get the following expressions for the coordinate vectors $\hat{r}_P$, $\hat{\theta}_P$, and $\hat{z}_P$ in the associated Cartesian system, for $P = (r_0, \theta_0, z_0)$.

(i) For $\hat{r}_P$, we use the path (3.1):

$$\vec{R}(t) = t \cos \theta_0 \hat{i} + t \sin \theta_0 \hat{j} + z_0 \hat{k}.$$

Then $\dfrac{d\vec{R}}{dt} = \cos \theta_0 \hat{i} + \sin \theta_0 \hat{j}$, and $\left| \dfrac{d\vec{R}}{dt} \right| = 1$.

Hence, for our unit tangent vector at the point $P = (r_0, \theta_0, z_0)$, we have

$$\hat{r}_P = \frac{d\vec{R}}{dt} \Big/ \left| \frac{d\vec{R}}{dt} \right| = \cos \theta_0 \hat{i} + \sin \theta_0 \hat{j}.$$

(ii) For $\hat{\theta}_P$, we use the path (3.2):

$$\vec{R}(t) = r_0 \cos t \hat{i} + r_0 \sin t \hat{j} + z_0 \hat{k}.$$

Then $\dfrac{d\vec{R}}{dt} = -r_0 \sin t\hat{i} + r_0 \cos t\hat{j}$, and $\left|\dfrac{d\vec{R}}{dt}\right| = r_0$.

Hence, for our unit tangent vector at the point (r_0, θ_0, z_0), we have

$$\hat{\theta}_P = \dfrac{d\vec{R}}{dt} \bigg/ \left|\dfrac{d\vec{R}}{dt}\right| = -\sin\theta_0\hat{i} + \cos\theta_0\hat{j}.$$

(iii)   For $\hat{z}_P$, we use the path (3.3):

$$\vec{R}(t) = -r_0 \sin\theta_0\hat{i} + r_0 \cos\theta_0\hat{j} + t\hat{k}.$$

Then $\dfrac{d\vec{R}}{dt} = \hat{k}$, and $\left|\dfrac{d\vec{R}}{dt}\right| = 1$.

Hence, for our unit tangent vector at the point (r_0, θ_0, z_0), we have

$$\hat{z}_P = \dfrac{d\vec{R}}{dt} \bigg/ \left|\dfrac{d\vec{R}}{dt}\right| = \hat{k}.$$

Thus, at the point $P = (r, \theta, z)$ *with* $r > 0$, *we have the expressions*

(4.1)
$$
\begin{aligned}
\hat{r}_P &= \cos\theta\hat{i} + \sin\theta\hat{j}, \\
\hat{\theta}_P &= -\sin\theta\hat{i} + \cos\theta\hat{j}, \\
\hat{z}_P &= \hat{k}
\end{aligned}
$$

in the associated Cartesian system. These formulas express the cylindrical coordinate vectors at P *as vectors in the associated Cartesian system, but with scalar components given in terms of the cylindrical coordinates at* P. *Moreover, these formulas define* $\hat{r}$ *and* $\hat{\theta}$ *as one-variable vector functions of* θ.

The formulas (4.1) are used in §7.5 below.

Coordinate frames. From the formulas for $\hat{r}_P, \hat{\theta}_P$, and $\hat{z}_P$, we see that $\hat{r}_P \cdot \hat{\theta}_P = \hat{r}_P \cdot \hat{z}_P = \hat{\theta}_P \cdot \hat{z}_P = 0$. Thus, at any given point P with $r > 0$, $\hat{r}_P$, $\hat{\theta}_P$, and $\hat{z}_P$ form a frame. This frame is called the *coordinate frame* at P. The directions of the unit vectors $\hat{r}_P$ and $\hat{\theta}_P$ in the coordinate frame at P depend upon the value of θ at P. The coordinate frame for cylindrical coordinates is important for the same reason that the moving frame is important in §6.7, namely, vectors of interest at the point P (for example, velocity and acceleration of a moving point at P) can be expressed in terms of the coordinate frame at P. In §7.6, we shall see how a problem with cylindrical symmetry may be simplified and clarified by the use of cylindrical coordinate frames.

Of course, the concepts of *coordinate surface, coordinate curve, coordinate vector,* and *coordinate frame* can be also applied to Cartesian coordinates. For Cartesian coordinates, we have: the coordinate surfaces are planes perpendicular to the axes; the coordinate curves are lines parallel to the axes; the coordinate vectors at every point are $\hat{i}, \hat{j},$ and $\hat{k}$; the coordinate vectors form a frame; and the directions of the coordinate frame remain unchanged as we move from one point to another.

Definition. A coordinate system is said to be *orthogonal* if, at every point P for which coordinate vectors are defined, the coordinate vectors at P form a frame. Cartesian systems and cylindrical systems are two examples of orthogonal coordinate systems.

Remark. It can be shown that Cartesian systems are the only orthogonal systems for which all coordinate curves are straight lines. (We saw above that the θ-curves in a cylindrical system are circles.) For this reason, non-Cartesian orthogonal systems are sometimes referred to as *curvilinear coordinate systems.* It can also be shown that Cartesian systems are the only orthogonal systems for which all coordinate frames have the same constant orientation in space.

§7.5 Paths, Velocity, and Acceleration

Cylindrical coordinates can be used to describe a given path in E^2. The first step is to obtain the cylindrical coordinates of the "moving point" as scalar functions, r(t), θ(t), and z(t), of a parameter t. Using the unit vectors $\hat{i},\hat{j},\hat{k}$ in the associated Cartesian system, we can express the position vector of the moving point:

(5.1) $\bar{R}(t) = r(t) \cos θ(t)\hat{i} + r(t) \sin θ(t)\hat{j} + z(t)\hat{k}$.

Example 1. In our helix example in Chapter 6, we have $\bar{R}(t) = 2 \cos t\hat{i} + 2 \sin t\hat{j} + 3t\hat{k}$ *on* $[0,4π]$. We see that the *cylindrical parametric equations* for this curve are:

$$\left.\begin{array}{l} r(t) = 2 \\ \theta(t) = t \\ z(t) = 3t \end{array}\right\}, \quad (0 \le t \le 4\pi).$$

The second step in using cylindrical coordinates with a path $\vec{R}(t)$ in $\mathbf{E}^3$ is to get an expression for $\vec{R}(t)$ which uses a cylindrical coordinate frame $\hat{r}, \hat{\theta}$, and $\hat{z}$ in place of the Cartesian frame $\hat{i}, \hat{j}$, and $\hat{k}$. That is to say, we seek to express the position vector $\vec{R}(t)$, for any chosen value of t ("any chosen moment t"), *in terms of the cylindrical frame at the point which $\vec{R}(t)$ locates*. This coordinate frame is called the *local cylindrical frame* at the point $\vec{R}(t)$.

We accomplish the second step by using (2.2) and (4.1) to calculate:

$$\vec{R}(t) \cdot \hat{r}(t) = r(t)(\cos^2 \theta(t) + \sin^2 \theta(t)) = r(t),$$
$$\vec{R}(t) \cdot \hat{\theta}(t) = r(t)(-\cos \theta(t) \sin \theta(t) + \sin \theta(t) \cos \theta(t)) = 0,$$
$$\vec{R}(t) \cdot \hat{z}(t) = z(t).$$

Hence, by the frame identity, we have:

(5.2) $\vec{R} = r\hat{r} + z\hat{z}$, *where* r *and* z *are scalar functions of* t, $\hat{r}$ *is a vector function of* t, *and* $\hat{z}$ *is a constant vector*, [5–1].

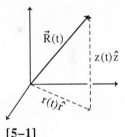

[5–1]

Example 1 (*continued*). For our helix we have $\vec{R}(t) = 2\hat{r} + 3t\hat{z}$.

Given a path $\vec{R}(t)$ in the form (5.2), our next goal is to get expressions for the *velocity* $\vec{v} = \dot{\vec{R}}$ and the *acceleration* $\vec{a} = \ddot{\vec{R}}$, at any given moment, *in terms of the derivatives of* $r(t), \theta(t),$ *and* $z(t)$ *at that moment and in terms of the local cylindrical frame at that moment.*

We begin by differentiating (5.2). This gives

$$\vec{v} = \dot{\vec{R}} = \dot{r}\hat{r} + r\dot{\hat{r}} + \dot{z}\hat{z} + z\dot{\hat{z}}.$$

From (4.1), we have

(5.3) $\dot{\hat{r}} = -(\sin \theta)\dot{\theta}\hat{i} + (\cos \theta)\dot{\theta}\hat{j},\ \dot{\hat{\theta}} = -(\cos \theta)\dot{\theta}\hat{i} - (\sin \theta)\dot{\theta}\hat{j},\ and\ \dot{\hat{z}} = \vec{0}.$

It remains to express $\dot{\hat{r}}$ in terms of the cylindrical frame. We do this by using the frame identity in the same way that we did to get (5.2).

From (4.1) and (5.3), we have

$$\dot{\hat{r}} \cdot \hat{r} = -\sin\theta(\cos\theta)\dot{\theta} + \cos\theta(\sin\theta)\dot{\theta} = 0,$$
$$\dot{\hat{r}} \cdot \hat{\theta} = (\sin^2\theta)\dot{\theta} + (\cos^2\theta)\dot{\theta} = \dot{\theta},$$
$$\dot{\hat{r}} \cdot \hat{z} = 0.$$

Hence, by the frame identity, we have

(5.4) $$\dot{\hat{r}} = \dot{\theta}\hat{\theta}.$$

This gives us our desired expression for $\vec{v}$:

(5.5) *Velocity in cylindrical coordinates:* $\quad \vec{v} = \dot{r}\hat{r} + r\dot{\theta}\hat{\theta} + \dot{z}\hat{z}.$

To get the acceleration, we differentiate again and repeat the frame identity procedure. From (5.5) we have

$$\vec{a} = \ddot{\vec{R}} = \dot{\vec{v}} = \ddot{r}\hat{r} + \dot{r}\dot{\hat{r}} + \dot{r}\dot{\theta}\hat{\theta} + r\ddot{\theta}\hat{\theta} + r\dot{\theta}\dot{\hat{\theta}} + \ddot{z}\hat{z}.$$

From (5.4), we have $\dot{\hat{r}} = \dot{\theta}\hat{\theta}$, and by a frame identity analysis we get

$$\dot{\hat{\theta}} \cdot \hat{r} = -\dot{\theta},$$
$$\dot{\hat{\theta}} \cdot \hat{\theta} = 0,$$
$$\dot{\hat{\theta}} \cdot \hat{z} = 0,$$

and hence

(5.6) $$\dot{\hat{\theta}} = -\dot{\theta}\hat{r}.$$

Using (5.4) and (5.6) in $\vec{a}$, we get our desired expression for $\vec{a}$:

(5.7) *Acceleration in cylindrical coordinates:*

$$\vec{a} = (\ddot{r} - r\dot{\theta}^2)\hat{r} + (r\ddot{\theta} + 2\dot{r}\dot{\theta})\hat{\theta} + \ddot{z}\hat{z}.$$

If we have cylindrical parametric equations for the motion of a particle, we can use (5.5) and (5.7) to calculate $\vec{v}$ and $\vec{a}$.

Example 1 (*continued*). We have $r(t) = 2$, $\theta(t) = t$, and $z(t) = 3t$. Therefore,

$$
\begin{array}{lll}
r = 2 & \dot{r} = 0 & \ddot{r} = 0 \\
\theta = t & \dot{\theta} = 1 & \ddot{\theta} = 0 \\
z = 3t & \dot{z} = 3 & \ddot{z} = 0
\end{array}
$$

and this gives the result
$$
\begin{cases}
\vec{v} = 2\hat{\theta} + 3\hat{z}, \\
\vec{a} = -2\hat{r}.
\end{cases}
$$

The expressions (5.5) and (5.7) for $\vec{v}$ and $\vec{a}$ are especially useful for analyzing rotating mechanical systems. The content of the expression (5.5) for $\vec{v}$ is intuitively natural; the $\hat{r}$, $\hat{\theta}$, and $\hat{z}$ components of the velocity of a moving particle are just what geometry would lead us to expect. The content of the expression (5.7) for $\vec{a}$ is less obvious. In particular, the term $2\dot{r}\dot{\theta}\hat{\theta}$, known as the *Coriolis term*, is related to special effects in rotating physical systems. The Coriolis term can be used, for example, along with Newton's second law, to analyze and predict the motion of a Foucault pendulum, to explain the slight tendency of a draining vortex to be counterclockwise in the northern hemisphere and clockwise in the southern hemisphere, and to help explain prevailing wind patterns in the earth's atmosphere. In all three cases, the rotating system is the surface of the earth.

§7.6 Motion under a Central Force

Let O be a fixed point in space. Let a particle P of mass m move in space under the influence of a force $\vec{F}$, where, at each moment, the direction of $\vec{F}$ is along the line determined by O and P and where $\vec{F}$ is the only force acting on P. (There is no other restriction on the force $\vec{F}$; $\vec{F}$ may be attractive or repulsive, and its magnitude may vary with position and time in an arbitrary and irregular way.) Under these circumstances, we say that the motion of the particle is a *central force motion with center at* O.

Assume that we have a particular central force motion of this kind. Let $\vec{R}(t)$ be the position vector of P with respect to O. Then, by our definition,

(6.1) $$\vec{F}(t) = g(t)\vec{R}(t)$$

for some scalar function g(t). By Newton's second law, we have

$$\vec{F} = g(t)\vec{R} = m\vec{a} = m\ddot{\vec{R}}.$$

Hence $$\ddot{\vec{R}} = \frac{g(t)}{m}\vec{R}.$$

It follows by vector algebra that

(6.2) $$\vec{R} \times \ddot{\vec{R}} = \frac{g(t)}{m}(\vec{R} \times \vec{R}) = \vec{0}.$$

By (4.4) of Chapter 5, we know that

$$\frac{d}{dt}(\vec{R} \times \dot{\vec{R}}) = \dot{\vec{R}} \times \dot{\vec{R}} + \vec{R} \times \ddot{\vec{R}} = \vec{R} \times \ddot{\vec{R}}.$$

Hence from (6.2) we have

(6.3) $$\frac{d}{dt}(\vec{R} \times \dot{\vec{R}}) = \vec{0}.$$

By §5.7, (6.3) implies that

(6.4) $$\vec{R} \times \dot{\vec{R}} = \vec{C}, \text{ for some constant vector } \vec{C}.$$

By the definition of cross product, it follows from (6.4) that $\vec{R} \cdot \vec{C} = 0$, and that $\dot{\vec{R}} \cdot \vec{C} = 0$, for all values of t. The equation $\vec{R} \cdot \vec{C} = 0$ tells us that the particle P must remain in a fixed plane M which contains O and is perpendicular to $\vec{C}$. The equation $\dot{\vec{R}} \cdot \vec{C} = 0$ tells us that throughout the motion, the velocity of P can be represented by an arrow which lies in the plane M.

We now introduce a system of cylindrical coordinates which has O as its origin and has the plane M as the plane z = 0. Using coordinate vectors, we see that

$$\vec{R} = r\hat{r}, \qquad\qquad \text{by (5.2),}$$

and $$\dot{\vec{R}} = \dot{r}\hat{r} + r\dot{\theta}\hat{\theta}, \qquad \text{by (5.5).}$$

Therefore, $$\vec{C} = \vec{R} \times \dot{\vec{R}} = r^2\dot{\theta}\hat{z}.$$

Since $\vec{C}$ is a constant vector, we have that

(6.5) $\qquad\qquad\qquad |\vec{C}| = r^2\dot{\theta}$ is a constant scalar.

For our central force motion, the constant vector $m\vec{C} = mr^2\dot{\theta}\hat{z}$ is known to physicists as the *angular momentum of* P *about the point* O. (Physicists use a more general definition of angular momentum which applies, as well, to other motions besides central force motions.) We have shown that this angular momentum about O is a constant vector (in direction and magnitude) during any central force motion with center at O.

Since P moves in the plane M during our central force motion, the moving position-arrow from O to P sweeps out area in M. By elementary calculus, the area A(d) swept out between t = a and t = d (see [6–1]) must be

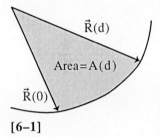

$\vec{R}(d)$

Area = A(d)

$\vec{R}(0)$

[6–1]

$$A(d) = \int_a^d \tfrac{1}{2}r^2\dot{\theta}\,dt\,.$$

If we apply the fundamental theorem of integral calculus, we get

(6.6) $\qquad\qquad\qquad \dfrac{dA}{dt} = \dfrac{1}{2}r^2\dot{\theta} = \dfrac{1}{2}|\vec{C}|\;.$

Thus the arrow from O *to* P *sweeps out area in* M *at a constant rate.* This result is known as *Kepler's second law*. (We state it here as a mathematical fact about central force motions. Kepler's original statement was an empirical assertion about the observed motions of planets around the sun.)

§7.7 Motion under an Inverse-square Central Force

A special case of central force motion arises when we have

(7.1) $\qquad\qquad\qquad \vec{F}(t) = \dfrac{k}{\vec{R}\cdot\vec{R}}\hat{R} = \dfrac{k}{|\vec{R}|^2}\hat{R}\,.$

where k is a given positive or negative constant, $\vec{R}$ is the position-vector of the moving particle, and $|\vec{R}|^2$ is the square of the distance of the particle from O. Such a force is called an *inverse-square central force*. The force acts directly towards, or directly away from, O, and its magnitude is inversely proportional to the square of its distance from O. Inverse-square central forces are important in elementary physics. The gravitational force caused by a uniform spherical mass with center at O is observed to be an inverse-square central force (*Newton's law of gravity*), and so is the electrostatic force caused by a uniform spherical electric charge with center at O (*Coulomb's law of electrostatic attraction and repulsion*).

(7.1) is a special case of (6.1), since $\dfrac{k}{|\vec{R}|^2}\hat{R}$ can be written as $\dfrac{k}{|\vec{R}|^3}\hat{R}$,

and $\dfrac{k}{|\vec{R}|^3}$ can then be viewed as the coefficient $g(t)$ in (6.1). ($\dfrac{k}{|\vec{R}|^3}$ depends on

t because $\vec{R}$ depends upon t.)

A dramatic early achievement of modern science was Isaac Newton's deduction, from his laws of motion and his law of gravity, that the *orbit* (curve traced out by the motion) of a particle of constant mass under an inverse-square central force must be a conic section in a plane through O. In order to achieve this deduction, Newton invented calculus and spent several years of intensive research. We shall use vector algebra (which Newton did not have) and differential calculus to make this deduction. With the benefit of three centuries of further mathematical experience, our deduction will occupy less than two pages of text.

(In Chapter 28, we shall see how to extend this result from a moving particle to a moving uniform solid sphere. Newton invented integral calculus in order to achieve this last step — which completed a theoretical deduction of Kepler's empirical observation that the orbits of planets are ellipses (*Kepler's first law*).)

We begin our deduction for a moving particle of constant mass m by applying Newton's second law of motion to (7.1). We have

$$m\ddot{\vec{R}} = \frac{k}{|\vec{R}|^3}\hat{R},\qquad \text{and this gives}$$

(7.2) $$\ddot{\vec{R}} = \frac{b}{|\vec{R}|^3}\hat{R},\qquad \text{where } b = \frac{k}{m}.$$

From §7.6, we know that there must be a constant vector $\vec{C}$ such that, for every t, (i) $\vec{R}\times\dot{\vec{R}} = \vec{C}$, and (ii) $\vec{R}$ and $\ddot{\vec{R}}$ lie in the plane M through O perpendicular to $\vec{C}$. As described in §7.6, we introduce cylindrical coordinates r, θ, and z, with origin at O and $\hat{z}$ in the direction of $\vec{C}$. Then, $\vec{R} = r\hat{r}$ and, by (6.5) and (6.6),

$$\vec{R}\times\dot{\vec{R}} = \vec{C} \Rightarrow \vec{C} = r^2\dot{\theta}\hat{z} \Rightarrow |\vec{C}| = r^2\dot{\theta} \Rightarrow$$

(7.3) $$\dot{\theta} = \frac{c}{r^2}, \text{ where } c = |\vec{C}|.$$

We now use (7.2) and (7.3) to prove Newton's result.

(7.4)   **_Theorem._** *The orbit of a particle of constant mass under an inverse-square central force is all or part of a conic section.*

   Proof of (7.4). $\vec{R} \times \dot{\vec{R}} = \vec{C} \Rightarrow \ddot{\vec{R}} \times (\vec{R} \times \dot{\vec{R}}) = \ddot{\vec{R}} \times \vec{C} = \frac{d}{dt}(\dot{\vec{R}} \times \vec{C})$,

by (4.4) in Chapter 5. By (7.2), we also have

$$\ddot{\vec{R}} \times \vec{C} = \frac{b}{r^3}\vec{R} \times \vec{C} = \frac{b}{r^2}\hat{r} \times r^2\dot{\theta}\hat{z} = -b\dot{\theta}\hat{\theta} = -b\dot{\hat{r}} \quad \text{by (5.4).}$$

Thus $\frac{d}{dt}(\dot{\vec{R}} \times \vec{C}) = -b\dot{\hat{r}}$.

Integrating, we have $$\dot{\vec{R}} \times \vec{C} = -b\hat{r} + \vec{A} = -\frac{b}{r}\vec{R} + \vec{A},$$

where the vector $\vec{A}$ is a constant of integration. $\vec{A}$ depends on the specific motion and must lie in M. We now take the scalar triple product

$$\vec{R} \cdot (\dot{\vec{R}} \times \vec{C}) = -\frac{b}{r}\vec{R}^2 + \vec{R} \cdot \vec{A} = -br + ra\cos\varphi,$$

where φ is the angle between $\hat{r}$ and $\vec{A}$ and $a = |\vec{A}|$.
By (6.1) of Chapter 3, we also have

$$\vec{R} \cdot (\dot{\vec{R}} \times \vec{C}) = (\vec{R} \times \dot{\vec{R}}) \cdot \vec{C} = \vec{C}^2 = c^2.$$

Therefore $c^2 = -br + r\,a\cos\varphi.$

Dividing by b and transposing, we have

$$r = -\frac{c^2}{b} + \frac{a}{b}r\cos\varphi.$$

We now change to Cartesian coordinates with positive x axis in the direction of $\vec{A}$ and positive z axis in the direction of $\hat{z}$. (This new Cartesian system is obtained by rotating the Cartesian axes associated with r, θ, and z about the z axis, and it has r, φ, z as its associated cylindrical coordinates.) The last equation above becomes

$$\sqrt{x^2 + y^2} = -\frac{c^2}{b} + \frac{a}{b}x.$$

Squaring, we have

$$x^2 + y^2 = \frac{a^2}{b^2} x^2 - 2 \frac{ac^2}{b^2} x + \frac{c^4}{b^2}.$$

This is a second-degree equation, and hence, by (1.4) and (1.5) of Chapter 4, it determines a conic section. In a principal case, this conic section will be an ellipse, a parabola, or a hyperbola, depending upon whether $\frac{a^2}{b^2}$ is < 1, $=1$, or >1. For certain values of a, b, and c, special-case conic sections can also occur. This proves (7.4).

As a corollary to the proof of (7.4) we obtain the result on hodographs mentioned in §5.3.

(7.5) **Corollary.** *The hodograph of the motion of a particle of constant mass under an inverse-square central force must be all or part of a circle.*

Proof of (7.5). From the proof of (7.4), we have

$$r = -\frac{c^2}{b} + \frac{a}{b} r \cos \varphi.$$

Dividing by r, we get

$$1 = -\frac{c^2}{br} + \frac{a}{b} \cos \varphi.$$

Transposing and differentiating with respect to t, we have

$$\frac{c^2}{br^2} \dot{r} = \frac{a}{b} \sin \varphi \, \dot{\varphi} = \frac{ac}{br^2} \sin \varphi, \text{ by (7.3) with } \dot{\varphi} \text{ for } \dot{\theta}.$$

Hence, dividing by $\frac{c^2}{br^2}$, we have

(7.6) $\dot{r} = \frac{a}{c} \sin \varphi.$

We now take new polar coordinates $\tilde{r}$ and $\tilde{\theta}$ in M with the same origin O and with the positive x axis in the direction of $\hat{z} \times \vec{A}$. We then have $\tilde{r} = r$ and $\tilde{\theta} = \varphi - \frac{\pi}{2}$. Then $\cos \tilde{\theta} = \cos(\varphi - \frac{\pi}{2}) = \sin \varphi$, and (7.6) becomes

$$\dot{\tilde{r}} = \frac{a}{c} \cos \tilde{\theta}.$$

Hence $2 \tilde{r} \dot{\tilde{r}} = 2 \tilde{r} \frac{a}{c} \cos \tilde{\theta},$

and this gives $\frac{d}{dt}(\tilde{r}^2) = 2\vec{K} \cdot \vec{R}$, where $|\vec{K}| = \frac{a}{c}$ and $\vec{K}$ has the direction $\tilde{\theta} = 0$.

But $\tilde{r}^2 = \vec{R}^2$. Hence $\frac{d}{dt}(\tilde{r}^2) = \frac{d}{dt}(\vec{R} \cdot \vec{R}) = 2\vec{R} \cdot \dot{\vec{R}}$, and we have

$$2\vec{R} \cdot \dot{\vec{R}} = 2\vec{K} \cdot \vec{R} \Rightarrow 2\vec{R} \cdot (\dot{\vec{R}} - \vec{K}) = 0$$

$$\Rightarrow \frac{2\ddot{\vec{R}}|\vec{R}|^3}{b} \cdot (\dot{\vec{R}} - \vec{K}) = 0 \quad (\text{by } (7.2)) \Rightarrow$$

$$2\ddot{\vec{R}} \cdot (\dot{\vec{R}} - \vec{K}) = 0 \Rightarrow 2\dot{\vec{R}} \cdot \ddot{\vec{R}} = 2\vec{K} \cdot \ddot{\vec{R}} \Rightarrow \frac{d}{dt}(\dot{\vec{R}}^2) = \frac{d}{dt}(2\vec{K} \cdot \dot{\vec{R}}).$$

Integrating this last equation, we get $\dot{\vec{R}}^2 = 2\vec{K} \cdot \dot{\vec{R}} + d$, where the scalar d is a constant of integration and depends on the specific motion. To find the hodograph, we view $\dot{\vec{R}}$ as a new position vector $\vec{V}$. We then have

$$\vec{V}^2 = 2\vec{K} \cdot \vec{V} + d.$$

In the associated Cartesian system, with $\vec{V} = x\hat{i} + y\hat{j}$ and $\vec{K} = \frac{a}{c}\hat{i}$, this equation in $\vec{V}$ becomes

$$x^2 + y^2 = 2\frac{a}{c}x + d.$$

Completing the square for x we have

$$(x - \frac{a}{c})^2 + y^2 = (\frac{a}{c})^2 + d,$$

and we see that this equation determines a circle with center at $(\frac{a}{c}, 0)$ and radius $= \sqrt{(\frac{a}{c})^2 + d}$. The case $(\frac{a}{c})^2 + d < 0$ cannot occur, since, for any given motion, the vector $\vec{V}$ must exist and must trace out some curve. This proves (7.5).

Applications of corollary (7.5) are given in the problems.

§7.8 Problems

§7.1

1. A point P has the cylindrical coordinates r = 3, $\theta = \frac{3\pi}{4}$, z = −1. What are the coordinates of P in the associated Cartesian system?

§7.4

1. A point P has the cylindrical coordinates r = 4, $\theta = \frac{2\pi}{3}$, z = 1.

(a) What are the coordinates of P in the associated Cartesian system?

(b) Give expressions in terms of $\hat{i}$, $\hat{j}$, and $\hat{k}$ for the coordinate vectors $\{\hat{r},\hat{\theta},\hat{z}\}$ at the point P.

(c) What are the scalar components of the vector $3\hat{j}$ in this frame?

2. A particle moves so that its cylindrical coordinates are given by:

$$\left.\begin{array}{l} r = \sqrt{2}t \\ \theta = \dfrac{\pi}{4} \\ z = t \end{array}\right\} (0 \le t \le 1), \text{ where } t \text{ is time.}$$

(a) Find parametric equations in the associated Cartesian coordinate system for the same path with time t as parameter.

(b) Give the components of the particle's velocity $\vec{v}$ in the cylindrical coordinate frame at $t = \dfrac{1}{2}$.

3. Consider a cylindrical coordinate system and its associated Cartesian system. Find the scalar components of $3\hat{j} - 4\hat{k}$ in the cylindrical coordinate frame at the point whose position-vector is $\hat{i} - 2\hat{j} + 2\hat{k}$.

4. A point P has the cylindrical coordinates $(2,\frac{\pi}{4},3)$. Let $\vec{V} = -y\hat{i} + x\hat{j} + z\hat{k}$, where x, y, and z are the associated Cartesian coordinates of P. Find the scalar components of $\vec{V}$ in the cylindrical coordinate frame at P.

5. A point P has the cylindrical coordinates $(2,\frac{\pi}{3},1)$. Let $\vec{V} = x\hat{i} + y\hat{j} + z\hat{k}$, where x, y, and z are the associated Cartesian coordinates of P. Find the scalar components of $\vec{V}$ in the cylindrical coordinate frame at P.

§7.5

1. The path of a moving particle is given in cylindrical coordinates by $r(t) = a(1 + \sin t)$, $\theta(t) =$ $1 - e^{-t}$, $z(t) = \cos t$. At the point where $t = 0$, find the components of the velocity and the acceleration in the cylindrical coordinate frame.

2. The path of a moving particle is given by $\vec{R}(t) = t\cos t\,\hat{i} + t\sin t\,\hat{j} + t\hat{k}$, $(0 \le t)$.

(a) Find the coordinates of the particle in the associated cylindrical system as functions of t.

(b) Find expressions for the velocity and acceleration of the particle in the local cylindrical coordinate frame.

3. You are given parametric equations for a curve in cylindrical coordinates: $r = t^2$, $\theta = 2t$, $z = t^3$. Express the unit tangent vector $\hat{T}$ as a function of t in the coordinate frame $\{\hat{r},\hat{\theta},\hat{z}\}$ at time t.

4. The path of a particle is given in cylindrical coordinates by the parametric equations $r(t) = t$, $\theta(t) = t^2$, $z(t) = t^3$, where t is time and $0 \le t \le 5$.

(a) Give the velocity $\vec{v}(t)$ in the coordinate frame at time t.

(b) Give the acceleration $\vec{a}(t)$ in the coordinate frame at time t.

(c) Give the speed $\frac{ds}{dt}$ at time t.

(d) Express, as a definite integral, the arc length of the particle's path. (*Hint.* Use (4.8) of Chapter 6.)

5. You are given parametric equations for a curve in cylindrical coordinates; $r = t^2$, $\theta = e^t$, $z = 0$, $(-\infty < t < \infty)$, where t is time. Find, as functions of t, the components of the unit tangent vector $\hat{T}$ at time t in the coordinate frame $\{\hat{r},\hat{\theta},\hat{z}\}$ at time t.

6. A particle moves so that its cylindrical coordinates are given as functions of time by $r = t$, $\theta = t$, $z = t$, $(0 \le t)$. Find

(a) the speed at time $t = 1$;

(b) the curvature at time $t = 1$.

7. A particle moves so that its cylindrical coordinates are given as functions of time by $r = t$, $\theta = t^2$, $z = t$, $(0 \le t)$.
    (a) Express velocity $\vec{v}$ and acceleration $\vec{a}$ at time $t = 1$ in the coordinate frame at time $t = 1$.
    (b) Find the curvature at time $t = 1$.

8. A particle moves so that its cylindrical coordinates are given as functions of time by $r = t^2$, $\theta = t$, $z = t$, $(0 \le t)$.
    (a) Express velocity $\vec{v}$ and acceleration $\vec{a}$ at time $t = 1$ in the coordinate frame at time $t = 1$.
    (b) Find the curvature at time $t = 1$.

9. A curve satisfies the equations $r = e^{a\theta}$ and $z = 0$ in cylindrical coordinates.
    (a) Use θ itself as parameter, and find the components of $\hat{T}$ and $\hat{N}$ in the frame $\{\hat{r}, \hat{\theta}, \hat{z}\}$ as functions of θ.
    (b) Find κ as a function of θ.

10. A path is given in cylindrical coordinates by the parametric equations $\left. \begin{array}{l} r = r(t) \\ \theta = \theta(t) \\ z = z(t) \end{array} \right\} (0 \le t)$. Find a formula for curvature κ as a function of t, in terms of the functions $r(t)$, $\theta(t)$, $z(t)$, and their derivatives.

11. A curve satisfies the equations $r = \theta$ and $z = 0$ in cylindrical coordinates. We are also given that $\frac{d\theta}{dt} = 2$. Find the scalar components of $\hat{T}$ and $\hat{N}$ (as functions of θ) in the coordinate frame $\{\hat{r}, \hat{\theta}, \hat{z}\}$, and find curvature κ as a function of θ. (*Hint.* Find $\vec{v}$ and $\vec{a}$ first. Find $\hat{N}$ by taking a vector product of $\hat{T}$ and $\hat{z}$.)

12. The motion of a particle is given in cylindrical coordinates by the parametric equations $r = r(t)$, $\theta = \theta(t)$, and $z = z(t)$. Find expressions giving the components of $\ddot{\theta}$ and $\dddot{\theta}$ in the cylindrical

coordinate frame $\{\hat{r}, \hat{\theta}, \hat{z}\}$ in terms of r, θ, and z and their first, second, and third derivatives with respect to t.

§7.6

1. In §7.6, we assume that the force $\vec{F}(t) = g(t)\vec{R}(t)$, on a particle P with position vector $\vec{R}(t)$, is directed along a line through a fixed point O. In most physical applications of this general kind, however, we will have a force $\vec{G}(t) = g(t)(\vec{R}_1(t) - \vec{R}_0(t))$ exerted by a particle P_0 with position vector $\vec{R}_0(t)$ upon a particle P_1, with position vector $\vec{R}_1(t)$. This force is directed along a line from P_1 through P_0, and the motion of P_0 is affected by the equal and opposite reactive force $(-\vec{G}(t))$ to the force which P_0 exerts upon P_1. Let m_0 be the mass of P_0, and let m_1 be the mass of P_1. Show that for all t,

    $$\frac{d^2}{dt^2}(\vec{R}_1(t) - \vec{R}_0(t)) = \frac{d^2\vec{R}}{dt^2},$$

    where $\vec{R}(t)$ is the position vector of a particle P with respect to a fixed center of force, and where (i) P is acted on by the central force $\vec{F}(t) = \vec{G}(t)$, and (ii) the mass of P is $\frac{m_0 m_1}{m_0 + m_1}$. (*Hint.* Use Newton's second law to find $\frac{d^2\vec{R}_1}{dt^2}$ and $\frac{d^2\vec{R}_0}{dt^2}$, separately, in terms of $\vec{G}(t)$. Then combine to get $\frac{d^2}{dt^2}(\vec{R}_1(t) - \vec{R}_0(t))$.)

    This remarkable and extremely simple result shows us that the motion of P_1 *relative to* P_0 can be calculated exactly as if P_0 were fixed in space, provided we leave $\vec{G}(t)$ unchanged and reduce the mass of P_1 to $\frac{m_0 m_1}{m_0 + m_1}$ when we apply Newton's second law. This result permits us, for example, to calculate the orbit of the earth relative to the sun as if the sun were fixed in space, provided that we make a slight reduction in the mass of the earth for the purpose of applying Newton's second law. (See also problem 1 for §7.7.)

§7.7

1. Newton's *law of gravity* asserts that the gravitational force exerted by a particle P_0 of mass m_0 and position vector $\vec{R}_0$, upon a particle P_1 of mass m_1 and position vector $\vec{R}_1$, is given by

$$\frac{\gamma m_0 m_1}{|R_0 - R_1|^3}(\vec{R}_0 - \vec{R}_1),$$

where γ is a certain physical constant.

Consider now the motion of P_1 relative to the center of mass CM of the two particles P_1 and P_0. By (6.2) of §5.6, CM may be viewed as a fixed point in space. Hence the motion of P_1 *relative to* CM may be viewed as a genuine central-force problem. Show that the gravitational force exerted by P_0 on P_1 may be expressed equivalently (at all times) as the gravitational force that would be exerted on P_1 by an imaginary particle P_I held fixed at the point CM and of mass $\left(\dfrac{m_0}{m_0 + m_1}\right)^2 m_0$. (*Hint.* This is a simple calculation, using the definition of CM.)

Using §7.7 and problem 1 of §7.6, and ignoring the effects of the moon and other planets, we see that the orbit of the earth *relative to* the sun is a true ellipse. Now, from the present problem, we see that the orbit of the earth relative to its common center of mass with the sun (this center of mass is a fixed point in space) must also be a true ellipse.

(*Note.* Our analysis in §7.6 and §7.7 has been carried out for small *particles* which we treat as points in space. Is it correct to apply the results of this analysis to extended, massive objects like the earth and the sun? In Chapter 28, we show that it is correct to do so.)

2. In an inverse-square motion, the circular hodograph is helpful for making an approximate qualitative analysis of the effects of certain perturbations on the motion. For example, the effect of a gentle, uniform, and constant external force (in addition to the inverse-square central force) in some fixed direction parallel to the plane of the orbit is to cause, by Newton's second law, a small, constant, additional acceleration in that direction. This results in a gradual shift of the circular hodograph in that same direction.

(a) Assume that the direction of such a force is also perpendicular to the major axis of an elliptical orbit. What changes in the shape of the orbit can be expected? What changes, if any, will occur in directions of the elliptic axes? (*Hint.* Use Kepler's second law to relate points on the hodograph to points on the orbit.)

(b) What happens to the hodograph when the circle of the hodograph ceases to enclose the origin?

Chapter 8

Scalar Fields and Scalar Functions

§8.1   Scalar Fields and Vector Fields
§8.2   Visualizing a Scalar Field
§8.3   Directional Derivatives
§8.4   Scalar Functions
§8.5   Partial Functions
§8.6   Partial Derivatives
§8.7   Limits and Continuity
§8.8   Counterexamples
§8.9   Problems

§8.1   Scalar Fields and Vector Fields

In the early development of calculus, the term "function" was understood to mean a numerical correspondence, from input numbers to output numbers, given by a mathematical formula of some kind. The word "function" was treated as simply another name for such a formula. As time passed, the word "function" came increasingly to mean an abstract numerical correspondence from inputs to outputs; it no longer referred to a particular formula. Thus mathematicians agreed to say, for example, that the *distinct* formulas $(x+1)^2-x^2$ and $2x+1$ define the *same* function. Still later, the meaning of the word "function" was broadened to include not just a numerical input-output correspondence defined by a mathematical formula, but also to include numerical input-output correspondences defined in other formal or empirical ways. Thus an empirically observed numerical correspondence between values of temperature (inputs) and values of pressure (outputs) for a given rigid container of a gas at equilibrium came to be called a "function," even in cases where no precise formula was known for the correspondence. Finally, in the twentieth century, the requirement that a function be a numerical correspondence was dropped, and the word "function" has come to be used for input-output correspondences for which the inputs or the outputs (or both) are objects of a non-numerical kind.

Scalar fields are functions in this broad sense. The outputs are *real numbers*, but the inputs are *points* in E^3 or E^2. Moreover, we do not require that there be a mathematical formula which defines the scalar field.

Definition. Let D be a subset of E^3. (D may be a bounded part of E^3, an unbounded part of E^3, or all of E^3.) Let f be a function which associates with each point P in D (as input) a corresponding real number f(P) (as output). Such a function f is called a *scalar field on* D *in* E^3, and D is called the *domain* of the scalar field f. Scalar fields with domains in E^2 are defined similarly.

In the natural sciences, scalar fields are widely used as mathematical models. For example, let D represent a region of physical space occupied by some solid material whose temperature varies from one point to another. The temperature of the material at a given moment can be represented by a scalar field f on D, where for any point P in D, f(P) is the temperature (in suitable units) at location P. Scalar fields can be similarly used to represent other physical scalar quantities which may vary from point to point in space, such as density, pressure, and electric potential.

A particular scalar field may be specified by a mathematical expression of some kind, or, in applied mathematics, it may simply be indicated by a single symbol (such as f) and treated as a function whose value at any given point can be obtained, as needed, by physical measurement. We shall often specify a scalar field f mathematically by using Cartesian coordinates and then giving f(P) as a numerical function of the coordinates of P. (See §8.4.)

Scalar fields can be combined by algebraic operations to form new scalar fields:

(1.1)   *Let* f *and* g *be given scalar fields on* D, *and let* c *be a given real number. Define* $h_1, h_2, h_3,$ *and* h_4 *by*

$$h_1(P) = cf(P)$$
$$h_2(P) = f(P) + g(P)$$
$$h_3(P) = f(P)g(P)$$
$$h_4(P) = f(P)/g(P).$$

Then $h_1, h_2,$ *and* h_3 *are scalar fields on* D, *and* h_4 *is a scalar field on the set of all points* P *in* D *for which* $g(P) \neq 0$.

Vector fields are also functions in this broader sense. The inputs are *points* in E^3 or E^2, and the outputs are *vectors* in E^3 or E^2.

Definition. Let D be a subset of $\mathbf{E}^3$. (D may be a bounded part of $\mathbf{E}^3$, an unbounded part of $\mathbf{E}^3$, or all of $\mathbf{E}^3$.) Let $\vec{F}$ be a function which associates with each *point* P in D (as input) a corresponding *vector* $\vec{F}(P)$ in $\mathbf{E}^3$ (as output). Such a function $\vec{F}$ is called a *vector field on* D *in* $\mathbf{E}^3$, and D is called the *domain* of the vector field $\vec{F}$. Similarly, *vector fields in* $\mathbf{E}^2$ are defined to be functions which take *points* in $\mathbf{E}^2$ as inputs and give *vectors* in $\mathbf{E}^2$ as outputs.

In the natural sciences, vector fields are widely used as mathematical models for *fluid flows* and *fields of force*. For example, a fluid-flow vector field in some given domain might have, as its output vector, a fluid velocity vector $\vec{v}$, a mass-flow vector $\vec{m}$, or (as a flow of energy) a heat-flow vector $\vec{H}$. (See §2.2.) Similarly, a vector field of force might have, as its output, an electric field vector $\vec{E}$, a magnetic field vector $\vec{B}$, or a gravitational field vector $\vec{G}$. (Fields of force are defined, and considered in more detail, in Chapter 18.)

Scalar and vector fields can be combined by the operations of vector algebra to form new scalar and vector fields:

(1.2) *Let* f *be a given scalar field on* D, *and let* $\vec{F}$ *and* $\vec{G}$ *be given vector fields on* D. *Define* h, $\vec{H}_1$, $\vec{H}_2$, *and* $\vec{H}_3$ *by*

$$h(P) = \vec{F}(P) \cdot \vec{G}(P)$$
$$\vec{H}_1(P) = f(P)\vec{F}(P)$$
$$\vec{H}_2(P) = \vec{F}(P) + \vec{G}(P)$$
$$\vec{H}_3(P) = \vec{F}(P) \times \vec{G}(P).$$

Then h *is a scalar field on* D, *and* $\vec{H}_1$, $\vec{H}_2$, *and* $\vec{H}_3$ *are vector fields on* D.

In this chapter, we begin a systematic study of *scalar fields* (as defined above) and *scalar functions* (defined in §8.4). The chapters which follow are chiefly concerned with scalar fields, although vector fields are occasionally mentioned. A systematic study of vector fields begins in Chapter 18.

§8.2   Visualizing a Scalar Field

Definition. Let c be a given real number. The *level set for* c of a scalar field f on D is the set of points P in D such that f(P) = c. When D is in $\mathbf{E}^3$, a level set is often called a *level surface*; when D is in $\mathbf{E}^2$, a level set is often called a *level curve.* (Note that "level surface" or "level curve" for f does *not* mean a "flat" or "horizontal" surface or curve. It means a set of points on which f has a constant ("level") value.)

The general form of a scalar field in $\mathbf{E}^3$ or $\mathbf{E}^2$ can be indicated or sketched by using one or more *level sets.*

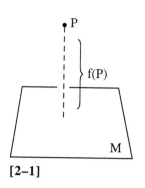

[2–1]

Examples. In the following examples, we use geometrically defined scalar fields to illustrate *level sets.*

(1) *Let* M *be a fixed plane in* $\mathbf{E}^3$. *We define a scalar field* f *on* $\mathbf{E}^3$ *by*

$$f(P) = \text{distance from P to M} \quad [2-1].$$

Here, the level set for 4 consists of two planes parallel to M, each at a distance 4 from M.

(2) *Let* M *be a given plane in* $\mathbf{E}^3$ *and let* Q *be a fixed point in* M. *We define a scalar field* g *on* $\mathbf{E}^3$ *by*

$$g(P) = \text{distance from Q to the projection of P on M} \quad [2-2].$$

Here, the level set for 4 consists of the cylinder of radius 4 whose axis goes through Q and is perpendicular to M.

(3) *Let* Q *be a fixed point in* $\mathbf{E}^2$. *We define a scalar field* h *on* $\mathbf{E}^2$ *by*

$$h(P) = \text{distance from Q to P} \quad [2-3].$$

Here, the level set for 4 consists of the circle of radius 4 with center at Q.

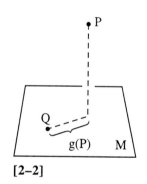

[2–2]

In applications, a level set for a temperature field is called an *isotherm,* for a pressure field is called an *isobar,* and for a potential field is called an *equipotential surface* (in $\mathbf{E}^3$) or an *equipotential curve* (in $\mathbf{E}^2$). (Isobars and isotherms appear as curves on weather maps.)

The general form of a scalar field in $\mathbf{E}^2$ can also be indicated by means of a *graph.* A graph for f is obtained by placing the domain D of f in a chosen plane M in $\mathbf{E}^3$, by taking a unit normal vector $\hat{n}$ to M, and by associating with each point P in D a point Q in $\mathbf{E}^3$ such that $\overrightarrow{PQ} = f(P)\hat{n}$. The set of points Q obtained in this way is called a *graph* of f, and can usually be pictured as a surface in $\mathbf{E}^3$.

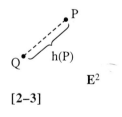

[2–3]

Example 3 (continued). In $\mathbf{E}^2$, we have h(P) = distance from Q to P. In $\mathbf{E}^3$, the *graph* of h is a one-sheeted cone with vertex at Q and axis perpendicular to M.

In flat geographic relief maps, the quantity

h(P) = altitude of the earth's surface above sea-level at map-point P

can be viewed as a scalar field in the plane of the map. Selected level curves of h appear on the map and are referred to as *contour-lines*. In this case, a graph of h would be, in effect, a scale model of the earth's surface over the plane of the map.

§8.3 Directional Derivatives

Let f be a given scalar field. Let P be a fixed point in the domain of f. Let $\hat{u}$ be a fixed unit vector. We wish to consider the rate, per unit distance, at which the value of f increases when a variable point moves through P in the direction $\hat{u}$. We make the following definition.

Definition. Let f be a scalar field on a domain D in $\mathbf{E}^3$. Let $\hat{u}$ be a fixed unit vector in $\mathbf{E}^3$. Let P be a fixed point in D, and let Q be another point in D such that $\overrightarrow{PQ}$ is parallel to $\hat{u}$ and such that the line segment PQ is contained in D. Let $\Delta f = f(Q)-f(P)$, and let $\Delta s = \overrightarrow{PQ}\cdot\hat{u}$. (Thus if $\overrightarrow{PQ}$ has direction $\hat{u}$, $\Delta s = d(P,Q)$, and if $\overrightarrow{PQ}$ has direction $-\hat{u}$, $\Delta s = -d(P,Q)$.) [3–1]. Then (for our given choice of P and $\hat{u}$) we consider the limit

$$\lim_{\Delta s \to 0} \frac{\Delta f}{\Delta s}.$$

If this limit exists, we write

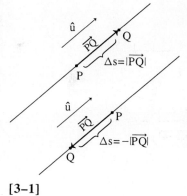

[3–1]

(3.1)
$$\left.\frac{df}{ds}\right|_{\hat{u},P} = \lim_{\Delta s \to 0} \frac{\Delta f}{\Delta s},$$

and we call this value the *directional derivative of* f *in the direction* $\hat{u}$ *at* P.

Evidently, if we take $\hat{v} = -\hat{u}$, we get

$$\left.\frac{df}{ds}\right|_{\hat{v},P} = -\left.\frac{df}{ds}\right|_{\hat{u},P}.$$

For the case of a scalar field f in $\mathbf{E}^2$, the definition of directional derivative is the same, except that D is a subset of $\mathbf{E}^2$, and $\hat{u}$ is a vector in $\mathbf{E}^2$.

Examples. *If* f *represents the spatially varying density of a solid material in a region* D, *then* $\dfrac{df}{ds}\Big|_{\hat{u},P}$ *is the instantaneous rate at which density increases (per unit distance) as we move in the direction* $\hat{u}$ *through* P *to nearby points along the line through* P *given by* $\hat{u}$. *Similarly for the temperature in a region* D.

In §8.4 below, we use scalar functions to describe scalar fields. In Chapter 9, we show how to use such a scalar function to calculate the directional derivative of a given scalar field, for a given direction $\hat{u}$ and a given point P.

§8.4   Scalar Functions

Definitions. An *ordered pair* of real numbers is a sequence of two real numbers, with repetition allowed. We use the notation (a,b) to denote the ordered pair formed of a and then b in that order. Two ordered pairs, such as (5,3) and (3,5), consisting of the same numbers in distinct orders, are treated as distinct ordered pairs.

Similarly, an *ordered triple* of real numbers is a sequence of three real numbers, with repetition allowed. We use the notation (a,b,c) to denote an ordered triple of real numbers. As with ordered pairs, the same numbers in distinct orders are treated as distinct ordered triples. For example, (2,1,2) and (2,2,1) are different as ordered triples.

Similarly, for any positive integer n, an *ordered* n-*tuple* of real numbers is a sequence of n real numbers with repetition allowed. We use the notation $(a_1,a_2,...,a_n)$ to denote an ordered n-tuple of real numbers.

Definition. We have defined **R** to be the set of all real numbers. We now define $\mathbf{R}^2$ to be the set of all ordered pairs of real numbers, $\mathbf{R}^3$ to be the set of all ordered triples of real numbers, and $\mathbf{R}^n$ (for any positive integer n) to be the set of all ordered n-tuples of real numbers.

If we introduce a system of Cartesian coordinates in $\mathbf{E}^3$, we obtain a natural, one-to-one correspondence between points in $\mathbf{E}^3$ and ordered triples in $\mathbf{R}^3$, where each point in $\mathbf{E}^3$ is associated with the corresponding ordered triple (in $\mathbf{R}^3$) of its coordinates. Similarly, if we introduce a system of Cartesian coordinates in $\mathbf{E}^2$, we obtain a one-to-one correspondence between points in $\mathbf{E}^2$ and ordered pairs in $\mathbf{R}^2$.

In one-variable calculus, we consider functions which take members of some set D in **R** as inputs and produce members of **R** as outputs. Such functions are called *real-valued functions of one real variable* or, more briefly, *functions of one variable* or *one-variable functions*.

Definition. Consider a function which takes, as inputs, ordered triples in some set D in $\mathbf{R}^3$ and produces members of **R** as outputs. Such a function is called a *scalar-valued function of three scalar variables* or, more briefly, a *function of three variables*. Similarly, a function from some set D in $\mathbf{R}^2$ into **R** is called a *scalar-valued function of two scalar variables* or, more briefly, a *function of two variables*; and a function from some set D in $\mathbf{R}^n$ into **R** is called a *scalar-valued function of* n *scalar variables* or a *function of* n *variables*. (The word "real" is sometimes used for the word "scalar" in these definitions.) For each such function, the set D is called the *domain* of that function.

Functions of more than one variable are referred to, collectively, as *multivariable scalar functions* or as *functions of several variables*. When it is clear from context that scalar-valued functions are intended, we sometimes refer to multivariable scalar functions as, simply, *multivariable functions*. We also sometimes speak of scalar functions of one or more variables as, simply, *functions*.
We use the letters

$$f, g, h, \ldots, F, G, H, \ldots$$

to denote multivariable functions. We use variable-symbols such as

$$x, y, z, \ldots$$

to represent input numbers for a multivariable function, and we use variable symbols such as

$$w, z, \ldots$$

to represent the output value of a multivariable function. We speak of the input variables as the *independent variables* and of the output variable as the *dependent variable* for the given function. We write

$$w = f(x, y, z)$$

to mean that w is the output value of f for the ordered triple (x, y, z) as input. We sometimes also use the notation $f(x, y, z)$ (rather than simply f) to denote a function. In this way, we indicate, at the same time, the variable-symbols that we intend to use for the independent variables of f.

There are a variety of ways in which a multivariable function can be introduced or defined.

(1) *A multivariable function* f *may be defined by choosing a coordinate system for the domain (in* $\mathbf{E}^3$ *or* $\mathbf{E}^2$*) of a given scalar field* g *and by taking the output value* f(a,b,c) *to be the output value* g(P) *for the point* P *whose coordinates are* (a,b,c). In this case, we say that the multivariable function f *represents* the scalar field g in the given coordinate system. Thus, in example 3 in §8.2, if we take Q to be the origin of a Cartesian system, then the scalar field h, where h(P) = distance from Q to P, is represented by the function h(x,y) = $\sqrt{x^2 + y^2}$. (We shall often use the same symbol (in this case h) to denote both a scalar field and a multivariable function which represents it in some chosen coordinate system.) Note that different choices of Cartesian coordinate system will usually produce different multivariable functions to represent the same given scalar field. Non-Cartesian coordinate systems can also be used to represent scalar fields. Thus, in example 3 of §8.2, if we take Q to be the origin in a system of *polar coordinates* r and θ, then the scalar field h is represented by the function f(r,θ) = r.

(2) *Multivariable scalar functions may also be used to represent the scalar components of a given vector field in a given coordinate system.* For example, if $\vec{F}(P)$ is a given vector field on a domain D in $\mathbf{E}^3$ with Cartesian coordinates, then we have a vector-valued function $\vec{F}(x,y,z)$ and three scalar functions L(x,y,z), M(x,y,z), and N(x,y,z), all with domain D, such that for any point P in D with coordinates (a,b,c),

(4.1) $\vec{F}(P) = \vec{F}(a,b,c) = L(a,b,c)\hat{i} + M(a,b,c)\hat{j} + N(a,b,c)\hat{k}$.

Conversely, any choice of three scalar functions L, M, and N, on a domain D with Cartesian coordinates *defines* a corresponding vector field $\vec{F}$ on D for which (4.1) is true. For example,

$$\vec{F}(x,y,z) = \frac{x}{(x^2+y^2+z^2)^{1/2}}\hat{i} + \frac{y}{(x^2+y^2+z^2)^{1/2}}\hat{j} + \frac{z}{(x^2+y^2+z^2)^{1/2}}\hat{k}$$

defines a unit vector at all points of $\mathbf{E}^3$ except the point with coordinates (0,0,0).

(3) *A multivariable function may be defined explicitly, in terms of known functions and operations, by an algebraic expression.* In this case, we must be careful to indicate the full set of independent variables, and their intended order. Thus, for example, the expression x^2+2v can be used to define the 2-variable function

$$f(x,v) = x^2 + 2v,$$

or the 2-variable function

$$g(v,x) = x^2 + 2v,$$

or a 3-variable function such as

$$h(x,z,v) = x^2 + 2v.$$

In each of these equations, the symbols on the left give a name for the function as well as information about intended independent variables and their intended order.

(4) *A multivariable function may also be introduced by an implicit definition.* An *implicit definition* defines a single multivariable function or, in some cases, several multivariable functions at the same time. It consists of one or more equations or relationships among variables, where the intended independent and dependent variables have been designated and where, for each dependent variable, the desired order of independent variables has been indicated. For example, the equation $w^2 - x^2 - y^2 - z^2 = 0$, with w as designated dependent variable and x,y,z as designated independent variables, and with (x,y,z) as the intended order of independent variables, implicitly defines $w = f(x,y,z)$ as one of two possible continuous functions: $f(x,y,z) = \sqrt{x^2 + y^2 + z^2}$ and $f(x,y,z) = -\sqrt{x^2 + y^2 + z^2}$. (Continuity for a multivariable function is defined in §8.7 below.)

(5) *A multivariable function may be introduced as a solution to a differential equation, in which partial derivatives (to be defined in §8.6) for the desired function appear.*

(6) *A multivariable function may be defined as an empirically given, numerical relationship.* For example, consider a small enclosed container of gas at equilibrium, where the pressure p and temperature T are each observed to have nearly constant values throughout the gas and where the volume of the container can be varied by means of a piston. Let V be the volume of the container. If we alter either V or T (or both) and then allow the gas to come again to equilibrium,

we observe, in general, a new value of p which depends only upon the new values of V and T. This dependence can be represented by a function of two variables

$$p = h(T,V).$$

Here the multivariable function h does not arise initially as a scalar field. (It is often helpful, however, to view a multivariable function of this kind as a scalar field on some imagined geometrical space. We can do so for $h(V,T)$ by taking $\mathbf{E}^2$ with Cartesian coordinates and naming these Cartesian coordinates V and T. For the case of a gas, the resulting geometrical domain of h is known to scientists and engineers as *VT space.*)

§8.5 Partial Functions

The geometric form of a *multivariable function* f can be visualized by viewing f as a *scalar field* on a space which has Cartesian coordinates. Then, for the case of a function of two variables, a graph can be used, and for the case of functions of either two or three variables, level sets can be used.

Example. Let $f(x,y) = x^2 - y^2$. The graph of f will be the surface satisfying $z = x^2 - y^2$. This surface is a hyperbolic paraboloid as described in §4.6. The level curve $x^2 - y^2 = c$ will be a hyperbola for $c \neq 0$ and a pair of intersecting straight lines for $c = 0$.

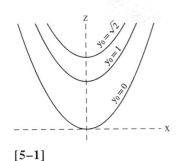

[5–1]

Another way to picture a function of several variables is to use *partial functions*. A *partial function* is a one-variable function obtained from a function of several variables by assigning constant values to all but one of the independent variables and by allowing the remaining independent variable to vary. Partial functions may be studied and graphed as functions of one variable.

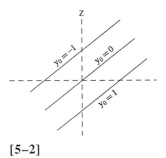

[5–2]

Examples.
(1) *For* $z = f(x,y) = x^2 + y^2$, *the partial functions in* x *have the form* $z = f(x,y_0) = x^2 + y_0^2$ *and are as in* [5–1]. The partial functions in y have the form $z = f(x_0,y) = x_0^2 + y^2$ and are similar. The graphs of these partial functions are parabolas. The graph of f itself is a paraboloid as described in §4.6.
(2) *For* $z = f(x,y) = x - y$, *the partial functions in* x *(where* $z = x - y_0$) *are as in* [5–2]. [5–3] *shows the partial functions in* y. The graphs of these partial functions are straight lines. The graph of f itself is a plane with $-\hat{i} + \hat{j} + \hat{k}$ as a normal vector.

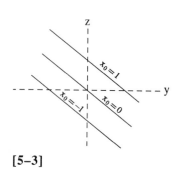

[5–3]

(3) *For* w = f(x,y,z) = x² + y² + z², *the partial functions in* x *are*
$f(x,y_0,z_0) = x^2 + y_0^2 + z_0^2$ [5–4]. Note that with a function of three variables, getting a partial function in x requires choosing a constant value y_0 *and* a constant
value z_0. The graphs of these partial functions are parabolas. (The graph of f
itself is a hypersurface in four-dimensional Euclidean space. See Chapter 34.)

[5–4]

§8.6 Partial Derivatives

Definition. Let f(x,y) be a function of two variables on D in $\mathbf{R}^2$, and let P =
(a,b) be in D. The *partial derivative of* f *with respect to* x *at* P is the derivative
at a (in the sense of one-variable calculus) of the partial function f(x,b). It is
written $\left.\dfrac{\partial f}{\partial x}\right|_P$. We can also write this partial derivative as $\left.\dfrac{\partial f}{\partial x}\right|_{(a,b)}$. Indeed, if we
let f also denote the scalar field represented by the function f(x,y) in Cartesian
coordinates, the partial derivative $\left.\dfrac{\partial f}{\partial x}\right|_P$ can also be written as the directional
derivative $\left.\dfrac{df}{ds}\right|_{\hat{i},P}$.

Similarly, the partial derivative of the function f with respect to y at P =
(a,b) is the derivative at b of the partial function f(a,y). We write it $\left.\dfrac{\partial f}{\partial y}\right|_P$.
$\left.\dfrac{\partial f}{\partial y}\right|_P$ can also be written as the directional derivative $\left.\dfrac{df}{ds}\right|_{\hat{j},P}$.

Since $\left.\dfrac{\partial f}{\partial x}\right|_P$ and $\left.\dfrac{\partial f}{\partial y}\right|_P$ are the derivatives of the partial functions f(x,b)
and f(a,y), we have, from the definition of *derivative* in elementary calculus, that
for P = (a,b):

(6.1)
$$\left.\frac{\partial f}{\partial x}\right|_P = \lim_{\Delta x \to 0} \frac{f(a+\Delta x,b)-f(a,b)}{\Delta x},$$

and

$$\left.\frac{\partial f}{\partial y}\right|_P = \lim_{\Delta y \to 0} \frac{f(a,b+\Delta y)-f(a,b)}{\Delta y}.$$

If the partial derivative of f with respect to x exists at every point in the
domain D of f, then the partial derivative itself is a function of two variables on
D. We write this new function as $\dfrac{\partial f}{\partial x}$. Similarly for $\dfrac{\partial f}{\partial y}$.

Evidently, $\frac{\partial f}{\partial x}$ may be calculated by taking the partial function in x, treating the independent variables other than x as constants, and applying the differentiation rules of one-variable calculus.

Example 1. Let $f(x,y) = x^2 + y^3 + e^{x^2 y}$. To find the function $\frac{\partial f}{\partial x}$, we treat y as a constant and differentiate. We get

$$\frac{\partial f}{\partial x} = 2x + 2xy\, e^{x^2 y}.$$

Similarly, treating x as a constant, we get

$$\frac{\partial f}{\partial y} = 3y^2 + x^2\, e^{x^2 y}.$$

From one-variable calculus, we know that $\left.\frac{\partial f}{\partial x}\right|_{(a,b)}$ must be the slope of the tangent line at $x = a$ to the graph of the partial function $f(x,b)$. It may also be interpreted as the instantaneous rate of change of the function f at (a,b) with respect to changes in x, when y is held at the constant value b. For example, consider $f(x,y) = x^2 + y^2$ at the point (1,10). Then, $\left.\frac{\partial f}{\partial x}\right|_{(1,10)} = 2$, while $\left.\frac{\partial f}{\partial y}\right|_{(1,10)} = 20$. Thus, at (1,10) the value of the function $f(x,y)$ is ten times more sensitive to a slight change in y than it is to a slight change in x.

Notation. A wide variety of notations for partial derivatives are used by mathematicians and scientists. The student should become familiar with the following. Let $z = f(x,y)$. Then $\frac{\partial f}{\partial x}$ is also written $\frac{\partial}{\partial x} f$ or $\frac{\partial z}{\partial x}$ or f_x. Similarly, $\frac{\partial f}{\partial y}$ is $\frac{\partial}{\partial y} f$ or $\frac{\partial z}{\partial y}$ or f_y. The notation f_x is analogous to the notation f' for the derivative of a one-variable function. Thus $f_x(P)$ is another way of writing $\left.\frac{\partial f}{\partial x}\right|_P$.

Since the partial derivatives of $f(x,y)$ may themselves be viewed as the functions f_x and f_y of two variables, we can seek *their* partial derivatives. Various notations for these *second-order* partial derivatives are as follows.

Let $z = f(x,y)$. Then the two partial derivatives of f_x are written

$$\frac{\partial}{\partial x}\left(\frac{\partial f}{\partial x}\right),\ \frac{\partial^2 f}{\partial x^2},\ (f_x)_x,\ f_{xx},\ \frac{\partial}{\partial x}\left(\frac{\partial z}{\partial x}\right),\ \text{or}\ \frac{\partial^2 z}{\partial x^2},$$

and

$$\frac{\partial}{\partial y}\left(\frac{\partial f}{\partial x}\right),\ \frac{\partial^2 f}{\partial y \partial x},\ (f_x)_y,\ f_{xy},\ \frac{\partial}{\partial y}\left(\frac{\partial z}{\partial x}\right),\ \text{or}\ \frac{\partial^2 z}{\partial y \partial x}.$$

Similarly, for the two partial derivatives of f_y, we have

$$\frac{\partial}{\partial x}\left(\frac{\partial f}{\partial y}\right),\ \frac{\partial^2 f}{\partial x \partial y},\ (f_y)_x,\ f_{yx},\ \frac{\partial}{\partial x}\left(\frac{\partial z}{\partial y}\right),\ \text{or}\ \frac{\partial^2 z}{\partial x \partial y},$$

and

$$\frac{\partial}{\partial y}\left(\frac{\partial f}{\partial y}\right),\ \frac{\partial^2 f}{\partial y^2},\ (f_y)_y,\ f_{yy},\ \frac{\partial}{\partial y}\left(\frac{\partial z}{\partial y}\right),\ \text{or}\ \frac{\partial^2 z}{\partial y^2}.$$

Third-order and higher-order partial derivatives are indicated in the same way. Sometimes, for clarity, we refer to f_x and f_y as the *first-order* partial derivatives of f.

Example 1 (continued). For $f(x,y) = x^2 + y^3 + e^{x^2 y}$, the second-order partial derivatives, as functions, are:

$$f_{xx} = \frac{\partial}{\partial x}(2x + 2xye^{x^2 y}) = 2 + 2ye^{x^2 y} + 4x^2 y^2 e^{x^2 y},$$

$$f_{xy} = \frac{\partial}{\partial y}(2x + 2xye^{x^2 y}) = 2xe^{x^2 y} + 2x^3 ye^{x^2 y},$$

$$f_{yx} = \frac{\partial}{\partial x}(3y^2 + x^2 e^{x^2 y}) = 2xe^{x^2 y} + 2x^3 ye^{x^2 y},$$

$$f_{yy} = \frac{\partial}{\partial y}(3y^2 + x^2 e^{x^2 y}) = 6y + x^4 e^{x^2 y}.$$

The fact that $f_{xy} = f_{yx}$ in this example is not an accident. We shall later see (in §9.8) that for *every* function f, $f_{xy} = f_{yx}$ provided that the second-order partial derivatives of f are continuous functions.

In the case of a function $g(x,y,z)$ of three variables, the definitions, notations, and calculations for partial derivatives are similar to those for a function of two variables. At the point $P = (a,b,c)$, we have, for example, the definition

(6.2)
$$\left.\frac{\partial g}{\partial y}\right|_P = \lim_{\Delta y \to 0} \frac{g(a,b+\Delta y,c) - g(a,b,c)}{\Delta y}.$$

Example 2. *For* $w = g(x,y,z) = 3x^2yz + xyz^2$, *find* g_x *and* g_{xy}. We have

$$g_x = \frac{\partial g}{\partial x} = 6xyz + yz^2,$$

and

$$g_{xy} = \frac{\partial^2 g}{\partial y \partial x} = 6xz + z^2.$$

In contrast to the *partial* derivatives of a multivariable function, the derivative $\frac{df}{dx}$ of a one-variable function f is sometimes called a *total* derivative. We shall often ignore this distinction and refer to the nth-order partial derivatives of a multivariable function f as, simply, the nth-order *derivatives* of f.

Example 3. *Let* $f(x,y,z) = xy + z^2$. *Find the values of all derivatives of all orders at* $P = (1,2,3)$,

$$\frac{df}{dx} = f_x(x,y,z) = y; \quad \text{hence} \quad \left.\frac{\partial f}{\partial x}\right|_P = f_x(1,2,3) = 2.$$

$$\frac{\partial f}{\partial y} = f_y(x,y,z) = x; \quad \text{hence} \quad \left.\frac{\partial f}{\partial y}\right|_P = f_y(1,2,3) = 1.$$

$$\frac{\partial f}{\partial z} = f_z(x,y,z) = 2z; \quad \text{hence} \quad \left.\frac{\partial f}{\partial z}\right|_P = f_z(1,2,3) = 6.$$

$$\frac{\partial^2 f}{\partial x^2} = f_{xx}(x,y,z) = 0; \quad \text{hence} \quad \left.\frac{\partial^2 f}{\partial x^2}\right|_P = f_{xx}(1,2,3) = 0.$$

$$\frac{\partial^2 f}{\partial y \partial x} = f_{xy}(x,y,z) = 1; \quad \text{hence} \quad \left.\frac{\partial^2 f}{\partial y \partial x}\right|_P = f_{xy}(1,2,3) = 1.$$

$$\frac{\partial^2 f}{\partial x \partial y} = f_{yx}(x,y,z) = 1; \quad \text{hence} \quad \left.\frac{\partial^2 f}{\partial x \partial y}\right|_P = f_{yx}(1,2,3) = 1.$$

$$\frac{\partial^2 f}{\partial z^2} = f_{zz}(x,y,z) = 2; \quad \text{hence} \quad \left.\frac{\partial^2 f}{\partial z^2}\right|_P = f_{zz}(1,2,3) = 2.$$

The five remaining second-order derivatives are 0, and all derivatives of third or higher order must be 0.

§8.7   Limits and Continuity

Definition. Let f be a scalar field on domain D. Let P be a point in D, and let L be a real number. We define

$$\lim_{Q \to P} f(Q) = L$$

to mean that we can force $f(Q)$ to be as close as we like to L merely by *taking* Q (in D) within a correspondingly small non-zero distance of P. More precisely, $\lim_{Q \to P} f(Q) = L \Leftrightarrow$ *there is a one-variable function* $\varepsilon(t)$*, defined on an interval of real numbers* [0,d] *for some* $d > 0$*, such that:* (i) $\varepsilon(t)$ *decreases steadily to 0 as* t *approaches 0; and* (ii) *for all* Q *in* D*, if* $0 < |\overrightarrow{QP}| \le d$*, then* $|f(Q) - L| \le \varepsilon(|\overrightarrow{QP}|)$. This definition of limit (for a scalar field) coincides formally with the definition of limit described in §5.2 (for a one-variable scalar function). As in the one-variable case, $\varepsilon(t)$ is called a *funnel function* for the limit of f at P.

A limit theorem on algebraic combinations of limits can be proved for scalar fields exactly as for one-variable scalar functions. (See problem 1 for §5.2.)

(7.1)   **Theorem.** *Let* f *and* g *be scalar fields. If* $\lim_{Q \to P} f(Q) = k$ *and*

$\lim_{Q \to P} g(Q) = \ell$*, then* $\lim_{Q \to P} \big(f(Q) + g(Q)\big) = k + \ell$*,* $\lim_{Q \to P} \big(f(Q)g(Q)\big) = k\ell$*, and,*

when $\ell \ne 0$*,* $\lim_{Q \to P} \dfrac{f(Q)}{g(Q)} = \dfrac{k}{\ell}$.

Definition. Let f be a scalar field with domain D and let P be a point in D. We say that f is *continuous at* P if $\lim_{Q \to P} f(Q) = f(P)$. We say that f is *continuous on* S if: (i) S is a set contained in D, and (ii) f is continuous at every point in S. We say that f is *continuous* if f is continuous on D.

As with one-variable functions, it follows, from the theorem above on algebraic combinations of limits, that corresponding combinations of continuous scalar fields must be continuous. We have:

(7.2)   **Theorem.** *Let* f *and* g *be continuous scalar fields on a domain* D*. Let* c *be a scalar constant. Then each of* $f + g$*,* fg*, and* cf *is a continuous scalar field on* D*, and* $\dfrac{f}{g}$ *is a continuous scalar field on* D'*, where* D' *is the set of points* P *in* D *for which* $g(P) \ne 0$*.*

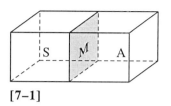

[7–1]

Most of the scalar fields that occur in mathematics are continuous. Discontinuous scalar fields can, however, easily arise in a carefully defined mathematical model.

Example 1. [7–1] *shows a block of steel* (S) *and a block of aluminum* (A) *in contact at the plane interface* M. In our model, we assume that the steel is a substance of constant density s and that the aluminum is a substance of constant density a, with a < s. We now treat the two blocks as a single solid object and let D be the region of space occupied by the two blocks together. We define a scalar field δ, called the *density*, on D as follows. At each P in D, δ(P) = $\lim\limits_{r \to 0} \dfrac{M_r}{V_r}$, where M_r is the mass of the material from D that is contained in a sphere of radius r with center at P, and V_r is the volume of this material. Then, as the reader may verify,

$$\delta(P) = a \text{ for all points in A but not on the interface M,}$$
$$\delta(P) = s \text{ for all points in S but not on the interface M,}$$

and
$$\delta(P) = \frac{a+s}{2} \text{ for all points on the interface M.}$$

Evidently, δ is continuous at points on D which are not on M, and δ is not continuous at points on M. Hence δ is not a continuous scalar field. We speak of M as a *surface of discontinuity* for the field δ.

Limits and *continuity* are defined for a multivariable *function* f by treating the independent variables of f as *Cartesian coordinates* and considering the scalar field (in E^3 or E^2) represented by f. We say that the function f *has a limit* or is *continuous* according as this scalar field has a limit or is continuous. For example, for a function f with domain in R^3, we say that f *has limit* L at (a,b,c) if the scalar field represented by f has limit L at the point whose coordinates are (a,b,c), and we say that f is *continuous* if this scalar field is continuous.

In particular, we have the following:

(7.3)   *Theorem. If a scalar function* f *is defined from given functions by an expression in which the given functions are combined under operations of addition, multiplication, and division, and if the given functions are continuous, then* f *is continuous.*

Example 2. f(x,y) = xy + sin x cos y must be continuous, since x, y, sin x, and cos y are continuous.

The proof of (7.3) uses (7.2) and is analogous to the proof of the corresponding theorem of one-variable calculus.

It follows from this theorem that every polynomial function in several variables is continuous on its domain and that every rational function in several variables (defined as the quotient of two polynomial functions) is continuous on its domain.

Definition. If a scalar function f on a domain D is obtained from a given function g by substituting given functions $h_1, h_2, \ldots$ for the independent variables of g, we say that f has been obtained by *composition* from g, $h_1, h_2, \ldots$.

Example 3. $f(x,y) = \sin(x+y)$ is obtained by composition from $g(z) = \sin z$ and $h(x,y) = x+y$. Similarly, $k(x,y) = \sin(xy + \cos x)$ is obtained by composition from $f(x,y) = \sin(x+y)$, $h_1(x,y) = xy$, and $h_2(x) = \cos x$.

The following more general version of (7.3) can be proved.

(7.4)   *If a scalar function f can be obtained from given functions by composition, and if the given functions are continuous, then f must be continuous.*

§8.8   Counterexamples

A *counterexample* is an example which shows that a particular general statement of the form:

"Whenever A holds, then B must hold" is not true. It does so by describing a situation in which A holds but B does not hold.

First counterexample. We show that a multivariable function which possesses each of its first-order derivatives at every point in its domain need not be continuous on its domain.

Let f be the scalar field defined as follows in a system of polar coordinates with origin at O. If P has polar coordinates r and θ, we set

$$f(P) = \sin 2\theta \quad \text{for } P \neq O;$$
$$f(P) = 0 \quad \text{for } P = O.$$

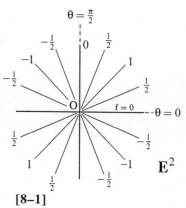

[8–1]

f has level curves as indicated in [8–1]. These level curves show that $\lim_{P \to O} f(P)$ cannot exist (since every possible funnel function is violated). Hence f is *not continuous* at O.

Let the 2-variable function $f(x,y)$ represent f in the associated Cartesian system of the given polar system. Then

$$f(x,y) = \begin{cases} \dfrac{2xy}{x^2+y^2} & \text{for } (x,y) \neq (0,0), \\ 0 & \text{for } (x,y) = (0,0). \end{cases}$$

(For $x \neq 0$, the formula can be obtained from $f(x,y) = \sin 2\theta = \sin 2(\tan^{-1} \frac{y}{x}) = $

$2 \sin (\tan^{-1} \frac{y}{x})\cos(\tan^{-1} \frac{y}{x}) = \dfrac{2y}{\sqrt{x^2+y^2}} \dfrac{x}{\sqrt{x^2+y^2}} = \dfrac{2xy}{x^2+y^2}$. For $x = 0$ and $y \neq 0$,

the formula is also correct, since $f(x,y) = 0$ and $\dfrac{2xy}{x^2+y^2} = 0$.)

The reader may verify that every *partial function* of f is continuous. (In particular, $f(x,0) = 0$ and $f(0,y) = 0$.) Furthermore, f possesses each of its first-order derivatives at every point. Note that the derivatives $\frac{\partial f}{\partial x}$ and $\frac{\partial f}{\partial y}$ are not themselves continuous (as 2-variable functions) at $O = (0,0)$; for $\frac{\partial f}{\partial x}\big|_O = 0$, but

$\frac{\partial f}{\partial x}\big|_P$ increases without limit as P approaches O along the line $y = 2x$, as may easily be shown. In (2.12) of §9.2, however, we shall see that if a function g has first-order derivatives which *are* continuous throughout the domain of g (this was *not* true for f above), then g itself must always be continuous on its domain.

Second counterexample. We show that a scalar field which possesses a directional derivative in every direction at a given point P need not be continuous at P.

Let f be the scalar field defined as follows in a system of polar coordinates with origin at O. If P has polar coordinates r and θ, where $0 \leq \theta < 2\pi$, we set

$$f(P) = \begin{cases} r^2 \tan \dfrac{\theta}{4} & \text{for } P \neq O, \\ 0 & \text{for } P = O. \end{cases}$$

The reader may verify: (i) that for every $\hat{u}$, $\frac{df}{ds}\big|_{\hat{u},O} = 0$, and (ii) that f is not continuous at O, since for every funnel function ε, there is a line through O on which f violates ε. See the problems.

§8.9 Problems

§8.1

1. Consider the scalar field f in example 1 of §8.2. Introduce a system of Cartesian coordinates and give an algebraic expression for f(P) as a function of the coordinates (x,y,z) of P. Try to choose the coordinate system so that the expression is as simple as possible.

2. Do the same as problem 1 for g in example 2 of §8.2.

3. Let Q_1 and Q_2 be fixed points in E^2 such that $d(Q_1,Q_2) = 2c$. Define the scalar field f by f(P) = $d(P,Q_1)$ + $d(P,Q_2)$. Do the same as in problem 1 for this f.

4. Let Q be a fixed point in E^2, and let L be a fixed line such that the distance from Q to L is 2c > 0. Define f by f(P) = d(P,Q) – d(P,L), where d(P,L) is the distance from the point P to the line L. Do the same as in problem 1 for this f.

5. Let Q_1 and Q_2 be fixed points in E^2 such that $d(Q_1,Q_2)$ = 2c. Define f by f(P) = $(d(P,Q_1)$ - $d(P,Q_2))^2$. Do the same as in problem 1 for this f.

§8.2

1. In your chosen coordinate system for problem 3 of §8.1, find (for the scalar field f) an equation for the level curve for 2a. Simplify as far as possible.

2. As in problem 1, find, for the scalar field in problem 4 of §8.1, an equation for the level curve for 0, and simplify as far as possible.

3. As in problem 1, find, for the scalar field in problem 5 of §8.1, as equation for the level curve for $4a^2$, and simplify as far as possible.

§8.3

1. For the scalar field f in example 1 of §8.2 and for a point P not on M, find $\left.\dfrac{df}{ds}\right|_{\hat{u},P}$:

(a) when $\hat{u}$ is normal to M and directed away from M (at P);
(b) when $\hat{u}$ is parallel to M;
(c) when $\hat{u}$ is normal to M and directed towards M (at P);
(d) when the angle between $\hat{u}$ and the plane M is $\pi/4$ and $\hat{u}$ is directed away from M (at P).

§8.5

1. Sketch the family of level curves for f(x,y) = ye^x.

2. (a) Sketch the family of partial functions in x for f(x,y) = ye^x.
   (b) Sketch the family of partial functions in y for f(x,y) = ye^x.

3. In each of the following cases, give a general description of (i) the level surface for 0 and (ii) the level surface for 1. With each description, give a sketch on the x, y, z axes. Be sure to label the axes.
   (a) f(x,y,z) = $x^2 + y^2 + z^2$
   (b) f(x,y,z) = $x^2 + y^2 - z^2$
   (c) f(x,y,z) = $x^2 - y^2 - z^2$
   (d) f(x,y,z) = xyz
   (*Hint.* See (6.2a) in §4.6.)

4. Give formulas for:
   (a) the two partial functions of f(x,y) = x^2-2xy at $(x_0,y_0) = (1,3)$.
   (b) the three partial functions of f(x,y,z) = $x^2-y^2-z^2$ at $(x_0,y_0,z_0) = (2,-1,3)$.

5. Let z = f(x,y) = $\dfrac{x}{y}$:
   (a) In a single graph, sketch the level curves for z = –1,0,1,2.
   (b) In a single graph, sketch the partial functions $f(x,y_0)$ for y_0 = –1, 0, 1, 2.
   (c) In a single graph, sketch the partial functions $f(x_0,y)$ for x_0 = –1, 0, 1, 2.
   (d) Find all points (x,y) at which the sensitivity of f(x,y) to changes in x is equal to the sensitivity of f to changes in y. (We define the

sensitivity of f to changes in x to be $\left|\dfrac{\partial f}{\partial x}\right|$.

Similarly for sensitivity of f to changes in y.)

6. Let $z = f(x,y) = xy^2$
  (a) In a single graph, sketch the level curves for z = 1, z = 0, z = –1.
  (b) In a single graph, sketch the partial functions $f(x,y_0)$ for $y_0 = 1,0,–1$.
  (c) In a single graph, sketch the partial functions $f(x_0,y)$ for $x_0 = 1,0,–1$.

§8.6

1. Find partial derivatives (first-order) for:
  (a) $f(x,y) = \dfrac{y}{x^2 + y^2}$;
  (b) $f(x,y) = x\cos(xy)$;
  (c) $f(x,y,z) = z\sin^{-1}\left(\dfrac{y}{x}\right)$.

2. Find first- and second-order derivatives for:
  (a) $f(x,y) = ye^x$;
  (b) $f(x,y) = \ln\sqrt{x^2 + y^2}$;
  (c) $f(x,y) = \tan^{-1}\left(\dfrac{y}{x}\right)$;
  (d) $f(x,y) = e^{-x^2-y}$.

3. Consider the determinant $\begin{vmatrix} e & f \\ g & h \end{vmatrix}$, where $|e| > |f| > |g| > |h|$. To changes in which element will the value of the determinant be most sensitive?

4. Let a and b be two sides of a triangle and let γ be the included angle (measures in radians). Let z be the area of the triangle.
  (a) Find $\dfrac{\partial z}{\partial a}$, $\dfrac{\partial z}{\partial b}$, $\dfrac{\partial z}{\partial \gamma}$.
  (b) To changes in which of the three quantities, a, b, and γ, is z most sensitive when $\gamma = \dfrac{\pi}{2}$ and a > b? Justify your answer.

  (c) Let $\gamma = \dfrac{\pi}{4}$. For what values of a and b will it be the case that z is most sensitive to changes in γ?

5. (a) Evaluate all first- and second-order partial derivatives of $z = e^{2x+y^2}$.
  (b) Find $\dfrac{\partial^2 w}{\partial y\, dx}$ for $w = \cos(\dfrac{xy}{z})$.

6. Let $f(x,y,z) = y\cos(e^{xz}) - \sin(xy^2z)$. Find f_{xx}, f_{xy}, f_{yz}, and f_{zy}.

7. Let
$$q = \begin{vmatrix} x & y & 1 \\ z & 2 & 5 \\ 0 & 3 & 1 \end{vmatrix}.$$
To changes in which of the three elements x, y, and z will changes in the value of q be most sensitive
  (a) when (x,y,z) = (0,0,0)?
  (b) when (x,y,z) = (10,15,20)?
  (c) when (x,y,z) = (15,15,15)?

8. Consider a solid cylindrical block in the shape of a soup can. Let a be its radius and let b be its height. Let V be its volume and let S be its total surface area (top and bottom included).
  (a) When is V more sensitive to changes in a than to changes in b, and when is it more sensitive to changes in b than to changes in a?
  (b) Consider S in place of V, and answer the same questions as in (a).

9. In Chapter 5, we introduced vector functions $\vec{R}(t)$ of one variable and considered their derivatives. We can also introduce vector functions of two or more variables. For example,
$$\vec{R}(u,v) = \ell n(u+v)\hat{i} - \cos(2uv)\hat{j} + \dfrac{1}{uv}\hat{k}.$$
Partial derivatives for such functions can be defined in the same way as for scalar functions. Thus for a vector function $\vec{R}(u,v)$ of 2 variables,

$$\left.\frac{\partial \vec{R}}{\partial u}\right|_{(u_0,v_0)} = \lim_{\Delta u \to 0} \frac{\vec{R}(u_0+\Delta u,v_0)-\vec{R}(u_0,v_0)}{\Delta u}.$$

In the example just given, find

$$\frac{\partial \vec{R}}{\partial u}, \ \frac{\partial \vec{R}}{\partial v}, \ \frac{\partial^2 \vec{R}}{\partial u^2}, \ \text{and} \ \frac{\partial^2 \vec{R}}{\partial u \partial v}.$$

§8.7

1. Use funnel functions to prove that if $\lim_{x \to c} f(x) = k$, $g(k) = \ell$, and $\lim_{x \to k} g(x) = \ell$, then $\lim_{x \to c} g(f(x)) = \ell$. (Results of this kind also hold for multivariable functions.)

2. (a) Prove the first conclusion in (7.1) by contructing an appropriate funnel function for $\lim_{Q \to P} (f(Q) + g(Q))$ from given funnel functions for $\lim_{Q \to P} f(Q)$ and $\lim_{Q \to P} g(Q)$.

(b) Prove the second conclusion in (7.1) similarly.

(c) Prove the third conclusion in (7.1) similarly.

3. Let h, g_1, and g_2 be continuous on $\mathbf{E}^2$. Prove that $f(x,y) = h(g_1(x,y),g_2(x,y))$ is continuous.

§8.8

1. In the first counterexample:
   (a) Sketch the partial functions in y of f for $x_0 = 0,1,$ and 2.
   (b) Show that every partial funciton of f is continuous.
   (c) Show that f possesses each of its first-order derivatives at every point.

2. In the first counterexample, show that $\left.\frac{df}{dx}\right|_P$ increases without limit as P approaches O along the line $y = 2x$.

3. In the second counterexample
   (a) prove (i);
   (b) prove (ii).

Chapter 9

Linear Approximation; the Gradient

§9.1   Linear Approximation for One-variable Functions

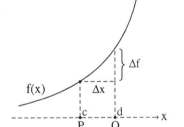

[1–1]

Consider the graph of the function $f(x) = x^2$ in the xy plane, as shown in [1–1]. Let P be a fixed point on the x axis at $x = c$. Let Q be a nearby variable point on the x axis at $x = c + \Delta x$, where $\Delta x \neq 0$ and Δx may be positive or negative. At Q, the value of f is $f(c+\Delta x) = (c+\Delta x)^2 = c^2 + 2c\Delta x + (\Delta x)^2$. Hence, as we move from P to Q, the value of f increases by the amount $\Delta f = f(c+\Delta x) - f(c) = c^2 + 2c\Delta x + (\Delta x)^2 - c^2 = 2c\Delta x + (\Delta x)^2$. Hence the average rate of increase for f, per unit increase in x, as we move from P to Q is $\frac{\Delta f}{\Delta x} = 2c + \Delta x$. If we now let variable point Q approach fixed point P (that is to say, if we now let Δx approach 0), then the value of the average rate $\frac{\Delta f}{\Delta x}$ $(= 2c + \Delta x)$ approaches the limit value 2c. Thus

$$\lim_{\Delta x \to 0} \left(\frac{\Delta f}{\Delta x}\right) = 2c.$$

This limit value of the average rate of increase is called the *instantaneous rate of increase for* f *at* c, or the *derivative of* f *at* c, and is written $\frac{df}{dx}\Big|_c$. In recent

182

decades, most treatments of elementary calculus have presented the derivative in this way. These treatments view the derivative as, primarily, an instantaneous rate of change.

There is, however, another, older and more traditional way to define and view the derivative. In this older approach, we begin by looking for a simple, linear-equation formula which will give us, for each value of Δx, an *estimate* of the corresponding value of Δf, and which will also have the special property that the *relative accuracy* of this estimate will improve as Q approaches P. More specifically, we will seek a formula of the form

$$(1.1) \qquad \Delta f_{app} = A_c \Delta x \, ,$$

where Δf_{app} is the estimated value of Δf (both Δf_{app} and Δf depend on Δx), where A_c is a constant coefficient whose value depends on c, and where the error in this estimated value becomes smaller and smaller in proportion to Δx as Q approaches P. If there is such a constant A_c, we will call it a *differential coefficient for* f *at* c, because it can be used to estimate the *difference* Δf. To make the idea of *relative accuracy* precise, we define

$$(1.2) \qquad \varepsilon_c(Q) = \frac{\Delta f - \Delta f_{app}}{\Delta x} = \frac{\Delta f - A_c \Delta x}{\Delta x} \, .$$

Thus $\varepsilon_c(Q)$ is the ratio of the *error* in the estimate for Q to the value of Δx. Evidently, we seek an A_c such that

$$(1.3) \qquad \lim_{Q \to P} \varepsilon_c(Q) = 0$$

We now observe that $\lim_{Q \to P} \varepsilon_c(Q) = 0 \Leftrightarrow \lim_{\Delta x \to 0} \left(\frac{\Delta f - A_c \Delta x}{\Delta x} \right) = 0 \Leftrightarrow$

$\left[\lim_{\Delta x \to 0} \left(\frac{\Delta f}{\Delta x} \right) - A_c \right] = 0 \Leftrightarrow \lim_{\Delta x \to 0} \left(\frac{\Delta f}{\Delta x} \right) = A_c \Leftrightarrow A_c = \frac{df}{dx} \Big|_c$. In other words, f

has a differential coefficient A_c at c $\Leftrightarrow$ f has a derivative at c; and the value of this differential coefficient at c is the same as the value of the derivative at c.

(1.4) **Definition.** Formula (1.1), with $A_c = \frac{df}{dx} \Big|_c$, becomes

$$\Delta f_{app} = \frac{df}{dx} \Big|_c \Delta x \, .$$

This formula is called the *linear approximation formula for* f *at* c.

The linear approximation formula provides a useful and practical short-cut for estimating the value of Δf for given values of c and Δx.

Example 1. *Let* $f(x) = x^2$; *let P be given by* $x = c = 2$, *and let Q be given by* $\Delta x = 0.1$. The differential coefficient $A_2 = \dfrac{df}{dx}\Big|_2 = 2x\big|_2 = 4$, and we get

$$\Delta f_{app} = 4\Delta x = 4(0.1) = 0.4 .$$

The exact value of Δf is, of course, $f(2.1) - f(2) = (2.1)^2 - 2^2 = 4.41 - 4 = 0.41$. Thus the error in this estimate is 0.01, and $\varepsilon_2(Q) = \dfrac{0.01}{0.1} = \dfrac{1}{10}$.

Consider now a new point Q given by $\Delta x = 0.01$. We then get

$$\Delta f_{app} = 4(0.01) = 0.04 .$$

The exact value of Δf is $f(2.01) - f(2) = (2.01)^2 - 2^2 = 4.0401 - 4 = 0.0401$. Hence the error in the estimate is 0.0001, and $\varepsilon_2(Q) = \dfrac{0.0001}{0.01} = \dfrac{1}{100}$. As expected, $\varepsilon_2(Q)$ is decreasing as Q approaches P.

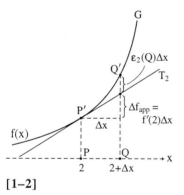

[1–2]

Our example is illustrated in [1–2], where P′ and Q′ are the points on the graph G of $f(x) = x^2$ whose projections on the x axis are the fixed point P and the nearby variable point Q. T_2 is the straight line through P′ with slope $= A_2 = \dfrac{df}{dx}\Big|_c = 4$. (The slope of T_2 in the figure is distorted to make the figure easier to view.) In elementary calculus, T_c (defined in this way for x = c) is called the *tangent line* to G for x = c. Note that T_c is the graph of the function

(1.5) $t_c(x) = f(c) + \Delta f_{app} = f(c) + A_c\Delta x = f(c) + A_c(x-c) .$

The function $t_c(x)$ is sometimes called the *linearization of* f *at* x = c. For c = 2 and $f(x) = x^2$, we have $t_2(x) = 4 + 4(x-2)$, and T_2 thus has the linear equation $y = 4 + 4x - 8 = 4x - 4$.

It is obvious from the figure that the product $\varepsilon_c(Q)\Delta x \ (= \Delta f_{app} - \Delta f)$ must approach 0 as Q approaches P. It is less obvious (but also true) that $\varepsilon_c(Q) \ (= \dfrac{\Delta f - A_c\Delta x}{\Delta x})$ must, by itself, approach 0 as Q approaches P. It is the requirement that $\varepsilon_c(Q)$ approach 0 which provides the gradual "touching" between line T_c and graph G at P' and which makes the differential coefficient A_c unique.

Here is a another example of how linear approximation can be used to estimate Δf for a given f, c, and Δx.

Example 2. Let $f(x) = x^{1/2}$, let $c = 81$, and let Q be chosen with $\Delta x = 0.1$. Then we have $f'(x) = \dfrac{1}{2x^{1/2}}$ and $f'(c) = \dfrac{1}{18}$. Hence, by (1.1),

$$\Delta f \approx \Delta f_{app} = \frac{1}{18}(0.1) = \frac{1}{180} = 0.005556 .$$

Since $f(c) = 9$, we have the estimate

$$f(Q) = (81.1)^{1/2} \approx 9.005556 .$$

(This estimate is correct to four decimal places; the exact value is $(81.1)^{1/2} \approx 9.005554$.) The accuracy of Δf_{app} in the example above is impressive, but obviously, for practical usefulness, we need to have some way of getting a *guaranteed* level of precision for a linear approximation without knowing ahead of time the exact value that we hope to approximate. In the problems, we find such a guarantee for the example above. More specifically, we use a simple calculation with the second derivative $f''(c)$ to show that our linear approximation must be accurate to within ±0.000004.

We summarize our discussion above in the following definition.

(1.6) **Definition.** Let $f(x)$ be a one-variable function on domain D. Let $x = c$ be in D and let P, Q, and Δx be defined as in [1–1] above. We treat c as a fixed value and Δx as a variable value. Without referring to the derivative $f'(c)$, we say that f *has linear approximation at* c if, for our chosen c, there exist a constant A_c (whose value depends on c) and a function $\varepsilon_c(Q)$ (whose values depend on c as well as on Q), such that

    (i)   for all Q in D, $\Delta f = A_c \Delta x + \varepsilon_c(Q)\Delta x$,

and (ii) $\lim\limits_{Q \to P} \varepsilon_c(Q) = 0$.

A_c is called the *differential coefficient for* f *at* c, and ε_c is called the *relative error function for* f *at* c.

The following facts are implied by this definition.

(1.7)   *Theorem.* *If* f *has linear approximation at* c, *then* f *is continuous at* c.

      Proof. This is immediate, since, by definition, f is continuous at P $\Leftrightarrow$ $\lim\limits_{Q\to P} \Delta f = 0$.

(1.8)   *Theorem.* f *has linear approximation at* c $\Leftrightarrow$ f *has a derivative at* c.

(1.9)   *Theorem.* *If* f *has linear approximation at* c *with differential coefficient* A_c, *then* $A_c = \dfrac{df}{dx}\Big|_c$.

      ***Proof for* (1.8) *and* (1.9):** Given in preceding discussion.

      Remark For one-variable functions, as we have summarized in (1.8) and (1.9) above, the concepts of *differential coefficient* and *derivative* are equivalent. We shall now turn to linear approximation for two- and three-variable functions. In the multivariable cases, we shall see that the concepts of differential coefficient and derivative are no longer equivalent and that the concept of differential coefficient is more fundamental and more theoretically useful than the concept of derivative.

§9.2   Linear Approximation for Multivariable Functions

      Two-variable functions. Let f(x,y) be a two-variable function on domain D.

(2.1)   *Definition.* Let P = (a,b) be a fixed point in D. Let Q = (a+Δx,b+Δy) be another point in D. Let Δf = f(a+Δx,b+Δy)–f(a,b). We say that f *has linear approximation at* P if, for our chosen input P, there exist constants A and B (whose values depend on P) and two functions $\varepsilon_1(Q)$ and $\varepsilon_2(Q)$ (whose values depend on P as well as on Q) such that:

        (i)   for all Q in D, $\Delta f = A\Delta x + B\Delta y + \varepsilon_1(Q)\Delta x + \varepsilon_2(Q)\Delta y$,

and   (ii) $\lim\limits_{Q\to P} \varepsilon_1(Q) = 0$ and $\lim\limits_{Q\to P} \varepsilon_2(Q) = 0$.

      We shall see below that if f has linear approximation at P, then the constants A and B are unique. (We shall see in the problems, however, that if f has linear approximation at P, then the functions $\varepsilon_1(Q)$ and $\varepsilon_2(Q)$ need not be unique.) The following facts are also implied by (2.1).

(2.2)   **Theorem.** *If* f *has linear approximation at* P, *then* f *is continuous at* P.

   Proof. This is true since $\lim\limits_{Q \to P} \Delta f = 0$ is equivalent to the definition of continuity for f at P.

(2.3)   **Theorem.** *If* f *has linear approximation at* P, *then* f *has first-order partial derivatives at* P *and these derivatives have the values* $\dfrac{\partial f}{\partial x}\Big|_P = A$ *and*

$\dfrac{\partial f}{\partial y}\Big|_P = B$.

   Proof. Set $\Delta y = 0$ in (2.1). Then divide by Δx and let $\Delta x \to 0$. We get $\dfrac{\partial f}{\partial x}\Big|_P = \lim\limits_{\Delta x \to 0} \dfrac{\Delta f}{\Delta x} = A$ by definition. Similarly, we get $\dfrac{\partial f}{\partial y}\Big|_P = B$.

(2.4)   **Theorem.** *It is possible for* f *to have first-order partial derivatives at every point in its domain but not to have linear approximation at a certain point* P *in its domain.*

   Proof. The first example in §8.8 gives a function f which has partial derivatives at every point but is not continuous at a certain point P. By (2.2), f does not have linear approximation at P.
   If f *has* linear approximation at P, as given by (2.1), we *define*

(2.5) $\Delta f_{app} = A\Delta x + B\Delta y$.

Note that Δf_{app} depends on P (as well as on Δx and Δy), since A and B depend on P. By (2.3), (2.5) can also be written

(2.6) $\Delta f_{app} = \dfrac{\partial f}{\partial x}\Big|_P \Delta x + \dfrac{\partial f}{\partial y}\Big|_P \Delta y$.

(2.6) is called the *linear approximation formula* for f at P.
   Consider the graph of f in E^3. Let $P = (a,b)$ (in the xy plane) be a given fixed point in D, as above, and let $Q = (a+\Delta x, b+\Delta y)$ be a given variable point in D (in the xy plane). Let $f(P) = c$. Then $P' = (a,b,c)$ in E^3 is the point on the graph of f given by P. Now define a new function t_P on D by

$$t_P(Q) = f(P) + \Delta f_{app}$$

for all Q in D. The function t_P is the approximation to the function f given by linear approximation at P. t_P is sometimes called the *linearization of f at P*. We find the graph of t_P as follows. Let $Q' = (x, y, z)$ be the point on this graph given by $Q = (a+\Delta x, b+\Delta y)$ in the xy plane. Then $(x,y,z) = (a+\Delta x, b+\Delta y, f(P)+\Delta f_{app})$. This implies that $\Delta x = x-a$, $\Delta y = y-b$, and $\Delta f_{app} = z - f(P)$. From (2.5), we now have

$$z - f(P) = \Delta f_{app} = A\Delta x + B\Delta y = A(x-a) + B(y-b).$$

Hence,

(2.7) $$z - Ax - By = f(P) - Aa - Bb.$$

The coordinates of $Q' = (x, y, z)$ must satisfy (2.7). (2.7) is the equation of a plane M_P in E^3; thus M_P is the graph of the function t_P. For our given fixed point P, we see that the point $P' = (a,b,f(P))$ lies in the plane M_P, since its coordinates satisfy (2.7).

(2.8) **Definition.** For the point $P = (a,b)$, the plane M_P described above is called the *tangent plane* to the graph of f at the point $P' = (a,b,f(a,b))$ on the graph of f. (See [2–1].)

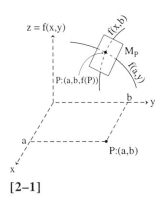

z = f(x,y)

$P:(a,b,f(P))$

M_P

f(x,b)

f(a,y)

b → y

a

x

P:(a,b)

[2–1]

For the two-variable function f, M_P is the geometrical counterpart to the tangent line T_c described in §9.1. When we use linear approximation at $P = (a,b)$ to find Δf_{app} for some point Q near P, we are, in effect, using the tangent plane M_P as an approximation to the graph of f at and near the point $P' = (a,b,f(a,b))$ on the graph of f. Thus, *saying that a 2-variable function $f(x,y)$ has linear approximation at P is the same as saying, informally, that there is a plane M_P which serves as a useful and good approximation to the graph of f in the immediate vicinity of input P.* Note that if a function f has first-order partial derivatives at P but does not have linear approximation at P (example 1 in §8.8 gives an example of such a function), then the plane given by (2.7), with values of the partial derivatives for A and B, does not serve as a good approximation to the graph of f in the vicinity of input P.

Since $A = \left.\dfrac{\partial f}{\partial x}\right|_P$ and $B = \left.\dfrac{\partial f}{\partial y}\right|_P$, the equation of the tangent plane may be written

(2.9) $$z - \left.\frac{\partial f}{\partial x}\right|_P x - \left.\frac{\partial f}{\partial y}\right|_P y = f(P) - \left.\frac{\partial f}{\partial x}\right|_P a - \left.\frac{\partial f}{\partial x}\right|_P b.$$

We note from (2.9) that an upwards normal vector to this plane is given by

$$(2.10) \qquad \vec{N}_P = -\frac{\partial f}{\partial x}\Big|_P \hat{i} - \frac{\partial f}{\partial y}\Big|_P \hat{j} + \hat{k}.$$

We say that of $\vec{N}_P$ is a *normal vector to the graph of* f *at* P'.

Example 1. Let $f(x,y) = x^2 + y^2$. *Find the tangent plane* M_P *to the graph of* f *for* $P = (2,3)$ *in the domain of* f.

Solution. This plane must touch the graph at $P' = (2,3,13)$. We assume that f has linear approximation at P. Then $\frac{\partial f}{\partial x} = 2x$ and $\frac{\partial f}{\partial y} = 2y$. Hence $\frac{\partial f}{\partial x}\Big|_P = 4$ and $\frac{\partial f}{\partial y}\Big|_P = 6$. By (2.10),

$$\vec{N}_P = -4\hat{i} - 6\hat{j} + \hat{k}.$$

The equation for M_P is

$$z - 4x - 6y = 13 - 8 - 18 = -13$$

or

$$4x + 6y - z = 13.$$

Three-variable functions. Linear approximation for three-variable functions is exactly analogous to linear approximation for two variable functions. The *linear approximation formula for* f *at* P is

$$(2.11) \qquad \Delta f_{app} = \frac{\partial f}{\partial x}\Big|_P \Delta x + \frac{\partial f}{\partial y}\Big|_P \Delta y + \frac{\partial f}{\partial z}\Big|_P \Delta z.$$

The graph of a three-variable function must be given in 4-dimensional Euclidean space ($\mathbf{E}^4$). We do not consider such graphs in this text. After we have developed definitions and techniques necessary to do geometry in $\mathbf{E}^4$, we can define, for a three-variable function f which has linear approximation at P, a "tangent hyperplane" at the corresponding point P' on the graph of f. This "hyperplane" has the same properties and significance as a tangent plane (in $\mathbf{E}^3$) to the graph of a two-variable function. The "tangent hyperplane" at P' is usually referred to as the *tangent space* for f at P'.

A major theoretical question remains: how can we be sure that a two-variable or three-variable function *has* linear approximation at a given point in its domain? The answer is given by one of the central and fundamental theorems of multivariable calculus: the *linear approximation theorem*. We state and illu-strate the theorem here. The proof is given in §9.3 below.

(2.12) *The linear approximation theorem:* *Let f be a two- or three-variable function on domain* D. *Let* P *be a point in* D. *Then* f *will have linear approxi-mation at* P *provided that there is some disk* U *(for two variable* f) *or solid sphere* U *(for three-variable f) in* D *with center at* P *and positive radius, such that the first-order partial derivatives of f exist on* U *and are continuous at the point* P.

If a two- or three-variable function has linear approximation at a point P, we can use (2.6) or (2.11) to get approximate values for Δf in the same way that we used linear approximation in §9.1 to get approximate values for Δf for a one-variable function f.

Example 2. *Find an approximate value for the function* $f(x,y,z) = x^3 + y^2 z$ *at* Q = (1.1, 0.9, 2.2).

Solution. We take P = (1,1,2). Then $\Delta x = 0.1$, $\Delta y = -0.1$, and $\Delta z = 0.2$. We see that the first-order derivatives of f are polynomials in x,y,z. Hence these derivatives are continuous everywhere. By (2.12), f has linear approximation at every point.   We have $\left.\dfrac{\partial f}{\partial x}\right|_P = 3x^2\Big|_P = 3$, $\left.\dfrac{\partial f}{\partial y}\right|_P = 2yz\Big|_P = 4$, and $\left.\dfrac{\partial f}{\partial z}\right|_P = y^2\Big|_P = 1$. By (2.11),

$$\Delta f_{app} = 3\Delta x + 4\Delta y + \Delta z = 0.3 - 0.4 + 0.2 = 0.1 .$$

Since f(P) = 1+2 = 3, our estimate gives

$$f(Q) \approx 3.1. \quad \text{(The exact value is 3.133.)}$$

As with the example in §9.1, a simple calculation with second derivatives can be used to get a guaranteed level of precision for this estimate. (See §12.6 and exam-ple 2 in §9.6.)

Definition. Let a function f have domain D. We say that f *has linear approxi-mation on* D *if, for every point* P *in* D, f *has linear approximation at* P.

Example 3. *Let* f *be the two-variable function* $f(x,y) = x^3 + xy^2$. *(a) Show that* f *has linear approximation on its domain.* *(b) Let* P = (1,1). *Then* $f(P) = 2$. *Find the normal vector* $\vec{N}_P$ *to the graph of* f *at* P' = (1,1,2), *and find the equation of the tangent plane* M *at* P'.

Solution. (a) f has first-order derivatives $\dfrac{\partial f}{\partial x} = 3x^2 + y^2$ and $\dfrac{\partial f}{\partial y} = 2xy$.

These derivatives are polynomial functions and hence continuous at every point in the domain of f. Hence by (2.12), f has linear approximation on its domain. (b)

We have $\dfrac{\partial f}{\partial x}\Big|_P = 3x^2 + y^2\Big|_P = 4$ and $\dfrac{\partial f}{\partial y}\Big|_P = 2xy\Big|_P = 2$. By (2.10), a normal

vector at P' is

$$\vec{N}_P = -4\hat{i} - 2\hat{j} + \hat{k}$$

Hence the equation of the tangent plane must be

$$-4x - 2y + z = -4(1) - 2(1) + 2 = -4$$

or

$$4x + 2y - z = 4.$$

§9.3 Proof of the Linear Approximation Theorem

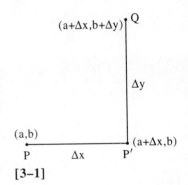

(a+Δx,b+Δy) Q

Δy

(a,b)

P   Δx   P'

(a+Δx,b)

[3–1]

We give the proof of (2.12) for the case of a two-variable function $f(x,y)$.

Let P be as assumed in the theorem. Let U be a disk with center at P such that the first-order derivatives of f are defined on U. Let P = (a,b). Let Q = (a+Δx,b+Δy) be some other point in U. Take P' = (a+Δx,b). (See [3–1].) Since Q is in U, and U is a disk with center at P, P' must be in U also. Let $\Delta f_1 = f(P') - f(P)$ and $\Delta f_2 = f(Q) - f(P')$. Then $\Delta f = \Delta f_1 + \Delta f_2$.

Since, by hypothesis, f_x and f_y exist in U, the partial functions of f must be differentiable, as functions of one variable, in U. Hence we can apply the mean-value theorem of elementary calculus to the partial function f(x,b) on [a,a+Δx]. The mean-value theorem gives a value c such that a < c < a+Δx and $\Delta f_1 = f_x(c,b)\Delta x$. Similarly, we apply the mean-value theorem to f(a+Δx,y) as partial function of y on [b,b+Δy], we get a value d such that b < d < b+Δy and $\Delta f_2 = f_y(a+Δx,d)\Delta y$. Hence

$$\Delta f = \Delta f_1 + \Delta f_2 = f_x(c,b)\Delta x + f_y(a+Δx,d)\Delta y.$$

Since, by hypothesis, f_x and f_y are continuous at P, we have $f_x(c,b) = f_x(a,b) + \varepsilon_1(Q)$ and $f_y(a+\Delta x,d) = f_y(a,b) + \varepsilon_2(Q)$, where $\lim\limits_{Q\to P} \varepsilon_1(Q) = \lim\limits_{Q\to P} \varepsilon_2(Q) = 0$.

Substituting for $f_x(c,b)$ and $f_y(a+\Delta x,d)$, we get

$$\Delta f = f_x(a,b)\Delta x + f_y(a,b)\Delta y + \varepsilon_1(Q)\Delta x + \varepsilon_2(Q)\Delta y.$$

By definition (2.1), the function f has linear approximation at P. This completes the proof. The proof for a three-variable function $f(x,y,z)$ is the same, except that an additional application of the mean-value theorem is required for the third variable of the function f.

§9.4   Finding Directional Derivatives; the Gradient Vector

Linear approximation provides a method for calculating directional derivatives. Let f be a scalar field in $\mathbf{E}^3$; let P be a point in the domain of f, and let the unit vector $\hat{u}$ specify a direction at P. We wish to calculate the directional derivative $\dfrac{df}{ds}\Big|_{\hat{u},P}$.

We introduce Cartesian coordinates and a three-variable function $f(x,y,z)$ which represents the scalar field f in that coordinate system. If this function f has linear approximation at P, we can calculate $\dfrac{df}{ds}\Big|_{\hat{u},P}$ as follows. Let Q be a point which is near P, such that $\overrightarrow{PQ}$ is parallel to $\hat{u}$. Let $\overrightarrow{PQ} = \Delta x\hat{i} + \Delta y\hat{j} + \Delta z\hat{k}$, and let $\Delta s = \overrightarrow{PQ}\cdot\hat{u}$. Then we have $\overrightarrow{PQ} = (\Delta s)\hat{u}$. (See [4–1].)

Since f has linear approximation at P, we know that

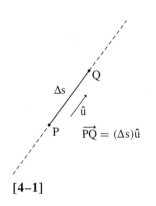

[4–1]

$$(4.1) \quad \Delta f = \frac{\partial f}{\partial x}\Big|_P \Delta x + \frac{\partial f}{\partial y}\Big|_P \Delta y + \frac{\partial f}{\partial z}\Big|_P \Delta z + \varepsilon_1(Q)\Delta x + \varepsilon_2(Q)\Delta y + \varepsilon_3(Q)\Delta z$$

where $\varepsilon_1(Q)$, $\varepsilon_2(Q)$, and $\varepsilon_3(Q)$ each have limit 0 as $Q \to P$.

Define $\vec{G}_P$ to be the vector

$$(4.2) \qquad\qquad \vec{G}_P = \frac{\partial f}{\partial x}\Big|_P \hat{i} + \frac{\partial f}{\partial y}\Big|_P \hat{j} + \frac{\partial f}{\partial z}\Big|_P \hat{k},$$

and define $\vec{\varepsilon}(Q)$ to be the vector

(4.3)
$$\vec{\varepsilon}(Q) = \varepsilon_1(Q)\hat{i} + \varepsilon_2(Q)\hat{j} + \varepsilon_3(Q)\hat{k} \ .$$

Then (4.1) can be rewritten

(4.4)
$$\Delta f = \vec{G}_P \cdot \overrightarrow{PQ} + \vec{\varepsilon}(Q) \cdot \overrightarrow{PQ}, \text{ where } \lim_{Q \to P} \vec{\varepsilon}(Q) = \vec{0}.$$

Thus
$$\Delta f = \vec{G}_P \cdot (\Delta s)\hat{u} + \vec{\varepsilon}(Q) \cdot (\Delta s)\hat{u} \text{ , and hence}$$
$$\frac{\Delta f}{\Delta s} = \vec{G}_P \cdot \hat{u} + \vec{\varepsilon}(Q) \cdot \hat{u} \ .$$

Letting $Q \to P$, we have $\Delta s \to 0$ and $\vec{\varepsilon}(Q) \to \vec{0}$. Thus

(4.5)
$$\frac{df}{ds}\bigg|_{\hat{u},P} = \lim_{\Delta s \to 0} \left(\frac{\Delta f}{\Delta s}\right) = \vec{G}_P \cdot \hat{u} \ .$$

Hence we calculate our desired directional derivative by finding the vector $\vec{G}_P$ *as in (4.2) and then taking the dot product of* $\vec{G}_P$ *with* $\hat{u}$.

Example. *Let a given scalar field* f *in* $\mathbf{E}^3$ *be represented by the function* f(x,y,z) = $x^2y + y^2z + z^2x$ *in Cartesian coordinates. Find the directional derivative of* f *at the point* P = (1,0,3) *in the direction given by the vector* $\vec{A} = \hat{i} - 2\hat{j} + 2\hat{k}$.

Solution. First we find the vector $\vec{G}_P$. $\vec{G}_P$ must have, as its components,

$$\frac{\partial f}{\partial x}\bigg|_P = 2xy + z^2\bigg|_P = 9,$$

$$\frac{\partial f}{\partial y}\bigg|_P = x^2 + 2yz\bigg|_P = 1,$$

$$\frac{\partial f}{\partial z}\bigg|_P = y^2 + 2xz\bigg|_P = 6.$$

Thus $\vec{G}_P = 9\hat{i} + \hat{j} + 6\hat{k}$.

Next, we find $\hat{u}$.

$$\hat{u} = \hat{A} = \frac{1}{|\vec{A}|}\vec{A} = \frac{1}{3}\hat{i} - \frac{2}{3}\hat{j} + \frac{2}{3}\hat{k} \ .$$

Finally we calculate

$$\frac{df}{ds}\bigg|_{\hat{u},P} = \vec{G}_P \cdot \hat{u} = 9(\tfrac{1}{3}) + 1(-\tfrac{2}{3}) + 6(\tfrac{2}{3}) = 3 - \tfrac{2}{3} + 4 = 6\tfrac{1}{3} \ .$$

Thus, for the given point P and direction $\hat{u}$,

$$\frac{df}{ds}\bigg|_{\hat{u},\,P} = 6\tfrac{1}{3} \ .$$

We summarize the method above in the following theorem.

(4.6)   *Gradient theorem. Let P be a point in the domain D of a scalar field f in E^3. Let f(x,y,z) be a function which represents the scalar field f in a given Cartesian coordinate system. Assume that f(x,y,z) has linear approximation on D. Then at each point P in D*

> (i)   *there exists a vector $\vec{G}_P$ such that for each direction $\hat{u}$ at P,*

(4.7) $$\frac{df}{ds}\bigg|_{\hat{u},P} = \vec{G}_P \cdot \hat{u} \ ; \ and$$

> (ii)   *the vector $\vec{G}_P$ is given by*

(4.8) $$\vec{G}_P = \frac{\partial f}{\partial x}\bigg|_P \hat{i} + \frac{\partial f}{\partial y}\bigg|_P \hat{j} + \frac{\partial f}{\partial z}\bigg|_P \hat{k} \ .$$

Definition and notation. The vector $\vec{G}_P$ is called the *gradient at P* of the scalar field f (and of the three-variable function f). We customarily write $\vec{G}_P$ as $\vec{\nabla}f\big|_P$. $\vec{\nabla}f$ (without subscript) will be the vector-valued function which, for each point P in D as input, gives the corresponding output value $\vec{\nabla}f\big|_P$. The symbol $\vec{\nabla}$ is called *del*. This symbol also has a more general use which is described in Chapters 25, 26, and 27.

Example (continued). For the scalar field represented by f(x,y,z) = $x^2y + y^2z + z^2x$, we have

$$\vec{\nabla}f = (2xy + z^2)\hat{i} + (x^2 + 2yz)\hat{j} + (y^2 + 2xz)\hat{k} \ .$$

§9.5 Facts about the Gradient Vector

Further useful facts about the gradient vector can be deduced from the discussion and results above. (For proofs, see the problems.)

(1) Consider the geometrical significance of the gradient vector, as expressed in the gradient theorem. We observe the following:

(5.1) *For a point* P *in the domain of the scalar field* f, *the direction of* $\vec{\nabla}f\big|_P$ *is the direction* $\hat{u}$ *for which* $\frac{df}{ds}\big|_{\hat{u},P}$ *has its maximum value, and the magnitude of* $\vec{\nabla}f\big|_P$ *is that maximum value.*

(5.2) *The formula* (4.7) *shows that as the direction* $\hat{u}$ *at* P *is varied, the resulting value of* $\frac{df}{ds}\big|_{\hat{u},P}$ *must vary in a highly regular and smooth way. In particular, we see that if* $\hat{u}$ *is orthogonal to* $\vec{\nabla}f\big|_P$, *then* $\frac{df}{ds}\big|_{\hat{u},P}$ *must be 0.*

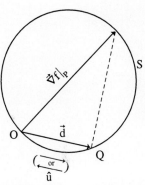

The variation of the directional derivative with the angle θ between $\hat{u}$ and $\vec{\nabla}f\big|_P$ is shown in [5–1], where S is a sphere of diameter $\big|\vec{\nabla}f\big|_P\big|$, and $\vec{d}$ is the vector $\left(\frac{df}{ds}\big|_{\hat{u},P}\right)\hat{u}$. By elementary vector geometry, $\vec{d} = \overrightarrow{OQ}$ where Q must be on the surface of S. For $\frac{\pi}{2} \le \theta \le \pi$, we have $\frac{df}{ds}\big|_{\hat{u},P} \le 0$, and Q continues to lie on the surface of S.   [5–1]

(2) Let f be a given scalar field, and let P be a given point in the domain of f. Since the definition of *directional derivative* does not depend on a chosen coordinate system, (5.1) and (5.2) show us that *we can visualize the vector* $\vec{\nabla}f\big|_P$, *without reference to a particular coordinate system, as the unique vector whose magnitude and direction are given by* (5.1).

(3) Let f be a scalar field. Let P be a point in the domain of f. Consider the following four statements:

(5.3) *There is some Cartesian system in which* f *is represented by a function which has linear approximation at* P.

(5.4) *For every choice of Cartesian system,* f *is represented by a function which has linear approximation at* P.

(5.5)   *There exist a vector $\vec{G}_P$ and a vector function $\vec{\varepsilon}(Q)$ (defined for all Q in the domain of* f*) such that* $f(Q) - f(P) = \vec{G}_P \cdot \vec{PQ} + \vec{\varepsilon}(Q) \cdot \vec{PQ}$ *and*
$$\lim_{Q \to P} \vec{\varepsilon}(Q) = \vec{0} \, .$$

(5.6)   *There exist a vector $\vec{G}_P$ and a scalar function $\varepsilon(Q)$ (defined for all Q in the domain of* f *) such that* $f(Q) - f(P) = \vec{G}_P \cdot \vec{PQ} + \varepsilon(Q)\left|\vec{PQ}\right|$ *and*
$$\lim_{Q \to P} \varepsilon(Q) = 0 \, .$$

From (2) above and the discussion in §9.4, we can show that these four statements are equivalent (that is to say, if any one of them is true about f and P, then the other three must also be true). Since (5.5) and (5.6) do not mention any chosen coordinate system, either (5.5) or (5.6) gives us a *coordinate-free vector-algebra definition* of what it means for a *scalar field* f *to have linear approximation at* P.

(4) Let f be a scalar field on D in $\mathbf{E}^3$. Let $\{\hat{u}, \hat{v}, \hat{w}\}$ be a frame. By the frame identity (§3.4), we have

$$\vec{\nabla}f\Big|_P = (\vec{\nabla}f\Big|_P \cdot \hat{u})\hat{u} + (\vec{\nabla}f\Big|_P \cdot \hat{v})\hat{v} + (\vec{\nabla}f\Big|_P \cdot \hat{w})\hat{w} \, .$$

Hence by the gradient theorem:

(5.6) $$\vec{\nabla}f\Big|_P = \frac{df}{ds}\Big|_{\hat{u},P} \hat{u} + \frac{df}{ds}\Big|_{\hat{v},P} \hat{v} + \frac{df}{ds}\Big|_{\hat{w},P} \hat{w} \, .$$

We shall later use (5.6) to get expressions for the gradient in cylindrical coordinates and in other non-Cartesian coordinate systems. (For cylindrical coordinates, see §11.5.)

(5) The following version of the gradient theorem holds for two-variable functions and scalar fields in $\mathbf{E}^2$.

(5.7)   *Let P be a point in the domain* D *of a scalar field* f *in* $\mathbf{E}^2$. *Let* f(x,y) *be a function which represents the scalar field* f *in a given Cartesian system. Assume that* f(x,y) *has linear approximation on* D. *Then at each point P in D,*

(i)   *there exists a vector $\vec{G}_P$ such that for each direction $\hat{u}$ at P,*

$$\frac{df}{ds}\bigg|_{\hat{u},P} = \vec{G}_P \cdot \hat{u}, \quad and$$

(ii)   *the vector* $\vec{G}_P$ *is given by*

$$\vec{G}_P = \frac{\partial f}{\partial x}\bigg|_P \hat{i} + \frac{\partial f}{\partial y}\bigg|_P \hat{j}.$$

We emphasize that for a two-variable function f(x,y), $\vec{\nabla}f\big|_P$ is a vector in the xy plane. Students are sometimes tempted to think that because $\vec{\nabla}f\big|_P$ points in the direction of most rapid increase for f, $\vec{\nabla}f\big|_P$ must be a vector in $\mathbf{E}^3$ pointing upwards along the steepest slope of a graph of f at a point P′ on the graph whose projection on the xy plane is P. This is not true. P is a point in the xy plane. $\vec{\nabla}f\big|_P$ gives the direction *in the* xy *plane* along which the steepest upwards slope of the graph at P′ occurs.

(6)   (2.6) and (2.11) were linear approximation formulas for two-variable functions and for three-variable functions. From (5.5), we see that both these formulas can be expressed by the same *vector linear approximation formula*:

(5.8) $\Delta f_{app} = \vec{\nabla}f\big|_P \cdot \overrightarrow{PQ}$.

§9.6 Examples and Applications

Example 1. A scalar field f *is represented in Cartesian coordinates by the function* $f(x,y,z) = x^2 + 2xyz + z^2$. By the linear approximation theorem and the gradient theorem, f has a gradient vector at P = (1,1,1). We ask the following questions.

(i)   *At the point* (1,1,1), *in what direction is the field* f *increasing most rapidly, and how fast is it increasing in that direction?*

(ii)   *At the point* (1,1,1), *how fast is the field* f *increasing in the direction of the vector* $\vec{A} = \hat{i} - \hat{j}$?

We first find $\vec{\nabla}f = f_x\hat{i} + f_y\hat{j} + f_z\hat{k} = (2x + 2yz)\hat{i} + 2xz\hat{j} + (2xy + 2z)\hat{k}$.

Then for P = (1,1,1), we have $\vec{\nabla}f\big|_P = 4\hat{i} + 2\hat{j} + 4\hat{k}$. Hence the *direction* of most rapid increase is given by the direction of $\vec{\nabla}f\big|_P$. This direction, as a unit-vector, is $\frac{1}{6}(4\hat{i} + 2\hat{j} + 4\hat{k}) = \frac{2}{3}\hat{i} + \frac{1}{3}\hat{j} + \frac{2}{3}\hat{k}$. The *rate* of this maximum increase must be

$$\left|\vec{\nabla}f\big|_P\right| = \left|4\hat{i} + 2\hat{j} + 4\hat{k}\right| = \sqrt{16 + 4 + 16} = 6. \text{ This answers (i).}$$

From the vector $\vec{A} = \hat{i} - \hat{j}$, we get $\hat{A} = \left(\frac{\sqrt{2}}{2}\right)\hat{i} - \left(\frac{\sqrt{2}}{2}\right)\hat{j}$. Hence

$$\frac{df}{ds}\bigg|_{\hat{A},P} = \vec{\nabla}f\big|_P \cdot \hat{A} = 4\left(\frac{\sqrt{2}}{2}\right) - 2\left(\frac{\sqrt{2}}{2}\right) = \sqrt{2}. \text{ This answers (ii).}$$

Example 2. *Find an approximate value for* $f(x,y) = x^2 y^3$ *at the point* $Q = (2.99, 1.02)$.

Solution. Take $P = (3,1)$. Then we have $\overrightarrow{PQ} = -(0.01)\hat{i} + (0.02)\hat{j}$; $\vec{\nabla}f = 2xy^3\hat{i} + 3x^2y^2\hat{j}$; and $\vec{\nabla}f\big|_P = 6\hat{i} + 27\hat{j}$. Using (5.8) above, we get

$$\Delta f_{app} = \vec{\nabla}f\big|_P \cdot \overrightarrow{PQ} = -0.06 + 0.54 = 0.48.$$

Since $f(P) = 9$, our estimate for $f(Q)$ is

$$f(Q) \approx f(P) + \Delta f_{app} = 9 + 0.48 = 9.48.$$

As in the case of one-variable linear approximation, a higher-order directional derivative can be used to get guaranteed bounds on this estimate. We consider the calculation of second-order directional derivatives in §12.6.

Example 3. *In a given Cartesian system, let* $\vec{u} = \hat{i}, \vec{v} = \hat{i} + \hat{j}$, *and* $\vec{w} = \hat{i} + \hat{j} + \hat{k}$. *For a scalar field* f *and a point* P, *we are given the information that* $\frac{df}{ds}\big|_{\hat{u},P} = 1$, $\frac{df}{ds}\big|_{\hat{v},P} = 2$, $\frac{df}{ds}\big|_{\hat{w},P} = -1$. *Assume that* f *has linear approximation at* P. *Find an expression for* $\vec{\nabla}f\big|_P$.

Let $\vec{\nabla}f\big|_P = a_1\hat{i} + a_2\hat{j} + a_3\hat{k}$, where a_1, a_2, and a_3 are the unknown scalar components of $\vec{\nabla}f\big|_P$. By the gradient theorem, we have

$$1 = \frac{df}{ds}\bigg|_{\hat{u},P} = \vec{\nabla}f\big|_P \cdot \hat{u} = (a_1\hat{i} + a_2\hat{j} + a_3\hat{k}) \cdot i = a_1. \text{ Thus } a_1 = 1.$$

Similarly,

$$2 = \frac{df}{ds}\bigg|_{\hat{v},P} = \vec{\nabla}f\big|_P \cdot \hat{v} = \vec{\nabla}f\big|_P \cdot \left(\frac{\hat{i}+\hat{j}}{|\hat{i}+\hat{j}|}\right)$$

$$= (\hat{i} + a_2\hat{j} + a_3\hat{k}) \cdot \left(\frac{\sqrt{2}}{2}\hat{i} + \frac{\sqrt{2}}{2}\hat{j}\right) = \frac{\sqrt{2}}{2}(1 + a_2).$$

Hence $a_2 = \frac{2}{\sqrt{2}}(2 - \frac{\sqrt{2}}{2}) = 2\sqrt{2} - 1$.

Finally,

$$-1 = \frac{df}{ds}\bigg|_{\hat{w},P} = \vec{\nabla}f\big|_P \cdot \hat{w} = \vec{\nabla}f\big|_P \cdot \left(\frac{\hat{i}+\hat{j}+\hat{k}}{|\hat{i}+\hat{j}+\hat{k}|}\right)$$

$$= (\hat{i} + (2\sqrt{2} - 1)\hat{j} + a_3\hat{k}) \cdot \left(\frac{\sqrt{3}}{3}\hat{i} + \frac{\sqrt{3}}{3}\hat{j} + \frac{\sqrt{3}}{3}\hat{k}\right)$$

$$= \frac{\sqrt{3}}{3}(1 + 2\sqrt{2} - 1 + a_3) = \frac{\sqrt{3}}{3}(2\sqrt{2} + a_3).$$

Hence $a_3 = -\sqrt{3} - 2\sqrt{2}$.

These results give us

$$\vec{\nabla}f\big|_P = \hat{i} + (2\sqrt{2} - 1)\hat{j} - (\sqrt{3} + 2\sqrt{2})\hat{k}.$$

This example illustrates a general fact: *if a three-variable function* f *has linear approximation at* P, *then we can find* $\vec{\nabla}f\big|_P$ *if we know the values of* $\frac{df}{ds}$ *at* P *in at least three different non-coplanar directions.* (See Chapter 30.)

§9.7 Differentiability

In one variable calculus, having a *derivative*, having *linear approximation*, and being *differentiable* all mean the same thing. In (2.4), we saw that a multivariable function f on D can have partial derivatives on D without having linear approximation on D. How then should we define "differentiable" for a multivariable function or scalar field? In multivariable calculus, mathematicians have chosen to define "differentiable" to mean *having linear approximation*.

Definition. Let f be a multivariable function or scalar field on D. Let P be a point in D. We say that f is *differentiable at* P if f has linear approximation at P. We say that f is *differentiable* if f is differentiable at every point P in D.

These definitions have the following consequences.

(7.1)   *If f is differentiable, then f is continuous* (by (2.2)).

(7.2)   f *can be continuous without being differentiable.* For example, the function $f(x,y) = \sqrt{x^2 + y^2}$ is continuous but has no gradient vector or directional derivatives at the origin. (The first example in §9.9 is of technical interest. It shows that it is possible for a function to be continuous at P and to have all its directional derivatives at P without being differentiable at P.)

Definition. Let f be a function of one or more variables on D. We say that f is *in class* C^1 (or, simply, *is in* C^1) if f has all its first-order derivatives on D and if these derivatives are continuous on D. Let f be a scalar field on D. We say that f *is in* C^1 if f can be represented in Cartesian coordinates by a function in C^1.

By the linear approximation theorem (2.11), we have:

(7.3)   *If f is in* C^1, *then f is differentiable.*

In example 2 of §9.9, we see that f *can be differentiable without being in* C^1.

Definition. We say that a function f on D is *twice differentiable*, if f has first-order partial derivatives on D, and these derivatives are differentiable.

Let f be a scalar field on D. We say that f is *twice differentiable* if f can be represented in Cartesian coordinates by a twice differentiable function.

By (7.1), we have:

(7.4)   *If* f *is twice differentiable, then* f *is in* C^1.

The following is also true (see the problems):

(7.5)   f *can be in* C^1 *without being twice differentiable.*

Definition. We say that a function f on D is *in class* C^2 (or, simply, *is in* C^2) if f has second-order derivatives on D and these second-order derivatives are continuous. Let f be a scalar field on D. We say that f *is in* C^2 if f can be represented in Cartesian coordinates by a function in C^2.

By (7.3), we have:

(7.6)   *If* f *is in* C^2, *then* f *is twice differentiable.*

One can also show (see the problems):

(7.7)   *There exists a twice differentiable function which is not in* C^2.

We can summarize (7.1)–(7.7) as follows. *The continuous functions include the differentiable functions, which include the* C^1 *functions, which include the twice differentiable functions, which include the* C^2 *functions; but none of the converse inclusions is true.*
Analogous versions of the equivalence between (5.3) and (5.4) can be proved for C^1 scalar fields, twice differentiable scalar fields, and C^2 scalar fields. It follows, as in §9.5, that these properties of scalar fields are coordinate-free.

Remark. Functions in C^1 are traditionally known as *continuously differentiable* functions. Functions in C^2 are traditionally known as *continuously twice differentiable functions*. For each n > 2, an analogous concept of n-*times differentiable* and an analogous class C^n can be defined. Results analogous to (7.4)–(7.7) can then be stated and proved. (See the problems.)

§9.8 Equality of Cross-derivatives

We present and prove a useful technical fact about derivatives.

(8.1)   *Theorem. (Equality of cross-derivatives for a two-variable function.)* Let f(x,y) be a function of two variables on domain D. Let P = (a,b) be a point in D. Assume that there is a disk in D with center at P such that f on D is in C^2. Then

$$f_{xy}(P) = f_{yx}(P) .$$

Proof. Let U be the given disk. Since the second-order derivatives of f are continuous on U, it follows, by the linear-approximation theorem, that the first-order derivatives of f must have linear approximation on U. Hence, by (2.2), the first-order derivatives of f must be continuous on U. Hence, by the same argument, f itself must be continuous on U.

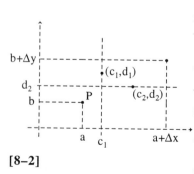

(a,b+Δy)   (a+Δx,b+Δy)

Δy

P   Δx

(a,b)   (a+Δx,b)

[8–1]

Let $\Delta x \neq 0$ and $\Delta y \neq 0$ have fixed values. (See [8–1].) Define the one-variable functions:

$$F(x) = f(x,b+\Delta y) - f(x,b),$$
$$G(y) = f(a+\Delta x,y) - f(a,y).$$

F is differentiable, since $F'(x) = f_x(x,b+\Delta y) - f_x(x,b)$. Similarly for G.
Let $\delta^* = F(a+\Delta x) - F(a)$.
Then $\delta^* = f(a+\Delta x,b+\Delta y) - f(a+\Delta x,b) - f(a,b+\Delta y) + f(a,b)$.
Similarly, $G(b+\Delta y) - G(b) = \delta^*$. ($\delta^*$ is sometimes called the *second difference* of f at P for Δx and Δy.)
By the mean-value theorem, $\delta^* = F'(c_1)\Delta x$ for some c_1 in the interval $(a,a+\Delta x)$. Since $F'(x) = f_x(x,b+\Delta y) - f_x(x,b)$, we have

$$F'(c_1) = f_x(c_1,b+\Delta y) - f_x(c_1,b) .$$

Since f_x is continuous and has first-order derivatives, we can also apply the mean-value theorem to $f_x(c_1,y)$. We get $F'(c_1) = f_{xy}(c_1,d_1)\Delta y$ for some d_1 in the interval $(b,b+\Delta y)$.

Thus we have $\delta^* = f_{xy}(c_1,d_1)\Delta x\Delta y$. By an exactly similar argument, beginning with $\delta^* = G(b+\Delta y) - G(b)$, and using the mean-value theorem twice, we get $\delta^* = f_{yx}(c_2,d_2)\Delta x\Delta y$ for some d_2 in the interval $(b,b+\Delta y)$ and some c_2 in the interval $(a,a+\Delta x)$. (See [8–2].)

b+Δy

(c₁,d₁)

d₂
b   P   (c₂,d₂)

a   c₁   a+Δx

[8–2]

Thus $f_{xy}(c_1,d_1)\Delta x\Delta y = f_{yx}(c_2,d_2)\Delta x\Delta y,$

and $f_{xy}(c_1,d_1) = f_{yx}(c_2,d_2).$

Here, for our fixed a and b, the value of each of c_1, d_1, c_2, and d_2, depends on Δx and Δy. Taking $Q = (a+\Delta x, b+\Delta y)$, we have, by the continuity of f_{xy} and f_{yx},

$$\lim_{Q \to P} f_{xy}(c_1, d_1) = f_{xy}(P),$$

and
$$\lim_{Q \to P} f_{yx}(c_2, d_2) = f_{yx}(P).$$

Since, for each Q, $f_{xy}(c_1, d_1) = f_{yx}(c_2, d_2)$, we must have

$$f_{xy}(P) = f_{yx}(P).$$

This completes the proof.

(8.2) **Corollary.** *(Equality of cross-derivatives for a three-variable function.) Let $f(x,y,z)$ be a function of three variables on domain D. Let P = (a,b,c) be a point in D. Assume that there is a solid sphere in D with center at P on which f is in C^2. Then*

$$f_{xy}(P) = f_{yx}(P), \; f_{xz}(P) = f_{zx}(P), \; and \; f_{yz}(P) = f_{zy}(P).$$

Proof. To get $f_{xy}(P) = f_{yx}(P)$, we hold $z = c$ constant and apply the theorem. Similarly for $f_{xz}(P) = f_{zx}(P)$ (with $y = b$), and for $f_{yz}(P) = f_{zy}(P)$ (with $x = a$).

Example. Consider $f(x,y,z) = \cos(xy) + y^2 z^3$. It is easy to see, by familiar differentiation rules, that the second-order derivatives of f must all be continuous. (Indeed, derivatives of all orders must exist and be continuous.) Calculating the cross-derivatives we get

$$f_x = -y \sin(xy),$$
$$f_y = -x \sin(xy) + 2yz^3,$$
$$f_z = 3y^2 z^2,$$

and then

$$f_{xy} = -\sin(xy) - xy \cos(xy),$$
$$f_{xz} = 0,$$
$$f_{yx} = -\sin(xy) - xy \cos(xy),$$
$$f_{yz} = 6yz^2,$$
$$f_{zx} = 0,$$
$$f_{zy} = 6yz^2.$$

We see that the cross-derivatives are equal as predicted and required by (8.1).

Provided that we have continuous derivatives as needed, the theorem implies not only that $f_{xy} = f_{yx}$, but that $f_{xyx} = f_{yxx} = f_{xxy}$, that $f_{xxyy} = f_{xyxy} = f_{xyyx} = \ldots$, and so forth. For example, to show that $f_{xyx} = f_{yxx}$, we note that $f_{xy} = f_{yx} \Rightarrow (f_{xy})_x = (f_{yx})_x$; and to show that $f_{xxyy} = f_{xyxy}$, we note that $(f_x)_{xy} = (f_x)_{yx} \Rightarrow ((f_x)_{xy})_y = ((f_x)_{yx})_y$.

Thus in taking an nth-order derivative (provided we have the required continuity of nth-order derivatives), the only information we need is the number of times $\frac{\partial}{\partial x}$ is applied, the number of times $\frac{\partial}{\partial y}$ is applied, and the number of times $\frac{\partial}{\partial z}$ is applied. We do not need to know the order in which these applications occur.

§9.9 Counterexamples

First counterexample. Consider the two statements about a scalar field f:

(a) f *is differentiable.*

(b) f *is continuous, and for every point P and direction $\hat{u}$,*
$\left. \dfrac{df}{ds} \right|_{\hat{u},P}$ *exists.*

We have seen in §9.7 that (a) implies (b). In the following example, (b) is true but (a) is not true.

Take a polar coordinate system in E^2. Let the point O be the origin of the system. Define a scalar field f by

$$f(P) = \begin{cases} r\sin 3\theta & \text{if } P \neq O \text{ and } P \text{ has coordinates } (r,\theta); \\ 0 & \text{if } P = O \end{cases}$$

The reader may verify that f is continuous at O and that for each direction $\hat{u} = \cos\theta\hat{i} + \sin\theta\hat{j}$ (in the associated Cartesian system), the directional derivative of f exists at O with $\left. \dfrac{df}{ds} \right|_{\hat{u},O} = \sin 3\theta$. Hence (b) holds for f. The directional derivative at O has 1 as its maximum value and takes this maximum value in each of the three directions given by $\theta = \dfrac{\pi}{6}, \dfrac{5\pi}{6}, \dfrac{3\pi}{2}$. Hence, by (5.1) and (5.2), f cannot have a gradient at O and is therefore not differentiable at O. Thus (a) fails for f. Note that f is represented in the associated Cartesian system by the function

$$f(x,y) = \begin{cases} \dfrac{3yx^2 - y^3}{x^2 + y^2} & \text{if } x \neq 0 \text{ or } x \neq 0, \\[2ex] 0 & \text{if } x = 0 \text{ and } y = 0. \end{cases}$$

This formula for f shows that on the set of all points other than O, f is differentiable (by (2.11), since the partial derivatives of f exist and are continuous at all points other than O). O is the only point at which f fails to be differentiable.

Second counterexample. Consider the two statements about a multi-variable function h:

(a)　　h *is in* C^1.
(b)　　h *is differentiable*.

We have seen in (2.12) that (a) implies (b). In the following example, (b) is true but (a) is not true.

The reader may verify that the one-variable function

$$g(x) = \begin{cases} x^2 \sin\dfrac{1}{x} & \text{for } x \neq 0, \\[2ex] 0 & \text{if } x = 0, \end{cases}$$

is differentiable and hence has linear approximation at every point, but that the derivative g′ is not continuous at x = 0. We define a two-variable function h by setting

$$h(x,y) = g(x) \text{ for all } x \text{ and } y.$$

Since g has linear approximation at every point, h has linear approximation at every point, as the reader may also verify. Hence h is differentiable at every point. On the other hand, the first-order derivative $h_x(x,y) = g'(x)$ cannot be continuous at points where x = 0. Thus (b) holds for h, but (a) fails.

Geometrically, the graph of h has a tangent plane for every point (x,y). For each point (0,y), this tangent plane has the unit normal vector $\hat{k}$. If we fix y, however, take x > 0, and let x approach 0, the unit normal vector for the tangent plane for (x,y) oscillates without approaching the limit $\hat{k}$.

§9.10 Problems

§9.1

1. Show that the estimate obtained in example 2 in §9.1 is accurate to within ± 0.000004. (*Hint.* Use the second derivative, evaluated at $x = 81$ and $x = 100$, to get bounds on the rate at which the first derivative is changing between 81 and 81.1. Then use the mean value theorem.)

2. Let $\vec{R}(t)$, $[a \leq t \leq b]$, be a vector function of t.
   (i) For a given c, $[a \leq c \leq b]$, give an appropriate definition for "$\vec{R}(t)$ has linear approximation at c."
   (ii) Show that $\vec{R}$ has linear approximation at c if and only if $\vec{R}$ has a derivative at c.

§9.2

1. Consider the function $f(x,y,z) = x^2 y + y^2 z$. Estimate the change in the value of f when we move from $P = (1,1,2)$ to the nearby point $(0.9,1.1,2.1)$.

2. We wish to determine the volume of a cylinder of radius about 2 cm and height about 3 cm. We plan to measure the diameter and height, getting each to the same accuracy in centimeters. Estimate what this accuracy should be if the calculated volume is not to be in error by more than 0.5 cm^3.

3. Find the equation of the tangent plane to $z = x^2 + y^2$ at the point $(3,4,25)$.

4. Find the tangent plane to the surface $z = \tan^{-1}\left(\frac{y}{x}\right)$ at $(1,1,\frac{\pi}{4})$. Find the line normal to the tangent plane through this point.

5. Use linear approximation to estimate how much an error of 2% in each of the factors a,b,c may affect the product abc.

6. Find the equation of the plane tangent to $z = e^{x+y^2}$ at $(1,1,e^2)$.

7. Consider the graph of the function $z = e^{3xy}$. Find the equation of the tangent plane to this surface at the point given by $x = 1$, $y = -2$.

8. The dimensions of a rectangular box are measured as 3, 4 and 12 cm. If the measurements may be in error by ± 0.01, ± 0.01, and ± 0.03 cm, respectively, calculate the length of the interior diagonal and estimate the possible error in this length.

9. A perfectly rectangular block has edges of approximate length 5 cm., 8 cm., and 3 cm. We plan to measure each edge to the same accuracy in centimeters. What should this accuracy be in order to guarantee that the calculated volume will not be in error by more than about 1 mm^3?

10. Let f have linear approximation at P as described in (2.1). Show that the functions ε_1 and ε_2 are not unique. (*Hint.* Increase $\varepsilon_1(Q)$ by a small amount $\delta_1(Q)$ and decrease $\varepsilon_2(Q)$ by $\delta_2(Q)$. Then look for a necessary relationship between $\delta_1(Q)$ and $\delta_2(Q)$.)

11. Let f be the function in the first counterexample in §8.8. By (2.2), f cannot have linear approximation at 0, and hence the hypothesis for (2.12) must fail at 0. Show directly, without using (2.12), that this hypothesis fails for the given f.

§9.4

1. Given $f(x,y,z) = x^2+y^2$, find the directional derivative of f at $Q = (1,2,2)$ in the direction of the origin.

2. At the point $P = (1,2)$ in the xy plane, a scalar field represented by $f(x,y)$ has $\frac{df}{ds} = \sqrt{2}$ in the direction towards the point $(2,3)$ and $\frac{df}{ds} = -1$ in the direction towards $(1,0)$. Find the gradient of the scalar field at P.

3. Let $f = x^3 - e^x z + y^2$. Find the value of the directional derivative of f in the direction of $\hat{T}$ at the point $\vec{R}(1)$ for the curve given by $\vec{R}(t) = (t-1)\hat{i} + t^2\hat{j} + t^3\hat{k}$, $t > 0$.

4. f is a given scalar field. At a certain point P, the directional derivative of f in the direction $\hat{u}$ is $\left.\dfrac{df}{ds}\right|_{\hat{u},P} = -2$, where $\hat{u} = \dfrac{1}{3}\hat{i} + \dfrac{2}{3}\hat{j} - \dfrac{2}{3}\hat{k}$. Given that $\left.\dfrac{df}{dy}\right|_P = 3$ and $\left.\dfrac{df}{dz}\right|_P = 6$, what is $\left.\dfrac{df}{dx}\right|_P$?

5. At a given point P, a scalar field f in Cartesian coordinates has the following directional derivatives:

$\dfrac{df}{ds} = 2$ in the direction $\hat{i}$,

$\dfrac{df}{ds} = -1$ in the direction $\hat{j}$,

$\dfrac{df}{ds} = 2\sqrt{3}$ in the direction of $\hat{i} + \hat{j} + \hat{k}$.

At the point P, what is the directional derivative of f in the direction of $\hat{i} + \hat{j} - \hat{k}$?

(*Hint.* Remember to use unit vectors when unit vectors are called for.)

6. Given $f = x^3 + 2x^2y + yz^2$, find the directional derivative of f at Q = (1,2,2) in the direction of the origin.

7. Let $f = \dfrac{x^2 - y^2}{x^2 + y^2}$ define a scalar field in $\mathbf{E}^2$ at all points except (0,0). In what directions is $\dfrac{df}{ds} = 0$ at (1,1)?

8. A certain function $f(x,y,z)$ has continuous partial derivatives. At a given point P, it has values for certain directional derivatives as follows:

$$\left.\frac{\partial f}{\partial s}\right|_{\hat{A},P} = \left.\frac{\partial f}{\partial s}\right|_{\hat{B},P} = \left.\frac{\partial f}{\partial s}\right|_{\hat{C},P} = 1.$$

where $\vec{A} = \hat{i}$, $\vec{B} = \hat{i} + \hat{j}$, and $\vec{C} = \hat{i} + \hat{j} + \hat{k}$. Find $\left.\dfrac{\partial f}{\partial s}\right|_{\hat{k},P}$.

9. At the point (1,1,0), $f(x,y,z)$ has $\dfrac{df}{ds} = 1$ in the direction towards the point (1,1,1), $\dfrac{df}{ds} = \sqrt{2}$ in the direction towards (1,0,-1), and $\dfrac{df}{ds} = 0$ in the direction towards (2,2,1). What is the value of $\dfrac{df}{ds}$ in the direction towards (0,0,0)?

10. Find the directional derivative of the function $f(x,y,z) = xy^2z^3$ at the point (1,1,1) in the direction of the vector $2\hat{i} + \hat{j} - 2\hat{k}$.

11. Consider the function $u(x,y) = x^2 + 2y^2$ at (1,1), in what directions is the directional derivative equal to zero?

12. A function $f(x,y,z)$ is known to have the following values: $f(1,1,1) = 1$, $f(2,1,1) = 2$, $f(2,2,1) = 3$, $f(2,2,2) = 4$. Give an approximate value for each of the directional derivatives

$$\left.\frac{\partial f}{\partial s}\right|_{\hat{i}}, \quad \left.\frac{\partial f}{\partial s}\right|_{\hat{i}+\hat{j}}, \quad \left.\frac{\partial f}{\partial s}\right|_{\hat{i}+\hat{j}+\hat{k}}$$

at the point (1,1,1).

13. (a) Find the gradient of $f(x,y,z) = x^3 + y^2z - xyz$.

   (b) Find the directional derivative of f in the direction of the gradient at the point (1,1,1).

   (c) Find the directional derivative of f at (1,1,1) in the direction of the vector $\hat{i} - 2\hat{j} + 2\hat{k}$.

   (d) Let $\vec{A}$ and $\vec{B}$ be vectors such that the directional derivative of f at (1,1,1) in the direction of $\vec{A}$ is $\sqrt{5}$ and in the direction of $\vec{B}$ is 0. What is the directional derivative of f at (1,1,1) in the direction of $\vec{A} \times \vec{B}$?

14. At the point P = (2,3) in a plane with Cartesian coordinates, a certain function f(x,y) has the directional derivatives $\frac{df}{ds}\big|_{\hat{i},P} = 4$ and $\frac{df}{ds}\big|_{\hat{j},P} = -3$.
    (a) In what direction at P is the function decreasing most rapidly?
    (b) In what directions will the directional derivative of f at P be 0.
    (c) What is the directional derivative of f at P in the direction of the vector $\hat{i} - \hat{j}$?.

15. Let f(x,y,z) = x²y + y²z, and let P be the point (1,1,−1.5).
    (a) At the point P, is the value of f *least* sensitive to change in x, to change in y, or to change in z? Give reasons for your answer.
    (b) Use linear approximation to estimate the change in f when we move from P to the point (1.1, 1.2,−1.4).
    (c) For what unit vector û will the directional derivative $\frac{df}{ds}\big|_{\hat{u},P}$ be a maximum?

§9.5

1. Prove (5.1).

2. In figure [5–1], show that Q must always lie on the sphere S, regardless of the direction of û.

3. Prove that the statements (5.3), (5.4), (5.5), and (5.6) are equivalent. (*Hint.* Prove each of the four implications (5.3) ⇒ (5.4) ⇒ (5.5) ⇒ (5.6) ⇒ (5.3).)

4. A scalar field f on **E**²-with-Cartesian-coordinates is represented by the function f(x,y) and has linear approximation at the point P = (a,b). Let P, P′, and M_P be as in (2.8), and let S be the graph of f.
    (a) For f(x,y) = x² + y², a = 2, and b = 1, find a vector $\vec{v}$ in **E**³ whose direction is the direction in which the value of z = f(x,y) increases most rapidly as we move through P′ in the plane M_P. This direction is called the *direction of steepest ascent on* S. (See the comment after (5.7).)
    (b) Find a general formula for V in terms of $\vec{\nabla}f$, a, b, and $\hat{k}$.

§9.6

1. Let f(x,y,z) = x²y−xyz.
    (a) Find the value of the directional derivative of f at the point (1,1,1) in the direction of $2\hat{i} + 2\hat{j} - \hat{k}$.
    (b) Find the equation of the plane tangent to the surface f(x,y,z) = 0 at (1,1,1).

2. At a given point P, the directional derivative of f becomes a maximum when the direction used is given by the vector $\hat{i} + \hat{j} - \hat{k}$. In this direction, $\frac{df}{ds} = 2\sqrt{3}$. What is the value at P of $\frac{df}{ds}$ in the direction of $\hat{i} + \hat{j}$?

3. Find an equation for the tangent plane to the graph of the function f(x,y) = x² + 3xy − y² at (1,−1,−3).

4. Consider the function $f(x,y) = \frac{2xy}{(x^2+y^2)}$ at the point P = (x,y) = (1,2).
    (a) Find the gradient of f at P.
    (b) Find the direction in which the directional derivative at P is maximum.
    (c) Find the value of this maximum directional derivative.
    (d) Find the directional derivative of f at P in the direction of the vector $\hat{i} + \hat{j}$.

5. In problem 4 find the equation of the plane tangent to the graph of f at the point on the graph given by (x,y) = (1,2).

6. Let f(x,y,z) = xy³ − e^{xz} + yz². Find the value of the directional derivative of f in the direction given by the unit tangent vector to the curve $\vec{R}(t) = (t-1)\hat{i} + t^2\hat{j} + t^2\hat{k}$ at the point $\vec{R}(1)$.

§9.7

1. Prove (7.5). (*Hint.* Define $f(x,y) = g(x)$, where $g(x)$ is defined to be an indefinite integral of the function $h(x) = x \sin\frac{1}{x}$.)

2. Let $f(x,y) = g(x) = \begin{cases} x^2 \sin(\frac{1}{x}) & \text{for } x \neq 0, \\ 0 & \text{for } x = 0. \end{cases}$

   Prove that f is differentiable but not in C^1. (This example is considered further in §9.9.)

3. Let $g(x)$ be a function of one variable. Show that

   $$g \text{ is in } C^n \Leftrightarrow \int g\,dx \text{ in is } C^{n+1}.$$

4. Let $g(x)$ be a function of one variable. Show that

   $$g \text{ is } n \text{ times differentiable} \Leftrightarrow$$
   $$\int g\,dx \text{ is } (n+1) \text{ times differentiable.}$$

5. Show that f in (7.2) has no gradient vector or directional derivatives at the origin.

§9.9

1. For the function g in the second counter-example, show that g is differentiable for all x but that $g'(x)$ is not continuous for all x.

2. In the first counter-example, express the directional derivative at O as a function of θ.

Chapter 10

The Chain Rule

§10.1 Physical Variables and Their Derivatives

In algebra and calculus, *variable symbols* such as x, y, u, t,... may be used to represent real numbers. Sometimes, a variable symbol of this kind will also be used to represent a given *function*. (For example, we may say "Let y = x^2; then y′ = 2x.") At other times, a variable symbol of this kind will be used to represent the *observed value*, at a given moment, of some measurable physical quantity in a given (real or imagined) physical system or experiment. (For example, we may say "Consider an object falling from rest in a vacuum at sea level. At a given moment, let s = distance fallen, v = speed, and t = elapsed time of fall.") When a variable symbol is used in this latter way, we say that it serves as a *physical variable*.

Physical variables of a given physical system may be mathematically related in various and different ways. For example, for the falling object, with feet and seconds as units, we have: s = $16t^2$, v = 32t, s = $\frac{1}{64}v^2$. A given physical variable (such as s) need not be associated with only one function, but may instead appear as the output of more than one function (such as s = $16t^2$ and s = $\frac{1}{64}v^2$). In such a case, if we take or use a derivative, we must make clear which specific function is being differentiated. Thus, for s = $16t^2$ and s = $\frac{1}{64}v^2$, we cannot simply use the notation s′ for a derivative of s. We must either introduce a function symbol (saying, for example, "Let s = f(v) = $\frac{1}{64}v^2$; consider the derivative $f'(v) = \frac{1}{32}v''$), or we must use a derivative notation which exhibits the desired independent variable (writing, for example, $\frac{ds}{dv} = \frac{1}{32}v$.)

Physical variables may also be related to each other by multivariable functions. To illustrate this, we consider the case of a small container of a gas at equilibrium. *Equilibrium* means that the gas has been allowed to reach a state in which the values of all macroscopic physical variables are steady and unchanging. A *macroscopic variable* for a gas is a variable (like p = pressure, or T = temperature) whose value is determined by the average behavior of many molecules. In the case of a gas, we shall limit our attention to macroscopic physical variables.

The equilibrium state of a given gas can be changed by adding or removing heat or by changing the shape of the container so that the volume occupied by the gas is increased or reduced. (We assume that the total mass of the gas remains unchanged.)

Physical variables include (among others): *temperature* T (which we can measure with a thermometer), *pressure* p (which we can measure with a manometer), *volume* V (which we can calculate from measured dimensions of the container), and *internal energy* U (which we can measure by choosing a "standard" state of the gas to which we assign energy U = 0 and by keeping track of the amount of heat that enters or leaves the gas and of the amount of work that the gas does on its environment, or has done on it by its environment, when the state of the gas is changed from the standard state to the currently observed state.)

Extensive study of these and other physical variables in a gas system of this kind reveals the following experimental facts:

(1.1)   *The value of* p *(for a given kind and quantity of gas) is found to depend on, and to be determined by, the values of* V *and* T. Hence we have p = h(V,T), where h is a function of two variables.

(1.2)   *In fact, the value of every physical variable (for our given system) depends on and is determined by the values of* V *and* T.

(1.3)   *The value of* T *is found to depend on and to be determined by the values of* V *and* U. Hence we have T = t(V,U), where t is a function of two-variables.

(1.4)   (1.2) and (1.3) imply that *the value of every physical variable depends on and is determined by the values of* V *and* U. In particular, there is a two-variable function g (different from the function h in (1.1)) such that p = g(V,U) = h(V,t(V,U)).

If we are using the function $p = h(V,T)$ and the function $p = g(V,U)$ at the same time, then the derivative notation $\frac{\partial p}{\partial V}$ will be ambiguous, since it does not show which function's derivative is intended. We can make our intended meaning clear in one of two possible ways: (i) we can use a function-symbol and write $\frac{\partial h}{\partial V}$ or $\frac{\partial g}{\partial V}$; or (ii) we can write $\left(\frac{\partial p}{\partial V}\right)_T$ or $\left(\frac{\partial p}{\partial V}\right)_U$. In the latter case, we show the intended function by writing the remaining input variable of that function as a subscript. Thus $\left(\frac{\partial p}{\partial V}\right)_T = \frac{\partial h}{\partial T}$, and $\left(\frac{\partial p}{\partial V}\right)_U = \frac{\partial g}{\partial T}$.

We shall often use this parenthesis notation to clarify such ambiguities. For example, if w, u, and v are physical variables in some physical system or experiment, and if w is a function of the independent variables u and v, then $\left(\frac{\partial w}{\partial u}\right)_v$ denotes the partial derivative of w with respect to u for w as a function of u and v. We speak of this derivative as "the derivative of w with respect to u when v is held constant." Similarly, $\left(\frac{\partial v}{\partial y}\right)_{x,z}$ denotes the partial derivative of v with respect to y for v as a function of x, y, and z. We speak of this derivative as "the derivative of v with respect to y when x and z are held constant." The notation $\left(\frac{\partial w}{\partial u}\right)_v$ may also be written $\left(\frac{\partial}{\partial u}\right)_v w$. Similarly, the notation $\left(\frac{\partial w}{\partial y}\right)_{x,z}$ may also be written $\left(\frac{\partial}{\partial y}\right)_{x,z} w$.

Example 1. *Consider a hypothetical physical system with physical variables* w, u, v, x, *and* y, *where these variables always satisfy the equations*

$$w = u + v,$$
$$u = xy,$$
$$v = x + y.$$

Hence w must also always satisfy the equation

$$w = u + x + y = u + x + \frac{u}{x}.$$

Evidently, in this example, we have

$$\left(\frac{\partial w}{\partial u}\right)_v = 1,$$

but
$$\left(\frac{\partial w}{\partial u}\right)_x = 1 + \frac{1}{x}.$$

Derivatives as physical variables. It is usually possible to measure the value of a specified derivative of a physical variable by direct physical experiment (see below). Hence this derivative is itself a physical variable which may be related in various ways to other physical variables.

Example 2. We return to our container of gas. For a given equilibrium state, the derivative $\left(\frac{\partial p}{\partial T}\right)_V$ *can be measured by holding the volume constant, causing the temperature to change by* ΔT; *then finding the corresponding change in pressure* Δp, *and calculating* $\frac{\Delta p}{\Delta T}$. We can, in principle, achieve any desired degree of precision in this measurement by taking ΔT sufficiently small. Thus $\left(\frac{\partial p}{\partial T}\right)_V$ is a physical variable. Moreover, by (1.2) and (1.4) this physical variable can be viewed as a function of V and T or as a function of V and U. If we wish, we can treat it as a function of V and U, and go on to consider a new derivative such as $\left(\frac{\partial}{\partial V}\right)_U \left(\frac{\partial p}{\partial T}\right)_V$. This last derivative, in its turn, is a physical variable since it can be directly measured by an appropriate experimental procedure, as follows. Let α be the physical variable $\left(\frac{\partial p}{\partial T}\right)_V$. To measure $\left(\frac{\partial \alpha}{\partial V}\right)_U$, we cause a change ΔV in the volume, then add or remove enough heat to leave U unchanged (and thus compensate for any work done by or on the gas during the change ΔV.) Finally, by measuring α in the new state of the gas, we find the change $\Delta \alpha$. $\frac{\Delta \alpha}{\Delta V}$ is then an approximation for the desired derivative. As before, we can achieve any desired degree of accuracy by taking ΔV sufficiently small and making sufficiently accurate measurements of $\Delta \alpha$. (Derivatives often provide important physical variables. For example, the physical variable defined as $\beta = -\frac{\alpha}{p}$ is known to engineers as "the coefficient of change of pressure at constant volume.")

Remarks. (1) In example 2, we have assumed that observed values for $\frac{\Delta p}{\Delta T}$ converge to some limit as ΔT approaches 0 and V is held constant. We usually incorporate this assumption into a stronger and more general statement, saying, "we assume that the function p = h(V,T) is differentiable." If, in fact, the values of $\dfrac{\Delta p}{\Delta T}$ do not appear to approach a limit, we may question or reject this assumption of differentiability.

(2) In a parenthesis notation such as $\left(\dfrac{\partial v}{\partial y}\right)_{x,z}$, the variable whose derivative is being taken (v in this case) is called the *dependent physical variable,* and the remaining variables ($\{x,y,z\}$ in this case) are called *independent physical variables.* The parenthesis notation is used only when it is known or assumed that for every independent physical variable, it is possible to vary the value of that variable while the value of each of the other independent variables is held constant. In this case, we sometimes say that the given set of independent variables is *functionally independent.* For our gas system, the set $\{V,T\}$ is functionally independent, but the set $\{p,V,T\}$ is not, since by (1.2), p cannot be varied while V and T are held constant. In example 1 above, we have assumed that the sets $\{u,v\}$, $\{u,x\}$, $\{x,y\}$ are functionally independent. The set $\{u,x,y\}$, however, cannot be functionally independent, since u must remain constant if x and y are held constant (because $u = xy$).

§10.2 The Chain Rule in Multivariable Calculus

The *multivariable chain rule* is a powerful and widely used theorem. Its theoretical, computational, and practical uses extend across all of science and technology. The chain rule can be used (i) to differentiate a given multivariable function when that function is expressed as a combination of other multivariable functions, and (ii) to deduce numerical and algebraic relationships among the *derivatives* of given physical variables from initially given relationships among the variables themselves.

(2.1) **Theorem.** *Let* w, u, v, x, *and* y *be physical variables and let* f, g_1 *and* g_2 *be two-variable functions such that* $w = f(u,v)$, $u = g_1(x,y)$, *and* $v = g_2(x,y)$. *Define* h *by* $w = h(x,y) = f(g_1(x,y),g_2(x,y))$. *Let* $P = (a,b)$ *be a point in the domain of* h. *(Thus* $Q = (g_1(a,b),g_2(a,b))$ *must be a point in the domain of* f.) *Assume that* g_1 *and* g_2 *are differentiable at* P, *and that* f *is differentiable at* Q. *It follows that*

(i) h *is differentiable at* P,

and   (ii) $\left.\dfrac{\partial h}{\partial x}\right|_P = \left.\dfrac{\partial f}{\partial u}\right|_Q \left.\dfrac{\partial g_1}{\partial x}\right|_P + \left.\dfrac{\partial f}{\partial v}\right|_Q \left.\dfrac{\partial g_2}{\partial x}\right|_P$,

$\left.\dfrac{\partial h}{\partial y}\right|_P = \left.\dfrac{\partial f}{\partial u}\right|_Q \left.\dfrac{\partial g_1}{\partial y}\right|_P + \left.\dfrac{\partial f}{\partial v}\right|_Q \left.\dfrac{\partial g_2}{\partial y}\right|_P$.

Conclusion (ii) is sometimes expressed with g_1 written as u and g_2 written as v:

$$\frac{\partial h}{\partial x}\bigg|_P = \frac{\partial f}{\partial u}\bigg|_Q \frac{\partial u}{\partial x}\bigg|_P + \frac{\partial f}{\partial v}\bigg|_Q \frac{\partial v}{\partial x}\bigg|_P ,$$

and

$$\frac{\partial h}{\partial y}\bigg|_P = \frac{\partial f}{\partial u}\bigg|_Q \frac{\partial u}{\partial y}\bigg|_P + \frac{\partial f}{\partial v}\bigg|_Q \frac{\partial v}{\partial y}\bigg|_P .$$

Conclusion (ii) can also be written

(2.2) $$h_x(P) = f_u(Q)g_{1x}(P) + f_v(Q)g_{2x}(P)$$

and $$h_y(P) = f_u(Q)g_{1y}(P) + f_v(Q)g_{2y}(P) .$$

(2.1) asserts, in effect, that two simple arithmetic relationships (given in (2.2)) must hold among the eight real numbers:

$$h_x(P),\ h_y(P),\ f_u(Q),\ f_v(Q),\ g_{1x}(P),\ g_{1y}(P),\ g_{2x}(P),\ \text{and } g_{2y}(P).$$

(2.1) implies, as a corollary, that if f, g_1, and g_2 are differentiable *functions* (on their domains), then the same simple relationships must hold among the derivative *functions* (on those domains).

(2.3) **Corollary.** *Let* w, u, v, x, y, f, g_1, g_2, *and* h *be as in the theorem. If* f, g_1 *and* g_2 *are differentiable, then*
  (i)   h *is differentiable, and*
  (ii) $h_x(x,y) =$
     $$f_u\big(g_1(x,y),g_2(x,y)\big)g_{1x}(x,y) + f_v\big(g_1(x,y),g_2(x,y)\big)g_{2x}(x,y),$$
and $h_y(x,y) =$
     $$f_u\big(g_1(x,y),g_2(x,y)\big)g_{1y}(x,y) + f_v\big(g_1(x,y),g_2(x,y)\big)g_{2y}(x,y),$$
for all (x,y) *in the domain of* h.

Conclusion (ii) is sometimes written in the form

(2.4) $$\frac{\partial h}{\partial x} = \frac{\partial f}{\partial u}\frac{\partial g_1}{\partial x} + \frac{\partial f}{\partial v}\frac{\partial g_2}{\partial x}, \text{ and } \frac{\partial h}{\partial y} = \frac{\partial f}{\partial u}\frac{\partial g_1}{\partial y} + \frac{\partial f}{\partial v}\frac{\partial g_2}{\partial y} ,$$

or as $h_x = f_u g_{1x} + f_v g_{2x}$, and $h_y = f_u g_{1y} + f_v g_{2y} .$

Using physical variable notation instead of function notation, we can write (2.4) as:

(2.5)
$$\left(\frac{\partial w}{\partial x}\right)_y = \left(\frac{\partial w}{\partial u}\right)_v\left(\frac{\partial u}{\partial x}\right)_y + \left(\frac{\partial w}{\partial v}\right)_u\left(\frac{\partial v}{\partial x}\right)_y$$

and
$$\left(\frac{\partial w}{\partial y}\right)_x = \left(\frac{\partial w}{\partial u}\right)_v\left(\frac{\partial u}{\partial y}\right)_x + \left(\frac{\partial w}{\partial v}\right)_u\left(\frac{\partial v}{\partial y}\right)_x .$$

We sometimes suppress the parenthesis notation and simply write (2.5) as:

(2.6)
$$\frac{\partial w}{\partial x} = \frac{\partial w}{\partial u}\frac{\partial u}{\partial x} + \frac{\partial w}{\partial v}\frac{\partial v}{\partial x} ,$$

and
$$\frac{\partial w}{\partial y} = \frac{\partial w}{\partial u}\frac{\partial u}{\partial y} + \frac{\partial w}{\partial v}\frac{\partial v}{\partial y} .$$

We only do this, however, when the precise character of the derivatives has been made clear, so that we do not need the help of the parenthesis notation.

Example. Let $w = f(u,v) = u^2+v^2$, $u = g_1(x,y) = xy$, and $v = g_2(x,y) = x+y$. Then corollary (2.3) asserts, for instance, that:

$$\left(\frac{\partial w}{\partial x}\right)_y = (2u)(y) + (2v)(1) = 2xy^2 + 2x + 2y.$$

We can check this result by substituting first and then differentiating.
We get $\quad\quad\quad w = x^2y^2 + x^2 + 2xy + y^2,$
and $\quad\quad\quad\quad \left(\frac{\partial w}{\partial x}\right)_y = 2xy^2 + 2x + 2y .$

Note that the chain rule allows us to get specific numerical results as well as algebraic relationships.

Example (continued). What is the value of $\left(\frac{\partial w}{\partial x}\right)_y$ when $u = 1$ and $v = 2$? We have:

$$u = 1 \text{ and } v = 2 \Rightarrow xy = 1 \text{ and } x + y = 2 \Rightarrow x^2 - 2x + 1 = 0 \Rightarrow$$
$$x = 1 \text{ and } y = 1 \Rightarrow \left(\frac{\partial w}{\partial x}\right)_y = 2 + 2 + 2 = 6.$$

Remark. The mathematical content of §10.2 is contained in (2.1) and (2.3). The remainder of §10.2 introduced a variety of notations for the chain rule. These notations occur because the chain rule is widely used in a variety of forms and ways.

§10.3 Proof of the Chain Rule

In (2.1), we write the constant values $\dfrac{\partial f}{\partial u}\Big|_Q, \dfrac{\partial f}{\partial v}\Big|_Q, \dfrac{\partial g_1}{\partial x}\Big|_P, \dfrac{\partial g_1}{\partial y}\Big|_P, \dfrac{\partial g_2}{\partial x}\Big|_P,$

$\dfrac{\partial g_2}{\partial y}\Big|_P$ as $\bar{f}_u, \bar{f}_v, \bar{g}_{1x}, \bar{g}_{1y}, \bar{g}_{2x}, \bar{g}_{2y}$ respectively. For any point $P' = (x,y)$ in

the domain of h, let $\Delta x = x - a$ and $\Delta y = x - b$. Let $c = g_1(a,b)$ and $d = g_2(a,b)$. Let $\Delta g_1 = g_1(x,y) - c$ and $\Delta g_2 = g_2(x,y) - d$. By the differentiability of g_1 and g_2 at P, there must exist error coefficients $\varepsilon_1(x,y)$, $\varepsilon_2(x,y)$, $\varepsilon_3(x,y)$, and $\varepsilon_4(x,y)$ such that for every (x,y) in the domain of h:

$$\Delta g_1 = (\bar{g}_{1x} + \varepsilon_1(x,y))\Delta x + (\bar{g}_{1y} + \varepsilon_2(x,y))\Delta y,$$

and
$$\Delta g_2 = (\bar{g}_{2x} + \varepsilon_3(x,y))\Delta x + (\bar{g}_{2y} + \varepsilon_4(x,y))\Delta y,$$

where each of the functions $\varepsilon_1, \varepsilon_2, \varepsilon_3$ and ε_4 has limit 0 as $(x,y) \to (a,b)$. By the differentiability of f at Q, there must exist error coefficients $\varepsilon_5(u,v)$ and $\varepsilon_6(u,v)$ such that for every $(u,v) = (c+\Delta u, d+\Delta v)$ in the domain of f:

$$\Delta f = f(u,v) - f(c,d) = (\bar{f}_u + \varepsilon_5(u,v))\Delta u + (\bar{f}_v + \varepsilon_6(u,v))\Delta v$$

where ε_5 and ε_6 have limit 0 as $(u,v) \to (c,d)$.

Substituting Δg_1 and Δg_2 respectively for Δu and Δv in the expression for Δf, we get

$$\begin{aligned}
\Delta f &= \Delta h \\
&= (\bar{f}_u + \varepsilon_5(g_1(x,y), g_2(x,y)))[(\bar{g}_{1x} + \varepsilon_1(x,y))\Delta x + (\bar{g}_{1y} + \varepsilon_2(x,y))\Delta y] \\
&\quad + (\bar{f}_v + \varepsilon_6(g_1(x,y), g_2(x,y)))[(\bar{g}_{2x} + \varepsilon_3(x,y))\Delta x + (\bar{g}_{2y} + \varepsilon_4(x,y))\Delta y].
\end{aligned}$$

Here $\varepsilon_5(g_1(x,y), g_2(x,y))$ and $\varepsilon_6(g_1(x,y), g_2(x,y))$ have limit 0 as $(x,y) \to (a,b)$, as the reader may verify by combining funnel functions for the limits for ε_5 and ε_6 at (c,d) with funnel functions for the continuity of g_1 and g_2 at (a,b).

Multiplying out this last expression for Δh, we get

$$\Delta h = (\bar{f}_u \bar{g}_{1x} + \bar{f}_v \bar{g}_{2x})\Delta x + (\bar{f}_u \bar{g}_{1y} + \bar{f}_v \bar{g}_{2y})\Delta y + \varepsilon_1^*(x,y)\Delta x + \varepsilon_2^*(x,y)\Delta y,$$

where
$$\varepsilon_1^*(x,y) = \varepsilon_1 \bar{f}_u + \varepsilon_5 \bar{g}_{1x} + \varepsilon_1 \varepsilon_5 + \varepsilon_3 \bar{f}_v + \varepsilon_6 \bar{g}_{2x} + \varepsilon_3 \varepsilon_6,$$

and $\qquad \varepsilon_2^*(x,y) = \varepsilon_2 \bar{f}_u + \varepsilon_5 \bar{g}_{1y} + \varepsilon_2 \varepsilon_5 + \varepsilon_4 \bar{f}_v + \varepsilon_6 \bar{g}_{2y} + \varepsilon_4 \varepsilon_6,$

and where we have used the abbreviations $\varepsilon_1 = \varepsilon_1(x,y),\dots,\varepsilon_5 = \varepsilon_5(g_1(x,y),g_2(x,y))$, ... We see that $\displaystyle\lim_{(x,y)\to(a,b)} \varepsilon_1^*(x,y) = 0$ and $\displaystyle\lim_{(x,y)\to(a,b)} \varepsilon_2^*(x,y) = 0.$

Thus h has linear approximation at (a,b). Hence, by definition, (i) h is differentiable at P, and (ii) the first order derivatives of h at P must be

$$\left.\frac{\partial h}{\partial x}\right|_P = \bar{f}_u \bar{g}_{1x} + \bar{f}_v \bar{g}_{2x}$$

and

$$\left.\frac{\partial h}{\partial y}\right|_P = \bar{f}_u \bar{g}_{1y} + \bar{f}_v \bar{g}_{2y}.$$

This proves the theorem.

Remark. If we make the stronger assumption in (2.1) that f, g_1, and g_2 are in C^1, the proof of (2.1) becomes shorter and simpler. It follows directly, by (7.3) and (7.4) in §8.7, that (i) h is in C^1 and (ii) the chain rule formulas hold.

§10.4 Terminology for the Chain Rule

In using the chain rule as given in (2.1) and (2.3), we shall refer to the function f as the *primary function* and to the functions g_1 and g_2 as the *secondary functions*. Analogous statements and proofs of the chain rule can be given for other cases where the primary function is of one, two, or three variables, and the secondary functions are either all of one, all of two, or all of three variables. We give several illustrations.

Let $w = f(u,v,t)$, $u = g_1(x,y)$, $v = g_2(x,y)$, and $t = g_3(x,y)$. Then, for example,

$$\left(\frac{\partial w}{\partial x}\right)_y = \frac{\partial f}{\partial u}\frac{\partial g_1}{\partial x} + \frac{\partial f}{\partial v}\frac{\partial g_2}{\partial x} + \frac{\partial f}{\partial t}\frac{\partial g_3}{\partial x}.$$

Similarly, let $w = f(u,v)$, $u = g_1(x,y,z)$, and $v = g_2(x,y,z)$. Then, for example,

$$\left(\frac{\partial w}{\partial y}\right)_{x,z} = \frac{\partial f}{\partial u}\frac{\partial g_1}{\partial y} + \frac{\partial f}{\partial v}\frac{\partial g_2}{\partial y}.$$

In applying the chain rule, with w, u, v, x, and y as in (2.1), we speak of w as the *dependent* variable, of {u,v} as the set of *intermediate* variables, and of {x,y} as the set of *independent* or "final" variables. (Similarly for other cases of the chain rule.) It is sometimes useful to indicate these sets of variables explicitly by writing, for example:

$$w \rightarrow \{u,v\} \rightarrow \{x,y\} \qquad \text{(theorem (2.1)),}$$
$$w \rightarrow \{u,v,t\} \rightarrow \{x,y\} \qquad \text{(the next-to-last illustration),}$$
$$w \rightarrow \{u,v\} \rightarrow \{x,y,z\} \qquad \text{(the last illustration).}$$

When we apply the chain rule, it is important to be clear and explicit as to the full sets of intermediate and independent variables being used. Most mistakes with the chain rule are caused by vagueness or confusion concerning these sets.

[4–1] illustrates the chain rule for the case of $w \rightarrow \{u_1,u_2,u_3\} \rightarrow \{x_1,x_2,x_3\}$, with f as primary function and g_1, g_2, g_3 as secondary functions. To find $\frac{\partial w}{\partial x_2}$ ($= \left(\frac{\partial w}{\partial x_2}\right)_{x_1,x_3}$), we draw *all* possible paths to x_2 through the interme-

diate variables. Then with each path we associate a product of the form $\frac{\partial f}{\partial u_i} \frac{\partial g_i}{\partial x_2}$.

We get the final expression for $\frac{\partial w}{\partial x_2}$ as the sum of these products.

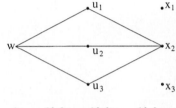

$$\frac{\partial w}{\partial x_2} = \frac{\partial f}{\partial u_1}\frac{\partial u_1}{\partial x_2} + \frac{\partial f}{\partial u_2}\frac{\partial u_2}{\partial x_2} + \frac{\partial f}{\partial u_3}\frac{\partial u_3}{\partial x_2}$$

[4–1]

When functions of one-variable appear in applications of the chain rule, one-variable notations are used for their derivatives. Thus, for example, in the case

$$w \rightarrow \{u,v\} \rightarrow \{x\} ,$$

the primary function is a two-variable function, and the secondary functions are one-variable functions. Here we write

$$\frac{dw}{dx} = \frac{\partial w}{\partial u}\frac{du}{dx} + \frac{\partial w}{\partial v}\frac{dv}{dx}.$$

Similarly, in the case

$$w \rightarrow \{u\} \rightarrow \{x,y,z\} ,$$

the primary function is a one-variable function, and the secondary function is a three-variable function. Hence we write

$$\frac{\partial w}{\partial x} = \frac{dw}{du}\frac{\partial u}{\partial x}, \quad \frac{\partial w}{\partial y} = \frac{dw}{du}\frac{\partial u}{\partial y}, \quad \text{and} \quad \frac{\partial w}{\partial z} = \frac{dw}{du}\frac{\partial u}{\partial z}.$$

Further variations occur when the same physical variable appears both as an intermediate and as an independent variable in an application of the chain rule. We consider these variations in §10.5.

When we apply the chain rule to given primary and secondary functions, the final expression that we obtain may include both intermediate and final variables. This does not lessen the usefulness of the expression. The rule always produces a valid numerical relationship between the value of the stated derivative and the values of other derivatives and variables. (Keep in mind that all of these derivatives and variables can be viewed as physical variables.)

Example. In the example in §10.2, we have $w = u^2 + v^2$, $u = xy$, and $v = x + y$. Application of the chain rule gave us, initially,

$$\left(\frac{\partial w}{\partial x}\right)_y = 2uy + 2v.$$

This expression is a correct relationship among the physical variables u, v, y, and $\left(\frac{\partial w}{\partial x}\right)_y$, even though it mixes the variables u and v (which we are treating as intermediate) with the variable y (which we are treating as independent).

§10.5 Formal Derivatives; Variant Forms

Formal derivatives.

Example 1. Let w, x, y, *and* u *be physical variables. Assume that these variables are interrelated by the functions* f *and* g, *where* $w = f(x,y)$ *and* $y = g(x,u)$ *and where we have formulas for the functions* f *and* g. *Then* w *can also be viewed as a function of the variables* x *and* u, *since, by substitution,* $w = f(x,g(x,u))$. *We wish to use the chain rule to find the derivatives* $\left(\frac{\partial w}{\partial x}\right)_u$ *and* $\left(\frac{\partial w}{\partial u}\right)_x$. We take

$$w \to \{x,y\} \to \{x,u\},$$

(That is to say, we take $\{x,y\}$ as intermediate variables and $\{x,u\}$ as independent variables.) As primary function, we have $w = f(x,y)$. As secondary functions, we have $x = h(x,u)$ and $y = g(x,u)$, where h is the two-variable function defined by

$$h(x,u) = x.$$

Thus $\frac{\partial h}{\partial x} = 1$ and $\frac{\partial h}{\partial u} = 0$. Applying the chain rule, we get

(5.1)
$$\left(\frac{\partial w}{\partial x}\right)_u = f_x h_x + f_y g_x = f_x \cdot 1 + f_y g_x$$
$$= f_x + f_y g_x,$$

and

(5.2)
$$\left(\frac{\partial w}{\partial u}\right)_x = f_x h_u + f_y g_u = f_x \cdot 0 + f_y g_u$$
$$= f_y g_u.$$

In a problem of this kind, the derivatives of the initially given functions are called *formal derivatives* because (i) we have expressions (or formulas) for these initially given functions, and (ii) we can use differentiation rules to calculate these derivatives directly. (In example 1, the formal derivatives are f_x, f_y, g_x, and g_u.)

Example 1 (*continued*). If we happen to have the following formulas for our initially given functions:

$$f(x,y) = x^2 y,$$
$$g(x,u) = u^2 + 2x,$$

then, for our formal derivatives, we have

$$f_x = 2xy, \quad f_y = x^2, \quad g_x = 2, \quad g_u = 2u.$$

Substituting in (5.1) and (5.2), we find

$$\left(\frac{\partial w}{\partial x}\right)_u = 2xy + 2x^2 \quad (= 2x(u^2+2x) + 2x^2 = 2xu^2 + 6x^2)$$

and
$$\left(\frac{\partial w}{\partial u}\right)_x = 2x^2 u.$$

In many chain-rule applications, our goal will be to express the derivatives of a newly defined function in terms of the formal derivatives of given functions. (In (5.1) and (5.2), we did exactly this.)

Variant forms. In the above example, the variable x occurs both as an intermediate variable and as a final variable. When the sets of intermediate and final variables have a member in common, we say that we have a *variant form* of the chain rule. It is easy to see that the proof of the chain rule remains valid for such variant forms.

Example 2. *We are given the physical variables* w, x, y, *and* z *related by the given functions*

$$w = f(x,y,z) \ \text{ and } \ z = h(x,y).$$

Hence w *can be viewed as*

$$w = f(x,y,h(x,y)).$$

We wish to find $\left(\frac{\partial w}{\partial x}\right)_y$ *and* $\left(\frac{\partial w}{\partial y}\right)_x$ *in terms of the formal derivatives of* f *and* h. We apply the chain rule for

$$w \rightarrow \{x,y,z\} \rightarrow \{x,y\}.$$

The secondary functions are $x = x(x,y) = x$, $y = y(x,y) = y$, and $z = z(x,y) = h(x,y)$. As in the previous example, we now have

$$\left(\frac{\partial w}{\partial x}\right)_y = f_x \cdot 1 + f_y \cdot 0 + f_z h_x$$
$$= f_x + f_z h_x \, ;$$

and
$$\left(\frac{\partial w}{\partial y}\right)_x = f_x \cdot 0 + f_y \cdot 1 + f_z h_y$$
$$= f_y + f_z h_y \, .$$

Example 3. *Let* $w = f(x,y,z) = xyz$ *and* $y = g(x,z) = e^{xz}$. *Use the chain rule to find the value of* $\left(\frac{\partial w}{\partial z}\right)_x$ *when* $x = 1$ *and* $z = 1$. We apply the chain rule for $w \rightarrow \{x,y,z\} \rightarrow \{x,z\}$. As in the previous example, we have

$$\left(\frac{\partial w}{\partial z}\right)_x = f_x \cdot 0 + f_y g_z + f_z \cdot 1 = f_y g_z + f_z$$
$$= xz(xe^{xz}) + xy \, .$$

Since $(x = 1$ and $z = 1) \Rightarrow y = e$, we have

$$\left(\frac{\partial w}{\partial z}\right)_x = e + e = 2e.$$

§10.6 Higher Derivatives by the Chain Rule

The chain rule can also be used to find expressions for *higher-order* derivatives in terms of formal derivatives of given functions. Calculations are straightforward, though sometimes tedious. (In these examples, we assume existence and continuity of higher order derivatives as needed.)

Example 1. Let $w = f(x)$ *and* $x = g(u,v)$. *Consider* w *as a function* $w = h(u,v)$ *of* u *and* v. *Then* $h(u,v) = f(g(u,v))$. *(Thus we have* $w \to \{x\} \to \{u,v\}$.) *Find* $\dfrac{\partial^2 h}{\partial u^2}$ *in terms of the formal derivatives of* f *and* g.

By a first application of the chain rule, we have

$$\frac{\partial h}{\partial u} = f_x g_u .$$

Then, by the product rule, we have

$$\frac{\partial^2 h}{\partial u^2} = \frac{\partial}{\partial u}(f_x g_u) = \left(\frac{\partial}{\partial u} f_x\right) g_u + f_x \left(\frac{\partial}{\partial u} g_u\right),$$

where, in each occurrence, $\dfrac{\partial}{\partial u}$ means $\left(\dfrac{\partial}{\partial u}\right)_v$.

By a second application of the chain rule, we get $\dfrac{\partial}{\partial u} f_x = f_{xx} g_u$, and we finally have

$$\frac{\partial^2 h}{\partial u^2} = f_{xx} g_u^2 + f_x g_{uu} .$$

Example 2. We check the result of example 1 for the special case $f(x) = x^3$ *and* $x = g(u,v) = u^2 v^2$.

If we first substitute and then differentiate, we get

$$h(u, v) = u^6 v^6 \quad \text{and} \quad \frac{\partial^2 h}{\partial u^2} = 30u^4 v^6.$$

If, instead, we apply the result of example 1 and then substitute at the end, we get

$$f_{xx} = 6x, \ g_u = 2uv^2, \ f_x = 3x^2, \text{ and } g_{uu} = 2v^2.$$

Hence
$$\frac{\partial^2 h}{\partial u^2} = 6x(2uv^2)^2 + 3x^2 2v^2$$
$$= 6u^2 v^2 (4u^2 v^4) + 3u^4 v^4 (2v^2)$$
$$= 24u^4 v^6 + 6u^4 v^6 = 30u^4 v^6.$$

Example 3. *Let* $w = f(x,y)$ *and* $y = g(x,t)$. *Consider* w *as a function* $w = h(x,t)$ *of* x *and* t. *(Thus we have* $w \to \{x,y\} \to \{x,t\}$.) *Find* $\dfrac{\partial^2 h}{\partial x \partial t}$ *in terms of the formal derivatives of* f *and* g.

By a first application of the chain rule (with x and t as independent variables), we have

$$\frac{\partial h}{\partial t} = f_x \cdot 0 + f_y g_t = f_y g_t.$$

By the product rule, we have

$$\frac{\partial^2 h}{\partial x \partial t} = \frac{\partial}{\partial x}(f_y g_t) = (\frac{\partial}{\partial x} f_y) g_t + f_y (\frac{\partial}{\partial x} g_t)$$

where $\dfrac{\partial}{\partial x}$ means $\left(\dfrac{\partial}{\partial x}\right)_t$.

By a second application of the chain rule to find $\dfrac{\partial}{\partial x} f_y$ (again with x and t as independent variables), we have

$$\frac{\partial^2 h}{\partial x \partial t} = (f_{yx} \cdot 1 + f_{yy} g_x) g_t + f_y g_{tx}$$
$$= f_{yx} g_t + f_{yy} g_x g_t + f_y g_{tx}.$$

Remark. Chapter 11 will present a formal algorithm, known as the *elimination method*, for applying the chain rule. This algorithm provides a simpler and more routine approach to the examples given above in §10.5 and §10.6.

§10.7 Problems

§10.2

1. Let $w = f(u)$ and $u = g(x,y,z)$. Find $\left(\frac{\partial w}{\partial z}\right)_{x,y}$ in terms of derivatives of f and g.

2. Let $q = f(x,y,z)$, $x = x(u,v,w)$, $y = y(u,v,w)$, and $z = z(u,v,w)$. Find $\left(\frac{\partial q}{\partial v}\right)_{u,w}$ in terms of derivatives of f, x, y, and z.

3. Let $w = xy$, $x = u^2 + v^2$, and $y = uv$.
   (a) Do not substitute for x and y. Instead, use the chain rule to find expressions for $\left(\frac{\partial w}{\partial u}\right)_v$ and $\left(\frac{\partial w}{\partial v}\right)_u$ in terms of the physical variables x, y, u, and v.
   (b) Find the values of x, y, and w when $u = 1$ and $v = 2$.
   (c) From (a), find the value of $\left(\frac{\partial w}{\partial u}\right)_v$ and $\left(\frac{\partial w}{\partial v}\right)_u$ when $u = 1$ and $v = 2$.

4. Let $z = x^2 + y^2$, $x = uvw$, and $y = uv + vw + wu$.
   (a) Use the chain rule to find an expression for $\left(\frac{\partial z}{\partial u}\right)_{v,w}$ in terms of u, v, w, x, and y.
   (b) Find the value of $\left(\frac{\partial z}{\partial u}\right)_{v,w}$ when $u = 1$, $v = 2$, and $w = 3$.

5. Let $w = x^2 + xy + y^2$, $x = v$, and $y = \sin(uv)$. Use the chain rule to find the value of $\left(\frac{\partial w}{\partial v}\right)_u$ when $u = \pi$ and $v = \frac{1}{2}$.

6. Let $w = e^{2x+3y} \cos 4z$, $x = \ln t$, $y = \ln(t^2+1)$, and $z = t$.
   (a) Use the chain rule to find an expression for $\frac{dw}{dt}$.
   (b) Find the values of x, y, z, and w when $t = \pi$.

7. If $w = f(u) + g(v)$, $u = x + ct$, $v = x - ct$, and f and y have continuous second derivatives, show that
$$\frac{\partial^2 w}{\partial t^2} = c^2 \frac{\partial^2 w}{\partial x^2}.$$

8. Let $w = u^2 + v^2$, $u = x + y$, and $v = xy$. Find the values of $\left(\frac{\partial w}{\partial x}\right)_y$ and $\left(\frac{\partial w}{\partial y}\right)_x$ for $(x,y) = (1,1)$, and for $(x,y) = (-2,0)$.

9. Let $w = f(u,v)$, $u = g(x,y)$, and $v = h(x,y)$. You are given the following information:

$g(1,2) = 3$	$g_x(1,2) = 2$
$h(1,2) = 5$	$g_y(1,2) = -2$
$f_u(3,5) = 12$	$h_x(1,2) = -3$
$f_v(3,5) = -1$	$h_y(1,2) = 7$

For each of the following, state whether the requested value can be found from the given information, and, if it can be found, find it
   (a) $f(3,5)$
   (b) $\left(\frac{\partial w}{\partial x}\right)_y$ for $x = 1$, $y = 2$.
   (c) $\left(\frac{\partial w}{\partial y}\right)_x$ for $x = 1$, $y = 2$.

10. Let $w = f(u,v) = u^2 + v^2$, $u = x + y$, and $v = x - y$. Calculate $\left(\frac{\partial w}{\partial x}\right)_y$:
   (i) by using the chain rule.
   (ii) by first substituting for u and v in f and then calculating the partial derivative directly.
Show that these two calculations lead to the same result.

11. Let $w = uv$, $u = \frac{x}{y}$, $v = x + y$.
   (a) Give expressions for w in terms of:
$$\{x,y\}, \{u,y\}, \{u,x\}, \{v,x\}, \text{ and } \{v,y\}.$$
   (b) Find $\left(\frac{\partial w}{\partial y}\right)_u$ and $\left(\frac{\partial w}{\partial y}\right)_v$.

12. Let $w = f(x,y,z)$, and let $y = g(x,z)$. Then w can be considered as a function of x and z. Find $\left(\frac{\partial w}{\partial x}\right)_z$ and $\left(\frac{\partial w}{\partial z}\right)_x$ in terms of the first-order derivatives of f and g.

§10.3

1. Assuming that f, g_1, and g_2 are in C^1, give a shorter and simpler proof, as suggested in the final remark in §10.3.

2. Outline a proof that if h is a function obtained by composition from differentiable functions, then h must be differentiable.

§10.5

1. The temperature T at a point (x,y,z) in space at time t is given by the function $T(x,y,z,t)$. A moving particle P follows a path given by the position vector $\vec{R}(t) = x(t)\hat{i} + y(t)\hat{j} + z(t)\hat{k}$. Find an expression, in terms of the derivatives of the functions T, x, y, and z, for the rate at which the temperature experienced by P is increasing.

2. We are given $w = f(u,v)$, $v = g(u,x)$, and $u = h(x)$. Find $\frac{dw}{dx}$ in terms of formal derivatives of f, g, and h.

3. We are given $w = f(x,t)$, $x = g(u,v)$, and $v = h(u,t)$. Find $\left(\frac{\partial w}{\partial t}\right)_u$ and $\left(\frac{\partial w}{\partial u}\right)_t$ in terms of the formal derivatives of f, g, and h.

4. We are given $w = f(x,y,z)$, $x = g(y,z,t)$, and $y = h(z,t)$. Use the chain rule to find $\left(\frac{\partial w}{\partial z}\right)_t$ and $\left(\frac{\partial w}{\partial t}\right)_z$ in terms of the formal derivatives of f, g, and h.

5. Let $w = xyz^2$, $x = \frac{yz^2}{t}$, $y = zt^2$. When $z = 1$ and $t = 2$, find the values of:
   (a) w, x, and y.
   (b) $\left(\frac{\partial w}{\partial z}\right)_t$.
   (c) $\left(\frac{\partial w}{\partial t}\right)_z$.

6. Let $w = u^2 + v^2$, $v = ux$, and $u = 2-x$. Use the chain rule to find the value of $\frac{dw}{dx}$ when $x = 0$.

7. (a) Given $w = f(u,y)$ and $u = g(x,y)$, express $\left(\frac{\partial w}{\partial y}\right)_x$ in terms of the formal derivatives f_u, f_y, g_x, and g_y.

   (b) Given f and g as in (a), express $\left(\frac{\partial w}{\partial x}\right)_y$ in terms of the formal derivatives f_u, f_y, g_x, and g_y.

8. You are given $w = M(x,y)$, $x = f(u,v)$, $y = g(u,v)$, $u = h(t)$, and $v = k(t)$.
   (a) Find $\frac{dx}{dt}$ and $\frac{dy}{dt}$ in terms of the derivatives of f, g, h, and k.
   (b) Find $\frac{dw}{dt}$ in terms of the derivatives of M, f, g, h, and k.
   (c) Let $M(x,y) = x^3 + y$, $f(u,v) = 2u^2 - v^2$, $g(u,v) = u + 2v$, $u = t^2$, $v = t^3$. Find $\frac{dw}{dt}$ for $t = 1$.

9. Given $w = f(u,v)$, $u = xy$, $v = \frac{x}{y}$; express the product $\left(\frac{\partial w}{\partial x}\right)_y \left(\frac{\partial w}{\partial y}\right)_x$ in terms of $\frac{\partial f}{\partial u}$, $\frac{\partial f}{\partial v}$, u and v.

10. You are given $w = f(x,y,z)$ and $z = g(x,y)$.
    (a) Express $\left(\frac{\partial w}{\partial x}\right)_y$ in terms of the formal partial derivatives of f and g.
    (b) Check your result by applying it to the specific case $w = xy + yz + zx$ and $z = xy$.

§10.6

1. Let $w = f(u,v)$, $u = g(x,y)$, and $v = h(x,y)$.

   (a) Using the chain rule several times, find an expression for $\dfrac{\partial^2 w}{\partial x^2}$ in terms of the formal derivatives of f, g, and h.

   (b) Do the same for $\dfrac{\partial^2 w}{\partial x \partial y}$.

2. Let $w = \sin x + \cos y$ and $y = \ln(x+t)$. Use the chain rule to find expressions for the following in terms of x, y, and t.

   (a) $\left(\dfrac{\partial w}{\partial x}\right)_t$.

   (b) $\left(\dfrac{\partial}{\partial x}\right)_t \left(\dfrac{\partial w}{\partial x}\right)_t$.

   (c) $\left(\dfrac{\partial}{\partial t}\right)_x \left(\dfrac{\partial w}{\partial x}\right)_t$.

   (d) $\left(\dfrac{\partial}{\partial x}\right)_t \left(\dfrac{\partial}{\partial x}\right)_t \left(\dfrac{\partial w}{\partial x}\right)_t$

   (e) Using equality of cross derivatives, check that the result of (c) agrees with the final formula in example 3 in §10.6.

3. In problem 2 of §10.5, find $\dfrac{d^2 w}{dx^2}$ in terms of the formal derivatives of f, g, and h.

4. In each of the following cases, express the desired derivatives in terms of the formal derivatives of f, g, and h.

   (a) Let $w = f(x,u,v)$, $u = g(x,y)$, $v = h(x,y)$. Find $\left(\dfrac{\partial w}{\partial x}\right)_y$ and $\left(\dfrac{\partial w}{\partial y}\right)_x$.

   (b) Let $w = f(u,v)$, $u = g(t)$, $v = h(t)$. Find $\dfrac{d^2 w}{dt^2}$.

5. Let $w = f(u,v)$ and $u = g(v)$. Using the chain rule, find an expression for $\dfrac{d^2 w}{dv^2}$ in terms of formal derivatives of f and g.

Chapter 11

Using the Chain Rule

§11.1 Differential Notation

Let $f(x,y)$ be a differentiable function of x and y. By §9.2, the linear approximation formula for f is

$$(1.1) \qquad \Delta f_{app} = f_x(x,y)\Delta x + f_y(x,y)\Delta y.$$

If we replace Δx and Δy by the notations dx and dy, we have

$$(1.2) \qquad \Delta f_{app} = f_x(x,y)dx + f_y(x,y)dy.$$

The expression (1.2) is called the *differential* of f. It is simply the linear approximation formula for f, with different symbols for the increments Δx and Δy. The symbols dx and dy are known as *differential increments*.

Similarly, let $g(x,y,z)$ be a differentiable function of x, y, and z. Then, from the linear approximation formula for g, we get the *differential* of g

$$(1.3) \qquad \Delta g_{app} = g_x dx + g_y dy + g_z dz.$$

The notation Δf_{app} is sometimes written df. Thus from (1.2) and (1.3), we have

and
$$df = f_x dx + f_y dy$$
$$dg = g_x dx + g_y dy + g_z dz \, .$$

Examples.
(1) $f(x,y) = x^2 + xy \Rightarrow df = (2x+y)dx + xdy$.
(2) $g(x,y,z) = e^x\cos yz \Rightarrow dg = e^x\cos yz\, dx - ze^x\sin yz\, dy - ye^x\sin yz\, dz$.
(3) $h(x,y,z) = z \Rightarrow dh = dz$.

Let an equation between two functions be given; for example, $f(x,u) = g(v,x)$. If we take the differentials of the two sides of this equation and set these differentials equal, this new equation is called the *differential* of the given equation. Thus the differential of

$$f(x,u) = g(v,x)$$

will be

$$f_x dx + f_u du = g_v dv + g_x dx.$$

Examples.
(4)   The differential of

$$xy = u^2 + 2y \text{ is}$$
$$ydx + xdy = 2udu + 2dy.$$

(5)   The differential of

$$w = x^2 + y^2 \text{ is}$$
$$dw = 2xdx + 2ydy\,.$$

The operation of forming the differential of a function, or of an equation, is a strictly formal, symbolic process. All the variables which appear in a given function or equation are treated in the same way, even though we may possess additional and separate information about their interdependence.

Example 6. We are given the two equations

$$w = x^2 + y^2,$$
$$y = e^x\,.$$

We then get the two differentials

$$dw = 2xdx + 2ydy,$$
$$dy = e^x dx\,.$$

In §11.2, we give a simple formal algorithm for applying the chain rule. The individual steps in this algorithm use differentials. A proof that this algorithm yields correct results appears in §11.7.

§11.2 The Elimination Method

We now give a general algorithm for carrying out applications of the chain rule. This algorithm uses differentials and is known as the *elimination method*. In this section, we give, with examples, a purely formal and mechanical description of the elimination method.

Assume the following: (i) that we have physical variables related by one or more equations where each side of each equation is an expression for a differentiable function of zero or more of these physical variables; (ii) that we have designated one physical variable as *dependent* and some of the other physical variables as *independent* (there may also remain some *undesignated* variables); and (iii) that each numerical assignment of values to the *independent* variables (from some specified domain of possible "input" values for the independent variables), determines (in a sense made precise at the end of §11.4) a corresponding assignment of "output" values (not necessarily unique) to the *dependent* and *undesignated* variables such that these assignments satisfy all the equations in (i). Under these assumptions, the *elimination method* enables us to find expressions for the first-order derivatives of the *dependent* variable, with respect to the various *independent* variables, by the following three steps.

Step 1. We take the differential of each given equation. The new equations will be linear equations in the differential increments of the various physical variables, where the coefficient of each differential increment will be an expression in zero or more of the physical variables.

Example 1. *Let the given equations be*

$$\left. \begin{array}{l} w = f(x,y,z) \\ z = h(x,y) \end{array} \right\}.$$

We wish to find $\left(\frac{\partial w}{\partial x}\right)_y$ *and* $\left(\frac{\partial w}{\partial y}\right)_x$ *in terms of the formal derivatives of* f *and* h. Let w be the dependent variable, and let x and y be the independent variables; then z remains as the undesignated variable. Step 1 yields

(2.1) $dw = f_x dx + f_y dy + f_z dz,$

(2.2) $dz = h_x dx + h_y dy .$

Step 2. By algebraic calculation, we eliminate the increments of the undesignated variables and express the dependent increment as a linear combination of independent increments. (If Step 2 cannot be carried out, this implies that assumption (iii) above is incorrect.) Each of the coefficients in this linear combination will be an expression in zero or more of the physical variables (including dependent, undesignated, and independent variables).

Example 1 (continued). Applying Step 2, we use (2.2) to substitute for dz in (2.1). We get

$$dw = f_x dx + f_y dy + f_z(h_x dx + h_y dy)$$
$$= (f_x + f_z h_x)dx + (f_y + f_z h_y)dy .$$

Step 3. The coefficient of each independent increment will then be the corresponding first-order derivative of the dependent variable with respect to the variable of that independent increment.

Example 1 (continued). We have

$$\left(\frac{\partial w}{\partial x}\right)_y = f_x + f_z h_x ,$$
$$\left(\frac{\partial w}{\partial y}\right)_x = f_y + f_z h_y .$$

(This result appears earlier as example 2 in §10.5.)

This method is called the *elimination method*, because the algebraic calculation in Step 2 "eliminates" the undesignated increments.

We further illustrate the method by applying it to several other examples from §10.5.

Example 2. Given $w = f(x,y)$ *and* $y = g(x,u),$ *find* $\left(\frac{\partial w}{\partial x}\right)_u$ *and* $\left(\frac{\partial w}{\partial u}\right)_x$ *in terms of the formal derivatives of* f *and* g. (This is example 1 in §10.5.) We designate w as dependent, and x and u as independent. We leave y undesignated. Then:

Step 1. Taking differentials, we have

$$dw = f_x dx + f_y dy,$$
$$dy = g_x dx + g_u du .$$

Step 2.   Eliminating dy, we get

$$dw = (f_x + f_y g_x)dx + f_y g_u du .$$

Step 3.   Hence $\left(\frac{\partial w}{\partial x}\right)_u = f_x + f_y g_x$, and $\left(\frac{\partial w}{\partial u}\right)_x = f_y g_u.$

Example 3. *Given* $w = xyz$ *and* $y = e^{xz}$, *find* $\left(\frac{\partial w}{\partial z}\right)_x$ *and* $\left(\frac{\partial w}{\partial x}\right)_z$ *when* $x = 1$ *and* $z = 1$. (This is example 3 in §10.5.) We designate w as dependent and x and z as independent. We leave y undesignated.

Step 1. $dw = yzdx + xzdy + xydz$
$$\qquad\qquad = edx + dy + edz \quad \text{(since } [x = 1 \text{ and } z = 1] \Rightarrow y = e),$$
$$\qquad dy = ze^{xz}dx + xe^{xz}dz$$
$$\qquad\qquad = edx + edz.$$

Step 2.   Eliminating dy, we have

$$dw = edx + edx + edz + edz$$
$$\qquad = 2edx + 2edz .$$

Step 3.   Hence $\left(\frac{\partial w}{\partial x}\right)_z = 2e$ and $\left(\frac{\partial w}{\partial z}\right)_x = 2e$ when $x = 1$ and $z = 1$.

Simplifying Step 2. If there is an independent variable for which we do *not* happen to seek a partial derivative of the dependent variable, then the differential increment of that independent variable can be set equal to zero in *Step 2* of the elimination method. Thus in example 2, if we only seek $\left(\frac{\partial w}{\partial u}\right)_x$, we have

Step 1. $dw = f_x dx + f_y dy,$
$$\qquad dy = g_x dx + g_u du .$$

Step 2.   Setting dx = 0, we get

$$dw = f_y dy,$$
$$dy = g_u du .$$

We then have $dw = f_y g_u du .$

Step 3. Hence $\left(\frac{\partial w}{\partial u}\right)_x = f_y g_u$.

Remarks. (1) In each of the examples above, the elimination method obtains, by means of a brief, formal calculation, the same result as is obtained in §10.5 by a direct application of the chain rule. Furthermore, the original statements of the chain rule itself are immediately given by the elimination method. Thus, for (2.1) in §10.2, we begin with

$$w = f(u,v),$$
$$u = g_1(x,y),$$
$$v = g_2(x,y).$$

where w is dependent; x and y are independent, and u and v are undesignated. Applying the elimination method, we have

Step 1.
$$dw = f_u du + f_v dv$$
$$du = g_{1x}x + g_{1y}dy$$
$$dv = g_{2x}dx + g_{2y}dy$$

Step 2. $\quad dw = (f_u g_{1x} + f_v g_{2x})dx + (f_u g_{1y} + f_v g_{2y})dy.$

Step 3.
$$\frac{\partial w}{\partial x} = f_u g_{1x} + f_v g_{2x} = \frac{\partial w}{\partial u}\frac{\partial u}{\partial x} + \frac{\partial w}{\partial v}\frac{\partial v}{\partial x},$$
$$\frac{\partial w}{\partial y} = f_u g_{1y} + f_v g_{2y} = \frac{\partial w}{\partial u}\frac{\partial u}{\partial y} + \frac{\partial w}{\partial v}\frac{\partial v}{\partial y}.$$

(2) The phrases "dependent variable" and "independent variable" have the same meaning, in applications of the elimination method, that they have in applications of the chain rule. "Undesignated variables," however, refers to those variables which, in a direct application of the chain rule (as in §10.5), would be *intermediate but not independent.*

§11.3 Implicit Differentiation in Elementary Calculus

Implicitly defined functions are briefly described in comment (3) in §8.4. We consider them now in more detail.

Definitions. In elementary calculus, let φ be a one-variable function on domain D. An equation of the form $y = \mathcal{F}[x]$ (where $\mathcal{F}[x]$ is an algebraic formula in the variable x) is called an *explicit definition* for φ if, for all a in D, $\mathcal{F}[a] = \varphi(a)$.

(Here, as in §4.1, $\mathcal{F}[a]$ is the real number obtained by evaluating $\mathcal{F}[x]$ with the real number a substituted for x.)

In elementary calculus, we say that a continuous one-variable function $y = \varphi(x)$ on domain D is *implicitly defined* by an equation of the form $\mathcal{F}[x,y] = 0$ (where $\mathcal{F}[x,y]$ is an algebraic formula in the variables x and y) if (i) for every a in D, $\mathcal{F}[a,\varphi(a)] = 0$, and (ii) for every a in D, $\mathcal{F}[a,b] \neq 0$ for values of b which are nearby to $\varphi(a)$ but different from $\varphi(a)$. (Condition (ii) is made more precise at the end of this section.)

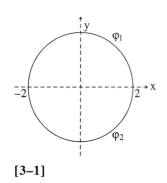

[3–1]

Example 1. The equation $x^2+y^2-4 = 0$ *implicitly defines two continuous one-variable functions,* $y = \varphi_1(x)$ *and* $y = \varphi_2(x)$, *whose graphs are as in* [3–1]. These functions are also given by the explicit definitions: $y = \varphi_1(x) = (4-x^2)^{1/2}$ and $y = \varphi_2(x) = -(4-x^2)^{1/2}$.

Example 2. The equation $y^3-y-x = 0$ *implicitly defines three continuous one-variable functions,* $y = \varphi_1(x)$, $y = \varphi_2(x)$, *and* $y = \varphi_3(x)$. See [3–2] for the graphs of these functions. There are no simple explicit definitions for these three functions.

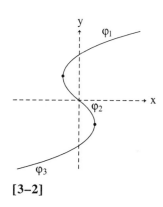

[3–2]

In elementary calculus, we can find an *explicit* expression for the derivative of an *implicitly* defined function by using the elementary calculus chain rule. This process is called *implicit differentiation.*

Example 1 (*continued*). In the case of $x^2+y^2-4 = 0$, we differentiate this equation to get

$$2x + 2y\frac{dy}{dx} = 0 \quad \text{(by the chain rule of elementary calculus),}$$

and this gives

$$\frac{dy}{dx} = -\frac{x}{y}$$

Example 2 (*continued*). For $y^3-y-x = 0$, this process gives

$$3y^2\frac{dy}{dx} - \frac{dy}{dx} - 1 = 0 \quad \text{(by the chain rule of elementary calculus),}$$

and we get

$$\frac{dy}{dx} = \frac{1}{3y^2-1}.$$

In both examples, the dependent variable y appears in the expression for $\frac{dy}{dx}$. Furthermore, for each implicitly defined function φ whose domain contains c, and for each corresponding value of y = φ(c), the expression for $\frac{dy}{dx}$ gives us the correct value of $\frac{dy}{dx}\bigg|_c$.

Example 2 (*continued*). In the case $y^3 - y - x = 0$, let c = 0. Then c lies in the domain of each of the three implicitly defined functions (see the figure above).

For $φ_1$, c = 0 ⇒ y = $φ_1(c)$ = 1, and $\dfrac{dφ_1}{dx}\bigg|_c = \dfrac{1}{3y^2 - 1} = \dfrac{1}{2}$.

For $φ_2$, c = 0 ⇒ y = $φ_2(c)$ = 0, and $\dfrac{dφ_2}{dx}\bigg|_c = \dfrac{1}{3y^2 - 1} = -1$.

For $φ_3$, c = 0 ⇒ y = $φ_3(c)$ = -1, and $\dfrac{dφ_3}{dx}\bigg|_c = \dfrac{1}{3y^2 - 1} = \dfrac{1}{2}$.

These examples illustrate the special advantage of implicit differentiation in elementary calculus: *even though we may have no explicit definition for an implicitly defined function y = φ(x), we can always find an explicit expression for the derivative $\frac{dφ}{dx}$.*

Remark. Condition (ii) in the definition of *implicit definability* (at the beginning of this section) can be made precise as follows: let y_0 be a real number and let r > 0. Then the *deleted neighborhood at* y_0 *of radius* r is the set of all real numbers b such that $0 < |b - y_0| \le r$. Condition (ii) asserts that for every a in D, there must be some r > 0 and some deleted neighborhood at φ(a) of radius r such that $\mathscr{F}[a,b] \ne 0$ for all values of b in that deleted neighborhood.

§11.4 Implicit Differentiation in Multivariable Calculus

Implicitly defined functions also occur in multivariable calculus, and we can use the multivariable chain rule to find explicit expressions for the derivatives of implicitly defined functions.

Definition. We say that a continuous two-variable function φ on domain D is *implicitly defined* by an equation of the form $\mathcal{F}[x,y,z] = 0$, where $\mathcal{F}[x,y,z]$ is an algebraic formula in the variables x, y, and z, if (i) for every (a,b) in D, $\mathcal{F}[a,b,\varphi(a,b)] = 0$, and (ii) for every (a,b) in D, $\mathcal{F}[a,b,c] \neq 0$ for values of c nearby to $\varphi(a,b)$ but different from $\varphi(a,b)$. ((ii) can be made fully precise as in the *remark* at the end of §11.3. We sometimes abbreviate conditions (i) and (ii) into the single statement: for every (a,b) in D, $z = \varphi(a,b)$ is an *isolated solution* to the equation $\mathcal{F}[a,b,z] = 0$.)

Notation. Since an algebraic formula of the form $\mathcal{F}[x,y,z]$ explicitly defines a function F such that for all a,b,c, $F(a,b,c) = \mathcal{F}[a,b,c]$, we shall henceforth write $\mathcal{F}[x,y,z] = 0$ as F(x,y,z) = 0. It is customary in multivariable calculus to use functions in this way to represent algebraic formulas.

When φ is implicitly defined by F(x,y,z) = 0, the graph of φ must occur as part of the level surface of F for 0. An equation of the form G(x,y,z) = H(x,y,z) can always be written in the standard form F(x,y,z) = 0, by taking F = G – H. Thus we may also speak of a function φ being *implicitly defined* by an equation of the form G(x,y,z) = H(x,y,z).

Example 1. *Let* $F(x,y,z) = x^2+y^2+z^2-1$. *Then the equation* F(x,y,z) = 0 *implicitly defines two continuous functions on the domain* $x^2+y^2 \leq 1$. *These functions also have the explicit definitions:* $z = \sqrt{1 - x^2 - y^2}$ *and* $z = -\sqrt{1 - x^2 - y^2}$.

Let F be a differentiable function of three variables. Given the equation F(x,y,z) = 0, we can ask two basic questions:

(I) Does the equation F(x,y,z) = 0 implicitly define one or more continuous functions $z = \varphi(x,y)$, and if so, what are their largest possible domains?

(II) If the equation F(x,y,z) = 0 implicitly defines a continuous function $z = \varphi(x,y)$, can we find *explicit* expressions for the first-order derivatives of φ in terms of the formal derivatives of F?

In the present section, we consider question (II). We return to question (I) in §11.7. In multivariable calculus, the process of calculating the derivatives of φ from the derivatives of F is known as *implicit differentiation*.

We can supply a positive answer to question (II) in two different ways: (a) we can make a direct application of the chain rule, or (b) we can use the elimination method. (a) gives better theoretical understanding; (b) gives a shorter and simpler calculation. We show both ways, doing (a) first and then (b).

(a) To apply the *chain rule* directly, we take w as dependent variable, $\{x,y,z\}$ as intermediate variables, $\{x,y\}$ as independent variables, $F(x,y,z)$ as the primary function, and $x(x,y) = x$, $y(x,y) = y$, and $z(x,y) = \varphi(x,y)$ as the secondary functions, where φ is the implicitly defined function whose derivatives we wish to find. As a function of the independent variables $\{x,y\}$, w must be the constant function $w = F(x,y,\varphi(x,y)) = 0$. Hence we must have $\left(\frac{\partial w}{\partial x}\right)_y = 0$ and $\left(\frac{\partial w}{\partial y}\right)_x = 0$. By the chain rule, this gives

$$0 = \left(\frac{\partial w}{\partial x}\right)_y = F_x \cdot 1 + F_y \cdot 0 + F_z \frac{\partial \varphi}{\partial x}.$$

(4.1)   Hence
$$\frac{\partial \varphi}{\partial x} = -\frac{F_x}{F_z}.$$

Similarly,
$$0 = \left(\frac{\partial w}{\partial y}\right)_x = F_x \cdot 0 + F_y \cdot 1 + F_z \frac{\partial \varphi}{\partial y}.$$

(4.2)   Hence
$$\frac{\partial \varphi}{\partial y} = -\frac{F_y}{F_z}.$$

These expressions for $\frac{\partial \varphi}{\partial x}$ and $\frac{\partial \varphi}{\partial y}$ give us our desired derivatives for all x, y, z such that $z = \varphi(x,y)$, $F(x,y,z) = 0$, and $F_z(x,y,z) \neq 0$.

Example 1 (*continued*). Let $\varphi(x,y)$ *be implicitly defined by* $F(x,y,z) = x^2+y^2+z^2-1 = 0$. *Find* $\frac{\partial \varphi}{\partial x}$ *and* $\frac{\partial \varphi}{\partial y}$ *at the point* $(x,y) = (\frac{1}{3},\frac{2}{3})$. (Here $\frac{\partial \varphi}{\partial x}$ can also be written $\left(\frac{\partial z}{\partial x}\right)_y$, and $\frac{\partial \varphi}{\partial y}$ can be written $\left(\frac{\partial z}{\partial y}\right)_x$.) From the preceding general formulas for $\frac{\partial \varphi}{\partial x}$ and $\frac{\partial \varphi}{\partial y}$, we have

$$\frac{\partial \varphi}{\partial x} = -\frac{F_x}{F_z} = -\frac{2x}{2z} = -\frac{x}{z}$$

and
$$\frac{\partial \varphi}{\partial y} = -\frac{F_y}{F_z} = -\frac{y}{z}.$$

For $x = \frac{1}{3}$ and $y = \frac{2}{3}$, $F(x,y,z) = 0$ gives us $z = \pm\frac{2}{3}$. This indicates that we have two implicitly defined functions, φ_1 and φ_2, whose domains contain $(\frac{1}{3},\frac{2}{3})$, where $\varphi_1(\frac{1}{3},\frac{2}{3}) = \frac{2}{3}$ and $\varphi_2(\frac{1}{3},\frac{2}{3}) = -\frac{2}{3}$. From our general expressions for $\frac{\partial\varphi}{\partial x}$ and $\frac{\partial\varphi}{\partial y}$, we get

$$\frac{\partial\varphi_1}{\partial x} = -\frac{1}{2}, \qquad \frac{\partial\varphi_1}{\partial y} = -1 \quad \text{(for } z = \frac{2}{3}\text{)};$$

and
$$\frac{\partial\varphi_2}{\partial x} = \frac{1}{2}, \qquad \frac{\partial\varphi_2}{\partial y} = 1 \quad \text{(for } z = -\frac{2}{3}\text{)};$$

(The reader may check these results by using the explicit expressions for φ_1 and φ_2 at the beginning of this example and differentiating directly.)

(b) We can also use the *elimination method* for implicit differentiation. We apply the steps of the method directly, as described in §11.2. The elimination method is usually the simplest and easiest way to carry out implicit differentiation.

Example 1 (continued). *Assume that* $x^2+y^2+z^2 = 1$ *defines* $z = \varphi(x,y)$ *implicitly. We seek formulas for* $\frac{\partial\varphi}{\partial x} = \left(\frac{\partial z}{\partial x}\right)_y$ *and* $\frac{\partial\varphi}{\partial y} = \left(\frac{\partial z}{\partial y}\right)_x$. (This is the same implicit differentiation problem as before.) We designate z as dependent variable, and x and y as independent variables. Applying the elimination method, we get

Step 1. $2x\,dx + 2y\,dy + 2z\,dz = 0.$

Step 2.   Solving for dz, $dz = -\frac{x}{z}dx - \frac{y}{z}dy$.

Step 3.   Hence $\left(\frac{\partial z}{\partial x}\right)_y = -\frac{x}{z}$, and $\left(\frac{\partial z}{\partial y}\right)_x = -\frac{y}{z}$.

Example 2. We are given the equations $y = f(x,u)$ *and* $u = g(x,v)$. *These are the same as* $y - f(x,u) = 0$ *and* $u - g(x,v) = 0$. *Assume that the second equation defines* x *implicitly as a function of* u *and* v. *Then the two equations together also define* y *implicitly as a function of* u *and* v. *We wish to find* $\frac{\partial\varphi}{\partial y} = \left(\frac{\partial z}{\partial y}\right)_x$ *and* $\left(\frac{\partial y}{\partial v}\right)_u$ *in terms of the formal derivatives of* f *and* g. *To* accomplish this, we designate y as *dependent* and u and v as *independent*, and we leave x *undesignated.* Applying elimination, we have

Step 1. $dy = f_x dx + f_u du,$
$du = g_x dx + g_v dv.$

Step 2.   In order to eliminate dx and then solve for dy in terms of du and dv, we first find

$$dx = \frac{1}{g_x} du - \frac{g_v}{g_x} dv$$

from the second equation in *Step 1*. Substituting this in the first equation in *Step 1*, we have

$$dy = \frac{f_x}{g_x} du - \frac{f_x g_v}{g_x} dv + f_u du$$

$$= \left(f_u + \frac{f_x}{g_x}\right) du - \frac{f_x g_v}{g_x} dv .$$

Step 3.   Hence $\left(\frac{\partial y}{\partial u}\right)_v = f_u + \frac{f_x}{g_x}$, and $\left(\frac{\partial y}{\partial v}\right)_u = -\frac{f_x g_v}{g_x}$. We have found the desired partial derivatives in terms of the formal derivatives of the given functions f and g.

Example 3. *We are given* $y = xu$ *and* $u = x^2+v^2$, *and we wish to find* $\left(\frac{\partial y}{\partial u}\right)_v$. (This is a specific instance of the preceding example.) We take y as *dependent*, and u and v as *independent* and x as *undesignated.*

Step 1. $dy = x \, du + u \, dx.$
$du = 2x \, dx + 2v \, dv.$

Step 2.   We set $dv = 0$ (see "Simplifying Step 2" at the end of §11.2) and eliminate dx:

$$dx = \frac{1}{2x} du, \text{and}$$
$$dy = x \, du + \frac{u}{2x} du = \left(x + \frac{u}{2x}\right) du.$$

Step 3.   Hence $\left(\frac{\partial y}{\partial u}\right)_v = x + \frac{u}{2x}$. Note that the undesignated variable x appears in this solution. We have *eliminated* the increment dx, but the variable x remains.

(For a check, $x = \pm\sqrt{u - v^2}$ Thus $y = u(\pm\sqrt{u - v^2})$, and

$$\left(\frac{\partial y}{\partial u}\right)_v = \pm\sqrt{u - v^2} + \frac{u}{2(\pm\sqrt{u-v^2})} = x + \frac{u}{2x}.)$$

The idea of implicit differentiation can be extended to situations where there is more than one given equation and where there may be more than one dependent variable. Consider, for example, two equations of the form

$$F(x,y,u,v) = 0,$$

and $$G(x,y,u,v) = 0,$$

where F and G are differentiable functions and u and v are dependent variables. We say that continuous two-variable functions φ and ψ on a common domain D are *simultaneously implicitly defined* by the given equations, if, for every (a,b) in D, the values $u = \varphi(a,b)$ and $v = \psi(a,b)$ occur as an isolated solution to the system of simultaneous equations

$$F(a,b,u,v) = 0,$$
$$G(a,b,u,v) = 0$$

in the unknowns u and v.

Example 4. *Assume that the two equations $u^2-v^2+x^2+y^2 = 2$ and $u+v+x+y = 0$ implicitly define u and v as functions of x and y. Find $\left(\frac{\partial u}{\partial x}\right)_y$ when $x = 1, y = 1$.*

Step 1. $2u\, du - 2v\, dv + 2x\, dx + 2y\, dy = 0,$
$\quad\quad\quad du + dv + dx + dy = 0.$

Step 2.   Letting $dy = 0$, we get the simultaneous system

$$\left(\begin{matrix} 2u\, du - 2v\, dv = -2x\, dx \\ du + dv = -dx \end{matrix}\right).$$

Treating du and dv as unknowns, and solving for du by Cramer's rule, we have

$$du = \frac{\begin{vmatrix} -2x\,dx & -2v \\ -dx & 1 \end{vmatrix}}{\begin{vmatrix} 2u & -2v \\ 1 & 1 \end{vmatrix}} = \frac{(-2x-2v)dx}{2u+2v}$$

$$= -\left(\frac{x+v}{u+v}\right)dx \ .$$

Step 3.   Hence $\left(\frac{\partial u}{\partial x}\right)_y = -\frac{x+v}{u+v}$.

From the original equations

$$x = 1, \ y = 1 \Rightarrow \left[\begin{matrix} u^2 - v^2 = 0 \\ u + v = -2 \end{matrix}\right] \Rightarrow u = -1, \ v = -1.$$

Therefore

$$\left(\frac{\partial u}{\partial x}\right)_y\bigg|_{(1,1)} = -\frac{1-1}{-1-1} = 0.$$

The reader will note, in these examples, the computational power provided by the use of the elimination method for implicit differentiation. If we can evaluate the initial formal derivatives, then we can always find the desired implicit derivatives by solving an appropriate system of linear equations.

Remark.   Returning to the description of the elimination method at the beginning of §11.2, we see that assumption (iii) can be made precise as follows: "(iii) that our given equations define the dependent and undesignated variables, simultaneously implicitly, as functions of the independent variables."

§11.5 Examples and Applications

As illustrations, we give geometric, algebraic, and physical applications of the chain rule.

Example 1. *If we move along a smooth directed curve* C *in the domain of a differentiable scalar field* f, *at what rate, at a given point* P *on* C, *is the observed value of the scalar field increasing, per unit of arc length along* C?

(5.1)   **Solution.**   *The desired rate of increase is, simply, the directional derivative of* f *at* P *in the direction of the unit tangent vector to the curve at* P. We show this by the following theoretical argument. (See [5–1].)

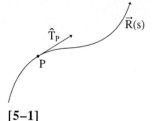

[5–1]

Let $\vec{R}(s)$ be the path for C with arc length s as parameter. We introduce Cartesian coordinates. Then f is represented by a function $f(x,y,z)$, and $\vec{R}(s) = x(s)\hat{i} + y(s)\hat{j} + z(s)\hat{k}$ for certain functions $x(s)$, $y(s)$, and $z(s)$. We define

$$h(s) = f(x(s),y(s),z(s)) \,.$$

For each value of s, $h(s)$ is the value of f at the point $\vec{R}(s)$ on C. Let s_P be the value of s at P. We wish to calculate $h'(s_P)$. By the chain rule, we have

$$h'(s_P) = \frac{\partial f}{\partial x}\Big|_P \frac{dx}{ds}\Big|_{s_P} + \frac{\partial f}{\partial y}\Big|_P \frac{dy}{ds}\Big|_{s_P} + \frac{\partial f}{\partial z}\Big|_P \frac{dz}{ds}\Big|_{s_P}$$

$$= \vec{\nabla}f\Big|_P \cdot \frac{d\vec{R}}{ds}\Big|_{s_P} = \vec{\nabla}f\Big|_P \cdot \hat{T}_P$$

(since the unit tangent vector $\hat{T}_P = \dfrac{d\vec{R}}{ds}\Big|_{s_P}$, see §6.3)

$$= \frac{df}{ds}\Big|_{\hat{T}_P,P} \quad \text{(by the gradient theorem).}$$

Example 2. Let a scalar field f be represented in cylindrical coordinates by the function $f(r,\theta,z)$. Find a formula for the gradient vector $\vec{\nabla}f$ in cylindrical coordinates.

Solution. Let r_P be the value of r at P. For any point P with $r_P > 0$,

(5.2)
$$\vec{\nabla}f\Big|_P = \frac{\partial f}{\partial r}\Big|_P \hat{r}_P + \frac{1}{r_P}\frac{\partial f}{\partial \theta}\Big|_P \hat{\theta}_P + \frac{\partial f}{\partial z}\Big|_P \hat{z}_P$$

Derivation. Let $\{\hat{u},\hat{v},\hat{w}\}$ be any frame at a point P in the domain of a differentiable scalar field f in E^3. Then, by (5.6) in Chapter 9,

$$\vec{\nabla}f\Big|_P = \frac{df}{ds}\Big|_{\hat{u},P} \hat{u} + \frac{df}{ds}\Big|_{\hat{v},P} \hat{v} + \frac{df}{ds}\Big|_{\hat{w},P} \hat{w} \,.$$

If we apply this to the cylindrical coordinate frame at P, we get

(5.3)
$$\vec{\nabla}f\Big|_P = \frac{df}{ds}\Big|_{\hat{r},P} \hat{r}_P + \frac{df}{ds}\Big|_{\hat{\theta},P} \hat{\theta}_P + \frac{df}{ds}\Big|_{\hat{z},P} \hat{z}_P \,.$$

It is immediate from the definition of directional derivative that

(5.4)
$$\frac{df}{ds}\Big|_{\hat{r},P} = \frac{\partial f}{\partial r}\Big|_{P} \quad \text{and} \quad \frac{df}{ds}\Big|_{\hat{z},P} = \frac{\partial f}{\partial z}\Big|_{P}.$$

We now show that $\dfrac{df}{ds}\Big|_{\hat{\theta},P} = \dfrac{1}{r_P}\dfrac{\partial f}{\partial \theta}\Big|_{P}.$

Let z_P be the value of z at P. Consider the circle through P in the plane $z = z_P$, with center on the z axis. This circle has radius r_P and $\hat{\theta}$ as tangent vector at P. The rate of change of f at P, per unit of arc length along this circle in the direction of increasing θ is given by

$$\lim_{\Delta s \to 0}\left(\frac{\Delta f}{\Delta s}\right) = \lim_{\Delta\theta \to 0}\left(\frac{\Delta f}{r_P\Delta\theta}\right) = \frac{1}{r_P}\frac{\partial f}{\partial\theta}\Big|_P.$$

Hence, by example 1, we have

(5.5)
$$\frac{df}{ds}\Big|_{\hat{\theta},P} = \frac{1}{r_P}\frac{\partial f}{\partial\theta}\Big|_{P}.$$

Substituting (5.4) and (5.5) in (5.3), we get (5.2), the *gradient formula for cylindrical coordinates*:

Example 3. *In cylindrical coordinates, a scalar field in* $\mathbf{E}^3$ *is given by* $f(r,\theta,z) = r\theta$, *for* $r > 0$ *and* $0 \le \theta \le \pi$. *Find the gradient of f at the point* P_0 *whose cylindrical coordinates are* (r_0,θ_0,z_0).

Solution. By (5.2), $\vec{\nabla}f = \theta_0\hat{r}_{P_0} + \hat{\theta}_{P_0}$.

Example 4. *In the science of* thermodynamics, *the chain rule can be used to derive mathematical identities which can be used to predict certain experimental observations.* We give one example.

Consider a given quantity of a certain gas at equilibrium in a small container whose volume can be varied. (See the discussion of such a gas in §10.1.) For a certain three-variable function F, the physical variables p, V, and T always satisfy the equation $F(p,V,T) = 0$. (We can use $F(p,V,T) = p - h(V,T)$, where h is the function for p described in §10.1. This equation $F(p,V,T) = 0$ is called the *equation of state* for the given gas. In the hypothetical case of a "perfect" gas, we can take $F(p,V,T) = p - \dfrac{kT}{V}$ for some constant k.) Each equilibrium state yields values for p, V, and T and therefore, if F is differentiable, we have values for the derivatives F_p, F_V, and F_T.   We make the following assumption:

(5.6)   *F is differentiable, and the first-order derivatives of F are non-zero for all equilibrium states.* Assumption (5.6) is, in effect, a *natural* law; it makes a statement about nature. It also implies, by implicit differentiation as in §11.4, that the derivatives $\left(\dfrac{\partial p}{\partial V}\right)_T$, $\left(\dfrac{\partial V}{\partial T}\right)_p$, $\left(\dfrac{\partial T}{\partial p}\right)_V$ exist for each equilibrium state.   By implicit differentiation, we know that $\left(\dfrac{\partial p}{\partial V}\right)_T = -\dfrac{F_V}{F_p}$, $\left(\dfrac{\partial V}{\partial T}\right)_p = -\dfrac{F_T}{F_V}$, $\left(\dfrac{\partial T}{\partial p}\right)_V = -\dfrac{F_p}{F_T}$. Hence these three derivatives must satisfy the identity

(5.7) $$\left(\frac{\partial p}{\partial V}\right)_T \left(\frac{\partial V}{\partial T}\right)_p \left(\frac{\partial T}{\partial p}\right)_V = \left(-\frac{F_V}{F_p}\right)\left(-\frac{F_T}{F_V}\right)\left(-\frac{F_p}{F_T}\right) = -1 .$$

For a given equilibrium state, each of these derivatives can itself be measured by its own appropriate and distinct physical procedure (see §10.1). The identity (5.7) has predictive power and can be checked empirically (for any equilibrium state) by measuring these derivatives and verifying that their product is −1. The identity also yields the useful fact that in any given equilibrium state, either all three derivatives are negative or else one is negative and two are positive. We have deduced these predictions, directly and logically, from the fact of nature assumed in (5.6).

The chain rule can also be used to calculate *higher derivatives* for implicitly defined functions.

Example 5. *Let z = φ (x,y) be defined implicitly by F(x,y,z) = 0, where F is C^2. Find $\dfrac{\partial^2 \varphi}{\partial x^2}$ in terms of the formal derivatives of F.*

Solution. By implicit differentiation,

(5.8) $$\frac{\partial \varphi}{\partial x} = -\frac{F_x}{F_z} .$$

Then $\dfrac{\partial^2 \varphi}{\partial x^2} = \left(\dfrac{\partial}{\partial x}\right)_y \dfrac{\partial \varphi}{\partial x} = \dfrac{-F_z \left(\frac{\partial}{\partial x}\right)_y F_x + F_x \left(\frac{\partial}{\partial x}\right)_y F_z}{F_z^2}$, by the quotient rule. By the

chain rule, we also have

$$\left(\frac{\partial}{\partial x}\right)_y F_x = \left(\frac{\partial}{\partial x}\right)_y F_x(x, y, \varphi(x, y)) = F_{xx} + F_{xz} \frac{\partial \varphi}{\partial x} = F_{xx} - \frac{F_{xz} F_x}{F_z} \quad \text{(by (5.8))};$$

and, similarly,

$$\left(\frac{\partial}{\partial x}\right)_y F_z = F_{zx} + F_{zz}\left(-\frac{F_x}{F_z}\right) = F_{zx} - \frac{F_{zz} F_x}{F_z}.$$

We thus get $\dfrac{\partial^2 \varphi}{\partial x^2} = \dfrac{-F_z F_{xx} + F_{xz} F_x + F_x F_{zx} - \dfrac{F_{zz} F_x^2}{F_z}}{F_z^2}$, and, by the equality of cross de-

rivatives for F, this gives

(5.9) $$\frac{\partial^2 \varphi}{\partial x^2} = -\frac{F_{xx}}{F_z} + \frac{2 F_{xz} F_x}{F_z^2} - \frac{F_{zz} F_x^2}{F_z^3}.$$

In a similar way, $\dfrac{\partial^2 \varphi}{\partial xy}$ and $\dfrac{\partial^2 \varphi}{\partial y^2}$ can be calculated. Indeed, nth-order

derivatives, for $n > 2$, can be calculated on the assumption that F is C^n, but these higher-order calculations may be lengthy.

§11.6 Normal Vectors to Surfaces

The following fact will be frequently used in later chapters to help us find a normal vector to a given surface S at a given point P.

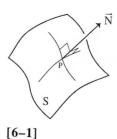

[6–1]

(6.1) *Let f be a C^1 scalar field in E^3, let S be a level surface of f, and let P be a point on S such that $\vec{\nabla}f\big|_P \neq \vec{0}$. It follows that the vector $\vec{\nabla}f\big|_P$ is normal to the surface S in the following precise sense:*
  (i) *For every directed smooth curve C which lies in S and goes through P, $\vec{\nabla}f\big|_P$ is orthogonal to the unit tangent vector of C at P. See [6–1].)*
  (ii) *Every unit vector orthogonal to $\vec{\nabla}f\big|_P$ is the unit tangent vector at P of some directed smooth curve which lies in S and goes through P.*

It immediately follows from (i) and (ii) that if $\vec{\nabla}f\big|_P \neq \vec{0}$, then any other non-zero vector $\vec{N}$ which is normal to S at P (in the sense of (i) and (ii) above) must be a scalar multiple of $\vec{\nabla}f\big|_P$.

Proof of **(6.1)**. We use the chain rule to prove (i). Let S be the level surface of f for c. We use Cartesian coordinates. Let P be a point on S for which $\vec{\nabla}f\big|_P \neq \vec{0}$. Let $\vec{R}(t) = x(t)\hat{i} + y(t)\hat{j} + z(t)\hat{k}$, t in [a,b], be some smooth path which lies entirely in S and goes through P. For each t in [a,b], let $P_t = \vec{R}(t)$ be the point (on the path) given by t. Define $h(t) = f(P_t)$. Since S is a level surface, we have $h(t) = c$ for all t in [a,b]; hence we know that $\frac{dh}{dt} = 0$. But by the chain rule, we have

$$0 = \frac{dh}{dt} = \frac{d}{dt}\big(f(x(t), y(t), z(t))\big) =$$

$$f_x \frac{dx}{dt} + f_y \frac{dy}{dt} + f_z \frac{dz}{dt} = \vec{\nabla}f \cdot \frac{d\vec{R}}{dt} = \vec{\nabla}f \cdot \frac{d\vec{R}}{ds}\frac{ds}{dt} = 0.$$

Since $\vec{R}(t)$ is smooth, $\frac{ds}{dt} \neq 0$. Hence

$$\vec{\nabla}f \cdot \frac{d\vec{R}}{ds} = \vec{\nabla}f \cdot \hat{T} = 0 \qquad \text{(since, from §6.2, } \hat{T} = \frac{d\vec{R}}{ds}\text{)}.$$

In particular, at the point P, $\vec{\nabla}f\big|_P \cdot \hat{T}_P = 0$. This proves (i).

The proof of (ii) uses the *implicit function theorem* described in §11.7 below. We do not give the proof here.

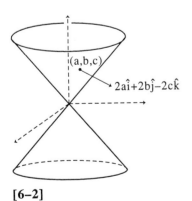

Example 1. Let f be represented in Cartesian coordinates by $f(x,y,z) = x^2 + y^2 - z^2$, and let S be the level surface given by $f(x,y,z) = 0$. Then S is the two-sheeted cone $z^2 = x^2 + y^2$ with vertex at the origin. See [6–2]. We have $\vec{\nabla}f = 2x\hat{i} + 2y\hat{j} - 2z\hat{k}$, and $\vec{\nabla}f\big|_P \neq \vec{0}$ for all points P except the origin. Hence, by (6.1), for all points P = (a,b,c) on S except the origin, $\vec{\nabla}f\big|_P = 2a\hat{i} + 2b\hat{j} - 2c\hat{k}$ must be normal to S at P.

The following special case of (6.1) is often useful.

[6–2]

(6.2)   *Corollary. Let* $g(x,y)$ *be a two-variable function in* C^1. *Let* S *be the graph of* g *in Cartesian coordinates in* E^3. *Let* Q *be a point in the domain of* g *in the* xy *plane, and let* P *be the point corresponding to* Q *on* S. *Let*

$$\vec{N} = -\frac{\partial g}{\partial x}\bigg|_Q \hat{i} - \frac{\partial g}{\partial y}\bigg|_Q \hat{j} + \hat{k}$$

Then $\vec{N}$ *is normal to the graph* S *of* g *at* P *in the sense of* (i) *and* (ii) *in* (6.1).

Proof of (6.2). The graph of g is the level surface for 0 of the function $f(x,y,z) = z - g(x,y)$. Then $\vec{\nabla}f\big|_P = -\frac{\partial g}{\partial x}\bigg|_Q \hat{i} - \frac{\partial g}{\partial y}\bigg|_Q \hat{j} + \hat{k} \neq \vec{0}$, since $\hat{k} \neq \vec{0}$. By (6.1), $\vec{\nabla}f\big|_P$ is normal to S at P.

Example 2. Let $g(x,y) = x^2 + y^2$. The graph of g is a paraboloid about the z axis. By (6.2), at any point $P = (a,b,g(a,b))$ on this graph, we have $\vec{N} = -2a\hat{i} - 2b\hat{j} + \hat{k}$ as a normal vector to the graph. See [6–3].

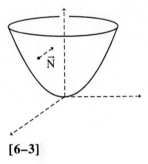

[6–3]

Remarks. (1) Note that the formula for $\vec{N}$ in (6.2) always gives a vector which is *non-zero* and *directed upwards*.

(2) Note that two-dimensional versions of (6.1) and (6.2) hold for a scalar field f in E^2. In particular: if $\vec{\nabla}f\big|_P \neq \vec{0}$, then $\vec{\nabla}f\big|_P$ is normal to the level curve through P; and $\vec{N} = -\frac{dg}{dx}\bigg|_c \hat{i} + \hat{j}$ is an upward directed normal to the graph of g at $P = (c,g(c))$.

(3) Note that the formula in (6.2) for a normal vector to a graph at a given point P on the graph is the same as the formula (2.10) in §9.2 for an upward normal vector to the tangent plane to a graph of a differentiable two-variable function at a given point P′ on the graph.

§11.7 Proof that the Elimination Method is Valid

(7.1)   *Theorem. If the assumptions of the elimination method hold for a given set of equations (as stated in the definition of the method in §11.2), then the derivative expressions produced for that set of equations by the elimination method must be correct.*

Proof. We prove the theorem for a particular case. It will be evident that the proof can be adapted to other cases.

Let w, u, and x be given physical variables. Let $F(w,u,x) = 0$ and $G(w,u,x) = 0$ be given equations in these variables, where F and G are differentiable functions. Let w be dependent, x be independent, and u be undesignated. Assume that the given equations define w and u implicitly as continuous functions of x. Let c be a given value of x. Let $\Delta x = x - c$ be a change in the value of x. Let Δu and Δw be the resulting changes in the values of u and w. Then by the continuity of u and w, if $\Delta x \to 0$, we must have $\Delta u \to 0$ and $\Delta w \to 0$. Moreover, for these values of Δx, Δu, and Δw, we must have $\Delta F = 0$ and $\Delta G = 0$. Hence, by linear approximation for the differentiable functions F and G, we have

$$0 = (F_w + \varepsilon_1)\Delta w + (F_u + \varepsilon_2)\Delta u + (F_x + \varepsilon_3)\Delta x, \text{ and}$$
$$0 = (G_w + \varepsilon_4)\Delta w + (G_u + \varepsilon_5)\Delta u + (G_x + \varepsilon_6)\Delta x,$$

where $\lim_{x \to c} \varepsilon_i = 0$ for $i = 1, 2, \ldots 6$.

We abbreviate the above equations as

$$0 = \widetilde{F}_w \Delta w + \widetilde{F}_u \Delta u + \widetilde{F}_x \Delta x, \text{ and}$$
$$0 = \widetilde{G}_w \Delta w + \widetilde{G}_u \Delta u + \widetilde{G}_x \Delta x.$$

If we use Cramer's rule to eliminate Δu from these last equations and solve for Δw, we get

(7.2) $$\Delta w = \frac{\widetilde{G}_x \widetilde{F}_u - \widetilde{F}_x \widetilde{G}_u}{\widetilde{G}_u \widetilde{F}_w - \widetilde{F}_u \widetilde{G}_w} \Delta x.$$

Hence

(7.3) $$\frac{dw}{dx} = \lim_{x \to c} \frac{\Delta w}{\Delta x} = \lim_{x \to c} \left(\frac{\widetilde{G}_x \widetilde{F}_u - \widetilde{F}_x \widetilde{G}_u}{\widetilde{G}_u \widetilde{F}_w - \widetilde{F}_u \widetilde{G}_w} \right)$$

$$= \frac{G_x F_u - F_x G_u}{G_u F_w - F_u G_w}, \quad \text{provided } G_u F_w - F_u G_w \neq 0.$$

But this last expression for $\frac{dw}{dx}$ is also the expression for $\frac{dw}{dx}$ produced by the elimination method. Hence the elimination method gives the correct result.

Remark. The following result is also true. If we make the additional assumption that $F(w,u,x)$ and $G(w,u,x)$ are in C^1, if $F(d,e,c) = 0$ and $G(d,e,c) = 0$, and if the *denominator* $G_uF_w - F_uG_w$ (in the expression above for $\frac{dw}{dx}$) is not zero at $P_0 = (d,e,c)$, then there must be some interval (a,b) with $a < c < b$ such that the given equations implicitly define w as a differentiable function of x on (a,b). This result is an instance of a more general result: *if a given set of equations is formed from continuously differentiable functions, has a designated dependent variable, and has designated independent variables* $x,\dots,$ *then this set of equations defines one or more functions implicitly in some neighborhood of a specified point* $(c,\dots),$ *provided that the denominators which appear when the elimination method is applied are not zero at the point* $(c,\dots).$ (Here, $(c,\dots)$ is a specific n-tuple of values for the designated independent variables $(x,\dots).)$ This result is known as the *implicit function theorem*. The proof of this theorem is not given in this text. The implicit function theorem answers question (I) in §11.4, since it shows us that a function will be implicitly defined in any region of values (for all the variables) for which all the elimination-method denominators are non-zero.

§11.8   Problems

§11.1

1. Find the differentials of the following equations:
    (a) $xy + yz = xz$
    (b) $z = xyz + y^2x$
    (c) $u + vx = ue^{xy}$
    (d) $\sin^2(xy) = x^2 + y^2$

§11.2

1. Use the elimination method to solve:
    (a) Problem 2 for §10.5
    (b) Problem 3 for §10.5
    (c) Problem 4 for §10.5

2. Use the elimination method to help solve:
    (a) Problem 5 for §10.5. (Numerical values may be inserted after Step 1.)
    (b) Problem 6 for §10.5

§11.4

1. The equation $xze^{y-1} - y^2z^3 = 0$ implicitly defines a function $z = z(x,y)$ in the vicinity of the point $(1,1,1)$. Find the value of $\frac{\partial z}{\partial y}$ when x = 1, y = 1, and z = 1.

2. The equation $F(x,y,z) = x^2yz + xyz^2 = 4$ defines z implicitly as a function of x and y. Find $\left(\frac{\partial z}{\partial x}\right)_y$ as a function of x, y, and z.

3. Let A,B,C be the angles of a triangle, and let a,b,c be the opposite sides. Then the value of   A   is determined by the values of a, b, and c. Find $\frac{\partial A}{\partial a}$ and $\frac{\partial A}{\partial b}$. Your answer may be expressed in terms of a, b, c, and A. (*Hint.* Use law of cosines.)

4. Let $w = f(x,y,z)$, and let $y = g(x,z)$ define z implicitly as a function of x and y. We are given that

$$g(1,3) = 2 \qquad f(1,2,3) = 7$$
$$g_x(1,3) = 2 \qquad f_x(1,2,3) = 5$$
$$g_z(1,3) = 3 \qquad f_y(1,2,3) = 1$$
$$f_z(1,2,3) = 6 .$$

Find the value of $\left(\frac{\partial w}{\partial x}\right)_y$ when $x = 1$ and $y = 2$.

5. You are given $x = f(y,v)$ and $v = g(y,u)$, where $v = g(y,u)$ defines y implicitly as a function of u and v.

(a) Give an expression for $\left(\frac{\partial y}{\partial u}\right)_v$ in terms of g_y and g_u.

(b) Give an expression for $\left(\frac{\partial x}{\partial u}\right)_v$ in terms of f_y, g_y, g_u.

6. u and v are defined implicitly as functions of x and y by the equations

$$x = u^2 - v^2$$
$$y = uv .$$

Find expressions for $\left(\frac{\partial u}{\partial x}\right)_y$ and $\left(\frac{\partial u}{\partial y}\right)_x$ as functions of u and v.

7. Let $w = f(x,y,z)$, and let the equation $x = g(y,z)$ define y implicitly as a function of x and z. Find $\left(\frac{\partial w}{\partial x}\right)_z$ in terms of the formal derivatives of f and g.

8. The equation $F(x,y,z) = \cos(xy) - \sin(xz) = 0$ defines z implicitly as a function of x and y. Find $\left(\frac{\partial z}{\partial x}\right)_y$ as a function of x, y, and z.

9. The following equation defines z implicitly as a function of x and y:

$$xz^2 + yz = 3.$$

Find all possible values of $\left(\frac{\partial z}{\partial x}\right)_y$ when $x = 1$ and $y = 2$.

10. You are given that $y = x^3u$ and $u = x^2 - v^2$. Find expressions for $\left(\frac{\partial x}{\partial u}\right)_v$ and $\left(\frac{\partial y}{\partial v}\right)_u$ in terms of x, u, and v.

11. Polar and Cartesian coordinates are related by the equations

$$x = r \cos \theta ,$$
$$y = r \sin \theta .$$

These equations define r and θ implicitly as functions of x and y. Find $\left(\frac{\partial r}{\partial x}\right)_y$ and $\left(\frac{\partial \theta}{\partial y}\right)_x$ in terms of r and θ.

12. u and v are defined implicitly as functions of x and y by the equations

$$x = u^2 + v^2 ,$$
$$y = uv .$$

Find $\left(\frac{\partial v}{\partial x}\right)_y$ when $u = \sqrt{2}$ and $v = 1$.

13. Let $w = f(u,v)$, $u = g(x,y)$, and $v = h(x,y)$. Assume that the second equation defines x implicitly as a function of y and u. Hence all three equations can be viewed as defining w as a function of y and u. Find $\left(\frac{\partial w}{\partial y}\right)_u$ in terms of the formal derivatives of f, g, and h.

14. The equation $x^2 - 4uvx + y^2 = 0$ can be viewed as defining x implicitly as a function of u, v, and y.

(a) Find an expression for $\left(\frac{\partial x}{\partial v}\right)_{u,y}$.

(b) What can you conclude as to the value (or values) of this derivative when $u = v = y = 1$?

15. The equation $x^2 + ux - v = 0$ defines x implicitly as a function of u and v.

(a) Use the elimination method to find $\frac{\partial x}{\partial u}$ and $\frac{\partial x}{\partial v}$.

(b) Find $\frac{\partial^2 x}{\partial v^2}$.

16. You are given $x = f(y,v)$ and $v = g(y,u)$, where $v = g(y,u)$ defines y implicitly as a function of u and v.

    (a) Give an expression for $\left(\frac{\partial y}{\partial v}\right)_u$ in terms of $\frac{\partial g}{\partial y}$.

    (b) Give an expression for $\left(\frac{\partial x}{\partial v}\right)_u$ in terms of $\frac{\partial f}{\partial y}$, $\frac{\partial g}{\partial y}$, and $\frac{\partial f}{\partial v}$.

17. A real-valued function $z = f(x,y)$ is implicitly defined by the equation $x^3 z + y^2 - z^4 = 0$. What are the possible values of z, of $\frac{\partial f}{\partial x}$, and of $\frac{\partial f}{\partial y}$, when $x = 0$ and $y = 1$?

18. We have $w = f(u,v,x)$ and $u = g(v,x)$ for given functions f and g.

    (a) From these equations, we see that we can choose v and x as final independent variables and treat w as a function of v and x. Find an expression for $\left(\frac{\partial w}{\partial v}\right)_x$ in terms of the formal derivatives f_u, f_v, f_x, g_v, and g_x.

    (b) Given the two equations above, assume that the equation $u = g(v,x)$ defines x implicitly as a function of u and v. Then we can choose u and v as final independent variables and treat w as a function of u and v. Find an expression for $\left(\frac{\partial w}{\partial v}\right)_u$ in terms of the formal derivatives f_u, f_v, f_x, g_v, and g_x.

19. Let x be implicitly defined as a function of u and v by the equation

$$x^2 + ux - v^2 = 0.$$

    (a) Find expressions for $\frac{\partial x}{\partial u}$ and $\frac{\partial x}{\partial v}$ in terms of x, u, and v.

    (b) What are the values of these derivatives when $u = 3$, $v = 2$?

20. We are given $w = f(x,y,z)$ and $y = g(x,z)$. Assume that $y = g(x,z)$ defines z implicitly as a function of x and y.

    (a) Find an expression for $\left(\frac{\partial w}{\partial y}\right)_x$ in terms of the formal derivatives of f and g.

    (b) Apply your result in (a) to find $\left(\frac{\partial w}{\partial y}\right)_x$ when $f(x,y,z) = x^2 + y^2 + z^2$ and $g(x,z) = x^2 - z^2$.

21. Given $z = x^4 + y^4$, find $\left(\frac{\partial x}{\partial y}\right)_z$ and $\left(\frac{\partial x}{\partial z}\right)_y$ for $x = 1$, $y = 1$, $z = 2$.

22. You are given $u = \sin(xy)$ and $x = \cos(yv)$, where the latter equation defines y implicitly as a function of x and v. Find $\left(\frac{\partial y}{\partial x}\right)_v$ and $\left(\frac{\partial u}{\partial x}\right)_v$ in terms of u, v, x, and y.

23. Let $w = u + v$, $u = \frac{x}{y}$, $v = x - y$. Find $\left(\frac{\partial w}{\partial y}\right)_v$ and $\left(\frac{\partial w}{\partial y}\right)_u$, and express each of these as a function of x and y.

24. If $f(x,y,z) = 0$, show that $\left(\frac{\partial x}{\partial y}\right)_z \left(\frac{\partial y}{\partial z}\right)_x \left(\frac{\partial z}{\partial x}\right)_y = -1$.

25. Given $x^3 - 2u^2 vx + y^4 = 0$.

    (a) Find an expression for $\left(\frac{\partial x}{\partial v}\right)_{u,y}$.

    (b) What can you conclude as to the value (or values) of this partial derivative when $v = u = y = 1$?

26. u is defined implicitly as a function of y and v by the equations

$$u = x^2 + y^2,$$
$$y = \sin(xv).$$

Find the values of $\left(\frac{\partial u}{\partial v}\right)_y$ and $\left(\frac{\partial u}{\partial y}\right)_v$ when $v = \pi$ and $x = 1$.

27. The equation $x^3 - 2u^2 vx^2 - yu^2 x + 2y^3 = 0$ defines x implicitly as a function of u, v, and y.

    (a) Find an expression for $\left(\frac{\partial x}{\partial v}\right)_{u,y}$.

    (b) What can you conclude as to the possible numerical values of this partial derivative when $v = u = y = 1$?

28. Cylindrical and Cartesian coordinates are related by the equations:

$$x = r \cos \theta,$$
$$y = r \sin \theta,$$
$$z = z.$$

Consider r, θ, and z as functions of x, y, z, and from these equations, find the first-order partial derivatives of r, θ, and z, with respect to x, y, and z, where these derivatives are themselves expressed as functions of r, θ, and z.

29. Let $w = f(x,y,z)$, and let x, y, and z be related to r, θ, and z as in problem 28. Express $\frac{\partial f}{\partial x}$, $\frac{\partial f}{\partial y}$, and $\frac{\partial f}{\partial z}$ in terms of r, θ, z, and the first-order partial derivatives of w as a function of r, θ, and z.
(*Hint.* Apply the chain rule to w with $\{r,\theta,z\}$ as intermediate variables and $\{x,y,z\}$ as final variables.)

30. Let x be implicitly defined as a function of u and v by the equation

$$x^2 + ux - vx = 0.$$

(a) Find an expression for $\frac{\partial x}{\partial u}$ in terms of x, u, and v.

(b) Find the numerical value (or values) of $\frac{\partial x}{\partial u}$ when $(u,v) = (1,-1)$.

31. Let $w = f(x,y,z)$, and let $x = g(y,z)$ define z implicitly as a function of x and y. We are given that

$$g(2,3) = 1, \qquad f(1,2,3) = 7,$$
$$g_y(2,3) = 6, \qquad f_x(1,2,3) = 2,$$
$$g_z(2,3) = 1, \qquad f_y(1,2,3) = 3,$$
$$f_z(1,2,3) = 2.$$

Find the value of $\left(\frac{\partial w}{\partial x}\right)_y$ when $x = 1$ and $y = 2$.

32. $F(x,y) = 0$ defines $y(x)$ implicitly. Find an expression for $\frac{d^2y}{dx^2}$ in terms of the formal partial derivatives of F. You may assume equality of cross derivatives for F.

33. $F(x,y,z)$ is a given differentiable function. Assume that $F(x,y,z) = 0$ defines x as a function of y and z. Find $\frac{\partial^2 x}{\partial y \partial z}$ in terms of the derivatives of F.

34. You are given: $\quad x = \cos(yz),$
$$z = u - v,$$
$$x = uv.$$

Find $\left(\frac{\partial y}{\partial u}\right)_v$ when $u = \frac{1}{2}$, $v = 0$, and $y = \pi$.

§11.5

1. The surfaces $z = f(x,y)$ and $z = g(x,y)$ intersect in a curve C. Find an expression, in terms of the derivatives of f and g, for a vector tangent to C at a point P on C where $x = a$ and $y = b$.

2. A scalar field f is represented by the function $f(r,\theta,z) = rz \sin \theta$ in cylindrical coordinates. Find the gradient in terms of the coordinate frame $\{\hat{r},\hat{\theta},\hat{z}\}$.

3. Let A,B,C be the angles of a triangle. Let a,b,c be the opposite sides. Consider b as a function of the independent variables A,a,c. Find $\frac{\partial b}{\partial A}$, $\frac{\partial b}{\partial a}$, and $\frac{\partial b}{\partial c}$.

4. Prove the following:
Theorem. Let f be a differentiable scalar field; let g be a differentiable one-variable function; and let P be a point in the domain D of f such that $f(P)$ is in the domain of g. Let h be the scalar field defined by

$$h(Q) = g(f(Q)) \quad \text{for all} \ Q \ \text{such that}$$
$$f(Q) \ \text{is in the domain of g.}$$

Then

$$\left. \vec{\nabla} h \right|_P = g'(f(P)) \left. \vec{\nabla} f \right|_P.$$

§11.6

1. Find a normal vector to the surface $2z^2 - 3x^2y + 5xy^2 = 0$ at the point $(-1,1,2)$.

2. (a) Let $f(x,y,z) = x^2 - y^2 + z^2 - 12$. Find the gradient of f at the point $(3,1,2)$.

   (b) Find the equation of the tangent plane to the surface $x^2 - y^2 + z^2 = 12$ at the point $(3,1,2)$.

3. Let $f(x,y,z) = x^2 + y^2 + z^2 - 1$, and let $P = (2,-1,2)$.

   (a) Find the gradient of f at P.

   (b) Find the directional derivative of f at P in the direction of the origin.

   (c) Find the equation of the tangent plane at P to the surface $x^2 + y^2 + z^2 = 9$.

   (d) Let Q be the point at distance 0.03 from P in the direction of the origin. Estimate the value of f at Q.

4. At the point $(1,1,1)$, find equations for the *tangent plane* and *normal line* to the hyperboloid given by $x^2 + y^2 - z^2 = 1$.

5. At the point $(3,2)$, find equations for the *tangent line* and *normal line* to the ellipse given by $2x^2 + 3y^2 = 30$.

6. At the point $(3,2,2)$, find equations for the *tangent plane* and *normal line* to the hyperboloid given by $x^2 - y^2 - z^2 = 1$.

§11.7

1. Prove (7.2).

2. Prove (7.3).

Chapter 12

Maximum-Minimum Problems

§12.1 Global Problems and Local Problems

Maximum-minimum problems are among the most common and important problems of applied mathematics. We shall refer to maximum-minimum problems as, more briefly, *max-min problems*.

In a multivariable max-min problem, we are given a particular scalar field (or scalar function) f on a specified domain D, and we want to determine: (a) the "maximum" and "minimum" values that f takes on (for points in D); and (b) the particular points in D at which f takes on those "maximum" and "minimum" values. In multivariable calculus, as in one-variable calculus, there are two forms of max-min problems: *global problems* and *local problems*.

(1.1)   In a *global max-min problem* (for f on D), we want to find the points P in D (if any) such that the value of f at P is *at least as great* as the value of f at any other point in D. Such points, if they exist, are called *global maximum points* (for f on D), and the corresponding value of f is called the *global maximum value of* f *on* D. *Global minimum points* and the *global minimum value of* f *on* D are defined similarly. (We know, of course, from one-variable calculus, that such global points and global values need not always exist. For example, the one-variable function $f(x) = \frac{1}{x}$ with the positive x axis as its domain has no global maximum or global minimum points or values.)

(1.2)   In a *local max-min problem* (for f on D), we want to find the points P in D (if any) such that the value of f at P is *greater than* the value of f at any other point in D "nearby" to P. (The idea of *nearby point* will be made precise in (1.5) below.) Such points, if they exist, are called *local maximum points*, and a value of f for such a point is called the *local maximum value* at that point. *Local minimum points* and *local minimum values* (for f on D) are defined similarly.

We illustrate these definitions with a one-variable example. Recall from elementary calculus that for real numbers a and b with $a < b$, [a,b] denotes the interval of real numbers x   such that $a \leq x \leq b$, and that (a,b) occasionally denotes the interval of real numbers x   such that $a < x < b$. (Use of the latter notation should not be confused with the more common use of (a,b) to denote a point with coordinates a and b.)

Example 1. *Let* f *be a one-variable function. Let the graph of* f *be as sketched in* [1–1]. *The domain of* f *is the interval* [a,c].
For the global max-min problem, we have:

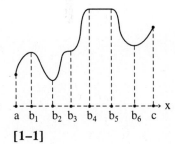

[1–1]

    The global maximum points are all points in $[b_4, b_5]$.
    The global maximum value is $f(b_4)$.
    There is a single global minimum point, the point b_2.
    The global minimum value is $f(b_2)$.
For the local max-min problem, we have:

    There are two local maximum points, the points b_1 and   c, with corresponding local maximum values $f(b_1)$ and $f(c)$.
    There are three local minimum points, the points a, b_2, and b_6, with corresponding local minimum values $f(a)$, $f(b_2)$, and $f(b_6)$.

(1.3)   We shall also use the following terminology. We define the *global extreme points* (for f on D) to be the points which are either global maximum points or global minimum points. We define the *global extreme values* (for f on D) to be the values of f at its global extreme points. (There can be at most two such global extreme values.) We define the *local extreme points* (for f on D) to be the points which are either local maximum points or local minimum points. Finally, we define the *extreme points* (for f on D) to be the points which are either global extreme points or local extreme points.

Example 1 (*continued*).
The global extreme points are b_2, $[b_4,b_5]$.
The local extreme points are a, b_1, b_2, b_6, and c.
The extreme points are a, b_1, b_2, $[b_4,b_5]$, b_6, and c.

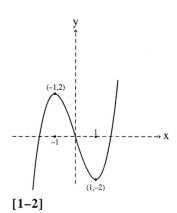

[1–2]

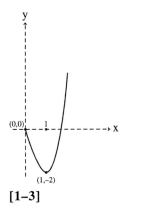

[1–3]

Example 2. *Let* $f(x) = x^3 - 3x$ *on* $D = (-\infty, \infty)$. (See [1–2].) Here $x = -1$ is a local-maximum point and $x = 1$ is a local-minimum point. The corresponding values are $f(-1) = 2$ and $f(1) = -2$. There are no global extreme points.

Example 3. *Let* $f(x) = x^3 - 3x$ *on* $D = [0, \infty)$. (See [1–3].) Here $x = 1$ is both a local-minimum point and a global minimum point, and $x = 0$ is a local-maximum point. There is, however, no global maximum point.

It remains to make precise the meaning of "nearby" in (1.2). We do so as follows.

(1.4) *Definitions.* Let x be a given real number. A set U of real numbers is called a *neighborhood of* x if $U = [x-r, x+r]$ for some $r > 0$. Thus U is the set of reals y such that $|x - y| \leq r$. r is called the *radius* of the neighborhood U.

Let P be a given point in E^2. A set U of points in E^2 is called a *neighborhood of* P if, for some $r > 0$, U is the disk of radius r with center at P. Thus U is the set of points Q in E^2 such that $d(P, Q) \leq r$.

Let P be a given point in E^3. A set U of points is E^3 is called a *neighborhood of* P if, for some $r > 0$, U is the ball of radius r with center at P. Thus U is the set of points Q in E^3 such that $d(P, Q) \leq r$.

Note that, by definition, the radius of a neighborhood is always positive.

(1.5) *Definitions.* Let f be a scalar field (or function) on domain D. Let P be a point in D. We say that:

(i)   P is a *local maximum point* for f if there is some neighborhood U of P such that for all points Q in both U and D,

$$Q \neq P \Rightarrow f(P) > f(Q).$$

(ii)   P is a *local minimum point* for f if there is some neighborhood U of P such that for all points Q in both U and D,

$$Q \neq P \Rightarrow f(P) < f(Q).$$

Clearly, solutions to the global or local max-min problems for f on D may change if we alter the domain D but keep the same scalar field f. We have the following: Let f be a scalar field (or function) with domain D. Let D' be a subset of D.

(1.6)   *If* P *in* D′ *is an extreme point for* f *on* D, *then* P *must also be an extreme point for* f *on* D′.

The converse of (1.6), however, fails.

(1.7)   *If* P *in* D′ *is an extreme point for* f *on* D′, *it does not necessarily follow that* P *is an extreme point for* f *on* D. See examples 2 and 3 above.

In the remaining sections of this chapter, we present methods for finding the extreme points (and corresponding values) of a given scalar field (or function). In most of our examples, we shall try to solve both the global max-min problem and the local max-min problem for the given field or function and given domain (that is to say, we shall try to find the location and nature of all the extreme points for that problem).

Remarks. (1) Several other definitions are sometimes used in connection with max-min problems. Given f on D, we say that P is a *strong global maximum point* for f, if for all points Q in D, $Q \neq P \Rightarrow f(P) > f(Q)$. We say that P is a *weak local maximum point* for f, if there is some neighborhood U of P such that for all points Q in both U and D, $f(P) \geq f(Q)$. *Strong global minimum point* and *weak local minimum point* are defined similarly. In example 1 above, b_2 is a strong global minimum point, the interval $[b_4,b_5]$ consists of weak local maximum points, and the points in the interval (b_4,b_5) are also weak local minimum points. We shall make infrequent use of the concepts of strong global and weak local points.

(2) In many applications, local extreme points represent points of stability for a physical system. Let a particle be acted on by a conservative field of force which occupies a region D in E^3 and which varies from point to point in D. A point Q in D is called a point of *stable equilibrium* if, when the particle is placed at Q, any slight displacement of the particle away from Q subjects the particle to a "restoring force" (caused by the field) in the direction of Q. Thus, if the particle is "nearly at rest" near Q, it must remain permanently near Q. If f is a scalar field of *potential energy* for the given particle at various positions in the given force field, then a point Q of stable equilibrium must be a point where the value of f is smaller than at all immediately nearby points. That is to say, Q is a point of stable equilibrium $\Leftrightarrow$ Q is a local minimum point for f.

§12.2 The Max-Min Existence Theorem

Let f be a scalar field (or scalar function) on domain D. The *max-min existence theorem* ((2.5) below) tells us that if (i) f is continuous on D, and (ii) the domain D has a special geometric property called "compactness," then there must be at least one global maximum point for f on D and at least one global minimum point for f on D. The existence of these global extreme points implies, in turn, that there is a maximum value for f on D which f actually reaches, and a minimum value for f on D which f actually reaches.

We first define *compactness*, then restate the theorem, and then consider some of its applications. The proof of the theorem is outlined in the problems.

We begin with some geometrical definitions. Let f be a field (or function) on domain D. Then D is contained in $\mathbf{E}^3$ or $\mathbf{E}^2$ or $\mathbf{R}$. Let P be a point in the space ($\mathbf{E}^3$ or $\mathbf{E}^2$ or $\mathbf{R}$) that contains D. (P is not necessarily in D.)

(2.1)   Definitions.

(i) We say that P is an *interior point of* D if there is some neighborhood U of P such that U is entirely contained in D. (It follows that if P is an interior point of D, then P must be in D.)

(ii) We say that P is a *boundary point of* D if every neighborhood U of P contains at least one point in D and at least one point not in D. (A boundary point of D may or may not itself be a member of D.)

(iii) We say that P is a *boundary point in* D if P is a boundary point of D which happens also to be in D.

Example 1.   (a) *Let* D *be the interval* $1 < x \le 2$ *in* **R**. Then the interior points of D are the points in the interval $1 < x < 2$); the boundary points of D are the points 1 and 2; 2 is a boundary point *in* D; but 1 is not a boundary point in D.

(b)   *Let* U *be a neighborhood of radius* r *of a point* P. Let Q be any point such that $d(P,Q) < r$. Then Q is an interior point of U. (To see this, take $r' = r - d(P,Q)$. Then the neighborhood of Q of radius r′ must lie entirely in U. (See [2–1]).)

These definitions imply the following (by elementary logic).

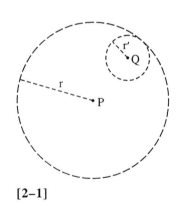

[2–1]

(2.2)   Theorem. *Let* D *be the domain of a scalar field (or function)* f. *Then*

(i)   every point in D *is either an interior point of* D *or a boundary point in* D, *and*

(ii)   no point can be both an interior point of D *and a boundary point of* D.

(2.3)   Definition. Let D be the domain of a field (or function) f. We say that

(i)   D is *open* if D contains no boundary point of D (and hence every point in D is an interior point of D);

(ii)   D is *closed* if D contains every boundary point of D.

Example 2. (a) *If* D *is all of* $\mathbf{E}^2$, *then* D *is both open and closed, since* D *has no boundary points.*

(b) *Let* U *be a neighborhood of a point* P. Since every boundary point of U is in U, U must be closed.

(c) *Given a point* P *and a positive real number* r, *let* V *be the set of all points* Q *such that* d(P,Q) < r. Then every point in V is an interior point of V, and V must be an open set.

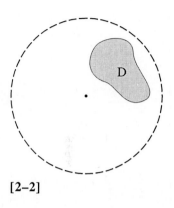

Definition. As in §1.6, we say that a set D (in $\mathbf{R}$, $\mathbf{E}^2$, or $\mathbf{E}^3$) is *bounded* if D is contained in some sufficiently large neighborhood of some point. (See [2–2].) (For D in $\mathbf{E}^3$, this simply means that D can be enclosed is some sufficiently large sphere. For D in $\mathbf{E}^2$, it means that D can be enclosed in some sufficiently large circle. For D in $\mathbf{R}$, it means that there are points a and b such that every point in D lies between a and b.)

[2–2]

Sets which are both closed and bounded are especially important in maximum-minimum problems, and we give these sets a special name.

(2.4) *Definition.* A set D (in $\mathbf{R}$, $\mathbf{E}^2$, or $\mathbf{E}^3$) is said to be *compact* if D is closed and bounded.

Example 3. The following examples are all from $\mathbf{E}^2$ with Cartesian coordinates.

(a) *The set* D *of all points* (x,y) *such that* $0 \le x \le 1$ *and* $0 \le y \le 1$. This set is compact, since it forms a square, and therefore bounded, region which includes all of its own boundary points.

(b) *The set of all points* (x,y) *such that* $x^2+y^2 \le 1$. This set is compact, since it forms a circular, and therefore bounded, disk which includes all of its own boundary points.

(c) *The set* D *of all points* (x,y) *such that* $x^2+y^2 = 1$. This set is compact. Here D has the special feature that every point in D is, in fact, a boundary point of D and that D has no other boundary points.

(d) *The set* D *of all points* (x,y) *such that* $0 \le x$ *and* $0 \le y$. This set is not compact. D is closed, but D is not bounded.

(e) *The set* D *of all points* (x,y) *such that* $x^2+y^2 < 1$. This set is not compact. Here D is bounded but not closed, since the boundary points on $x^2+y^2 = 1$ are not in D.

(f) *The set* $D = \mathbf{E}^2$ is not compact. D is closed but not bounded.

(g) *The line* x = 1 is not compact. Here D is closed but not bounded.

(h) *If* D *is a finite set of points in* $\mathbf{E}^2$, *then* D *is compact.*

We can now restate the theorem.

(2.5)   *Maximum-minimum existence theorem.* Let f *be a continuous scalar field (or function) whose domain is compact. Then* f *must have at least one global maximum point and at least one global minimum point.*

Example 4. In E^2-*with-Cartesian-coordinates, let* f *be a scalar field represented by the function* $f(x,y) = x^2 y$ *and have, as its domain* D, *the circle* $x^2 + y^2 = 1$. It is easy to show, by elementary calculus or by methods to be described in Chapter 13, that the only possible extreme points are the six points $\left(\pm \frac{\sqrt{6}}{3}, \frac{\sqrt{3}}{3} \right)$, $\left(\pm \frac{\sqrt{6}}{3}, -\frac{\sqrt{3}}{3} \right)$, and $(0, \pm 1)$. At the points $\left(\pm \frac{\sqrt{6}}{3}, \frac{\sqrt{3}}{3} \right)$, f has the

value $\frac{2}{9}\sqrt{3}$, at the points $\left(\pm \frac{\sqrt{6}}{3}, -\frac{\sqrt{3}}{3} \right)$, f has the value $-\frac{2}{9}\sqrt{3}$; and at the

points $(0, \pm 1)$, f has the value 0. By example 3c above, D is *compact.* Moreover, f is *continuous* because it is represented by a polynomial function (§8.7). By the max-min existence theorem (2.5), there must be at least one global maximum point and at least one global minimum point. Hence $\left(\pm \frac{\sqrt{6}}{3}, \frac{\sqrt{3}}{3} \right)$ must be global maxi-

mum points, and $\frac{2}{9}\sqrt{3}$ is the global maximum value. Similarly, $\left(\pm \frac{\sqrt{6}}{3}, -\frac{\sqrt{3}}{3} \right)$

must be global minimum points, and $-\frac{2}{9}\sqrt{3}$ is the global minimum value. It is

easy to show that these four points are also local extreme points, as we see in the problems, and that the points $(0, \pm 1)$ are local extreme points.

Further uses of the existence theorem appear in the remaining sections of this chapter and in Chapter 13.

§12.3 Critical Points

In this section and in §12.4, we consider the problem of finding and identifying those extreme points which happen to be *interior points* of D. In Chapter 13, we consider the problem of finding and identifying those extreme points which happen to be *boundary points* in D. In the present section, we assume that f is differentiable.

(3.1) **Definition.** Let f be a differentiable scalar field (or function) on domain D. We say that P in D is a *critical point for* f if $\vec{\nabla}f\big|_P = \vec{0}$. Since we can visualize the gradient vector without reference to a particular coordinate system, we can also visualize critical points of a differentiable scalar field without reference to a coordinate system: a critical point is simply a point where, in every direction, the directional derivative is zero. (Otherwise, we must have $\vec{\nabla}f\big|_P \neq \vec{0}$.)

Our approach to finding interior points which are extreme points is based on the following theorem, which is proved at the end of this section.

(3.2) **The critical-point theorem.** *Let* f *be a differentiable scalar field or scalar function on domain* D. *Let* P *be an interior point of* D *and also an extreme point of* f. *Then* P *must be a critical point for* f.

How do (3.1) and (3.2) help us to solve max-min problems? (3.1) helps us to find the set of critical points. (3.2) provides us with the knowledge that every interior extreme point must be included among those critical points. (3.2) does not tell us which critical points are extreme points. We shall use two further means to make this identification and to discover the specific nature of these extreme points. One of these is the *max-min existence theorem* (see example 4 in §12.2). The other is *second-order derivative tests* (described in §12.4).

Example. *Assume Cartesian coordinates in* E^2, *and let a scalar field* f *be represented by* $f(x,y) = x^3 + xy + y^3$ *on* $D = E^2$. To find the critical points of f, we set

$$\vec{\nabla}f = (3x^2 + y)\hat{i} + (x + 3y^2)\hat{j} = \vec{0},$$

and solve the system of equations

$$\left.\begin{array}{r}3x^2 + y = 0 \\ x + 3y^2 = 0\end{array}\right\}.$$

The solutions are $(-\frac{1}{3}, -\frac{1}{3})$ and $(0,0)$. Since D has no boundary points, these are the only possible extreme points. In §12.4, we get further information about the points $(-\frac{1}{3}, -\frac{1}{3})$ and $(0,0)$. In particular we shall see that $(-\frac{1}{3}, -\frac{1}{3})$ is a local maximum point, while $(0,0)$ is not an extreme point.

In order to prove the critical point theorem (3.2), we first prove the following lemma.

(3.3) **Lemma.** *Let* f *be a differentiable scalar field (or function) on domain* D. *Let* P *be an interior point of* D *and also an extreme point for* f. *Then in every direction* û *at* P, *the directional derivative* $\left.\dfrac{df}{ds}\right|_{\hat{u},P}$ *must be 0.*

Proof. Assume that $\left.\dfrac{df}{ds}\right|_{\hat{u},P} > 0$ for some direction û. Let L be the line through P parallel to û. Then

$$\lim_{\Delta s \to 0} \frac{\Delta f}{\Delta s} > 0,$$

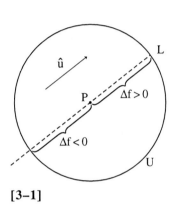

[3–1]

where Δf and Δs are defined as in §8.3. (Recall that $\Delta s = \overrightarrow{PQ} \cdot \hat{u}$, where Q is a variable point on L.) From the definition of limit, there must be some neighborhood U of P such that U is contained in D and for all points Q on L and in U, $Q \neq P \Rightarrow \dfrac{\Delta f}{\Delta s} > 0$. (See [3–1].) Let r be the radius of U. It follows that $0 < \Delta s \leq r \Rightarrow \Delta f > 0$, and that $-r \leq \Delta s < 0 \Rightarrow \Delta f < 0$. Hence, in every neighborhood of P, f must take on values > f(P) and values < f(P). Thus, by the definition of *extreme point*, P cannot be an extreme point for f. The proof for $\left.\dfrac{df}{ds}\right|_{\hat{u},P} < 0$ is similar. This completes the proof of (3.3).

The critical point theorem (3.2) now follows from (3.3) by the comment in (3.1) above.

§12.4 Derivative Tests at Critical Points

We now present a method for exploring the specific nature of each interior critical point (of a given field or function f) as a possible extreme point. This method is known as the *second-order derivative test for a local extreme point*. It requires the additional assumption that f be in C^2. In multivariable calculus, the derivative test for a function uses the second order derivatives of that function. In order to apply the test to a scalar field, we introduce Cartesian coordinates, represent the field as a function, and apply the derivative test to that function.

   We give two forms of the test: the *test for two-variable functions* and the *test for three-variable functions*.   The correctness of the two-variable test is proved in §12.6; the three-variable test is verified in Chapter 34.

Definition. Let f be a scalar field or function with domain D. Let D′ be a subset of D. Then $f|_{D'}$ denotes the new scalar field or function which coincides with f on D′ and which has D′ as its entire domain. $f|_{D'}$ is called the *restriction* of f to the new and smaller domain D′.

Definition. Let f be a scalar field or function with domain D. Let P be an interior point of D. P is called a *saddle point* of f if there exist straight lines L_1 and L_2 through P such that: P is a relative-maximum point for $f|_{L_1'}$, and P is a relative-minimum point for $f|_{L_2'}$, where L_1' is the portion of L_1 contained in D, and L_2' is the portion of L_2 contained in D.

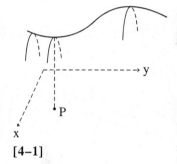

[4–1]

   Evidently, a saddle point of f cannot be an extreme point for f. In the case of a two-variable C^2 function f, we shall see in §21.8 that the graph of f, over a sufficiently small neighborhood of a saddle point, has the "saddle-shaped" form illustrated in [4–1]. (Of course, in this figure, the saddle point itself is the point P down in the domain of f.)
   When we apply a derivative test, we hope to discover whether a given critical point is a local maximum point, a local minimum point, or a saddle point (it may also be none of these).
   Recall from elementary calculus, the derivative test for a one-variable function:

(4.1)   *If f is a twice differentiable one-variable function, and if c is a critical point for f, then* $[f''(c) > 0 \Rightarrow c$ *is a local minimum point] and* $[f''(c) < 0 \Rightarrow c$ *is a local maximum point].*

   The derivative tests in multivariable calculus are based on (4.1). The underlying idea is straightforward. Let f be a given scalar field (or function) in E^3 or E^2 and in C^2. For a given point in the domain of f and a given direction $\hat{u}$, we define $\left.\dfrac{d^2f}{ds^2}\right|_{\hat{u},P}$, *the second-order directional derivative of f at P in the direction* $\hat{u}$, in the following natural way: we take a number line L with origin at P and positive axis in the direction $\hat{u}$. Our given scalar field (or function) f determines a one-variable function $\tilde{f}(s)$ on L. This function will be in C^2. $\tilde{f}'(0)$ is the directional derivative $\left.\dfrac{df}{ds}\right|_{\hat{u},P}$ as previously defined in §8.3. We now define

$\left.\dfrac{d^2f}{ds^2}\right|_{\hat{u},P}$ to be $\tilde{f}''(0)$. To carry out our derivative tests in $\mathbf{E}^3$ or $\mathbf{E}^2$, we use (4.1) as follows:

(4.2)   **Theorem.** *If* P *is a critical point of* f, *then*

      (i)   *if for every direction* $\hat{u}$, $\left.\dfrac{d^2f}{ds^2}\right|_{\hat{u},P} < 0$, *then* P *must be a local*

maximum point for f;

      (ii)   *if for every direction* $\hat{u}$, $\left.\dfrac{d^2f}{ds^2}\right|_{\hat{u},P} > 0$, *then* P *must be a local*

minimum point for f;

      (iii)   *if* $\left.\dfrac{d^2f}{ds^2}\right|_{\hat{u},P} > 0$ *for some direction* $\hat{u}$ *and* $\left.\dfrac{d^2f}{ds^2}\right|_{\hat{u},P} < 0$ *for some*

other direction $\hat{v}$, *then* P *must be a saddle point for* f.

    Proof. Immediate from (4.1).

    For (4.2) to be useful, we must represent a given scalar field f as a function f in some coordinate system and then express $\left.\dfrac{d^2f}{ds^2}\right|_{\hat{u},P}$ in terms of the derivatives of this function. With Cartesian coordinates in $\mathbf{E}^2$, we get

(4.3) $$\left.\dfrac{d^2f}{ds^2}\right|_{\hat{u},P} = f_{xx}(P)u_1^2 + 2f_{xy}(P)u_1u_2 + f_{yy}(P)u_2^2 ,$$

where $\hat{u} = u_1\hat{i} + u_2\hat{j}$. (See (6.3) in §12.6.) Similarly in $\mathbf{E}^3$, we find

(4.4) $$\left.\dfrac{d^2f}{ds^2}\right|_{\hat{u},P} = f_{xx}(P)u_1^2 + f_{yy}(P)u_2^2 + f_{zz}(P)u_3^2 +$$

$$2f_{xy}(P)u_1u_2 + 2f_{xz}(P)u_1u_3 + 2f_{yz}(P)u_2u_3,$$

where $\hat{u} = u_1\hat{i} + u_2\hat{j} + u_3\hat{k}$.

    In §12.6, we carry out an algebraic analysis of (4.3), and in §34.5, we use linear algebra to make a corresponding analysis of (4.4). We obtain the following convenient and useful results.

The two-variable test.

Definition. Let f(x,y) be a C^2 function of two variables. We use the second-order derivatives of f (and determinant notation) to define the following two-variable functions. For each P in the domain of f,

$$H_1(P) = f_{xx}(P),$$

and
$$H_2(P) = \begin{vmatrix} f_{xx}(P) & f_{xy}(P) \\ f_{yx}(P) & f_{yy}(P) \end{vmatrix}$$

$$= f_{xx}(P)f_{yy}P) - (f_{xy}(P))^2 \qquad \text{(by the equality of cross-derivatives).}$$

H_1 and H_2 are called the *Hessian functions* for f.

(4.5)   *Theorem. (Derivative test for a two-variable function.) Let f be a two-variable C^2 function. Let P be an interior critical point for f. Then*
   (i)   [$H_2(P) > 0$ and $H_1(P) > 0$] $\Rightarrow$ P is a local minimum point for f,
   (ii)   [$H_2(P) > 0$ and $H_1(P) < 0$] $\Rightarrow$ P is a local maximum point for f,
   (iii) $H_2(P) < 0 \Rightarrow$ P is a saddle point for f.

Example 1. Let $f(x,y) = x^3+xy+y^3$. In the example in §12.3, we saw that $P_1 = (-\frac{1}{3},-\frac{1}{3})$ and $P_2 = (0,0)$ are the critical points for f. Calculating the Hessian functions, we have

$$H_1(x,y) = f_{xx}(x,y) = 6x,$$

and
$$H_2(x,y) = \begin{vmatrix} 6x & 1 \\ 1 & 6y \end{vmatrix} = 36xy - 1.$$

[4-2]

For P_1 this gives $H_2(P_1) = 4-1 = 3$ and $H_1(P_1) = -2$. Hence, by the derivative test, P_1 is a local maximum point. P_1 is not a global maximum point, however, since f has values greater than $f(P_1) = \frac{1}{27}$; for example, $f(1,1) = 3$. For P_2, we have $H_2(P_2) = -1 < 0$. Hence, by the derivative test, P_2 is a saddle point. The graph of f is indicated in [4-2], which shows two curves lying in the graph of f: the curve where $f(x,-x) = -x^2$, and the curve where $f(x,x) = 2x^3+x^2 = 2x^2(x+\frac{1}{2})$.

Example 2. If $H_2(P) = 0$, it is still possible for P to be an extreme point, or saddle point, and it is also possible for P to be neither of these. In each of the following examples, P = (0,0) is a critical point and $H_2(P) = 0$:

(a) $f(x,y) = x^4 + y^4$; here P is a local minimum point.

(b) $f(x,y) = x^3 y$; here P is a saddle point.

(c) $f(x,y) = x^3 + y^3$; here P is neither a local extreme point nor a saddle point.

There is, however, a partial converse to the two-variable derivative test:

(4.6)   *If $H_1(P)$ and $H_2(P)$ are not both zero, then the only way for an interior critical point P to be a local extreme point or a saddle point is for one of the three Hessian conditions in (4.5) to hold.*

The three variable test.

Definition. Let $f(x,y,z)$ be a C^2 function of three variables representing a scalar field f. We use the second-order derivatives of f (and determinant notation) to define the following three-variable functions. For each P in the domain of f,

$$H_1(P) \ = \ f_{xx}(P),$$

$$H_2(P) \ = \ \begin{vmatrix} f_{xx}(P) & f_{xy}(P) \\ f_{yx}(P) & f_{yy}(P) \end{vmatrix},$$

and

$$H_3(P) \ = \ \begin{vmatrix} f_{xx}(P) & f_{xy}(P) & f_{xz}(P) \\ f_{yx}(P) & f_{yy}(P) & f_{yz}(P) \\ f_{zx}(P) & f_{zy}(P) & f_{zz}(P) \end{vmatrix}.$$

H_1, H_2, and H_3 are called the *Hessian functions* for f.

(4.7)   **Theorem.** *(Derivative test for a three-variable function.) Let f be a three-variable C^2 function. Let P be an interior critical point for f. Then*

(i) $[H_2(P) > 0, H_1(P) > 0, and H_3(P) > 0] \Rightarrow P$ *is a local minimum point for f;*

(ii) $[H_2(P) > 0, H_1(P) < 0, and H_3(P) < 0] \Rightarrow P$ *is a local maximum point for f;*

(iii) $H_2(P) < 0 \Rightarrow P$ *is a saddle point for f.*

(4.7) is an abbreviated statement of the three-variable test. There are other cases which also yield information about P. For example, if either $[H_1(P) \geq 0$ and $H_3(P) < 0]$ or $[H_1(P) \leq 0$ and $H_3(P) > 0]$, then P must be a saddle point. We can however, make the following general statement:

(4.8)   *if* $H_1(P)$, $H_2(P)$, *and* $H_3(P)$ *are not all* 0, *then the only way for an interior critical point* P *to be a local extreme point is for one of the first two Hessian conditions in* (4.7) *to hold.*

Example 3. Let $f(x,y,z) = x^3 + x^2 + y^2 + z^2 - xy + xz$. The reader may verify that $P_1 = (0,0,0)$ and $P_2 = (-\frac{1}{3}, -\frac{1}{6}, \frac{1}{6})$ are the critical points, and that

$$H_1(x,y,z) = 6x + 2,$$
$$H_2(x,y,z) = 12x + 3,$$
and $$H_3(x,y,z) = 24x + 4$$

are the Hessian functions. Since $H_1(P_1) > 0$, $H_2(P_1) > 0$, and $H_3(P_1) > 0$, P_1 must be a local minimum point. We also note that P_1 is not a global minimum point, since for example, $f(-2,0,0) = -4$, while $f(P_1) = 0$. Since $H_2(P_2) = -1 < 0$, P_2 must be a saddle point.

Remark. As we have noted, derivative tests may not identify all the interior extreme points. Other algebraic or geometric techniques, tailored to the given function or field f, may be needed. The max-min existence theorem is often helpful in these cases.

§12.5 Complete Solutions to Maximum-Minimum Problems

In a max-min problem, we have a given scalar field or function f on a specified domain D, and we seek to locate and identify extreme points of f. In §12.3, we have described a method for identifying a subset of the set of all *interior* points of D, where this subset (the interior *critical* points of f) must include among its members all *interior* extreme points of f. Similarly, in the next chapter (§13.5), we describe a method for identifying a subset of the set of all *boundary* points of D, where this subset must include among its members all *boundary* extreme points of f. Using these two methods together, we can look for a complete answer to the problem of finding all extreme points for f on D. Our general procedure will be as follows.

Step I. We divide D into two subsets: D_i = the set of interior points of D, and D_b = the set of those boundary points of D which happen to lie in D.

Step II_i. We try to find a "small" (usually *finite*) subset S_i of D_i with the property that every extreme point (of f on D) which lies in D_i must necessarily lie in S_i.

Step II_b. We try to find a "small" (usually *finite*) subset S_b of D_b with the property that every extreme point (of f on D) which lies in D_b must necessarily lie in S_b.

Steps III_i *and* III_b. We examine each individual point in S_i or S_b and decide whether or not it is a global or local, maximum or minimum point.

The following example illustrates the above steps for the simple case of a one-variable, differentiable function f.

Example. Consider the function $f(x) = x^3 - 3x$ on the domain $D = [-2,3]$.

Step I. We see that D_i is the interval $(-2,3)$ and that $D_b = \{-2,3\}$. Thus D_b consists of two points.

Step II_i. From elementary calculus, we know that an interior point c of D can be an extreme point *only if* $f'(c) = 0$. Thus we take S_i to be the set of c in D_i such that $f'(c) = 0$. In our example, $f'(x) = 3x^2 - 3$. Thus $3x^2 - 3 = 0 \Rightarrow x^2 - 1 = 0 \Rightarrow x = \pm 1$. Hence $S_i = \{-1, 1\}$.

Step II_b. Since D_b is already finite, we take $S_b = D_b = \{-2,3\}$.

Steps III_i *and* III_b. We have four points to consider: $-2, -1, 1, 3$. Evaluating f at each point, we find the values: $f(-2) = -2$, $f(-1) = 2$, $f(1) = -2$, $f(3) = 18$. D is compact, since it is closed and bounded; and f on D is continuous, since f is a polynomial. Hence, by the existence theorem, f on D must have at least one global maximum point and at least one global minimum point. From steps II_i and II_b, we know that any global maximum or global minimum point must occur among the four points: $-2, -1, 1, 3$. From the values of f just obtained, we conclude: (i) there are two global minimum points, $x = -2$ and $x = 1$; (ii) the global minimum value of f is -2; (iii) there is a unique global maximum point, $x = 3$; and (iv) the global maximum value of f is 18.

Since $x = -2$ and $x = 1$ are the only global minimum points, we see that each must also be a local minimum point. (Consider a neighborhood of radius 1 for each point.) Similarly, $x = 3$ must also be a local maximum point. Finally, by applying the same analysis to $g = f|_{[-2,1]} = $ the restriction of f to the new and smaller domain $[-2,1]$, we see that $x = -1$ is the unique global maximum point for g, and hence that x = −1 must be a local maximum point for f (consider the neighborhood of radius 1 for x = −1). Graphs of f and g are shown in [5–1] and [5–2]. Note that the logical analysis above does not depend on or use the geometrical information given in these figures.

§12.6 Proof for the Two-variable Derivative Test

For the *two-variable derivative test* (4.1), we assume that f is a two-variable function with continuous second-order derivatives. Let H_1 and H_2 be the *Hessian functions* for f as defined in §12.4. (4.1) then asserts:

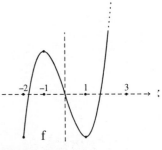

[5–1]

[5–2]

(6.1)   *For an interior point of the domain of* f:

  (i) $[\vec{\nabla}f\big|_P = \vec{0}$ *and* $H_1(P) > 0$ *and* $H_2(P) > 0] \Rightarrow$ P *is a local minimum point.*

  (ii) $[\vec{\nabla}f\big|_P = \vec{0}$ *and* $H_1(P) < 0$ *and* $H_2(P) > 0] \Rightarrow$ P *is a local maximum point.*

  (iii) $[\vec{\nabla}f\big|_P = \vec{0}$ *and* $H_2(P) < 0] \Rightarrow$ P *is a saddle point.*

To prove that the test is correct, we recall definitions and a theorem from elementary calculus. In particular, let g be a one-variable function. We say that g is an *increasing function* on the interval [a,b] if, for all c and d in [a,b], $c < d \Rightarrow g(c) < g(d)$.

(6.2)   **Theorem.** *If a one-variable function* g *is differentiable on* [a,b], *and if* $g' > 0$ *on* (a,b), *then* g *must be an increasing function on* [a,b]. (The mean-value theorem is used to prove this theorem.)

A corollary is immediate.

(6.3)   **Corollary.** *If* g *has a second derivative* g″, *then* $g'' > 0$ *on* (a,b) $\Rightarrow g'$ *is increasing on* [a,b].

This corollary leads us to the *one-variable derivative test*. We state this test for local minimum points and in the following form.

(6.4)   *If* g *has a second derivative* g″, *then* $[[g'(c) = 0$ *and for some* $r > 0$, $g'' > 0$ *on* (c−r,c+r)] $\Rightarrow g(c) < g(x)$ *for all* $x \neq c$ *in* [c−r,c+r]]. (*Thus* c *is a local minimum point for* g.)

(6.4) is true, since $[g'(c) = 0$ and $g'' > 0$ on $(c-r,c+r)] \Rightarrow [g' < 0$ on $(c-r,c)$ and $g' > 0$ on $(c,c+r)] \Rightarrow [g$ decreasing on $[c-r,c]$ and g increasing on $[c,c+r]] \Rightarrow g(c) < g(x)$ for all $x \neq c$ in $[c-r,c+r]$.

We shall use (6.4) to prove (6.1). (Simpler and more familiar versions of the one-variable test may also be deduced from (6.4). See the problems.)

Proof of (6.1).

Proof of (i): Let $\hat{u} = u_1\hat{i} + u_2\hat{j}$ be a fixed direction. Let $f_{(\hat{u})}(Q)$ be another notation for $\left.\dfrac{df}{ds}\right|_{\hat{u},Q}$. Then $f_{(\hat{u})}(Q) = \left.\dfrac{df}{ds}\right|_{\hat{u},Q} = \left.\vec{\nabla}f\right|_Q \cdot \hat{u} = f_x(Q)u_1 + f_y(Q)u_2$. Let $g(Q) = f_{(\hat{u})}(Q)$. Then $\left.\dfrac{dg}{ds}\right|_{\hat{u},Q}$ is the "second-order directional derivative" of f in direction $\hat{u}$ at Q. We write it more briefly as $\left.f''\right|_{\hat{u},Q}$. We have

$$(6.5) \qquad \left.f''\right|_{\hat{u},Q} = \left.\vec{\nabla}g\right|_Q \cdot \hat{u} = \left.\vec{\nabla}f_{(\hat{u})}\right|_Q \cdot \hat{u}$$

$$= \left[(f_{xx}(Q)u_1 + f_{yx}(Q)u_2)\hat{i} + (f_{xy}(Q)u_1 + f_{yy}(Q)u_2)\hat{j}\right] \cdot (u_1\hat{i} + u_2\hat{j})$$

$$= f_{xx}(Q)u_1^2 + 2f_{xy}(Q)u_1u_2 + f_{yy}(Q)u_2^2.$$

We abbreviate this last expression as $Au_1^2 + 2Bu_1u_2 + Cu_2^2$, keeping in mind that A, B, and C depend on Q.

Let P be a point such that $\left.\vec{\nabla}f\right|_P = \vec{0}$ and such that $H_1(P) > 0$ and $H_2(P) > 0$. H_1 and H_2 are continuous functions, since the second-order derivatives of f are continuous. Hence there is a neighborhood U of P such that $H_1 > 0$ on U and $H_2 > 0$ on U. By the definitions of H_1 and H_2, we have $A > 0$ on U and $AC - B^2 > 0$ on U.

By completing the square, we get the identity

$$(6.6) \qquad Au_1^2 + 2Bu_1u_2 + Cu_2^2 = Au_1^2 + 2Bu_1u_2 + (\tfrac{B^2}{A})u_2^2 + (C - \tfrac{B^2}{A})u_2^2$$

$$= (\sqrt{A}\,u_1 + \tfrac{B}{\sqrt{A}}u_2)^2 + (C - \tfrac{B^2}{A})u_2^2.$$

Now $\qquad\qquad\qquad |\hat{u}| = 1 \Rightarrow [u_1 \neq 0 \text{ or } u_2 \neq 0],$

and $\qquad\qquad [AC - B^2 > 0 \text{ and } A > 0] \Rightarrow C - \dfrac{B^2}{A} > 0.$

Thus $u_2 \neq 0 \Rightarrow f''|_{\hat{u},Q} > 0$ for all Q in U, since $f''|_{\hat{u},Q}$ is the sum of a non-negative term and a positive term, while $u_2 = 0 \Rightarrow u_1 \neq 0 \Rightarrow f''|_{\hat{u},Q} = Au_2^2 > 0$ for all Q in U. Thus, for every direction $\hat{u}$, $f''|_{\hat{u},Q} > 0$ for all Q in U.

For each direction $\hat{u}$, let L be the line through P parallel to $\hat{u}$. Apply (6.4) to $f|_{L'}$, where L' is the portion of L in U. This gives f(P) < f(Q) for every $Q \neq P$ in U and on L. Since this is true for every $\hat{u}$, we have f(P) < f(Q) for all $Q \neq P$ in U. Thus P is a local minimum point. This proves (i).

Proof of (ii). Define g = –f. Then the Hessian functions for g must both be positive at P. By (i), g has a local minimum at P. Hence f has a local maximum at P. This proves (ii).

Proof of (iii). Assume first that $H_1(P) > 0$. Then in (6.5) and (6.6) for $f''|_{\hat{u},Q}$, the coefficient $C - \dfrac{B^2}{A}$ is negative. Consider two cases:

(1) $\hat{u} = \hat{i}$. Then $u_2 = 0$ and $f''|_{\hat{u},Q} = A > 0$.

(2) $\hat{u} = \hat{v}$, where $\bar{v} = \dfrac{B}{\sqrt{A}}\hat{i} - \sqrt{A}\hat{j}$. In this case, $u_1 = \dfrac{B}{\sqrt{A}}$, and $u_2 = -\sqrt{A}$, and $\sqrt{A}u_1 + \dfrac{B}{\sqrt{A}}u_2 = 0$. Hence $f''|_{\hat{u},P} = (C - \dfrac{B^2}{A})u_2^2 < 0$. It follows from (6.4) that on a line L through P parallel to $\hat{u}$, f has a local minimum at P in case (1) and a local maximum at P in case (2). By definition, P must be a saddle point.

For $H_1(P) < 0$, the proof is similar, using g = –f.
For $H_1(P) = f_{xx}(P) = 0$, either $f_{yy}(P) \neq 0$ or $f_{yy}(P) = 0$.
If $f_{yy}(P) \neq 0$, the proof is as for $H_1(P) \neq 0$, interchanging A and C. If $f_{yy}(P) = 0$, we have

$$H_2(P) = -(f_{xy}(P))^2 < 0.$$

Assume $f_{xy}(P) > 0$. Then $f''|_{\hat{u},Q} = 2f_{xy}(P)u_1u_2$ must be positive if $u_1u_2 > 0$ and negative if $u_1u_2 < 0$. Since u_1u_2 is positive for a direction such as $\frac{3}{5}\hat{i} + \frac{4}{5}\hat{j}$ and negative for a direction such as $\frac{3}{5}\hat{i} - \frac{4}{5}\hat{j}$, we again have a saddle point. The proof for $f_{xy}(P) < 0$ is similar. This proves (iii).

§12.7   Problems

§12.1

1. Let f(x) = sin x with [0,3π] as domain. Describe the global and local extreme points and values for f.

2. Let g(x) = cos x with [0,2π] as domain. Describe the global and local extreme points and values for g.

3. Let f(x) be continuous and have the domain [a,b]. Assume that $f'(x) < 0$ for all x in (a,b). What can you conclude about the global and local extreme points and values for f ?

4. Let f(x) be defined on the interval [–a,a]. Assume that $f'(x) < 0$ for $-a < x < 0$ and $f'(x) > 0$ for $0 < x < a$. What can you conclude about the global and local extreme points and values for f ?

§12.2

1. Prove (2.5), the max-min existence theorem.
   (*Hint*. Use the following steps.
   (i) Define a *limit point* P of a set S of points to be a point such that every neighborhood of P contains at least one member of S different from P.
   (ii) Show that if D is a compact set, then every infinite subset S of D has a limit point in D. Prove this by dividing D into a finite number of smaller pieces, each of which has "diameter" ≤ half the diameter of D. At least one of these pieces must contain infinitely many members of S. Repeating this division process again and again, we narrow down to the desired limit point.
   (iii) *Assume f on D has no maximum value.* Then we can find an infinite set S of points in D such that the set of values on S has no bound in the set of all values of f on D.
   (iv) Use (ii) and the definition of continuity for f to get a contradiction.)

2. For any set S in E^3, define *Int*(S) to be the set of interior points of S. What can you conclude about *Int*(*Int*(S))?

3. Let S be a bounded set in E^3. Let *Bd*(S) be the set of boundary points of S.
   (a) Show that *Bd*(S) cannot be empty.
   (b) Show that *Bd*(S) is closed.
   (c) Show that *Bd*(S) is compact.

4. Let S be a bounded set. Define *Int*(S) and *Bd*(S) as in problems 2 and 3. What can you conclude about
   (a) *Bd*(*Bd*(S))
   (b) *Bd*(*Int*(S))

5. Give an example of a bounded set S for which *Int* (*Bd* (S)) is not empty.

6. Let D_1 and D_2 be compact sets. Let $D_1 \cap D_2$ be the set of points which lie in *both* D_1 and D_2. Let $D_1 \cup D_2$ be the set of points which lie in *either* D_1 or D_2 or both.
   (a) Show that $D_1 \cap D_2$ must be compact.
   (b) Show that $D_1 \cup D_2$ must be compact.

7. Which of the following sets are closed, which are open, which are neither, and which are compact. We assume Cartesian coordinates.
   (i) The set of all points (x,y) in E^2 such that $0 \le x \le 1$ and $x^2 \le y \le x$.
   (ii) The set of all points (x,y) in E^2 such that $y = x^2$.
   (iii) The set of all points (x,y,z) in E^3 such that $1 < x^2 + y^2 + z^2 < 4$.
   (iv) The set of all points (x,y,z) in E^3 such that $1 \le x^2 + y^2 + z^2 < 4$.

8. Assume that the six points given in example 4 are the only possible extreme points. Show that each of these points is a local extreme point. (*Hint*. Use the max-min existence theorem.)

9. Give details for proving (2.2).

§12.3

1. For each of the following statements, say whether it is *always true* or *sometimes false*, and give reasons. In each case, f is a differentiable scalar field with domain D.

    (i) If P is an interior point of D and a global extreme point, then P must be a local extreme point.

    (ii) If P is an interior point of D and a weak local extreme point, then P must be a critical point.

    (iii) If P is an interior point of D and a strong global extreme point, then P must be a critical point.

§12.4

1. For each of the following functions, use the critical point theorem and an appropriate derivative test or other argument to identify and find the extreme points and values. In each case, the domain is largest possible.

    (a) $f(x,y) = 3x + 12y - x^3 - y^3$

    (b) $f(x,y) = \cos x + \cos y$

    (c) $f(x,y) = 5xy - 7x^2 - y^2 + 3x - 6y + 2$

    (d) $f(x,y) = 5x^2 + 6xy + 2y^2 - 4x - 2y + 7$

    (e) $f(x,y) = x^2 + xy + 3x + 2y + 5$

    (f) $f(x,y) = 2x^2 - xy - 3y^2 - 3x + 7y$

    (g) $f(x,y) = e^{-(x^2+y^2+z^2)}$

    (h) $f(x,y) = \sec(xy)$

    (i) $f(x,y,z) = x^2 + 2y^2 + 2z^2 + 2xy + xz + yz$

    (j) $f(x,y,z) = x^2 + 2y^2 + 2z^2 + 4xy + xz + yz$

    (k) $x^3 - y^3 - 2xy + z^2$

    (l) $3x^2 + 5xy + 2y^2 - x - y + 3$.

    (m) $x + \frac{1}{2}x^2y - \frac{1}{6}y^3$

    (n) $x^2 + 2xy + y^3$

    (o) $x^2 + 2y^2 + 3z^2$

    (p) $x^2 - 2y^2 + 3z^2$

    (q) $x^2 - 2y^2 - 3z^2$

2. Prove (4.6) for the case where f is given by a second degree polynomial.

3. Prove the results stated in example 2.

4. In example 3, show that P_1 and P_2 are the critical points and that H_1, H_2, and H_3 are as given.

Chapter 13

Constrained Maximum-Minimum Problems

§13.1 Constrained Problems

Let a scalar field f give the temperature at each point in a region D of the atmosphere. Let a curve C be the trajectory of a particle moving in D. We can think of C itself as a new *restricted domain* D′ for f, and we can look for extreme points and extreme values of f on D′. In so doing, we might hope to find, for example, the maximum and minimum atmospheric temperatures to which the particle is subjected as it moves along C.

In this chapter, we consider three forms of max-min problems on restricted domains:

    (i) f is given in E^2, and the restricted domain D′ is a curve C in E^2.

    (ii) f is given in E^3, and the restricted domain D′ is a surface S in E^3.

    (iii) f is given in E^3, and the restricted domain D′ is a curve C in E^3.

Example 1. *In* E^2 *with Cartesian coordinates, a scalar field* f *is given by* $f(x,y) = x^2y$ *on* $D = E^2$. D′ *is the circle of radius* 1 *with center at the origin.* (This example appears in Chapter 12 as example 4 in §12.2.) D′ can be described by the equation $x^2 + y^2 = 1$. We refer to this equation as a *constraint*. This constraint presents D′ as a *level curve* of another scalar field, $g(x,y) = x^2 + y^2$.

Example 2. *In* E^3 *with Cartesian coordinates, a scalar field* f *is given by* $f(x,y,z) = xyz$ *on* $D = E^3$. D′ *is the surface* S *described by the equation*

274

x+2y+2z = 108. We again refer to this equation as a *constraint*. This constraint gives D′ as a *level surface* of the scalar field g(x,y,z) = x+2y+2z.

Example 3. In $\mathbf{E}^3$ *with Cartesian coordinates, a scalar field f is given by* $f(x,y,z) = x^2+y^2+z^2$ *on* $D = \mathbf{E}^3$. D′ *is the line through (6,0,2) parallel to* $\hat{i} - \hat{j}$. Here, D′ can be described as the intersection of two chosen surfaces. For example, D′ is the intersection of the plane given by the equation z = 2 with the plane given by the equation x+y = 6. We refer to two such equations as *constraints*. These constraints give D′ as the *intersection of level surfaces* of the scalar fields $g_1(x,y,z) = z$ and $g_2(x,y,z) = x+y$. (Various other pairs of planes could be used to give the same D′.)

We refer to problems of the three kinds described and illustrated above as *constrained max-min problems* and to the restricted domains in these problems as *constrained domains*.

In applied mathematics, a variety of problems can be formulated as constrained max-min problems. In physical applications, constraints usually express existing physical restrictions or limitations. Consider the following.

Example 4. *The U.S. Postal Service will accept a rectangular package for mailing provided that the sum of its girth and its length is not greater than 108 inches. What is the maximum volume for an acceptable package?*

We first formulate this problem as a constrained max-min problem. In $\mathbf{E}^3$-with-Cartesian coordinates, we take the scalar function f(x,y,z) = xyz, where x, y, and z are the length, width, and height of the package in inches; f(x,y,z) then gives the volume. The *girth* of the package is defined to be y+z+y+z = 2y+2z. Hence the condition for acceptability is that x+2y+2z ≤ 108. This condition, together with the condition that x ≥ 0, y ≥ 0, and z ≥ 0, defines a restricted domain D′ for f in the first octant of our coordinate system. D′ is a solid pyramid, bounded, closed, and hence compact; f, moreover, is continuous. Hence, by the max-min existence theorem, there must be a global maximum point, and a global maximum value, for f on D′. We note that any global maximum point (x,y,z) for f on D′ must satisfy the equation x+2y+2z = 108, since the volume f(x,y,z) for a point (x,y,z) with x+2y+z < 108 could be slightly increased by slightly increasing x while still remaining in D′. Hence we can further restrict the domain of f to that portion of the plane x+2y+2z = 108 which lies in the first octant.

This suggests that we formulate the problem of example 4 as the constrained max-min problem in example 2. The constrained problem in example 2 is not exactly the same as example 4, since the constrained domain in example 2 is the entire plane $x+2y+2z = 108$, while the final domain for example 4 is the portion of that plane contained in the first octant (where $x \geq 0$, $y \geq 0$, and $z \geq 0$). Nevertheless, if we can solve example 2 and find an extreme point inside the first octant, then that extreme point will also be an extreme point for example 4 (by (1.6) in §12.1). It will follow, as we shall see, that this extreme point is, in fact, a global extreme point for example 4.

In each of examples 1, 2, and 3, the constrained domain consists entirely of boundary points of the original domain D of f. Since there are no interior points, the interior-point methods of Chapter 12 cannot be directly applied, and other methods must be used. In the remainder of this chapter, we describe two general methods for solving constrained max-min problems: the *Lagrange method* and the *parametric method*. We shall also see, at the end of the chapter, how these constrained-problem methods can be used to carry out a boundary-point analysis (*Steps* IIb and IIIb in §12.5) for an unconstrained maximum-minimum problem in which both interior points and boundary points occur.

§13.2 The Lagrange Method for One Constraint

For the case of a single constraint (examples 1 and 2 of §13.1), we assume:

(2.1) (i) f *and* g *are* C^1.

(ii) D′ *is the set of all points P such that* $g(P) = c$.

(iii) $\vec{\nabla}g\big|_P \neq \vec{0}$ *at all points P in* D′.

For f in $\mathbf{E}^2$, (2.1) ensures that D′ will be a curve with a normal line given by $\vec{\nabla}g\big|_P$, at every point P of D′; while for f in $\mathbf{E}^3$, (2.1) ensures that D′ will be a surface with a normal line given by $\vec{\nabla}g\big|_P$ at every point P of D′. (2.1) also implies that the constrained domain D′ will consist entirely of boundary points.

The Lagrange method uses an idea analogous to the idea of *critical point* in §12.3 and a theorem analogous to the *critical point theorem* in §12.3.

Definition. We define P in D′ to be a *constrained critical point for* f *on* D′ if there is some scalar value λ^P such that

(2.2) $\vec{\nabla}f\big|_P = \lambda^P \vec{\nabla}g\big|_P \,.$

λ^P is called the *Lagrange multiplier* for the constrained critical point P. Note, from (2.1), that P is a constrained critical point $\Leftrightarrow$ either $\vec{\nabla}f\big|_P = \vec{0}$ or $\vec{\nabla}f\big|_P$ is normal to D' at P.

We now state and prove:

(2.3)   **The constrained critical point theorem for a single constraint.** *Let* f, g, *and* c *be as assumed in* (2.1). *If* P *is an extreme point for* f *on* D', *then* P *must be a constrained critical point for* f *on* D'.

Proof. Assume that P is an extreme point for f on D'. Suppose that P is not a constrained critical point. Then we would have $\vec{\nabla}f\big|_P \neq \vec{0}$. Furthermore, if D' is a curve (in $\mathbf{E}^2$), $\vec{\nabla}f\big|_P$ would not be normal to that curve; and if D' is a surface (in $\mathbf{E}^3$), then, by the definition of normal vector in §11.6, $\vec{\nabla}f\big|_P$ would not be normal to that surface. By a similar argument to that for the critical point theorem in §12.3, it would follow that in every neighborhood of P, there would be a point Q_1 in D' such that $f(Q_1) > f(P)$ and a point Q_2 in D' such that $f(Q_2) < f(P)$. (See the problems.) This would imply that P could not be an extreme point, contrary to our assumption that P is an extreme point. Thus our supposition that P is not a constrained critical point must be false. This completes the proof.

In the *Lagrange method*, we first find the constrained critical points and then go on to identify the nature of these points as possible extreme points for f on D'. We therefore begin by finding all points P which satisfy the system of equations:

(2.4)
$$\left(\begin{array}{c} \vec{\nabla}f\big|_P = \lambda\vec{\nabla}g\big|_P \\ g(P) = c \end{array} \right) \text{ for some } \lambda .$$

Example 1 *(continued from §13.1). We have* $f(x,y) = x^2y$, $g(x,y) = x^2 + y^2$, *and* c = 1. Let P have coordinates (x,y). Then (2.4) becomes

$$\left(\begin{array}{c} 2xy\hat{i} + x^2\hat{j} = \lambda(2x\hat{i} + 2y\hat{j}) \\ x^2 + y^2 = 1 \end{array} \right).$$

This gives us a system of three scalar equations in the three unknowns x, y, and λ:

$$\begin{pmatrix} 2xy = 2\lambda x \\ x^2 = 2\lambda y \\ x^2 + y^2 = 1 \end{pmatrix}.$$

We solve this particular system as follows. From the first equation,

$$2xy - 2\lambda x = 0 \Rightarrow 2x(y-\lambda) = 0 \Rightarrow [x = 0 \text{ or } y = \lambda].$$

We consider the two possibilities: $x = 0$ and $y = \lambda$. We have

$$x = 0 \Rightarrow [y = \pm 1 \text{ and } \lambda = 0], \text{ and}$$

$$y = \lambda \Rightarrow x^2 = 2y^2 \Rightarrow 3y^2 = 1 \Rightarrow \lambda = y = \pm\frac{\sqrt{3}}{3} \Rightarrow x = \pm\frac{\sqrt{6}}{3}.$$

Thus the constrained critical points must be

$$(0,1), (0,-1), (\tfrac{1}{3}\sqrt{6},\tfrac{1}{3}\sqrt{3}), (\tfrac{1}{3}\sqrt{6},-\tfrac{1}{3}\sqrt{3}), (-\tfrac{1}{3}\sqrt{6},\tfrac{1}{3}\sqrt{3}), (-\tfrac{1}{3}\sqrt{6},-\tfrac{1}{3}\sqrt{3}).$$

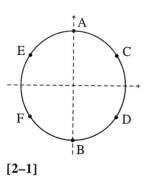

[2–1]

We now complete our application of the Lagrange method by exploring the nature of these points as possible extreme points. In [2–1], we show these points as A, B, C, D, E, F. The values of f at these points are

$$0, 0, \tfrac{2}{9}\sqrt{3}, -\tfrac{2}{9}\sqrt{3}, \tfrac{2}{9}\sqrt{3}, -\tfrac{2}{9}\sqrt{3}.$$

From these values, we see that C and E, $(\pm\frac{1}{3}\sqrt{6},\frac{1}{3}\sqrt{3})$, must be *global maximum points* and that D and F, $(\pm\frac{1}{3}\sqrt{6},-\frac{1}{3}\sqrt{3})$, must be *global minimum points*.

Since these are the only global extreme points (by (2.3)), it follows that they must also be local extreme points. By applying the existence theorem to the restriction of f to the arc EAC, we see that A must be a local minimum point, and by applying the theorem to the arc FBD, we see that B must be a local maximum point. (See the problems.) Thus A, D, and F are *local minimum points* of f, and B, C, and E are *local maximum points*.

Example 2 (*continued from §13.1*). *We have* f(x,y,z) = xyz, g(x,y,z) = x + 2y + 2z, *and* c = 108. Let P have coordinates (x,y,z). Then (2.4) becomes

$$\left(\begin{array}{c} yz\hat{i} + xz\hat{j} + xy\hat{k} = \lambda(\hat{i} + 2\hat{j} + 2\hat{k}) \\ x + 2y + 2z = 108 \end{array} \right).$$

Expressing the vector equation as three scalar equations, we have

$$\left(\begin{array}{c} yz = \lambda \\ xz = 2\lambda \\ xy = 2\lambda \\ x + 2y + 2z = 108 \end{array} \right).$$

The specific form of the first three equations suggests that we can solve this system by considering two cases:

Case 1. $\lambda = 0$. By the first three equations, at least two of x, y, and z must be 0. The fourth equation then shows that the possible points are

$$P = (0,0,54), \quad P = (0,54,0)), \quad \text{and} \quad P = (108,0,0).$$

Case 2. $\lambda \neq 0$. By the first three equations, none of x, y, and z is 0. Hence $xz = xy \Rightarrow z = y$, and $xy = 2yz \Rightarrow x = 2z$. It follows from the fourth equation that $x = 36$ and $y = z = 18$. We have

$$P_4 = (36,18,18)$$

as our fourth and last point. (For P_4, the first equation then gives $\lambda = 18^2 = 324$.)

We complete our application of the Lagrange method by examining the four points P_1, P_2, P_3, P_4 to see which, if any, are extreme or relative-extreme points for f. The values of f at these points are

$$f(P_1) = f(P_2) = f(P_3) = 0, \quad \text{and} \quad f(P_4) = 11,664.$$

From the formula for f, it is easy to see that in every neighborhood of P_1, f takes on both positive and negative values. Similarly for P_2 and P_3. Hence none of P_1, P_2, and P_3 can be an extreme point. Since f is continuous, since $f(P_4) > 0$ and $f = 0$ on the boundaries of the triangular portion of D' contained in the first quadrant, and since this triangular portion is compact, it follows, by the max-min existence theorem, that P_4 is a local maximum point. P_4 is not a global maximum since for $Q = (-10,-41,100)$, for example, $g(Q) = 108$ and $f(Q) = 41,000 > f(P_4)$. (Of course D' itself is not compact.)

From our discussion of example 4 in §13.1, it follows that P_4 must be a global maximum point for example 4. Hence for example 4 (the postal service problem) the maximum volume is 11,664 in^3 = 6.75 ft^3.

In the Lagrange method, it is useful to have an easily remembered algorithmic format for finding the system of scalar equations to be solved.

(2.5) ***Standard format for the one-constraint Lagrange method.*** (We give this format for the case where D′ is in E^3. The format for D′ in E^2 is exactly similar.)

Step 1. *Define a new function* h *of four variables by*

$$h(x,y,z,\lambda) = f(x,y,z) - \lambda g(x,y,z) .$$

Step 2. *Take the first-order derivatives of* h *with respect to* x, y, *and* z, *and set each equal to 0. Then add the constraint equation* to form the system

$$\begin{pmatrix} h_x(x,y,z,\lambda) = 0 \\ h_y(x,y,z,\lambda) = 0 \\ h_z(x,y,z,\lambda) = 0 \\ g(x,y,z) = c \end{pmatrix} .$$

Step 3. *Solve the equations obtained in step 2 as a system of simultaneous equations in the unknowns* x, y, z, λ. *The solutions* (x,y,z) *give the constrained critical points.*

Step 4. *Identify, from among these constrained critical points, the extreme points for* f *on* D′. (The existence theorem may be useful in this step, as we have seen in examples 1 and 2.)

Example 1 (*continued*). *We apply this standard format to example 1.* Here, f and g are two-variable functions. The standard steps give us:

Step 1. Define $h(x,y,\lambda) = f(x,y) - \lambda g(x,y) = x^2 y - \lambda(x^2 + y^2)$.

Step 2. Take the first-order derivatives of h and set each equal to 0 then add the constraint equation.

$$\begin{pmatrix} h_x = 2xy - 2\lambda x = 0 \\ h_y = x^2 - 2\lambda y = 0 \\ x^2 + y^2 = 1 \end{pmatrix} .$$

Step 3. Solve these equations to get the constrained critical points. This was done above.

Step 4. Examine the constrained critical points. This was done above.

In carrying out the multiplier method, there is no single, standard strategy for the calculations in step 3. In some cases, it is helpful to solve first for λ; in others we may not need to find all the values for λ. In some cases, inventive use of algebra can lead to quick and elegant solutions; in others, solutions by numerical approximation may be necessary.

§13.3 The Lagrange Method for Two Constraints

For the case of two constraints (example 3 in §13.1), we assume:

(3.1) (i)   f, g_1, and g_2 are in C^1.

(ii) D′ is the set of all points P such that both $g_1(P) = c_1$ and $g_2(P) = c_2$.

(iii) $\vec{\nabla}g_1\big|_P \times \vec{\nabla}g_2\big|_P \neq \vec{0}$ at all points P in D′.

(3.1) ensures that D′ is a curve in $\mathbf{E}^3$ with a normal plane, at every point P in D′, given by $\vec{\nabla}g_1\big|_P \times \vec{\nabla}g_2\big|_P$ as normal vector. (3.1) also implies that D′ will consist entirely of boundary points.

For the Lagrange method, we again use analogues of *critical point* and of the *critical point theorem.*

Definition. We define P in D′ to be a *constrained critical point for* f *on* D′ if there exist scalar values λ_1^P and λ_2^P such that

(3.2)
$$\vec{\nabla}f\big|_P = \lambda_1^P \vec{\nabla}g_1\big|_P + \lambda_2^P \vec{\nabla}g_2\big|_P .$$

λ_1^P and λ_2^P are called *Lagrange multipliers* for the constrained critical point P. Note, from (3.1), that P is a constrained critical point $\Leftrightarrow$ either $\vec{\nabla}f\big|_P = \vec{0}$ or $\vec{\nabla}f\big|_P$ is normal to D′ at P.

We now state and prove:

(3.3) *The constrained critical point theorem for two constraints.* Let f, g_1, g_2, c_1, and c_2 be as assumed in (3.1). If P is an extreme point for f on D′, then P must be a constrained critical point for f on D′.

Proof. Assume that P is an extreme point for f on D′. Suppose that P were not a constrained critical point. Then, by the same argument as for (2.3), this would imply that P would not be an extreme point. (See the problems.) This proves the theorem.

In the *Lagrange method* for two constraints, we first find the constrained critical points, and then go on to identify the nature of these points as possible extreme points for f on D'. We therefore begin by finding all points which satisfy the system of equations:

(3.4)
$$\left(\begin{array}{c} \vec{\nabla}f\big|_P = \lambda_1 \vec{\nabla}g_1\big|_P + \lambda_2 \vec{\nabla}g_2\big|_P \\ g_1(P) = c_1 \\ g_2(P) = c_2 \end{array} \right) \quad \text{for some } \lambda_1 \text{ and } \lambda_2.$$

Example 3 (*continued from §13.1*).   We have $f(x,y,z) = x^2+y^2+z^2$, $g_1(x,y,z) = z$, $g_2(x,y,z) = x+y$, $c_1 = 2$ *and* $c_2 = 6$. Let P have coordinates (x,y,z). (3.4) becomes

$$\left(\begin{array}{c} 2x\hat{i} + 2y\hat{j} + 2z\hat{k} = \lambda_1\hat{k} + \lambda_2(\hat{i}+\hat{j}) \\ z = 2 \\ x+y = 6 \end{array} \right).$$

This gives us a system of five scalar equations in the five unknowns x, y, z, λ_1, and λ_2:

$$\left(\begin{array}{c} 2x = \lambda_2 \\ 2y = \lambda_2 \\ 2z = \lambda_1 \\ z = 2 \\ x+y = 6 \end{array} \right).$$

The solution is immediate, yielding $x = 3$, $y = 3$, $z = 2$, $\lambda_1 = 4$ and $\lambda_2 = 6$. We thus have (3,3,2) as the unique constrained critical point. Finally, using the max-min existence theorem on a finite segment of the constrained domain D', we can show that $P = (3,3,2)$ must be a global minimum point. We do so by noting that for any Q in D', f(Q) is the square of the distance from the origin to Q. If P were not a global minimum, there would have to be some point Q with $f(Q) < f(P)$. Since D' is a straight line, there would also be a point Q' on D', on the other side of Q from P, with $f(Q') > f(Q)$. Applying the max-min existence theorem to f on the compact line segment from P to Q', we see that there would have to be a minimum point on the interior of this segment and that this minimum point would have to be a constrained critical point, contrary to our result that P is the only constrained critical point.

The standard format for two constraints is exactly similar to (2.5).

(3.5) ***Standard format for the two-constraint Langrange method.***
      Step 1. Define a new function h of five variables by
$$h(x,y,z, \lambda_1,\lambda_2) = f(x,y,z) - \lambda_1 g_1(x,y,z) - \lambda_2 g_2(x,y,z) \ .$$
      Step 2. Take the first-order derivatives of h with respect to x, y, and z, and set each equal to 0. Then add the constraint equations to form a system of five equations.
      Step 3. Solve the equations in step 2 as a system of simultaneous equations in the unknowns x, y, z, λ_1, λ_2. The solutions (x,y,z) give the constrained critical points.
      Step 4. Identify, from among these points, the extreme points for f on
D'.

§13.4 Significance of Multipliers; Derivative Tests

Significance of multipliers. The Langrange multipliers λ, λ_1, and λ_2 have a simple and useful mathematical significance. As in (2.1), let f, g, and c define a constrained max-min problem in E^2 with Cartesian coordinates. (2.1) then gives us the system of equations

(4.1)
$$\left(\begin{matrix} f_x(x,y) = \lambda g_x(x,y) \\ f_y(x,y) = \lambda g_y(x,y) \\ g(x,y) = c \end{matrix} \right).$$

By (2.3), a point P is a constrained critical point $\Leftrightarrow$ its coordinates satisfy (4.1) for some λ. We now add to this system the equation

$$w = f(x,y)$$

with the new variable w. We abbreviate our new system as

(4.2)
$$\begin{matrix} \text{(i)} \\ \text{(ii)} \\ \text{(iii)} \\ \text{(iv)} \end{matrix} \left(\begin{matrix} w = f \\ f_x = \lambda g_x \\ f_y = \lambda g_y \\ g = c \end{matrix} \right).$$

From §13.2, we see that (4.2) is satisfied by the values w, x, y, c, and $\lambda \Leftrightarrow$ P = (x,y) is a constrained critical point, w is the value of f at P, and λ is the Lagrange multiplier for P. We now assume that (4.2) simultaneously implicitly defines w, x, y, and λ as functions of the independent variable c. Using the elimination method, we can find expressions for $\frac{dx}{dc}$, $\frac{dy}{dc}$, $\frac{dw}{dc}$, and $\frac{d\lambda}{dc}$. These expressions will tell us, for any given constrained critical point P = (x,y) and for the values at P of λ and w, the approximate changes which will occur in the position of P and in the values of λ and w when a slight change is made in the value c. The calculations for $\frac{dx}{dc}$, $\frac{dy}{dc}$, and $\frac{d\lambda}{dc}$ are lengthy and use the second order derivatives of f and g. The calculation for $\frac{dw}{dc}$, however, is direct and simple. We get

$$dw = f_x dx + f_y dy \qquad \text{(from (i))}$$
$$= \lambda(g_x dx + g_y dy) \qquad \text{(from (ii) and (iii))}$$
$$g_x dx + g_y dy = dc \qquad \text{(from (iv))}.$$

Therefore, $dw = \lambda dc$, and we have

(4.3) $$\frac{dw}{dc} = \lambda \,.$$

By a similar argument and calculation, we can show that (4.3) holds for a one-constraint problem in $\mathbf{E}^3$. The result for a max-min problem with two constraints is similarly simple. We find

(4.4) $$\frac{\partial w}{\partial c_1} = \lambda_1 \quad \text{and} \quad \frac{\partial w}{\partial c_2} = \lambda_2. \text{ (See the problems.)}$$

Example 1. *In §13.2, we saw that* P_4 = (36,18,18) *is the global maximum point for the postal service problem (example 4 in §13.1) with* λ = *324.* This tells us that if the postal service were to increase the value of c from 108 to 109 inches, then the maximum allowed value would increase by approximately 324 in³.

Example 2. *Similarly, for example 1 in §13.1, we found in §13.2 that* $\lambda = \frac{\sqrt{3}}{3}$ *for the global maximum points. Hence* $\frac{\sqrt{3}}{3}$ is the instantaneous rate of increase of the global maximum value of f per unit increase in the square of the circle's radius.

Derivative tests. Second-order derivative tests can be used to identify the extreme points from among the constrained critical points. The principle underlying these tests is the same as for unconstrained problems. At a constrained critical point, we consider the second-order directional derivatives in all the directions permitted by the constraints. If these derivatives are all positive, we have a local minimum point, and if they are all negative, we have a local maximum point.

In the case of a single constraint in E^2, there are only the directions along one line of motion to consider, and a second order directional derivative at P along that line can be shown to have the same sign as

$$(4.5) \qquad H_c(P) = - \begin{vmatrix} 0 & -g_x(P) & -g_y(P) \\ -g_x(P) & f_{xx}(P) & f_{xy}(P) \\ -g_y(P) & f_{yx}(P) & f_{yy}(P) \end{vmatrix}.$$

(See the problems.) *If* $H_c(P) > 0$, *then* P *must be a local minimum; if* $H_c(P) <$ 0, *then* P *must be a local maximum.* An example of this test is given in §13.5.

The case of two constraints in E^3 is similarly limited to one line of motion, but an expression analogous to (4.5) requires a 5×5 determinant. See the problems for Chapter 34.

In the case of one constraint in E^3, the permitted directions are all the directions in the plane normal to $\vec{\nabla}g\big|_P$. The corresponding derivative test is similar in nature to the two-variable test for unconstrained problems (§12.4) and requires a 4×4 as well as a 3×3 determinant. See the problems for Chapter 34.

§13.5 A Further Example

The following problem can either be solved by elementary calculus or be set up as a constrained problem and solved by the multiplier method. The multiplier solution is simpler and more informative.

Example. A straight irrigation canal is to have a rectangular cross-section of fixed area A. *Its sides and bottom will be thin walls of concrete of a*

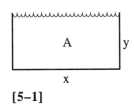

[5–1]

given, constant thickness.   What dimensions should the cross section have in order to minimize the amount of concrete used?

Let x be width and y be depth. (See [5–1].) We wish to minimize $f(x,y) = x+2y$ under the constraint that $g(x,y) = xy = A$. We apply the multiplier method.

Step 1. $h(x,y,\lambda) = x + 2 - \lambda(xy)$.

Step 2. $h_x = 1 - \lambda y = 0$,

$h_y = 2 - \lambda x = 0$,

$xy = A$.

Step 3.   From the first two equations, we have

$$\left. \begin{array}{c} 1 - \lambda y = 0 \\ 2 - \lambda x = 0 \end{array} \right\} \Rightarrow \left. \begin{array}{c} x - \lambda xy = 0 \\ 2y - \lambda xy = 0 \end{array} \right\} \Rightarrow x = 2y.$$

Without further calculation, we see that we have a constrained critical point $\Leftrightarrow$ the width of the canal is equal to twice its depth. It now follows, by the third equation, that $x = \sqrt{2A}$ and $y = \frac{1}{2}\sqrt{2A}$. We also get $\lambda = \frac{1}{y} = \sqrt{2/A}$. By (4.3), λ is the instantaneous rate of increase in the critical value of $f(x,y)$, per unit increase in the fixed cross-sectional area A.

Step 4.   The point $(\sqrt{2A}, \frac{1}{2}\sqrt{2A})$ is a minimum and relative-minimum point for f under the given constraint. We can show this in any of three ways:

(a) By the existence theorem: If x becomes large, then y must become small and $f(x,y) = x + 2y$ must become large; if x becomes small, then y must become large and $f(x,y) = x + 2y$ must become large. By the existence theorem applied to an appropriate finite closed interval of values of x, there must be a minimum point. Since $(\sqrt{2A}, \frac{1}{2}\sqrt{2A})$ is the only critical point, it must be the unique global minimum point.

(b) By considering $f(x,y) = x + 2y = x + \frac{2A}{x}$ as a one-variable max-min problem: We can use the one-variable derivative test to identify $x = \sqrt{2A}$ as a unique local minimum point. The max-min existence theorem then tells us that it is a unique global minimum point.

(c) By using the *Lagrange-multiplier derivative test* described in §13.4: For our present problem, we have

$$g_x = y \Rightarrow g_x(P) = \tfrac{1}{2}\sqrt{2A},$$

$$g_y = x \Rightarrow g_y(P) = \sqrt{2A},$$

$$f_{xx} = 0 \text{ and } f_{yy} = 0,$$

$$f_{xy} = f_{yx} = -\lambda_P = -\sqrt{2/A}.$$

Hence

$$H_c(P) = -\begin{vmatrix} 0 & -\tfrac{1}{2}\sqrt{2A} & -\sqrt{2A} \\ -\tfrac{1}{2}\sqrt{2A} & 0 & -\sqrt{2/A} \\ -\sqrt{2A} & -\sqrt{2/A} & 0 \end{vmatrix} = \sqrt{2A}\sqrt{2A}\sqrt{2/A} = 2\sqrt{2A} > 0.$$

The test tells us that P must be a local minimum point.

§13.6 The Parametric Method for Constrained Problems

In a constrained max-min problem, the constrained domain D′ of f consists entirely of boundary points. In the *parametric method*, we use one or more parameters to form a parametric representation (parametric equations) for this set of boundary points. We then express the value of the given field or function f as a function h of these parameters and solve the corresponding unconstrained max-min problem for h. As we shall see, the number of parameters for the function h will be smaller than the number of variables of f, and most (if not all) of the parameter points in the domain of h will be interior points. This permits us to use the interior point methods of §12.3 and §12.4 to solve the max-min problem for h. The parameter values that we obtain as extreme points for h then give us, by means of the parametric equations, the points of D′ that are extreme points for f.

Example 1. Consider example 1 in §13.1. In E^2 with Cartesian coordinates, f is given by the function $f(x,y) = x^2 y$ and has as its domain D′ the circle $x^2 + y^2 = 1$. We locate the extreme points for f by using the parametric representation $\vec{R}(t) = \cos t\,\hat{i} + \sin t\,\hat{j}$ $(-\infty < t < \infty)$ for D′. We have $h(t) = f(\vec{R}(t)) = \cos^2 t \sin t$ on the domain $(-\infty, \infty)$. (We are allowing each point in D′ to be represented by more than one point in the domain of h. This is convenient in the present example, but not essential. We could have used $0 \le t < 2\pi$.)

The max-min problem for h is a one-variable problem, and the domain of h consists entirely of interior points. We therefore proceed to find the critical points for h. We have

$$
\begin{aligned}
h'(t) &= -2 \cos t \sin^2 t + \cos^3 t \\
&= \cos^3 t - 2 \cos t\,(1 - \cos^2 t) \\
&= \cos^3 t + 2 \cos^3 t - 2 \cos t \\
&= \cos t\,(3 \cos^2 t - 2) = 0\,.
\end{aligned}
$$

Hence the critical points for h must be those values of t which satisfy one or the other of the following equations:

$$
\cos t = 0,
$$
and
$$
3\cos^2 t - 2 = 0.
$$

These are the values of t such that

$$
\cos t = 0,
$$
or
$$
\cos t = \pm \tfrac{1}{3}\sqrt{6}\,.
$$

Since $x = \cos t$, we see that all extreme points for f must occur at points where

$$
x = 0,\; x = \tfrac{1}{3}\sqrt{6},\; x = -\tfrac{1}{3}\sqrt{6}.
$$

Thus the possible extreme points in the domain of f are

$$
(0,1),\, (0,-1),\, (\tfrac{1}{3}\sqrt{6},\tfrac{1}{3}\sqrt{3}),\, (\tfrac{1}{3}\sqrt{6},-\tfrac{1}{3}\sqrt{3}),\, (-\tfrac{1}{3}\sqrt{6},\tfrac{1}{3}\sqrt{3}),\, (-\tfrac{1}{3}\sqrt{6},-\tfrac{1}{3}\sqrt{3}).
$$

If we wish, we can now use the one-variable derivative test, with $h''(t) = \sin t(2 - 9 \cos^2 t)$, to identify the extreme points for f on D'.

In Chapter 21, we shall see that, just as a curve can be represented by a continuous vector function $\vec{R}(t)$ in a single parameter t, a surface in $\mathbf{E}^3$ can be represented by a continuous vector function $\vec{R}(u,v)$ in two parameters u and v. If the constrained domain of a given scalar field f in $\mathbf{E}^3$ is such a surface, we can then form

$$
h(u,v) = f(R(u,v))
$$

and proceed to solve the unconstrained problem for h as in the following example.

Example 2. *Consider example 2 in §13.1. In* E^3 *with Cartesian co-ordinates,* f *is given by the function* $f(x,y,z) = xyz$ *and has as its domain* D' *the plane* x+2y+2z = 108. *We wish to find the extreme points (if any) for* f.

We get a parametric expression for the plane by using x and y themselves as parameters. We have

$$\vec{R}(x,y) = x\hat{i} + y\hat{j} + \tfrac{1}{2}(108 - x - 2y)\hat{k}$$

Then $\qquad h(x,y) = f(\vec{R}(x,y)) = \tfrac{1}{2}xy(108 - x - 2y) = 54xy - \tfrac{1}{2}x^2y - xy^2.$

It remains to find the extreme points (if any) of h. The domain of h is all of E^2. Hence every point of this domain is an interior point, and we can use the methods of §12.3 and §12.4. We find the critical points of h by setting $\vec{\nabla}h = \vec{0}$. This gives:

$$\frac{\partial h}{\partial x} = 54y - xy - y^2 = y(54 - x - y) = 0$$

and $\qquad \frac{\partial h}{\partial y} = 54x - \tfrac{1}{2}x^2 - 2xy = x(54 - \tfrac{1}{2}x - 2y) = 0$

We solve these equations as follows:

$$(0,0) \text{ is a solution.}$$

$[x = 0 \text{ and } y \neq 0] \;\Rightarrow 54 - x - y = 0 \Rightarrow 54 - y = 0 \Rightarrow y = 54.$

$[x \neq 0 \text{ and } y = 0] \;\Rightarrow 54 - \tfrac{1}{2}x - 2y = 0 \Rightarrow 54 - \tfrac{1}{2}x = 0 \Rightarrow x = 108.$

$[x \neq 0 \text{ and } y \neq 0] \;\Rightarrow \left\{ \begin{matrix} 54 - x - y = 0 \\ 54 - \tfrac{1}{2}x - 2y = 0 \end{matrix} \right\} \Rightarrow x = 36, y = 18.$

Thus the critical points of h are

$$(0,0), \ (0,54), \ (108,0), \ (36,18) \,.$$

The corresponding points in the domain of f are $P_1(0,0,54)$, $P_2 = (0,54,0)$, $P_3 = (108,0,0)$, and $P_4 = (36,18,18)$. The values of f at these points are

$$0, \ 0, \ 0, \ 11{,}664.$$

Applying the derivative test, we have $h_{xx} = -y$, $h_{yy} = -2x$, $h_{xy} = h_{yx} = 54-x-2y$. Hence,

$$H_1(x,y) = -y,$$
$$H_2(x,y) = 2xy - (54-x-2y)^2.$$

Evaluating H_1 and H_2 at the four critical points, we get $H_2(0,0) = H_2(0,54) = H_2(108,0) = -54^2 < 0$, while $H_2(36,18) = 36^2-18^2 > 0$, and $H_1(36,18) = -18 < 0$. Thus, by the two-variable derivative test, the points $(0,0)$, $(0,54)$, and $(108,0)$ are saddle points for h, while $(36,18)$ is a local maximum point for h. Hence $P_4 = (36,18,18)$ is a local maximum point for f and is the only extreme point for f. It is easy to find a point Q in D such that $f(Q) > f(P_4)$. (See §13.2.) Hence P_4 is not a global maximum point.

Example 3. Consider example 3 in §13.1. In $\mathbf{E}^3$ with Cartesian coordinates, f is given by the function $f(x,y,z) = x^2+y^2+z^2$ and has, as its domain D', the intersection of the planes $z = 2$ and $x+y = 6$. We have a path for D' from its description in §13.1.

$$\vec{R}(t) = 6\hat{i} + 2\hat{k} + t(\hat{i} - \hat{j}), \quad (-\infty < t < \infty)$$
$$= (t+6)\hat{i} - t\hat{j} + 2\hat{k},$$

and the parametric equations for D' are

$$\left. \begin{aligned} x &= 6+t \\ y &= -t \\ z &= 2 \end{aligned} \right\} \quad t \text{ in } (-\infty,\infty).$$

Substituting in $f(x,y,z)$, we have

$$h(t) = f(\vec{R}(t)) = (6+t)^2 + t^2 + 4$$
$$= 2t^2 + 12t + 40.$$

We now solve the max-min problem for h. The domain of h, $(-\infty,\infty)$, consists entirely of interior points. We find the critical points for h. We have

$$h'(t) = 4t+12 = 0 \Rightarrow t = -3.$$

This is the only critical point. It gives us the corresponding point on D':

$$P = (3,3,2).$$

Using the one-variable derivative test, we find

$$h''(t) = 4 \Rightarrow h''(-3) = 4 > 0.$$

Thus $t = -3$ is a local minimum point for h, and $P = (3,3,2)$ is a local minimum point for f. The value of f at P is

$$f(P) = h(-3) = 22 \; .$$

As in §13.2, we can now use the max-min existence theorem to show that $t = -3$ is, in fact, a global minimum point for h. It follows that P = (3,3,2) is a global minimum point for f, and that $f(P) = 22$ is the global minimum value of f.

§13.7 Boundary-point Analysis for Unconstrained Problems

In solving a general (unconstrained) maximum-minimum problem, the analysis of boundary points (Step IIb in §12.5) can often be formulated as the solution of one or more constrained max-min problems. These constrained problems can then be solved by multiplier methods or by parametric methods. This approach to boundary-point analysis is justified by the following fact.

(7.1)   *Theorem. Let f have domain D and let D_b be the set of those boundary points of D that are in D. If P in D_b is an extreme point of f on D, then P must also be an extreme point of $f|_{D_b}$.*

Proof. The theorem is immediate from (1.6) and (1.7) in §12.1.

Thus, in searching among the boundary points of D for extreme or relative-extreme points of f on D, it is enough to consider the extreme and relative-extreme points of $f|_{D_b}$. These latter points can often be found by solving a constrained problem. (The converse of the theorem, however, does not always hold. It is possible for P to be a relative-extreme point for $f|_{D_b}$ without also being a relative-extreme point for f on D. We shall see this in the example.)

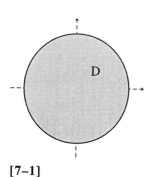

[7–1]

Example. *In* $\mathbf{E}^2$ *with Cartesian coordinates let* $f(x,y) = x^3+y^2$ *on the region bounded by the circle* $x^2+y^2 = 1$. The domain D of f consists of the shaded region in [7–1] together with its boundary. Evidently D is compact. Hence, by the existence theorem, there must be at least one global maximum point and at least one global minimum point for f.

We begin with the interior points of D. We find the interior critical points by setting $\vec{\nabla}f = 3x^2\hat{i} + 2y\hat{j} = \vec{0}$. We have $3x^2 = 0 \Rightarrow x = 0$, and $2y = 0 \Rightarrow y = 0$. Hence $O = (0,0)$ is the only interior critical point. The derivative test gives no further information, since $H_1(O) = 0$ and $H_2(O) = 0$. We note, however, that $f(O) = 0$, that $f(Q) < 0$ for all points on the x axis with $x < 0$, and that $f(Q) > 0$ for all points on the x axis with $x > 0$. Hence P cannot be an extreme point for f. It follows that the global maximum and minimum points for f must lie on the circular boundary D_b.

We now use the Lagrange-multiplier method to solve the constrained max-min problem for f on D_b. We have

$$h(x,y,\lambda) = x^3 + y^2 - \lambda(x^2+y^2),$$
$$h_x = 3x^2 - 2\lambda x = 0,$$
$$h_y = 2y - 2\lambda y = 0,$$
$$x^2+y^2 = 1.$$

Then $2y - 2\lambda y = 0 \Leftrightarrow 2y(1-\lambda) = 0 \Leftrightarrow y = 0$ or $\lambda = 1$, and $\lambda = 1 \Leftrightarrow 3x^2-2x = 0 \Leftrightarrow x(3x-2) = 0 \Leftrightarrow x = 0$ or $x = \frac{2}{3}$. Finally, using $x^2+y^2 = 1$, we have

$$y = 0 \Rightarrow x = \pm1,$$
$$x = 0 \Rightarrow y = \pm1,$$
$$x = \tfrac{2}{3} \Rightarrow y = \pm\tfrac{1}{3}\sqrt{5}.$$

We thus have six constrained critical points on D_b:

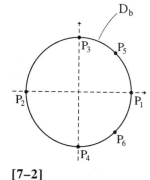

[7–2]

$$P_1 = (1,0), \ P_2 = (-1,0), \ P_3 = (0,1), \ P_4 = (0,-1), \ P_5 = (\tfrac{2}{3},\tfrac{1}{3}\sqrt{5}),$$
and
$$P_6 = (\tfrac{2}{3},-\tfrac{1}{3}\sqrt{5}). \text{ (See [7–2].)}$$

Evaluating f at each point, we find

$$f(P_1) = f(P_3) = f(P_4) = 1, \ f(P_2) = -1, \ f(P_5) = f(P_6) = \tfrac{23}{27}.$$

P_1, P_3, and P_4 must therefore be global maximum points on D_b, and P_2 must be a global minimum point on D_b. Hence these points must also be the desired global extreme points for f on D. By (7.1), these must be the only global extreme points for f on D. (It follows that each global extreme point must also be a local extreme point for f on D.)

What about P_5? Applying the existence theorem to the arc $P_1P_5P_3$ of D_b, we see that P_5 must be a local minimum point for f on D_b. Is P_5 also a local minimum point for f on D? We note that $\vec{\nabla}f\big|_{P_5}$ is non-zero and outwardly normal to D_b at P_5. This implies that immediately inside the circle D_b on the radial segment $\overline{OP_5}$, f has values $<\frac{23}{27}$. Hence P_5 cannot be a local minimum point for f on D. By a similar argument, P_6 is a local minimum point for f on D_b but not for f on D. We have thus identified P_1, P_2, P_3, and P_4 as the set of all extreme and relative-extreme points for f on D.

Remark. In the example above, we could have used the parametric method for D_b instead of the Lagrange-multiplier method.

§13.8 Problems

§13.2 & §13.3

For problems 1–24, use the Lagrange method.

1. Where on the surface $x^2+y^2+4z^2 = 2$ does the function $f(x,y,z) = x+2z$ take on maximum and minimum values?

2. A parcel delivery service requires that the dimensions of a rectangular box be such that the length plus twice the width plus twice the height be no more than twelve feet ($x+2y+2z \leq 12$). Find the volume in cubic feet of the largest volume box that the service will deliver.

3. What are the maximum and minimum values attained by $f(x,y,z) = xyz$ on the ellipsoid $2x^2+2y^2+z^2 = 2$.

4. A convex pentagon consists of three sides of a rectangle attached to the two equal sides of an isosceles triangle. The perimeter of the pentagon is P. Find the maximum possible area of the pentagon.

5. Consider the function $f(x,y,z) = x-y$ with domain restricted to the spherical surface $x^2+y^2+z^2 = 8$. Find the maximum and minimum points.

6. An open rectangular box has length x, width y, and depth z. Assume that the volume of the box is fixed. Find the relative proportions of x, y, and z that will minimize the interior surface area of the box.

7. (a) Find the maximum value of $f(x,y) = xy$ on that portion of the line $2x+y = 1$ which lies in the first quadrant.
    (b) Explain why this must be the maximum value.

8. You wish to construct an open-topped box with a square base. The box is to have a fixed volume. Let x be the length of one side of the base, and let y be the altitude. Use the Lagrange multiplier method to determine the ratio $\frac{y}{x}$ that will minimize the total external surface area of the base and side.

9. Find and analyze (giving reasons) the constrained critical points for the function $f(x,y) = 2x + y$ on the circle $x^2 + y^2 = 5$.

10. Find the minimum value of $f(x,y,z) = x^2 + y^2 + z^2$ on the plane $x + 2y + 3z = 6$.

11. The equation $4x^2 + y^2 = 4$ defines an ellipse C through the points $(0,2)$, $(0,-2)$, $(1,0)$, and $(-1,0)$. Let $f(x,y) = x + y^2$.
    (a) Find the maximum and minimum values of f on the curve C. (Be sure to consider all possible constrained critical points.)
    (b) Give the point or points at which the maximum value occurs and the point or points at which the minimum value occurs.
    (c) There is a local extreme point for f on C which is not a global extreme point for f on C. What is this point, and what kind of local extreme point is it?

12. Assume
    $$x > 0, \, y > 0,$$
    $$f(x,y) = x^4 + x^2 y^2,$$
    $$x^2 y = \sqrt{2}.$$
    (a) Use the Lagrange multiplier method to find all extreme and relative extreme points of f under these conditions.
    (b) For each such point, state its extremal nature (maximum or minimum, local or global), and give your reasoning.

13. Consider $f(x,y) = x^3 y$. If we limit our attention to points on the circle $x^2 + y^2 = 4$, f has six critical points. Find these critical points and show them on a figure. Indicate on the figure which are maximum points, which are minimum points, and which, if any, are not extreme points.

14. (a) Using Lagrange multipliers, find constrained local maxima and minima for the function $f(x,y,z) = x$ subject to the constraint $x^2 + y^2 + z^2 = 1$.
    (b) Do the same for the function $f(x,y,z) = x^2$ with the same constraint. (After differentiating, be

sure to consider all logically possible cases for coordinate values to be 0 or not 0.)

15. (a) Using Lagrange multipliers, find constrained local maxima and minima for the function $f(x,y,z) = x - y$ subject to the constraint $x^2 + y^2 + z^2 = 1$.
    (b) Do the same for the function $f(x,y,z) = x^2 - y^2$ with the same constraint $x^2 + y^2 + z^2 = 1$. (After differentiating, be sure to consider all logically possible cases for coordinate values to be 0 or not 0.)

16. Use Lagrange multipliers to find the closest point to the origin on the intersection of the two planes $x + y - z = 3$ and $x - y + z = 1$.

17. The surfaces $x^2 + y^2 + z^2 = 6$ and $x + y + z = 0$ intersect in a circle. Find the extreme points on this circle (if any) for the function $f(x,y,z) = xyz$.

18. Use Lagrange multipliers to find all extreme points for the function $f(x,y,z) = x + y + z$ on the intersection of the cylinder $x^2 + y^2 - 1 = 0$ with the plane $z - 2 = 0$.

19. Let P be the point $(0,1,1)$, and let S be the sphere $x^2 + y^2 + z^2 = 4$. Find the point on S which is closest to P and the point on S which is farthest from P.

20. Find the maximum and minimum values of the function $f(x,y,z) = xy + xz$ on the curve of intersection of the cylinder $x^2 + y^2 = 1$ and the hyperbolic cylinder $xz = 1$.

21. A cylindrical metal can is to contain 1 liter of water and is to have no top. Find the dimensions of the can which uses a minimum amount of metal.

22. The plane $x - y + 4z = 1$ intersects the elliptical cylinder given by the equation $2x^2 + 4y^2 = 3$. Find the points on the curve of intersection where the coordinate z has maximum and minimum values, and give these values.

23. (a) Use a Lagrange multiplier to find the point (x,y,z) closest to the origin on the graph of the function $z = x + y - 3$. (To simplify work, take

$f(x,y,z) = \frac{1}{2}(x^2+y^2+z^2)$ as the function to be minimized.

(b) Consider the general problem of finding points (x,y,z) which minimize distance from the origin on the graph of the function $z = g(x,y)$. Give a system of equations in x, y, z, $\frac{\partial g}{\partial x}$, and $\frac{\partial g}{\partial y}$ whose solution will give such points.

24. A light ray travels from point A to point B crossing a plane boundary between two different media. The path of the ray lies in a plane perpendicular to this boundary plane. In the first medium, its speed is c_1 and its direction makes an angle θ_1 with a line normal to the plane boundary. Similarly for c_2 and θ_2 in the second medium. Use a Lagrange multiplier to show that the trip is made in minimum time when *Snell's Law* holds:

$$\frac{\sin \theta_1}{\sin \theta_2} = \frac{c_1}{c_2}.$$

25. In example 1, use the existence theorem to prove that A is a local minimum point and that B is a local maximum point.

26. Complete the proof of (2.3).

27. Complete the proof of (3.3).

§13.4

1. Complete the calculation for (4.4).

2. Derive the derivative test (4.5).

3. Apply (4.5) to the case of example 1 in §13.2.

§13.6

Use the parametric method to solve the following problems.

1. Where on the ellipse $x^2+2y^2 = 1$ does the function $f(x,y) = xy$ take on maximum and minimum values?

2. Consider the curve $xy = 2$ in the first quadrant. Find where, on this curve, the function $f(x,y) = x+2y$ takes on a minimum value.

3. Consider the scalar field given by $f(x,y) = xy^2$ in the xy plane. If we limit our attention to points on the circle $x^2+y^2 = 9$, f has six critical points. Find these points and show them on a figure. Indicate on the figure which are maximum points and which are minimum points.

4. Consider $f(x,y) = x^3y$. If we limit our attention to points on the circle $x^2+y^2 = 4$, f has six critical points. Find these critical points and show them on a figure. Indicate on the figure which are maximum points, which are minimum points, and which, if any, are not extreme points.

5. (a) Find the minimum value of the function $f(x,y) = 2x+y$ on the curve $x^2y = 1$ in the first quadrant. Give the point or points at which this minimum value occurs.

    (b) How do you know that this is a minimum value?

6. (a) Find the minimum value of the function $f(x,y) = x+y^2$ on the curve $xy^2 = 1$ in the first quadrant. Give the point or points at which this minimum occurs.

    (b) How do you know that this is a minimum value?

7. Where on the line $x+3y = 2$ does the function $f(x,y) = xy^2$ take on maximum and minimum values?

8. Let M be the plane given by the equation $x+2y+2z = 18$. Find the coordinates (x,y,z) of the point P in M which is closest to the origin. Do so by minimizing $|\overrightarrow{OP}|^2 = x^2+y^2+z^2$ under the constraint that P must lie in M.

9. Use the parametric method to find the maximum area of a rectangle in the xy plane such that
    (i) the rectangle has its sides parallel to the axes,
    (ii) the rectangle has one vertex at the origin and the diagonally opposite vertex on the curve $x^2+y = 1$,
    (iii) the rectangle lies entirely in the first quadrant.

10. Consider the function $f(x,y) = x^2y$.
   (a) Locate all constrained critical points of f on the line $x+y = 3$. (Be careful to consider all possible cases.)
   (b) What is the nature (as maxima or minima) of the points in (a)?

11. Consider $f(x,y) = xy+y$.
   (a) Find all constrained critical points of f on the line $x+y = 3$.
   (b) What is the nature (as max or min) of the point or points found in (a)? Explain your reasoning.

12. Find the point in the first quadrant on the hyperbola $xy = 1$ closest to $(2,-2)$.

§13.7

1. In $\mathbf{E}^3$ with Cartesian coordinates, let $f(x,y) = x^3+xy+y^3$ on the region bounded by the x axis and the lower semi-circle of $x^2+y^2 = 1$. The domain D of f consists of this region together with its boundary. Give a complete solution to the max-min problem for f on D.

Chapter 14

Multiple Integrals

§14.1 Definite Integrals in One-variable Calculus

We begin with a review of the definition of *definite integral* in elementary calculus. In elementary calculus, the definite integral, like the derivative, is defined as a limit.

(1.1)   *Definition.* Let a and b be real numbers with a < b. Let f be a one-variable function whose domain includes [a,b]. Then the *definite integral of f over* [a,b] is written

$$\int_a^b fdx \quad (\text{or } \int_a^b f(x)dx)$$

and is defined to be the scalar value given by

$$\lim_{\substack{n \to \infty \\ \max \Delta x_i \to 0}} \sum_{i=1}^{n} f(x_i^*)\Delta x_i,$$

if this limit exists, where, for each n, the interval [a,b] is subdivided into n subintervals and then, for each i, $1 \le i \le n$, Δx_i is the length of the i-th subinterval and x_i^* is an arbitrary *sample point* chosen in the i-th subinterval. For each n,

the sum $\sum_{i=1}^{n} f(x_i^*)\Delta x_i$ is called a *Riemann sum*. The meaning of $\lim_{\substack{n\to\infty \\ \{\max \Delta x_i \to 0\}}}$

was explained and made precise in §5.8 as the unique limit (if it exists) of *proper sequences* of Riemann sums. This unique limit exists and has the value ℓ, we say that *the integral* $\int_a^b fdx$ *exists and has the value* ℓ.

Examples.

(1)   Let f be a continuous function such that $f(x) \geq 0$ for all x in [a,b]. Consider the graph of f in E^2 with Cartesian coordinates. Then   is the area of the region in E^2 whose upper boundary is the curve $y = f(x)$ for $a \leq x \leq b$, whose lower boundary is the segment [a,b] on the x axis, and whose side boundaries (if any) are given by lines perpendicular to the x axis at $x = a$ and $x = b$. See [1–1].

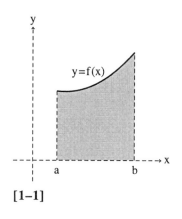

[1–1]

(2)   Let $f(t) = \dfrac{dx}{dt}$ be the velocity at time t of a point moving along the x axis. Then the *net displacement* of the point along the x axis between times $t = a$ and $t = b$ is $\int_a^b fdt$, and the *total distance* moved by the point is $\int_a^b |f|\,dt$.

(3)   Let $\delta(x)$ be the density (per unit length) at the point x on a thin, straight wire between $x = a$ and $x = b$ on the x axis. Then $\int_a^b \delta(x)dx$ is the total *mass* of the wire.

(1.2)   The definite integral satisfies the following algebraic laws. These laws follow directly from definition (1.1).

(I) *If [a,c] is included in the domain of f, and $a < b < c$, then* $\int_a^c fdx =$ $\int_a^b fdx + \int_b^c fdx$.

(II) *If [a,b] is included in the domains of both f and g, then* $\int_a^b (f+g)dx = \int_a^b fdx + \int_a^b gdx$.

(III) *If [a,b] is included in the domain of f, then for any constant k,* $\int_a^b kfdx = k\int_a^b fdx$.

(IV) *Let m_1 and m_2 be given constants such that $m_1 \leq f(x) \leq m_2$ for all x in [a,b]. Then*

$$m_1(b-a) \le \int_a^b fdx \le m_2(b-a) \,,$$

and hence for some M *in* $[m_1, m_2]$, $\int_a^b fdx = M(b-a)$.

For certain functions f and intervals $[a,b]$, the limit which defines $\int_a^b fdx$ may not exist. When this happens, we say that *the integral* $\int_a^b fdx$ *does not exist.* The following theorem gives a condition under which we can be sure that the integral does exist.

(1.3) *Existence theorem for the definite integral. Let* f *be continuous on* $[a,b]$. *Then the definite integral* $\int_a^b fdx$ *exists.*

Theorem (1.3) is a central theorem of elementary calculus. It is also a corollary to the *existence theorem for multiple integrals*. The latter theorem will be stated in §14.4 and proved in §14.10.

Remark. For a one-variable function f, we know: f differentiable on $[a,b]$ $\Rightarrow$ f is continuous on $[a,b]$ $\Rightarrow$ $\int_a^b f(x)dx$ exists. The converse implications, however, are not always true. It is possible for $\int_a^b f(x)dx$ to exist even if f is not continuous on $[a,b]$, and it is possible for f to be continuous on $[a,b]$ even if f is not differentiable on $[a,b]$. See (4.4) below.

§14.2 Scalar-valued Multiple Integrals

We begin by defining certain subregions of $\mathbf{E}^2$ and $\mathbf{E}^3$ which will play the same role for integrals in multivariable calculus that intervals of the form $[a,b]$ play for integrals in one-variable calculus.

Definition. A region R in $\mathbf{E}^2$ is said to be *elementary* if there is a piecewise smooth, simple, closed curve C such that R consists of the points on C together with the points inside C. See [2–1]. It follows from the definition of *compactness* that every elementary region must be compact. (Recall that a set is *compact* if it is bounded and includes all its boundary points.)

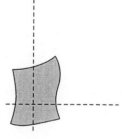

elementary

[2–1]

Definition. A region in $\mathbf{E}^2$ is said to be *regular* if it is either elementary or can be divided into a finite number of elementary regions by a regular subdivision. A *regular subdivision in* $\mathbf{E}^2$ is a subdivision in which the elementary regions form a connected whole and in which neighboring elementary regions may only touch along a single segment of common boundary curve of positive length. A more formal definition of *regular subdivision in* $\mathbf{E}^2$, similar to the formal definition of polyhedron in §1.6, will be given in §23.1.

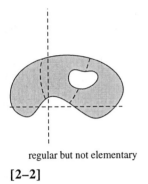

regular but not elementary

[2–2]

Examples. The region enclosed by (and including) a circle is elementary (and hence regular), and so is the region enclosed by (and including) a polygon. The annular ring enclosed by (and including) two concentric circles is regular but not elementary (see [2–2]).

It is immediate that (1) every elementary region in $\mathbf{E}^2$ is also a regular region, and (2) every regular region in $\mathbf{E}^2$ is compact.

In Chapter 21, we define *smooth surface, piecewise smooth surface, closed surface, and simple, closed surface.* "Smooth," "piecewise smooth," "closed" and "simple, closed" will have meanings for *surfaces* closely analogous to their meanings for *curves.* For the time being, it will be enough for the reader to have a general idea of these concepts.

Definition. A region R in $\mathbf{E}^3$ is said to be *elementary* if there is a piecewise smooth, simple, closed surface S such that R consists of the points on S together with the points inside S. It follows that every elementary region in $\mathbf{E}^3$ must be compact.

Examples. This definition includes such familiar pieces of $\mathbf{E}^3$ as solid spheres and solid cubes. An elementary region in $\mathbf{E}^3$ can be visualized as a "solid" piece of space which is compact, has no internal cavities, is pierced by no "doughnut holes," and has an exterior surface which can be subdivided into a finite number of "smooth" pieces.

Definition. A region in $\mathbf{E}^3$ is said to be *regular* if it is either elementary or can be divided into a finite number of elementary regions by a regular subdivision. A *regular subdivision in* $\mathbf{E}^3$ is a subdivision in which the elementary regions form a connected whole and in which neighboring elementary regions may only touch across a single piece of common boundary surface of positive area. A more formal definition of *regular subdivision in* $\mathbf{E}^3$ is given in §23.1.

Examples. The region enclosed by (and including) a tetrahedron is elementary. The region enclosed by (and including) a torus is regular but not elementary. The region between (and including) two concentric spheres is regular but not elementary.

It is immediate that every elementary region in $\mathbf{E}^3$ is regular and that every regular region in $\mathbf{E}^3$ is compact.

Every elementary region in $\mathbf{E}^2$ has a unique and well-defined area. Similarly, every elementary region in $\mathbf{E}^3$ has a unique and well-defined volume. We do not prove these facts in this text. It follows that every regular region in $\mathbf{E}^2$ has a unique and well-defined area and that every regular region in $\mathbf{E}^3$ has a unique and well-defined volume.

(2.1)   *Definition.* Let R be a regular region in $\mathbf{E}^2$. Let f be a scalar field whose domain includes R. Then the *double integral of f over R* is written

$$\iint_R f dA$$

and is defined to be the scalar value (real number) given by

$$\lim_{\left\{\substack{n\to\infty \\ \max d_i \to 0}\right\}} \sum_{i=1}^n f(P_i^*)\Delta A_i \,,$$

if this limit exists, where, for each n, the region R is subdivided into n elementary subregions, and then, for each i, $1 \le i \le n$, ΔA_i is the area of the i-th subregion, P_i^* is a *sample point* chosen in the i-th subregion, and d_i is the maximum distance between two points in the i-th subregion. This maximum distance is called the *diameter* of the subregion. A sum of this kind is called a *Riemann sum* for f on R. If the unique limit in this definition exists and has the value L, we say that the double integral $\iint_R f dA$ *exists and has the value* L.

The particular choice of a subdivision of R into subregions and the particular choice of the sample points are referred to, together, as a *decomposition* of R. For a given decomposition, the maximum value of the d_i is called the *mesh* of the decomposition. For a given f on R, the form and value of a Riemann sum are determined by the decomposition for that sum. *Proper sequences* of decompositions were defined in §5.8. A sequence of Riemann sums is said to be a *proper sequence* if it is determined by a proper sequence of decompositions.

(2.2)   *Definition.* Let R be a regular region in $\mathbf{E}^3$. Let f be a scalar field whose domain includes R. Then the *triple integral of* f *over* R is written

$$\iiint_R f dV$$

and is defined to be the scalar value given by

$$\lim_{\substack{n \to \infty \\ \max d_i \to 0}} \sum_{i=1}^{n} f(P_i^*) \Delta V_i \, ,$$

if this limit exists, where, for each n, the region R is subdivided into n elementary subregions, and then, for each i, $1 \le i \le n$, ΔV_i is the volume of the i-th subregion, P_i^* is a *sample point* chosen in the i-th subregion, and d_i is the maximum distance between two points in the i-th subregion. This maximum distance is again called the *diameter* of the subregion. A sum of this kind is called a *Riemann sum* for f on R. If the unique limit in this definition exists and has the value L, we say that the triple integral $\iiint_R$ fdV *exists and has the value* L. The concepts of *decomposition* and *mesh* are defined as for regions in $\mathbf{E}^2$.

The scalar field f is called the *integrand* of the double or triple integral described above.   Double and triple integrals are collectively referred to as *multiple integrals.*

Notation. When Cartesian axes are chosen in $\mathbf{E}^2$, and the integrand f can be represented by a function f(x,y) of two variables, then the double integral is often written as $\iint_R f(x,y)dA$ or as $\iint_R f(x,y)dxdy$. In the latter case, dxdy is a reminder that we are integrating in a plane which has Cartesian coordinates x and y. dxdy also suggests that the i-th subregion for a Riemann sum could be taken (except near the boundary of R) to be rectangular in shape, with sides of length Δx_i and Δy_i parallel to the axes, and with area $\Delta A_i = \Delta x_i \Delta y_i$. Similarly, in $\mathbf{E}^3$ with Cartesian coordinates, when f can be represented as a function f(x,y,z) of three variables, we write $\iiint_R f(x,y,z)dV$ or $\iiint_R f(x,y,z)dxdydz$ for the triple integral.

The following fact about elementary subregions in $\mathbf{E}^2$ and $\mathbf{E}^3$ is useful in theoretical derivations.

(2.3)   **Lemma.** *Let* R *be an elementary region (in* $\mathbf{E}^2$ *or* $\mathbf{E}^3$*), where* R *has diameter* d. *Then* R *can be subdivided into a finite number of elementary subregions, each of which has diameter* $< \frac{1}{2}$d.

For the proof of (2.3), see the problems. In §14.10, (2.3) is used in the proof of the *existence theorem for multiple integrals*. The existence theorem is stated as (4.2) in §14.4 below.

§14.3 Applications of Scalar-valued Integrals

Examples. (1) *Let* R *be a regular region in* $\mathbf{E}^2$, *and let* f *have the constant value* 1. *Then*

(3.1) $\iint_R f dA = $ area of R . This integral is usually written $\iint_R dA$.

Similarly, let R *be a regular region in* $\mathbf{E}^3$, *and take* f *to have the constant value* 1. *Then*

(3.2) $\iiint_R f dV = $ volume of R . This integral is usually written $\iiint_R dV$.

(2)   *Consider* $\mathbf{E}^3$ *with Cartesian coordinates. Then the* xy *plane in* $\mathbf{E}^3$ *may be viewed as* $\mathbf{E}^2$ *with Cartesian coordinates. Let* R *be a regular region in this plane. Let* f(x,y) *be a continuous function on* R *such that* f(x,y) $\geq$ 0 *for all points* (x,y) *in* R. *Then* $\iint_R f(x,y) dA$ *is the volume of the solid, cylinder-like region* R' *in* $\mathbf{E}^3$ *whose upper surface is given by the graph of* f(x,y) *for* (x,y) *in* R, *whose lower surface is* R, *and whose lateral (cylinder-like) surfaces (if any) are formed of straight lines perpendicular to* R *from the boundary of* R. *(See* [3–1].*)* Thus, for f $\geq$ 0, $\iint_R f dA$ is "volume under a surface and above R" in exactly the same way that $\int_b^a f dx$ in one-variable calculus is "area under a curve and above [a,b]" for the case where f(x) $\geq$ 0 on [a,b].

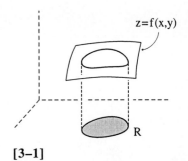

[3–1]

(3)   *For every point* P *in* R, *let* δ(P) *be the density at* P *of a substance* S *which occupies a region* R *in* $\mathbf{E}^3$. *Then*

(3.3) $\iiint_R \delta dV$ *gives the total* mass *of* S.

(4)   Multiple integrals can also be used to define and express the *average value* of a scalar field over a region R as $\dfrac{\iiint_R f dV}{\iiint_R dV}$ (for $\mathbf{E}^3$) or $\dfrac{\iint_R f dA}{\iint_R dA}$

(for $\mathbf{E}^2$). For example, for the substance S in (3),

(3.4) $\dfrac{\iiint_R \delta dV}{\iiint_R dV}$ $\left(= \dfrac{M}{V} \right.$, where M is total mass in R and V is volume of R.$\left.\right)$

is the average value of the density of S. (To see why we use the word "average," imagine that R is subdivided into n subregions, each having the same volume $\Delta V = \dfrac{1}{n} V$. Then

$$\frac{M}{V} \approx \frac{1}{V}(\delta(P_1)\Delta V_1 + \cdots + \delta(P_n)\Delta V_n) = \frac{1}{n}(\delta(P_1) + \cdots \delta(P_n)),$$

and this last expression is clearly an *average* in the usual sense.)

Notation. The average value of f over R is sometimes written as $\bar{f}$ *over* R. Thus (3.4) can be written: $\bar{\delta} = \dfrac{M}{V}$

(5)   *In some cases, the integrand of a scalar-valued multiple integral will be given by a scalar-valued expression of vector algebra.* For example, let $\vec{A}(P)$ be the position vector of P with respect to a reference point O. Then $f(P) = \left|\vec{A}(P)\right| = (\vec{A} \cdot \vec{A})^{1/2}$ is a scalar field, and

$$\frac{\iiint_R f dV}{\iiint_R dV} = \frac{\iiint_R (\vec{A} \cdot \vec{A})^{1/2} dV}{\iiint_R dV}$$

gives the *average distance* from O of points in R.

(6)   A substance S occupies a regular region R in $\mathbf{E}^3$ and has its density given by the scalar field δ. Let L be a straight line through a point Q, and let the scalar field r_L give the distance $r_L(P)$ of each point P from the line L. See [3–2]. The *moment of inertia of S about the axis L* is defined to be

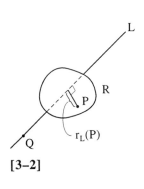

[3–2]

(3.5) $I_L = \iiint_R \delta r_L^2 \, dV.$

If δ is constant throughout S, if the total mass of S is M, and if the volume of D is V, then we have $\delta = \frac{M}{V}$. In this special case,

$$(3.6) \qquad\qquad I_L = \frac{M}{V} \iiint_R r_L^2 \, dV.$$

Remarks. (a) The concepts and expressions defined in the above examples are useful in theoretical derivations, even if we do not have mathematical expressions for the integrals or know how to evaluate the integrals numerically.

(b) We learn in elementary physics that the rotation of a rigid body about a given axis L must satisfy the equation $\tau_L = I_L \ddot{\theta}$, where I_L is the moment of inertia about L (as defined in (3.5) above), τ_L is the torque about L, and $\ddot{\theta}$ is the angular acceleration about L in the direction of the torque. This fact follows from Newton's laws of motion. Thus the equation $\tau_L = I_L \ddot{\theta}$ is the analogue, for rigid body rotation, of the equation $\vec{F} = m\vec{a}$ for the motion of a particle.

§14.4 Laws; the Existence Theorem

(4.1) Multiple integrals satisfy *algebraic laws* analogous to those satisfied by one-variable integrals. We state these laws for triple integrals. (The statements for double integrals are similar.) These laws follow directly from (2.2) (the definition of triple integral) and are analogous to the laws for one-variable integrals in (1.2) above.

(I) *If a regular region* R *is included in the domain of* f, *and if* R *is subdivided into two regular regions* R_1 *and* R_2, *then*

$$\iiint_R f dV = \iiint_{R_1} f dV + \iiint_{R_2} f dV.$$

(Note that the only overlap between R_1 and R_2 will be a common portion of boundary curve.)

(II) *If a regular region* R *is included in the domains of both* f *and* g, *then*

$$\iiint_R (f + g) dV = \iiint_R f dV + \iiint_R g dV.$$

(III) *If a regular region* R *is included in the domain of* f, *then for any given constant* k,

$$\iiint_R kf dV = k \iiint_R f dV.$$

(IV) *Let* m_1 *and* m_2 *be given constants such that* $m_1 \le f(P) \le m_2$ *for all* P *in a regular region* R, *and let* V *be the volume of* R, *then*

$$m_1 V \le \iiint_R f dV \le m_2 V.$$

(For a double integral, the volume V of R is replaced by the area A of R.)

The following theorem gives a condition under which we can be sure that a multiple integral exists.

(4.2) **Existence theorem for scalar-valued multiple integrals.** *Let* f *be a continuous scalar field on a regular region* R *in* $\mathbf{E}^2$. *Then the double integral* $\iint_R f dA$ *exists. Similarly, let* f *be a continuous scalar field on a regular region* R *in* $\mathbf{E}^3$. *Then the triple integral* $\iiint_R f dV$ *exists.*

A proof for (4.2) is given in §14.10.

With an appropriate definition of *piecewise continuous* for a scalar field or function f, we get the following corollary to (4.2):

(4.3) **Corollary.** *The existence theorem (4.2) continues to hold if "piecewise continuous" is substituted for "continuous."*

The definition of "piecewise continuous" is natural but technically subtle.

(4.4) **Definition.** A scalar field or function f is said to be *piecewise continuous* on a regular region R if (i) R is included in the domain of f, and (ii) it is possible to subdivide R into finitely many regular subregions $R_1,...,R_n$ such that for each subregion R_i, f as defined on the interior of R_i can be extended to a field or function f_i which is defined and *continuous* on the entire subregion R_i including its boundary. f_i need not agree with f on this boundary. (See the problems.) If f is piecewise continuous on R, then the value of the integral of f on R is obtained by taking the integral of f_i over R_i for each i and summing the values of these integrals.   It is easy to show that the max-min existence theorem holds for piecewise continuous scalar fields.

If f is continuous, then law (IV) can be strengthened to the following:

(4.5)   (V) *The mean-value law.* *Let* V *be the volume of a regular region* R, *and let* f *be continuous over* R. *Then there must be a point* P *in* R *such that*

$$\iiint_R f \, dV = f(P)V.$$

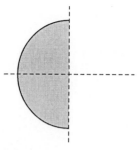

[5–1]

See the problems.

§14.5 Evaluating Multiple Integrals from the Definition

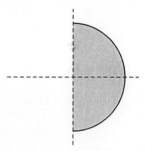

In Chapter 15, we give a systematic method for evaluating multiple integrals. This is the *method of iterated integrals*, which uses the fundamental theorem of integral calculus for one-variable integrals.   Sometimes, however, without using this systematic method, we can find an exact or approximate value for a multiple integral by considering geometric symmetries in f and R, or by using geometric figures whose areas or volumes we already know.

[5–2]

Examples. The first four examples use Cartesian coordinates in $\mathbf{E}^2$.

(1)   *Let* R *be a disk of radius* 1 *centered at the origin. Let* $f(x,y) = x$. We see that $\iint_R f \, dA = 0$, since the integral over [5–1] is the negative of the integral over [5–2]. (See (I) in §14.4.)

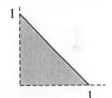

[5–3]

(2)   *Let* $f(x,y) = 1 - x - y$, *where* R *is* [5–3]. Then, by application 2 in §14.3, $\iint_R (1 - x - y) dA$ can be viewed as the volume of a pyramid with base R and altitude = 1. Hence $\iint_R f \, dA = \frac{1}{3}$ (altitude)(area of base) = $\frac{1}{6}$.

(3)   *Let* $f(x,y) = \sqrt{1 - x^2 - y^2}$, *where* R *is* [5–4]. Then, again by (2) in §14.3, $\iint_R \sqrt{1 - x^2 - y^2} dA$ can be viewed as the volume of a hemisphere of a radius 1. Hence $\iint_R f \, dA = \frac{1}{2} \cdot \frac{4}{3} \pi = \frac{4\pi}{6} = \frac{2\pi}{3}$.

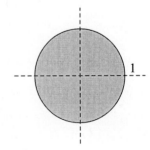

[5–4]

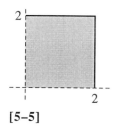

[5–5]

(4)   *Let* $f(x,y) = 2 + \dfrac{xy}{100}$, *where R is* [5–5]. In this case we cannot get the exact value of $\iint_R f dA$ from the definition, but we can get bounds on the value. Because, in R, $2 \le 2 + \dfrac{xy}{100} \le 2.04$, we have 2(area of R) $\le \iint_R f dA \le 2.04$ (area of R). (See (IV) in (§14.4).) Therefore $8 \le \iint_R f dA \le 8.16$. (In fact, $\iint_R f dA = 8.04$. We will see how to calculate this in the next chapter.)

(5)   As in (1), we can sometimes use symmetries in f and R to find a value for a triple integral. *Let R be a solid sphere of radius* 1 *centered at the origin in* $\mathbf{E}^3$ *with Cartesian coordinates.* Let $f(x,y,z) = x$. Then $\iiint_R f dV = 0$, since the integral over the x < 0 hemisphere must be the negative of the integral over the x > 0 hemisphere.

(6)   In symmetry arguments, we sometimes use Riemann sums based on specific decompositions of the region R. *Consider, for example,* $\iint_R x^3 y^3 dA$, *where R in* $\mathbf{E}^2$ *is the disk of radius* a *with center at the origin.* For a Riemann sum, we take a decomposition of R in which each subregion in the left half (x < 0) is the congruent mirror-image of a corresponding subregion in the right half (x > 0), and in which each sample point (–a,b) on the left is the mirror image of a corresponding sample point (a,b) on the right. (See [5–6].) Then the two corresponding terms in the Riemann sum must be $(-a)^3 b^3 \Delta A$ for the left and $a^3 b^3 \Delta A$ for the right, where ΔA is the area of each of the two subregions. The sum of these two terms is 0. As all of the subregions can be paired in this way, the value of the Riemann sum must be zero. Since a Riemann sum of this special kind can be constructed with an arbitrarily large number of arbitrarily small subregions, the value of the limit of a proper sequence of such special sums must be zero. (Recall the definition of *proper sequence* from §5.8.) Now $\iiint_R x^3 y^3 dA$ exists (by the existence theorem in §14.4), and *every* proper sequence must have the value of this integral as its limit. Hence the value of this integral must be zero. We refer to this method of showing an integral to be zero by the use of appropriate special Riemann sums as the *cancellation method*.

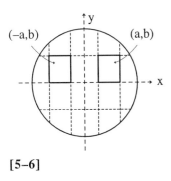

[5–6]

(7)   Riemann sums can also be used in other ways to find short cuts to the value of a multiple integral. Let R in $\mathbf{E}^2$ be the same region as in (6). What is $\iint_R (x^2 + y^2) dA$? We can express this integral as an elementary-calculus integral by using Riemann sums in which the disk is divided into m concentric

annular rings, and then each ring is divided by radial segments into further sub-regions. (See [5–7].) For the j-th ring ($1 \leq j \leq m$), we choose a value r_j^* such that the area of the ring is $2\pi r_j^* \Delta r_j$ (see the remark below), and for each subregion in that ring, we choose a sample point $P^* = (x^*, y^*)$ such that $\left| \overrightarrow{OP^*} \right| = r_j^*$ and hence $f(P^*) = (x^*)^2 + (y^*)^2 = (r_j^*)^2$. Hence the entire Riemann sum has the value

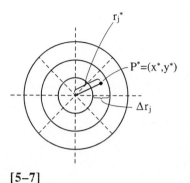

$$\sum_{j=1}^{m} (r_j^*)^2 \, 2\pi r_j^* \Delta r_j.$$ But this last sum is also a Riemann sum for the elementary-

[5–7]

calculus integral $\int_0^a 2\pi r^3 dr = \frac{2\pi a^4}{4} = \frac{\pi a^4}{2}$. Hence $\iint_R (x^2 + y^2) dA = \frac{\pi a^4}{2}$.

    (8) *Consider the integrals $\iint_R x^2 dA$ and $\iint_R y^2 dA$, where R is the region in (7). By geometric symmetry, these two integrals must be equal. By (II) in §14.4*, we have $\iint_R (x^2 + y^2) dA = \iint_R x^2 dA + \iint_R y^2 dA$. Hence from (7) we get

$$\iint_R x^2 dA = \iint_R y^2 dA = \frac{\pi a^4}{4}.$$

    Remark. In example (7), we can choose r_j^* with the desired property by taking

(5.1) $$r_j^* = \tfrac{1}{2}(r_j^{(o)} + r_j^{(i)}),$$

where $r_j^{(o)}$ is the outer radius of the j-th ring, and $r_j^{(i)}$ is the inner radius. (See the problems.)

    It is not necessary, however, to know (5.1) in order to do the example in (7). We need only observe that since $2\pi r_j^{(i)} \Delta r_j$ gives too small an area for the j-th ring, and $2\pi r_j^{(o)} \Delta r_j$ gives too large an area, then (by the continuity of $g(r) = 2\pi r$) there must be an intermediate value r_j^* which gives the correct area.

§14.6 Vector-valued Multiple Integrals

In Chapter 8, we noted that a vector field $\vec{F}$ in $\mathbf{E}^2$ is like a scalar field in $\mathbf{E}^2$, except that it associates with each point P in its domain (in $\mathbf{E}^2$) a certain vector $\vec{F}(P)$ in $\mathbf{E}^2$. Similarly, a vector field $\vec{F}$ in $\mathbf{E}^3$ associates with each point P in its domain (in $\mathbf{E}^3$) a certain vector $\vec{F}(P)$ in $\mathbf{E}^3$. We also saw that a *vector field* $\vec{F}$ may be represented in Cartesian coordinates by a *vector-valued function* which takes the coordinates of a point P as input and gives the $\hat{i}, \hat{j}, \hat{k}$ representation of the corresponding vector $\vec{F}(P)$ as output. (See §8.4.)

In the foregoing definitions of multiple integrals (in §14.2), the scalar field f can be replaced by a vector field $\vec{F}$. An integral so defined then has a certain *vector* (rather than scalar) as its value, provided that the Riemann sums have a limit. For example, for a regular region R (in $\mathbf{E}^2$) included in the domain of a vector field $\vec{F}$, the *vector-valued triple integral of* $\vec{F}$ *over* R is written

$$\iiint_R \vec{F} dV$$

and is defined to be the vector value given by

$$\lim_{\substack{n \to \infty \\ \max d_i \to 0}} \sum_{i=1}^{n} \vec{F}(P_i^*) \Delta V_i \, ,$$

where each Riemann sum is a *vector sum*, and where n, d_i, P_i^*, and ΔV_i are as before.

Continuity for a vector field $\vec{F}$ is defined exactly as for a scalar field, and the *existence theorem for vector-valued multiple integrals* has the same form as our previous theorem (4.2) for scalar-valued integrals:

(6.1)   ***Existence theorem for vector-valued integrals.*** *Let* $\vec{F}$ *be a continuous vector field or vector function on a regular region* R *in* $\mathbf{E}^2$. *Then the double integral* $\iint_R \vec{F} dA$ *exists. Similarly let* $\vec{F}$ *be a continuous vector field or function on a regular region* R *in* $\mathbf{E}^3$. *Then the triple integral* $\iiint_R \vec{F} dV$ *exists.*

A definition of *piecewise continuous* analogous to (4.4) can be given, and a corollary analogous to (4.3) follows:

(6.2) **Corollary.** *The existence theorem (6.1) continues to hold if "continuous" is replaced by "piecewise continuous."*

(6.3)   The *algebraic laws* for vector-valued integrals are as follows:

(I), (II), and (III) are as stated for scalar integrals in §14.4. The fourth law for scalar integrals does not apply. Instead, we give an alternative form of (III) in which a constant vector is factored out of a vector integral. In this case, the remaining integral is a scalar integral. We state this law for $\mathbf{E}^3$. (Other alternative forms of (III) are considered in the problems.)

(III′)   *Let* $\vec{A}$ *be a constant vector, let* f *be a scalar field whose domain includes* R, *and let* $\vec{F} = f\vec{A}$. *Then*

$$\iiint_R \vec{F}dV = \iiint_R f\vec{A}dV = \left(\iiint_R fdV\right)\vec{A}.$$

These laws imply that if a vector field $\vec{F}$ is expressed in a Cartesian coordinate system as a vector-valued function

$$\vec{F}(x,y,z) = L(x,y,z)\hat{i} + M(x,y,z)\hat{j} + N(x,y,z)\hat{k},$$

then a vector-valued integral of $\vec{F}$ can be expressed in terms of corresponding scalar-valued integrals of the scalar functions L, M, and N. We have

$$
\begin{aligned}
(6.4) \quad \iiint_R \vec{F}dV &= \iiint_R (L\hat{i} + M\hat{j} + N\hat{k})dV \\
&= \iiint_R L\hat{i}dV + \iiint_R M\hat{j}dV + \iiint_R M\hat{k}dV \quad \text{(by II)} \\
&= \left(\iiint_R LdV\right)\hat{i} + \left(\iiint_R MdV\right)\hat{j} + \left(\iiint_R NdV\right)\hat{k} \quad \text{(by III′).}
\end{aligned}
$$

The laws above, like the laws in §14.4 and the laws in §14.1, are immediate consequences of our definitions of integrals as limits of Riemann sums.

§14.7 Applications of Vector-valued Integrals

Vector-valued multiple integrals can be used to give general definitions of certain important mathematical and physical quantities.

Examples.

(1) Let D be a regular region in $\mathbf{E}^3$ with volume V. Let O be a fixed reference point, and let $\vec{R}(P)$ be the position vector of P. Let $\vec{C}$ be the average value of $\vec{R}$ as defined by

(7.1)
$$\vec{C} = \frac{\iiint_D \vec{R}\,dV}{V} = \frac{\iiint_D \vec{R}\,dV}{\iiint_D dV}.$$

Let $\overline{P}$ be the point whose position is given by the vector $\vec{C}$. $\overline{P}$ is called the *centroid* of D. (The *centroid* of a regular region in $\mathbf{E}^2$ is defined similarly.) It is easy to show that the point $\overline{P}$ given by (7.1) does not depend upon our choice of the reference point O. (See the problems.)

To compute $\vec{C}$, we usually introduce Cartesian coordinates with origin at O. Then

$$\vec{R} = x\hat{i} + y\hat{j} + z\hat{k},$$

and
$$\vec{C} = \frac{\iiint_D x\,dV}{V}\hat{i} + \frac{\iiint_D y\,dV}{V}\hat{j} + \frac{\iiint_D z\,dV}{V}\hat{k}.$$

The coordinates of the centroid are written $\overline{x}$, $\overline{y}$, and $\overline{z}$, and we have

(7.2) $$\overline{x} = \frac{1}{V}\iiint_D x\,dV, \quad \overline{y} = \frac{1}{V}\iiint_D y\,dV, \quad \overline{z} = \frac{1}{V}\iiint_D z\,dV,$$

where volume V can be obtained as $\iiint_D dV$. (The same formulas for $\overline{x}$ and $\overline{y}$, with volume V replaced by area A, give the Cartesian coordinates of the *centroid* of a region D in $\mathbf{E}^2$.) Thus, in Cartesian coordinates, the coordinates $\overline{x}$, $\overline{y}$, and $\overline{z}$ of the centroid are the *average* values of the x, y, and z coordinates over D. (This is only true for Cartesian coordinates. It is not true, for example, in polar or cylindrical coordinates. See §16.5.)

(2) Let the scalar field $\delta(P)$ give the density (at P) of a substance S which occupies a regular region D in $\mathbf{E}^3$ and has total mass M. Let $\vec{C}_m$ be the "weighted average" of $\vec{R}$ as defined by

(7.3)
$$\vec{C}_m = \frac{\iiint_D \delta\vec{R}dV}{M} = \frac{\iiint_D \delta\vec{R}dV}{\iiint_D \delta dV}.$$

The point whose position is given by the vector $\vec{C}_m$ is called the *center of mass* of S. Using Cartesian coordinates, we write its coordinates as $\bar{x}_m, \bar{y}_m, \bar{z}_m$. With $\vec{R} = x\hat{i} + y\hat{j} + z\hat{k}$, we have

(7.4)
$$\vec{C}_m = \bar{x}_m\hat{i} + \bar{y}_m\hat{j} + \bar{z}_m\hat{k}$$
$$= \left(\frac{1}{M}\iiint_D \delta x dV\right)\hat{i} + \left(\frac{1}{M}\iiint_D \delta y dV\right)\hat{j} + \left(\frac{1}{M}\iiint_D \delta z dV\right)\hat{k}$$

In the special case where the density is constant throughout D, we have

$$\vec{C}_m = \frac{\iiint_D \delta\vec{R}dV}{\iiint_D \delta dV} = \frac{\delta\iiint_D \vec{R}dV}{\delta\iiint_D dV} = \frac{\iiint_D \vec{R}dV}{\iiint_D dV},$$

and we see that the center of mass of the substance S is the same as the centroid of the region D.

(3)   A particle of mass m is located at the point O. A substance S occupies a regular region D and has its density given by a scalar field δ on D. We assume that the point O lies outside the region D. Newton's law of gravity suggests that the total gravitational force $\vec{G}$ on S due to m (in both direction and magnitude) will be given by the vector-valued integral

(7.5)
$$\vec{G} = \iiint_D \frac{-\gamma m\delta}{|\vec{R}|^3}\vec{R}dV = \iiint_D \frac{-\gamma m\delta}{|\vec{R}|^2}\hat{R}dV,$$

where $-\gamma m\delta$ replaces the constant k in the formula for inverse-square central force in §7.7. The appropriateness of this expression is evident if we consider Riemann sums and use the physical fact that the total force on a substance can be obtained by vector addition of forces at various points (see §2.6).

§14.8 An Example

In almost all fields of science and technology, vector- and scalar-valued integrals can be used to give technical definitions that are concise and useful. We illustrate with a simple example. Let D be a given geographical region (for example, the continental United States). Assume, for simplicity, that D can be treated as a regular flat region in E^2, and that population density (per unit area) can be represented, for the purposes of a mathematical model, by a continuous scalar field δ. Let the vector field $\vec{R}(P)$ give the position vector of any point P in D with respect to a fixed reference point O. We can then make the following demographic definitions:

$$area \text{ of } D = \iint_D dA ,$$

$$population \text{ of } D = \iint_D \delta dA ,$$

$$\text{position vector of the } geographical\ center \text{ of } D = \frac{\iint_D \vec{R} dA}{\iint_D dA} , \text{ and}$$

$$\text{position vector of the } center\ of\ population \text{ of } D = \frac{\iint_D \delta \vec{R} dA}{\iint_D \delta dA} .$$

Let Q be any fixed point in D (for example, the city of Washington, DC in the United States). Let $\vec{R}_Q$ be the position vector of Q. Then

$$\frac{\iint_D \delta |\vec{R} - \vec{R}_Q| dA}{\iint_D \delta dA} = average\ distance\ from \text{ Q for inhabitants of D.}$$

§14.9 Uniform Continuity

In this section, we state a theoretical fact about continuous scalar fields which will be used in later proofs.

Recall that a scalar field f is continuous at a given point $P \Leftrightarrow \lim_{Q \to P} f(Q) = f(P)$. Recall also that $\lim_{Q \to P} f(Q) = f(P) \Leftrightarrow$ there is a funnel function $\varepsilon(t)$ (on $[0,d]$ for some d) such that $\varepsilon(t)$ decreases to 0 as $t \to 0$ and such that for all Q, $0 < |\vec{PQ}| \le d \Rightarrow |f(P) - f(Q)| \le \varepsilon(|\vec{PQ}|)$. In general, for a continuous f, this funnel function $\varepsilon(t)$ will depend upon the point P. (In the case of $f(x) = \frac{1}{x}$ on $(0,\infty)$, for example, the funnel function needed for continuity at x becomes increasingly steep as x is taken closer to 0.) In some cases, however, there may be a single funnel function which can be used at all points P in the domain of f.

Definition. A scalar field f is said to be *uniformly continuous* on a domain R if there is a funnel function ε(t) on [0,d] which can be used at every point P in R to give the continuity of f at P. We refer to ε(t) as a *uniform funnel function for* f *on* R.

In the problems, we develop a proof for the following fundamental theorem about continuous scalar fields on compact domains. The proof is similar to the proof of the max-min existence theorem.

(9.1)   *Uniform continuity theorem. Let a scalar field be continuous on a compact domain* R. *Then* f *is uniformly continuous on* R.

(9.2)   *Corollary. Let* f *be continuous on a compact domain* R. *Let* d* *be the diameter of* R (= $\max\limits_{P,Q \text{ in } R} |\overrightarrow{PQ}|$). *Then* f *on* R *has a uniform funnel* ε*(t) *on* [0,d*].

Proof of (9.2). Let ε(t) on [0,d] be some uniform funnel for f on R. Define ε*(t) on successive intervals [0,d], [d,2d], [2d,3d],... by

$$\varepsilon^*(t) = n\varepsilon(d) + \varepsilon(t - nd) \text{ for } t \text{ in } [nd,(n+1)d] .$$

See the problems for details.

Remark. In §14.10, we use (9.1) to prove the existence theorem for multiple integrals. An appropriate version of (9.2) also holds for scalar fields which are piecewise continuous on a compact domain. See the problems.

§14.10 Existence for Multiple Integrals

Recall that a multiple integral is said to *exist* if there is a real number L such that every proper sequence of Riemann sums (for that integral) has the limit L. L is then called the *value* of the multiple integral. We now prove the fundamental *existence theorem for scalar-valued multiple integrals* as stated in §14.4.

(10.1) *Existence theorem for multiple integrals.* *Let* f *be a scalar field which is continuous on a regular region* R *in* E^3. *Then* $\iiint_R f dV$ *exists. Similarly, if* f *is a scalar field which is continuous on a regular region* R *in* E^2, *then* $\iint_R f dA$ *exists.*

Proof. We give the proof for the case of a double integral.

The proof is in two parts. In part 1, we show that if S_n and S'_n (n = 1,2,3,...) are the values of the n-th Riemann sums in two given proper sequences for $\iint_R f dA$, then it must be true that $\lim_{n\to\infty}(S_n - S'_n) = 0$. In part 2, we show that there exist a real number L and a particular proper sequence of Riemann sums for $\iint_R f dA$ with values S''_n (n = 1,2,3,...) such that $\lim_{n\to\infty} S''_n = L$. Parts 1 and 2 together imply that every proper sequence of Riemann sums for $\iint_R f dA$ must have this same limit L. It then follows, by definition, that $\iint_R f dA$ exists and has the value L.

Proof of part 1. Let d_n be the maximum diameter (of a subregion) occurring in either of the two Riemann sums, S_n and S'_n. By the definition of proper sequence, $\lim_{n\to\infty} d_n = 0$. For a given n, let $S_n = \sum_{i=1}^{n} f(P_i^*) \Delta A_i$ and $S'_n = \sum_{j=1}^{n} f(Q_j^*) \Delta B_j$, where ΔA_i is the area of a subregion R_i of R, and ΔB_j is the area of a subregion T_j of R.

Let U_{ij} be the set of points in R common to both R_i and T_j. (U_{ij} is not necessarily an elementary region.) For each i,j, U_{ij} has a well-defined area $\Delta C_{ij} \geq 0$.

Consider the sum $\sum_{i,j} f(P_i^*) \Delta C_{ij}$, where $\sum_{i,j}$ ranges over all pairs (i,j) with $1 \leq i \leq n$, $1 \leq j \leq n$. By grouping terms, we see that this sum has the value S_n. Similarly, $\sum_{i,j} f(Q_j^*) \Delta C_{ij}$ has the value S'_n. Finally, for all i,j, define

$$\Phi_{ij} = \begin{cases} \left| f(P_i^*) - f(Q_j^*) \right| & \text{for } \Delta C_{ij} \neq 0, \\ 0 & \text{for } \Delta C_{ij} = 0. \end{cases}$$

Then $\left| S_n - S_n' \right| \leq \sum_{i,j} \Phi_{ij} \Delta C_{ij} \leq \mu \cdot (\text{area of } R)$, where μ is the maximum value of Φ_{ij} for all i,j with $\Delta C_{ij} \neq 0$. Using the uniform continuity of f, we can make μ as small as we like by increasing n so that d_n becomes sufficiently small. (Note that $\Delta C_{ij} \neq 0 \Rightarrow d(P_i^*, Q_j^*) \leq 2d_n$). Hence $\lim_{n \to \infty} (S_n - S_n') = 0$.

Proof of part 2. We can use lemma (2.3) in §14.2 to construct the sequence S_n'' (n = 1,2,3, ...) inductively.

Stage 1. We define $S_1'' = f(P)(\text{area of } R)$, where the sample point P is chosen to minimize the value of f on R.

Stage 2. We use (2.3) to divide R into smaller elementary subregions. Let n_2 be the number of these subregions. We form the Riemann sum S_{n_2}'' by choosing, in each subregion, a sample-point which minimizes the value of f over that subregion.

. . . .

Stage j+1. We use the lemma to divide each of the subregions in stage j into still smaller elementary subregions. Let n_{j+1} be the total number of these new, smaller subregions. We form the Riemann sum $S_{n_{j+1}}''$ by choosing, in each new subregion, a sample-point which minimizes the value of f on that subregion.

. . . .

It follows that the values of these successive Riemann sums are non-decreasing ($S_{n_j}'' \leq S_{n_{j+1}}''$). Moreover, these values are all < MA, where M is the maximum value of f on R and A is the area of R. Hence, by the *least-upper-bound principle* (which asserts that a bounded, non-decreasing sequence of values must approach some limit), these values must approach a limit L. Since the maximum diameter of the subregions at stage j+1 is less than one-half the maximum diameter of the subregions at stage j, it is easy to fill in a full proper sequence $(S_1'', S_2'', S_3'', ...)$ of Riemann sums which has the limit L. (See the problems for further details.)

This completes the proof. Apart from obvious notational changes and the use of *volume* in place of *area*, the proof for triple integrals is the same. The proof of (10.1) holds good if f is only piecewise continuous. This proves (4.3).

Let S be a Riemann sum. Recall from §14.2 that m_S is the *mesh* of S if m_S is the maximum diameter of the subregions appearing in the decomposition of S. Let f be a continuous scalar field on a regular region R. The following corollary gives useful information as to how fast a proper sequence of Riemann sums for f on R must approach the limit $\iiint_R f dV$.

(10.2) *Corollary. Let $\varepsilon(t)$ on $[0,d]$ be a funnel function which gives the uniform continuity of f on R. Let V = volume of R. Then for a Riemann sum whose mesh is m_S with $m_S \leq d$,*

$$\left| S - \iiint_R f dV \right| \leq \varepsilon(m_S) V.$$

Let $\Delta_1, \Delta_2, \ldots$ be an infinite sequence of decompositions of a regular region R. (We use the abbreviated notation $\{\Delta_i\}$ for this sequence.) For each i, let n_i be the number of subregions appearing in Δ_i, and let δ_i be the mesh of Δ_i. Recall that $\{\Delta_i\}$ is called a *proper sequence* if $\lim_{i\to\infty} \delta_i = 0$ and if, for each i, $n_i = i$. If $\{\Delta_i\}$ is a proper sequence of decompositions, then we say that the sequence of Riemann sums determined by $\{\Delta_i\}$ is a *proper sequence* of sums.

Definition. A sequence of decompositions $\{\Delta_i\}$ is called a *proper sequence in the broad sense* if $\lim_{i\to\infty} \delta_i = 0$ and $\lim_{i\to\infty} n_i = \infty$ (but $n_i = i$ is not necessarily true). If $\{\Delta_i\}$ is a proper sequence in the broad sense, then we say that the sequence of Riemann sums determined by $\{\Delta_i\}$ is a *proper sequence* of sums *in the broad sense.*

The following result follows from (10.2).

(10.3) *Corollary. For continuous f and regular R, every sequence of partial sums in the broad sense converges to $\iiint_R f dV$.*

Corollaries (10.2) and (10.3) also hold for double integrals and one-variable integrals. We shall use the corollaries in §15.8. The proofs of (10.2) and (10.3) are given as problems. Appropriate versions of (10.2) and (10.3) also hold when f is piecewise continuous.

§14.11 Problems

§14.2

1. Use the max-min existence theorem to prove that every elementary region has a diameter.

2. Prove (2.3) for elementary regions in E^2.

§14.3

1. The region R in E^3 is occupied by solid matter whose density at each point is $\delta = x^2+y^2+z^2$. In each case, express the answer or answers in terms of appropriate multiple integrals with appropriate integrands. Integrands should be expressed as functions of x, y, and z, but do not need to be simplified:
   (a) Give the total mass.
   (b) Give the moment of inertia about the z axis.
   (c) Give the moment of inertia about the x axis.
   (d) Give the moment of inertia about the line determined by the points (0,0,0) and (1,1,1).

2. A thin metal plate occupies a region R in the upper half of the xy plane. The density per unit area of the plate is $\delta = ye^{xy}$ at each point. In each case, express the answer or answers in terms of appropriate multiple integrals with appropriate integrands. Integrands should be expressed as functions of x and y but do not need to be simplified.
   (a) Give the moment of inertia about the z axis.
   (b) Give the moment of inertia about the y axis.
   (c) Give the moment of inertia about the line y = 3.

3. Let S be a solid sphere of radius a with center at P_0. Let d be the average distance from P_0 of points in S. Express d in terms of appropriate multiple integrals.

§14.4

1. For each of the following functions, decide whether or not it is *piecewise continuous* as defined in (4.4).

(a) $f(x) = \frac{1}{x}$ for $x > 0$, and $= 0$ for $x \le 0$.

(b) $f(x) = \sin\frac{1}{x}$ for $x > 0$, and $= 0$ for $x \le 0$.

(c) $f(x) = x \sin\frac{1}{x}$ for $x > 0$, and $= 0$ for $x \le 0$.

2. Prove that the max-min existence theorem continues to hold if we replace *continuity* with *piecewise continuity* as defined in (4.4).

3. Show that (4.3) follows from (4.2).

4. Give an example to show that part (ii) of (4.4) is required for (4.3). (*Hint.* See problem 1 for an appropriate example.)

5. Prove the mean-value law for multiple integrals ((V) in §14.4.) (*Hint.* Use (IV).)

§14.5

1. Use the definition of multiple integral to evaluate each of the following. Solutions may include using Riemann sums to suggest equivalent one-variable (elementary calculus) definite integrals.
   (a) R is a disk in the xy plane with center at the origin and radius a. Find
   $$\iint_R (x^3 + xy^2)dA.$$
   (b) R is a solid sphere in xyz space with center at the origin and radius a. Find
   $$\iiint_R (x(y+z)^2 + 3)dV.$$
   (c) R is a solid cone with base in the xy plane of radius 2 with center at the origin and with vertex at (0,0,2). Find the volume of R. (*Hint.* Think of R as a collection of circular disks parallel to the xy plane with centers on the z axis. Express the total volume of these disks as a Riemann sum for a one-variable integral. Evaluate this integral.)

2. A thin square metal plate has total mass M and edge-length a. It occupies a region R in the xy plane with its edges parallel to the coordinate

axes and center-point at the origin. The plate has constant density (per unit area) $\delta = \dfrac{M}{a^2}$.

(a) Find the moment of inertia of the plate about the y axis. (*Hint.* Divide the plate into narrow vertical strips. Use a Riemann sum to get a one-variable definite integral, and evaluate.) Give your answer in terms of M and a.

(b) Find the moment of inertia about the z axis. (*Hint.* Use strips as in (a). Use a one-variable integral to get the contribution of each strip to the desired moment. Then use another one-variable integral to combine these contributions.) Give your answer in terms of M and a.

3. In problem 3 for §14.3, find the value of d by using one or more one-variable integrals and evaluating. (*Hint.* Think of S as a collection of thin, concentric, spherical shells.)

4. Evaluate $\iint_R (x^2 - y^2)dA$, where R is the square with vertices $(1,0)$, $(0,1)$, $(-1,0)$, $(0,-1)$.

5. Verify that r_j^* as defined in (5.1) has the desired property for the example in (7).

§14.6

1. Let the vector field $\hat{F} = xy\hat{i} + (x+1)\hat{j}$ on R, where R is a disk of radius 2 with center at the origin. Use Riemann sums and/or one variable integrals to evaluate $\iint_R \vec{F}dA$.

2. Show that for a constant vector field $\vec{A}$ on the domain D of a vector field $\vec{F}$

(a) $\iiint_R \vec{A} \times \vec{F}dV = \vec{A} \times \iiint_R \vec{F}dV$,

(b) $\iiint_R \vec{A} \cdot \vec{F}dV = \vec{A} \cdot \iiint_R \vec{F}dV$.

§14.7

1. (a) For problem 1 of §14.3, express the coordinates of the center of mass in terms of appropriate multiple integrals.

   (b) Do the same for problem 2 of §14.3.

2. Prove that the point located by the vector $\vec{C}$ in (7.1) is independent of the choice of a reference point for the position vector $\vec{R}$.

3. For R as in problem 1c of §14.5, find the centroid of R.

4. R is a hemisphere of radius a with center at the origin and lying above the xy plane. Find the centroid of R. (*Hint.* See the hint for problem 1c for §14.5.)

5. Two masses M_1 and M_2 occupy non-overlapping regions R_1 and R_2. Let P_1 and P_2 be the centers of mass of R_1 and R_2, respectively. Show that the center of mass of M_1 and M_2 considered as a single mass in space must be the same as it would be if the entire mass of M_1 were concentrated at P_1, and the entire mass of M_2 were concentrated at P_2. Do not assume constant density.

6. Show whether or not the result of problem 5 holds for three masses M_1, M_2, M_3 in non-overlapping regions R_1, R_2, and R_3.

7. (a) Does the result of (7.2) depend on the Cartesian coordinate system used? Justify your answer.

   (b) Is the same true for (7.4)? Justify.

§14.9

1. Let R be a given set (in $\mathbf{E}^2$ or $\mathbf{E}^3$.) Let $\mathscr{F}$ be a collection of open neighborhoods. If every point in R is contained in at least one of the neighborhoods in $\mathscr{F}$, we say that $\mathscr{F}$ *covers* R. The *Heine-Borel theorem* asserts that *if R is compact, then every collection of open neighborhoods which covers R must have a finite subcollection which also covers R.* Prove this theorem. (*Hint.* Assume that R is compact, that $\mathscr{F}$ covers R, but

that no finite subcollection of $\mathcal{F}$ covers R. Use subneighborhoods whose centers have rational coordinates and whose radii are rational to obtain an infinite sequence $U_1, U_2, \ldots$ of members of $\mathcal{F}$ which covers R. For $n = 1, 2, \ldots$, define P_n to be a point in R not covered by the finite collection $\{U_1, U_2, \ldots, U_n\}$. Use compactness of R (see problem 1 for §12.2) to show that there must be a point P^* in R such that every neighborhood which contains P^* must also contain P_n for infinitely many values of n. This leads to a contradiction.)

2. Use the Heine-Borel theorem to prove (9.1).

3. Show that (9.2) holds with *piecewise continuity* in place of *continuity*.

§14.10

1. For part 2 of the proof of (10.1), explain how to define the full proper sequence $S_1'', S_2'', \ldots$.

2. Prove (10.2).

3. Deduce (10.3) from (10.2).

Chapter 15

Iterated Integrals

§15.1 Iterated Integrals

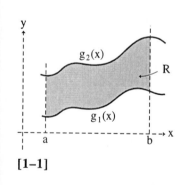

[1–1]

Let f be a continuous two-variable function. Let g_1 and g_2 be continuous one-variable functions, and let a and b be real numbers. We assume: (i) that $a < b$, (ii) that $[a,b]$ is contained in the domain of g_1 and in the domain of g_2, and (iii) that for all x, $a < x < b \Rightarrow g_1(x) < g_2(x)$. Let R be the set of all points (x,y) in the xy plane such that $a \leq x \leq b$ and $g_1(x) \leq y \leq g_2(x)$. See [1–1]. Finally, we assume that the region R is contained in the domain of f.

Consider the operation on f indicated by the following formula:

$$(1.1) \qquad \int_a^b \left[\int_{g_1(x)}^{g_2(x)} f(x,y)dy \right] dx .$$

The *first stage* of this operation consists of the one-variable integral

$$(1.2) \qquad \int_{g_1(x)}^{g_2(x)} f(x,y)dy .$$

In evaluating this integral, the variable x is treated as a constant. For each value of the constant x in $[a,b]$, we get a corresponding value for the integral (1.2). Thus (1.2) defines a function k on $[a,b]$:

(1.3)
$$k(x) = \int_{g_1(x)}^{g_2(x)} f(x,y)dy .$$

The existence of the integral (1.2) for each x in [a,b] follows from the continuity of f and the existence theorem in §14.1. The continuity of k can be proved from the continuity of f, g_1, and g_2. (See the problems.)

The *second stage* of the operation indicated in (1.1) consists of the one-variable integral

(1.4)
$$\int_a^b k(x)dx .$$

This integral exists, by the existence theorem in §14.1, and yields a real number ℓ as its output.

Definition. A two-stage operation of the kind described above, using two successive one-variable integrals on a given function of two variables, is called a *double iterated integral*. The region R, as described above, is called the *region of integration* for the iterated integral. The final output ℓ is called the *value* of the iterated integral. We say that the *order of integration* for the iterated integral is y *then* x.

We also include, as *double iterated integrals*, operations of the form

(1.5)
$$\int_c^d \left[\int_{h_1(y)}^{h_2(y)} f(x,y)dx \right] dy$$

on an analogous region R as in [1–2], where h_1 and h_2 are given continuous one-variable functions, and c and d are given real numbers. In this case, we say that the order of integration is x *then* y.

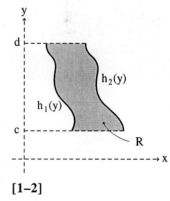

[1–2]

Notation. The following notations are commonly used for double iterated integrals:

$$\int_a^b \int_{g_1(x)}^{g_2(x)} f(x,y)dydx \quad \text{or} \quad \int_a^b dx \int_{g_1(x)}^{g_2(x)} f(x,y)dy$$

for
$$\int_a^b \left[\int_{g_1(x)}^{g_2(x)} f(x,y)dy \right] dx ,$$

and
$$\int_c^d \int_{h_1(y)}^{h_2(y)} f(x,y)dxdy \quad \text{or} \quad \int_c^d dy \int_{h_1(y)}^{h_2(y)} f(x,y)dx$$

for
$$\int_c^d \left[\int_{h_1(y)}^{h_2(y)} f(x,y)dx \right] dy .$$

In the following example, we evaluate a given double iterated integral by successively evaluating the two one-variable integrals.

Example 1. *We are given* $\int_0^1 \int_{x^2}^x xydydx$. (We note that $a = 0 < b = 1$ and that $g_1(x) = x^2 < g_2(x) = x$ for $0 < x < 1$.)

First stage:

$$\int_{x^2}^x xydy = \frac{1}{2}xy^2 \Big]_{y=x^2}^{y=x} = \frac{1}{2}(x^3 - x^5) .$$

Second stage:

$$\int_0^1 \frac{1}{2}(x^3 - x^5)dx = \frac{1}{2}\left(\frac{x^4}{4} - \frac{x^6}{6} \right)\Big]_0^1 = \frac{1}{2}\left(\frac{1}{4} - \frac{1}{6} \right) = \frac{1}{24} .$$

Let f be a continuous three-variable function. Let h_1 and h_2 be continuous two-variable functions; let g_1 and g_2 be continuous one-variable functions, and let a and b be real numbers. We assume (i) that $a < b$, (ii) that $[a,b]$ is contained in the domain of g_1 and in the domain of g_2, (iii) that for all x, $a < x < b \Rightarrow g_1(x) < g_2(x)$, and (iv) that for all (x,y) such that $a < x < b$ and $g_1(x) < y < g_2(x)$, (x,y) is in the domains of h_1 and h_2 and $h_1(x,y) < h_2(x,y)$. Let R be the set of all points (x,y,z) such that $a \le x \le b$, $g_1(x) \le y \le g_2(x)$, and $h_1(x,y) \le z \le h_2(x,y)$. (See [1–3].) Finally, we assume that the region R is contained in the domain of f.

Consider the operation on f indicated by the following formula:

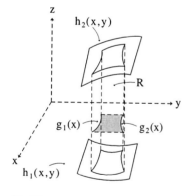

[1–3]

(1.6)
$$\int_a^b \left[\int_{g_1(x)}^{g_2(x)} \left[\int_{h_1(x,y)}^{h_2(x,y)} f(x,y,z)dz \right] dy \right] dx .$$

Here the first (innermost) stage of the operation is a one variable integral in which both x and y are treated as constants. This first stage produces a two-variable function $k(x,y)$. The second stage is a one-variable integral of $k(x,y)$ in which x is treated as a constant. This second stage produces a one-variable function $\ell(x)$.

Finally, the third stage is a one-variable integral of $\ell(x)$ which produces a single real number m as output.

Definition. A three-stage operation of the kind just described, using three successive one-variable integrals on a given function of three variables, is called a *triple iterated integral.* The output m is called the *value* of the iterated integral. The order of integration is z *then* y *then* x. There are five other possible orders of integration, depending upon the shape of R: z *then* x *then* y, y *then* z *then* x, etc. Given a function $f(x,y,z)$, we also include, as *triple iterated integrals*, cases in which these other orders of integration are used.

    Notation. As with double iterated integrals, the following notations are commonly used for a triple iterated integral:

$$\int_a^b \int_{g_1(x)}^{g_2(x)} \int_{h_1(x,y)}^{h_2(x,y)} f(x,y,z)dzdydx \quad \text{or} \quad \int_a^b dx \int_{g_1(x)}^{g_2(x)} dy \int_{h_1(x,y)}^{h_2(x,y)} f(x,y,z)dz$$

for

$$\int_a^b \left[\int_{g_1(x)}^{g_2(x)} \left[\int_{h_1(x,y)}^{h_2(x,y)} f(x,y,z)dz \right] dy \right] dx .$$

    Example 2. We are given $\int_0^1 \int_0^x \int_0^{xy} (xy + 2z)dzdydx$. (We note that $a = 0 < b = 1$, that $g_1(x) = 0 < g_2(x) = x$ for $0 < x < 1$, and that $h_1(x,y) = 0 < h_2(x,y) = xy$ for $0 < x < 1$ and $0 < y < x$. It is easiest to verify this by sketching, or visualizing, the region of integration R.) A figure for this particular example appears as [4–2] in §15.4.

    First stage:

$$\int_0^{xy} (xy + 2z)dz = xyz + z^2 \Big]_{z=0}^{z=xy} = x^2y^2 + x^2y^2 = 2x^2y^2 .$$

    Second stage:

$$\int_0^x 2x^2y^2 dy = \frac{2}{3}x^2y^3 \Big]_{z=0}^{z=xy} = \frac{2}{3}x^5 .$$

    Third stage:

$$\int_0^1 \frac{2}{3}x^5 dx = \frac{2}{18} \Big]_0^1 = \frac{1}{9} .$$

Remark. We normally use the fundamental theorem of integral calculus to evaluate the successive one-variable integrals in an iterated integral, as in the examples above. Numerical approximation techniques may also be used for the final stage of an iterated integral.

§15.2 Evaluating Double Integrals on Simple Regions; Fubini's Theorem

(2.1) *Definition.* Consider E^2 with Cartesian coordinates. We say that a region R is *simple* if it has one of the two following forms.

(a)   The projection of R on the x axis is an interval [a,b], and there are continuous functions g_1 and g_2 on [a,b] such that if $a < x < b$, then $g_1(x) < g_2(x)$, and R is the collection of points (x,y) such that $a \leq x \leq b$ and $g_1(x) \leq y \leq g_2(x)$. In this case, we say that R is y-*simple* or "vertically simple."

(b)   The projection of R on the y axis is an interval [c,d], and there are continuous functions h_1 and h_2 on [c,d] such that if $c < y < d$, then $h_1(y) < h_2(y)$, and R is the collection of points (x,y) such that $c \leq y \leq d$ and $h_1(y) \leq x \leq h_2(y)$. In this case, we say that R is x-*simple*, or "horizontally simple."

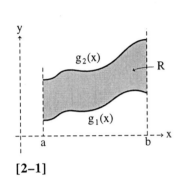

[2–1]

We observe that *a region is y-simple if and only if it can serve as the region of integration for an iterated double integral with order y then x, and that a region is x-simple if and only if it can serve as the region of integration for an iterated integral with order x then y.*

Examples of simple regions are given in [2–1] and [2–2]. We note that every simple region is elementary, but not every elementary region is simple.

As we shall see below, it is often possible for a given region R to be both y-simple and x-simple. In this case, we can have iterated integrals over R with both possible orders of integration.

The importance of iterated integrals lies in the fact that the evaluation of an iterated integral is a problem in one-variable calculus, and in the fact that iterated integrals can be used to evaluate multiple integrals. This is described for double integrals in the following theorem.

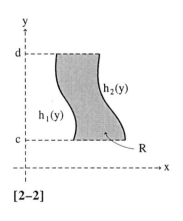

[2–2]

(2.2) *Evaluation theorem for double integrals.* *In E^2 with Cartesian coordinates, let R be a y-simple region (as described above), and let f be a continuous two-variable function whose domain includes R. Then*

$$\iint_R fdA = \int_a^b \int_{g_1(x)}^{g_2(x)} f(x,y)dydx.$$

Similarly, let R *be an* x-simple region (as described above) contained in the domain of f. Then

$$\iint_R fdA = \int_c^d \int_{h_1(y)}^{h_2(y)} f(x,y)dxdy.$$

(2.2) is commonly known as *Fubini's theorem for double integrals.*

Example. *In* E^2 *with Cartesian coordinates, let* R *be the region contained between the curves* $y = x^2$ *and* $y = x,$ *for* $0 \le x \le 1$. *(See [2–3].)* *Find the coordinates of the centroid of* R.

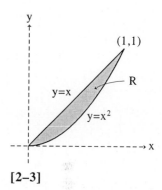

[2–3]

From Chapter 14, the answer is given by the double integrals

$$\overline{x} = \frac{1}{A}\iint_R xdA \quad \text{and} \quad \overline{y} = \frac{1}{A}\iint_R ydA,$$

where $A = \iint_R dA$, and $\overline{x}$ and $\overline{y}$ are the desired coordinates of the centroid. Since R is y-simple, we can use Fubini's theorem (2.2).

$$A = \int_0^1 \int_{x^2}^x dydx = \int_0^1 y\ \Big]_{y=x^2}^{y=x} dx = \int_0^1 (x - x^2)dx = \frac{x^2}{2} - \frac{x^3}{3}\Big]_0^1 = \frac{1}{2} - \frac{1}{3} = \frac{1}{6},$$

$$\overline{x} = 6\int_0^1 \int_{x^2}^x xdydx = 6\int_0^1 xy\ \Big]_{y=x^2}^{y=x} dx = 6\int_0^1 (x^2 - x^3)dx$$

$$= 6\left(\frac{x^3}{3} - \frac{x^4}{4}\right)\Big]_0^1 = \frac{6}{12} = \frac{1}{2},$$

$$\overline{y} = 6\int_0^1 \int_{x^2}^x ydydx = 6\int_0^1 \frac{y^2}{2}\Big]_{y=x^2}^{y=x} dx = 3\int_0^1 (x^2 - x^4)dx$$

$$= 3\left(\frac{x^3}{3} - \frac{x^5}{5}\right)\Big]_0^1 = 1 - \frac{3}{5} = \frac{2}{5}.$$

Thus the centroid is $\left(\frac{1}{2}, \frac{2}{5}\right)$.

Note that the region R is also x-simple, with $h_1(y) = y$ and $h_2(y) = y^{1/2}$. Hence, as the reader may verify, we can reach the same answer by using iterated integrals with the order x *then* y.

The idea behind the proof of (2.2) can be intuitively summarized as follows. (The full proof is given in §15.8.) Assume that R is y-simple with side boundaries $x = a$ and $x = b$, and with $y = g_2(x)$ and $y = g_1(x)$ as upper and lower boundaries.   Assume that f is continuous on R.   Let $k(x) = \int_{g_1(x)}^{g_2(x)} f(x, y)dy$.

Then k, g_1, and g_2 are continuous on [a,b]. We wish to show that $\iint_R fdA =$

$\int_a^b k(x)dx$. We divide R into strips perpendicular to the x axis, then $\iint_R fdA$ is

the sum of the double integrals of f over the strips. The integral over the i-th

strip is approximately $\left[\int_{g_1(x_i^*)}^{g_2(x_i^*)} f(x_i^*, y)dy \right] \Delta x_i = k(x_i^*)\Delta x_1$. Hence $\iint_R fdA \approx$

$\sum_i k(x_i^*)\Delta x_i \approx \int_a^b k(x)dx$.   Here x_i^* is a sample   x   value for the i-th strip.

§15.3 Evaluating Double Integrals on Regular Regions

Law I in §14.4 implies that if a regular region R in E^2-with-Cartesian-coordinates can be subdivided into a finite number of *simple* regions $R_1, R_2, \ldots, R_n$, then, for a two-variable function f whose domain includes R, we will have

$$\iint_R fdA = \iint_{R_1} fdA + \iint_{R_2} fdA + \cdots + \iint_{R_n} fdA .$$

By Fubini's theorem in §15.2 above, we can use iterated integrals to evaluate $\iint_{R_i} fdA$ for each i, and hence we can evaluate $\iint_R fdA$. The following theorem is true, but is not proved in this text.

(3.1)   *Subdivision principle for* E^2. (a) *For a regular region* R *in* E^2-*with-Cartesian-coordinates, it is always possible to subdivide* R *into a finite number of simple regions.*

(b) *It is not, however, always possible to divide such a region* R *into finitely many simple subregions such that either all the subregions are y-simple or all the subregions are x-simple.*

The proof of (3.1) requires methods and examples which are more complex than those treated in this text. In contrast to (b), we note that in virtually all examples which arise in practical applications, it *is* possible to divide R into subregions all of which are y-simple and it is also possible to divide R   into subregions all of which are x-simple.

[3–1] and [3–2] illustrate this subdivision principle. (In these examples, all the subregions are y-simple and use only lines parallel to the y axis.)

Sometimes, we will subdivide a region which is already simple in order to simplify the resulting iterated integrals. For example, in a y-simple region, if the function g_1 is defined by different formulas for different parts of the interval [a,b], we will subdivide so that each iterated integral uses only one of those formulas (and similarly for the function g_2).

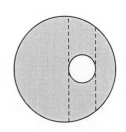

(4 regions)

[3–1]

Example. *Find the centroid of the region R enclosed by the triangle with vertices at (0,0), (3,0), and (1,4).* (See [3–3].) By definition, the centroid is at $(\bar{x}, \bar{y})$, where $\bar{x} = \frac{1}{A}\iint_R x\,dA$ and $\bar{y} = \frac{1}{A}\iint_R y\,dA$, with $A = \iint_R dA$. Evidently, from elementary geometry,

$$A = \tfrac{1}{2}\text{base} \times \text{altitude} = \tfrac{1}{2}3\cdot 4 = 6.$$

(8 regions)

[3–2]

For $\bar{x}$ and $\bar{y}$, we subdivide R into the y-simple regions R_1 and R_2 as in the figure. We have

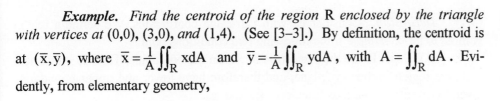

$$\begin{aligned}
\bar{x} &= \tfrac{1}{6}\iint_R x\,dA = \tfrac{1}{6}\iint_{R_1} x\,dA + \tfrac{1}{6}\iint_{R_2} x\,dA \\
&= \tfrac{1}{6}\int_0^1\int_0^{4x} x\,dy\,dx + \tfrac{1}{6}\int_1^3\int_0^{6-2x} x\,dy\,dx \\
&= \tfrac{1}{6}\int_0^1 xy \Big]_{y=0}^{y=4x} dx + \tfrac{1}{6}\int_1^3 xy \Big]_{y=0}^{y=6-2x} dx \\
&= \tfrac{1}{6}\int_0^1 4x^2\,dx + \tfrac{1}{6}\int_1^3 (6x - 2x^2)\,dx \\
&= \tfrac{1}{6}\tfrac{4}{3}x^3 \Big]_0^1 + \tfrac{1}{6}(3x^2 - \tfrac{2}{3}x^3)\Big]_1^3 = \tfrac{2}{9} + \tfrac{1}{6}(27 - 18 - 3 + \tfrac{2}{3}) = \tfrac{4}{3}.
\end{aligned}$$

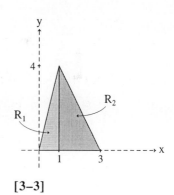

[3–3]

Similarly, $\bar{y} = \tfrac{1}{6}\int_0^1\int_0^{4x} y\,dy\,dx + \tfrac{1}{6}\int_1^3\int_0^{6-2x} y\,dy\,dx$. The reader may verify that this yields $\bar{y} = \tfrac{4}{3}$.

This example can also be done without subdivision, since R, as given, is also x-simple, so that we can use the order x *then* y. We then have

$$\bar{x} = \tfrac{1}{6}\int_0^4\int_{y/4}^{3-y/2} x\,dx\,dy,$$

$$\bar{y} = \frac{1}{6} \int_0^4 \int_{y/4}^{3-y/2} y \, dx \, dy \,.$$

The gain in avoiding subdivision, however, is offset (in this case) by increased algebraic complexity in the evaluation of the new iterated integrals.

§15.4 Evaluating Triple Integrals on Simple Regions

(4.1) *Definition.* Consider $\mathbf{E}^3$ with Cartesian coordinates. We define a region R in $\mathbf{E}^3$ to be *zy-simple* if the following two conditions hold:

(1) Let R′ be the projection of R on the xy plane. Then R′ must be a y-simple region in the xy plane, and therefore have lower and upper boundaries given by certain continuous functions g_1 and g_2.

(2) There must exist two continuous two-variable functions $h_1(x,y)$ and $h_2(x,y)$ on R′ such that for every point (x,y) in R′ but not on the boundary of R′, $h_1(x,y) < h_2(x,y)$; and such that R is the collection of all points (x,y,z) such that (x,y,0) is in R′ and $h_1(x,y) \le z \le h_2(x,y)$.

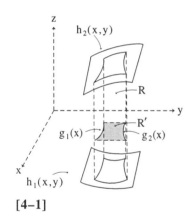

Clearly, a region is zy-simple if and only if it can serve as the region of integration for a triple iterated integral with the order z *then* y *then* x. [4–1] illustrates a zy-simple region in $\mathbf{E}^3$. Analogous versions of *zx-simple, yz-simple, yx-simple, xz-simple,* and *xy-simple* can be defined for each of the five other possible orders of integration. We say that a region R is *simple* (in $\mathbf{E}^3$) if it is either zy-, zx-, yz-, yx-, xz-, or xy-simple.

The following theorem enables us to use iterated integrals to evaluate triple integrals on simple regions.

[4–1]

(4.2) *Fubini's theorem for triple integrals. In $\mathbf{E}^3$ with Cartesian coordinates, let R be a zy-simple region (for the order z then y then x, as described above.) Let f be a continuous three-variable function whose domain includes R. Then*

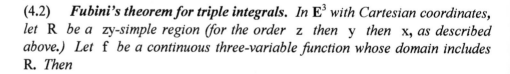

$$\iiint_R f dA = \int_a^b \int_{g_1(x)}^{g_2(x)} \int_{h_1(x,y)}^{h_2(x,y)} f(x,y,z) dz \, dy \, dx \,.$$

Analogous statements hold for each of the other five orders of integration.

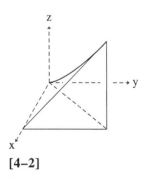

[4–2]

Example. Let f(x,y,z) = xy + 2z. Let R consist of all points (x,y,z) such that $0 \le x \le 1$, $0 \le y \le x$, and $0 \le z \le xy$, [4–2]. Then R has a triangular region as its projection R′ in the xy plane, where the vertices of R′ are (0,0), (1,0), and

(1,1). R is evidently zy-simple (for the order z *then* y *then* x) with $g_1(x) = 0$, $g_2(x) = x$, $h_1(x,y) = 0$, and $h_2(x,y) = xy$. It follows from the evaluation theorem that

$$\iiint_R f dA = \int_0^1 \int_0^x \int_0^{xy} (xy + 2z) dz dy dx.$$

This iterated integral is evaluated as example 2 in §15.1.

> *Proof of* (4.2). Let R' be as in the definition of zy *simple.*
> By steps exactly analogous to the construction for the proof of (2.2) in §15.8, we can show that

$$\iiint_R f dV = \iint_{R'} \left[\int_{h_1(x,y)}^{h_2(x,y)} f(x,y,z) dx \right] dA \ .$$

Then applying (2.2) to the double integral over R', we have our desired result.

§15.5 Evaluating Triple Integrals on Regular Regions; Rods and Slices

An analogue to (3.1) holds for $\mathbf{E}^3$.

(5.1) *Subdivision principle for* $\mathbf{E}^3$. *For a regular region R in* $\mathbf{E}^3$*-with-Cartesian coordinates, it is always possible to subdivide R into a finite number of simple regions.*

Hence, by law (I) in §14.4 and Fubini's theorem in §15.4, we can express a given triple integral of a regular region R as a sum of iterated integrals over simple subregions of R. It may be difficult, however, to visualize a suitable subdivision of a regular region in $\mathbf{E}^3$ into simple regions. It is sometimes easier and more efficient, in evaluating triple integrals, to use one of the following methods.

The method of rods. In this method, we express the given triple integral as an iteration of a one-variable integral followed by a double integral:

$$\iiint_R f dV = \iint_{R'} \left[\int_{h_1(x,y)}^{h_2(x,y)} f(x,y,z) dz \right] dA.$$

Here, the regular region R' is the projection of R on the xy plane. R' need not be a simple region; the method requires only that R be the set of all points (x,y,z) such that (x,y) is in R' and $h_1(x,y) \leq z \leq h_2(x,y)$. The result of the *first* integral, $\int_{h_1(x,y)}^{h_2(x,y)} f(x,y,z)dz$, is expressed as a function of x and y, and

$$\left[\int_{h_1(x,y)}^{h_2(x,y)} f(x,y,z)dz \right] \Delta A$$ gives, approximately, the triple integral of f on a *rod*

of cross-section area ΔA at (x,y) between the upper and lower surfaces of R. The *second* integral, as a double integral, can then be evaluated by any suitable means including, if necessary, the subdivision of R' into simple regions. In effect, this double integral combines all the rods together to get the desired triple integral over the full three-dimensional region R.

The method of slices. In this method, we express the given triple integral as an iteration of a double integral followed by a one-variable integral:

$$\iiint_R f dV = \int_{c_1}^{c_2} \left[\iint_{R_z} f(x,y,z)dA \right] dz \ .$$

Here, for each value of z, R_z is the projection on the xy plane of the intersection of R with a plane normal to the z axis at z. (R_z need not be a simple region.) c_1 and c_2 are the minimum and maximum values of z in R. The method requires that the value of the double integral be a continuous function of z. The value of the double integral may be obtained by any suitable means. The result of the double integral, $\iint_{R_z} f(x,y,z)dA$, is expressed as a function of z, and

$$\left[\iint_{R_z} f(x,y,z)dA \right] \Delta z$$ gives, approximately, the triple integral of f on a *slice* of

R of thickness Δz, perpendicular to the z axis at z. In effect, the final one-variable integral combines all the slices together to get the desired triple integral over the full region R.

We illustrate these methods with two examples.

Example 1. We use the slice method to derive the volume of a sphere of radius a with center at the origin. We take a slice R_z normal to the z axis. It is a disk of radius $\sqrt{a^2 - z^2}$ and hence has area $\pi(a^2-z^2)$. (See [5–1].) The final one-variable integral must be

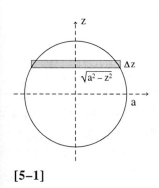

[5–1]

$$\int_{-a}^{a} \pi(a^2 - z^2)dz = \pi a^2 z - \frac{z^3}{3}\Bigg]_{-a}^{a} = 2\pi a^3 - \frac{2}{3}\pi a^3 = \frac{4}{3}\pi a^3.$$

Example 2. *We wish to find the volume of the region* R *consisting of all points which lie between the plane* $z = 0$ *and the plane* $z + y = 2$, *and which project onto the annular ring* $1 \le x^2 + y^2 \le 4$ *in the* xy *plane. We call this ring* R′. *Using the rod method, we find that the volume of* R *must be*

$$\iint_{R'} \left[\int_0^{2-y} dz \right] dA = \iint_{R'} (2-y)dA = \iint_{R'} 2dA - \iint_{R'} ydA$$

The last integral is 0 by cancellation, and $\displaystyle\iint_{R'} 2dA = 2(\text{area of } R') = 2(4\pi - \pi) = 6\pi$.

Remark. The advantages of these methods are (i) that they can often be applied directly, without considering whether or not the region R is simple, and (ii) that if a further subdivision is required, it can be carried out as a two-dimensional rather than three-dimensional subdivision. Furthermore, evaluation of the double integral can be approached as a new and separate problem — to be solved perhaps by the use of a different coordinate system or by direct argument from the definition of integral.

§15.6   Setting up Iterated Integrals

The following suggestions may be helpful for finding the correct limits of integration for an iterated integral in Cartesian coordinates.

Double iterated integral, Cartesian coordinates.

    (a)   For an iterated integral $\displaystyle\iint_R f(x,y)dydx$ over a y-simple region R

[6–1]

[6–1], we can find the limits of integration as follows:

        (i)   Hold x fixed, let y increase. As the point (x,y) moves, it traces a vertical line.

        (ii)   Find the y-value where this vertical line enters the region R and the y-value where it leaves R. These y-values depend upon x and give the limits for the dy integral.

        (iii)   Then choose the limits for the dx integral to include all vertical lines which pass through any part of R.

[6-2]

(b)   For an iterated integral $\iint_R f(x,y)dxdy$ over an x-simple region R [6-2], we do the same as (a) with x and y interchanged, and with "horizontal" in place of "vertical."

Example 1. *Find the area between the parabola* $y^2 = x$ *and the line* $y = x-2$. This area is $\iint_R dA$ where R is as shown in [6-3].

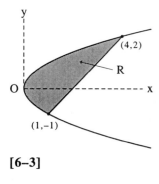

[6-3]

If we choose the order of integration $\iint_R dydx$, we simplify evaluation by dividing R into two subregions. By the rules above, we get:

$$\iint_R dA = \int_0^1 \int_{-\sqrt{x}}^{\sqrt{x}} dydx + \int_1^4 \int_{x-2}^{\sqrt{x}} dydx = \frac{4}{3} + \frac{19}{6} = \frac{9}{2}.$$

If we choose the order $\iint_R dxdy$, we get

$$\iint_R dA = \int_{-1}^2 \int_{y^2}^{y+2} dxdy = \frac{9}{2}.$$

Triple iterated integral, Cartesian coordinates.

For an iterated integral $\iiint_R f(x,y,z)dzdydx$ over a zy-simple region R as described in §15.4, we can find the limits of integration as follows:

(i)   Hold x and y fixed, let z increase. As the point (x,y,z) moves, it traces a vertical line.

(ii)   Find the z-value where this line enters R and the z-value where it leaves R. These values depend on x and y and give the limits for the dz integral.

(iii)   Supply the remaining limits to include all vertical lines which pass through any part of R. We find these limits by analyzing the projection R′ of R onto the xy plane as a y-simple region of integration for a double iterated integral.

Analogous steps hold for each of the other five orders of integration.

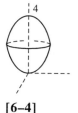

[6-4]

Example 2. *Set up an iterated integral for the volume of the region* R *lying above* $z = x^2 + y^2$ *and below* $z = 4-x^2-y^2$, [6-4].

Here the projection on the xy plane is the disk $x^2 + y^2 \le 2$. We get

$$\iiint_R dV = \int_{-\sqrt{2}}^{\sqrt{2}} \int_{-\sqrt{2-x^2}}^{\sqrt{2-x^2}} \int_{x^2+y^2}^{4-x^2-y^2} dz\,dy\,dx.$$

(This iterated integral has the value 4π, as the reader may verify. The third-stage integral is difficult. The volume of R can be more easily found by the use of cylindrical coordinates, as we shall see in §16.5. Easiest of all is to use circular slices normal to the z-axis; see the problems.)

§15.7 Examples

It is useful to think of the solution of a multiple integral problem as occurring in three steps.

Step 1. Describe carefully the multiple integral that we wish to evaluate, including a description of the integrand and of the region over which the multiple integral is taken.

Step 2. Set up one or more iterated integrals for the purpose of evaluating the multiple integral from step 1. This requires that we choose a coordinate system for the region R; express the integrand in terms of these coordinates; subdivide R, if necessary, to obtain simple regions; choose, on each simple region, an order of integration for the iterated integral; and find correct limits of integration for the one-variable integrals that make up each iterated integral.

Step 3. Evaluate the iterated integrals obtained in step 2.

Step 1 uses the concepts of Chapter 14. Step 3 makes straightforward use of the techniques of integration of one-variable calculus. Step 2 is usually the most difficult step; it requires skill at analyzing geometrical figures for the purpose of division into simple regions and for the purpose of setting up iterated integrals with appropriate limits of integration. This skill is acquired and sharpened by practice.

Example 1. The axes of two cylinders (of circular cross-section) of radius 1 intersect and are perpendicular. Find the volume of the intersection of the two cylinders.

Step 1. Take the two cylinder axes as the y and z axes. (See [7–1].) The region of intersection R is zy-simple (for the order z then y then x). The projected region R′ in the xy plane is the disk $x^2 + y^2 \le 1$.

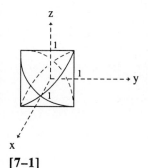

[7–1]

Step 2. We set up $\iiint_R dxdydz$. For the limits for dz, we take a line through a point (x,y) in the xy plane and find the z-values where it enters and leaves R. These values are determined by the cylindrical surface $x^2 + z^2 = 1$. Hence we have $z = -\sqrt{1-x^2}$ and $z = \sqrt{1-x^2}$ as our desired z-values. For the limits for dy, we use the disk R' given by $x^2 + y^2 \le 1$ and get $y = -\sqrt{1-x^2}$ and $y = \sqrt{1-x^2}$ as our desired y-values. Finally, the limits for dx must be x = −1 and x = 1. We have

$$\text{Volume} = \iiint_R dV = \int_{-1}^1 \int_{-\sqrt{1-x^2}}^{\sqrt{1-x^2}} \int_{-\sqrt{1-x^2}}^{\sqrt{1-x^2}} dzdydx.$$

Step 3. Evaluating, we have

$$= \int_{-1}^1 \int_{-\sqrt{1-x^2}}^{\sqrt{1-x^2}} 2\sqrt{1-x^2}\,dydx = \int_{-1}^1 4(1-x^2)dx$$

$$= 4x - \frac{4}{3}x^3 \Big]_{-1}^1 = 4 - \frac{4}{3} + 4 - \frac{4}{3} = \frac{16}{3}.$$

The volume of the intersection is $\frac{16}{3}$. (An alternative approach is to use square slices normal to the x axis. This leads directly to the integral $\int_{-1}^1 4(1-x^2)dx$; see the problems.)

Example 2. *Let R be a solid cube of total mass* M, *constant density* δ, *and edge-length* a. (Thus M = δa³.) *Find the moment of inertia of R with respect to a line L which passes through the center of the cube and is perpendicular to two opposite faces of the cube. Express this moment in terms of* M *and* a.

Step 1. We use Cartesian axes with origin at the center of the cube and with the z axis as L. (See [7–2].) From §14.3, the desired moment of inertia is

$$I_L = \iiint_R \delta r_L^2 dV = \frac{M}{V} \iiint_R r_L^2 dV.$$

Since $r_L = \sqrt{x^2 + y^2}$, we have

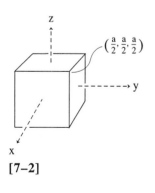

[7–2]

$$I_L = \frac{M}{a^3} \iiint_R (x^2 + y^2) dV.$$

Step 2. We can choose any order of integration. We choose y *then* x *then* z. Letting $b = \frac{a}{2}$, we get, as the reader may verify,

$$I_L = \frac{M}{a^3} \int_{-b}^{b} \int_{-b}^{b} \int_{-b}^{b} (x^2 + y^2) dy dx dz.$$

Step 3. $I_L = \frac{2bM}{a^2} \int_{-b}^{b} \int_{-b}^{b} (x^2 + \frac{b^2}{3}) dx dz$

$$\left(\text{since } x^2 y + \frac{y^3}{3} \Big]_{-b}^{b} = 2b(x^2 + \frac{b^2}{3}) \right)$$

$$= \frac{8b^4 M}{a^2} \int_{-b}^{b} dz \quad \left(\text{since } \frac{x^3}{3} + \frac{b^2 x}{3} \Big]_{-b}^{b} = \frac{4b^3}{3} \right)$$

$$= \frac{16b^5}{3a^2} M \qquad \left(\text{since } z \Big]_{-b}^{b} = 2b \right)$$

$$= \frac{1}{6} Ma^2 \qquad \left(\text{since } b = \frac{a}{2} \right).$$

The desired moment of inertia is $I_L = \frac{1}{6} Ma^2$.

§15.8 Proving Fubini's Theorem

Let R, a, b, g_1, g_2, f, and k be as in the assumptions for (2.2). We wish to show that $\iint_R f dA = \int_a^b \int_{g_1(x)}^{g_2(x)} f(x,y) dy dx$.

(1) Let L = b–a. By §14.9, we know that g_1 and g_2 are uniformly continuous on [a,b]. Let $\varepsilon_1(t)$ and $\varepsilon_2(t)$ on [0,L] be uniform funnels for g_1 and g_2 on [a,b]. Then $\varepsilon(t) = \varepsilon_1(t) + \varepsilon_2(t)$ on [0,L] is a uniform funnel for g_1 and also for g_2. In the problems, we show that $\varepsilon(t)$ can be strictly increasing. It follows that on any subinterval of [a,b] of length < δ, the difference between the maximum and minimum values g_1 is less than $\varepsilon(\delta)$ and similarly for g_2.

(2) Our first goal is to obtain a Riemann sum for f on R by dividing R into vertical strips and then, insofar as possible, dividing each strip into rectangular subregions of the same width as the strip. For any given δ, 0 < δ < L, we can

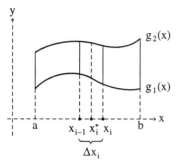

[8–1]

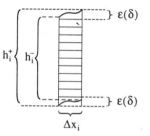

[8–2]

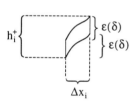

[8–3]

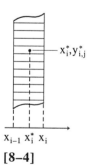

[8–4]

get our strips by taking a finite decomposition of [a,b] of mesh $< \delta$, with $a = x_0 < x_1 < \ldots < x_n = b$, $\Delta x_i = x_i - x_{i-1}$, and x_i^* a sample point in $[x_{i-1}, x_i]$ for $1 \le i \le n$, where n is the number of subintervals of the decomposition. (See [8–1].) Then, we divide the i-th strip of R into subregions as follows. Let $\max_i g_2$, $\min_i g_2$, $\max_i g_1$, $\min_i g_1$ be the maximum and minimum values of g_2 and g_1 on the interval $[x_{i-1}, x_1]$. Let $h_i^+ = \max_i g_2 - \min_i g_1$ and $h_i^- = \min_i g_2 - \max_i g_1$. If $h_i^+ > 2\varepsilon(\delta)$, then $h_i^- > 0$ and we can insert horizontal dividing lines of length Δx_i at $y = \min_i g_2 - \varepsilon(\delta)$ and at $y = \min_i g_1 + \varepsilon(\delta)$, and we can go on, if necessary, to insert further horizontal dividing lines of length Δx_i between those two lines until we have divided the i-th strip into subregions all of height $\le \varepsilon(\delta)$. (See [8–2].) Let n_i be the number of subregions obtained. If $h_i^+ \le 2\varepsilon(\delta)$, there may be no way to insert two or more dividing lines of length Δx_i, and we leave the i-th strip without further subdivisions, and $n_i = 1$. (See [8–3].) (Evidently, for each i, $n_i = 1$ or $n_i \ge 3$.) Let $R_{i,j}$ be the j-th subregion from the bottom in the i-th strip. For each j, we choose a sample point in $R_{i,j}$ on the line $x = x_i^*$. Let $(x_i^*, y_{i,j}^*)$ be the coordinates of this point. (See [8–4].) Finally let $\Delta A_{i,j}$ be the area of $R_{i,j}$. For each i, we define

$$S_{\delta,i} = \sum_{j=1}^{n_i} f(x_i^*, y_{i,j}^*) \Delta A_{i,j},$$

and

$$S_\delta = \sum_{i=1}^{n} S_{\delta,i}.$$

S_δ is evidently a Riemann sum of mesh $< \delta + 2\varepsilon(\delta)$ for $\iint_R f dA$, and S_δ must have the limit $\iint_R f dA$ as $\delta \to 0$.

(3) For each i, we modify $S_{\delta,1}$ as follows. If $n_i = 1$, we replace $\Delta A_{i,1}$ by $(g_2(x_i^*) - g_1(x_i^*))\Delta x_i$. (See [8–5].) If $n_i = 1$, we replace $\Delta A_{i,1}$ by $(\min_i g_1 + \varepsilon(\delta) - g_1(x_i^*))\Delta x_i$ and $\Delta A_{i,n_i}$ by $(g_2(x_i^*) - \max_i g_2 + \varepsilon(\delta))\Delta x_i$. For $1 < j < n_i$, the subregions of $S_{\delta,1}$ remain unchanged. (See [8-6].) Let $S_{\delta,i}^r$ be the value of this modified sum. ("r" reminds us that all the subregions in this new sum are rectangular.)

These modifications (of the single subregion if $n_i = 1$, of the top and bottom subregions if $n_i > 1$) must affect the sum $S_{\delta,1}$ by less than $M\varepsilon(\delta)\Delta x_i +$ $M\varepsilon(\delta)\Delta x_1 = [2M\varepsilon(\delta)]\Delta x_i$, where M is the maximum value of $|f|$ on R. Let

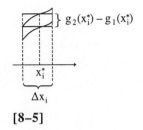

$$S_\delta^r = \sum_{i=1}^n S_{\delta,i}^r .$$

[8–5]

Then we have $\left|S_\delta - S_\delta^r\right| < [2M\varepsilon(\delta)]L$, and S_δ^r must have the same limit as S_δ when $\delta \to 0$; this limit is $\iint_R f\,dA$.

(4) In the subdivision for $S_{\delta,i}$, let $y_{i,1},\ldots,y_{i,n_i-1}$ be the positions of the inserted dividing lines, let $y_{i,0} = g_1(x_i^*)$, and let $y_{i,n_i} = g_2(x_i^*)$. Then let $\Delta y_{i,j} = y_{i,j} - y_{i,j-1}$, for $1 \le j \le n_i$. $S_{\delta,i}^r$ can now be expressed

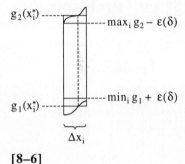

(8.1)
$$\left[\sum_{j=1}^{n_i} f(x_i^*, y_{i,j}^*)\Delta y_{i,j}\right]\Delta x_i ,$$

[8–6]

and we see that the sum in (8.1) is a Riemann sum of mesh $\le 2\varepsilon(\delta)$ for $k(x_i^*) =$ $\int_{g_1(x_i^*)}^{g_2(x_i^*)} f(x_i^*,y)dy$.

Define $S_\delta^k = \sum_{i=1}^n k(x_i^*)\Delta x_i$. Then S_δ^k is a Riemann sum of mesh $< \delta$ for the final one-variable integral $\int_a^b k(x)dx$ of the iterated integral $\int_a^b \int_{g_1(x)}^{g_2(x)} f(x,y)dydx$.

We consider

$$S_\delta^r - S_\delta^k = \sum_{i=1}^n \left[\sum_{j=1}^{n_i} f(x_i^*,y_{i,j}^*)\Delta y_{i,j} - k(x_i^*)\right]\Delta x_i .$$

By §14.9, k is uniformly continuous and must have a uniform funnel $\eta(t)$ on $[0,L]$. Let $H = \max_{[a,b]} g_2 - \min_{[a,b]} g_1$. By (10.2) in §14.10,

$$\left| \sum_{j=1}^{n_i} f(x_i^*, y_{i,j}^*) \Delta y_{i,j} - k(x_i^*) \right| \Delta x_i < [H\eta(2\varepsilon(\delta))] \Delta x_i .$$

Hence $\left| S_\delta^r - S_\delta^k \right| \le [H\eta(2\varepsilon(\delta))] L$. Thus S_δ^r must have the same limit as S_δ^k as

$\delta \to 0$, and this limit must be the value of the iterated integral for f.
The results of (3) and (4) together give our desired result.

§15.9   Problems

§15.1

1. Evaluate:

(a) $\displaystyle\int_1^{\ln 8} \int_0^{\ln y} e^{x+y} dxdy$

(b) $\displaystyle\int_1^2 \int_y^{y^2} dxdy$

2. Evaluate the iterated double integrals for $\bar{x}$ and $\bar{y}$ given at the end of §15.3.

3. Evaluate:

(a) $\displaystyle\int_0^1 \int_x^{2x} \int_0^{xy} x^2 yzdzdy\,dx$

(b) $\displaystyle\int_0^1 \int_0^y \int_x^y dzdxdy$

4. Describe geometrically, in words and with sketches, the regions of integration for:

(a) problem 3a

(b) problem 3b

5. For k, g_1, g_2, and f as in (1.3), prove that the continuity of k follows from the continuity of f, g_1, and g_2. (*Hint.* Use uniform continuity and (10.2) in §14.10.)

§15.2

1. Write as iterated integrals with order of integration reversed:

(a) $\displaystyle\int_0^1 \int_{\sqrt{y}}^1 dxdy$

(b) $\displaystyle\int_{-2}^1 \int_{x^2+4x}^{3x+2} dydx$

2. Evaluate $\displaystyle\int_0^2 \int_y^2 e^{x^2} dxdy$.

(*Hint.* You may wish to change the order of integration.)

3. By an iterated double integral, find the area of the region bounded by $x = y-y^2$ and the line $x+y = 0$.

4. Find the volume of the solid that lies under the surface $z = 4x^2+y^2$ and above the region in the xy-plane bounded by the curves $x = 0$, $y = 0$, $2x+y = 2$.

5. In the example in §15.2, find $\bar{x}$ and $\bar{y}$ by treating R as an x-simple region and using appropriate iterated integrals for A, $\bar{x}$, and $\bar{y}$.

§15.3

1. In the example in §15.3, verify that the indicated sum of two iterated double integrals gives the result $\bar{y} = \dfrac{4}{3}$.

2. Evaluate directly the two iterated integrals which appear at the end of §15.3; use the order x then y.

3. Evaluate $\iint_R xy\,dA$ when R is the triangular region enclosed by the lines x+y = 4, x–y = 0, and 3y–x = 0.

§15.4

1. By an iterated triple integral, find the volume of the tetrahedron bounded by the plane $\frac{x}{a}+\frac{y}{b}+\frac{z}{c}=1$ (with a,b,c > 0) and by the coordinate planes.

2. (a) Set up an iterated triple integral for the volume contained in the first octant inside the cylinder $x^2+y^2 = 1$, and below the plane x+y+2z = 2.
   (b) Evaluate this iterated integral.

§15.5

1. Set up an iterated integral giving the volume of the region bounded below by $z = x^2+y^2$ and above by z = y+2.

2. (a) Express the integral in the example in §15.4 as the sum of two iterated integrals, each on an xy-simple region.
   (b) Evaluate. (You may use example 2 in §15.1 as a check.)

§15.6

1. A sheet of metal occupies the region bounded by the curves y = x+2 and $y = x^2$. At each point, the density (per unit area) of the sheet is equal to the square of distance from the y axis. Find the mass of the sheet.

2. A sheet of metal of constant density = 1 (per unit area) occupies the region defined by the inequalities $-y^4 \le x \le y^2$, $-1 \le y < 1$.
   (a) Find its moment of inertia about the x axis.
   (b) Find its moment of inertia about the y axis.

3. Assuming constant density, find the coordinates of the center of mass of a body that occupies the region bounded by the parabolic cylinder $x = y^2$ and the planes z = 0 and x+z = 1.

4. A square piece of sheet metal of constant density per unit area has sides of length a and has mass M.
   (a) Find its moment of inertia about an axis perpendicular to its plane through one corner.
   (b) Find its moment of inertia about an axis along one edge of the square.

5. (a) Use a triple integral to find the volume of a solid pyramid whose base is a square of side a, whose altitude is b, and whose vertex projects onto the center of its base.
   (b) Locate the centroid of the pyramid.
   (c) Assuming constant density and total mass M, find the moment of inertia of the pyramid about a line L which goes through the vertex and is perpendicular to the base.

6. Use circular slices to evaluate the volume of R in example 2.

§15.7

1. Use square slices normal to the x axis to find the volume in example 1.

§15.8

1. Let ε(t) be a uniform funnel for a continuous function g on [a,b]. Show that there must, in fact, be a strictly increasing uniform funnel ε'(t) for g on [a,b].

2. Consider the function $f(x,y) = x^2 + 2y^2$ defined on the square domain $0 \le x \le 1, 0 \le y \le 1$. Give a formula for a uniform funnel function ε(t) for f on this domain.

Chapter 16

Integrals in Polar, Cylindrical, or Spherical Coordinates

§16.1 Cylindrical Coordinates, Polar Coordinates, and Spherical Coordinates

The natural symmetries of a given problem may suggest that we use a non-Cartesian coordinate system. We saw this in our study of vectors. We shall see it now with multiple integrals, where, in both theoretical and numerical calculations, the use of an appropriate non-Cartesian coordinate system may offer striking advantages. Among the most commonly used non-Cartesian coordinate systems are *polar coordinates* in $\mathbf{E}^2$, *cylindrical coordinates* in $\mathbf{E}^3$, and *spherical coordinates* in $\mathbf{E}^3$.

Cylindrical coordinates. These are described in §7.1 and §7.2.

Polar coordinates. Recall that in *polar coordinates*, we have an origin O, a fixed ray L from O, and a direction of positive rotation about O. See ([1–1].) A point P can be located by giving its distance r from O and an angle θ of rotation about O which will carry L to OP. r and θ are called *polar coordinates* of the point P.

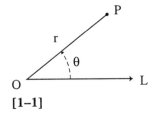
[1–1]

342

As we have seen, Cartesian coordinates are often used along with a given system of polar coordinates. It is customary to do this by taking O also to be the origin for the Cartesian coordinates, by taking L to be the positive x axis, and by taking the ray at $\theta = \frac{\pi}{2}$ to be the positive y axis. (See [1–2].) We then have the equations

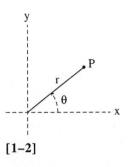

$$x = r \cos\theta, \qquad\qquad y = r \sin\theta,$$

$$r = \sqrt{x^2 + y^2},$$

$$\theta = \left\{ \begin{array}{l} \tan^{-1}(\tfrac{y}{x}) + 2n\pi, \text{ for } x > 0; \quad \cot^{-1}(\tfrac{x}{y}) + 2n\pi, \text{ for } y > 0 \\ \tan^{-1}(\tfrac{y}{x}) + (2n+1)\pi, \text{ for } x < 0; \quad \cot^{-1}(\tfrac{x}{y}) + (2n+1)\pi, \text{ for } y < 0 \end{array} \right\}$$

$$n = 0, \pm 1, \pm 2, \ldots.$$

[1–2]

relating Cartesian and polar coordinates. This Cartesian system in $\mathbf{E}^2$ is called the *associated Cartesian system* for the given system of polar coordinates. If we start with a given system of Cartesian coordinates in $\mathbf{E}^2$, then the polar system with $\theta = 0$ as the positive x axis and $\theta = \frac{\pi}{2}$ as the positive y axis is called the *associated polar system* for the given Cartesian system.

The equations above show, for example, that if we have the values $r = \sqrt{2}$ and $\theta = \pi/4$ for polar coordinates of a point, then we find the Cartesian coordinates for the same point (in the associated Cartesian system) to be $x = \sqrt{2} \cos(\pi/4) = 1$, and $y = \sqrt{2} \sin(\pi/4) = 1$.

Spherical coordinates. Our first step for defining a *spherical coordinate* system is the same as for defining a cylindrical coordinate system. We choose a plane M, an origin O in M, a ray L from O in M, and a number axis Z perpendicular to M at O. (See [1–3].) Then, for any point P, we get the spherical coordinates of P by (i) taking the direct distance ρ of P from O, (ii) taking the angle φ which $\overrightarrow{OP}$ makes with the positive direction of Z (we restrict φ to the interval $[0,\pi]$), and (iii) taking an angle θ between L and the projection of $\overrightarrow{OP}$ on M. The direction of increasing θ is given by a right-hand rule with respect to the positive direction of Z. If P is on Z, the value of θ is arbitrary. If P is O, the value of φ is arbitrary in $[0,\pi]$. ρ, φ, and θ are called the *spherical coordinates* of the point P. Cartesian coordinates are often used along with spherical coordinates. This is usually done by taking O to be the origin of Cartesian coordinates, Z to be the z axis, L to be the positive x axis, and the ray perpen-

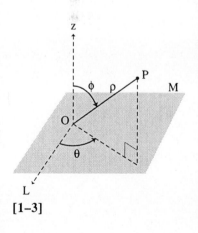

[1–3]

dicular to Z at $\theta = \frac{\pi}{2}$ to be the positive y axis. This Cartesian system is called the *associated Cartesian system* for the given system of spherical coordinates, and this spherical system is also called the *associated spherical system* for that Cartesian system. Evidently, the spherical coordinates (ρ, φ, θ) and the associated Cartesian coordinates (x,y,z) are related by the equations

(1.1)   for $x = \rho \sin \varphi \cos \theta$, $y = \rho \sin \varphi \sin \theta$, $z = \rho \cos \varphi$,

$$\rho = \sqrt{x^2 + y^2 + z^2} \ ,$$

$$\varphi = \cos^{-1}\left(\frac{z}{\sqrt{x^2+y^2+z^2}}\right) \text{ for } \rho > 0,$$

$$\theta = \begin{cases} \tan^{-1}(\frac{y}{x}) + 2n\pi, \text{ for } x > 0; \quad \cot^{-1}(\frac{x}{y}) + 2n\pi, \text{ for } y > 0 \\ \tan^{-1}(\frac{y}{x}) + (2n+1)\pi, \text{ for } x < 0; \quad \cot^{-1}(\frac{x}{y}) + (2n+1)\pi, \text{ for } y < 0 \end{cases}$$

$$n = 0, \pm 1, \pm 2, \ldots .$$

Using the z axis and the xy plane, we can also introduce an *associated system of cylindrical coordinates.* We then have

$$r = \rho \sin \varphi, \qquad\qquad \rho = \sqrt{r^2 + z^2}$$

$$\theta = \theta, \qquad\qquad \varphi = \cot^{-1}(\tfrac{z}{r}) \text{ for } r > 0$$

$$z = \rho \cos \varphi, \qquad\qquad \theta = \theta.$$

(Our choice of symbols for spherical coordinates is usual in mathematics. Physicists, however, sometimes use φ for θ, θ for φ, and r for ρ.)

A surface of constant ρ is a spherical surface. If we view the surface of the earth as such a sphere, with the axis z taken through the north and south poles, then the curves of constant φ on this surface are "parallels" of latitude, and the curves of constant θ are "meridians" of longitude.

§16.2   Iterated Integrals in Polar Coordinates

Given a double integral $\iint_R f(x,y)dA$ in Cartesian coordinates, we will sometimes be led, by the form of f or by the shape of R, to use polar coordinates in the integral. To find a general procedure for doing this, we view $\iint_R f(x,y)dA$ as the limit of a proper sequence (see §5.8) of Riemann sums of the form $\sum_i f(x_i^*,y_i^*)\Delta A_i$, where, except for *boundary subregions* (subregions that

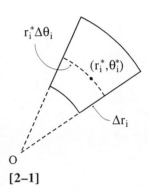

[2–1]

touch the boundary of R), we use elementary subregions of R of the following special kind: each subregion has, as boundary, *segments of two rays* from O and *arcs of two circles* centered at O. (See [2–1].) We now delete, from these Riemann sums, the terms for all boundary subregions. The sums that remain after these deletions are called *interior Riemann sums*, and we write them $\sum_{i\ (int)} f(x_i^*,y_i^*)\Delta A_i$. It is easy to show (see the problems for §17.7) that $\iint_R f(x,y)dA$ is also the limit of the remaining sequence of interior Riemann sums. (For these interior sums, every subregion is of the special kind shown in [2–1].)

Let (r_i^*,θ_i^*) be the polar coordinates of a chosen sample point in the i-th interior subregion. Let Δr_i be the change in r from one curved edge to the other of this subregion, and let $\Delta\theta_i$ be the angle between the straight edges of this subregion. Then the distance between the straight edges is approximately $r_i^*\Delta\theta_i$. Since the subregion is approximately rectangular, we have ΔA_i is approximately $r_i^*\Delta r_i\Delta\theta_i$. In fact, we can choose a sample point for each i so that $\Delta A_i = r_i^*\Delta r_i\Delta\theta_i$ (see the remark in §14.5). Thus $\iint_R f(x,y)dA$ is the limit of a sequence of interior Riemann sums of the form

$$\sum_{i(int)} f(r_i^* \cos\theta_i^*, r_i^* \sin\theta_i^*)r_i^*\Delta r_i\Delta\theta_i \ .$$

It follows that each of these Riemann sums is also an interior Riemann sum for a double integral with integrand $f(r \cos\theta, r \sin\theta)r$ over a corresponding region $\hat{R}$ in a plane with Cartesian coordinates where the rectangular axes in this plane are labeled r and θ instead of x and y. We call this new plane the rθ *plane*. We therefore have

(2.1) $$\iint_R f(x,y)dA = \iint_{\hat{R}} f(r \cos\theta, r \sin\theta)rdrd\theta .$$

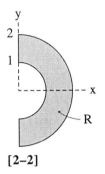

[2–2]

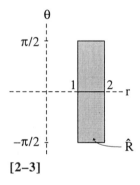

[2–3]

Here drdθ indicates integration in the rθ plane. (2.1) *thus expresses the original double integral over the region R in the xy plane as a double integral over a corresponding region* $\hat{R}$ *in the rθ plane. We say that (2.1) expresses the integral* $\iint_R f(x,y)dA$ *in polar coordinates.*

Example 1. *Consider the double integral* $\iint_R xdA$ *over the region R in* [2–2]. Expressing this in polar coordinates, we get $\iint_{\hat{R}} (r\cos\theta)rdrd\theta = \iint_{\hat{R}} r^2 \cos\theta drd\theta$, where $\hat{R}$ is the region in the rθ plane shown in [2–3]. When a double integral is expressed in polar coordinates, it may (depending on the shape of $\hat{R}$ and on the form of f(r cos θ, r sin θ)) become easier to evaluate. This is the case in our example, where

$$\iint_{\hat{R}} r^2 \cos\theta\, drd\theta = \int_1^2 dr \int_{-\pi/2}^{\pi/2} r^2 \cos\theta d\theta$$

$$= \int_1^2 2r^2 dr = \frac{2}{3}r^3 \Big|_1^2 = \frac{16}{3} - \frac{2}{3} = \frac{14}{3} \ .$$

(If we remain with Cartesian coordinates, we can get $\iint_R xdA = \frac{14}{3}$ after a lengthier calculation.)

Example 2. *Find the area of a circle of radius* a.

This area is simply $\iint_R dxdy$, where R is a circle of radius a centered at O. Going to polar coordinates, we have $\iint_{\hat{R}} rdrd\theta$, where $\hat{R}$ is the rectangle in rθ space bounded by the straight lines r = 0, r = a, θ = 0, and θ = 2π. We therefore have the iterated integral

$$\int_0^{2\pi} d\theta \int_0^a rdr = \int_0^{2\pi} \frac{a^2}{2} d\theta = \pi a^2 .$$

In the examples above, we have used the region $\hat{R}$ to find the limits of integration for an iterated integral in polar coordinates. From a theoretical point of view, we do this whenever we express a double integral in polar coordinates. In the next section, however, we see that we can often find the correct limits of integration for r and θ by direct reference to R, without using R̂ . Both of the examples above can be done in this way.

§16.3   Setting up Integrals in Polar Coordinates

We may not need to draw (or even to visualize) the rθ plane. The correct limits of integration for the iterated polar-coordinate integral may be directly evident from the region R in the xy plane. We must always, however, be sure that the additional factor r is put into the polar integrand.

(3.1)   *Thus, for a double integral over a region* R *where the integral is expressed in polar coordinates, it is often possible to integrate in the order* r *then* θ *(as in the examples above). In this case, we can find the correct limits as follows, using the region* R *in the* xy *plane (and ignoring the region* R̂ *in the* r *θ plane):*

(i) *Hold* θ *fixed, let* r *increase. As the point* (r,θ) *moves, it traces a ray going out from the origin.*

(ii) *Integrate from the* r-*value where the ray enters* R *to the* r-*value where it leaves* R. *This gives the limits for* r. *(These limits will depend on the fixed value of* θ.)

(iii) *Choose the limits for* θ *to include all rays which pass through any part of* R.

Example. *Set up an iterated integral in polar coordinates for the area of the region* R *in* [3–1].

For vertical line, we have x = −1 ⇔ r cos θ = −1 ⇔ r = $\frac{-1}{\cos\theta}$ ⇔ r = −sec θ. Moreover, the points P and Q have the polar coordinates $(2,\frac{2\pi}{3})$ and $(2,\frac{4\pi}{3})$. If we use the order r *then* θ, we have

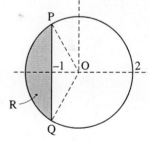

[3–1]

$$\int_{2\pi/3}^{4\pi/3} d\theta \int_{-\sec\theta}^{2} r\,dr = \int_{2\pi/3}^{4\pi/3}(2-\tfrac{1}{2}\sec^2\theta)d\theta = 2\theta - \tfrac{1}{2}\tan\theta \; \Big]_{2\pi/3}^{4\pi/3} = \frac{4\pi}{3}-\sqrt{3}.$$

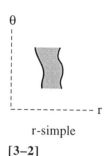

r-simple

[3–2]

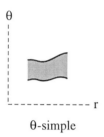

θ-simple

[3–3]

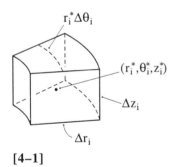

[4–1]

(Of course, the area of R in this example is also easily found by subtracting the area of the triangle OPQ $(=\sqrt{3})$ from the area of the circular sector OPQ $(=\frac{4\pi}{3})$.)

The procedure (3.1) for finding limits of integration assumes, in effect, that the region $\hat{R}$ is r-simple. It will usually be geometrically obvious (without directly considering $\hat{R}$) whether or not the procedure can be applied. If the region $\hat{R}$ does not have r-simple form [3–2] or θ-simple form [3–3], then we must divide $\hat{R}$ (as in §15.3) into subregions which do have r-simple or θ-simple form.

§16.4   Iterated Integrals in Cylindrical Coordinates

Given a triple integral $\iiint_R f(x,y,z)dV$, we may be led to use cylindrical coordinates. The method and argument are exactly as for double integrals and polar coordinates. We view $\iiint_R f(x,y,z)dV$ as the limit of Riemann sums of the form, $\sum_i f(x_i^*,y_i^*,z_i^*)\Delta V_i$, where, except along the boundary of R, we use elementary subregions of R of the following special kind: each subregion is bounded by portions of six coordinate surfaces, two of the form r = constant, two of the form θ = constant, and two of the form z = constant. (See [4–1].) For the i-th interior subregion, let Δr_i, $\Delta\theta_i$, and θz_i be the differences in coordinate values between opposite coordinate surfaces. For each i, let (r_i^*,θ_i^*,z_i^*) be the cylindrical coordinates of a chosen sample point (x_i^*,y_i^*,z_i^*) in the i-th subregion. Then ΔV_i, the volume of the i-th interior subregion, is approximately $(\Delta r_i)(r_i^*\Delta\theta_i)(\Delta z_i)=r_i^*\Delta r_i\Delta\theta_i\Delta z_i$, since the i-th subregion is approximately a rectangular solid in shape. In fact, it is possible to choose the sample point so that $\Delta V_i=r_i^*\Delta r_i\Delta\theta_i\Delta z_i$. (See (5.1) in §14.5.) Then $\iiint_R fdV$ is the limit of a sequence of interior Riemann sums of the form

$$\sum_{i(int)} f(r_i^*\cos\theta_i^*,r_i^*\sin\theta_i^*,z_i^*)r_i^*\Delta r_i\Delta\theta_i\Delta z_i.$$

It follows that each of these Riemann sums is also an interior Riemann sum for a triple integral with integrand $f(r\cos\theta,r\sin\theta,z)r$ over a corresponding region $\hat{R}$

in a three-dimensional Euclidean space with Cartesian coordinates, whose rectangular axes are labeled r, θ, and z instead of x, y, and z. This three-dimensional space is called rθz *space*. Hence we have

(4.1) $$\iiint_R f(x,y,z)dxdydz = \iiint_{\hat{R}} f(r\cos\theta, r\sin\theta, z)rdrd\theta dz ,$$

where drdθdz indicates integration in rθz space. (4.1) *thus expresses the original triple integral over the region* R *in* xyz *space as a new equivalent triple integral over a corresponding region* $\hat{R}$ *in* rθz *space.*

  Example 1. Consider example 1 in §15.7, where we found the volume of the intersection of two perpendicular cylinders of radius 1. Take the axes of the cylinders to be the x and z axes. (In §15.7, we used y and z.) Then, by symmetry, the volume is $8\iiint_R dxdydz$, where R is the wedge-shaped region (with x,z ≥ 0) contained between the plane z = 0, the cylinder $x^2 + y^2 = 1$, and the plane x = z. (See [4–2].) The total volume = $8\iiint_R dxdydz = 8\iiint_{\hat{R}} rdrd\theta dz$, where $\hat{R}$ is the region in rθz space defined by $0 \le z \le r\cos\theta$, $0 \le r \le 1$, and $-\frac{\pi}{2} \le \theta \le \frac{\pi}{2}$. Thus we have expressed the original triple integral in cylindrical coordinates. When a triple integral is expressed in cylindrical coordinates, it may (depending on the shape of $\hat{R}$ and the form of f(r cos θ, r sin θ, z)) become easier to evaluate by iterated integrals. This is the case in our example, where

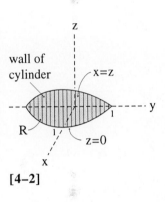

[4–2]

$$\text{Volume} = 8\iiint_{\hat{R}} rdrd\theta dz$$

$$= 8\int_{-\pi/2}^{\pi/2} d\theta \int_0^1 dr \int_0^{r\cos\theta} rdz$$

(see §16.5 for the best way to find these limits of integration)

$$= 8\int_{-\pi/2}^{\pi/2} d\theta \int_0^1 (r^2 \cos\theta)dr = 8\int_{-\pi/2}^{\pi/2} (\tfrac{1}{3}\cos\theta)d\theta = \tfrac{16}{3} .$$

  Example 2. Use cylindrical coordinates to find the volume of a sphere of radius a. We take the solid sphere of radius a with center at O as our region R. In Cartesian coordinates, the equation of the spherical surface is $x^2 + y^2 + z^2 = a^2$. In cylindrical coordinates, this becomes $r^2 + z^2 \equiv a^2$. Then $\iiint_R dV =$

$\iiint_{\hat{R}} r dr d\theta dz$, where $\hat{R}$ is a half-cylinder of radius a in $r\theta z$ space bounded above and below by the surface $r^2 + z^2 = a^2$ and on three sides by the planes r = 0, $\theta = 0$, and $\theta = 2\pi$. Hence we can go to the iterated integral:

$$\text{Volume} = \int_0^{2\pi} d\theta \int_0^a dr \int_{-\sqrt{a^2-r^2}}^{\sqrt{a^2-r^2}} r dz$$

(see §16.5 for the best way to find these limits of integration)

$$= \int_0^{2\pi} d\theta \int_0^a 2r\sqrt{a^2-r^2}\ dr$$

$$= -\int_0^{2\pi} d\theta \left(\left[(a^2-r^2)^{3/2} \cdot \tfrac{2}{3} \right]_0^a \right)$$

$$= \int_0^{2\pi} \tfrac{2}{3} a^3 d\theta = \frac{4\pi a^3}{3}.$$

Example 3. *Find the moment of inertia* I *of a solid circular cylinder of constant density, about its axis, where the cylinder has radius* a, *length* ℓ, *and mass* M.

Let the density be δ. Take the axis of the cylinder along the z axis and its base in the xy plane. Then we seek $\iiint_R (x^2 + y^2)\delta dV$, where $M = \pi a^2 \ell \delta$. Thus $\delta = \dfrac{M}{\pi a^2 \ell}$. Expressing this integral in cylindrical coordinates, we get I = $\iiint_R (x^2 + y^2)\delta dV = \iiint_{\hat{R}} \delta r^2 r dr d\theta dz$, where, in $r\theta z$ space, $\hat{R}$ is the region $0 \le r \le a, 0 \le z \le d, 0 \le \theta \le 2\pi$. Thus

$$I = \delta \int_0^{2\pi} \int_0^a \int_0^\ell r^3 dz dr d\theta$$

(see §16.5 for the best way to find these limits of integration)

$$= \delta \int_0^{2\pi} d\theta \int_0^\ell dz \int_0^a r^3 dr = \delta \int_0^{2\pi} d\theta \int_0^\ell \frac{a^4}{4} dz = \frac{2\pi \delta \ell a^4}{4} = \frac{2\pi a^4 \ell M}{4\pi a^2 \ell} = \frac{Ma^2}{2}.$$

In the examples above, we have used the region $\hat{R}$ to find the limits of integration in cylindrical coordinates. From a theoretical point of view, we do this whenever we express a triple integral in cylindrical coordinates. In the next section, however, we see that we can often find the correct cylindrical limits of integration by direct reference to R, without using $\hat{R}$. Each of the three examples above can be done in this way.

§16.5   Setting up Integrals in Cylindrical Coordinates

(5.1)   *For an iterated integral expressed in cylindrical coordinates, if we wish to integrate in the order z then r then θ (this is the case where $\hat{R}$ is zr-simple), we can find the limits of integration as follows (using the region R in xyz space and ignoring the region $\hat{R}$ in rθz space):*

(i)   *Hold r and θ fixed, let z increase. This gives us a vertical line.*

(ii)   *Integrate from the z-value where the vertical line enters R to the z-value where it leaves R. This gives the limits on z. (These limits will depend on the fixed values of r and θ.)*

(iii)   *Supply the remaining limits, as for a double integral in polar coordinates, to include all vertical lines which pass through any part of R. (We are therefore, in step (iii), integrating over the projection R′ of R onto the xy plane.)*

Example 1. *Use cylindrical coordinates to find the volume of the region* R *lying above* $z = x^2 + y^2$ *and below* $z = 4 - x^2 - y^2$. (The same region occurs in example 2 of §15.6.) The two surfaces are $z = r^2$ and $z = 4 - r^2$. R′ is the disk $r^2 \le 2$ in the plane z = 2.

We have:

$$\text{Volume} = \int_0^{2\pi} d\theta \int_0^{\sqrt{2}} dr \int_{r^2}^{4-r^2} r\,dz = 2\pi \int_0^{\sqrt{2}} (4r - 2r^3)\,dr = 4\pi .$$

The procedure (5.1) assumes, in effect, that the region $\hat{R}$ is zr-simple. It will usually be geometrically obvious (without directly considering $\hat{R}$) whether or not the procedure (5.1) can be applied. If the procedure cannot be applied, this shows that $\hat{R}$ is not zr-simple. If the region $\hat{R}$ is not simple, then we must divide $\hat{R}$ (as in §15.5) into simple subregions or use the other methods of §15.5.

Each of the examples in §16.4 used the order z *then* r *then* θ. The reader may check that the integration limits used in each of those examples can be easily obtained from the given R in xyz space by the method just described, without reference to $\hat{R}$.

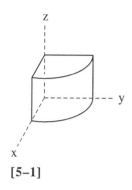

[5-1]

The order z *then* r *then* θ is the most commonly occurring order for iterated integrals in cylindrical coordinates. The order z *then* θ *then* r occasionally appears in applications. The other possible orders are less common.

Example 2. *Find the centroid of the region* D *defined by* $0 \le \theta \le \frac{\pi}{2}$, $0 \le r \le 1$, $0 \le z \le 1$. See [5-1].

Since $\iiint_R dV = \frac{\pi}{4}$, we have

$$\vec{C} = \frac{\iiint_D \vec{R}\, dV}{\iiint_D dV} = \frac{4}{\pi} \iiint_D \vec{R}\, dV. \quad \text{(See §14.7.)}$$

Then $\vec{R} = x\hat{i} + y\hat{j} + z\hat{k}$ in the associated Cartesian system, and

$$\vec{C} = \left(\frac{4}{\pi}\iiint_D x\, dV\right)\hat{i} + \left(\frac{4}{\pi}\iiint_D y\, dV\right)\hat{j} + \left(\frac{4}{\pi}\iiint_D z\, dV\right)\hat{k}$$
$$= \bar{x}\hat{i} + \bar{y}\hat{j} + \bar{z}\hat{k}.$$

Evidently, by geometrical symmetry, $\bar{z} = \frac{1}{2}$ and $\bar{x} = \bar{y}$. In order to evaluate $\bar{x} = \frac{4}{\pi}\iiint x\, dV$, we go to cylindrical coordinates:

$$\bar{x} = \frac{4}{\pi}\iiint r \cos\theta\, r\, dz\, dr\, d\theta$$
$$= \frac{4}{\pi}\int_0^{\pi/2} d\theta \int_0^1 dr \int_0^1 r^2 \cos\theta\, dz$$
$$= \frac{4}{\pi}\int_0^{\pi/2} d\theta \int_0^1 r^2 \cos\theta\, dr = \frac{4}{3\pi}\int_0^{\pi/2} \cos\theta\, d\theta = \frac{4}{3\pi}.$$

Thus $\bar{x} = \frac{4}{3\pi}$, $\bar{y} = \frac{4}{3\pi}$, and $\bar{z} = \frac{1}{2}$. If we wish, we can now get cylindrical coordinates for this centroid. These coordinates are

$$r_c = \sqrt{\bar{x}^2 + \bar{y}^2} = \frac{4\sqrt{2}}{3\pi} = 0.600,$$
$$\theta_c = \tan^{-1}\left(\frac{\bar{y}}{\bar{x}}\right) = \frac{\pi}{4},$$
$$z_c = \bar{z} = \frac{1}{2}.$$

Note that r_c does *not* turn out to be the average value of r in **D**. This average value is

$$\frac{\iiint_D r\,dV}{V} = \frac{4}{\pi}\int_0^{\pi/2}d\theta\int_0^1 dr\int_0^1 r^2 dz = \frac{4}{\pi}\cdot\frac{\pi}{2}\cdot\frac{1}{3} = \frac{2}{3} = 0.667\,.$$

Indeed, if we alter example 2 by taking $0 \le \theta \le 2\pi$, we get $r_c = 0$ but still have $\frac{2}{3}$ as the average value of r.

§16.6 Iterated Integrals in Spherical Coordinates

Given a triple integral $\iiint_R f(x,y,z)\,dV$, we may be led, by the form of f or the shape of R, to express the integral in spherical coordinates. The method and argument are exactly as for triple integrals in cylindrical coordinates. We view $\iiint_R f(x,y,z)\,dV$ as the limit of Riemann sums of the form $\sum_i f(x_i^*,y_i^*,z_i^*)\Delta V_i$, where except along the boundary of R, we use elementary subregions of R of the following special kind: each subregion is bounded by portions of six coordinate surfaces, two of the form $\rho = $ constant, two of the form $\varphi = $ constant, and two of the form $\theta = $ constant. (See [6–1].) For the i-th interior subregion, let $\Delta\rho_i$, $\Delta\varphi_i$, and $\Delta\theta_i$ be the differences in coordinate values between opposite coordinate surfaces. For each i, let $(\rho_i^*,\varphi_i^*,\theta_i^*)$ be the spherical coordinates of a chosen sample point (x_i^*,y_i^*,z_i^*) in the i-th subregion. Then ΔV_i, the volume of the i-th interior subregion, is approximately $(\Delta\rho_i)(\rho_i^*\Delta\varphi_i)(\rho_i^*\sin\varphi_i^*\Delta\theta_i) = (\rho_i^*)^2\sin\varphi_i^*\Delta\rho_i\Delta\varphi_i\Delta\theta_i$, since the i-th subregion is approximately a rectangular solid in shape. In fact, it is possible to choose the sample point so that

$$\Delta V_i = (\rho_i^*)^2\sin\varphi_i^*\Delta\rho_i\Delta\varphi_i\Delta\theta_i\,.$$

(See the problems.) Then $\iiint_R f\,dV$ is the limit of a sequence of interior Riemann sums of the form

$\rho_i^*\sin\varphi_i^*\Delta\theta$

$\Delta\rho_i$

$\rho_i^*\Delta\varphi_i$

$(\rho_i^*,\varphi_i^*,\theta_i^*)$

[6–1]

$$\sum_{i(\text{int})} f(\rho_i^* \sin \varphi_i^* \cos \theta_i^*, \rho_i^* \sin \varphi_i^* \sin \theta_i^*, \rho_i^* \cos \theta_i^*)(\rho_i^*)^2 \sin \varphi_i^* \Delta\rho_i \Delta\varphi_i \Delta\theta_i \,.$$

But each of these Riemann sums is also an interior Riemann sum for a triple integral with integrand $f(\rho \sin \varphi \cos \theta, \rho \sin \varphi \sin \theta, \rho \cos \theta)\rho^2 \sin \varphi$ over a corresponding region $\hat{R}$ in a three-dimensional Euclidean space with Cartesian coordinates, whose axes are labeled ρ, φ, and θ instead of x, y, and z. This three-dimensional space is called $\rho\varphi\theta$ *space*. Hence we have

$$(6.1) \quad \iiint_R f(x,y,z)dxdydz$$

$$= \iiint_{\hat{R}} f(\rho \sin \varphi \cos \theta, \rho \sin \varphi \sin \theta, \rho \cos \varphi)\rho^2 \sin \varphi d\rho d\varphi d\theta \,,$$

where $d\rho d\varphi d\theta$ indicates integration in $\rho\varphi\theta$ space. (6.1) *thus expresses the original triple integral over the region* R *in* xyz *space as a new, equivalent triple integral over a region* $\hat{R}$ *in* $\rho\varphi\theta$ *space.*

Example 1. *Find the moment of inertia, about an axis through its center, of a solid sphere of radius* a *and constant density* δ . Using Cartesian coordinates and taking the chosen axis to be the z axis, we get

$$I = \iiint_R (x^2 + y^2)\delta dV \,,$$

where R is the given solid sphere with center at the origin. Going to spherical coordinates, we have, from (6.1):

$$\iiint_{\hat{R}} (\rho^2 \sin^2 \varphi \cos^2 \theta + \rho^2 \sin^2 \varphi \sin^2 \theta)\delta\rho^2 \sin \varphi d\rho d\varphi d\theta$$

$$= \iiint_{\hat{R}} \delta\rho^4 \sin^3 \varphi \, d\rho d\varphi d\theta \,,$$

where $\hat{R}$ is the region in $\rho\varphi\theta$ space given by $0 \le \rho \le a, 0 \le \varphi \le \pi, -\pi < \theta \le \pi$. We have now expressed the triple integral for I in spherical coordinates. Next, we set up and evaluate the appropriate iterated integral.

As an iterated integral,

$$I = \iiint_{\hat{R}} \delta\rho^4 \sin^3\varphi \, d\rho \, d\varphi \, d\theta \qquad \text{becomes}$$

$$\int_{-\pi}^{\pi} d\theta \int_0^{\pi} d\varphi \int_0^a \{\delta\rho^4 \sin^3\varphi\} d\rho \qquad \text{(see §16.7 for the best way to find}$$

these limits of integration)

$$= \delta \int_{-\pi}^{\pi} d\theta \int_0^{\pi} \left(\frac{a^5}{5} \sin^3\varphi\right) d\varphi$$

$$= \frac{\delta a^5}{5} \int_{-\pi}^{\pi} d\theta \int_0^{\pi} (1 - \cos^2\varphi) \sin\varphi \, d\varphi$$

$$= \frac{2\pi\delta a^5}{5} \left[\frac{\cos^3\varphi}{3} - \cos\varphi\right]_0^{\pi}$$

$$= \frac{2\pi\delta a^5}{5} (1 - \tfrac{1}{3} + 1 - \tfrac{1}{3}) = \frac{8}{15}\pi a^5 \delta .$$

Let M be the total mass of the sphere, then $M = \frac{4}{3}\pi a^3 \delta$, and $\delta = \frac{3M}{4\pi a^3}$. Hence,

in terms of M, we have $I = \frac{8}{15}\pi a^5 \frac{3M}{4\pi a^3} = \frac{2}{5}Ma^2$.

Example 2. *Consider a thick hemispherical shell, of constant density* δ, *formed by removing a hemisphere of radius* 1 *from a concentric hemisphere of radius* 2. *What are the coordinates* $\overline{x}_m, \overline{y}_m, \overline{z}_m$ *of its center of mass in a Cartesian coordinate system where the center of the spheres is at the origin and the hemispheres lie above the xy plane?* (See [6–2].) Since the density is constant, the coordinates of the center of mass are the same as the coordinates $(\overline{x}, \overline{y}, \overline{z})$ of the centroid. By symmetry, $\overline{x} = \overline{y} = 0$. From §14.7,

$$\overline{z} = \frac{\iiint_R z\delta dV}{\iiint_R \delta dV} = \frac{\iiint_R z \, dx \, dy \, dz}{\iiint_R dx \, dy \, dz}, \qquad\qquad [6\text{--}2]$$

where R is the hemispherical shell.

We express both integrals in spherical coordinates and then evaluate by iterated integrals. Here the region $\hat{R}$ is the rectangular block in $\rho\varphi\theta$ space: $1 \le \rho \le 2, 0 \le \varphi \le \frac{\pi}{2}, -\pi \le \theta \le \pi$. For the numerator, we get

$$\iiint_R z\,dx\,dy\,dz = \iiint_{\hat{R}} \rho\cos\varphi\,\rho^2\sin\varphi\,d\rho\,d\varphi\,d\theta$$

$$= \iiint_{\hat{R}} \rho^3\sin\varphi\cos\varphi\,d\rho\,d\varphi\,d\theta$$

$$= \int_{-\pi}^{\pi} d\theta \int_0^{\pi/2} d\varphi \int_1^2 \rho^3\sin\varphi\cos\varphi\,d\rho$$

(see §16.7 for the best way to find these limits of integration)

$$= \int_{-\pi}^{\pi} d\theta \int_0^{\pi/2} \frac{15}{4}\sin\varphi\cos\varphi\,d\varphi$$

$$= \int_{-\pi}^{\pi} \frac{15}{8}\,d\theta = \frac{15\pi}{4}.$$

For the denominator, we get

$$\iiint_R dx\,dy\,dz = \iiint_{\hat{R}} \rho^2\sin\varphi\,d\rho\,d\varphi\,d\theta$$

$$= \int_{-\pi}^{\pi} d\theta \int_0^{\pi/2} d\varphi \int_1^2 \rho^2\sin\varphi\,d\rho$$

$$= \int_{-\pi}^{\pi} d\theta \int_0^{\pi/2} \frac{7}{3}\sin\varphi\,d\varphi$$

$$= \int_{-\pi}^{\pi} \frac{7}{3}\,d\theta = \frac{14\pi}{3}.$$

(Since the denominator is simply the volume of the shell, we could also have computed the denominator by subtracting the volume of the smaller solid hemisphere from the volume of the larger, getting $\frac{1}{2}\cdot\frac{4}{3}\pi 8 - \frac{1}{2}\cdot\frac{4}{3}\pi = \frac{14\pi}{3}$.) We therefore have

$$\bar{z} = \frac{\dfrac{15\pi}{4}}{\dfrac{14\pi}{3}} = \frac{45}{56} = 0.804.$$

Thus we find that the center of mass, in our Cartesian system, is (0,0,0.804).

§16.7   Setting up Integrals in Spherical Coordinates

(7.1)   *For an iterated integral expressed in spherical coordinates, where we hope to integrate in the order ρ then φ then θ, we can find the limits of inte-*

gration as follows, using the region R *in* xyz *space* (*and ignoring the region* $\hat{R}$
in ρφθ *space*).

(i)   *Hold* φ *and* θ *fixed, let* ρ *increase. This gives a ray going out
from the origin. Integrate from the* ρ-*value where the ray enters* R *to the* ρ-
value where it leaves R. *This gives the limits on* ρ. (*These limits will depend on
the fixed values for* φ *and* θ.)

(ii)   *Hold* θ *fixed, let* φ *increase from 0 to* π. *This gives a fan-
shaped family of rays from* (i). *Integrate over those* φ-*values for which the rays
pass through* R. *This gives the limits on* φ. (*These limits depend on the fixed
value of* θ.)

(iii)   *Choose the limits for* θ *to include all the fans from* (ii) *which
intersect the region* R.

Example. *Use spherical coordinates to find the volume contained inside
the cone* $z^2 = x^2 + y^2$, *inside the sphere of radius 1 with center at origin, and
inside the half-space* z ≥ 0. (*See* [7-1].)

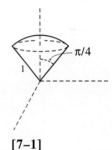

[7-1]

We have:

$$\iiint_R dV = \iiint_{\hat{R}} \rho^2 \sin\varphi \, d\rho d\varphi d\theta$$

$$= \int_{-\pi}^{\pi} d\theta \int_0^{\pi/4} d\varphi \int_0^1 \rho^2 \sin\varphi \, d\rho$$

$$= \frac{1}{3} \int_{-\pi}^{\pi} d\theta \int_0^{\pi/4} \sin\varphi d\varphi$$

$$= \frac{2\pi}{3} \int_0^{\pi/4} \sin\varphi \, d\varphi$$

$$= \frac{1}{3}\pi(2 - \sqrt{2}) \ .$$

(Note that we can evaluate the dθ integral independently of the dρ and dφ integrals, because θ does not appear in the integrands or limits of integration of the
dρ and dφ integrals.)

The procedure (7.1) assumes, in effect, that the region $\hat{R}$ is ρφ-simple.
Usually, it will be geometrically obvious (without directly considering $\hat{R}$) whether
or not the procedure (7.1) can be applied. If the procedure cannot be applied, this
shows that $\hat{R}$ is not ρφ-simple. If the region $\hat{R}$ is not simple, then we must
divide $\hat{R}$ (as in §15.5) into simple subregions or use other methods of §15.5.

§16.8   The Parallel-axis Theorem for Moments of Inertia

The following theorem is useful for calculating moments of inertia. Its proof is a good example of the use of multiple integrals in a theoretical calculation.

(8.1)   *The parallel-axis theorem. A body, not necessarily of constant density, occupies a region R in $\mathbf{E}^3$. Let L be a given fixed straight line. We wish to find* I_L*, the moment of inertia about* L. *We can do so as follows:*

Let $\overline{P}$ *be the center of mass. Let* $\overline{L}$ *be a line through* $\overline{P}$ *parallel to* L. *Let* M *be the total mass and let* $I_{\overline{L}}$ *be the moment of inertia about* $\overline{L}$. *If* d *is the distance from* $\overline{P}$ *to* L, *then*

$$I_L = I_{\overline{L}} + Md^2.$$

The theorem tells us that in order to calculate the moment of inertia of a body about any axis L, all we need to know is, (a) the mass M; (b) the moment of inertia $I_{\overline{L}}$ about a parallel axis $\overline{L}$ through the center of mass $\overline{P}$, and (c) the distance between L and $\overline{L}$.

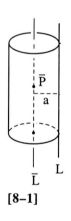

[8–1]

Example 1. In §16.4, we calculated the moment of inertia of a constant-density solid cylinder about its own axis to be $\frac{1}{2}Ma^2$, where a is the radius of the cylinder. What is the moment of inertia of the cylinder about an axis L lying in the surface of the cylinder? (See [8–1].) Here, by symmetry, we see that the center of mass lies on the cylinder's own axis, and this axis, $\overline{L}$, is parallel to L. Moreover, we have d = a. By the theorem,

$$I_L = I_{\overline{L}} + Ma^2.$$

From §16.4, $I_{\overline{L}} = \frac{1}{2}Ma^2$. Hence

$$I_L = \frac{1}{2}Ma^2 + Ma^2 = \frac{3}{2}Ma^2.$$

Proof of (8.1). The theorem follows from the definition of moment of inertia in §14.3 and from the basic laws for multiple integrals in §14.4. We introduce Cartesian coordinates as follows. Take L to be the z axis, with any chosen point on L as origin. Let $(\bar{x}_m, \bar{y}_m, \bar{z}_m)$ be the point $\overline{P}$. Then $d = \sqrt{\bar{x}_m^2 + \bar{y}_m^2}$. Let $\delta(x,y,z)$ be the density. Then $M = \iiint_R \delta dV$, and, by the definitions of $\bar{x}_m$ and $\bar{y}_m$, $M\bar{x}_m = \iiint_R x\delta dV$ and $M\bar{y}_m = \iiint_R y\delta dV$. By definition, we have

$$I_L = \iiint_R (x^2 + y^2)\delta dV,$$

and

$$I_{\overline{L}} = \iiint_R \left((x - \bar{x}_m)^2 + (y - \bar{y}_m)^2\right)\delta dV.$$

Hence

$$I_{\overline{L}} = \iiint_R \left(x^2 - 2\bar{x}_m x + \bar{x}_m^2 + y^2 - 2\bar{y}_m y + \bar{y}_m^2\right)\delta dV.$$

Since $\bar{x}_m$ and $\bar{y}_m$ are constants, we can use our basic algebraic laws for multiple integrals to get

$$I_{\overline{L}} = \iiint_R (x^2 + y^2)\delta dV - 2\bar{x}_m \iiint_R x\delta dV - 2\bar{y}_m \iiint_R y\delta dV$$
$$+ (\bar{x}_m^2 + \bar{y}_m^2)\iiint_R \delta dV.$$

Therefore,

$$I_L - I_{\overline{L}} = 2M\bar{x}_m^2 + 2M\bar{y}_m^2 - M(\bar{x}_m^2 + \bar{y}_m^2)$$
$$= M(\bar{x}_m^2 + \bar{y}_m^2) = Md^2.$$

Hence

$$I_L = I_{\overline{L}} + Md^2, \text{ as desired.}$$

Example 2. *Let* R *be a solid cube of total mass* M, *constant density* δ, *and edge-length* a. *What is the moment of inertia of* R *with respect to a line* L *which contains one edge of the cube?*

Let $\overline{L}$ be a line parallel to L through the center of the cube. In example 2 in §15.7, we saw that the moment of inertia of R with respect to $\overline{L}$ is $\frac{1}{6}Ma^2$. Since the center point of the cube is, by symmetry, its center of mass, we can apply the parallel axis theorem with $d = \frac{\sqrt{2}}{2}a$ (see [8–2]). Thus $d^2 = \frac{a^2}{2}$, and

$$I_L = \frac{1}{6}Ma^2 + \frac{1}{2}Ma^2 = \frac{2}{3}Ma^2.$$

[8–2]

§16.9 Problems

§16.1

1. For each of the following, give an equation for the same curve in the associated polar system. If possible, give r or r^2 as a function of θ.
   (a) $x^2 - y^2 = 1$ (c) $x^2 + 2y^2 = 1$
   (b) $x^2 - 2y^2 = 1$ (d) $y = x^2$

2. Let S be the surface given by $x^2 + y^2 - z^2 = 1$.
   (a) Give an equation for S in the associated cylindrical system.
   (b) Give an equation for S in the associated spherical system.

§16.2

1. Set up, but do not evaluate, the iterated integrals in polar coordinates giving the center of mass of the plane region bounded by $(x-1)^2 + y^2 = 1$ if the density (per unit area) is given by $\delta(x,y) = x^2 + y^2$. (Give expressions for $\bar{x}$ and $\bar{y}$.)

2. Find the area in the plane that lies inside the curve $r = a(1 + \cos\theta)$ (called a *cardioid*) and outside the circle $r = a$.

3. Find the coordinates $\bar{x}_m$ and $\bar{y}_m$ of the center of mass of the plane region contained in the circle $(x-1)^2 + y^2 = 1$ when the density (per unit area) is given by $\delta(x,y) = x^2 + y^2$. (*Hint*: Use polar coordinates. You may assume that $\int_0^{\pi/2} \cos^4 u\, du = (\frac{3}{16})\pi$ and that $\int_0^{\pi/2} \cos^6 u\, du = (\frac{5}{32})\pi$.)

4. Use Cartesian coordinates to evaluate the double integral in example 1.

§16.4

1. Express as an iterated integral in Cartesian coordinates, but do not evaluate
   $$\int_0^{\pi/2} \int_0^1 \int_0^{1-r} r^2 z \cos\theta\, dz\, dr\, d\theta \quad \text{(in cylindrical coordinates).}$$

2. A solid is bounded above by the cone $z = \sqrt{x^2 + y^2}$, below by the plane $z = 0$, and on the side by the surface $x^2 + y^2 = 4$. Assume that is has constant density and that has total mass M. (See [9-1].)

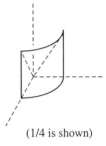

(1/4 is shown)

[9-1]

   (a) Give an expression in terms of multiple integrals, for its moment of inertia about the z axis.
   (b) Express these multiple integrals as iterated integrals in cylindrical coordinates.
   (c) Evaluate these integrals and find the moment of inertia in (a). (Be careful about order of integration.)

3. Two cylinders of radius 2 have the x axis and the y axis as axes. Let a third cylinder have radius 1 and the z axis as its axis. Express the volume of the common intersection of all three cylinders by an iterated integral in cylindrical coordinates. Do not evaluate. (*Hint*. For help in visualizing, see figure [7-1] in §15.7, but with the x and z axes interchanged.)

4. Use iterated integration in cylindrical coordinates to calculate the mass of a wedge-shaped solid bounded below by the plane $z = 0$, above by the plane $z = y$, and on the side by the cylinder $x^2 + y^2 = 4$, when the density is $\delta(x,y,z) = z/y$.

5. Find the center of mass of the volume bounded above by the sphere $x^2 + y^2 + z^2 = 2a^2$ and below by

the paraboloid $az = x^2+y^2$. Assume uniform density.

6. A torus of mass M is generated by rotating a circle of radius a about an axis in its plane at a distance b from its center (b > a). Let I be the moment of inertia about the axis of revolution. Assuming uniform density, express I in terms of iterated integrals. Do not evaluate.

7. Let R be the region bounded by the sphere $x^2 + y^2 + z^2 = 1$ (above) and by the cone $z = \sqrt{x^2 + y^2}$ (below). (See [7–1] in §16.7.)
   (a) Set up an iterated integral in cylindrical coordinates for the volume of this region.
   (b) Evaluate this integral.
   (In §16.7, this volume is expressed and calculated in spherical coordinates.)

8. Find the volume of the intersection of three mutually perpendicular cylinders, each of radius 1. (*Hint:* To visualize, take a 1/16th portion of the intersection of two cylinders, as pictured in [7–1] of §15.7, and intersect it with a third cylinder coming along the x axis.)

§16.6

1. R is the solid that remains when a solid sphere of radius 1 is removed from a concentric solid sphere of radius 2. The solid has constant density = 1/4.
   (a) Set up an iterated integral for the moment of inertia of R about the axis through the center of the sphere.
   (b) Evaluate this integral.

2. (a) Find the equation of the straight line x+y = 1 in polar coordinates.
   (b) The following iterated integral is given in cylindrical coordinates. Express it as an iterated integral in spherical coordinates. Do not evaluate.

$$\int_0^{\pi/2} \int_0^1 \int_0^{1-r} rz \, dz \, dr \, d\theta \,.$$

3. We wish to find the average distance from the origin to points inside the sphere $x^2+y^2+z^2 = 1$. (This problem has previously appeared as problem 3 for §14.3 and as problem 3 for §14.5.)
   (a) Give an expression for this desired quantity in terms of one or more multiple integrals.
   (b) Expressing these integrals in an appropriate spherical coordinate system, find the value of this quantity.

4. Find the moment of inertia about the z axis of a solid of uniform density and total mass M of the shape given in the example in §16.7 (the solid region inside the cone $z^2 = x^2+y^2$, above the xy plane, and inside the sphere $\rho = 1$.) Use spherical coordinates. You may assume the result of the example in §16.7.

5. Find the volume enclosed by the surface $\rho = a \sin \varphi$ (in spherical coordinates).

6. In figure [6–1], show that it is always possible to choose the sample point $(\rho_i^*, \varphi_i^*, \theta_i^*)$ so that $\Delta V_i = (\rho_i^*)^2 \sin \varphi_i^* \Delta \rho_i \Delta \varphi_i \Delta \theta_i$. (*Hint:* Choose ρ_i^* so that the area of the $\rho\theta$ cross-section through the sample is equal to $\rho_i^* \Delta \rho_i \Delta \theta_i$; then use the continuity of $\sin \varphi$ to choose φ_i^*. See the conclusion of the remark in §14.5.)

§16.8

1. A solid sphere S of radius a, total mass M, and uniform density, rotates around a line L tangent to (and fixed to) its surface. Use the parallel-axis theorem and the result of example 1 in §16.6 to find the moment of inertia of S about L.

Chapter 17

Curvilinear Coordinates and Change of Variables

§17.1 Curvilinear Coordinates

We consider a general procedure for introducing non-Cartesian coordinates u,v,w (in E^3) in place of Cartesian coordinates x,y,z. The introductions of cylindrical and of spherical coordinates are special cases of this procedure. We begin with *defining equations* of the general form

$$(1.1) \qquad \begin{aligned} x &= x(u,v,w), \\ y &= y(u,v,w), \\ z &= z(u,v,w), \end{aligned}$$

where the functions x, y, and z are C^2. For example, in the case of spherical coordinates, these equations are

$$\begin{aligned} x &= \rho \sin \varphi \cos \theta, \\ y &= \rho \sin \varphi \sin \theta, \\ z &= \rho \cos \varphi, \end{aligned}$$

and in the case of cylindrical coordinates, the equations are

$$\begin{aligned} x &= r \cos \theta, \\ y &= r \sin \theta, \\ z &= z. \end{aligned}$$

When we have introduced non-Cartesian coordinates in this way, it is customary to say that we have introduced a system of *curvilinear coordinates*.

The defining equations (1.1) enable us, by substitution, to express a scalar field f(P), given by a function f(x,y,z) in Cartesian coordinates, as a function f(x(u,v,w),y(u,v,w),z(u,v,w)) in the new coordinates. Similarly, an equation F(x,y,z) = 0 of a given surface in space can be replaced by an equation F(x(u,v,w),y(u,v,w),z(u,v,w)) = 0 for the same surface in terms of the new coordinates.

Defining equations of the form

(1.2)
$$x = x(u,v),$$
$$y = y(u,v)$$

can also be used, in E^2, to introduce curvilinear coordinates u and v in place of Cartesian coordinates x and y.

Example in E^2-with-Cartesian-coordinates. Let the equations $\begin{pmatrix} x = u - v \\ y = 2u + v \end{pmatrix}$ *define a new system of coordinates in the plane.* In this case, the function f(x,y) = $x^2 + y^2$ becomes f(x(u,v),y(u,v)) = $(u-v)^2 + (2u+v)^2$ in the new system. Similarly, the curve F(x,y) = $2x^2 - y = 0$ becomes F(x(u,v),y(u,v)) = $2(u-v)^2 - 2u - v = 0.$

When we introduce a curvilinear system in E^3-with-Cartesian-coordinates, the defining equations for new coordinates u,v,w can be conveniently summarized in vector form by giving the position vector

$$\vec{R} = x\hat{i} + y\hat{j} + z\hat{k}$$

as a vector-valued function of u, v, and w:

(1.3) $$\vec{R}(u,v,w) = x(u,v,w)\hat{i} + y(u,v,w)\hat{j} + z(u,v,w)\hat{k}.$$

We sometimes refer to this last vector equation as the *vector defining equation* for the new coordinate system. In the case of cylindrical coordinates, for example, we have

$$\vec{R}(r,\theta,z) = r\cos\theta\hat{i} + r\sin\theta\hat{j} + z\hat{k}.$$

Similarly, for E^2, we have $\vec{R}(u,v) = x(u,v)\hat{i} + y(u,v)\hat{j}$.

§17.2 Coordinate Curves, Surfaces, and Unit Vectors

In a curvilinear coordinate system given by $\vec{R}(u, v, w)$, if we hold $v = v_0$ and $w = w_0$ fixed and let u vary as a parameter, we get the curve $\vec{R}(t, v_0, w_0) = x(t, v_0, w_0)\hat{i} + y(t, v_0, w_0)\hat{j} + z(t, v_0, w_0)\hat{k}$ (where we have written u as t.) This curve is called the *coordinate curve in* u for the fixed values v_0 and w_0. Similarly, we get coordinate curves in v and in w. Coordinate curves in curvilinear coordinates are analogous to lines parallel to the axes in Cartesian coordinates.

In a curvilinear coordinate system given by $\vec{R}(u, v, w)$, if we hold $w = w_0$ fixed and let u and v vary as parameters, we get a surface called the *coordinate surface in* u *and* v for the fixed value $w = w_0$. Coordinate surfaces in curvilinear coordinates are analogous to planes perpendicular to the axes in Cartesian coordinates.

***Example in* E^2** (*continued from* §17.1). Given $x = u-v$, $y = 2u+v$, the coordinate curves in u are given by $\begin{cases} x = t - v_0 \\ y = 2t + v_0 \end{cases}$, and the coordinate curves in v are given by $\begin{cases} x = u_0 - t \\ y = 2u_0 + t \end{cases}$. In this example, the coordinate curves are straight lines. Some are shown in [2–1].

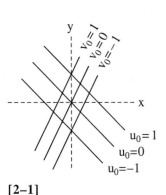

[2–1]

In working with a curvilinear coordinate system u,v,w in E^3, it is often helpful to introduce, at a given point P with curvilinear coordinates (u,v,w), the *unit coordinate vectors* $\hat{u}_P, \hat{v}_P, \hat{w}_P$. $\hat{u}_P$ is defined to be the unit tangent vector $\hat{T}$ to the coordinate curve in u at P. Thus, to find the unit coordinate vector $\hat{u}_P$ we let

$$\vec{R}(u, v, w) = x(u, v, w)\hat{i} + y(u, v, w)\hat{j} + z(u, v, w)\hat{k} \ ;$$

we take

$$\left.\frac{\partial \vec{R}}{\partial u}\right|_P = \left.\frac{\partial x}{\partial u}\right|_P \hat{i} + \left.\frac{\partial y}{\partial u}\right|_P \hat{j} + \left.\frac{\partial z}{\partial u}\right|_P \hat{k} \ ;$$

and we form $\hat{u}_P$ = the unit vector $\left.\frac{\partial \vec{R}}{\partial u}\right|_P \Big/ \left|\left.\frac{\partial \vec{R}}{\partial u}\right|_P\right|$. Similarly for $\hat{v}_P$ and $\hat{w}_P$.

(In Chapter 7, we obtained the coordinate unit vectors $\hat{r}_P, \hat{\theta}_P, \hat{z}_P$ for cylindrical coordinates.) In spherical coordinates, we begin with

(2.1) $\qquad \vec{R}(\rho,\varphi,\theta) = \rho \sin \varphi \cos \theta \hat{i} + \rho \sin \varphi \sin \theta \hat{j} + \rho \cos \varphi \hat{k}$.

Treating ρ and θ as constants, we get

$$\frac{\partial \vec{R}}{\partial \varphi} = \rho \cos \varphi \cos \theta \hat{i} + \rho \cos \varphi \sin \theta \hat{j} - \rho \sin \varphi \hat{k} .$$

Then

$$\left| \frac{\partial \vec{R}}{\partial \varphi} \right| = [\rho^2 \cos^2 \varphi \cos^2 \theta + \rho^2 \cos^2 \varphi \sin^2 \theta + \rho^2 \sin^2 \varphi]^{1/2} = \rho .$$

Therefore, we have the unit vector

(2.2) $\qquad \hat{\varphi}_P = \left.\frac{\partial \vec{R}}{\partial \varphi}\right|_P \left/ \left.\left| \frac{\partial \vec{R}}{\partial \varphi} \right|\right|_P\right. = \cos \varphi \cos \theta \hat{i} + \cos \varphi \sin \theta \hat{j} - \sin \varphi \hat{k}$

$\qquad\qquad\qquad\qquad\qquad\qquad$ (defined for $\rho > 0$ and $0 < \varphi < \pi$).

In the same way, we get

(2.3) $\qquad \hat{\rho}_P = \sin \varphi \cos \theta \hat{i} + \sin \varphi \sin \theta \hat{j} + \cos \varphi \hat{k}$ (defined for $\rho > 0$),

and

(2.4) $\qquad \hat{\theta}_P = -\sin \theta \hat{i} + \cos \theta \hat{j}$ $\qquad\qquad$ (defined for $\rho > 0$ and $0 < \varphi < \pi$).

$\qquad$ Similarly, for a system u,v in $\mathbf{E}^2$, we can get unit coordinate vectors $\hat{u}_P$ and $\hat{v}_P$.

Example in $\mathbf{E}^2$ (continued). Given $x = u-v$, $y = 2u+v$, we get the coordinate vectors $\hat{u}_P$ and $\hat{v}_P$ by the same method.

$\qquad$ We have $\vec{R}(u,v) = x\hat{i} + y\hat{j} = (u - v)\hat{i} + (2u + v)\hat{j}$.

Then $\qquad\qquad \dfrac{\partial \vec{R}}{\partial u} = \hat{i} + 2\hat{j} \quad\Rightarrow\quad \hat{u}_P = \dfrac{\dfrac{\partial \vec{R}}{\partial u}}{\left| \dfrac{\partial \vec{R}}{\partial u} \right|} = \dfrac{\sqrt{5}}{5}(2\hat{i} + \hat{j})$,

and
$$\frac{\partial \vec{R}}{\partial v} = -\hat{i} + \hat{j} \Rightarrow \hat{v}_P = \frac{\frac{\partial \vec{R}}{\partial v}}{\left|\frac{\partial \vec{R}}{\partial v}\right|} = \frac{\sqrt{2}}{2}(-\hat{i} + \hat{j}).$$

In this case, $\hat{u}_P$ and $\hat{v}_P$ are constant vectors (independent of P).

When a curvilinear coordinate system is used in $\mathbf{E}^3$ or $\mathbf{E}^2$, there may be points for which not all the unit coordinate vectors are uniquely defined. This occurs in the case of spherical coordinates at all points on the z axis. At a point on the positive z axis, for example, $\hat{\theta} = -\sin\theta\hat{i} + \cos\theta\hat{j}$ is not unique, since θ can have any value at that point. Points at which not all unit coordinate vectors are uniquely defined or at which, in the case of $\mathbf{E}^3$, the unit vectors lie in a common plane (or, in the case of $\mathbf{E}^2$, lie on a common line) are called *singular points* of the coordinate system. For the curvilinear coordinate systems customarily used, singular points, if they occur at all, occur as isolated points or as points that lie along one or more curves (for example, the origin in polar coordinates or the z axis in spherical or cylindrical coordinates). In the example in $\mathbf{E}^2$ above, with $x = u - v$ and $y = 2u + v$, there are no singular points.

Definition. If, at every non-singular point in a given curvilinear coordinate system, the unique unit coordinate vectors form a frame (that is to say, the unit coordinate vectors are mutually perpendicular), we say that the system of coordinates is *orthogonal.*

Polar coordinates, cylindrical coordinates, and spherical coordinates are examples of orthogonal coordinate systems. The system in the example in $\mathbf{E}^2$ above is not orthogonal, since $\hat{u}_P \cdot \hat{v}_P \neq 0$. A system u,v,w of curvilinear coordinates may be useful (for either theoretical or computational purposes) even if it is not orthogonal. As we have already seen with our uses of spherical and cylindrical coordinates for evaluating multiple integrals, a good choice of curvilinear coordinate system may be suggested by particular algebraic or geometrical symmetries occurring in a given problem.

Remark. In the case of the example in $\mathbf{E}^2$, it is clear from the defining equations that each given point has unique values for its new coordinates. In most curvilinear systems, however, there will be geometrical points with more than one set of coordinate values. For spherical coordinates, this is the case at the singular points, where all values of θ can occur for the same point, and it is also the case at all non-singular points, since a point with the coordinate values $\theta = \theta_0$ also has the coordinate values $\theta = \theta_0 + 2\pi$, $\theta = \theta_0 + 4\pi, \ldots$. At singular points, this difficulty may be unavoidable, but at non-singular points it can usually be avoided by limiting the permissible values of the curvilinear coordinates. Thus, in spherical coordinates, we can limit φ to $0 \leq \varphi \leq \pi$ and θ to $-\pi < \theta \leq \pi$. (This limitation on θ may not be natural in certain kinds of physical problem. For example, if we wish to use parametric equations in spherical coordinates to describe the motion of the moon around the earth, and if we wish to have continuous functions for the parametric equations, then we must allow all values of θ. More specifically, $\theta(t)$ must be allowed to keep on increasing as time t increases.)

§17.3 The Jacobian

In the same way that we used $r\theta z$ space with cylindrical coordinates and $\rho\varphi\theta$ space with cylindrical coordinates, we can introduce, for any system $\{u,v,w\}$ of curvilinear coordinates, a three-dimensional uvw *space*, with rectangular axes for u, v, and w. For a region R in our given xyz space, we can seek a corresponding region $\hat{R}$ in uvw space (see example 1 in §16.2).

Definition. Given a curvilinear coordinate system with $x = x(u,v,w)$, $y = y(u,v,w)$, $z = z(u,v,w)$, we define the *Jacobian of* x,y,z *with respect to* u,v,w to be the determinant

(3.1)
$$\begin{vmatrix} \dfrac{\partial x}{\partial u} & \dfrac{\partial x}{\partial v} & \dfrac{\partial x}{\partial w} \\[2mm] \dfrac{\partial y}{\partial u} & \dfrac{\partial y}{\partial v} & \dfrac{\partial y}{\partial w} \\[2mm] \dfrac{\partial z}{\partial u} & \dfrac{\partial z}{\partial v} & \dfrac{\partial z}{\partial w} \end{vmatrix}.$$

We abbreviate this determinant as $\dfrac{\partial(x,y,z)}{\partial(u,v,w)}$. Note that the elements of this determinant are scalar functions of u,v,w, and hence that the Jacobian itself is a scalar function of u,v,w. At any particular point $\hat{P}_0$ in uvw space, with coordinates (u_0,v_0,w_0), the Jacobian takes on the specific numerical value

$$\frac{\partial(x,y,z)}{\partial(u,v,w)}\bigg|_{\hat{P}_0} = \begin{vmatrix} \frac{\partial x}{\partial u}\big|_{\hat{P}_0} & \frac{\partial x}{\partial v}\big|_{\hat{P}_0} & \cdots \\ \frac{\partial y}{\partial u}\big|_{\hat{P}_0} & \cdots & \cdots \\ \cdots & \cdots & \cdots \end{vmatrix}.$$

In $\mathbf{E}^2$, the *Jacobian of* x,y *with respect to* u,v is defined similarly:

$$(3.2) \qquad\qquad \frac{\partial(x,y)}{\partial(u,v)} = \begin{vmatrix} \frac{\partial x}{\partial u} & \frac{\partial x}{\partial v} \\ \frac{\partial y}{\partial u} & \frac{\partial y}{\partial v} \end{vmatrix},$$

and at a particular point $\hat{P}_0 = (u_0, v_0)$ in uv space, we have

$$\frac{\partial(x,y)}{\partial(u,v)}\bigg|_{\hat{P}_0} = \begin{vmatrix} \frac{\partial x}{\partial u}\big|_{\hat{P}_0} & \frac{\partial x}{\partial v}\big|_{\hat{P}_0} \\ \frac{\partial y}{\partial u}\big|_{\hat{P}_0} & \frac{\partial y}{\partial v}\big|_{\hat{P}_0} \end{vmatrix}.$$

Example. *For polar coordinates* r, θ, *we have* $x = r\cos\theta$, $y = r\sin\theta$. Hence

$$\frac{\partial(x,y)}{\partial(r,\theta)} = \begin{vmatrix} \cos\theta & -r\sin\theta \\ \sin\theta & r\cos\theta \end{vmatrix} = r\cos^2\theta + r\sin^2\theta = r.$$

In §17.5, we verify that, for cylindrical coordinates, $\dfrac{\partial(x,y,z)}{\partial(r,\theta,z)} = r$, and that, for spherical coordinates, $\dfrac{\partial(x,y,z)}{\partial(\rho,\varphi,\theta)} = \rho^2 \sin\varphi$.

The Jacobian is a new kind of derivative with many uses in multivariable calculus. Its most important property is described in the following theorem.

(3.3)   *Jacobian theorem. Let a curvilinear coordinate system be introduced in* $\mathbf{E}^3$-*with-Cartesian-coordinates by the defining equations* $x = x(u,v,w)$, $y = y(u,v,w)$, $z = z(u,v,w)$, *where the functions in these equations are* C^2. *Let* P_0 *be a given point in our given* xyz *space, and let* $\hat{P}_0$ *be a corresponding point in* uvw *space. Let* $\hat{U}$ *be the neighborhood of* $\hat{P}_0$ *of radius* a *in* uvw *space, and let* U *be the corresponding region in* xyz *space.* (U *consists of all points* P *in* xyz *space for which* $\hat{P}$ *is in* $\hat{U}$.) $\hat{V}_a = \frac{4}{3}\pi a^3$ *is the volume of* $\hat{U}$ *in* uvw *space. Let* V_a *be the volume of* U, *in* xyz *space.*

$$\textit{Then} \quad \lim_{a \to 0} \frac{V_a}{\hat{V}_a} = \left| \frac{\partial(x,y,z)}{\partial(u,v,w)} \right|_{\hat{P}_0} \right| \qquad \text{(where the outermost bars indicate}$$

absolute value).

The analogous statement holds for Jacobians in $\mathbf{E}^2$, *with "volume" replaced by "area."*

The Jacobian theorem tells us that the Jacobian is an *amplification factor.* In $\mathbf{E}^3$, it gives us the ratio of the *volume* of a small region in xyz space to the *volume* of the corresponding region in uvw space. Similarly for *areas* in $\mathbf{E}^2$. This ratio itself, of course, may vary with the position of P_0. For example, in polar coordinates, $\frac{\partial(x,y)}{\partial(r,\theta)} = r$. This tells us that near $r = 10$, a small region of the xy plane will have approximately ten times the area of its corresponding region in the $r\theta$ plane.

Proof of the Jacobian theorem *(in outline).* We give the proof for $\mathbf{E}^3$; the proof for $\mathbf{E}^2$ is similar. Let $P_0 = (x_0, y_0, z_0)$ and $\hat{P}_0 = (u_0, v_0, w_0)$. Instead of a spherical neighborhood $\hat{U}$ of $\hat{P}_0$, we use a rectangular solid of dimensions $\Delta u, \Delta v, \Delta w$ with one corner at $\hat{P}_0$. Then $\hat{V} = \Delta u \Delta v \Delta w$. The corresponding region U in xyz space is the region bounded by the coordinate surfaces in xyz space: $u = u_0, u = u_0 + \Delta u, v = v_0, v = v_0 + \Delta v, w = w_0$, and $w = w_0 + \Delta w$. (See [3–1].) Because the defining equations are C^2, U is an approximate parallelepiped whose edges are segments of coordinate curves given by the changes $\Delta u, \Delta v, \Delta w$ in u, v, and w. Thus the three edges adjacent to the vertex P_0 in the figure are approximately given by the vectors $\vec{u}, \vec{v}, \vec{w}$ (these are not unit vectors), where

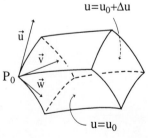

$$\vec{u} = \left. \frac{\partial \vec{R}}{\partial u} \right|_{\hat{P}_0} \Delta u = \left. \frac{\partial x}{\partial u} \right|_{\hat{P}_0} \Delta u \hat{i} + \left. \frac{\partial y}{\partial u} \right|_{\hat{P}_0} \Delta u \hat{j} + \left. \frac{\partial z}{\partial u} \right|_{\hat{P}_0} \Delta u \hat{k} \qquad [3-1]$$

$$= \Delta u \left(\left. \frac{\partial x}{\partial u} \right|_{\hat{P}_0} \hat{i} + \left. \frac{\partial y}{\partial u} \right|_{\hat{P}_0} \hat{j} + \left. \frac{\partial z}{\partial u} \right|_{\hat{P}_0} \hat{k} \right),$$

and similarly for $\vec{v}$ and $\vec{w}$. Hence the volume V of U is approximately

$$\left|[\vec{u},\vec{v},\vec{w}]\right| = \left|\begin{vmatrix} -\vec{u}- \\ -\vec{v}- \\ -\vec{w}- \end{vmatrix}\right| = \left|\begin{vmatrix} \vec{u} & \vec{v} & \vec{w} \end{vmatrix}\right| ,$$

where we have expressed the absolute value of the scalar triple product $[\vec{u},\vec{v},\vec{w}]$ first as the absolute value of the determinant whose rows are the $\hat{i},\hat{j},\hat{k}$ components of $\vec{u},\vec{v},$ and $\vec{w}$ and then as the absolute value of the determinant whose columns are the components of $\vec{u},\vec{v},$ and $\vec{w}$. Hence, by the definition of the Jacobian, we have

(3.4)
$$\left|[\vec{u},\vec{v},\vec{w}]\right| = \left|\frac{\partial(x,y,z)}{\partial(u,v,w)}\right|_{\hat{P}_0} \Delta u \Delta v \Delta w .$$

Let $a = \sqrt{(\Delta u)^2 + (\Delta v)^2 + (\Delta w)^2}$. It is possible to show that

(3.5)
$$\lim_{a \to 0}\left(\frac{V}{\left|[\vec{u},\vec{v},\vec{w}]\right|}\right) = 1 .$$

This is accomplished by finding, for P_0, a funnel function $\varepsilon(t)$ on [0,d] such that

$$0 < a < d \Rightarrow \left|1 - \frac{V}{\left|[\vec{u},\vec{v},w]\right|}\right| < \varepsilon(a).$$ We omit details. Since $\hat{V} = \Delta u \Delta v \Delta w$, it follows that

$$\lim_{a \to 0}\frac{V}{\hat{V}} = \lim_{a \to 0}\left(\frac{V}{\left|[\vec{u},\vec{v},\vec{w}]\right|} \cdot \frac{\left|[\vec{u},\vec{v},\vec{w}]\right|}{\hat{V}}\right)$$

$$= \lim_{a \to 0}\left(\frac{V}{\left|[\vec{u},\vec{v},\vec{w}]\right|}\right) \lim_{a \to 0}\frac{\left|[\vec{u},\vec{v},\vec{w}]\right|}{\hat{V}}$$

$$= \lim_{a \to 0}\left(\frac{\left|[\vec{u},\vec{v},\vec{w}]\right|}{\Delta u \Delta v \Delta w}\right) = \left|\frac{\partial(x,y,z)}{\partial(u,v,w)}\right|_{P_0}\Big| \qquad \text{(by (3.4)).}$$

§17.4 Change of Variables for Multiple Integrals

Integration by substitution is as important and useful in multivariable calculus as it is in elementary calculus. In multivariable calculus, this procedure is known as *change of variables*. We present it below as a way of changing the coordinate system in which a given multiple integral is expressed. The procedure is summarized in the following theorem.

(4.1)   ***Change-of-variables theorem* I.** *Let* f *be a continuous scalar field represented by the function* f(x,y,z) *in Cartesian coordinates in* $\mathbf{E}^3$. *Let R be a given regular region in the domain of* f. *Let curvilinear coordinates* u,v,w *be given by the* C^2 *defining equations* x = x(u,v,w), y = y(u,v,w), z = z(u,v,w). *Let* $\hat{R}$ *be a corresponding region in* uvw *space such that for each point P in R, there is a unique corresponding point* $\hat{P}$ *in* $\hat{R}$. *Then, if* $\dfrac{\partial(x,y,z)}{\partial(u,v,w)} \neq 0$ *on* $\hat{R}$, *the following* change-of-variables formula *is true:*

(4.2) $$\iiint_R f(x,y,z)dxdydz =$$

$$\iiint_{\hat{R}} f(x(u,v,w),y(u,v,w),z(u,v,w))\left|\frac{\partial(x,y,z)}{\partial(u,v,w)}\right|dudvdw,$$

where the vertical bars indicate absolute value.

   For $\mathbf{E}^2$, *under exactly similar assumptions, we have that the following* change of variables formula *is true.*

(4.3) $$\iint_R f(x,y)dxdy = \iint_{\hat{R}} f(x(u,v),y(u,v))\left|\frac{\partial(x,y)}{\partial(u,v)}\right|dudv.$$

The proof of (4.1) is given in §17.7 below. Let $J(\hat{P}) = \left.\dfrac{\partial(x,y,z)}{\partial(u,v,w)}\right|_{\hat{P}}$. By the use of more advanced methods, it is possible to show that:

(4.4)   ***Change-of-variables theorem* II.** *The formulas (4.2) and (4.3) are also true if we strengthen the condition on* f *to* C^1 *differentiability but allow the conditions that* J(P) ≠ 0 *on* $\hat{R}$, *and that each P in R determines a unique* $\hat{P}$ *to fail at singular points in* $\hat{R}$, *provided that either* J(P) ≥ 0 *on all of* $\hat{R}$ *or* J(P) ≤ 0 *on all of* $\hat{R}$, *and provided that the singular points occur only as isolated points or as points lying on some surface (for* $\mathbf{E}^3$) *or curve (for* $\mathbf{E}^2$).

We note the following features of this theorem:

(i)   The Jacobian notation makes the theorem easy to remember and use.

(ii)   Integration by substitution in a definite integral in elementary calculus can be viewed as a special one-dimensional case of these formulas.

(iii)   (4.4) includes, as special cases, our earlier results about change of variables for polar, cylindrical, and spherical coordinates.

Proof of (**4.1**).   The proof follows directly from the Jacobian theorem (3.3). We outline the proof here and complete the proof in §17.7. We begin with

(4.5) $\iiint_R f(x,y,z)dxdydz$ as the limit of a sequence of interior Riemann sums

of the form $\displaystyle\sum_{i(int)} f(x_i^*,y_i^*,z_i^*)\Delta V_i$,

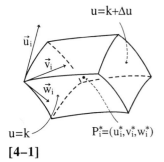

u=k+Δu

$\vec{u}_i$

$\vec{v}_i$

$\vec{w}_i$

u=k $P_i^*=(u_i^*,v_i^*,w_i^*)$

[4–1]

where R is divided into interior subregions by surfaces of the form u = constant, v = constant, w = constant, and where $P_i^* = (x_i^*,y_i^*,z_i^*)$ is a point in the i-th subregion. (See [4–1].) Then the i-th interior subregion is approximately a parallelepiped whose edges are segments of coordinate curves given by the changes Δu_i, Δv_i, Δw_i in the u,v,w coordinates. (This follows from the fact that the defining equations for the curvilinear coordinates are C^2.) By the Jacobian theorem,

(4.6) ΔV_i is approximately $\left|\dfrac{\partial(x,y,z)}{\partial(u,v,w)}\right|_{\hat{P}_i}\Delta u_i\Delta v_i\Delta w_i$, where $\hat{P}_i$ is the point in

uvw space corresponding to P_i^* .

We would like to conclude from (4.5) and (4.6) that

(4.7) $\iiint_R f(x,y,z)dxdydz$ is the limit of Riemann sums of the form

$$\sum_{i(int)} f(x(u_i^*,v_i^*,w_i^*),y(\cdots),z(\cdots))\left|\dfrac{\partial(x,y,z)}{\partial(u,v,w)}\right|_{\hat{P}_i}\Delta u_i\Delta v_i\Delta w_i ,$$

where $\hat{P}_i = (u_i^*,v_i^*,w_i^*)$. Since this last sum is an interior Riemann sum for the triple integral

$$\iiint_{\hat{R}} f(x(u,v,w),y(\cdots),z(\cdots))\left|\dfrac{\partial(x,y,z)}{\partial(u,v,w)}\right|dudvdw$$

in uvw space, our theorem will be proved if we can prove (4.7). We must be sure, however, that the small errors in (4.6) do not combine (when we use (4.6) to substitute for all the ΔV_i at the same time) to invalidate (4.7)). This is assured by *Duhamel's principle*, proved in §17.7.

Remark. Note from the proof of the Jacobian theorem that the Jacobian is > 0 when the vectors $\{\hat{u}, \hat{v}, \hat{w}\}$ are right-handed in xyz space, and < 0 when $\{\hat{u}, \hat{v}, \hat{w}\}$ are left-handed.

§17.5 Examples

Example 1. To verify that the theorem is consistent with our previous work on spherical coordinates, we calculate $\dfrac{\partial(x,y,z)}{\partial(\rho,\varphi,\theta)}$. We get

$$
\begin{vmatrix}
\sin\varphi\cos\theta & \rho\cos\varphi\cos\theta & -\rho\sin\varphi\sin\theta \\
\sin\varphi\sin\theta & \rho\cos\varphi\sin\theta & \rho\sin\varphi\cos\theta \\
\cos\varphi & -\rho\sin\varphi & 0
\end{vmatrix}
$$

$$
= \rho^2\cos^2\varphi\sin\varphi\cos^2\theta + \rho^2\sin^3\varphi\sin^2\theta
$$
$$
+ \rho^2\sin\varphi\cos^2\varphi\sin^2\theta + \rho^2\sin^3\varphi\cos^2\theta
$$
$$
= \rho^2\cos^2\varphi\sin\varphi + \rho^2\sin^3\varphi = \rho^2\sin\varphi.
$$

Thus our previous procedure for expressing triple integrals in spherical coordinates agrees with the change-of-variables theorem II. Since $\sin\varphi \geq 0$ for $0 \leq \varphi \leq \pi$, we have $\left|\rho^2\sin\varphi\right| = \rho^2\sin\varphi$.

Example 2. Similarly, for cylindrical coordinates we get

$$
\frac{\partial(x,y,z)}{\partial(r,\theta,z)} =
\begin{vmatrix}
\cos\theta & -r\cos\theta & 0 \\
\sin\theta & r\sin\theta & 0 \\
0 & 0 & 1
\end{vmatrix}
= r\cos^2\theta + r\sin^2\theta = r.
$$

Since $r \geq 0$, we have $|r| = r$, and (4.4) agrees with our previous procedure for expressing triple integrals in cylindrical coordinates.

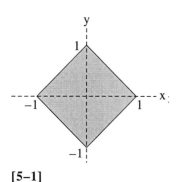

[5-1]

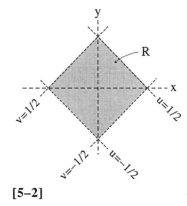

[5-2]

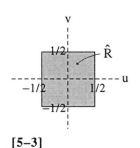

[5-3]

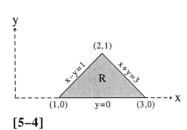

[5-4]

Example 3. Find $\iint_R (x+y)^2\,dxdy$ *where R is as in* [5-1]. This integral is evaluated easily by finding a new coordinate system in which both the integrand and the limits for the iterated integrals are especially simple. We try $\begin{pmatrix} x = u - v \\ y = u + v \end{pmatrix}$. Then the scalar field becomes $(x+y)^2 = (2u)^2 = 4u^2$, and the region R is bounded by the coordinate curves $u = -\frac{1}{2}$, $u = \frac{1}{2}$, $v = -\frac{1}{2}$, $v = \frac{1}{2}$. (See [5-2].) Thus $\hat{R}$ is as shown in [5-3]. Finally, the Jacobian $\dfrac{\partial(x,y)}{\partial(u,v)} = \begin{vmatrix} 1 & -1 \\ 1 & 1 \end{vmatrix} = 2$. Thus the theorem gives

$$\iint_R (x+y)^2\,dxdy = \iint_{\hat{R}} 4u^2 \left| \frac{\partial(x,y)}{\partial(u,v)} \right| dudv = \int_{-1/2}^{1/2} \int_{-1/2}^{1/2} 4u^2 \cdot 2\,dudv$$

$$= \int_{-1/2}^{1/2} \left(\frac{8u^3}{3} \right)\Bigg|_{-1/2}^{1/2} dv = \int_{-1/2}^{1/2} \frac{2}{3}\,dv = \frac{2}{3} \ .$$

(This result can be checked by doing the original integral in terms of x and y. The calculation is longer, however, and R must be broken into subregions for the iterated integration.)

Example 4. Let R be the triangular region in the xy plane with vertices at $(1,0)$, $(2,1)$, *and* $(3,0)$. *See* [5-4]. *Find*

$$I = \iint_R \sqrt{\frac{(x+y)}{(x-y)}}\ dA \ .$$

Since dA represents the area of a small piece in the xy plane, we use dxdy as an alternative notation for dA. Thus we write

$$I = \iint_R \sqrt{\frac{(x+y)}{(x-y)}}\ dxdy\,;$$

x and y are our given *old* variables. We introduce the *new* variables u and v by the equations

$$u = x + y,$$
$$v = x - y.$$

These equations give the new variables in terms of the old; we call these equations the *inverse defining equations* for the change of variables from x and y to u and v. If we solve these equations for x and y, we get

$$x = (\tfrac{1}{2})(u+v),$$

$$y = (\tfrac{1}{2})(u-v).$$

These are the *direct defining equations* for the change of variables from x and y to u and v. We now apply (4.1). The original integrand becomes $\sqrt{\dfrac{(x+y)}{(x-y)}} = \sqrt{\dfrac{u}{v}}$.

The Jacobian is $\begin{vmatrix} \dfrac{1}{2} & \dfrac{1}{2} \\ \dfrac{1}{2} & -\dfrac{1}{2} \end{vmatrix} = -(\tfrac{1}{2})$. Hence its absolute value is $\tfrac{1}{2}$. Thus we get

$\iint_R \sqrt{\dfrac{(x+y)}{(x-y)}}\, dxdy = \iint_{\hat{R}} \sqrt{\dfrac{u}{v}}(\tfrac{1}{2})dudv$, where $\hat{R}$ is the region in the uv plane which corresponds to the region R in the xy plane. The reader may verify that since R is bounded by the lines x–y = 1, x+y = 3, and y = 0, then $\hat{R}$ must be the triangular region in the uv plane bounded by the lines v = 1, u = 3, and u = v. See [5–5].

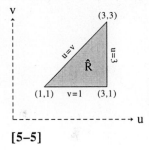

[5–5]

We can now set up and evaluate I as an iterated integral in the variables u and v. We have

$$I = (\tfrac{1}{2})\int_1^3 \int_1^u \left(\dfrac{u^{1/2}}{v^{1/2}} \right) dvdu$$

$$= (\tfrac{1}{2})\int_1^3 2u^{1/2}v^{1/2} \Big]_{v=1}^{v=u} du = \int_1^3 (u - u^{1/2})du$$

$$= \dfrac{u^2}{2} - \dfrac{2}{3}u^{3/2} \Big]_1^3 = \dfrac{14}{3} - 2\sqrt{3} = 1.20.$$

Note the simplicity of our iterated integration, once we have made the change of variables.

In the example above, we began with the inverse defining equations for u and v. We calculated the Jacobian by first *solving* the inverse equations (u = x+y and v = x–y) to get the direct equations $(x = (\tfrac{1}{2})(u+v)$ and $y = (\tfrac{1}{2})(u-v))$ and then differentiating the direct equations to get the Jacobian. Sometimes, in applications, we may begin with inverse equations but find that we cannot solve them to get direct equations. In such cases, the following fact will usually give us the Jacobian value we need.

(5.1)   **Theorem.** *In a change of variables between x,y, and u,v,* $\left.\dfrac{\partial(x,y)}{\partial(u,v)}\right|_{\hat{P}}$ *is the reciprocal of* $\left.\dfrac{\partial(u,v)}{\partial(x,y)}\right|_{P}$ *, provided that* $\left.\dfrac{\partial(u,v)}{\partial(x,y)}\right|_{P} \neq 0$.

This theorem is proved in §17.6 below.

Thus, in our example above, we could have first found

$$\frac{\partial(u,v)}{\partial(x,y)} = \begin{vmatrix} 1 & 1 \\ 1 & -1 \end{vmatrix} = -2 .$$

Hence, by (5.1), we would have had $\dfrac{\partial(x,y)}{\partial(u,v)} = -(\frac{1}{2})$ without finding or using the direct equations.

Similarly, in $\mathbf{E}^3$ we have:

(5.2)   *In a change of variables between x,y,z and u,v,w,* $\left.\dfrac{\partial(x,y,z)}{\partial(u,v,w)}\right|_{\hat{P}}$ *is the reciprocal of* $\left.\dfrac{\partial(u,v,w)}{\partial(x,y,z)}\right|_{P}$ *, provided that* $\left.\dfrac{\partial(u,v,w)}{\partial(x,y,z)}\right|_{P} \neq 0$.

Remark. An appropriate change of variables will often simplify both the given integrand *and* the given region of integration. (This happened in example 4 above.) When this occurs in an applied mathematics problem, it suggests that our previous choice of coordinates for the problem may not have been the best or most natural choice.

We conclude this section with a summary of suggested steps for using a change of variables to evaluate a multiple integral.

Step 1. Choose new coordinates and get either direct or inverse defining equations.

Step 2. Visualize or sketch $\hat{R}$.

Step 3. Express the given integrand in terms of the new coordinates.

Step 4. Find the Jacobian or the inverse Jacobian.

Step 5. Set up the integral (or integrals) on $\hat{R}$ *with appropriate limits of integration.*

Step 6. Evaluate.

§17.6 The Inverse Jacobian

We have noted that (5.1) and (5.2) are computationally useful in applications of the change-of-variable theorem. We now prove (5.1) for $\mathbf{E}^2$. The proof of (5.2) for $\mathbf{E}^3$, is similar. We begin with a detailed restatement of (5.1).

(6.1) **Theorem.** *Let* $x = f(u,v)$ *and* $y = g(u,v)$ *be defining equations for a curvilinear coordinate system in* $\mathbf{E}^2$. *Let* P *be a point in the* xy *space and let* $\hat{P}$ *be the corresponding point in the* uv *space. If* $\left.\dfrac{\partial(x,y)}{\partial(u,v)}\right|_{\hat{P}} \neq 0$, *then*

(i) u *and* v *are implicitly defined as functions of* x *and* y *in some neighborhood of* P, *and*

(ii) $\left.\dfrac{\partial(u,v)}{\partial(x,y)}\right|_{P} = \dfrac{1}{\left.\dfrac{\partial(x,y)}{\partial(u,v)}\right|_{\hat{P}}}$.

Proof. For (ii), we calculate $\dfrac{\partial(u,v)}{\partial(x,y)}$ by implicit differentiation. Using the elimination method we have

$$dx = f_u\,du + f_v\,dv$$
$$dy = g_u\,du + g_v\,dv\ .$$

By Cramer's rule, we get $du = \dfrac{\begin{vmatrix} dx & f_v \\ dy & g_v \end{vmatrix}}{D}$, and $dv = \dfrac{\begin{vmatrix} f_u & dx \\ g_u & dy \end{vmatrix}}{D}$,

where $$D = \begin{vmatrix} f_u & f_v \\ g_u & g_v \end{vmatrix} = \dfrac{\partial(x,y)}{\partial(u,v)}\ .$$

Hence $\dfrac{\partial u}{\partial x} = \dfrac{g_v}{D}, \quad \dfrac{\partial u}{\partial y} = \dfrac{-f_v}{D}, \quad \dfrac{\partial v}{\partial x} = \dfrac{-g_u}{D}, \quad \dfrac{\partial v}{\partial y} = \dfrac{f_u}{D}\ .$

This gives $\dfrac{\partial(u,v)}{\partial(x,y)} = \begin{vmatrix} \dfrac{\partial u}{\partial x} & \dfrac{\partial u}{\partial y} \\ \dfrac{\partial v}{\partial x} & \dfrac{\partial v}{\partial y} \end{vmatrix} = \dfrac{1}{D^2}\begin{vmatrix} g_v & -f_v \\ -g_u & f_u \end{vmatrix} = \dfrac{D}{D^2} = \dfrac{1}{D} = \dfrac{1}{\dfrac{\partial(x,y)}{\partial(u,v)}}\ .$

(i) now follows by the final remark in §11.7.

(6.1) can also be proved as a corollary to the Jacobian theorem (3.3).

§17.7 Duhamel's Principle

Near the end of §17.4, we question the validity of using the approximation given by the Jacobian theorem (3.3) to make a change of variables in a multiple integral. Might it be possible that the approximation errors introduced into the individual terms in a Riemann sum could accumulate in a way that would create an error in the limit of those sums (and hence in the value of the integral being calculated)? For (4.1), this question is settled by a theorem known as *Duhamel's principle* and sometimes called *the substitution principle for Riemann sums.*

Consider a given proper sequence of Riemann sums for a multiple integral of a continuous scalar field f over a compact region R, where the value of the integral is ℓ, and where for each n, the n-th sum in the proper sequence is itself a sum of n terms. Let S_n be the value of this sum. Then $\lim_{n \to \infty} S_n = \ell$. Let Δ_{in} be the i-th term $(1 \le i \le n)$ in the n-th sum. Then $S_n = \sum_{i=1}^{n} \Delta_{in}$. Thus, for example, in the case of a double integral $\iint_R f dA$, Δ_{in} will be the i-th term, $f(P_i^*)\Delta A_i$, in the n-th Riemann sum in a given proper sequence of sums for that integral.

Next, consider a second proper sequence of Riemann sums for a second multiple integral of a continuous scalar field g over a compact region R', where the value of the integral is ℓ', and the n-th sum is a sum of n terms. For each n, let Δ'_{in} be the i-th term $(1 \le i \le n)$ in the n-th sum, and let $S'_n = \sum_{i=1}^{n} \Delta'_{in}$ be the value of the n-th sum. Then $\lim_{n \to \infty} S'_n = \ell'$. Thus, for example, Δ'_{in} might be the i-th term, $g(Q_i^*)\Delta A'_i$, in the n-th Riemann sum in a proper sequence of Riemann sums for a double integral $\iint_{R'} g dA$.

(7.1)   *Theorem. (Duhamel's principle) Let proper sequences for each of two integrals be given as above. Assume that these sequences are related to each other as follows: (i) for every n and every* $i \le n$, $\Delta_{in} = 0 \Leftrightarrow \Delta'_{in} = 0$, *and* (ii) *there exists a funnel function* $\varepsilon(t)$ *on* $[1,\infty)$, *decreasing to 0 as* $t \to \infty$, *such that for every* n *and every* $i \le n$,

(7.2) $$\Delta'_{in} \ne 0 \Rightarrow \left| \frac{\Delta_{in}}{\Delta'_{in}} - 1 \right| < \varepsilon(n) \ .$$

Then the two integrals must have the same value. (Condition (7.2) is sometimes expressed as: $\lim\limits_{n\to\infty}\dfrac{\Delta_{in}}{\Delta'_{in}}=1$ uniformly.)

Proof of Duhamel's principle. Let ℓ and ℓ' be the values of the two given integrals. We have

$$S_n = \sum_{i=1}^{n}\Delta_{in} = \sum_{\substack{i=1\\(\Delta'_{in}\neq 0)}}^{n}\frac{\Delta_{in}}{\Delta'_{in}}\Delta'_{in}\,,$$

and

$$S'_n = \sum_{i=1}^{n}\Delta'_{in}\;.$$

Let $\varepsilon(t)$ be the given funnel function. Then

$$S_n - S'_n = \sum_{\substack{i=1\\(\Delta'_{in}\neq 0)}}^{n}\left(\frac{\Delta_{in}}{\Delta'_{in}}-1\right)\Delta'_{in}\,,$$

and

$$\left|S_n - S'_n\right| \le \sum_{\substack{i=1\\(\Delta'_{in}\neq 0)}}^{n}\left|\frac{\Delta_{in}}{\Delta'_{in}}-1\right|\left|\Delta'_{in}\right| \le \varepsilon(n)\sum_{i=1}^{n}\left|\Delta'_{in}\right|\;.$$

If g is a continuous scalar field on the compact region R', then $|g|$ must also be a continuous scalar field on R', and $\iint_{R'}|g|\,dA$ must exist. Hence a fundamental sequence for $\iint_{R'}|g|\,dA$ must converge to a limit. It follows that $\lim\limits_{n\to\infty}\sum_{i=1}^{n}\left|\Delta'_{in}\right|$ exists. Since $\lim\limits_{n\to\infty}\varepsilon(n)=0$, we have $\lim\limits_{n\to\infty}\left|S_n - S'_n\right|=0$. Thus $\lim\limits_{n\to\infty}S_n = \lim\limits_{n\to\infty}S'_n$ and $\ell = \ell'$ as desired.

Remark. For simplicity, we have stated and proved (7.1) for general Riemann sums. It is easy to formulate (7.1) for interior Riemann sums, and the proof is the same.

Example. We complete the proof of (4.1). We abbreviate $\left.\dfrac{\partial(x,y,z)}{\partial(u,v,w)}\right|_{\hat{P}}$ as

$J(\hat{P})$, and we write $t(\hat{P})$ for the point P in R corresponding to the point $\hat{P}$ in $\hat{R}$.

Let $S'_n = \displaystyle\sum_{i=1}^{n} \Delta'_{in}$ be the n-th sum in a proper sequence of interior Riemann sums

for $\iiint_{\hat{R}} f(t(\hat{P}))\left|J(\hat{P})\right|d\hat{V}$ based on rectangular decompositions in $\hat{R}$ with sample

points $\hat{P}^*_{in}$ in $\hat{R}$. Let $S_n = \displaystyle\sum_{i=1}^{n} \Delta_{in}$ be the n-th sum in the sequence of Riemann

sums for $\iiint_{R} f(P)dV$ based on corresponding decompositions of R, with corre-

sponding sample points $P^*_{in} = t(\hat{P}^*_{in})$. Assumption (i) in (7.1) holds, since

$\Delta'_{in} = 0 \Leftrightarrow f(t(\hat{P}^*_{in}))\left|J(\hat{P}^*_{in})\right| = 0 \Leftrightarrow f(t(\hat{P}^*_{in})) = 0 \Leftrightarrow f(P^*_{in}) = 0 \Leftrightarrow \Delta_{in} = 0$.

Since $J(\hat{P})$ is continuous and $\neq 0$ on $\hat{R}$, there exist positive m and M

such that for all $\hat{P}$ in $\hat{R}$, $m \leq \left|J(\hat{P})\right| \leq M$. Under these circumstances, it is

possible to show (see the problems) that the limit in (3.3) is uniform; that is to say,

there is a single funnel function $\varepsilon(t)$ which applies for all $\hat{P}$. This, in turn,

implies that there is a single funnel function $\varepsilon''(t)$ on $[1,\infty)$, with $\displaystyle\lim_{t\to\infty} \varepsilon''(t) = 0$,

such that for all n, and $i \leq n$,

$$\left|\frac{\Delta V_{in}}{\Delta \hat{V}_{in}} - \left|J(\hat{P}^*_{in})\right|\right| < \varepsilon''(n).$$

Hence $\qquad \left|\dfrac{\Delta V_{in}}{\left|J(\hat{P}^*_{in})\right|\Delta \hat{V}_{in}} - 1\right| < \dfrac{1}{\left|J(\hat{P}^*_{in})\right|}\varepsilon''(n) < \dfrac{1}{m}\varepsilon''(n)$.

Since $\qquad \dfrac{\Delta V_{in}}{\left|J(\hat{P}^*_{in})\right|\Delta \hat{V}_{in}} = \dfrac{f(P^*_{in})\Delta V_{in}}{f(t(\hat{P}^*_{in}))\left|J(\hat{P}^*_{in})\right|\Delta \hat{V}_{in}} = \dfrac{\Delta_{in}}{\Delta'_{in}}$,

assumption (ii) in (7.1) holds, with $\varepsilon(n) = \dfrac{1}{m}\varepsilon''(n)$. Applying (7.1) we conclude

that

$$\iiint_R f(P)dV = \iiint_{\hat{R}} f(t(\hat{P}))\left|J(\hat{P})\right|d\hat{V},$$

and the proof is complete.

§17.8 Problems

§17.2

1. In each change of variables from x and y to u and v, let P be a point with curvilinear coordinates (u_0, v_0) and Cartesian coordinates (x_0, y_0). Let $\hat{u}_P$ and $\hat{v}_P$ be the unit coordinate vectors at P for the curvilinear system. Find $\hat{i}, \hat{j}$-representations for $\hat{u}_P$ and $\hat{v}_P$ with scalar components expressed in terms of u_0 and v_0.

(a) $x = \frac{1}{2}(u+v), y = \frac{1}{2}(u-v)$

(b) $x = 3u \cos v, y = 2u \sin v$

(c) $u = x + y, v = x - 2y$

(d) $u = xy, v = xy^2$

In each case, state whether or not the curvilinear system is orthogonal.

2. (a) Assume that $x = x(u,v,w),$
$y = y(u,v,w),$
$z = z(u,v,w)$

define an orthogonal curvilinear coordinate system. Let a scalar field $f(x,y,z)$ be given. Find an expression for $\bar{\nabla}f$ in terms of $\hat{u}, \hat{v}, \hat{w}$, where the coefficients are expressed in terms of f_u, f_v, f_w and the partial derivatives of x,y,z with respect to u,v,w.

(b) Apply this formula to the case of spherical coordinates.

§17.3

1. Consider the change of variables $u = xy, v = xy^2$, from x and y to u and v. For what points in the xy plane will it be true that the area-amplification factor from the uv plane is exactly 2?

§17.5

1. Convert the integral $\iint_R \sin\left(\frac{x+y}{x-2y}\right)dA$ into an iterated integral in u and v coordinates, where $u = x+y, v = x-2y$, and R is as in [8–1].

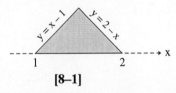

[8–1]

2. Let R be the square region in the xy plane with vertices $(1,0), (0,1), (-1,0),$ and $(0,-1)$. Consider the double integral $\iint_R (x+y)^2 dA$. Make the change of variables $u = x+y, v = x-y$, and then evaluate.

3. Find the area enclosed between the curves $xy = 1$, $xy = 3$, $xy^2 = 1$, and $xy^2 = 2$ by going to the coordinate system given by $u = xy, v = xy^2$.

4. Evaluate $\iint_R (x-y)^2(x+y)^4 dxdy$, where R is the square having its vertices $(1,0), (-1,0), (0,1), (0,-1)$, by making the change of variables $x = \frac{1}{2}(u+v)$, $y = \frac{1}{2}(u-v)$, and evaluating an iterated integral over the corresponding region in uv space.

5. The region R is bounded by the curves $y = x, y = -x$, and $x+y = \cos(x-y)$ as indicated in [8–2]. Find the area of R by making a suitable change of variables.

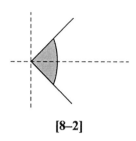

[8–2]

6. Evaluate $\iint_R \left(\dfrac{(x^2-y^2)}{xy}\right) dxdy$, where R is the region

in the first quadrant bounded by the curves xy = 1, xy = 3, $x^2-y^2 = 1$, $x^2-y^2 = 4$. (See [8–3].) Do so by introducing the variables u = xy, v = x^2-y^2, and evaluating an iterated integral over the corre-

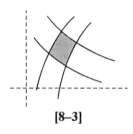

[8–3]

sponding region $\hat{R}$ in uv space. (*Hint*: As a first step, find $\dfrac{\partial(u,v)}{\partial(x,y)}$.)

7. Evaluate the double integral

$$\iint_R \sqrt{\frac{(x+y)}{(x-2y)}}\, dA ,$$

where R is the region enclosed between y = $\frac{1}{2}$(x–1), y = 0, and x+y = 4. (*Hint*. Use new coordinates u = x+y, v = x–2y.)

8. Find the volume of the region that is bounded by the plane z = 0, the paraboloid z = x^2+y^2, and the elliptical cylinder $\dfrac{x^2}{9}+\dfrac{y^2}{4}=1$. Use the "elliptical polar" coordinates given by x = 3u cos v, y = 2u sin v.

9. You are given the double integral $\iint_R \dfrac{y-x}{y+x} dxdy$, where R is the triangular region in the xy plane with vertices (2,0), (1,0), and (1,1).
   (a) Express this integral as an iterated integral in new variables u and v such that u = y+x and v = y–x.
   (b) Evaluate this iterated integral.

10. You are given the double integral $\iint_R (x + y)\cos(x^2 - y^2)dA$, where R is the square region in the xy plane with vertices (0,0), $(\frac{1}{2},\frac{1}{2})$, (1,0), and $(\frac{1}{2},-\frac{1}{2})$.
    (a) Let u and v be new variables such that u = x+y and v = x–y. Describe or sketch the region $\hat{R}$ in the uv plane that corresponds to the region R.
    (b) Make this change of variables to obtain an iterated integral in u and v.
    (c) Evaluate this iterated integral.

11. Evaluate $\iint_R \left(\dfrac{(x^2+y^2)}{xy}\right) dxdy$ where R is the region

in the first quadrant bounded by the curves xy = 1, xy = 3, $x^2-y^2 = 1$, $x^2-y^2 = 4$. (See [8–4].) Do so by introducing the variables u = xy, v = x^2-y^2, and

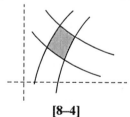

[8–4]

evaluating an iterated integral over the corresponding region $\hat{R}$ in uv space.

§17.7

1. For a given double integral $\iint_R f(x, y)dA$, consider a proper sequence of Riemann sums of the form

$\sum\limits_i f(x_i^*, y_i^*)\Delta A_i$ as described in §16.2, and then consider the corresponding sequence of *interior* sums $\sum\limits_{i(\text{int})} f(x_i^*, y_i^*)\Delta A_i$ as defined in §16.2. Use (7.1) to prove that, for a continuous f, these two sequences must have the same limit.

2. Complete the proof of (4.1) by showing that the limit in (3.3) is uniform.

Chapter 18

Vector Fields

§18.1   Vector Fields; Gradient Fields

Definitions. A region D in $\mathbf{E}^3$ or $\mathbf{E}^2$ is *connected* if, for every pair of points P and Q in D, there exists a piecewise smooth curve which connects P and Q and which lies entirely in D. (Thus a region D is connected if it consists of a "single piece" rather than of two or more "separated pieces.") As introduced in §8.1, a *vector field* $\bar{\mathbf{F}}$ *in* $\mathbf{E}^3$ is a function which takes points in some connected region D of $\mathbf{E}^3$ as *inputs* and gives vectors in $\mathbf{E}^3$ as *outputs*. D is called the *domain* of $\bar{\mathbf{F}}$. For a given P in D, we often think of the output $\bar{\mathbf{F}}(P)$ as the vector in $\mathbf{E}^3$ *assigned to* (or *located at*) the point P in $\mathbf{E}^3$ by the vector field $\bar{\mathbf{F}}$. The definition of *vector field in* $\mathbf{E}^2$ is the same, except that $\mathbf{E}^3$ is replaced by $\mathbf{E}^2$.

Vector fields are used in science and engineering as mathematical pictures of *force-fields*, of *fluid flows* (gas or liquid), and of *energy flows*. Vector fields also arise as *gradients* of scalar fields.

For an example of a force-field, let D be a region of space, and let p be a *test particle* of mass m_0 which can be placed at any chosen point P in D. The total effect on p of gravitational force from other matter (both inside and outside D) when p is placed at P can then be measured. At a given moment of time, for each point P in D, let $\bar{\mathbf{G}}(P)$ be the force upon p that would be measured if p

384

were placed at the point P at that moment. Thus $\vec{G}$ is a vector field on D. $\vec{G}$ provides an instantaneous "snapshot" of the gravitational forces in D that could act on p.

By Newton's law of gravity, the force that acts on any test particle has a magnitude proportional to the particle's mass, but a direction which does not depend on that mass. It follows that the vector field $\vec{F}$, defined by

$$\vec{F}(P) = \frac{1}{m_0}\vec{G}(P) \qquad\qquad \text{(for P in D),}$$

gives the force that would act on a test particle of mass 1 at P. The vector field $\vec{F}$ is called *the gravitational field* in D. From this $\vec{F}$ we can find the gravitational force that would act on any other particle. It will simply be $m\vec{F}(P)$, for a particle of mass m at P.

Vector fields for electric and magnetic forces are defined similarly. For an electric force field, the test particle p carries an electric charge; the force on p has (by laws of physics) magnitude proportional to the charge on p; and the direction of this force is reversed when the sign of the charge on p is reversed. For a magnetic force-field, the test-object p is a tiny magnetized needle whose direction, when pivoted at P, gives the direction of magnetic force at P, and whose resistance to rotation can be used to calculate the force that would act on a unit magnetic pole at P.

Vector fields are highly useful mathematical pictures of force-fields at points whose distances from sources of force are not too small. When these distances are submicroscopic, a simple vector-field picture is less accurate, and more complex mathematical models (such as quantum field theory) may be needed.

For another application of vector fields, consider a region D of physical space through which (or within which) a fluid (gas or liquid) is flowing. Provided that the pattern of flow is not turbulent (microscopically irregular), we can assign to each point P in D a vector $\vec{v}(P)$ which gives the average velocity of the fluid particles in a small neighborhood of P at a given moment. $\vec{v}(P)$ is called the *velocity* of the flow at P at the given moment. $\vec{v}$ is evidently a vector field and is known as the *velocity-field* of the flow at that moment. The vector field $\vec{v}$ gives an instantaneous "snapshot" of the fluid's motion.

In a non-turbulent fluid flow on D, let $\delta(P)$, for each point P, be the *density* of the fluid at P at a given moment. (Thus δ is a scalar field on D.) Then the vector $\delta(P)\vec{v}(P)$ gives the *mass-flow* at P at that moment. (See the discussion of mass-flow in §2.2.) The mass-flow vector can be abbreviated as

(1.1) $$\vec{m} = \delta\vec{v}.$$

$\vec{m}$ is a vector field on D and is called the *mass-flow field* or "mass-transport" field.

Vector fields can be used to picture flows of energy as well as flows of matter. For example the flow of heat energy in a solid material (at a given moment) can be expressed by a vector field $\vec{H}$ on D, where D is the interior of the solid and $\vec{H}$ is the *heat-flow vector* described in §2.2.

In addition to force-fields and flow-fields, a third kind of vector field arises when we take the *gradient* of a given differentiable scalar field. For example, if the scalar field T gives the temperature T(P), at a given moment, at each point P in a region D, then the gradient $\vec{\nabla}T$ is a vector field which gives the gradient vector $\vec{\nabla}T\big|_P$ at each point P in D at that moment.

Definition. If a vector field $\vec{F}$ on D happens to coincide with the gradient of some scalar field f on D, then $\vec{F}$ is called a *gradient field*, and f is called a *scalar potential for* $\vec{F}$.

We shall see in §18.5 that not every vector field is a gradient field. Moreover, a gradient field can have more than one scalar potential, since, for example, if f is a scalar potential for $\vec{F}$, then f+c, for any given scalar constant c, is also a scalar potential for f.

Other kinds of vector fields arise when we use vectors for geometrical purposes; for example, we have the vector field $\vec{R}$ which associates, with each point P, the *position vector* of P with respect to some chosen reference point, and, in cylindrical coordinates, we have the vector fields $\hat{r}$ and $\hat{\theta}$, which give unit coordinate vectors for each point P.

§18.2   Vector Fields in Cartesian Coordinates

Let $\vec{F}$ be a vector field on D in E^3. If we introduce Cartesian coordinates, then, for each P in D, we have

(2.1) $$\vec{F}(P) = L(P)\hat{i} + M(P)\hat{j} + N(P)\hat{k},$$

where the components L, M, and N are scalar fields on D defined by $L = \vec{F} \cdot \hat{i}$, $M = \vec{F} \cdot \hat{j}$, and $N = \vec{F} \cdot \hat{k}$. If we use the coordinates (x,y,z) of P, we get

(2.2) $$\vec{F}(x, y, z) = L(x, y, z)\hat{i} + M(x, y, z)\hat{j} + N(x, y, z)\hat{k},$$

where $\vec{F}$ is now expressed as a *vector function* of the three variables x, y, and z, and where L, M, and N are expressed as scalar functions of x, y, and z.

Similarly, let $\vec{F}$ be a vector field on D in E^2. If we introduce Cartesian coordinates, then we get

(2.3) $$\vec{F}(P) = L(P)\hat{i} + M(P)\hat{j}, \text{ and}$$

(2.4) $$\vec{F}(x, y) = L(x, y)\hat{i} + M(x, y)\hat{j}.$$

Vector fields are often defined or expressed in a chosen Cartesian system as vector functions of the form (2.2) or (2.4). We shall see below in §18.6 that it is sometimes simpler and more convenient to express a given vector field as a vector function in a non-Cartesian coordinate system.

Example 1. Let $\vec{F}(x, y, z) = (2xy + yz)\hat{i} + (x^2 + xz + z)\hat{j} + (xy + y)\hat{k}$. *Does $\vec{F}$ represent a gradient field?* We can show that the answer is "yes" by producing a scalar function f(x,y,z) such that $\vec{\nabla}f = \vec{F}$. If we happen to try

$$f(x, y, z) = x^2 y + xyz + yz,$$

we find that

$$\vec{\nabla}f = f_x\hat{i} + f_y\hat{j} + f_z\hat{k} = (2xy + yz)\hat{i} + (x^2 + xz + z)\hat{j} + (xy + y)\hat{k},$$

and we see that $\vec{F}$ is, in fact, a gradient field. (In Chapter 20, we develop systematic ways to look for a scalar potential for an alleged gradient field.)

Example 2. Let $\vec{G}(x, y, z) = -y\hat{i} + x\hat{j} + z\hat{k}$. *Is $\vec{G}$ a gradient field?* If we try various possibilities for a scalar potential, we do not find one. For example $f(x, y, z) = xy + \frac{1}{2}z^2$ gives $\vec{\nabla}f = y\hat{i} + x\hat{j} + z\hat{k} \neq \vec{G}$. In §18.5 below, we *prove* that $\vec{G} = -y\hat{i} + x\hat{j} + z\hat{k}$ is not a gradient field.

Example 3. Let $\vec{R}\,(P)$ *be the position vector of* P *with respect to the origin. Then* $\vec{R}\,(x,y,z) = x\hat{i} + y\hat{j} + z\hat{k}$. *Is* $\vec{R}$ *a gradient field?* Yes, the scalar function $f(x,y,z) = \frac{1}{2}(x^2 + y^2 + z^2)$ is a scalar potential for $\vec{R}$ as the reader may verify.

Remark. In this text, for the most part, we limit our attention to vector functions which carry pairs of scalars, as inputs, to output vectors in $\mathbf{E}^2$, or triples of scalars, as inputs, to output vectors in $\mathbf{E}^3$. Occasionally, however, we use vector functions which carry triples of scalars (as inputs) to output vectors in $\mathbf{E}^2$. This is the case in §18.8 below where, in order to represent a vector field in $\mathbf{E}^2$ which is changing with time, we use a vector function with triples of scalars as inputs and with vectors in $\mathbf{E}^2$ as outputs.

§18.3 Visualizing a Vector Field

We consider three different ways to visualize a vector field $\vec{F}$: (1) by *sample arrows*, (2) by *integral curves*, and (3) by *equipotential surfaces* or *curves*. ((3) only applies when $\vec{F}$ is a gradient field.)

(1) *Sample arrows.* We take a sample of points in D, and at each point P of the sample, we draw, or imagine, an arrow $\overrightarrow{PQ}$ such that $\overrightarrow{PQ} = \vec{F}(P)$. (This arrow has its tail at P.)

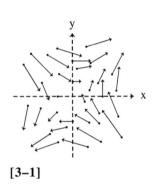

[3–1]

Example. In [3–1] *we show some sample arrows for* $\vec{F}(x,y) = y\hat{i} + x\hat{j}$ *in* $\mathbf{E}^2$. (In figures [3–1], [3–2], and [3–3], for easier visualizing, the arrows are shown with correct directions but one-half their correct lengths.)

(2) *Integral curves.*

Definition. Let $\vec{F}$ be a vector field on domain D. A directed smooth curve C in D is called an *integral curve of* $\vec{F}$ if, at every point P on C, (i) $\vec{F}(P) \neq \vec{0}$, and (ii) $\vec{F}(P)$ has the same direction as $\hat{T}_P$, the unit tangent vector to C at P.

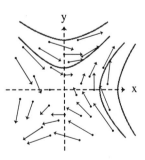

[3–2]

Example (*continued*). [3–2] *shows several integral curves for* $\vec{F} = y\hat{i} + x\hat{j}$ *in* $\mathbf{E}^2$. For this example, an equation for the integral curves is found by noting that, at any point P on an integral curve, we must have (by (ii) in the definition of integral curve)

(3.1)
$$\frac{dy}{dx}\bigg|_P = \frac{\vec{F}(P)\cdot\hat{j}}{\vec{F}(P)\cdot\hat{i}} = \frac{x}{y}, \text{ for } y \neq 0.$$

This differential equation has the solutions

$$y^2 - x^2 = a.$$

These hyperbolas are the desired integral curves of $\vec{F}$. The general problem of finding integral curves for a vector field in $\mathbf{E}^2$ is considered in §20.6.

   The following general fact can be proved from the elementary theory of differential equations.

(3.2)   *If $\vec{F}$ is a C^1 vector field on domain* $\mathbf{D}$ *(the concept of C^1 vector field is defined in (4.3) below), and if* P *is a point in* $\mathbf{D}$ *such that* $\vec{F}(P) \neq \vec{0}$, *then* $\vec{F}$ *has a unique integral curve* C *through* P *such that any other integral curve (for $\vec{F}$) through* P *must be a subset of* C. *(Thus* C *is a unique "longest" integral curve in* $\mathbf{D}$ *through* P.)

   In a force-field, integral curves are usually called *lines of force*. In a fluid flow, integral curves of the velocity field are usually called *stream-lines* or *lines of flow*. We note from (1.1) that the integral curves for the velocity field must be the same as the integral curves for the mass-flow field.

   (3) *Equipotential surfaces and curves.* If a vector field $\vec{F}$ is a gradient field, then $\vec{F} = \vec{\nabla}f$, for some scalar field f. The level surfaces (in $\mathbf{E}^3$) or level curves (in $\mathbf{E}^2$) of the scalar potential f provide another way to visualize $\vec{F}$. These surfaces or curves are called *equipotential surfaces* or *curves* for the gradient field $\vec{F}$. From §11.6, we know that if S is a level surface for f in $\mathbf{E}^3$, then at every point P on S, $\vec{\nabla}f\big|_P$ is normal to S at P, and hence, since $\vec{F}(P) = \vec{\nabla}f\big|_P$, $\vec{F}(P)$ must be normal to S at P. Similarly for $\vec{F}$ and f in $\mathbf{E}^2$: at every point P on a level curve C for the scalar potential f, we must have $\vec{F}$ (P) normal to C at P. It is easy to show (see (3.5) in §20.3) that if f_1 and f_2 are different scalar potentials for the same vector field $\vec{F}$, then they must have the same level surfaces of curves. It follows, for any *gradient field* $\vec{F}$, that

(3.3)   *At every point where $\vec{F}(P) \neq \vec{0}$, the integral curve through* P *must be normal to the equipotential surface or curve through* P. In particular, for a gradient field in $\mathbf{E}^2$, the integral curves and the equipotential curves form two mutually orthogonal families of curves. (Furthermore, we see, from the gradient's having the direction and magnitude of the maximum directional derivative, that as two equipotential surfaces or curves come closer together, the magnitude of the vector field $\vec{F}$ must increase.)

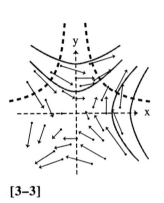

[3–3]

Example (*continued*). $\vec{F} = y\hat{i} + x\hat{j}$ *is a gradient field with* $f = xy$ *as a scalar potential.* Hence the equipotential curves of f are given by the equation

$$xy = c.$$

[3–3] shows two of these equipotential curves. Note how these curves are orthogonal to the integral curves and how the magnitude of $\vec{F}$ increases as two equipotential curves approach each other asymptotically.

§18.4   Limits, Continuity, Differentiability, and Algebraic Combinations of Vector Fields

In order to develop calculus for vector fields, we need the following definitions.

(4.1)   *Definition.* Let $\vec{F}$ be a vector field on domain D. Let P be a point in D. Let $\vec{A}$ be a given fixed vector. We define

$$\lim_{Q \to P} \vec{F}(P) = \vec{A}$$

to mean that as Q varies and P is held fixed, $\displaystyle\lim_{|\overrightarrow{QP}| \to 0} |\vec{A} - \vec{F}(Q)| = 0$. (More precisely stated: there exists a funnel function $\varepsilon(t)$ on [0,d], for some $d > 0$, such that for all Q in D,

$$0 < |\overrightarrow{QP}| \le d \Rightarrow |\vec{A} - \vec{F}(Q)| < \varepsilon(|\overrightarrow{QP}|).$$

The definitions for continuity follow in the normal way:

(4.2)   **Definition.** Let $\vec{F}$ be a vector field on domain D. Let P be a point in D. We say that $\vec{F}$ is *continuous at* P if $\lim\limits_{Q \to P} \vec{F}(Q) = \vec{F}(P)$.

Definition. Let $\vec{F}$ be a vector field on domain D. We say that $\vec{F}$ is *continuous* on D if, for every P in D, $\vec{F}$ is continuous at P.

From these definitions, it follows that if $\vec{F}$ is a vector field on D expressed in Cartesian coordinates as

$$\vec{F} = L\hat{i} + M\hat{j} + N\hat{k} ,$$

where L, M, and N are scalar fields on D, then, for a point P in D,

> $\vec{F}$ *is continuous at* P $\Leftrightarrow$
> *each of the scalar fields* L, M, *and* N *is continuous at* P.

To define differentiability and continuous differentiability for a vector field $\vec{F}$, we use Cartesian coordinates.

(4.3)   **Definition.** Let $\vec{F}$ on D be expressed in Cartesian coordinates as $\vec{F} = L\hat{i} + M\hat{j} + N\hat{k}$. Let P be a point in D. Then $\vec{F}$ is said to be *differentiable at* P, *differentiable on* D, C^1 *at* P, or C^1 *on* D according as all three of the scalar fields L, M, and N are differentiable at P, differentiable on D, C^1 at P, or C^1 on D.

Thus, for example, if

$$\vec{F} = L(x,y,z)\hat{i} + M(x,y,z)\hat{j} + N(x,y,z)\hat{k} ,$$

then $\vec{F}$ is C^1 on D $\Leftrightarrow$ the first-order partial derivatives of L, M, and N are all continuous on D.

Higher-order differentiabilities and continuous higher-order differentiabilities are defined for vector fields in the same way. It follows, as in §9.5, that the above definitions of differentiability, which use Cartesian coordinates, do not depend on the particular Cartesian system chosen.

Vector fields and scalar fields on a common domain can be combined, under the operations of vector algebra, to form new vector fields and scalar fields on that domain.

(4.4) *Definition.* Let k be a scalar field on domain D in $\mathbf{E}^3$, and let $\vec{F}$ and $\vec{G}$ be vector fields on D. Define a scalar field h and vector fields $\vec{H}_1$, $\vec{H}_2$, and $\vec{H}_3$ as follows. For P in D:

$$h(P) = \vec{F}(P) \cdot \vec{G}(P),$$
$$\vec{H}_1(P) = k(P)\vec{F}(P),$$
$$\vec{H}_2(P) = \vec{F}(P) + \vec{G}(P),$$
$$\vec{H}_3(P) = \vec{F}(P) \times \vec{G}(P).$$

These fields are customarily denoted as $h = \vec{F} \cdot \vec{G}$, $\vec{H}_1 = k\vec{F}$, $\vec{H}_2 = \vec{F} + \vec{G}$, $\vec{H}_3 = \vec{F} \times \vec{G}$. (Similar definitions and notations are used in $\mathbf{E}^2$.)

Limit theorems of the usual kind can be proved for these fields:

(4.5) *Limit theorems. Let* h, $\vec{H}_1$, $\vec{H}_2$, *and* $\vec{H}_3$ *be obtained from* k, $\vec{F}$, *and* $\vec{G}$ *on D as in (4.4). Let P be a given point in D. Assume that* $\lim_{Q \to P} k(Q) = \ell$, $\lim_{Q \to P} \vec{F}(Q) = \vec{L}_1$ *and* $\lim_{Q \to P} \vec{G}(Q) = \vec{L}_2$. *Then:*

(i) $\lim_{Q \to P} h(Q) = \vec{L}_1 \cdot \vec{L}_2,$

(ii) $\lim_{Q \to P} \vec{H}_1(Q) = \ell\vec{L}_1,$

(iii) $\lim_{Q \to P} \vec{H}_2(Q) = \vec{L}_1 + \vec{L}_2,$

(iv) $\lim_{Q \to P} \vec{H}_3(Q) = \vec{L}_1 \times \vec{L}_2.$

(i)-(iv) are proved by assuming given funnel functions for the limits of k, $\vec{F}$, and $\vec{G}$ at P and then using these assumed functions to construct new funnel functions for the limits of h, $\vec{H}_1$, $\vec{H}_2$, and $\vec{H}_3$ at P. See the problems.

Remark. In §2.6, we saw that the addition operation for vectors appears in the law of nature known as the *law of composition of forces*. Similarly, the addition operation for vector fields appears in the laws of nature known as *laws of superposition for force fields*. These laws, which hold for electric, magnetic, and gravitational fields, assert (for the case of electric fields, for example) that if source A, by itself, would create an electric field $\vec{E}_A$ in a certain region of space, and if source B, by itself, would create an electric field $\vec{E}_B$ in that region, then the electric field created by the sources A and B acting at the same time must be

$$\vec{E}_S = \vec{E}_A + \vec{E}_B .$$

(Analogous statements hold for magnetic fields and for gravitational fields.)

The superposition laws hold to an extraordinary degree of precision. (This precision is associated, for example, with the fact that a video receiver can select any one of hundreds of signals (channels) from a single ambient electromagnetic field.)

§18.5   Derivative Test for Gradient Fields

In this section, we present a simple diagnostic test for helping to decide whether or not a vector field in Cartesian coordinates is a gradient field. Let $\vec{F} = L\hat{i} + M\hat{j} + N\hat{k}$, where L, M, and N are C^1. If $\vec{F}$ is a gradient field, then, by definition, there is a scalar field f such that

$$\vec{F} = \vec{\nabla}f = f_x\hat{i} + f_y\hat{j} + f_z\hat{k} .$$

Since $\vec{F} \cdot \hat{i} = L$ and $\vec{\nabla}f \cdot \hat{i} = f_x$, we have $L = f_x$. Similarly, we have $M = f_y$ and $N = \hat{i}_z$. Since L, M, and N are C^1, f must, by definition, be C^2. Hence the theorem on equality of cross-derivatives (§9.8) applies to f, and we must have $f_{xy} = f_{yx}$, $f_{yz} = f_{zy}$, and $f_{zx} = f_{xz} f_{ZX} = f_{XZ}$. But these three equations assert that

(5.1) $L_y = M_x, M_z = N_y, \text{ and } N_x = L_z .$

We have thus proved the following.

(5.2)   *Derivative test for gradient fields* in $\mathbf{E}^3$. *Let $\vec{F}$ be a vector field such that $\vec{F} = L\hat{i} + M\hat{j} + N\hat{k}$ in a Cartesian system and such that L, M, and N are C^1. If either $L_y \neq M_x$ or $M_z \neq N_y$ or $N_x \neq L_z$, then $\vec{F}$ cannot be a gradient field.*

Similarly, we have

(5.3)   *Derivative test for gradient fields in* E^2. *Let* $\vec{F}$ *be a vector field such that* $\vec{F} = L\hat{i} + M\hat{j}$ *in a Cartesian system and such that* L *and* M *are* C^1. *If* $L_y \neq M_x$, *then* $\vec{F}$ *cannot be a gradient field.*

Thus (5.1) provides a *necessary condition* for a vector field in E^3 to be a gradient field. Is (5.1) also a *sufficient condition* for $\vec{F}$ to be a gradient field; that is to say, if (5.1) holds, can we always then conclude that $\vec{F}$ must be a gradient field? Example 4 in §18.7 shows that the answer to this question is negative.

Examples.   (1) Let $\vec{G} = L\hat{i} + M\hat{j} + N\hat{k} = -y\hat{i} + x\hat{j} + z\hat{k}$. Then $L_y = -1$, but $M_x = 1$. Hence (5.1) fails. By (5.2), $\vec{G}$ is *not* a gradient field.

(2) Let $\vec{F} = (2xy + yz)\hat{i} + (x^2 + xz + z)\hat{j} + (xy + y)\hat{k}$.

In this case,
$$L_y = 2x + z = M_x,$$

$$M_z = x + 1 = N_y,$$

and
$$N_x = y = L_z.$$

Hence (5.1) holds. Since (5.1) is not a sufficient condition, the question of whether or not $\vec{F}$ is a gradient field is not settled by the derivative test. If we can find an actual scalar potential for $\vec{F}$, then we will know for certain that $\vec{F}$ is a gradient field. In §14.2 above, we saw that $f = x^2y + xyz + yz$ is a scalar potential for $\vec{F}$. Hence we know that $\vec{F}$ is, in fact, a gradient field.

§18.6   Vector Fields in Non-Cartesian Coordinates

We can express a vector field $\vec{F}$ in an orthogonal curvilinear coordinate system by using the curvilinear coordinate frame at P (for any given point P) and giving the components of $\vec{F}(P)$ in that frame as functions of the curvilinear coordinates at P. (We must remember, of course, that the unit vectors in this frame are themselves vector fields which may vary in direction as P varies.) We illustrate for the case of *cylindrical* coordinates.

In *cylindrical coordinates*, we can express a vector field $\vec{G}$ as

(6.1) $\vec{G} = G_1\hat{r} + G_2\hat{\theta} + G_3\hat{z},$

where the scalar fields G_1, G_2 and G_3 are expressed as functions of r, θ, and z:

(6.2) $\qquad\qquad \vec{G}(r,\theta,z) = G_1(r,\theta,z)\hat{r} + G_2(r,\theta,z)\hat{\theta} + G_3(r,\theta,z)\hat{z}$;

and where $\hat{r}$ and $\hat{\theta}$ also vary with θ. (In §7.4, $\hat{r}$ and $\hat{\theta}$ were written as $\hat{r}_P$ and $\hat{\theta}_P$.)

An important case arises when $\vec{G}$ has the simple form

(6.3) $\qquad\qquad\qquad\qquad \vec{G} = g(r)\hat{r}$.

(Here $G_1 = g(r)$ is a function of r only, and G_2 and G_3 both have the constant value 0.)

$\vec{G}$ is then called a *uniform radial field*. The electric field of an infinitely long, thin, charged rod whose charge density per unit length is constant is an example of a uniform radial field, with $g(r) = \frac{c}{r}$, where the constant c is proportional to the linear density of charge on the rod. The expression for this same field in the associated Cartesian system is $\vec{G} = \frac{c(x\hat{i}+y\hat{j})}{x^2+y^2}$. This Cartesian expression is less simple and informative than the cylindrical expression $\vec{G} = \frac{c}{r}\hat{r}$.

Let g be a scalar field given by the function g(r,θ,z) in cylindrical coordinates. How can we find the gradient of g? In example 2 of §11.5, we saw that

(6.4) *Gradient in cylindrical coordinates.* $\qquad \vec{\nabla}g = \frac{\partial g}{\partial r}\hat{r} + \frac{1}{r}\frac{\partial g}{\partial \theta}\hat{\theta} + \frac{\partial g}{\partial z}\hat{z}$.

Similarly, for the case of g in $\mathbf{E}^2$ given as g(r,θ) in polar coordinates, we have

(6.5) *Gradient in polar coordinates.* $\qquad \vec{\nabla}g = \frac{\partial g}{\partial r}\hat{r} + \frac{1}{r}\frac{\partial g}{\partial \theta}\hat{\theta}$.

From (6.4) and (6.5), we see that a uniform radial field $\vec{G} = g(r)\hat{r}$ in $\mathbf{E}^3$ or $\mathbf{E}^2$ (with g(r) a piecewise continuous function) is necessarily a gradient field, since any antiderivative of g(r), such as $\int_1^r g(t)dt$, will be a scalar potential.

Similarly, in *spherical coordinates*, we can express a vector field $\vec{H}$ as

(6.6) $\vec{H} = H_1\hat{\rho} + H_2\hat{\phi} + H_3\hat{\theta}$,

where the scalar fields H_1, H_2, and H_3 are expressed as functions of ρ, ϕ, and θ:

(6.7) $\vec{H}(\rho,\phi,\theta) = H_1(\rho,\phi,\theta)\hat{\rho} + H_2(\rho,\phi,\theta)\hat{\phi} + H_3(\rho,\phi,\theta)\hat{\theta}$;

and where $\hat{\rho}$, $\hat{\phi}$, and $\hat{\theta}$ are vector fields which vary with ϕ and θ; see (2.2), (2.3) and (2.4) in §17.2.)

A simple and important case arises when $\vec{H}$ has the form

(6.8) $\vec{H} = h(\rho)\,\hat{\rho}$.

$\vec{H}$ is then called a *uniform central field*. The gravitational field due to a stationary particle M in space is an example of a uniform central field, with $h(\rho) = -\dfrac{b}{\rho^2}$, where the constant b is proportional to the mass of M. The expression for this same field in the associated Cartesian system is $\vec{H} = -\dfrac{b(x\hat{i}+y\hat{j}+z\hat{k})}{(x^2+y^2+z^2)^{3/2}}$. The spherical expression $\vec{H} = -\dfrac{b}{\rho^2}\hat{\rho}$ is simpler, more natural, and more convenient.

Let h be a scalar field given by the function $h(\rho,\phi,\theta)$ in spherical coordinates. By an argument similar to the argument in example 2 of §11.5, we can show

(6.9)   ***Gradient in spherical coordinates:*** $\vec{\nabla}h = \dfrac{\partial h}{\partial \rho}\hat{\rho} + \dfrac{1}{\rho}\dfrac{\partial h}{\partial \phi}\hat{\phi} + \dfrac{1}{\rho \sin \phi}\dfrac{\partial h}{\partial \theta}\hat{\theta}$.

From (6.9), we see that a uniform central field $\vec{H} = h(\rho)\hat{\rho}$ (with $h(\rho)$ a piecewise continuous function) is necessarily a gradient field, since any antiderivative of $h(\rho)$, such as $\displaystyle\int_1^{\rho} h(t)dt$, will be a scalar potential.

§18.7   Five Examples in E^2

The following examples illustrate some of the preceding concepts and ideas. In each case, we do the following: (a) we define a vector field in polar coordinates and give its domain; then (b) we express it in the associated Cartesian system; then (c) we visualize it with arrows and integral curves; then (d) if it *is* a gradient field, we give a scalar potential for the field and use equipotential curves to help visualize the field; and finally (e) we mention, if applicable, a physical example for which the vector field serves as a model.

Example 1.

(a) $\vec{F} = r\hat{r}$. D is all of E^2. (See [7–1].)

(b) $\vec{F} = x\hat{i} + y\hat{j}$, since $r\hat{r} = r(\cos\theta\hat{i} + \sin\theta\hat{j}) = r\cos\theta\hat{i} + r\sin\theta\hat{j}$.

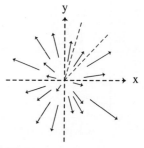

[7–1]

(c)   The integral curves are rays from the origin O, directed away from O. Arrows point away from O and increase in magnitude with distance from O. (For an arrow $\overrightarrow{PQ}$ at P, $\left|\overrightarrow{PQ}\right| = \left|\overrightarrow{OP}\right|$.)

(d)   The derivative test gives $L_y = 0 = M_x$ and leaves open the possibility that $\vec{F}$ is a gradient field. $f = \frac{1}{2}(x^2 + y^2) = \frac{1}{2}r^2$ is, in fact, a scalar potential for $\vec{F}$ (by (6.5) above, or by taking the gradient in Cartesian coordinates). This shows that $\vec{F}$ *is* a gradient field. (These results also follow directly from the comment after (6.5).) The equipotential curves are circles with center at O.

Example 2.

(a) $\vec{F} = \frac{1}{r}\hat{r}$. D is all of E^2 except for O. (See [7–2].)

(b) $\vec{F} = \dfrac{x\hat{i} + y\hat{j}}{x^2 + y^2}$, since $\frac{1}{r}\hat{r} = \frac{1}{r^2}r\hat{r}$.

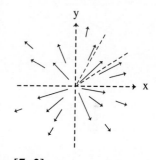

[7–2]

(c)   Integral curves are rays from O, with vertex O deleted, directed away from O. Arrows decrease in magnitude with increasing distance from O. (For an arrow $\overrightarrow{PQ}$ at P, $\left|\overrightarrow{PQ}\right| = \dfrac{1}{\left|\overrightarrow{OP}\right|}$.)

(d)   The derivative test gives $L_y = \dfrac{-2xy}{(x^2 + y^2)^2} = M_x$. This allows the possibility that $\vec{F}$ is a gradient field. $f = \frac{1}{2}\ln(x^2 + y^2) = \ln r$ is, in fact, a scalar potential for f. This shows that $\vec{F}$ *is* a gradient field. (This also follows from the comment after (6.5).) The equipotential curves are again circles with center at O.)

(e)   As mentioned in §18.6 above, a cross-section of the electrical field of an infinitely long charged rod with a charge density of 1 unit of positive charge per unit distance along the rod, has this form.

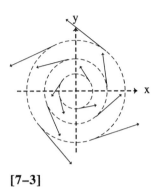

[7–3]

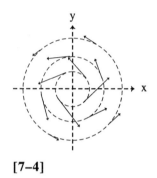

[7–4]

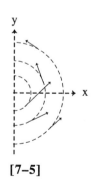

[7–5]

Example 3.

(a) $\vec{F} = r\hat{\theta}$. D is all of $\mathbf{E}^2$. (See [7–3].)

(b) $\vec{F} = -y\hat{i} + x\hat{j}$, since $\hat{\theta} = -\sin\theta\hat{i} + \cos\theta\hat{j}$.

(c)   Integral curves are circles with center at O, directed counter-clockwise. Arrows are tangent to these circles, where $|\overrightarrow{PQ}| = |\overrightarrow{OP}|$.

(d)   The derivative test gives $L_y = -1 \neq M_x = 1$. Thus $\vec{F}$ cannot be a gradient field.

(e)   As a velocity field, $\vec{F}$ describes the motion of the surface of a rigid turntable with center at O, rotating with the angular velocity 1 rad/sec.

Example 4.

(a) $\vec{F} = \frac{1}{r}\hat{\theta}$. D is all of $\mathbf{E}^2$ except for O. (See [7–4].)

(b) $\vec{F} = \dfrac{-y\hat{i}+x\hat{j}}{x^2+y^2}$, since $\frac{1}{r}\hat{\theta} = \frac{1}{r^2}(r\hat{\theta})$.

(c)   Integral curves are counterclockwise circles as in Example 3. Arrows are tangent to these circles, where $|\overrightarrow{PQ}| = \dfrac{1}{|\overrightarrow{OP}|}$.

(d)   The derivative test gives $L_y = \dfrac{y^2-x^2}{(x^2+y^2)^2} = M_x$. This allows the possibility that $\vec{F}$ is a gradient field. In Chapter 20, however, we show that $\vec{F}$ has no scalar potential and therefore is not a gradient field.

(e)   A cross-section of the magnetic field caused by a steady direct current in a long, straight, wire has the form $\vec{F} = \frac{c}{r}\hat{\theta}$, where c is proportional to the strength of the current. A rotating column of fluid whose velocity field has a cross-section of this form and (no vertical component) is called a *vortex.*.

Example 5.

(a) $\vec{F} = \frac{1}{r}\hat{\theta}$. D is the half-plane $x > 0$ (in the associated Cartesian system).

(b) $\vec{F} = \dfrac{-y\hat{i}+x\hat{j}}{x^2+y^2}$. Thus $\vec{F}$ has the same algebraic form as in (4). The domain, however, is different. (See [7–5].)

(c)   Integral curves are counter-clockwise semi circles with center at O. Arrows are as in (4).

(d)   As in (4), the derivative test gives $L_y = M_x \, L_y = M_x$. In this case, however, $f(r,\theta) = \theta$ is a scalar potential (by (6.5)), and $\vec{F}$ is hence a gradient field. In the associated Cartesian system, f on D can be expressed as $f = \tan^{-1}(\frac{y}{x})$. The equipotential curves are rays out from O, with vertex O deleted.

We see from examples 4 and 5 that whether or not $\vec{F} = \frac{1}{r}\hat{\theta}$ is a gradient field depends upon its domain. The scalar potential $f = \theta$ in example 5 does not serve as a potential for example 4, because a potential for a vector field must be a scalar field which is differentiable, and hence continuous, on the entire domain of the given vector field. $f = \theta$ cannot be continuous on the domain of example 4, since it must fail to be continuous on any circle with center at O.

§18.8   Time-dependent Fields and Steady-state Fields

So far, in our illustrations, we have treated both scalar fields and vector fields as instantaneous "snapshots," at a given moment, of a given physical situation. In many applications, however, we will prefer to have a "motion picture" of a force field or fluid flow which is evolving and changing in the course of time. We can accomplish this by introducing a parameter t (for time) into our algebraic expressions. Thus, for example, if we define a vector field in E^2 by

$$\vec{F}(P,t) = tr\,\hat{\theta} \quad (0 < a \le t \le b),$$

we have a rotating-turntable flow whose rate of angular rotation increases steadily from a rad/sec at time a to b rad/sec at time b. In the associated Cartesian coordinates, this same $\vec{F}(P,t)$ becomes

$$\vec{F}(x,y,t) = L(x,y,t)\hat{i} + M(x,y,t)\hat{j}$$
$$= -ty\hat{i} + tx\hat{j}.$$

Here $\vec{F}(x,y,t)$ is a function of three scalar variables and has vectors in E^2 as its values. Definitions of *limit, continuity,* and *differentiability* for such functions are similar to the definitions in §18.4.

Definition. In physical applications, a vector or scalar field in E^3 or E^2 which changes with a time parameter t is said to be a *time-dependent field*.

As models of evolving physical situations, most vector and scalar fields will be time-dependent. In some physical situations, however, we may wish to take, as a model, a vector or scalar field which does not change as time passes.

Definition. In a mathematical model of a physical process, a vector or scalar field which does not change as time passes is said to be a *steady-state field*. We can express it without using the parameter t. For example, the gravitational field caused by a stationary mass can be expressed as a steady-state vector field, and the velocity-field of a slowly flowing river whose level remains constant for a period of days might be represented for that period as a steady-state vector field.

Remark. It can be shown, from (3.2), that the path of a particular particle (the *trajectory* of that particle) in a steady-state fluid flow must follow a stream-line (integral curve) of the velocity field. On the other hand, it is not true for a steady-state force field that the path of a particle under that force must follow a line of force (integral curve). Indeed, the earth's orbit is almost perpendicular to the lines of force of the sun's gravitational field. (In the fluid-flow case, it is the velocity of the particle that is tangent to the integral curve. In the force-field case, it is the acceleration of the particle that is tangent to the integral curve.)

§18.9   Directional Derivatives for Vector Fields

Directional derivatives for scalar fields are defined in §8.3. We make the following analogous definition for vector fields.

Definition. Let $\vec{F}$ be a vector field on a domain D in E^3. Let $\hat{u}$ be a fixed unit vector in E^3. Let P be a fixed point in D, and let Q be another point in D such that $\overrightarrow{PQ}$ is parallel to $\hat{u}$ and such that the line segment PQ is contained in D. Let $\Delta\vec{F} = \vec{F}(Q) - \vec{F}(P)$, and let $\Delta s = \overrightarrow{PQ}\cdot\hat{u}$. Then, for our given P and $\hat{u}$, we consider the limit $\lim\limits_{\Delta s\to 0}\dfrac{\Delta\vec{F}}{\Delta s}$. If this limit exists, we write

$$(9.1) \qquad\qquad \left.\frac{d\vec{F}}{ds}\right|_{\hat{u}.P} = \lim_{\Delta s\to 0}\frac{\Delta\vec{F}}{\Delta s},$$

and we call this vector the *directional derivative of $\vec{F}$ in the direction $\hat{u}$ at P.*

As with the directional derivative of a scalar field, we have $\hat{v} = -\hat{u} \Rightarrow$ $\left.\dfrac{d\vec{F}}{ds}\right|_{\hat{v}.P} = -\left.\dfrac{d\vec{F}}{ds}\right|_{\hat{u}.P}$. Evidently, $\left.\dfrac{d\vec{F}}{ds}\right|_{\hat{u}.P}$ is simply the derivative (in the sense of §5.3) of the one-variable vector function $\vec{F}(s)$ obtained by using $\hat{u}$ to define an "s axis" through P.

We can use Cartesian axes to calculate $\left.\dfrac{d\vec{F}}{ds}\right|_{\hat{u}.P}$. Let $\vec{F} = L\hat{i} + M\hat{j} + N\hat{k}$, $\hat{u} = u_1\hat{i} + u_2\hat{j} + u_3\hat{k}$, and P be a given point in D. Then, for $\vec{F}$, $\hat{u}$, and P, we have

(9.2)
$$\begin{aligned}
\frac{d\vec{F}}{ds} &= \frac{dL}{ds}\hat{i} + \frac{dM}{ds}\hat{j} + \frac{dN}{ds}\hat{k} \\
&= (\hat{u}\cdot\vec{\nabla}L)\hat{i} + (\hat{u}\cdot\vec{\nabla}M)\hat{j} + (\hat{u}\cdot\vec{\nabla}N)\hat{k} \\
&= (u_1\left.\frac{\partial L}{\partial x}\right|_P + u_2\left.\frac{\partial L}{\partial y}\right|_P + u_3\left.\frac{\partial L}{\partial z}\right|_P)\hat{i} + (u_1\left.\frac{\partial M}{\partial x}\right|_P + u_2\left.\frac{\partial M}{\partial y}\right|_P + u_3\left.\frac{\partial M}{\partial z}\right|_P)\hat{j} \\
&\quad + (u_1\left.\frac{\partial N}{\partial x}\right|_P + u_2\left.\frac{\partial N}{\partial y}\right|_P + u_3\left.\frac{\partial N}{\partial z}\right|_P)\hat{k}.
\end{aligned}$$

Example 1. *Find* $\left.\dfrac{d\vec{F}}{ds}\right|_{\hat{u}.P}$ *for* $\vec{F} = x^2\hat{i} + y^2\hat{j} + z^2\hat{k}$, P $= (a,b,c)$, *and* $\hat{u} = u_1\hat{i} + u_2\hat{j} + u_3\hat{k}$. *Then* $\left.\vec{\nabla}L\right|_P = 2a\hat{i}$, $\left.\vec{\nabla}M\right|_P = 2b\hat{j}$, $\left.\vec{\nabla}N\right|_P = 2c\hat{k}$, *and*

$$\left.\frac{d\vec{F}}{ds}\right|_{\hat{u}.P} = 2au_1\hat{i} + 2bu_2\hat{j} + 2cu_3\hat{k}.$$

Example 2. *Find* $\left.\dfrac{d\vec{F}}{ds}\right|_{\hat{u}.P}$ *for* $\vec{F} = \vec{R} = x\hat{i} + y\hat{j} + z\hat{k}$, P $= (a,b,c)$, *and* $\hat{u} = u_1\hat{i} + u_2\hat{j} + u_3\hat{k}$. *Here,* $\vec{\nabla}L = \hat{i}$, $\vec{\nabla}M = \hat{j}$, $\vec{\nabla}N = \hat{k}$. *We have*

$$\left.\frac{d\vec{F}}{ds}\right|_{\hat{u}.P} = u_1\hat{i} + u_2\hat{j} + u_3\hat{k} = \hat{u}.$$

It is easy to show, from the definition, that the directional derivative for a given $\hat{u}$ is linear (it satisfies $\dfrac{d}{ds}(a\vec{F} + b\vec{G}) = a\dfrac{d\vec{F}}{ds} + b\dfrac{d\vec{G}}{ds}$), and that it therefore obeys the usual derivative laws for sums and products. (See §5.4.)

Example 3. *Find a general formula for* $\left.\dfrac{d\vec{F}}{ds}\right|_{\hat{u}.P}$ *for* $\vec{F} = (\vec{R} \cdot \vec{R})\vec{R}$ *, where*

$\vec{R} = x\hat{i} + y\hat{j} + z\hat{k}$.

We have
$$\frac{d\vec{F}}{ds} = (\vec{R} \cdot \vec{R})\frac{d\vec{R}}{ds} + \frac{d}{ds}(\vec{R} \cdot \vec{R})\vec{R} \qquad \text{(by a product rule)}$$

$$= (\vec{R} \cdot \vec{R})\frac{d\vec{R}}{ds} + (2\vec{R} \cdot \frac{d\vec{R}}{ds})\vec{R} \qquad \text{(by a product rule)}$$

$$= (\vec{R} \cdot \vec{R})\hat{u} + (2\vec{R} \cdot \hat{u})\vec{R} \qquad \text{(by example 2).}$$

Remark. Can we find a simple, coordinate-free formula for $\dfrac{d\vec{F}}{ds}$ analogous to the formula for a scalar field: $\left.\dfrac{d\vec{F}}{ds}\right|_{\hat{u}.P} = \left.\vec{\nabla}f\right|_{P} \cdot \hat{u}$? There is no analogous formula in the vector algebra that we have developed in this book so far. In Chapter 27, we will extend our vector algebra to include coordinate-free expressions for $\left.\dfrac{d\vec{F}}{ds}\right|_{\hat{u}.P}$. This extension will also provide us with a coordinate-free linear approximation formula for vector fields.

§18.10 Problems

§18.2

1. The vector field $\vec{m} = yz\hat{i} + zx\hat{j} + xy\hat{k}$ represents a mass flow. At what net rate, per second, is mass flowing upwards through the triangle whose vertices are $(0,0,0)$, $(1,0,0)$, and $(1,1,1)$? ($\vec{m}$ has the units gm/cm^{2}sec). (*Hint.* Integrate over the projected region in the xy plane. Find an appropriate form of Riemann sum for the desired mass flow. Then deduce from this the needed integrand.)

2. Same as problem 1, but so with vertices $(0,0,0)$, $(1,0,0)$, and $(1,1,0)$.

3. Find a scalar potential for the field $\vec{m}$ in problem 1.

§18.4

1. Complete the proof of (4.5) by constructing appropriate funnel functions at P from the assumed funnel functions for k, $\vec{F}$, and $\vec{G}$ at P.

§18.5

In each of the following, apply the derivative test for a gradient field. If the field passes the test, try to find, by "educated" guessing, a scalar potential for it.

1. $\vec{F} = (z^3 + 2xy)\hat{i} + (x^2 + 2y)\hat{j} + 3xz^2\hat{k}$

2. $\vec{F} = x\hat{i} + 2yz\hat{j} + y^2\hat{k}$

3. $\vec{F} = xz\hat{i} - y\hat{j} + \dfrac{x^2}{2}\hat{k}$

4. $\vec{F} = -y\hat{i} - x\hat{j} + x\hat{k}$

5. $\vec{F} = 2xyz\hat{i} + (x^2z + 2yz)\hat{j} + (x^2y + y^2)\hat{k}$

6. $\vec{F} = 4xy\hat{i} - x^2\hat{j} + 4z\hat{k}$

§18.6

1. Use (6.4) to find the gradient for each scalar field (given in cylindrical coordinates).
  (a) $f(r,\theta,z) = r(\theta^2 + z^2)$ $(-\pi < \theta < \pi)$.
  (b) $f(r,\theta,z) = r\theta + z$ $(-\pi < \theta < \pi)$.

2. Use (6.9) to find the gradient for each scalar field (given in spherical coordinates).
  (a) $f(\rho,\varphi,\theta) = \rho\varphi\theta$ $(-\pi < \theta < \pi)$
  (b) $f(\rho,\varphi,\theta) = \rho^2\varphi\sin\theta$ $(-\pi < \theta < \pi)$

3. Derive the general formula (6.9) for the gradient in spherical coordinates.

§18.7

1. In example 5 in §18.7, change the domain to $-0.9\pi \le \theta \le \pi$. Is $\vec{F}$ still a gradient field? Explain. If it is a gradient field, give a potential.

2. Let $\vec{F} = (r + \frac{1}{r})\hat{r}$ on all of $\mathbf{E}^2$ except for O. Express $\vec{F}$ in the associated Cartesian system. Sketch $\vec{F}$ with arrows and draw some integral curves. Give a general equation (in polar coordinates) for the integral curves. If $\vec{F}$ is a gradient field, give a scalar potential for it, sketch some equipotential for it, sketch some equipotential curves, and give a general equation for the equipotential curves.

3. Same as problem 2, but with $\vec{F} = \frac{1}{r}\hat{r} + \frac{1}{r}\hat{\theta}$ on the same domain.

§18.8

1. Find $\dfrac{d\vec{F}}{ds}\bigg|_{\hat{u}.P}$ for $\vec{F} = yz\hat{i} + xz\hat{j} + xy\hat{k}$, $P = (a,b,c)$, and $\hat{u} = u_1\hat{i} + u_2\hat{j} + u_3\hat{k}$.

2. For the vector field $\vec{F}(P) = \hat{r}_P$, find $\dfrac{d\vec{F}}{ds}\bigg|_{\hat{u}.P}$ in terms of $\hat{r}_P$, $\hat{\theta}_P$, $\hat{z}_P$, r, θ, z, and $\hat{u}$, where (r,θ,z) are cylindrical coordinates of P.

Chapter 19

Line Integrals

§19.1   Integrals on Curves

In the present chapter, we introduce two new kinds of definite integral: *scalar line integrals* and *vector line integrals*. For each kind, we proceed in two steps. First, we give a formal, conceptual *definition* of the integral. This definition will help us to think about, understand, and use the integral. Second, we describe methods for *evaluating* the integral.

The conceptual definitions are closely analogous to the definition of definite integral for a one-variable scalar-valued function (as given in elementary calculus), to the definition of definite integral for a one-variable vector-valued function (as given in §5.8), and to the definitions of multiple integral (as given in Chapter 14). The new integrals are different, however, in that (i) the *region of integration*, instead of being a bounded interval of real numbers or a general bounded region in $\mathbf{E}^3$ or $\mathbf{E}^2$, is now a *finite curve* C in $\mathbf{E}^3$ or $\mathbf{E}^2$, and (ii) the *integrand* is now a scalar or vector field whose domain contains C.

These integrals on curves are traditionally referred to as *line integrals*, where the word "line," in this case, refers to a general curve in $\mathbf{E}^3$ or $\mathbf{E}^2$ and not just to a straight line. The words "scalar" and "vector" in the phrases "scalar line integral" and "vector line integral" refer to the nature of the *integrand* (as a scalar or vector field), not to the *value* of the integral. We shall see that both scalar line integrals and vector line integrals, as defined in §20.2 and §20.4 below, have scalar values.

Other kinds of definite integrals on curves (besides the scalar-valued integrals defined below as *scale line integrals* and *vector line integrals*) can also be defined. We consider two of these in §19.6. Some of these other integrals on curves will have vector values.

§19.2   Scalar Line Integrals

A *scalar line integral* is much like a definite integral in elementary calculus. The only essential difference is that instead of having an interval of real numbers (usually on the x axis) as its region of integration, a scalar line integral uses a *finite curve* in $\mathbf{E}^3$ or $\mathbf{E}^2$.

(2.1)   *Definition.* Let C be a finite, rectifiable curve in $\mathbf{E}^3$ or $\mathbf{E}^2$, and let f be a scalar field whose domain contains C. We write the *scalar line integral of f on C* as

$$\int_C f\,ds\,,$$

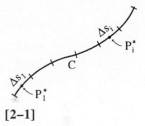

[2–1]

and we define it as follows. For a given positive integer n, divide C into n pieces whose arc-lengths are $\Delta s_1, \Delta s_2, \ldots, \Delta s_n$, and for each $i \le n$, choose a *sample point* P_i^* on the i-th piece of curve. (See [2–1].) Using this *decomposition* of C (see §5.8), form the *Riemann sum* $\sum_{i=1}^{n} f(P_i^*)\Delta s_i$. Finally, using *proper sequences* of such decompositions (see §5.8 for details), find if the following limit exists and is unique:

$$\lim_{\substack{n\to\infty \\ \max\Delta s_i \to 0}} \sum_{i=1}^{n} f(P_i^*)\Delta s_i \,.$$

If so, we write this limit as $\int_C f\,ds$; we say that the integral $\int_C f\,ds$ *exists* for the given f and C, and we call this integral the *scalar line integral* of f on C.

The uses and purposes of the scalar line integral are similar to those of the definite integral in elementary calculus. As we have noted, the definition is the same except that we have used, as our region, a curve instead of a segment along an axis.

Applications of scalar line integrals include the following.

(1)   If f(P) = 1, for all P, then $\int_C f\,ds = \int_C ds =$ the arc length of C.

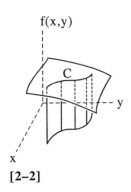

f(x,y)

C

y

x

[2–2]

(2)   If C is a curve in the xy plane and f is a scalar field ≥ 0 on the xy plane, then $\int_C f\,ds$ is the area of the surface in $\mathbf{E}^3$ formed by those points in $\mathbf{E}^3$ which project onto C and which lie between C and the graph of f. See [2–2].

(3)   A thin wire has the shape of a curve C in $\mathbf{E}^3$. f has C itself as its domain and gives at each point the density per unit arc length of the wire at that point. Then $\int_C f\,ds$ = the total mass of the wire.

(4)   Let f and C be given as in the definition. The *average value of* f *on* C can be defined as $\dfrac{\int_C f\,ds}{\int_C ds}$. (See the discussion of *average value* in §14.3.)

The following two laws hold for scalar line integrals and are immediate consequences of the formal definition (2.1).

(I)   *Let f and g be scalar fields whose common domain contains the curve* C. *Let* a *and* b *be given scalar constants. Then*

$$\int_C (af + bg)\,ds = a\int_C f\,ds + b\int_C g\,ds\,.$$

This is the *linearity law* for scalar line integrals.

(II)   *Let* f *be a scalar field whose domain contains the curve* C. *Let* C *be divided into two adjacent pieces,* C_1 *and* C_2. *Then*

$$\int_C f\,ds = \int_{C_1} f\,ds + \int_{C_2} f\,ds.$$

This is the *additivity law* for scalar line integrals and is sometimes written:

$$\int_{C_1+C_2} f\,ds = \int_{C_1} f\,ds + \int_{C_2} f\,ds\,,$$

where $C_1 + C_2$ denotes the curve obtained by combining the adjacent curves C_1 and C_2.

Under what conditions for f and C can we be sure that $\int_C f\,ds$ exists? The answer is given by the following:

(2.2) *Existence theorem. If* C   *is a finite, rectifiable curve and* f *is a continuous scalar field whose domain contains* C, *then* $\int_C f\,ds$ *exists.*

(2.2) can be proved by the methods of §14.10.

Remark. The following extended version of this existence theorem is also easy to prove.

(2.3)   *If* f *is a scalar field whose domain contains the finite, rectifiable curve* C, *and if* f *is piecewise continuous on* C, $\int_C f\,ds$ *exists.* Here, *piecewise continuous on* C is defined analogously to (4.4) of §14.4.

§19.3   Evaluating Scalar Line Integrals

We describe two methods for evaluating scalar line integrals: (i) *evaluation by definition* and (ii) *parametric evaluation.*

Evaluation by definition is based on the definition of the integral as the limit of Riemann sums. By considering specific Riemann sums, we try to recognize directly what the value of the integral must be.

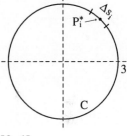

[3–1]

Example 1. *In* E^2 *with Cartesian coordinates, let* C *be the circle of radius* 3 *with center at the origin, and let* f *be given by the function* f(x,y) = $x^2 + y^2$. *(See* [3–1].) Here, for every point P on C, we have f(P) = 9. Hence every Riemann sum must have the form $\sum_i 9\Delta s_i$, and its value must be $\sum_i 9\Delta s_i =$

$$9\sum_i \Delta s_i = 9(\text{arc length of } C) = 9(6\pi) = 54\pi \,.$$ Hence, $\int_C f\,ds = 54\pi \,.$

The *symmetric-cancellation principle*, described for multiple integrals in Chapter 14, is often a useful aid in evaluating of scalar line integrals by definition.

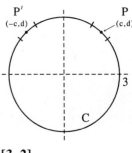

[3–2]

Example 2. *Let* C *be as in example* 1. *Let* f *be given by the function* f(x,y) = x^3. We note that for every point P = (c,d) on the right-hand semicircle of C, we have f(P) = c^3, and for each mirror image point P' = (–c,d) on the left-hand semicircle, we have f(P') = $(-c)^3 = -c^3$. (See [3–2].) If we form a Riemann sum: (i) by choosing the pieces of C on the left to be the mirror-images of the pieces of C chosen on the right, and (ii) by choosing the sample points on the left to be the mirror images of the sample points chosen on the right; then the terms in this Riemann sum must cancel in pairs, and the entire sum must have the value 0. We can define a proper sequence of such sums (as n → ∞), and hence we have $\int_C f\,ds = 0\,.$

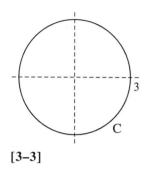

[3–3]

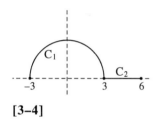

[3–4]

Laws I and II can also be helpful in evaluation by definition. Using (I), we can separately consider different terms in the integrand; using (II), we can separately consider different parts of C.

Example 3. *Let C be as in example 1. Let* f *be given by* $f(x,y) = x^3+y^2$. *(See [3–3].)* By (I), $\int_C f\,ds = \int_C x^3 ds + \int_C y^2 ds$. From example 2, $\int_C x^3 ds = 0$. For $\int_C y^2 ds$, we note that $\int_C y^2 ds = \int_C x^2 ds$ by the identical forms of their Riemann sums, and hence, by example 1, $\int_C y^2 ds = \frac{1}{2}\int_C (x^2 + y^2)ds = \frac{1}{2}(54\pi) = 27\pi$. Thus $\int_C f\,ds = 27\pi$.

Example 4. *In* $\mathbf{E}^2$ *with Cartesian coordinates, let C consist of two parts,* C_1 *and* C_2, *where* C_1 *is the upper semicircle of radius 3 with center at the origin, and* C_2 *is the straight segment from (3,0) to (6,0).* *(See [3–4].)* *Let* f *be given by* $f(x,y) = x^2+y^2$. By (II), we have $\int_C f\,ds = \int_{C_1} f\,ds + \int_{C_2} f\,ds$.

From example 1, we see that $\int_{C_1} f\,ds = 27\pi$ (since C_1 has half the length of the circle in example 1). Consider any Riemann sum for $\int_{C_2} f\,ds$. We see that it is identical with a Riemann sum for the integral $\int_3^6 x^2 dx$ as defined in elementary calculus, since, on C_2, $f(x,y) = f(x,0) = x^2$. Hence $\int_{C_2} f\,ds = \int_3^6 x^2 dx = \frac{x^3}{3}\Big]_3^6 = 63$. Thus we have $\int_C f\,ds = 27\pi + 63$.

Parametric evaluation is a more general and systematic (but sometimes longer) procedure. We first describe and illustrate the procedure. Then, at the end of this section, we justify it. For parametric evaluation, we assume not only that C is rectifiable, but also that C is smooth.

In parametric evaluation, we use a parametric path, in some coordinate system, for the given curve C, and we use a function f in that coordinate system for the given scalar field f. Let t be the parameter, with $a \le t \le b$. The notation $\int_C f\,ds$ suggests the following easily remembered evaluation rule:

(3.1)
$$\int_C f\,ds = \int_a^b f\frac{ds}{dt}\,dt = \int_a^b f\left|\frac{d\vec{R}}{dt}\right|dt\;.$$

For Cartesian coordinates in $\mathbf{E}^3$, we have

$$\vec{R}(t) = x(t)\hat{i} + y(t)\hat{j} + z(t)\hat{k},\quad a \le t \le b,$$

with
$$\frac{d\vec{R}}{dt} = \dot{x}(t)\hat{i} + \dot{y}(t)\hat{j} + \dot{z}(t)\hat{k},$$

and
$$f(P) = f(x,y,z),\text{ for }P = (x,y,z).$$

This gives:

(3.2) $$\int_C f\,ds = \int_a^b f\left|\frac{d\vec{R}}{dt}\right|dt = \int_a^b f(x(t), y(t), z(t))(\dot{x}(t)^2 + \dot{y}(t)^2 + \dot{z}(t)^2)^{1/2}\,dt\;.$$

Note that the parametric equations for C appear in (3.2) in three different places: (i) they give the limits of the final definite integral, (ii) they are used to express f as a function of the parameter, and (iii) derivatives of the parametric functions are used to get the expression for $\left|\dfrac{d\vec{R}}{dt}\right|$. The formula in (3.2) is lengthy but easy to summarize:

(a)   We find parametric equations for C: $\vec{R}(t) = x(t)\hat{i} + y(t)\hat{j} + z(t)\hat{k}$, $a \le t \le b$.

(b)   We find $\dfrac{d\vec{R}}{dt}$ as a function of t.

(c)   We find $\left|\dfrac{d\vec{R}}{dt}\right|$ as a function of t.

(d)   We express f as a function of the parameter t, $a \le t \le b$.

(e)   We take the product of (d) and (c).

(f)   We integrate this product with respect to t from a to b. (This final integral is a one-variable definite integral of elementary calculus.)

Example 5. In $\mathbf{E}^2$ with Cartesian coordinates, let C be the upper semicircle of radius 3 with center at the origin. Let f be given by f(x,y) = y. (See [3–5].) There is no evident way to use evaluation by definition. We therefore turn to parametric evaluation. Then

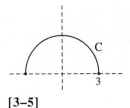

[3–5]

(a)   As our path for C, we take $\vec{R}(t) = 3\cos t\,\hat{i} + 3\sin t\,\hat{j}$, $0 \le t \le \pi$.

(b) $\dfrac{d\vec{R}}{dt} = -3 \sin t\,\hat{i} + 3 \cos t\,\hat{j}.$

(c) $\left|\dfrac{d\vec{R}}{dt}\right| = (9\sin^2 t + 9\cos^2 t)^{1/2} = 9^{1/2} = 3.$

(d)   We have $f(x(t),y(t)) = y(t) = 3 \sin t.$

(e) $f\left|\dfrac{d\vec{R}}{dt}\right| = (3 \sin t)3 = 9 \sin t.$

(f) $\displaystyle\int_C f\,ds = \int_0^{\pi} 9 \sin t\,dt = -9 \cos t\,\Big]_0^{\pi} = 18.$

As with evaluation by definition, we can use laws I and II to help with parametric evaluation.

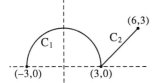

[3–6]

Example 6. *Again in* E^2 *with Cartesian coordinates, let* C *consist of* C_1 *and* C_2, *where* C_1 *is the upper semicircle of radius 3 with center at the origin and* C_2 *is the straight segment from (3,0) to (6,3).* *(See [3–6].)* *Let* f *be given by* $f(x,y) = x+y.$ *Then* $\displaystyle\int_C f\,ds = \int_{C_1} f\,ds + \int_{C_2} f\,ds$

For C_1: $\displaystyle\int_{C_1} f\,ds = \int_{C_1} x\,ds + \int_{C_1} y\,ds.$ By cancellation, we see that $\displaystyle\int_{C_1} x\,ds = 0.$ From example 5, we have $\displaystyle\int_{C_1} y\,ds = 18.$ Thus $\displaystyle\int_{C_1} f\,ds = 0 + 18 = 18.$

For C_2: we use parametric evaluation. We have:

(a) $\vec{R}(t) = t\hat{i} + (t-3)\hat{j},\ \ 3 \le t \le 6.$ (See §4.4.)

(b) $\dfrac{d\vec{R}}{dt} = \hat{i} + \hat{j}.$

(c) $\left|\dfrac{d\vec{R}}{dt}\right| = (1+1)^{1/2} = \sqrt{2}.$

(d) $f(x(t),y(t)) = x(t)+y(t) = t+(t-3) = 2t-3.$

(e) $f\left|\dfrac{d\vec{R}}{dt}\right| = \sqrt{2}(2t-3).$

(f) $\displaystyle\int_{C_2} f\,ds = \int_3^6 \sqrt{2}(2t-3)dt = \sqrt{2}(t^2-3t)\,\Big]_3^6 = 18\sqrt{2}.$

Finally, $\displaystyle\int_C f\,ds = \int_{C_1} f\,ds + \int_{C_2} f\,ds = 18 + 18\sqrt{2} = 18(1+\sqrt{2}).$

Justifying parametric evaluation. We must show that $\int_C f\, ds =$
$\int_a^b f\dfrac{ds}{dt}\, dt$. Let $\displaystyle\sum_{i=1}^{n} f(P_i^*)\Delta s_i$ be a Riemann sum for $\int_C f\, ds$. For each $i \le n$, let t_i^* be the parameter value which gives P_i^*, and let Δt_i be the parameter interval which corresponds to Δs_i. Now form the corresponding Riemann sum

$$\sum_i f(\vec{R}(t_i^*))\dfrac{ds}{dt}\Big|_{t_i^*}\Delta t_i.$$

This Riemann sum for $\int_a^b f\dfrac{ds}{dt}\, dt$ need not have exactly the same value as the previous sum for $\int_C f\, ds$, since it need not be true that $\Delta s_i = \dfrac{ds}{dt}\Big|_{t_i^*}\Delta t_i$ exactly. It will be true, however, that $\dfrac{ds}{dt}\Big|_{t_i^*}\Delta t_i$ is a sufficiently close approximation to Δs_i to guarantee that the Riemann sums for $\int_a^b f\dfrac{ds}{dt}\, dt$ will have the same limit as the Riemann sums for $\int_C f\, ds$. More specifically, the conditions for *Duhamel's principle* (see §17.7) can be shown to hold, since $\dfrac{\Delta s_i}{\dfrac{ds}{dt}\Big|_{t_i^*}\Delta t_i} = \dfrac{\dfrac{\Delta s_i}{\Delta t_i}}{\dfrac{ds}{dt}\Big|_{t_i^*}}$ approaches the limit 1 uniformly as $n \to \infty$. (The uniformity of this last limit follows from the smoothness of C.)

§19.4 Vector Line Integrals

(4.1) *Definition.* Let C be a finite, smooth, *directed* curve in $\mathbf{E}^3$ or $\mathbf{E}^2$, and let $\vec{F}$ be a vector field in $\mathbf{E}^3$ or $\mathbf{E}^2$ whose domain contains C. We write the *vector line integral of $\vec{F}$ on C* as

$$\int_C (\vec{F}\cdot\hat{T})\, ds$$

and we define it as follows. For a given positive integer n, divide C into n pieces whose arc-lengths are $\Delta s_1, \Delta s_2, ..., \Delta s_n$, and for each $i \le n$, choose a *sample point*

P_i^* on the i-th piece of curve. Then for each i, find the dot product $\vec{F}(P_i^*) \cdot \hat{T}(P_i^*)$, where $\hat{T}(P_i^*)$ is the unit tangent vector to C at P_i^*. Form the Riemann sum $\sum_{i=1}^{n} [\vec{F}(P_i^*) \cdot \hat{T}(P_i^*)]\Delta s_i$. Finally, using proper sequences of such decompositions (as in §5.8), see if the following limit exists and is unique:

$$\lim_{\substack{n \to \infty \\ \max \Delta s_i \to 0}} \sum_{i=1}^{n} [\vec{F}(P_i^*) \cdot \hat{T}(P_i^*)]\Delta s_i \ .$$

If so, we write this limit as $\int_C (\vec{F} \cdot \hat{T})ds$; we say that the integral $\int_C (\vec{F} \cdot \hat{T})ds$ *exists*; and we call this integral the *vector line integral of* $\vec{F}$ *on* C. We shall usually write the *vector line integral of* $\vec{F}$ *on* C as

$$\int_C \vec{F} \cdot d\vec{R} \ ,$$

where $d\vec{R}$ is an abbreviation for $\hat{T}ds$, and where, in a Riemann sum, we write $\Delta \vec{R}_i^*$ for $\hat{T}(P_i^*)\Delta s_i$.

Definition (4.1) can be expressed informally as follows. For the vector line integral $\int_C \vec{F} \cdot d\vec{R}$, we start with a finite, smooth, directed curve C as our region of integration and with a vector field $\vec{F}$ as our integrand. Because the curve C is smooth and directed, we can assign a unique unit tangent vector $\hat{T}(P)$ to each point P on C. Then we can use $\vec{F}$ to assign the scalar value $\vec{F}(P) \cdot \hat{T}(P)$ to each point P on C. Finally, defining $g(P) = \vec{F}(P) \cdot \hat{T}(P)$, we define the *value* of our vector line integral $\int_C \vec{F} \cdot d\vec{R}$ to be the value of the scalar line integral $\int_C gds$. This informal description can be summarized even more informally: $\int_C \vec{F} \cdot d\vec{R}$ *is the scalar line integral on* C *of the tangential component of* $\vec{F}$ *in the direction of* C.

In this text, we shall consider two common applications of vector line integrals to physical situations.

Application 1. If a particle p moves along a straight line in the direction $\hat{u}$ and is subjected to a *constant* force $\vec{F}$ (which need not be parallel to $\hat{u}$) then the *net work done by* $\vec{F}$ *on* p is defined (in physical theory) to be $(\vec{F} \cdot \hat{u})$s, where s is the distance moved by p. $(\vec{F} \cdot \hat{u})$s can also be written $\vec{F} \cdot \vec{s}$, where $\vec{s} = s\hat{u}$ is the *displacement* of p. These expressions suggest that when a vector field $\vec{F}$ represents a steady-state force field which varies continuously (in magnitude and direction) from point to point in space, and when a particle p, subject to the force $\vec{F}$, moves along a finite, smooth, directed curve C in the domain of $\vec{F}$ and moves from the initial point of C to the final point of C, then the *net work done by* $\vec{F}$ *on* p can be approximated by a Riemann sum for the integral $\int_C \vec{F} \cdot d\vec{R}$. This in turn suggests that the net work done by $\vec{F}$ on p can be *defined* as the vector line integral $\int_C \vec{F} \cdot d\vec{R}$. The use of line integrals, in this way, to define and express work done by a force (electromagnetic, gravitational, or nuclear) is a fundamental feature of modern physical theory.

Application 2. As we have seen in Chapter 18, vector fields can be used to represent a flow of matter or a flow of energy. For example, in a fluid-flow, the velocity $\vec{v}$ at each point can be represented as a vector field, and the mass-flow $\vec{m}$ at each point can be represented as a vector field. Let C be a directed smooth curve in the region of the flow. Then $\int_C \vec{v} \cdot d\vec{R}$ is called the *circulation along* C of the flow. It can be thought of as the average tangential component of $\vec{v}$, in the direction of C, multiplied by the arc length of C. Vector line integrals for circulation are a fundamental feature of theories of fluid-flow in modern engineering.

Applications (1) and (2) will be used in this text to motivate and illustrate our later study of vector fields. Notable applications of line integrals also occur in thermodynamics and in physical chemistry.

(4.2)   The following three laws follow from the definition of vector line integral.

(I) *Let* $\vec{F}$ *and* $\vec{G}$ *be vector fields whose common domain contains the finite, smooth, directed curve* C. *Let* a *and* b *be given scalar constants. Then*

$$\int_C (a\vec{F} + b\vec{G}) \cdot d\vec{R} = a \int_C \vec{F} \cdot d\vec{R} + b \int_C \vec{G} \cdot d\vec{R} \ .$$

This is the *linearity law* for vector line integrals.

(II) *Let* $\vec{F}$ *be a vector field whose domain contains the finite, smooth directed curve* C. *Let* C *be divided into two directed curves* C_1 *and* C_2, *where the directions for* C_1 *and* C_2 *are the same as for* C. *Then* $\int_C \vec{F} \cdot d\vec{R} = \int_{C_1} \vec{F} \cdot d\vec{R} + \int_{C_2} \vec{F} \cdot d\vec{R}$. This is the *additivity law* for vector line integrals.

(III) *Let* $\vec{F}$ *be a vector field whose domain contains a finite, smooth directed curve* C, *and let* –C *denote the same curve with its direction reversed. Then*

$$\int_{-C} \vec{F} \cdot d\vec{R} = -\int_C \vec{F} \cdot d\vec{R} \ .$$

This is the *direction-reversal law* for vector line integrals.

Our definition of vector line integral can be enlarged as follows.

(4.3) *Definition.* Let C be a finite, piecewise smooth, directed curve and let $C_1, C_2, \ldots, C_n$ be successive smooth, directed pieces which make up C. Let $\vec{F}$ be a vector field whose domain includes C. Then, provided that each of $\int_{C_1} \vec{F} \cdot d\vec{R}, \ldots, \int_{C_n} \vec{F} \cdot d\vec{R}$ exists, we define

$$\int_C \vec{F} \cdot d\vec{R} = \int_{C_1} \vec{F} \cdot d\vec{R} + \int_{C_2} \vec{F} \cdot d\vec{R} + \cdots + \int_{C_n} \vec{F} \cdot d\vec{R} \ .$$

(It is easy to show that this extended definition satisfies (I), (II), and (III), and that the value of $\int_C \vec{F} \cdot d\vec{R}$ does not depend on the way in which C is divided into smooth pieces.)

Under what conditions for $\vec{F}$ and C can we be sure that $\int_C \vec{F} \cdot d\vec{R}$ exists? The answer is given by the following

(4.4) *Existence theorem. If* C *is a finite, piecewise smooth, directed curve and* $\vec{F}$ *is a continuous vector field whose domain contains* C, *then* $\int_C \vec{F} \cdot d\vec{R}$ *exists.*

(4.4) is readily proved by the methods of §14.10. (4.4) can be extended to the case where $\vec{F}$ is piecewise continuous. The technical details are similar to those in (4.4) of §14.4.

Remark. (4.4) and the extended version of (4.4) are used in physical applications when we have, in a mathematical model, a path which has a sharp change in direction at some point or a vector field which has a discontinuity along some surface.

§19.5   Evaluating Vector Line Integrals

We describe two ways to evaluate a vector line integral: (i) *evaluation by definition*, and (ii) *parametric evaluation.*

In *evaluation by definition,* we consider specific Riemann sums and try to recognize directly what the value of the integral must be.

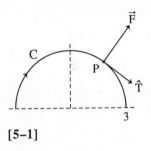

[5–1]

Example 1. In E^2 *with Cartesian coordinates, let* C *be the upper semi-circle of radius 3 with center at the origin, directed from left to right, and let* $\vec{F} = x\hat{i} + y\hat{j}$. *At every point* P *on* C, $\vec{F}(P)$ *is perpendicular to* C. *(Recall that* $\vec{F} = r\hat{r}$ *in polar coordinates.) See [5–1].* Hence, in a Riemann sum for $\int_C (\vec{F} \cdot \hat{T}) ds$, we must have $\vec{F}(P_i^*) \cdot \hat{T}(P_i^*) = 0$ for all i ≤ n. Thus every Riemann sum is 0, and $\int_C (\vec{F} \cdot \hat{T}) ds = \int_C \vec{F} \cdot d\vec{R} = 0$. (This result holds in general: if, at every point P on a smooth directed curve C, $\vec{F}(P)$ is perpendicular to C, then $\int_C \vec{F} \cdot d\vec{R} = 0$.)

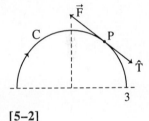

[5–2]

Example 2. In E^2 *with Cartesian coordinates, let* C *be as in example* 1, *and let* $\vec{F} = -y\hat{i} + x\hat{j}$. *In polar coordinates,* $\vec{F} = r\hat{\theta}$ *(see §18.7).* Hence, at every point P on C, $\vec{F}(P) = 3\hat{\theta}$ has the opposite direction to the unit tangent vector $\hat{T}(P) = -\hat{\theta}$. (See [5–2].) Thus at every P on C, $\vec{F}(P) \cdot \hat{T}(P) = -3$. Hence every Riemann sum = –3(arc length of C) = –9π, and $\int_C \vec{F} \cdot d\vec{R} = \int_C (\vec{F} \cdot \hat{T}) ds = -9\pi$.

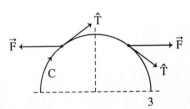

[5–3]

Example 3. Let C *be as in example 2, and let* $\vec{F} = x\hat{i}$. *(See [5–3].)* We can use cancellation, noting that for every point P = (c,d) on the right-hand side of C, with $\vec{F}(P) = c\hat{i}$ and $\hat{T}(P) = a\hat{i} - b\hat{j}$, there is a mirror image point P′ = (–c,d) on the left-hand side of C, with $\vec{F}(P') = -c\hat{i}$ and $\hat{T}(P') = a\hat{i} + b\hat{j}$. (See [5–3].) If we form a Riemann sum by choosing pieces of C on the left to be mirror images of pieces of C on the right and sample points on the left to be mirror images of sample points on the right, we then find, for any pair of mirror image terms with sample points P on the right and P′ on the left, that $\vec{F}(P) \cdot \hat{T}(P) = ac$ and $\vec{F}(P') \cdot \hat{T}(P') = -ac$. Thus the paired terms in the Riemann sum cancel, and the entire Riemann sum must be 0. Therefore $\int_C \vec{F} \cdot d\vec{R} = 0$.

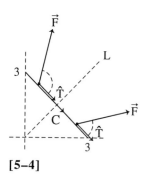

[5–4]

Example 4. *In* $\mathbf{E}^2$ *with Cartesian coordinates, let* C *be a straight segment directed from* (0,3) *to* (3,0). *Let* $\vec{F} = x\hat{i} + y\hat{j}$. We can arrange Riemann sums for $\int_C \vec{F} \cdot d\vec{R}$ so that each term can be paired with a mirror-image with respect to the line L in [5–4], so that these paired terms cancel. We thus have $\int_C \vec{F} \cdot d\vec{R} = 0$.

In *parametric evaluation* for a vector line integral, we use a parametric path, in some coordinate system, for the given directed curve C, we assume that C is smooth, and we express the given vector field $\vec{F}$ in that coordinate system. Let t be the parameter, with $a \le t \le b$. (Note that in some cases the direction of the directed curve C may be opposite to the direction given by the parameter t.) The notation $\int_C \vec{F} \cdot d\vec{R}$ suggests the following evaluation rule:

(5.1) $$\int_C \vec{F} \cdot d\vec{R} = \pm \int_a^b (\vec{F} \cdot \frac{d\vec{R}}{dt}) dt .$$

Here + is used when the direction of C agrees with the direction of t, and − is used when its direction is opposite to the direction of t.

For Cartesian coordinates in $\mathbf{E}^3$, we have

$$\vec{R}(t) = x(t)\hat{i} + y(t)\hat{j} + z(t)\hat{k}, \quad a \le t \le b,$$
$$\frac{d\vec{R}}{dt} = \dot{x}(t)\hat{i} + \dot{y}(t)\hat{j} + \dot{z}(t)\hat{k} ,$$

and $\vec{F}(P) = \vec{F}(x, y, z) = L(x, y, z)\hat{i} + M(x, y, z)\hat{j} + N(x, y, z)\hat{k}$ for $P = (x, y, z)$. This gives

(5.2) $$\int_C \vec{F} \cdot d\vec{R} = \pm \int_a^b (\vec{F} \cdot \frac{d\vec{R}}{dt}) dt$$
$$= \pm \int_a^b \left(L(x(t), y(t), z(t))\dot{x}(t) + M(x(t), y(t), z(t))\dot{y}(t) + N(x(t), y(t), z(t))\dot{z}(t) \right) dt .$$

The general formula (5.2) is lengthy, but the procedure is not difficult to carry out in practice:

(a) We find parametric equations for C: $\vec{R}(t) = x(t)\hat{i} + y(t)\hat{j} + z(t)\hat{k}$, $a \leq t \leq b$,

(b) We find $\dfrac{d\vec{R}}{dt}$ as a function of t.

(c) We express $\vec{F}$ as a function of t.

(d) We take the dot product of (c) and (b).

(e) We integrate this product with respect to t, from a to b.

(f) We multiply by -1 if the parameter direction is opposite to the curve direction.

Example 5. *We calculate the result in example 2 by parametric evaluation.* Recall that $\vec{F} = -y\hat{i} + x\hat{j}$ (in E^2) and that C is the upper semicircle of radius 3 with center at the origin, directed from left to right. (See [5–5].) We take the path:

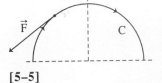

[5–5]

(a) $\vec{R}(t) = 3\cos t\,\hat{i} + 3\sin t\,\hat{j}$, $0 \leq t \leq \pi$, noticing that curve direction and parameter direction are opposite. We then have

(b) $\dfrac{d\vec{R}}{dt} = -3\sin t\,\hat{i} + 3\cos t\,\hat{j}$.

(c) $\vec{F} = -3\sin t\,\hat{i} + 3\cos t\,\hat{j}$.

(d) $\vec{F} \cdot \dfrac{d\vec{R}}{dt} = 9\sin^2 t + 9\cos^2 t = 9$.

(e) $\displaystyle\int_0^\pi 9\,dt = 9\pi$.

(f) Multiplying by -1, we get $\displaystyle\int_C \vec{F} \cdot d\vec{R} = -9\pi$.

Example 6. *Consider the directed curve* $C = C_1 + C_2$ *shown in* [5–6]. *Let* $\vec{F} = -y\hat{i} + x\hat{j}$. *We evaluate* $\displaystyle\int_{C_1} \vec{F} \cdot d\vec{R}$ *and* $\displaystyle\int_{C_2} \vec{F} \cdot d\vec{R}$ *separately in order to*

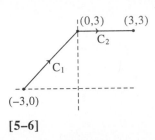

[5–6]

find $\displaystyle\int_C \vec{F} \cdot d\vec{R} = \int_{C_1} \vec{F} \cdot d\vec{R} + \int_{C_2} \vec{F} \cdot d\vec{R}$.

For C_1:

(a) $\vec{R}(t) = -3\hat{i} + t(\hat{i} + \hat{j}) = (t - 3)\hat{i} + t\hat{j}$, $0 \leq t \leq 3$; (here parameter direction agrees with curve direction).

(b) $\dfrac{d\vec{R}}{dt} = \hat{i} + \hat{j}$.

(c) $\vec{F} = -t\hat{i} + (t-3)\hat{j}$.

(d) $\vec{F} \cdot \dfrac{d\vec{R}}{dt} = -t + t - 3 = -3$

(e) $\displaystyle\int_0^3 -3\,dt = -9$. This is $\displaystyle\int_{C_1} \vec{F} \cdot d\vec{R}$.

For C_2:

(a) $\vec{R}(t) = 3\hat{j} + t\hat{i} = t\hat{i} + 3\hat{j}$, $0 \le t \le 3$; (parameter and curve directions agree).

(b) $\dfrac{d\vec{R}}{dt} = \hat{i}$.

(c) $\vec{F} = -3\hat{i} + t\hat{j}$.

(d) $\vec{F} \cdot \dfrac{d\vec{R}}{dt} = -3$.

(e) $\displaystyle\int_0^3 -3\,dt = -9$. This is $\displaystyle\int_{C_2} \vec{F} \cdot d\vec{R}$.

Hence $\displaystyle\int_C \vec{F} \cdot d\vec{R} = \int_{C_1} \vec{F} \cdot d\vec{R} + \int_{C_2} \vec{F} \cdot d\vec{R} = -9 - 9 = -18$.

Justifying parametric evaluation.   We must show that $\displaystyle\int_C \vec{F} \cdot d\vec{R} = \int_a^b \left(\vec{F} \cdot \dfrac{d\vec{R}}{dt}\right)dt$. This is proved in the same way as for scalar line integrals by constructing comparable Riemann sums and using Duhamel's principle. See the end of §19.3.

Remark.   An alternative notation for vector line integrals, based in Cartesian coordinates, is sometimes used. For E^3, $\vec{F}$ is expressed $L\hat{i} + M\hat{j} + N\hat{k}$, and $d\vec{R}$ is expressed as $dx\,\hat{i} + dy\,\hat{j} + dz\,\hat{k}$. $\displaystyle\int_C \vec{F} \cdot d\vec{R}$ then appears as:

$$\int_C (L\,dx + M\,dy + N\,dz),$$

and parametric evaluation is indicated by

$$\int_a^b \left(L\frac{dx}{dt} + M\frac{dy}{dt} + N\frac{dz}{dt}\right)dt .$$

Similarly, for E^2 with $\vec{F} = L\hat{i} + M\hat{j}$, we have $\int_C (Ldx + Mdy)$ and

$\int_a^b \left(L\frac{dx}{dt} + M\frac{dy}{dt} \right) dt$.

   Remark. In Chapter 20, we state and prove the *conservative field theorem*. The conservative field theorem sometimes gives a shorter and simpler way to evaluate a vector line integral than either of the methods described above. In particular, it simplifies the evaluation of examples 1, 3, and 4. In Chapter 26, we state and prove *Stokes's theorem*. Stokes's theorem sometimes gives an even shorter and simpler way to evaluate vector line integrals. In particular, it simplifies the evaluation of examples 2, 5, and 6.

§19.6 Other Integrals on Curves

   Other kinds of integrals on curves can be expressed by the use of appropriate vector-algebra notations. We give some examples below and in the problems. In each case we find that the notation itself suggests: (i) an appropriate formal definition in terms of Riemann sums, and (ii) an appropriate and correct method of parametric evaluation. In some cases, these new integrals have vector rather than scalar values.

   Example 1. *For a given finite, rectifiable curve C and for a vector field* $\vec{F}$ *whose domain contains C, we consider the new integral*

(6.1)
$$\int_C \vec{F} ds .$$

This integral is vector-valued and, as the notation suggests, is defined by

$$\lim_{\substack{n \to \infty \\ \max \Delta s_i \to 0}} \sum_{i=1}^{n} \vec{F}(P_i^*) \Delta s_i ,$$

where the P_i^* are sample points on C and the Δs_i are arc lengths of pieces of C. If $\vec{F}$ is continuous and C is smooth, this new integral can be parametrically evaluated by

(6.2)
$$\int_C \vec{F} ds = \int_a^b (\vec{F}\frac{ds}{dt})dt .$$

(In Cartesian coordinates, we have $\vec{F} = L\hat{i} + M\hat{j} + N\hat{k}$ and get $\int_C \vec{F} ds =$ $\left(\int_C L ds\right)\hat{i} + \left(\int_C M ds\right)\hat{j} + \left(\int_C N ds\right)\hat{k}$, where the scalar coefficients in the last three terms are scalar line integrals.)

If we use the position vector $\vec{R} = x\hat{i} + y\hat{j} + z\hat{k}$ (with respect to the origin) as the vector field $\vec{F}$, this new integral can be used to define and calculate the *centroid of a given curve* C. By the same reasoning as was used in Chapter 14 for the centroid of a region in E^3 or E^2, we define

$$(6.3) \qquad\qquad \vec{R}_0 = \int_C \vec{R} ds \bigg/ \int_C ds$$

to be (the position-vector for) the *centroid* of C. (For a thin wire in the shape of C and of constant density per unit arc length, this centroid will be the center of mass.)

Example 2. *For a given finite, smooth, directed curve* C *and a given vector field* $\vec{F}$ *whose domain contains* C, *we consider the new integral*

$$(6.4) \qquad\qquad \int_C (\vec{F} \times \hat{T}) ds,$$

where this integral may also be written as $\int_C \vec{F} \times d\vec{R}$. *This integral is vector-valued and, as the notation suggests, is defined by*

$$\lim_{\substack{n \to \infty \\ \max \Delta s_i \to 0}} \sum_{i=1}^{n} \left[\vec{F}(P_i^*) \times \hat{T}(P_i^*)\right] \Delta s_i.$$

If $\vec{F}$ is continuous, the integral can be parametrically evaluated by

$$(6.5) \qquad\qquad \int_C \vec{F} \times d\vec{R} = \pm \int_a^b \left(\vec{F} \times \frac{d\vec{R}}{dt}\right) dt.$$

This integral is used in the general theory of vector fields.

§19.7 Examples and Applications

Our previous illustrative examples of line integrals have been in $\mathbf{E}^2$. In our first example, we evaluate two vector line integrals in $\mathbf{E}^3$.

Example 1. *Let $\vec{F}$ be the vector field on $\mathbf{E}^3$ defined in Cartesian coordinates by*

$$\vec{F}(x,y,z) = (x+y)\hat{i} + (y-x)\hat{j} + z\hat{k}.$$

We consider two directed curves, C and C'. C consists two straight segments: C_1 from $(0,0,0)$ to $(1,0,0)$, and C_2 from $(1,0,0)$ to $(1,1,1)$. C' consists of a single straight segment from $(0,0,0)$ to $(1,1,1)$. The three segments form the sides of a triangle. (See [7-1].) We wish to calculate and compare $\int_C \vec{F} \cdot d\vec{R}$ and $\int_{C'} \vec{F} \cdot d\vec{R}$. To do this, we take each segment in turn and evaluate the integral for that segment parametrically.

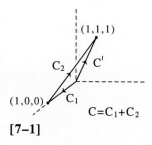

[7-1]

For C_1:
$$\vec{R}(t) = t\hat{i}, \quad 0 \le t \le 1.$$
$$\frac{d\vec{R}}{dt} = \hat{i}.$$
$$\vec{F} = t\hat{i} - t\hat{j} \quad \text{(putting } x = t, y = 0, z = 0 \text{ in } \vec{F}\text{)}$$
$$\vec{F} \cdot \frac{d\vec{R}}{dt} = t, \text{ and } \int_{C_1} \vec{F} \cdot d\vec{R} = \int_0^1 t\,dt = \frac{1}{2}.$$

For C_2:
$$\vec{R}(t) = \hat{i} + t\hat{j} + t\hat{k}, \quad 0 \le t \le 1.$$
$$\frac{d\vec{R}}{dt} = \hat{j} + \hat{k}.$$
$$\vec{F} = (1+t)\hat{i} + (t-1)\hat{j} + t\hat{k}.$$
$$\vec{F} \cdot \frac{d\vec{R}}{dt} = t - 1 + t = 2t - 1, \text{ and } \int_{C_2} \vec{F} \cdot d\vec{R} = \int_0^1 (2t-1)dt = 0.$$

For C':
$$\vec{R}(t) = t\hat{i} + t\hat{j} + t\hat{k}, \quad 0 \le t \le 1.$$
$$\frac{d\vec{R}}{dt} = \hat{i} + \hat{j} + \hat{k}.$$
$$\vec{F} = 2t\hat{i} + t\hat{k}.$$

$$\vec{F} \cdot \frac{d\vec{R}}{dt} = 2t + t = 3t, \text{ and } \int_{C'} \vec{F} \cdot d\vec{R} = \int_0^1 3t\,dt = \frac{3}{2}.$$

Thus $\int_C \vec{F} \cdot d\vec{R} = 0 + \frac{1}{2} = \frac{1}{2}$, while $\int_{C'} \vec{F} \cdot d\vec{R} = \frac{3}{2}$. (The fact that C and C' have the same initial point and the same final point but $\int_C \vec{F} \cdot d\vec{R} \neq \int_{C'} \vec{F} \cdot d\vec{R}$ shows that the vector field $\vec{F}$ has the property of *path dependence* (defined in §20.1.)

Example 2. *We illustrate the definition of centroid for a curve (in §19.6) by finding the centroid of a semicircle of radius* 1. *We choose Cartesian axes so that the semicircle lies in the upper half-plane of the* xy *plane, and has its center at the origin. Then*

$$\vec{R}_0 = \int_C \vec{R}\,ds \Big/ \int_C ds = \frac{1}{\pi}\left(\int_C x\,ds\,\hat{i} + \int_C y\,ds\,\hat{j} \right).$$

Using cancellation, we see that $\int_C x\,ds = 0$. For $\int_C y\,ds$, we have

$$\vec{R}(t) = \cos t\,\hat{i} + \sin t\,\hat{j}, \quad 0 \leq t \leq \pi,$$
$$\frac{ds}{dt} = (\sin^2 t + \cos^2 t)^{1/2} = 1, \text{ and}$$
$$y = \sin t.$$

Then $\int_C y\,ds = \int_0^\pi y\frac{ds}{dt}dt = \int_0^\pi \sin t\,dt = 2$, while $\int_C ds =$ arc length of $C = \pi$. Hence the centroid is at $(0,\frac{2}{\pi}) = (0,0.637)$.

Example 3. *Note that our basic notations and expressions for line integrals are coordinate-free and can therefore be directly used with non-Cartesian coordinate systems. Consider, the vector line integral* $\int_C \vec{F} \cdot d\vec{R}$ *in example 2 in* §19.5. (Recall that C was the upper semicircle of radius 3, center at 0, and that $\vec{F}$ was $-y\hat{i} + x\hat{j}$.) In *polar coordinates* we have:

Parametric path: $r(t) = 3$, $\theta(t) = \pi - t$, $0 \leq t \leq \pi$, where curve and parameter directions agree. (See [7–2].)

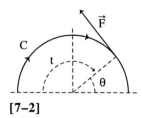

[7–2]

$$\frac{d\vec{R}}{dt} = \dot{r}\hat{r} + r\dot{\theta}\hat{\theta} = -3\hat{\theta} \quad \text{(see §7.4).}$$

$$\vec{F} = r\hat{\theta} = 3\hat{\theta}$$

$$\vec{F} \cdot \frac{d\vec{R}}{dt} = -9, \text{ and } \int_C \vec{F} \cdot d\vec{R} = \int_0^\pi -9dt = -9\pi .$$

Example 4. Let $\vec{F} = f(r)\hat{r}$ be a uniform radial field in cylindrical coordinates (as described in §18.6). How much net work does the field perform on a particle which follows a curve C from P to Q, where $P = (r_1, \theta_1, z_1)$ and $Q = (r_2, \theta_2, z_2)$? Assume that C is given in terms of $r(t)$, $\theta(t)$, and $z(t)$ for some parameter t, $a \le t \le b$.

Then $$\frac{d\vec{R}}{dt} = \dot{r}\hat{r} + r\dot{\theta}\hat{\theta} \quad \text{by §7.4,}$$

And $$\vec{F} \cdot \frac{d\vec{R}}{dt} = f(r)\hat{r} \cdot (\dot{r}\hat{r} + r\dot{\theta}\hat{\theta}) = f(r)\dot{r} = f(r)\frac{dr}{dt} .$$

Hence $\int_C \vec{F} \cdot d\vec{R} = \int_a^b f(r)\frac{dr}{dt} dt = \int_{r_1}^{r_2} f(r)dr$, by the change-of-variable theorem for definite integrals in elementary calculus. We see that the net work done by $\vec{F}$ depends only on the function $f(r)$ and on the values of the coordinate r at P and Q. It does not depend on the route followed by C to get from P to Q.

In the problems, we see that the same result holds for a *uniform central field* in spherical coordinates: if $\vec{F} = h(\rho)\hat{\rho}$, then $\int_C \vec{F} \cdot d\vec{R} = \int_{\rho_1}^{\rho_2} h(\rho)d\rho$, where ρ_1 and ρ_2 are the values of the coordinate ρ at P and Q.

§19.8 Problems

§19.3

1. Let $f(x,y,z) = xy(z+1)$. Find $\int_C f \, ds$:

    (a) when C is $x = \cos t, y = \sin t, z = 0,$
         $0 \le t \le 2\pi;$

    (b) when C is $x = \cos t, y = \sin t, z = t,$
         $0 \le t \le 2\pi.$

2. Find the average value of the function $f(x,y) = x^2 + y^2$ on the line segment from $(0,1)$ to $(1,0)$.

3. Find the average value of the function $f(x,y) = x^2$ on the circle $x^2 + y^2 = a^2$ in the xy-plane.

4. Explain and justify the final sentence in §19.3.

§19.5

1. Find the work done by the force field $\vec{F} = xy\hat{i} + z\hat{j} - yz\hat{k}$ in moving a test particle along the path $x = t, y = t^2, z = 2 - t, 0 \le t \le 1.$

2. Let $\vec{F}(x,y,z) = (y+z)\hat{i} + (x+z)\hat{j} + (x+y)\hat{k}$. Find $\int_C \vec{F} \cdot d\vec{R}$:

    (a) when C is as in problem 1a for §19.3;

    (b) when C is as in problem 1b for §19.3.

3. A particle moves along the helical path: $x = \cos t$, $y = \sin t, z = t, 0 \le t \le 2\pi.$ It moves in the force

field $\vec{F} = -y\hat{i} + x\hat{j} + z\hat{k}$. Find the work done on the particle by the field.

4. C is the perimeter of the triangle with vertices $(0,0,0)$, $(1,0,0)$, $(2,1,2)$. C is directed: $(0,0,0) \rightarrow (1,0,0) \rightarrow (2,1,2) \rightarrow (0,0,0)$. $\vec{F}(x,y,z) = x^2\hat{i} + xy\hat{j} + z\hat{k}$. Evaluate $\int_C \vec{F} \cdot d\vec{R}$.

5. Let $\vec{F} = \frac{1}{r}\hat{\theta}$ be a vector field on $\mathbf{E}^2$ minus the origin. Let C be the circle with radius a, center at $(b,0)$, counter-clockwise direction, and initial and final point at $(b+a,0)$. Show, by directly evaluating the integral, that

    (i)   if $b > a > 0$, then $\int_C \vec{F} \cdot d\vec{R} = 0$; and

    (ii) if $a > b > 0$, then $\int_C \vec{F} \cdot d\vec{R} = 2\pi$.

    (*Hint*: You may use the fact, from a table of integrals, that $\int_0^\pi \frac{dx}{A+B\cos x} = \frac{\pi}{\sqrt{A^2-B^2}}$, if $A > B > 0$.)

§19.6

1. Let $\vec{F} = r\hat{r}$ on all of $\mathbf{E}^2$. Let C be a circle of radius a in the plane $z = 0$, with center at the origin. Evaluate $\int_C \vec{F}ds$.

2. Let $\vec{F}$ and C be as in problem 1 with C directed counter clockwise. Evaluate $\int_C \vec{F} \times d\vec{R}$.

§19.7

1. Assume that a curve C is given in terms of $\rho(t), \varphi(t), \theta(t)$ for some parameter t, $a \le t \le b$. Find the components of $\frac{d\vec{R}}{dt}$ in the spherical coordinate frame. (See §7.5.)

2. A uniform central field is given, in spherical coordinates, by $\vec{F} = h(\rho)\hat{\rho}$ for some one-variable function h. Prove that for any directed, piecewise smooth curve C from $P = (\rho_1, \varphi_1, \theta_1)$ to $Q = (\rho_2, \varphi_2, \theta_2)$ in the domain of $\vec{F}$,

    $$\int_C \vec{F} \cdot d\vec{R} = \int_{\rho_1}^{\rho_2} h(\rho)d\rho.$$

Chapter 20

Conservative Fields

§20.1 Independence of Path; Conservative Fields

In this section, we present and illustrate several fundamental definitions.

Definition. Let $\vec{F}$ be a continuous vector field on domain D. We say that $\vec{F}$ has *independence of path on* D if the following statement is true: ' for *every pair* of piecewise smooth, directed curves C_1 and C_2 (in D) with a common initial point and a common final point (see [1–1]), we have $\int_{C_1} \vec{F} \cdot d\vec{R} = \int_{C_2} \vec{F} \cdot d\vec{R}$.

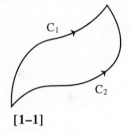

[1–1]

    Example 1. Let $\vec{F} = 4\hat{i} + 7\hat{j}$ *on* E^2. *Let* $Q = (x_1, y_1)$ *and* $P = (x_2, y_2)$. Let C be any smooth curve from Q to P. Let $\vec{R}(t) = x(t)\hat{i} + y(t)\hat{j}$, $a \le t \le b$, where $\vec{R}(a) = x_1\hat{i} + y_1\hat{j}$ and $\vec{R}(b) = x_2\hat{i} + y_2\hat{j}$, be a smooth path for C. Then, by parametric evaluation,

$$\int_C \vec{F} \cdot d\vec{R} = \int_a^b \left(4\frac{dx}{dt} + 7\frac{dy}{dt} \right) dt = \int_a^b 4\frac{dx}{dt} dt + \int_a^b 7\frac{dy}{dt} dt =$$

$$\int_{x_1}^{x_2} 4dx + \int_{y_1}^{y_2} 7dy = 4(x_2 - x_1) + 7(y_2 - y_1).$$

We note that this last expression continues to be true if C is a piecewise smooth curve from Q to P, since, for example, if $C = C_1 + C_2$, where C_1 and C_2 are smooth curves joined at (x′,y′), then $\int_{C_1} \vec{F} \cdot d\vec{R} = 4(x' - x_1) + 7(y' - y_1)$, $\int_{C_2} \vec{F} \cdot d\vec{R} =$

$4(x_2 - x') + 7(y_2 - y')$, and $\int_C \vec{F} \cdot d\vec{R} = \int_{C_1} \vec{F} \cdot d\vec{R} + \int_{C_2} \vec{F} \cdot d\vec{R} = 4(x_2 - x_1) +$ $7(y_2 - y_1)$, as before. Thus, for any given piecewise smooth C, the value of $\int_C \vec{F} \cdot d\vec{R}$ depends only on the coordinates of P and Q. Hence $\vec{F}$ has independence of path.

Definition. Let $\vec{F}$ be a continuous vector field on domain D. If $\vec{F}$ fails to have independence of path on D (that is to say, if there exist two piecewise smooth, directed curves C_1 and C_2 in D such that C_1 and C_2 have a common initial point and a common final point, but $\int_{C_1} \vec{F} \cdot d\vec{R} \neq \int_{C_2} \vec{F} \cdot d\vec{R}$), then we say that $\vec{F}$ on D has *path dependence*.

Example 2. *In example 1 of §19.7, for* $\vec{F} = (x + y)\hat{i} + (y - x)\hat{j} + z\hat{k}$ *on* E^3, *we considered two curves,* C *and* C', *from* (0,0,0) *to* (1,1,1) *and showed that* $\int_C \vec{F} \cdot d\vec{R} = \frac{1}{2}$ *but* $\int_{C'} \vec{F} \cdot d\vec{R} = \frac{3}{2}$. This tells us that $\vec{F}$ on E^3 has path dependence.

Note that path dependence of a vector field $\vec{F}$ does not imply that we *always* have $\int_{C_1} \vec{F} \cdot d\vec{R} \neq \int_{C_2} \vec{F} \cdot d\vec{R}$ when C_1 and C_2 are distinct curves with a common initial point and a common final point. It only implies that there is *some* pair of curves C_1 and C_2, with a common initial point and a common final point, such that $\int_{C_1} \vec{F} \cdot d\vec{R} \neq \int_{C_2} \vec{F} \cdot d\vec{R}$. (In the problems, however, we shall see that if $\vec{F}$ is path-dependent, then for *every* pair of points P and Q in D and for any given curve C_1 from Q to P, there must be some other curve C_2 from Q to P such that $\int_{C_1} \vec{F} \cdot d\vec{R} \neq \int_{C_2} \vec{F} \cdot d\vec{R}$.)

In example 4 of §19.7, we saw that for any smooth C in a uniform radial field $\vec{F} = f(r)\hat{r}$, the value of $\int_C \vec{F} \cdot d\vec{R} = \int_{r_1}^{r_2} f(r)dr$, where r_1 and r_2 are the values of the r coordinate at the initial and final points of C. By the same reasoning as in example 1, this result can be extended to any piecewise smooth curve with these initial and final points. Hence

(1.1) *Every uniform radial field has independence of path. The same is true, by a similar argument, for every uniform central field* $\vec{F} = h(\rho)\hat{\rho}$.

Recall that a piecewise smooth, directed curve is said to be *closed* if its initial and final points are the same. Let $\vec{F}$ be a vector field on D and let C be a piecewise smooth, directed, closed curve in D. (See [1–2].) In this case, we sometimes write $\int_C \vec{F} \cdot d\vec{R}$ as $\oint_C \vec{F} \cdot d\vec{R}$, where the circular mark on the integral sign serves as a reminder that C is closed.

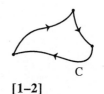

[1–2]

Definition. Let $\vec{F}$ be a vector field on D. We say that $\vec{F}$ is a *conservative field* if it is true that for every piecewise smooth, directed, closed curve C in D, $\oint_C \vec{F} \cdot d\vec{R} = 0$.

Example 3. *Let* $\vec{F} = 4\hat{i} + 7\hat{j}$ *on* $\mathbf{E}^2$ *as in example* 1. Let C be any piecewise smooth, directed, closed curve in $\mathbf{E}^2$. Applying the final expression obtained in example 1, we have $x_2 = x_1$, $y_2 = y_1$, and $\oint_C \vec{F} \cdot d\vec{R} = c(x_1 - x_1) + d(y_1 - y_1) = 0$. Thus $\vec{F}$ is a conservative field.

Example 4. *Similarly, from* (1.1), *we see that every uniform radial field must be conservative, and that every uniform central field must be conservative.*

In the definition of *conservative field*, the word "conservative" is chosen for the following reason. Let $\vec{F}$ be a physical force field for a moveable particle *p*. Assume that $\vec{F}$ is not a conservative field. Then there must be some closed, directed, piecewise smooth curve C such that $\oint_C \vec{F} \cdot d\vec{R} \neq 0$. Hence there must be some closed, directed curve C′ such that $\oint_{C'} \vec{F} \cdot d\vec{R} > 0$. (If $\oint_C \vec{F} \cdot d\vec{R} > 0$, take C′ = C; if $\oint_C \vec{F} \cdot d\vec{R} < 0$, take C′ = −C.) If the particle *p* is moved around the "loop" formed by C′, and in the direction of C′, then *p* returns to its initial position after a net positive amount of work is done on *p* by $\vec{F}$. If this cycle is repeated indefinitely, an arbitrarily large amount of work can be drawn from the field. In the absence of other changes in the physical universe, this cyclical process would violate the physical law of *conservation of energy*. A *conservative* force field, however, cannot be used in this way to violate the conservation-of-energy law.

It is easy to show that the property of being conservative is equivalent to the property of having independence of path.

(1.2) *Let $\vec{F}$ be a vector field on domain D. Then $\vec{F}$ is conservative $\Leftrightarrow$ $\vec{F}$ has independence of path.*

Proof of (1.2). We first prove $\Rightarrow$. Assume $\vec{F}$ is conservative. Let C_1 and C_2 be any two directed, piecewise smooth curves in D from a common initial point Q to a common final point P. Then $C_1+(-C_2)$ is a closed, directed, piecewise smooth curve. Since $\vec{F}$ is conservative, $\oint_{C_1+(-C_2)} \vec{F} \cdot d\vec{R} = 0 = \int_{C_1} \vec{F} \cdot d\vec{R} +$

$\int_{-C_2} \vec{F} \cdot d\vec{R} = \int_{C_1} \vec{F} \cdot d\vec{R} - \int_{C_2} \vec{F} \cdot d\vec{R}$. Thus $\int_{C_1} \vec{F} \cdot d\vec{R} = \int_{C_2} \vec{F} \cdot d\vec{R}$, and we have

shown that $\vec{F}$ has independence of path.

We next prove $\Leftarrow$. Assume that $\vec{F}$ has independence of path. Let C be any directed, piecewise smooth, closed curve in D. Let Q be the initial and final point of C. Let P be any other point on C. Then $C = C_1+C_2$, where C_1 goes from Q to P and C_2 goes from P to Q. Then C_1 and $-C_2$ both go from Q to P. By assumption, $\int_{C_1} \vec{F} \cdot d\vec{R} = \int_{-C_2} \vec{F} \cdot d\vec{R}$. Hence $\int_{C_1} \vec{F} \cdot d\vec{R} = -\int_{C_2} \vec{F} \cdot d\vec{R}$ (by (III) in §19.4),

and $\int_{C_1} \vec{F} \cdot d\vec{R} + \int_{C_2} \vec{F} \cdot d\vec{R} = 0$. Then $\int_{C_1+C_2} \vec{F} \cdot d\vec{R} = 0$ (by (II) in §19.4). Thus

$\oint_C \vec{F} \cdot d\vec{R} = 0$, and we have shown that $\vec{F}$ is conservative.

Example 5. *In E^2 with polar coordinates, let $\vec{F} = \frac{1}{r}\hat{\theta}$ on E^2-minus-the-origin. We show that $\vec{F}$ has path dependence.* Let C be a circle of radius a, with center at O, directed counter clockwise. See [1–3]. Then C is a smooth, directed, closed curve. For any point P on C, $\vec{F}(P) = \frac{1}{a}\hat{\theta}_P = \frac{1}{a}\hat{T}_P$. Forming a Riemann sum

$\oint_C \vec{F} \cdot d\vec{R}$, we get $\sum_i \vec{F}(P_i^*) \cdot \hat{T}(P_i^*)\Delta s_i = \sum_i \frac{1}{a}\hat{T}(P_i^*) \cdot \hat{T}(P_i^*)\Delta s_i = \sum_i \frac{1}{a}\Delta s_i =$

$\frac{1}{a}2\pi a = 2\pi \neq 0$. Thus $\oint_C \vec{F} \cdot d\vec{R} = 2\pi$ and $\vec{F}$ is not a conservative field. By (1.2),

$\vec{F}$ has path dependence.

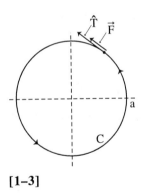

[1–3]

Notation. Let $\vec{F}$ be a conservative field. Let C be a piecewise smooth curve from Q to P. For such an $\vec{F}$ and C, we shall sometimes write $\int_C \vec{F} \cdot d\vec{R}$ as

$\int_Q^P \vec{F} \cdot d\vec{R}$, since the value of $\int_C \vec{F} \cdot d\vec{R}$ depends only on the points Q and P

(because $\vec{F}$ has independence of path).

§20.2 The Conservative-field Theorem

We now prove a central theorem of multivariable calculus. The theorem is of interest not only because it asserts a remarkable fact, but also because its proof will have a variety of applications.

(2.1) *Conservative-field theorem. Let $\vec{F}$ be a continuous vector field on a domain* D. *Then*

$$\vec{F} \text{ is conservative} \Leftrightarrow \vec{F} \text{ is a gradient field.}$$

***Proof of* (2.1).** We prove the theorem for the case of $\vec{F}$ in E^3. The proof for E^2 is similar.

We first prove $\Leftarrow$. Assume $\vec{F}$ is a gradient field. Then $\vec{F} = \vec{\nabla} f$ for some scalar field f on D. (See [2–1].) We show that $\vec{F}$ has independence of path. Let C be a smooth directed curve in D with initial point Q and final point P. Let $\vec{R}(t) = x(t)\hat{i} + y(t)\hat{j} + z(t)\hat{k}$, $a \le t \le b$, be a path for C in Cartesian coordi-

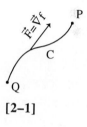

[2–1]

nates. Then $\int_C \vec{F} \cdot d\vec{R} = \int_C \vec{\nabla} f \cdot d\vec{R} = \int_a^b \vec{\nabla} f \cdot \dfrac{d\vec{R}}{dt} dt = \int_a^b (f_x \dfrac{dx}{dt} + f_y \dfrac{dy}{dt} + f_z \dfrac{dz}{dt})dt =$

$\int_a^b \dfrac{d}{dt} f(\vec{R}(t))dt$ (by the chain rule) $= f(\vec{R}(b)) - f(\vec{R}(a)) = f(P) - f(Q)$. By the same argument as was used at the end of example 1 in §20.1, this result can be extended from a smooth C to a piecewise smooth C. Since the value of $\int_C \vec{F} \cdot d\vec{R}$

depends only on the value of f at the initial and final points of C, $\vec{F}$ has independence of path. It follows, by (1.2), that $\vec{F}$ is conservative.

We next prove $\Rightarrow$. Assume that $\vec{F}$ is a conservative field. Then, by (1.2), $\vec{F}$ has independence of path. Choose a fixed point Q in D. We define a scalar field on D as follows. For every point P in D, let f(P) be the value of $\int_C \vec{F} \cdot d\vec{R}$

for some chosen piecewise smooth curve C from Q to P. Since $\vec{F}$ has independence of path, the value f(P) does not depend on which curve from Q to P we choose. We now introduce Cartesian coordinates and show that this scalar field f must be a scalar potential for $\vec{F}$. Let $\vec{F} = L\hat{i} + M\hat{j} + N\hat{k}$. Let $P = (x_1, y_1, z_1)$. Then $f(P) = \int_C \vec{F} \cdot d\vec{R}$. We calculate $f_x(P)$. Let $P' = (x_1 + \Delta x, y_1, z_1)$. (See

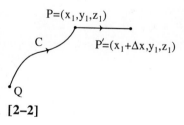

[2–2]

[2–2].) Then $f(P') = \int_{C+\Delta C} \vec{F} \cdot d\vec{R}$, where ΔC is the line-segment from P to P'.

Hence $f(P') - f(P) = \int_{\Delta C} \vec{F} \cdot d\vec{R}$. ΔC has the path $\vec{R}(t) = (x_1 + t)\hat{i} + y_1\hat{j} + z_1\hat{k}$,

$0 \le t \le \Delta x$. Hence $\dfrac{d\vec{R}}{dt} = \hat{i}$, and $\int_{\Delta C} \vec{F} \cdot d\vec{R} = \int_0^{\Delta x} \vec{F} \cdot \dfrac{d\vec{R}}{dt} dt = \int_0^{\Delta x} L(x_1 + t, y_1, z_1) dt$.

By definition $f_x(P) = \lim\limits_{\Delta x \to 0} \dfrac{f(P') - f(P)}{\Delta x} = \lim\limits_{\Delta x \to 0} \dfrac{\int_0^{\Delta x} L(x_1 + t, y_1, z_1) dt}{\Delta x}$.   We can

evaluate this limit by using the continuity of L together with either (i) l'Hopital's rule (from elementary calculus) or (ii) the mean-value theorem for definite integrals (from elementary calculus). (For (ii), see the problems. For a discussion of l'Hopital's rule, see the problems for §5.5.) Applying l'Hopital's rule, we differentiate numerator and denominator with respect to Δx. Differentiating the numerator, we get, by the fundamental theorem of integral calculus (see (3.2) below),

that $\dfrac{d}{d(\Delta x)} \int_0^{\Delta x} L(x_1 + t, y_1, z_1) dt = L(x_1 + \Delta x, y_1, z_1)$.   Differentiating   the

denominator, we get $\dfrac{d}{d(\Delta x)}(\Delta x) = 1$. This gives $f_x(P) = \lim\limits_{\Delta x \to 0} \dfrac{L(x_1 + \Delta x, y_1, z_1)}{1}$.

But $\lim\limits_{\Delta x \to 0} L(x_1 + \Delta x, y_1, z_1) = L(x_1, y_1, z_1)$ by the continuity of L. Hence

$f_x(P) = L(P)$. In the same way, we get $f_y P) = M(P)$ and $f_z(P) = N(P)$. Hence

$\vec{\nabla} f \big|_P = f_x(P)\hat{i} + f_y(P)\hat{j} + f_z(P)\hat{k} = L(P)\hat{i} + M(P)\hat{j} + N(P)\hat{k} = \vec{F}(P)$, for all P in D.

We have proved that f is a scalar potential for $\vec{F}$ and that $\vec{F}$ is therefore a gradient field. This completes the proof of the conservative field theorem.

The conservative-field theorem brings together two separate and distinct ideas (*conservative field* and *gradient field*) and shows that if a vector field has one of these properties, then it must also have the other. Henceforth, we shall use the phrases "conservative field," "gradient field," and "field with independence of path" interchangeably, since they apply to exactly the same vector fields. We shall most commonly use the phrase "conservative field."

Two important corollaries follow from the proof of the theorem.

(2.2)   *Corollary. Let $\vec{F}$ be a conservative field with scalar potential f. Then*

$$\int_Q^P \vec{F} \cdot d\vec{R} = f(P) - f(Q).$$

Proof of **(2.2)**. This fact is proved as part of the proof of $\Leftarrow$ in the theorem.

(2.3)   *Corollary. It is possible for a vector field to satisfy the derivative test* ((5.1) *in* §18.5) *throughout its domain without being a gradient field.* This corollary is sometimes stated as follows: *"Satisfying the derivative test is not a sufficient condition for a vector field to be a gradient field."*

Proof of (2.3). In example 4 of §18.7, we saw that $\vec{F} = \frac{1}{r}\hat{\theta} = \frac{-y\hat{i}+x\hat{j}}{x^2+y^2}$, on E^2-minus-the-origin, satisfies the derivative test. In example 5 of §20.1, we saw that $\vec{F}$ is not conservative. This implies, by the conservative-field theorem, that $\vec{F}$ is not a gradient field.

A third corollary is also obvious.

(2.4)   *Corollary. If* f *is a scalar potential for a continuous vector field, then* f *must be* C^1.

§20.3 Integration in a Conservative Field

Corollary (2.2) simplifies the evaluation of vector line integrals for a conservative field $\vec{F}$ with a known scalar potential f. Given any piecewise smooth, directed curve C, no matter how complex, we need only find f(P) – f(Q), where Q and P are the initial and final points of C, and we have the value of $\int_C \vec{F} \cdot d\vec{R}$. This simplifies the evaluation of $\int_C \vec{F} \cdot d\vec{R}$ in three ways: (i) we avoid the use of a specific curve C; (ii) we avoid any need for a formal integration procedure; and (iii) we replace a vector calculation with a scalar calculation.

Example. Find $\int_C \vec{F} \cdot d\vec{R}$, *where* (i) C *is a semicircular curve from* Q = (1,2,1) *to* P = (–1,3,–2), *where* C *has* PQ *as a diameter and lies above* PQ *in a plane perpendicular to the* xy *plane; and* (ii) $\vec{F} = yz\hat{i} + xz\hat{j} + xy\hat{k}$. **Solution.** C is difficult to visualize. We note, however, that f(x,y,z) = xyz is a scalar potential for $\vec{F}$. By (2.2), $\int_C \vec{F} \cdot d\vec{R} = xyz \Big]_{(1,2,1)}^{(-1,3,-2)} = 6 - 2 = 4$.

In elementary calculus, the fundamental theorem of integral calculus (the FTIC) is given in two equivalent forms (see §5.8):

(3.1)   *If* $f = \frac{dg}{dx}$, *then* $\int_a^b f(x)dx = g(b) - g(a)$.

(3.2) $\dfrac{d}{dx}\displaystyle\int_a^x f(u)du = f(x)$.

The conservative-field theorem gives us, as corollaries, two analogous multivariable versions of the FTIC. In these versions, $\vec{\nabla}$ plays the role of $\dfrac{d}{dx}$, and vector line integrals appear in place of elementary definite integrals.

(3.3)   *Corollary. If* $\vec{F} = \vec{\nabla}g$, *then* $\displaystyle\int_Q^P \vec{F}\cdot d\vec{R} = g(P) - g(Q)$. *(Analogue to (3.1).)*

(3.4)   *Corollary. For a conservative* $\vec{F}$ *and some chosen fixed* Q, *define a scalar field* g *by* $g(P) = \displaystyle\int_Q^P \vec{F}\cdot d\vec{R}$ *for each P.   Then* $\vec{\nabla}g = \vec{F}$. *(Analogue to (3.2).)*

(3.3) leads us to a proof of the following simple but useful fact.

(3.5)   *Let* f *be a differentiable scalar field on domain* D. *Then:*
  (i)   *for any scalar constant* c, $\vec{\nabla}(f + c) = \vec{\nabla}f$;
  (ii)   *if* g *is a differentiable scalar field on* D *such that* $\vec{\nabla}f = \vec{\nabla}g$ *on* D, *then there must be some scalar constant* c *such that for all P in D,* g(P) = f(P) + c.

Proof of (3.5). For (i), we have $\vec{\nabla}(f + c) = \vec{\nabla}f + \vec{\nabla}c = \vec{\nabla}f + \vec{0} = \vec{\nabla}f$. For (ii), assume $\vec{\nabla}f = \vec{\nabla}g$ on D. Let Q be a fixed point in D. Then $\vec{\nabla}f = \vec{\nabla}g \Rightarrow$ $\vec{\nabla}g - \vec{\nabla}f = \vec{0} \Rightarrow \vec{\nabla}(g - f) = \vec{0} \Rightarrow \displaystyle\int_Q^P \vec{\nabla}(g - f)\cdot d\vec{R} = 0$ for   every   P   in   D $\Rightarrow$ (g(P) − f(P)) − (g(Q) − f(Q)) = 0 for every P in D (by (3.3)) $\Rightarrow$ g(P) − f(P) = g(Q) − f(Q). Thus for all points P in D, g(P) − f(P) has the same constant value c = g(Q) − f(Q).
  (ii) in (3.5) can be restated: *"any two scalar potentials for the same vector field must differ by a constant."*

§20.4 Finding a Potential: the Line-integral Method

Because a scalar potential is so useful for evaluating a vector line integral, we wish to have systematic methods for finding a potential for a given conservative vector field. We describe two such methods: the *method of line integrals* (in this section) and the *method of indefinite integrals* (in §20.5). In addition to these two methods, there is also the less systematic method of an *inspired guess*. When successful, an inspired guess is the fastest method, since we can immediately confirm or disconfirm a guess by taking its gradient.

The *method of line integrals* is based on the proof of ⇒ for the conservative-field theorem. This proof tells us that if we are given a conservative $\vec{F}$ on D, we can choose a fixed *base point* Q in D and then define the value at P of a potential function f to be $f(P) = \int_C \vec{F} \cdot d\vec{R}$, where C is some systematically chosen, piecewise smooth curve from Q to P. We illustrate this method in several examples.

Example 1. *If* D *is convex, we can take* C *to be a line-segment* QP *from* Q *to* P, *since for every* P *in* D, *the segment* QP *must lie in* D. *Let* $\vec{F} = (x+y)\hat{i} + (x-y)\hat{j}$ *on* E^2. Take Q to be the origin. Let P = (c,d). Then C = QP is given by the path $\vec{R}(t) = tc\hat{i} + td\hat{j}$, $0 \le t \le 1$. (See [4–1].) We find $\int_C \vec{F} \cdot d\vec{R}$ by parametric evaluation:

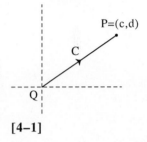

[4–1]

$$\frac{d\vec{R}}{dt} = c\hat{i} + d\hat{j},$$

$$\vec{F} = (tc + td)\hat{i} + (tc - td)\hat{j},$$

$$\vec{F} \cdot \frac{d\vec{R}}{dt} = tc^2 + tcd + tcd - td^2,$$

$$\int_0^1 \vec{F} \cdot \frac{d\vec{R}}{dt} dt = \int_0^1 (tc^2 + 2tcd - td^2)dt = (c^2 + 2cd - d^2)\frac{t^2}{2} \Big]_{t=0}^{t=1}$$

$$= \tfrac{1}{2}c^2 + cd - \tfrac{1}{2}d^2.$$

Putting x and y for c and d, we have

$$f(x,y) = \tfrac{1}{2}x^2 + xy - \tfrac{1}{2}y^2 \text{ as a scalar potential.}$$

Example 2. *It is sometimes convenient to take* C *to be a piecewise smooth curve formed of straight segments, where each segment is parallel to a coordinate axis. We take* $\vec{F}$, Q, *and* P *as in example 1.* Instead of a straight segment from Q to P, we use two successive line-segments: C_1 from (0,0) to (c,0), and C_2 from (c,0) to (c,d). (See [4–2].) We evaluate $\int_{C_1} \vec{F} \cdot d\vec{R}$ and $\int_{C_2} \vec{F} \cdot d\vec{R}$ by parametric evaluation.

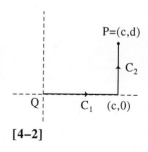

[4–2]

For C_1: $\vec{R}(t) = tc\hat{i}, \quad 0 \leq t \leq 1,$

$\dfrac{d\vec{R}}{dt} = c\hat{i},$

$\vec{F} = tc\hat{i} + tc\hat{j},$

$\displaystyle\int_0^1 \vec{F} \cdot \dfrac{d\vec{R}}{dt}\, dt = \int_0^1 tc^2 dt = \tfrac{1}{2}c^2 .$

For C_2: $\vec{R}(t) = c\hat{i} + td\hat{j}, \quad 0 \leq t \leq 1,$

$\dfrac{d\vec{R}}{dt} = d\hat{j},$

$\vec{F} = (c + td)\hat{i} + (c - td)\hat{j},$

$\displaystyle\int_0^1 \vec{F} \cdot \dfrac{d\vec{R}}{dt}\, dt = \int_0^1 (cd - td^2)dt = cd - \tfrac{1}{2}d^2 .$

Therefore $\displaystyle\int_C \vec{F} \cdot d\vec{R} = \int_{C_1} \vec{F} \cdot d\vec{R} + \int_{C_2} \vec{F} \cdot d\vec{R} = \tfrac{1}{2}c^2 + cd - \tfrac{1}{2}d^2$ and, as in example 1, $f(x,y) = \tfrac{1}{2}x^2 + xy - \tfrac{1}{2}y^2 .$

The line-integral method can also be applied when we are unsure whether $\vec{F}$ is a conservative field. If $\vec{F}$ is not conservative, then the scalar function that we obtain will not have $\vec{F}$ as its gradient. We see this in the next example.

Example 3. *Let* $\vec{F} = (x + y)\hat{i} + (y - x)\hat{j}$ *on* $\mathbf{E}^1$. If we apply the line-integral method with a straight segment from the origin, we have (with $P = (c,d)$):

$$\vec{R}(t) = tc\hat{i} + td\hat{j}, \quad 0 \leq t \leq 1,$$

$$\dfrac{d\vec{R}}{dt} = c\hat{i} + d\hat{j},$$

$$\vec{F} = (tc + td)\hat{i} + (td - tc)\hat{j},$$

$$\int_0^1 \vec{F} \cdot \dfrac{d\vec{R}}{dt}\, dt = \int_0^1 (tc^2 + tcd + td^2 - tcd)dt = \tfrac{1}{2}(c^2 + d^2) .$$

This gives $f(x,y) = \tfrac{1}{2}(x^2 + y^2)$, but $\vec{\nabla}f = x\hat{i} + y\hat{j}$, which is different from $\vec{F}$. Thus $\vec{F}$ is not conservative. We could already have discovered this by the derivative test, since $\dfrac{\partial}{\partial y}(x + y) = 1$ while $\dfrac{\partial}{\partial x}(y - x) = -1$.

It is sometimes possible, in using the line-integral method, to make an artful choice of a piecewise smooth curve C (from Q to P) so that the vector line integral for each piece of C can be evaluated by definition rather than by parametric evaluation, as we see in the next example.

Example 4. *Let* $\vec{F} = \frac{1}{r}\hat{\theta} = \frac{-y\hat{i}+x\hat{j}}{x^2+y^2}$ *on the half-plane* $x > 0$. (This is

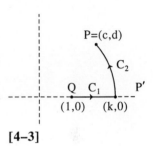

example 5 in §18.7, where we note that $\vec{F}$ is conservative.) Let Q be the point $(x,y) = (1,0)$. Let P be $(x,y) = (c,d)$. We use $C = C_1 + C_2$, where C_1 is a segment of the x axis from $(1,0)$ to $P' = (k,0)$, for $k = \sqrt{c^2+d^2}$, and C_2 is the circular arc, with center at 0, from P' to P. (See [4–3].) Evaluating by definition, we have $\int_{C_1} \vec{F}\cdot d\vec{R} = 0$, since $\vec{F}$ is perpendicular to C_1 at every point on C_1, and $\int_{C_2} \vec{F}\cdot d\vec{R} = \frac{1}{k}$(length of C_2), since $\vec{F}$ is tangent to C_2 at every point on C_2. **[4–3]**

Thus $\int_{C_2} \vec{F}\cdot d\vec{R} = \frac{1}{k}k\alpha = \alpha$, where α is the angle POP'. Since $\alpha = \tan^{-1}(\frac{d}{c})$,

we have $f(x,y) = \tan^{-1}(\frac{y}{x})$ as a scalar potential for $\vec{F}$. (In polar coordinates, this potential is more simply expressed as $g(r,\theta) = \theta$.)

§20.5 Finding a Potential: the Indefinite-integral Method

We now describe a second systematic method for finding a potential for a conservative field: the *method of indefinite integrals*. This method is often simpler and easier to use than the method of line integrals of §20.4. It requires, however, that the domain D be convex. We illustrate the procedure in several examples. We then describe it more formally and justify it.

Example 1. *We are given* $\vec{F} = L\hat{i} + M\hat{j} = (2x - 3y - 4)\hat{i} + (4y - 3x + 2)\hat{j}$ *on the domain* E^2, *which is convex.* We wish to find a scalar function $f(x,y)$ such that $f_x = L = 2x - 3y - 4$ and $f_y = M = 4y - 3x + 2$. We obtain f by "successive approximations" as follows.

Step 1. Since $L = f_x$, we express f as the indefinite integral $\int L dx = \int (2x - 3y - 4)dx$. In calculating this integral, we treat y as a constant. We get

$$f = \int (2x - 3y - 4)dx = x^2 - 3xy - 4x + C(y),$$

where the "constant of integration" C(y) may depend upon y, because, in finding f_x, we treat y as a constant. Thus

$$f = x^2 - 3xy - 4x + C(y) \text{ is our } first \ approximation.$$

Step 2. Using the result of step 1, we calculate f_y. This gives

$$f_y = -3x + C'(y).$$

But we also have $f_y = M = 4y - 3x + 2$.

Thus $-3x + C'(y) = 4y - 3x + 2$, and $C'(y) = 4y + 2$.

Taking an indefinite integral again, we have

$$C(y) = \int (4y + 2)dy = 2y^2 + 2y + D,$$

where D is a scalar constant. Thus

$$f = x^2 - 3xy - 4x + 2y^2 + 2y + D$$

is our *final version* of f.

The process is similar in $\mathbf{E}^3$.

Example 2. *We are given*

$$\vec{F} = L\hat{i} + M\hat{j} + N\hat{k} = (y - z)\hat{i} + (x + 2yz - 3z)\hat{j} + (y^2 - x - 3y + 4)\hat{k}$$

on the domain $\mathbf{E}^3$, *which is convex.*

Step 1. We have $f_x = y - z$. Integrating with respect to x, we get

$$f = \int (y - z)dx = yx - zx + C(y, z) \qquad (first \ approximation \text{ for } f),$$

where the "constant of integration" may be a function of y and z.

Step 2. We have $f_y = x + \dfrac{\partial}{\partial y} C(y, z)$ (from step 1)

$$\text{and }\; f_y = x + 2yz - 3z \qquad\qquad \text{(from the given } \vec{F}\text{)}.$$

Thus $\frac{\partial}{\partial y}C = 2yz - 3z$. Integrating with respect to y, we have $C = y^2 z - 3yz + D(z)$, where the "constant of integration" $D(z)$ may be a function of z. This gives

$$f = yx - zx + y^2 z - 3yz + D(z) \qquad \text{(second approximation)}.$$

Step 3. We have $f_z = -x + y^2 - 3y + D'(z)$ (from step 2)
$$\text{and }\; f_z = y^2 - x - 3y + 4 \qquad\qquad \text{(from the given } \vec{F}\text{)}.$$

Thus $D'(z) = 4$. Integrating with respect to z, we have $D = 4z + E$, where E is a scalar constant, and

$$f = xy - xz + y^2 z - 3yz + 4z + E \qquad \text{(final version of f)}.$$

Example 3. What happens if we use the indefinite integral method on a vector field which is not *conservative? We are given* $\vec{F} = -y\hat{i} + x\hat{j}$ *on the domain* E^2. (By the derivative test, $\vec{F}$ is not conservative, since $\frac{\partial L}{\partial y} = -1 \neq \frac{\partial M}{\partial x} = 1$.)

Step 1. We have $f_x = -y$. Integrating, we get
$$\text{and }\; f = -yx + C(y) \qquad\qquad \text{(first approximation)}.$$

Step 2. We have $f_y = -x + C'(y)$ (from step 1)
$$\text{and }\; f_y = x \qquad\qquad\qquad \text{(from the given } \vec{F}\text{)}.$$

Thus $C'(y) = 2x$, violating our requirement that C be a function of y only. Our procedure has led us to a contradiction. In the following proof, the derivative test condition (6.1) assures that there is no such contradiction.

Justifying the indefinite integral method. We give a proof a C^1 field $\vec{F} = L\hat{i} + M\hat{j}$ in E^2. The case for E^3 is treated in the problems. We first note the following facts about indefinite integration of a function $g(x,y)$.

(5.1) *It is always true that* $\frac{\partial}{\partial x}\int g\,dx = g$ (by the definition of indefinite integral).

(5.2)   *It is not always true that* $\int \frac{\partial g}{\partial x} dx = g$.

For example, if $g(x,y) = x+y$, then (by (5.1)) $\frac{\partial}{\partial x} \int g dx = \frac{\partial}{\partial x} (\frac{x^2}{2} + xy + D) = x+y$, but $\int \frac{\partial g}{\partial x} dx = \int dx = x + D \neq x + y$.

The justification for the indefinite-integral method is as follows. We are given that $\vec{F} = L\hat{i} + M\hat{j}$ and that $L_y = M_x$. We begin the indefinite integral procedure.

Step 1. We have $f_x = L$. Integrating with respect to x, we get

$$f = \overline{\int L dx} + C(y) \qquad \text{(first approximation).}$$

Here $\overline{\int L dx}$ denotes a specific, arbitrarily chosen, indefinite integral of L.

Step 2. $f_y = \frac{\partial}{\partial y} \overline{\int L dx} + C'(y)$ \qquad (from step 1)

 and $f_y = M$ \qquad\qquad\qquad (from the given $\vec{F}$).

This gives $C'(y) = M - \frac{\partial}{\partial y} \overline{\int L dx}$.

To rule out a possible contradiction, we must show that $C'(y)$ does not depend on x, that is to say, we must show that $\frac{\partial}{\partial x}(M - \frac{\partial}{\partial y} \overline{\int L dx}) = 0$. Computing, we get

$$\frac{\partial}{\partial x}(M - \frac{\partial}{\partial y} \overline{\int L dx}) = M_x - \frac{\partial}{\partial x} \frac{\partial}{\partial y} \overline{\int L dx}$$

$$= M_x - \frac{\partial}{\partial y} \frac{\partial}{\partial x} \overline{\int L dx} \quad \text{(eq. of cross-derivatives by } \vec{F} \text{ in } C^1)$$

$$= M_x - L_y \qquad\qquad \text{(by (7.1))}$$

$$= 0, \qquad\qquad\qquad \text{(since } M_x = L_y \text{).}$$

This completes the justification.

Remark. The proof above is valid if the domain of $\vec{F}$ is convex. This can be shown by a more detailed consideration of the domains of the functions which occur in the proof. This gives us, as a corollary, the following converse form of the derivative test. *If* $\vec{F} = L\hat{i} + M\hat{j}$ *(in* E^2*) is* C^1 *and has a convex domain D, and* if $L_y = M_x$ on D, *then* $\vec{F}$ *must be conservative.* A similar converse holds for a convex domain in $\mathbf{E}^3$. A stronger and more useful converse form of the derivative test is obtained in Chapter 26.

§20.6 Exact Differential Equations

This section presents the theory of *first-order differential equations in differential form.* Let $\vec{F} = L\hat{i} + M\hat{j}$ be a C^1 vector field on a domain D in $\mathbf{E}^2$, such that $\vec{F}(P) \neq \vec{0}$ for all P in D.

Definition. If C is a directed smooth curve in D such that at every point P on C, $\vec{F}(P) \cdot \hat{T}(P) = 0$, where $\hat{T}(P)$ is the unit tangent vector for C at P, we say that C is an *orthogonal curve* for $\vec{F}$. (See [6–1].) (Note that if C is an orthogonal curve for $\vec{F}$, then so is –C.)

If we form a vector line integral $\int_{\Delta C} \vec{F} \cdot d\vec{R}$, where ΔC is a portion of an orthogonal curve, then $\int_{\Delta C} \vec{F} \cdot d\vec{R} = \int_{\Delta C} (\vec{F} \cdot \hat{T}) ds = 0$. Recall that $\int_{\Delta C} \vec{F} \cdot d\vec{R}$ can be expressed as $\int_{\Delta C} (Ldx + Mdy)$. The fact that $\vec{F} \cdot \hat{T} = 0$ at every point on C is sometimes expressed by saying that C satisfies

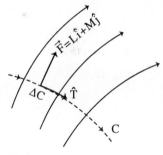

[6–1]

(6.1) $$Ldx + Mdy = 0.$$

The expression (6.1) is called a *first-order differential equation in differential form,* and an orthogonal curve for $\vec{F}$ is called a *solution* to (6.1). Given expressions for L and M, the problem of finding a useful expression for the family of orthogonal curves for $\vec{F} = L\hat{i} + M\hat{j}$ is known as the problem of *solving the differential equation* (6.1).

Definition. The differential equation (6.1) is said the be *exact* if $\vec{F} = M\hat{i} + N\hat{j}$ is a conservative field.

The problem of solving an exact differential equation is straightforward. By §18.3, the solutions will be the equipotential curves for $\vec{F}$. We first find a scalar potential f(x,y) for $\vec{F}$. Each solution is then given by the equation

$$f(x,y) = c$$

for some value of the scalar constant c.

    Example 1. *Consider the equation*

$$ydx + xdy = 0 \qquad\qquad on\ \mathbf{E}^2.$$

This equation is exact, since $L_y = M_x = 1$ and $\mathbf{E}^2$ is convex. By the indefinite integral method, we get

$$\begin{aligned}
f(x,y) &= xy + C(y) &&\textit{(first approximation)}\\
f_y &= x = x + C'(y)\\
C'(y) &= 0 \Rightarrow C(y) = E, &&\text{a scalar constant.}\\
\text{Thus}\qquad f(x,y) &= xy + E &&\textit{(final version)}.
\end{aligned}$$

The equipotential curves $xy = c$ are the solutions to the exact equation $ydx + xdy = 0$.

    If $\vec{F}$ is *not* conservative, it will still be true that $\vec{F}$ has orthogonal curves, because an orthogonal curve for $\vec{F}$ is the same as an integral curve for the vector field $\vec{G} = -M\hat{i} + L\hat{j}$. At every point P in D, $\vec{G}(P) \neq \vec{0}$, since $\vec{F}(P) \neq \vec{0}$. By (3.2) in §18.3, there must be an integral curve for $\vec{G}$ through P and hence a solution for (6.1) through P.

    How can we find these solutions when (6.1) is not exact (that is to say, when $\vec{F}$ is not conservative)? A commonly used approach is to try to find a scalar function   m(x,y)   on D such that (i) m(P) > 0 for all P in D and (ii) mLdx + mMdy = 0 *is* an exact equation. A solution to equation (ii) must be a solution to (6.1) and vice-versa, since the vector field $mL\hat{i} + mM\hat{j}$ has the same direction as $\vec{F}$ at every point. A scalar function m(x,y) which produces an exact equation   mLdx + mMdy = 0   is called an *integrating factor* for the given equation Ldx + Mdy = 0. If we have an integrating factor, we can find the solutions to the exact equation by the method of indefinite integrals (or by the method of line-integrals), and these solutions must also be the solutions to the original inexact equation.

Example 2. *Consider* $-ydx + xdy = 0$ *on the half-plane* $y > 0$. This equation is not exact. However, $m(x,y) = \dfrac{1}{y^2}$ serves as an integrating factor. It gives us

$$-\frac{1}{y}dx + \frac{x}{y^2}dy = 0.$$

This is exact, since $L_y = M_x = \dfrac{1}{y^2}$. Finding a potential by the method of indefinite integrals, we get

$$f(x,y) = -\frac{x}{y} + C(y) \qquad \text{(\textit{first approximation})}$$

$$f_y = \frac{x}{y^2} = \frac{x}{y^2} + C'(y)$$

$$C'(y) = 0 \Rightarrow C(y) = E.$$

Thus $\qquad\qquad f(x,y) = -\dfrac{x}{y} + E \qquad$ (*final version*).

Hence the solutions can be expressed:

$$\frac{x}{y} = c \quad \text{or} \quad x = cy .$$

(This is a simple example to illustrate the formal steps. In fact, we can immediately see, from the figure for example 3 in §18.7, that the orthogonal curves for $\vec{F} = -y\hat{i} + x\hat{j}$ must be rays from the origin.)

Remark. In the problems we outline a proof for:

(6.2) *Let* $\vec{F} = L\hat{i} + M\hat{j}$ *be* C^1 *and non-zero on a convex domain* D. *There must be a conservative continuously differentiable vector field* $\vec{G}$ *on* D *such that for all* P *in* D, $\vec{G}(P) \neq \vec{0}$, *and* $\vec{G}(P)$ *has the same direction as* $\vec{F}(P)$. *If we define the scalar field* m *by*

$$m(P) = \frac{\left|\vec{G}(P)\right|}{\left|\vec{F}(P)\right|},$$

then $m\vec{F} = \vec{G}$, *and* m *is an integrating factor for* $Ldx + Mdy = 0$.

This shows that if L and M are C^1, then an integrating factor for the equation $Mdx + Ndy = 0$ must always exist. Unfortunately, it is not always true that if we have simple expressions for L and M, then there will be a simple expression for an integrating factor m which will lead to an algebraically simple solution of the exact equation $mLdx + mMdy = 0$.

§20.7 Problems

§20.1

1. Show that if $\vec{F}$ on D is path dependent, then for *every* pair of points P and Q in D and for *any* given curve C_1 from Q to P, there must be *some* other curve C_2 from Q to P such that
$$\int_{C_1} \vec{F} \cdot d\vec{R} \neq \int_{C_2} \vec{F} \cdot d\vec{R}.$$

§20.2

1. The *mean-value theorem for definite integrals* in elementary calculus asserts that for a continuous one-variable function g, if $\int_a^b g(x)dx = L$, then there must be some c, $a \leq c \leq b$, such that $L = g(c)(b-a)$. Use this fact, in place of l'Hopital's rule, to evaluate $f_x(P)$ in the proof of (2.1).

2. Let $\vec{F} = \frac{1}{r}\hat{r} + \frac{1}{r}\hat{\theta}$ on all of $\mathbf{E}^2$ except O. Show that $\vec{F}$ satisfies the derivative test but is not conservative. (*Hint.* Use examples 2 and 4 of §18.7 and example 5 of §20.1.)

§20.3

1. Let $\vec{F} = yz\hat{i} + zx\hat{j} + xy\hat{k}$. Guess and verify a scalar potential for $\vec{F}$. Use this potential to calculate $\int_P^Q \vec{F} \cdot d\vec{R}$ when
   (a) P = (1,2,–1) and Q = (4,0,–3)
   (b) P = (–2,–1,4) and Q = (3,3,3)
   (c) P = (2,2,2) and Q = (4,2,1)

2. The force acting on a particle in the plane is given by $\vec{F} = 3\hat{i} - 4\hat{j}$. To which point on the ellipse $x^2 + 4y^2 = 4$ does $\vec{F}$ do the most work in moving a particle from the origin? To which point does it do the least work?

3. Find a scalar potential for the uniform central field $\vec{F} = \left(\frac{\rho+1}{\rho^2}\right)\hat{\rho}$.

4. Use the conservative field theorem to simplify the evaluation in:
   (a) example 1 in §19.5,
   (b) example 3 in §19.5,
   (c) example 4 in §19.5.

§20.4

1. (a) Which of the following vector fields is a gradient field?
   (i) $(2xyz^2 + 3)\hat{i} + (x^2z^2 + 2yz)\hat{j} + (2x^2yz + y^2)\hat{k}$
   (ii) $4xy\hat{i} - x^2\hat{j} + 4z\hat{k}$
   (b) Find a scalar potential for each gradient field.

2. In each case find a scalar potential for the given vector field.
   (a) $\vec{F} = (yz + y + z + 1)\hat{i} + (xz + x + z + 1)\hat{j} + (xy + y + x + 1)\hat{k}$
   (b) $\vec{F} = ye^{xy}\hat{i} + (xe^{xy} - z\sin(yz))\hat{j} - y\sin(yz)\hat{k}$

3. $\vec{F}$ is defined by

$$\vec{F} = yz\hat{i} + (xz + e^y z)\hat{j} + (xy + e^y + \tfrac{1}{2})\hat{k}.$$

If $\vec{F}$ is conservative, find a potential function. If not, give a reason.

§20.5

1. $\vec{F} = (ye^{xy} + ze^{xz})\hat{i} + (xe^{xy} + 2)\hat{j} + xe^{xz}\hat{k}$ is a conservative field. Find a scalar potential for $\vec{F}$, and use it to evaluate $\displaystyle\int_C \vec{F} \cdot d\vec{R}$ where C is x = sin t, y = sin t, z = cos t, $0 \le t \le \pi/2$.

2. In each of the following cases, decide whether $\vec{F}$ is conservative. If $\vec{F}$ is conservative, find a potential function. If $\vec{F}$ is not conservative, show why.

   (a) $\vec{F}(x,y,z) = (yz + e^x z)\hat{i} + xz\hat{j} + (xy + e^x + \tfrac{1}{z})\hat{k}$ for all points with $z > 0$.

   (b) $\vec{F}(x,y) = \left(x/(x^2+y^2)^2\right)\hat{i} + \left(y/(x^2+y^2)^2\right)\hat{j}$ for all points except the origin.

   (c) $\vec{F}(x,y) = \left(y/(x^2+y^2)^2\right)\hat{i} - \left(x/(x^2+y^2)^2\right)\hat{j}$ for all points except the origin.

3. Justify the indefinite integral method for the case of $\vec{F} = L\hat{i} + M\hat{j} + N\hat{k}$ in $\mathbf{E}^3$. In particular, indicate how and where this justification requires that the domain of $\vec{F}$ be convex. (*Hint.* Extend the justification for $\mathbf{E}^2$ in §20.5.)

§20.6

1. Find a general solution to each of the following differential equations.

   (a) $(x + y)dx + xdy = 0$

   (b) $(3x^2 y + 2xy)dx + (x^3 + x^2 + 2y)dy = 0$

   (c) $ydx + (y - x)dy = 0$ (*Hint.* See example 2.)

   (d) $(xy^2 + y)dx + xdy = 0$ (*Hint.* Use the fact that $ydx + xdy = 0$ is exact to find an integrating factor.)

2. Outline a proof for (6.2). (*Hint.* Use the integral curves and orthogonal curves for $\vec{F}$ to define a curvilinear coordinate system on D. In particular, choose a point P_0 as origin. Let C_0 be the integral curve through P_0 and let D_0 be the orthogonal curve through P_0 (directed by $-M\hat{i} + L\hat{j}$). Define the coordinates (u,v) of a point P in D by taking the integral curve C_P and the orthogonal curve D_P through P. Then u = distance from P_0 along C_0 of intersection of D_P with C_0, and v = distance from P_0 along D_0 of intersection of C_P with D_0. Define $\vec{R}(u,v) = $ position vector of point P with coordinates (u,v) with respect to P_0. Define f(u,v) = u. Show that

$$m(P) = \cfrac{1}{\left|\dfrac{\partial\vec{R}}{\partial u}\right|_P \cdot |\vec{F}(P)|}$$

is the desired integrating factor yielding $\vec{G} = \nabla f$.)

Chapter 21

Surfaces

§21.1 Parametric Representation of Surfaces

In our work so far, *points, lines, planes* and *curves* are mathematical objects in an idealized, non-physical universe, where they have their own precise mathematical meanings and properties. In mathematics, a *surface* is a geometrical figure of this same kind. Like lines, planes, and curves, surfaces are infinitely thin and precise figures in E^3. Sometimes, in mathematical applications, a mathematical surface will represent a particular "physical surface" (for example, the surface of the earth). On other occasions, a mathematical surface may be used in a more abstract and hypothetical way. For example, we might choose a point P in the region of a given, steady-state fluid flow and consider an "imaginary," stationary, spherical surface S of radius a and center at P. We might then ask for the net rate (in cm³/sec) at which fluid volume is flowing outward through S.

In our study of multivariable calculus, we have already used several different methods for specifying and describing surfaces:

(1) We can define a surface S by a *synthetic geometrical description.* For example: *let S be the spherical surface of radius 5 with center at O.*

(2) We can define a surface S by an *algebraic equation.* For example: *let S be the set of points which satisfy the equation $x^2 + y^2 + z^2 = 25$.*

(3) We can define a surface S as a *level surface* of a scalar field or function. For example: *let S be the level set for 25 of $f(x,y,z) = x^2 + y^2 + z^2$.*

(4) We can define a surface S as the *graph* of a two-variable function. For example: *let S be the graph of $f(x,y) = \sqrt{25 - x^2 - y^2}$.*

In this chapter, we develop a terminology and theory of surfaces which is analogous to the terminology and theory developed for curves in Chapter 6. We begin by showing how a surface S can be given by a *parametric representation*. Recall that a *parametric representation for a finite curve* C *in* $\mathbf{E}^3$ (or $\mathbf{E}^2$) consists of (i) a fixed reference point O in $\mathbf{E}^3$ (or $\mathbf{E}^2$), (ii) a closed interval $\hat{D}$ of real numbers which is the domain of possible values for the parameter t, and (iii) a continuous, one-variable, vector-valued function $\bar{R}(t)$ whose vector values are in $\mathbf{E}^3$ (or $\mathbf{E}^2$). Then the curve C is the set of points given by the position arrow for $\bar{R}(t)$ as t varies on $\hat{D}$. (See [1–1].) In Cartesian coordinates in $\mathbf{E}^3$, we write

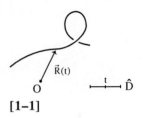

$$\bar{R}(t) = x(t)\hat{i} + y(t)\hat{j} + z(t)\hat{k} ,$$

[1–1]

where x(t), y(t) and z(t) are the coordinates of the point $\bar{R}(t)$ on C as functions of t. We speak of the equations x = x(t), y = y(t), and z = z(t) as *parametric equations for the curve* C.

Similarly, for a surface S, we have

(1.1)   **Definition.** A *parametric representation for a surface* S will consist of (i) a fixed reference point O in $\mathbf{E}^3$; (ii) a convex elementary region $\hat{D}$ in $\mathbf{E}^2$, where $\mathbf{E}^2$ has Cartesian axes for u and v and where possible pairs of values for the parameters u and v are given by the coordinates of points in $\hat{D}$; and (iii) a continuous, two-variable, vector-valued function $\bar{R}(u,v)$ whose vector values are in $\mathbf{E}^3$, where (iv) $\bar{R}(u,v)$ is injective on the interior of $\hat{D}$ (that is to say, distinct points in the interior of $\hat{D}$ are carried to distinct values of $\bar{R}$) and no value of $\bar{R}$ is given by both an interior point of $\hat{D}$ and a boundary point of $\hat{D}$. Then the surface S is the set of points given by the position arrow for $\bar{R}(u,v)$ as (u,v) varies over $\hat{D}$. (See [1–2].) In Cartesian coordinates, we write

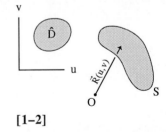

$$\bar{R}(u,v) = x(u,v)\hat{i} + y(u,v)\hat{j} + z(u,v)\hat{k} ,$$

[1–2]

where x(u,v), y(u,v), and z(u,v) are the coordinates of the point $\bar{R}(u,v)$ on S as functions of u and v. We speak of the equations x = x(u,v), y = y(u,v), z = z(u,v) as *parametric equations for the surface* S. (Condition (iv) in (1.1) is the analogue, for surfaces, of the condition for a curve that the curve be *simple*.) See the remark below.

(2.1) **Definition.** A figure S in E^3 is said to be a *parametric surface* if there is a parametric representation for S which satisfies (1.1).

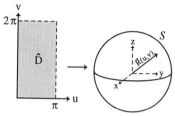

[1–3]

Example 1. Let $\hat{D}$ be the rectangular region $0 \leq u \leq \pi, 0 \leq v \leq 2\pi$, and let $\vec{R}(u,v) = 5\sin u \cos v \hat{i} + 5\sin u \sin v \hat{j} + 5\cos u \hat{k}$. If we think of the parameters u and v as the spherical coordinates φ and θ of a point on S, we see that S is the sphere of radius 5 with center at the origin, [1–3].

Example 2. Let $\hat{D}$ be the rectangular region $0 \leq u \leq 3, 0 \leq v \leq 2\pi$, and let

$$\vec{R}(u,v) = u\cos v \hat{i} + u\sin v \hat{j} + 2u\hat{k}.$$

S is a *cone* of radius 3, between $z = 0$ and $z = 6$, with vertex at the origin, [1–4]. Here the parameters u and v are the cylindrical coordinates r and θ of a point on S.

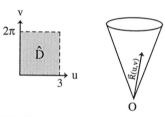

[1–4]

Example 3. If $\vec{R}(u,v)$ has the form $\vec{R}(u,v) = u\hat{i} + v\hat{j} + f(u,v)\hat{k}$, we can write u as x and v as y and think of $\hat{D}$ as a region in the xy plane of E^3. The surface S is then the graph of f(x,y) on $\hat{D}$. Let $\hat{D}$ be the circular region $u^2+v^2 \leq 9$, and let $\vec{R}(u,v) = u\hat{i} + v\hat{j} + 2(u^2+v^2)^{1/2}\hat{k}$. We can also write this as $\vec{R}(x,y) = x\hat{i} + y\hat{j} + 2(x^2+y^2)^{1/2}\hat{k}$, and we see that S is the graph of $f(x,y) = 2(x^2+y^2)^{1/2}$ over the disk $x^2 + y^2 \leq 9$, [1–5]. It is easy to show that this surface is the same as the surface of example 2.

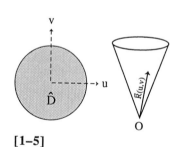

[1–5]

Example 4. Let $\hat{D}$ be the circular region $u^2+v^2 \leq 4$, and let $\vec{R}(u,v) = u\hat{i} + v\hat{j} + 2(u^2+v^2)\hat{k}$. We can write this as $\vec{R}(x,y) = x\hat{i} + y\hat{j} + 2(x^2+y^2)\hat{k}$, and we see that S is the portion between $z = 0$ and $z = 8$ of a circular paraboloid with vertex at the origin, [1–6].

In a parameterization of a surface S, each point in $\hat{D}$ is mapped to a corresponding point on S. In particular, the coordinate lines in $\hat{D}$ (u = constant and v = constant) will be mapped to corresponding curves on S. More specifically, let S be given by $\vec{R}(u,v)$ on $\hat{D}$. Let $\hat{P} = (a,b)$ be a given point in $\hat{D}$. Then $\hat{P}$ maps to the point $P = \vec{R}(a,b)$ on S. If we fix $v = b$ and allow u to vary, we have

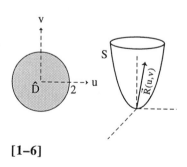

[1–6]

$$\vec{R}(u,b) = x(u,b)\hat{i} + y(u,b)\hat{j} + z(u,b)\hat{k}, \quad c \le u \le d,$$

where c and d are the minimum and maximum values of u for which (u,b) is in $\hat{D}$. $\vec{R}(u,b)$ is a path for a curve $C_{u,P}$ which lies in S and goes through P. $C_{u,P}$ is the image, under $\vec{R}(u,v)$, of the line-segment $v = b$ in $\hat{D}$. $C_{u,P}$ is called the u-*curve through* P *for the parameterization* $\vec{R}(u,v)$ of S. $C_{v,P}$, the v-*curve through* P *for* $\vec{R}(u,v)$, is similarly defined by fixing $u = a$ and let v vary, [1–7]. $C_{u,P}$ and $C_{v,P}$ are the *parameter curves through* P *for* $\vec{R}(u,v)$. The parameter curves can be viewed as the "coordinate curves" of a "coordinate system" on the surface S itself.

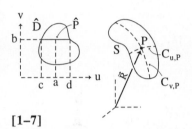

[1–7]

In example 1 above, the u-curves are circles of "latitude," and the v-curves are semicircles of "longitude" on the sphere S. In example 2, the u-curves are circles parallel to the xy plane and the v-curves are line segments orthogonal to those circles. In example 3, the x-curves are single portions of hyperbolas parallel to the xz plane, since an x-curve is the intersection of S with a plane perpendicular to the y axis; similarly, the y-curves are portions of hyperbolas parallel to the yz plane. In example 4, the x-curves and y-curves are portions of parabolas.

Remark. The requirement in (1.1) that $\hat{D}$ be convex guarantees that the parameter curves $C_{u,P}$ and $C_{v,P}$ will occur on S as uninterrupted curves. As we have noted, requirement (iv) in (1.1) is an exact counterpart to the assertion that a parameterization of a *curve* is *simple* (see Chapter 6). For curves, we did not require that all curve parameterizations be simple, because we wished to include applications to moving particles whose paths might be self-intersecting. For surfaces, we shall have no applications analogous to moving particles. Hence, for geometric convenience, we require that all parameterizations be "simple" in the sense of (iv).

§21.2 Elementary Surfaces; Smoothness

We begin with an informal and intuitive definition. A surface S will be called an *elementary surface* if S can be obtained (as a figure in E^3) by taking a finite, convex, elementary region $\hat{D}$ in E^2, treating it as if it were a thin sheet of elastic material, and twisting, bending, stretching, and/or compressing it. We do not allow $\hat{D}$ to be torn, nor do we allow any two

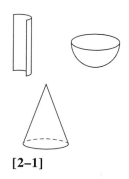

[2–1]

distinct points in $\hat{D}$ to come together as a single point of S. For example, we can take a flat rectangle $\hat{D}$ and curve it into a *half-cylinder*; we can take a flat disk and gradually push out its middle so that it becomes a hollow, bowl-shaped *hemisphere*, or we can take a flat disk and pull it out by its center point to get a tent-shaped *cone*, [2–1]. We make this informal idea precise with a definition.

(2.1)   *Definition.* A surface S is an *elementary surface* if S has a parametric representation $\vec{R}(u,v)$ on $\hat{D}$ such that $\vec{R}$ is *injective* on all of $\hat{D}$. (This is a strengthening of condition (iv) in (1.2).)

Example 1. The parameterization of the sphere in example 1 of §21.1 does not satisfy (2.1), since, for example, all points on the v axis in $\hat{D}$ map to the same point, (0,0,5), on S. In fact, it can be proved that there is no parameterization of S which satisfies (2.1). See (2.2). Hence S is not an elementary surface.

Example 2. The parameterization of the cone in example 2 of §21.1 does not satisfy (2.1), since for example, all points on the v axis in $\hat{D}$ map to the vertex point of P.

Example 3. On the other hand, the parameterization of the cone in example 3 of §21.1 does satisfy (2.1). Hence the cone is, in fact, an elementary surface.

Example 4. Similarly, the parameterization in example 4 of §21.1 satisfies (2.1), and hence this paraboloid is an elementary surface.

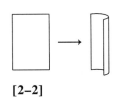

[2–2]

Example 5.   Using the cylindrical coordinates θ *and* z *as parameters, and taking* a > 0, *the injective parameterization* $\vec{R}(\theta,z) = a\cos\theta\,\hat{i} + a\sin\theta\,\hat{j} + z\hat{k}$ *on* $0 \leq \theta \leq \pi$, $0 \leq z \leq b$, *describes a half-cylinder of radius* a *and height* b, and we see that this half-cylinder is an elementary surface, [2–2].

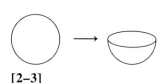

[2–3]

Example 6.   Using x *and* y *as parameters, the injective parameterization* $\vec{R}(x,y) = x\hat{i} + y\hat{j} + (1 - x^2 - y^2)^{1/2}\hat{k}$ *on* $x^2+y^2 \leq 1$ *gives a hemisphere of radius* 1, and we see that this hemisphere is an elementary surface, [2–3].

We make use of the following notation. Let B be a subset of $\hat{D}$.
Then $\bar{R}[B]$ will denote the set of points on S to which the points in B are
mapped by $\bar{R}$. It is possible to prove the following.

(2.2)   *Let $\bar{R}$ on $\hat{D}$ be a parameterization (satisfying (2.1)) for an elemen-*
tary surface S. Let $B(\hat{D})$ be the boundary of $\hat{D}$. Then $\bar{R}[B(\hat{D})]$ is a simple
closed curve in E^3. Let $\bar{R}_1$ on $\hat{D}_1$ be a different parameterization of the
same surface S (also satisfying (2.1), and let $B(\hat{D}_1)$ be the boundary of
$\hat{D}_1$. *Then $\bar{R}_1[B(\hat{D}_1)] = \bar{R}[B(\hat{D})]$.*

(2.2) shows that the set $\bar{R}[B(\hat{D})]$ is independent of the particular injective parameterization used for an elementary surface S. We call $\bar{R}[B(\hat{D})]$
the *surface-boundary* of S. The points of S not on the surface-boundary form
the *surface-interior* of S. The surface-boundary can be visualized as the
points "on the edge" of S. By (2.2), the surface-boundary of an elementary
surface is a simple, closed curve.

Example 7. The hemisphere and cone have circles as their surface-
boundaries. The surface-boundary of a half-cylinder is a simple, closed curve
consisting of two semicircles and two straight segments.

Smoothness. As with curves, we give a geometrical definition of
smoothness for surfaces.

Definition. Let P be a point on a given elementary surface S. We say that a
line L through P is *normal to S at P* if L and S satisfy the following two
conditions:
    (i)   For every directed smooth curve C which lies in S and goes
through P, L is orthogonal to the unit tangent vector of C at P.
    (ii) Every unit vector orthogonal to L is the unit tangent vector at P of
some directed smooth curve which lies in S and goes through P. (See [2–4].)
(Note that for a given S and P, there can be at most one normal line to S at P.)

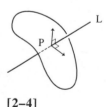

[2–4]

(2.3)   *Definition.* An elementary surface S will be called *smooth* if (i) with
every surface-interior point P, we can associate a unique straight line N_P such
that N_P is normal to S at P; (ii) this normal line varies from point to point on
S in a continuous way, that is to say, $\lim_{Q \to P} N_Q = N_P$, where Q is a variable

surface-interior point and where this limit means that the non-obtuse angle between N_Q and N_P goes to zero; (iii) with every *surface-boundary* point P_b, we can associate a unique straight line N_{P_b} (which we call the *boundary normal* at P_b) such that for interior points Q, $\lim\limits_{Q \to P_b} N_Q = N_{P_b}$; and (iv) the surface-boundary of S is a piecewise smooth, simple, closed curve.

The *half-cylinder* and *hemisphere* (examples 5 and 6) are smooth elementary surfaces, since, by elementary geometry, the desired normal lines and boundary normals exist. The *cone* is not a smooth elementary surface, since the vertex is a surface-interior point which has no normal line.

[2–5]

Example 8. Let S be a half-cone (see [2–5]). Here S is an elementary surface and the vertex V is a surface-boundary point. There is, however, no boundary-normal at V, since there can be no line L through V such that for all interior points Q, $\lim\limits_{Q \to V} N_Q = L$. Hence S is not smooth.

Note. In what follows, we shall refer to the surface-boundary of a surface S as, simply, the *boundary* of S.

§21.3 The Parametric Normal Vector $\vec{w}$

(3.1) *Definition.* We say that a parameterization $\vec{R}(u,v)$ on $\hat{D}$, of a surface S, is C^1 if each of the scalar coordinate functions $x(u,v)$, $y(u,v)$, and $z(u,v)$ is C^1 on $\hat{D}$.

In §21.1, the parameterizations of examples 1, 2, and 4 are in C^1, but the parameterization of example 3 is not, since $\frac{\partial z}{\partial u}$ and $\frac{\partial z}{\partial v}$ do not exist at $(u,v) = (0,0)$. In §21.2, the parameterization in example 5 is C^1, but the parameterization of example 6 is not, since $z(x,y) = (1 - x^2 - y^2)^{1/2}$ has no derivatives for $x^2 + y^2 = 1$.

Let $\vec{R}$ on $\hat{D}$ be in C^1, and let $\hat{P}$ be an interior point of $\hat{D}$. We define the vector $\vec{w}(\hat{P})$ by

(3.2)
$$\vec{w}(\hat{P}) = \frac{\partial \vec{R}}{\partial u}\Big|_{\hat{P}} \times \frac{\partial \vec{R}}{\partial v}\Big|_{\hat{P}} .$$

$\vec{w}(\hat{P})$ is defined for all interior points $\hat{P}$ of $\hat{D}$ and is continuous as a vector-valued function on $\hat{D}$. We sometimes write $\vec{w}(u,v)$ for $\vec{w}(\hat{P})$. $\vec{w}(\hat{P})$ is called the *parametric normal vector given by* $\hat{P}$. If $P = \vec{R}(\hat{P})$, we sometimes write $\vec{w}(\hat{P})$ as $\vec{w}_P$ and refer to $\vec{w}_P$ as the *parametric normal vector at the point P on S.*

$\left.\dfrac{\partial\vec{R}}{\partial u}\right|_{\hat{P}}$ and $\left.\dfrac{\partial\vec{R}}{\partial v}\right|_{\hat{P}}$ are velocity vectors at P for the u-curve and the v-curve through P. If $\vec{w}(\hat{P}) \neq \vec{0}$, then (3.2) implies that $\left.\dfrac{\partial\vec{R}}{\partial u}\right|_{\hat{P}} \neq \vec{0}$, $\left.\dfrac{\partial\vec{R}}{\partial v}\right|_{\hat{P}} \neq \vec{0}$, and that $\left.\dfrac{\partial\vec{R}}{\partial u}\right|_{\hat{P}}$ and $\left.\dfrac{\partial\vec{R}}{\partial v}\right|_{\hat{P}}$ are non-collinear vectors. This implies that $\vec{w}_P$ $(=\vec{w}(\hat{P}))$ is perpendicular to the plane through P determined by those non-collinear velocity vectors. (See [3–1].) This leads us to the first of two fundamental properties of the vector $\vec{w}$. (The second fundamental property will be described in §21.4.)

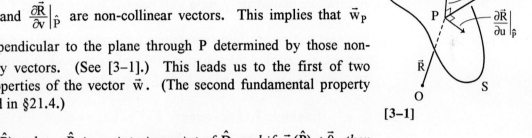

[3–1]

(3.3) *If* $P = \vec{R}(\hat{P})$, *where* $\hat{P}$ *is an interior point of* $\hat{D}$ *and if* $\vec{w}(\hat{P}) \neq \vec{0}$, *then* $\vec{w}_P$ *is normal to S at P.*

Proof. We must show (i) that $\vec{w}_P$ is normal at P to every smooth curve which lies in S and gives through P, and (ii) that there is such a smooth curve in S for every possible normal direction to $\vec{w}_P$.

To prove (i), let C be a smooth curve which lies in S and goes through P. Let t be the parameter for a smooth parameterization of C. For each value of t, let P_t be the corresponding point on C (and hence in S). When P_t is close enough to P, P_t must come from an interior point of $\hat{D}$. Let $\hat{P}_t = (u(t), v(t))$ be this interior point of $\hat{D}$. In the problems we show, using implicit differentiation, that $u(t)$ and $v(t)$ are C^1 functions. Let $\vec{V}_P$ be the velocity vector for C at P. $\vec{V}_P \neq \vec{0}$, since the parameterization of C is smooth. By the chain rule, we have

$$\vec{V}_P = \left.\frac{\partial\vec{R}}{\partial u}\right|_{\hat{P}} \left.\frac{\partial u}{\partial t}\right|_{\hat{t}} + \left.\frac{\partial\vec{R}}{\partial v}\right|_{\hat{P}} \left.\frac{\partial v}{\partial t}\right|_{\hat{t}}, \qquad \text{where} \quad P_{\hat{t}} = P.$$

Hence $\quad \vec{V}_P \cdot \vec{w}_P = \left.\dfrac{\partial u}{\partial t}\right|_{\hat{t}} \left[\left.\dfrac{\partial\vec{R}}{\partial u}\right|_{\hat{P}}, \left.\dfrac{\partial\vec{R}}{\partial u}\right|_{\hat{P}}, \left.\dfrac{\partial\vec{R}}{\partial v}\right|_{\hat{P}}\right] + \left.\dfrac{\partial v}{\partial t}\right|_{\hat{t}} \left[\left.\dfrac{\partial\vec{R}}{\partial v}\right|_{\hat{P}}, \left.\dfrac{\partial\vec{R}}{\partial u}\right|_{\hat{P}}, \left.\dfrac{\partial\vec{R}}{\partial v}\right|_{\hat{P}}\right].$

Since both scalar triple products are identically zero, we have

$$\vec{V}_P \cdot \vec{w}_P = 0 .$$

Hence if $\vec{w}_P \neq \vec{0}$, $\vec{w}_P$ must be perpendicular to $\vec{V}_P$. This completes the proof of (i). The proof of (ii) is easier and is given as a problem.

In view of (3.3), we can use $\vec{w}_P$ for the following purposes. Under the assumptions of (3.3), $\vec{w}_P$ gives us a *normal vector* to S at P. From this, in turn, we immediately get a *unit normal vector* to S at P: $\hat{w}_P = \dfrac{\vec{w}_P}{|\vec{w}_P|}$. We can also use $\vec{w}_P$ to find expressions for a *normal line* to S at P. Finally, we can use $\vec{w}_P$ to find the equation of the plane through P perpendicular to $\vec{w}_P$. We call this plane the *tangent plane* to S at P.

Example 1. *For example 2 of §21.1, we have*

$$\vec{R}(u, v) = u \cos v\hat{i} + u \sin v\hat{j} + 2u\hat{k} .$$

Then
$$\frac{\partial \vec{R}}{\partial u} = \cos v\hat{i} + \sin v\hat{j} + 2\hat{k} ,$$

$$\frac{\partial \vec{R}}{\partial v} = -u \sin v\hat{i} + u \cos v\hat{j} ,$$

and
$$\vec{w}(u, v) = -2u \cos v\hat{i} - 2u \sin v\hat{j} + u\hat{k} .$$

We note that at the origin in $\mathbf{E}^3$, with $u = 0$, we have $\vec{w}_P = \vec{0}$. This accords with the geometrical fact that there is no normal vector at the vertex of a cone.

Example 2. *For example 4 of §21.1, we have*

$$\vec{R}(u, v) = u\hat{i} + v\hat{j} + 2(u^2 + v^2)\hat{k} .$$

Then
$$\frac{\partial \vec{R}}{\partial u} = \hat{i} + 4u\hat{k} ,$$

$$\frac{\partial \vec{R}}{\partial v} = \hat{j} + 4v\hat{k} ,$$

and
$$\vec{w}(u, v) = -4u\hat{i} - 4v\hat{j} + \hat{k} .$$

Example 2 is a special case of the next example.

Example 3. *Let* $\vec{R}(x,y) = x\hat{i} + y\hat{j} + f(x,y)\hat{k}$, *where* f *is* C^1 *on* $\hat{D}$ *in the* xy *plane. Here* S *is the graph of* f. *Then*

$$\frac{\partial \vec{R}}{\partial x} = \hat{i} + f_x\hat{k},$$

$$\frac{\partial \vec{R}}{\partial y} = \hat{j} + f_y\hat{k},$$

and
$$\vec{w}(x,y) = \frac{\partial \vec{R}}{\partial x} \times \frac{\partial \vec{R}}{\partial y} = -f_x\hat{i} - f_y\hat{j} + \hat{k} .$$

Note that this expression for $\vec{w}$ turns out to be the same as our earlier formula (see (6.2) of §11.6) for an upward normal vector to the graph of f.

A general formula for $\vec{w}$, as given by (3.2), is:

$$(3.4) \qquad \vec{w} = \frac{\partial \vec{R}}{\partial u} \times \frac{\partial \vec{R}}{\partial v}$$

$$= \left(\frac{\partial y}{\partial u} \frac{\partial z}{\partial v} - \frac{\partial z}{\partial u} \frac{\partial y}{\partial v} \right)\hat{i} + \left(\frac{\partial z}{\partial u} \frac{\partial x}{\partial v} - \frac{\partial x}{\partial u} \frac{\partial z}{\partial v} \right)\hat{j} + \left(\frac{\partial x}{\partial u} \frac{\partial y}{\partial v} - \frac{\partial y}{\partial u} \frac{\partial x}{\partial v} \right)\hat{k}$$

$$= \frac{\partial(y,z)}{\partial(u,v)}\hat{i} + \frac{\partial(z,x)}{\partial(u,v)}\hat{j} + \frac{\partial(x,y)}{\partial(u,v)}\hat{k} .$$

When S is the graph of f(x,y), this formula becomes (as in example 3):

$$(3.5) \qquad \vec{w} = -f_x\hat{i} - f_y\hat{j} + \hat{k}.$$

Let S be a parametric surface. What conditions on a parameterization $\vec{R}(u,v)$ on $\hat{D}$ for S will guarantee that S is smooth? The following theorem supplies an answer to this question.

(3.6) Theorem. *Let* S *be given by* $\vec{R}(u,v)$ *on* $\hat{D}$. *If the following three conditions hold, then* S *must be smooth.*

(1)   R *is injective on* $\hat{D}$,

(2) $\vec{R}$ *is* C^1 *on* $\hat{D}$,

(3) $\vec{w}(\hat{P}) \neq \vec{0}$ *on* $\hat{D}$.

(i) and (ii) of (2.3) are immediate from (3.3). For (iii) and (iv), see §21.9. Corollaries to (3.6) can be formulated for cases where $\vec{w} = \vec{0}$ on the boundary of $\hat{D}$, and/or $\vec{R}$ is not injective on the boundary of $\hat{D}$, but where S must still be smooth. Example 1 of §21.1 is such a case.

Remark. Even if a parametric representation $\vec{R}(u,v)$ on $\hat{D}$ is not C^1 for all of S, it may be C^1 for portions of S. For these portions, $\vec{w}$ can be calculated and will have its usual properties. Thus, for example 3 of §21.1, we get $\vec{w}(\hat{P}) = -\dfrac{2x}{(x^2+y^2)^{1/2}}\hat{i} - \dfrac{2y}{(x^2+y^2)^{1/2}}\hat{j} + \hat{k}$, which is valid for all points in $\hat{D}$ except $(0,0)$.

§21.4 Surface Area

We have seen that the first fundamental property (3.3) of $\vec{w}_P$ concerns the *direction* of $\vec{w}_P$. The *second* fundamental property of $\vec{w}_P$ concerns the *magnitude* of $\vec{w}_P$. We begin with an informal statement of this second property.

(4.1) $\left|\vec{w}_P\right|$ *serves as a local amplification factor from area in* $\hat{D}$ *to surface area on S. That is to say, if* $P = \vec{R}(\hat{P})$, *and if a small region* $\hat{U}$ *around* $\hat{P}$ *maps to a corresponding region* U *around* P, *then*

$$\text{surface area }(U) \approx \left|\vec{w}_P\right|\ \text{area}(\hat{U}).$$

In the remainder of this section, we proceed as follows: (1) We define *surface area* for a smooth surface S. (2) We put (4.1) into precise mathematical form and prove it. (3) We show how to use $\vec{w}(u,v)$ to calculate surface area for a given parameterized surface. (4) We show how, in the case of a simple or familiar parameterization, (3.3) and (4.1) can sometimes be used to find $\vec{w}_P$ directly, without the lengthy Jacobian calculation indicated in (3.4).

Let S be a smooth parametric surface with a parameterization $\vec{R}$ on $\hat{D}$. Since $\hat{D}$ is an elementary region in E^2, $\hat{D}$ is compact. Since $\hat{D}$ is compact, S must be a compact (and hence bounded) set in E^3. (See the problems.)

(1) For given $n > 0$, divide $\hat{D}$ into n elementary pieces $\hat{D}_1,...,\hat{D}_n$. For each $i \le n$, let d_i be the *diameter* of $\hat{D}_i$ (the maximum distance between a pair of points in $\hat{D}_i$.) Let $S_i = \vec{R}[\hat{D}_i]$. Then the pieces $S_1,...,S_n$ constitute S. On each S_i, choose a *sample point* P_i^*. (See [4-1].) Let M_i be the tangent plane to S at P_i^*. Let A_i be the *area of the projection of* S_i *onto* M_i. Consider

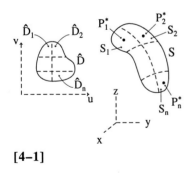

[4-1]

(4.2)
$$\lim_{\substack{n \to \infty \\ \max d_i \to 0}} \sum_{i=1}^{n} A_i \, .$$

It can be shown, by an argument similar to the proof of the existence theorem for multiple integrals in §14.10, that the plane areas A_i exist, that this limit exists, and that the value of the limit does not depend on the particular parameterization chosen. We call this value the *area of the surface* S.

(2) Next we turn to (4.1). We restate it as follows.

(4.3)   *Let* $\hat{P}$ *be an interior point of* $\hat{D}$, *and let* $P = \vec{R}(\hat{P})$. *Let* $\hat{D}(a,\hat{P})$ *be the set of all points* $\hat{Q}$ *in* $\hat{D}$ *such that* $d(\hat{Q},\hat{P}) \le a$. *Then*

$$\left| \vec{w}_P \right| = \left| \vec{w}(\hat{P}) \right| = \lim_{a \to 0} \frac{surface\ area(\vec{R}[\hat{D}(a,\hat{P})])}{area(\hat{D}(a,\hat{P}))} \, .$$

In other words, the magnitude $\left| \vec{w}(\hat{P}) \right|$ *gives us the instantaneous rate at which mapped surface area begins to appear around* P *per unit of area in* $\hat{D}$ *beginning to appear around* $\hat{P}$.

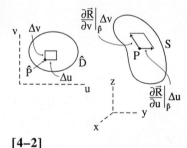

[4–2]

We outline a proof for (4.3). Consider a small rectangle at $\hat{P}$ in $\hat{D}$ with sides Δu and Δv, and hence with area $\Delta u \Delta v$. This rectangle maps to a region at P in S bounded by parametric curves. We call this region (in S) S*. Let F* be the parallelogram with adjacent sides given by the vectors $\left. \frac{\partial \vec{R}}{\partial u} \right|_{\hat{P}} \Delta u$ and $\left. \frac{\partial \vec{R}}{\partial v} \right|_{\hat{P}} \Delta v$. (Thus F* lies in the tangent plane to S at P.) (See [4–2].) It can be shown that $\frac{surface\ area(S^*)}{area(F^*)}$ approaches 1 as Δu and Δv become

small.   But $area(F^*) = \left| \left. \frac{\partial \vec{R}}{\partial u} \right|_{\hat{P}} \Delta u \times \left. \frac{\partial \vec{R}}{\partial v} \right|_{\hat{P}} \Delta v \right| = \left| \left. \frac{\partial \vec{R}}{\partial u} \right|_{\hat{P}} \times \left. \frac{\partial \vec{R}}{\partial v} \right|_{\hat{P}} \right| \Delta u \Delta v = \left| \vec{w}(\hat{P}) \right| \Delta u \Delta v$.

Then $\frac{surface\ area(S^*)}{\Delta u \Delta v} = \frac{surface\ area(S^*)}{area(F^*)} \cdot \frac{area(F^*)}{\Delta u \Delta v} = \frac{surface\ area(S^*)}{area(F^*)} \left| \vec{w}(\hat{P}) \right|$, and

hence $\frac{surface\ area(S^*)}{\Delta u \Delta v}$ approaches the limit $\left| \vec{w}(\hat{P}) \right|$ as Δu and Δv become

small. This is (4.3).

(3) From the definition of surface area (4.2) and from property (4.3) of $\vec{w}$, we can get an integral formula for the area of a smooth S given by $\vec{R}(u,v)$. In the definition of surface area, let $\hat{A}_i = Area(\hat{D}_i)$. Then $Area(S) \approx \sum_{i=1}^{n} A_i = \sum_{i=1}^{n} \frac{A_i}{\hat{A}_i}\hat{A}_i \approx \sum_{i=1}^{n}|\vec{w}(u_i,v_i)|\hat{A}_i$ by (4.3), where $(u_i,v_i) = \hat{P}_i^*$. Going to the limit as $n \to \infty$ and $\max d_i \to 0$, and using Duhamel's principle, we get

(4.4)   **Theorem.** *Let S have a* C^1 *parameterization* $\vec{R}$ *on* $\hat{D}$, *where* $\vec{w}(\hat{P}) \neq \vec{0}$ *on* $\hat{D}$ *and where* $\hat{D}$ *is a convex, elementary region in* $\mathbf{E}^2$. *Then*
  (i)   *S has finite surface area;*
  (ii)   *surface area* (S) $= \iint_{\hat{D}}|\vec{w}(u,v)|dudv$.

Example 1. *For the half-cylinder of radius* a *and height* b *in example 5 of §21.2, we had* θ *and* z *as parameters and*

$$\vec{R}(\theta,z) = a\cos\theta\hat{i} + a\sin\theta\hat{j} + z\hat{k} \text{ on } 0 \leq \theta \leq \pi, 0 \leq z \leq b .$$

Then
$$\frac{\partial\vec{R}}{\partial\theta} = -a\sin\theta\hat{i} + a\cos\theta\hat{j},$$

and
$$\frac{\partial\vec{R}}{\partial z} = \hat{k} .$$

Hence
$$\vec{w} = \frac{\partial\vec{R}}{\partial\theta} \times \frac{\partial\vec{R}}{\partial z} = a\cos\theta\hat{i} + a\sin\theta\hat{j},$$

and
$$|\vec{w}| = a$$

Thus   *surface area*(S) $= \iint_{\hat{D}}|\vec{w}|dudv = \int_0^{\pi}\int_0^{b} a\,du\,dv = \pi ab$.

Example 2. *We find the area of the paraboloid given by the graph of* $f(x,y) = x^2 + y^2$ *over the domain* $x^2 + y^2 \leq 1$. *By* (3.5), *we have* $\vec{w} = -2x\hat{i} - 2y\hat{j} + \hat{k}$ *on* $x^2 + y^2 \leq 1$. *Then surface area* $= \iint_{\hat{D}}\sqrt{4x^2 + 4y^2 + 1}dxdy$. We go to polar coordinates to evaluate this double integral.

$$surface\ area = \int_0^{2\pi}\int_0^1 (4r^2 + 1)^{1/2} r\,dr\,d\theta$$

$$= \frac{1}{8}\int_0^{2\pi}\int_{r=0}^{r=1} u^{1/2}du\,d\theta = \frac{1}{8}\int_0^{2\pi}\frac{2}{3}u^{3/2}\Big]_{r=0}^{r=1} d\theta$$

$$= \frac{1}{8} \int_0^{2\pi} \frac{2}{3}(4r^2+1)^{3/2} \bigg]_0^1 d\theta = \frac{2\pi}{8} \cdot \frac{2}{3}(5^{3/2}-1) = \frac{\pi}{6}(5^{3/2}-1).$$

(4) If S is given as the graph of a function f(x,y), then it is usually simplest to find $\vec{w}$ by (3.5). In other cases, we can sometimes use (3.3) and (4.3) to get an explicit expression for $\vec{w}(u,v)$ without the calculation suggested by (3.4). We give two examples.

Example 3. *Consider the parameterization of a sphere of radius* a *given by*

$$\vec{R}(\varphi,\theta) = a\sin\varphi\cos\theta\,\hat{i} + a\sin\varphi\sin\theta\,\hat{j} + a\cos\varphi\,\hat{k}$$

for $0 \le \varphi \le \pi, 0 \le \theta \le 2\pi$.

We wish to find $\vec{w}_P$ when P has the spherical coordinates (a,φ,θ).

By geometry, we know that the direction of $\vec{w}_P$ must be the direction of the coordinate vector $\hat{\rho}_P$. Let P on S come from $\hat{P}$ in $\hat{D}$. Consider a piece of S at P given by a small rectangle F^* at $\hat{P}$ in $\hat{D}$ with sides $\Delta\varphi$ and $\Delta\theta$. By the geometry of spherical coordinates, $\vec{R}[F^*]$ at P is approximately a rectangle with sides $a\Delta\varphi$ and $a\sin\varphi\,\Delta\theta$. Hence $\vec{R}[F^*]$ has the approximate area $a^2\sin\varphi\Delta\varphi\Delta\theta$. Thus, by (4.3),

$$\left|\vec{w}_P\right| = \frac{a^2\sin\varphi\Delta\varphi\Delta\theta}{\Delta\varphi\Delta\theta} = a^2\sin\varphi .$$

We therefore have

$$\vec{w}_P = a^2\sin\varphi\hat{\rho} .$$

(As a check, we can use (4.4) to get the area of S.

$$\int_0^{2\pi} \int_0^{\pi} a^2\sin\varphi\,d\varphi d\theta = 4\pi a^2 .)$$

Example 4. For the surface S in example 1 above, we can use an even simpler direct argument to get $\vec{w}_P$. First, the direction of $\vec{w}_P$ must be $\hat{r}_P$. Second, $\left|\vec{w}_P\right| = \frac{ad\theta dz}{d\theta dz} = a$. Hence $\vec{w}_P = a\hat{r}_P$.

Remark. The normal line of $\vec{w}_P$ is determined by the geometry of S and does not depend on the C^1 parameterization used. The magnitude $|\vec{w}_P|$, however, may vary from one parameterization to another. In this respect, as in others, the parametric normal vector $\vec{w}_P$ for surfaces is analogous to the velocity vector $\vec{v}_P$ for curves, with "normal" in place of "tangent."

§21.5 Finite Surfaces

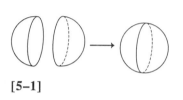

[5–1]

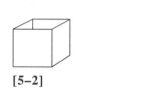

[5–2]

Let S be a connected set of points in $\mathbf{E}^3$ (see §18.1). We say that S is a *finite surface* if S can be divided into a finite collection of elementary surfaces where no two of these elementary surfaces intersect (or overlap) except along segments of common boundary, and where no segment of boundary is fully contained in more than two of these elementary surfaces. For example, a sphere is a finite surface, since it can be divided into two elementary hemispheres by a great circle, [5–1]; an open-topped cube is a finite surface, since it can be divided into five elementary square surfaces, [5–2]; a closed cylinder is an elementary surface, since it can be divided into four elementary pieces - two half-cylinders and two disks, [5–3]; and a Möbius strip (as in the figure) is a finite surface, since it can be divided into two elementary semi-circular bands, one of which has a half-twist, [5–4].

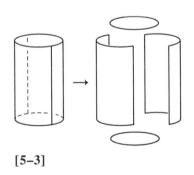

[5–3]

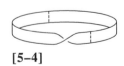

[5–4]

(5.1) ***Informal definition.*** A figure which can be assembled from a finite number of elementary surfaces in this way will be called a *finite surface*. In such a division of a finite surface S into elementary surfaces, a segment of joint boundary shared between two elementary pieces will be called a *seam* of that division, and a segment which remains as boundary to only one elementary piece will be called a *boundary segment* of S. The boundary segments, taken together, are called the *boundary* of the finite surface S. It follows that the boundary of a finite surface must consist of zero or more simple-closed, disjoint (non-intersecting) curves.

The informal definition of finite surface above will suffice for our present purposes. For the reader's interest, a formal definition of *finite surface*, from which the various statements above can be deduced, is included at the end of this section.

If S is a finite surface, then any division of S into elementary surfaces, as described above, is called a *decomposition* of S. Obviously, a given finite surface may have more than one decomposition. (For example, a given sphere may be decomposed into two hemispheres in infinitely many different ways.)

It is easy to show that the boundary of a finite surface (as defined above) does not depend on the decomposition used to define that surface.

(5.2) **Definition.** We say that a finite surface S is *piecewise smooth* if S has a decomposition in which every elementary surface is smooth. (In this case, by (2.3), every seam must be a piecewise smooth curve.)

(5.3) **Definition.** We say that a finite surface S is *smooth* if S has a decomposition into smooth elementary surfaces such that, for each point P on each seam, there is a neighborhood of P in E^3 whose intersection with S is a smooth elementary surface.

Example 1. A *sphere* is a smooth finite surface which can be decomposed into two hemispheres.

Example 2. Similarly, a *cylinder open at both ends* is a smooth finite surface, which can be decomposed into two half-cylinders, and which has two simple closed curves (which are circles) as its boundary.

Example 3. A *closed cylinder* and an *open-topped cube* are both piecewise smooth, finite surfaces. Neither is smooth. Moreover, the cylinder has no boundary; and the open cube has a single, simple, closed curve as its boundary. See [5–5].

[5–5]

Example 4. A *torus* is a smooth finite surface. It is not an elementary surface, but it can be assembled from four elementary surfaces with seams as in [5–6]. It has no boundary.

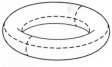

[5–6]

Example 5. A *Möbius strip*, as shown in [5–7], is a smooth finite surface. It has a single, simple, closed, smooth curve as its boundary. In §21.6, however, we shall see that it is not an elementary surface.

We conclude this section with a formal definition of *finite surfaces*.

[5–7]

Definition. A figure S in E^3 is a *finite surface* if we can designate a finite number of elementary surfaces in S, which we call *faces*, and a finite number of simple curves in S, which we call *edges* and whose endpoints we call *vertices*, such that
   (i) every point in S belongs to at least one face;
   (ii) every edge is part of the boundary of some face, and the boundary of each face is made up of edges;

(iii)   faces cannot intersect except in common edges or vertices;

(iv)   each edge is shared by at most two faces;

(v)   two faces with a common edge are said to be *adjacent*, and for every pair of faces S_1 and S_2, there is a chain of successively adjacent faces which connects S_1 to S_2;

(vi)   for every pair of faces S_1 and S_2 which share a common vertex, there is a chain of successively adjacent faces which connects S_1 to S_2 and whose members all share that vertex.

§21.6 Sidedness; Closed Surfaces

Consider a Möbius strip S as defined in §21.5. S is a smooth finite surface. Let P be a point on the mid-line of S. Let B be a bug standing on S at P. An erect arrow A is attached to B's back, so that wherever B stands, A is normal to S at that point. Let B leave P and follow the mid-line of S until B is again standing at the point P. We see that B is now on the "other side" of S, and that A points in a direction exactly opposite to its previous direction at P. We describe this circumstance by saying that the Möbius strip is a "one-sided surface."

We define *one-sidedness* for a piecewise smooth finite surface S as follows. We allow B to start at any surface-interior point of any smooth elementary piece of S. We allow B to move along any path on S, provided only that B does not touch the boundary of S. When B moves across a seam between two smooth pieces at a crossing point P, the arrow A switches from being normal to the first piece to being normal to the second according to the following *rule*: Let $\hat{T}_1$ and $\hat{T}_2$ be unit tangent vectors at P to the incoming and outgoing paths. Let $\hat{N}_1$ be the direction of A as it arrives at P. Let $\hat{E}$ be a unit tangent vector to the seam at P. Then the direction of A as it leaves P must be the direction of $[\hat{T}_1, \hat{N}_1, \hat{E}](\hat{E} \times \hat{T}_2)$. This informal idea of a moving bug can be made precise and formal by using a unit vector $\hat{n}$ normal to S for the direction of A, by requiring that $\hat{n}$ be a continuous function of position on S as B moves along a smooth path, and by requiring that $\hat{n}$ change abruptly across a seam in accord with the rule above.

(6.1)   *Definition.* A given piecewise smooth finite surface S is *one-sided* if it is possible for B to move from some starting point P on S and to find itself, eventually, again at P with the direction of its arrow reversed. If this is not possible (that is to say, if B always has its arrow in the same direction if and when it arrives back to its starting point), we say that the piecewise smooth, finite surface is two-sided.

The following fact can be proved.

(6.2)   *Every elementary smooth surface is two-sided.*

   Examples. A *sphere*, a *torus*, and a *closed cylinder* are each two-sided. An *open cylinder* and an *open-topped cube* are each two-sided.

   Let S be the piecewise smooth finite surface in [6–1]. S consists of eight congruent plane squares joined to form a piecewise smooth surface. S is one-sided.

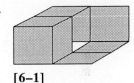

[6–1]

Definition. We say that a finite surface is *closed* if it has no boundary.

   We have already seen several examples of closed surfaces, including the *sphere*, the *closed cylinder*, and the *torus*. The following fundamental fact about closed surfaces in $\mathbf{E}^3$ can be proved.

(6.3)   *If a finite surface S in* $\mathbf{E}^3$ *is closed and piecewise smooth, then S divides the rest of* $\mathbf{E}^3$ *(that is, the points in* $\mathbf{E}^3$ *which are not on S) into two non-empty sets: namely, the* interior *to S, which is bounded, and the* exterior *to S, which is unbounded. Each of these sets is a connected set, but these two sets together do not form a connected set, since any curve which connects a point in the interior to S to a point in the exterior to S must intersect S.* (Note that in the context of (6.3) the words "closed" and "interior" are used with new meanings, distinct from, and unrelated to, the meanings introduced and used for these words in §12.2.)

   As a corollary to (6.3) we have:

(6.4)   *If a piecewise smooth surface is closed, then it must be two-sided.* (See the problems.)

   Remark. The concepts of sidedness and closure can also be applied to finite surfaces whose elementary pieces are smooth except at isolated surface-interior points. Thus the open cone in example 1 of §21.1 is two-sided and so is the closed circular cone obtained by adding a disk to the top of this open cone. Appropriate versions of (6.3) and (6.4) hold for such cases.

§21.7 Connectivity

The ideas of *surface* and *closed piecewise smooth surface* are used for the following geometric definitions.

(7.1) *Definition.* Let D be a connected region in E^2. We say that D is *simply connected* (or "one-connected") if, for every simple, closed, piecewise smooth curve C lying entirely in D, the interior of C (in the sense of §6.5) also lies entirely in D.

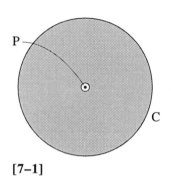

P

C

[7–1]

Example 1. In E^2, let D consist of all of E^2 except for a single point P, [7–1]. Then D is not simply connected: let C be the circle of radius 1 with center at P; then C lies entirely in D, but the interior of C does not lie entirely in D, because P is in the interior of C but is not in D.

(7.2) *Definition.* Let D be a connected region in E^3. We say that D is *simply connected* (or "one-connected") if, for every simple, closed, piecewise smooth curve C lying entirely in D, there is a finite, piecewise smooth, two-sided surface S such that (i) S has C as its complete boundary, and (ii) S lies entirely in D.

(The following geometrical fact is far from obvious. It can, however, be proved by more advanced methods:

(7.3) *For every closed, piecewise smooth curve C in E^3, there is a finite, piecewise smooth, two-sided surface S which has C as its complete boundary.* Several examples, for "knotted" curves, appear in the problems.)

Example 2. In E^3, let D consist of all of E^3 except for a single point P. Then D is simply connected because for any simple, closed, piecewise smooth curve C in D, we can construct a piecewise smooth surface S which has C as its boundary and which does not pass through P. (For example, if C is a circle of radius 1 with center at P, we can take S to be a hemisphere of radius 1 with center at P.

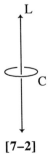

L

C

[7–2]

Example 3. In E^3, let D consist of all of E^3 except for a straight line L. Then D is not simply connected. For example, let C be a circle of radius 1 lying in a plane perpendicular to L with the center of C on L [7–2]. Then there is no finite, piecewise smooth surface S which has C as its boundary and which lies completely in D. Any such S must intersect L and hence have a point which is not in D.

(7.4) **Definition.** Let D be a connected set of points in E^3. We say that D is *two-connected* if for every piecewise smooth, finite, closed surface S lying entirely in D, the interior of S also lies entirely in D.

[7–3]

Example 4. Let D *be a solid sphere in* E^3, [7–3]. Then D is two-connected.

Example 5. Let D *be a solid sphere in* E^3 *with its center point removed*, [7–4]. Then D is simply connected but not two-connected.

[7–4]

Example 6. Let D *consist of a closed surface (for example a torus) together with its interior.* It follows from (6.3) that D is two-connected. (D need not be simply connected; see §21.9.)

[7–5]

§21.8 Curvature

We seek a natural definition of *curvature* for smooth surfaces. Let P be a fixed point on a smooth surface S, and let P′ be a variable point on S. We want the "curvature of S at P" to give us information about the rate and manner in which a normal line at P′ changes its direction as P′ moves through P. This information should tell us, for example, how "flat" or "rounded" S is at P and whether S is "cup-shaped" or "saddle-shaped" near P. We shall see that we need two numbers to give us this information in a simple and natural form.

To get useful definitions and facts, we must assume that S has a C^2 parameterization (that is to say, the functions x(u,v), y(u,v), and z(u,v) have continuous second-order derivatives) with $\vec{w}_P \neq \vec{0}$ at all points P on S.

Let P be a surface-interior point on S. Take a point Q in E^3 so that $\overrightarrow{PQ} = \vec{w}_P$. Let M be a plane containing P and Q, [8–1]. We call M a *normal plane* to S at P. Then M intersects S in a curve C_M which goes through P. Let κ_M be the curvature of C_M at P. If $\kappa_M \neq 0$, let $\hat{N}_M$ be the unit normal for C_M at P. (Then $\hat{N}_M$ must be collinear with $\vec{w}_P$.) Define

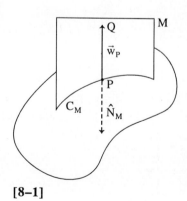

[8–1]

(8.1) $K_M = (\hat{N}_M \cdot \hat{w}_P)\kappa_M, \text{if } \kappa_M \neq 0; K_M = 0, \text{if } \kappa_M = 0$.

Thus $K_M = \pm\kappa_M$, depending on whether or not $\hat{N}_M$ and $\hat{w}_P$ have the same or opposite directions. K_M is called the *normal curvature* of S at P for the normal plane M and the normal direction $\hat{w}_P$. Obviously, K_M may vary as the plane M is rotated about the line of $\overrightarrow{PQ}$. In the problems, we shall show, from our assumption that S has a C^2 parameterization, that K_M must vary in an extraordinarily regular and symmetric way. The details of this regularity are known as *Euler's theorem*.

(8.2)   ***Euler's theorem.*** *Let S be a smooth surface given by a C^2 parameterization $\bar{R}$ as described above. Let P be a fixed interior point of S. Let $\hat{w}_P$, Q, M, and K_M be as described above. Then*

   (a) K_M *exists for all choices of* M

   (b) *As the plane M rotates about the line of $\overrightarrow{PQ}$, K_M has a maximum value K_1 and a minimum value K_2. (K_1 and K_2 are called the* principal curvatures *of S at P.) The values of K_1 and K_2 can be calculated from the first- and second-order derivatives of $\bar{R}$ at P.*

   (c) *If $K_1 \neq K_2$, then there are unique normal planes M_1 and M_2 at P such that $K_1 = K_{M_1}$ and $K_2 = K_{M_2}$. Moreover, the planes M_1 and M_2 are perpendicular to each other.*

   (d) *Assume $K_1 \neq K_2$, and let M_1 be as in (c). Let M be any normal plane to S at P, and let θ be the non-obtuse angle between M and M_1. Then*

$$K_M = K_1 \cos^2 \theta + K_2 \sin^2 \theta.$$

We see from Euler's theorem that the principal curvatures K_1 and K_2 give us full information about the possible normal curvatures of S at P. K_1 and K_2 are often encoded into two other numbers:

(8.3) $K = K_1 K_2$ is called the *total* (or "Gaussian") *curvature* of S at P.

(8.4) $H = \frac{1}{2}(K_1 + K_2)$ is called the *mean curvature* of S at P.

(The values of K_1 and K_2 can be easily recovered from K and H.) Several uses of K and H are described in the following examples.

   Example 1. *A soap film which is not a closed surface and which is constrained only at its edges must have H = 0 at every point.* Such a surface is called a *minimal surface.* Note that $H = 0 \Rightarrow K \leq 0$.

Example 2. *If a closed soap bubble is not acted on by external forces, then its surface is a sphere and K has a positive constant value at every point.* The increase in air pressure inside such a bubble (due to surface tension in the film) is then proportional to K.

Example 3. *Let S be a surface of constant total curvature.* (It can be proved that if S is closed, then S must be a sphere.) Assume that S is not closed and that S is formed of a bendable but not stretchable material. Then any surface into which S can be distorted must have the same total curvature at all points as before. (This can be illustrated with half a tennis ball or the rind of half an orange.)

Example 4. *It can be proved that if S has constant H and constant positive K, then S is all or part of a sphere.*

Remark. For the reader's possible interest, the formulas for calculating K_1 and K_2 at P from the derivatives of $\vec{R}$ at $\hat{P}$ are as follows: K_1 and K_2 are the roots of the quadratic equation

(8.5) $$K^2 - AK + B = 0 ,$$

where $$A = \frac{EN+GL-2FM}{EG-F^2} \quad \text{and} \quad B = \frac{LN-M^2}{EG-F^2} ,$$

where $$E = \vec{R}_u^2, \; F = \vec{R}_u \cdot \vec{R}_v, \; G = \vec{R}_v^2,$$

$$L = [\vec{R}_u, \vec{R}_v, \vec{R}_{uu}], \; M = [\vec{R}_u, \vec{R}_v, \vec{R}_{uv}], \text{ and } \; N = [\vec{R}_u, \vec{R}_v, \vec{R}_{vv}],$$

and where the indicated derivatives of $\vec{R}$ are evaluated at $\hat{P}$.

§21.9 Problems

§21.1

1. Show that the parameterization in example 3 gives the surface described in example 2.

For each of the following problems, give the indicated parameterization, including the region $\hat{D}$ of parameter values.

2. Use x and y to parameterize the planar region determined by the triangle with vertices at (0,0,0), (1,0,0), and (1,1,1).

3. Use θ and z to parameterize the cone of examples 2 and 3.

4. Parameterize the paraboloid of example 4
   (a) using r and θ;
   (b) using θ and z.

5. Use θ and z to parameterize the hyperboloid (of one sheet) $x^2 + y^2 - z^2 = 1$ between $z = -2$ and $z = 2$.

6. Parameterize the upper hemisphere of radius a with center at the origin,
   (a) using x and y;
   (b) using r and θ.

7. Use θ and z to parameterize the sphere of example 1. (*Hint.* Note that $z^2 + r^2 = 25$.)

8. Use parameters based on φ and θ to parameterize the ellipsoid $\dfrac{x^2}{a^2} + \dfrac{y^2}{b^2} + \dfrac{z^2}{c^2} = 1$. (*Hint.* Recall that the ellipse $\dfrac{x^2}{a^2} + \dfrac{y^2}{b^2} = 1$ can be parameterized by $\vec{R}(t) = a\cos t\,\hat{i} + b\sin t\,\hat{j}$.)

§21.3

1. For each of the following parameterizations, find the parametric normal vector $\vec{w}(u,v)$:
   (a) problem 2 for §21.1;
   (b) problem 3 for §21.1;
   (c) problem 4a for §21.1;
   (d) problem 5 for §21.1;
   (e) problem 6a for §21.1;
   (f) problem 6b for §21.1;
   (g) problem 8 for §21.1.

2. In the proof of (i) for (3.3), use implicit differentiation to show that the functions u(t) and v(t) must be in C^1.

3. Prove (ii) in the proof of (3.3).

4. Formulate and prove one or more corollaries to (3.6) for cases where S is smooth but $\vec{R}(u,v)$ is not injective on the boundary of $\hat{D}$.

§21.4

1. For each of the following parameterizations, find an integral for the surface area of the surface. Evaluate if possible.
   (a) problem 2 for §21.1;
   (b) problem 3 for §21.1;
   (c) problem 4a for §21.1;
   (d) problem 5 for §21.1;
   (e) problem 6a for §21.1; (*Hint.* Use polar coordinates to evaluate the integral.)
   (f) problem 6b for §21.1;
   (g) problem 8 for §21.1.

2. Let S be a surface with parameterization $\vec{R}(u,v)$ on $\hat{D}$, where $\hat{D}$ is an elementary region in $\mathbf{E}^2$. Show that S must be compact.

§21.6

1. (a) Outline a proof for (6.2).
   (b) Outline a proof for (6.4).
   (*Hint.* Let $\vec{R}(t) = x(t)\hat{i} + y(t)\hat{j} + z(t)\hat{k}$ be a parameterization of a closed path of the bug B showing that S is one-sided, and let u(t), v(t), w(t) be the components of the unit normal vector $\hat{A}$ attached to B. Then for (a), take the corresponding path in $\hat{D}$, and let this path shrink down to a single point. The proof for (b) is similar.)

§21.7

1. In each case, show by a sketch that there is a finite, smooth, two-sided surface having C as its complete boundary.
   (a) C is the boundary of a smooth Möbius strip.
   (b) C is the trefoil knot shown in [9–1].

[9–1]

§21.8

1. Prove Euler's theorem. (*Hint.* To express the normal curvature K_M, use the second-order directional derivative as described in (6.5) of §12.6.

2. As defined in example 1, a *minimal surface* is a C^2-surface for which $H = 0$ at all points.
    (a) Prove that $H = 0 \Rightarrow K \leq 0$.
    (b) Use (a) to prove that a minimal surface with a given fixed boundary is a surface of minimum area among all C^2-surfaces which have that boundary.

3. Give physical derivations for each of the facts asserted in example 2.   You may assume (i) that the pull of surface tension across any segment of curve in a given soap film is proportional to the length of that segment with a constant of proportionality which is the same for all curves in that film, and (ii) that the force of a constant gas pressure on a smooth piece of surface acts normal to the surface and is proportional to surface area.

4. Outline a proof for the fact given in example 3. (*Hint.* Begin with a suitable definition of what it means for a physical surface to be "bendable but not stretchable.")

5. Outline a proof for the fact given in example 4.

Chapter 22

Surface Integrals

§22.1 Integrals on Surfaces

In this chapter, we introduce two additional forms of definite integral: *scalar surface* integrals and *vector surface* integrals. As with previously considered integrals, we proceed in two steps. First, we give a formal, conceptual definition, which will help us to think about, understand, and use the integral. Second, we describe methods for evaluating the integral.

Our treatment of *surface integrals* is closely analogous to our treatment of *line integrals* in Chapter 19. Surface integrals differ from line integrals in that the *region of integration*, instead of being a finite curve in $\mathbf{E}^2$ or $\mathbf{E}^3$, is a finite surface in $\mathbf{E}^3$. As with line integrals, the *integrand* is a scalar or vector field whose domain contains the region of integration.

The words "scalar" and "vector" in the phrases "scalar surface integral" and "vector surface integral" refer to the nature of the integrand, not to the value of the integral. Both scalar surface integrals and vector surface integrals, as defined below in §22.2 and §22.4, have scalar values.

Other kinds of definite integrals on surfaces (besides the scalar-valued integrals defined below as *scalar surface integrals* and *vector surface integrals*) can also be defined. We consider two of these in §22.6. Some of these other integrals on surfaces will have vector values.

468

§22.2 Scalar Surface Integrals

Let S be a piecewise smooth, finite surface in $\mathbf{E}^3$. (S may be either one-sided or two-sided.) Let f be a scalar field in $\mathbf{E}^3$ whose domain contains S. We write the *scalar surface integral of* f *on* S as

$$\iint_S f d\sigma ,$$

and we define it as follows.

(2.1) **Definition.** For a sufficiently large positive integer n (n ≥ the minimum number of elementary pieces of S), divide S into n smooth elementary pieces $\Delta S_1, \ldots, \Delta S_n$. For each i, $0 < i \le n$, choose a *sample point* P_i^* on ΔS_i, and let $\Delta \sigma_i$ be the surface area of ΔS_i. (See [2–1].) Form the Riemann sum

$$\sum_{i=1}^{n} f(P_i^*) \Delta \sigma_i .$$

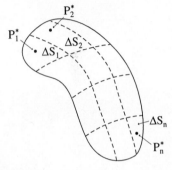

[2-1]

For each i, define d_i, the *diameter of* ΔS_i, as follows. d_i is the maximum value, over all pairs of points P and Q on ΔS_i, of the minimum arc length from P to Q as measured along a smooth curve in ΔS_i. (For the existence of this maximum value see the problems.) Finally (as in §5.8), see if the following limit exists and is unique:

$$\lim_{\substack{n \to \infty \\ \max d_i \to 0}} \sum_{i=1}^{n} f(P_i^*) \Delta \sigma_i .$$

If so, we write this limit as $\iint_S f d\sigma$; we say that the integral $\iint_S f d\sigma$ *exists* for the given f and S, and we call this integral the *scalar surface integral of* f on S.

Applications of scalar surface integrals include the following.

(1) If $f(P) = 1$ for all P in S, then $\iint_S f d\sigma = \iint_S d\sigma$ = surface area of S.

(2) A thin metal sheet has the shape of S. f has S as its domain and $f(P)$ gives the mass-density per unit surface-area at P. Then $\iint_S f d\sigma$ gives the total mass of the sheet.

(3) The *average value of* f *on* S can be defined as $\iint_S f d\sigma \Big/ \iint_S d\sigma$.

The following *linearity law* (I) and *additivity law* (II) hold for scalar surface integrals and are immediate consequences of (2.1).

(I)   *Let* f *and* g *be scalar fields whose common domain contains the surface* S. *Let* a *and* b *be given scalar constants. Then*

$$\iint_S (af + bg)d\sigma = a\iint_S fd\sigma + b\iint_S gd\sigma \ .$$

(II)   *Let* f *be a scalar field whose domain contains the surface* S. *Let* S *be divided into two piecewise smooth, finite surfaces* S_1 *and* S_2. *Then*

$$\iint_S fd\sigma = \iint_{S_1} fd\sigma + \iint_{S_2} fd\sigma \ .$$

In (II), $\iint_S fd\sigma$ is sometimes written as $\iint_{S_1 + S_2} fd\sigma$, where $S_1 + S_2$ denotes a surface which decomposes into the adjacent surfaces S_1 and S_2.

Under what conditions for f and S can we be sure that $\iint_S fd\sigma$ exists? An answer is given by the following *existence theorem.*

(2.2)   *If* S *is a piecewise smooth finite surface, and* f *is a continuous scalar field whose domain contains* S, *then* $\iint_S fd\sigma$ *exists.*

(2.2) can be proved by the methods of §14.10.

Remark. The following extended version of (2.2) can also be proved.

(2.3)   *If* f *is a scalar field whose domain contains the piecewise smooth finite surface* S, *and if* f *is piecewise continuous on* S, *then* $\iint_S fd\sigma$ *exists.*

Here "piecewise continuous on S" is defined analogously to (4.4) of §14.4.

§22.3 Evaluating Scalar Surface Integrals

We describe two ways to evaluate a scalar surface integral: (i) *evaluation by definition*, and (ii) *parametric evaluation.*

Evaluation by definition is based on the definition of the integral as the limit of Riemann sums. By considering specific Riemann sums, we recognize directly what the value of the integral must be.

Example 1. In E^3 *with Cartesian coordinates, let* S *be the open-ended cylinder of radius* a, *between* $z = 0$ *and* $z = b$, *with the* z *axis as its axis; and let* f *be given by* $f(x,y,z) = x^2 + y^2$. For every point P on S, we have $f(P) = a^2$. Hence every Riemann sum must have the form $\sum a^2 \Delta \sigma_i$. ([3–1].) The value of this sum is $\sum a^2 \Delta \sigma_i = a^2 \sum \Delta \sigma_i = a^2(\text{surface area of } S) = a^2(2\pi a b) = 2\pi a^3 b$.

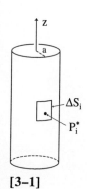

[3–1]

Symmetric cancellation can be helpful in evaluation by definition. In the following example, we use symmetric cancellation together with linearity.

Example 2. Let S *be as in example* 1. *Let* f *be given by the function* $f(x,y,z) = x^3 + y^3$. By linearity, $\iint_S (x^3 + y^3) d\sigma = \iint_S x^3 d\sigma + \iint_S y^3 d\sigma$. Let $g(x,y,z) = x^3$. For $\iint_S g d\sigma$, we note that for every point $P = (a,b,c)$ on S, we have a mirror-image point $P' = (-a,b,c)$ on S such that $g(P) = a^3$ and $g(P') = (-a)^3 = -a^3$. If we form a Riemann sum by (i) taking each ΔS with $x < 0$ to be the mirror-image of a ΔS with $x > 0$ and by (ii) taking the sample points with $x < 0$ to be mirror-images of the sample points with $x > 0$, then the terms in the Riemann sum must cancel in pairs, and the entire sum must have the value 0. Hence, in the limit, we have $\iint_S x^3 d\sigma = 0$. By a similar argument, we have $\iint_S y^3 d\sigma = 0$. Thus $\iint_S f d\sigma = 0$.

Just as linearity allows us to consider different terms in the integrand separately, additivity allows us to consider different pieces of S separately.

Example 3. In E^3 *with Cartesian coordinates, let* $S = S_1 + S_2 + S_3$, *where* S_1 *is the disk in the* xy *plane with radius* a *and center at the origin,* S_2 *is the cylindrical surface in example* 1 *above, and* S^3 *is the disk in the plane* $z = b$ *with radius* a *and center on the z axis.* ([3–2].) *Thus* S *is a closed cylinder.* We wish to find the average value of z^2 on S. As noted in §22.2, this value is given by $\oiint_S z^2 d\sigma / \oiint_S d\sigma$.

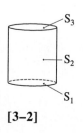

[3–2]

(For a surface integral over a *closed* surface, the notation $\iint$ is often replaced by $\oiint$. The latter notation serves as a reminder that the region of integration is a *closed* surface.) In the denominator, we have $\oiint_S d\sigma = \text{area}(S) = \text{area}(S_1) +$

area(S_2) + area(S_3) = $\pi a^2 + 2\pi ab + \pi a^2 = 2\pi a(a+b)$. In the numerator, $\oiint_S z^2 d\sigma =$ $\iint_{S_1} z^2 d\sigma + \iint_{S_2} z^2 d\sigma + \iint_{S_3} z^2 d\sigma$. Since $z^2 = 0$ on S_1, $\iint_{S_1} z^2 d\sigma = 0$. Since $z^2 = b^2$ on S_3, $\iint_{S_3} b^2 d\sigma = \pi a^2 b^2$. For S_2, we divide the interval $[0,b]$ on the z axis into successive intervals of lengths $\Delta z_1, \ldots, \Delta z_m$. Then for each j, $0 < j \le m$ we choose a sample point $(0,0,z_j^*)$ in the j-th interval. Finally, we divide S_2 into successive circular bands corresponding to the intervals on the z axis. We can now form a Riemann sum (for $\iint_{S_2}$) where the pieces from the j-th band contribute the total amount $(z_j^*)^2 2\pi a \Delta z_j$ to this sum. Hence the entire Riemann sum can be expressed as $\sum_{j=1}^{m} 2\pi a(z_j^*)^2 \Delta z_j$. We thus have

$$\iint_{S_2} z^2 d\sigma = \lim_{\substack{m \to \infty \\ \max \Delta z_j \to 0}} \sum_{j=1}^{m} 2\pi a(z_j^*)^2 \Delta z_j = 2\pi a \int_0^b z^2 dz = \tfrac{2}{3}\pi ab^3.$$

We now have $\oiint_S z^2 d\sigma = \pi a^2 b^2 + \tfrac{2}{3}\pi ab^3 = \pi ab^2(a + \tfrac{2}{3}b)$. Our desired average value of z^2 is

$$\frac{\oiint_S z^2 d\sigma}{\oiint_S d\sigma} = \frac{\pi ab^2(a+\tfrac{2}{3}b)}{2\pi a(a+b)} = \frac{b^2(a+\tfrac{2}{3}b)}{2(a+b)} \ .$$

Parametric evaluation is a more general and systematic procedure for evaluating a scalar surface integral $\iint_S f d\sigma$, when S is smooth and has a parameterization $\vec{R}$ on $\hat{D}$ such that $\vec{R}$ is C^1 with $\vec{w} \ne \vec{0}$ on the interior of $\hat{D}$.) (S need not be an elementary surface, provided that it is finite and smooth and we have a parameterization for it.)

Let S be a surface given by a smooth $\vec{R}(u,v)$ on $\hat{D}$, and let f be a scalar field whose domain contains S. We use $\vec{R}(u,v)$ to evaluate $\iint_S f d\sigma$. Recall (from (3.3) and (4.3) in Chapter 21) the basic properties of the parametric normal vector $\vec{w}$. The second of these properties suggests the following notational equation in differential form:

(3.1) $$d\sigma = |\vec{w}| du dv \ .$$

This in turn suggests

(3.2) $$\iint_S f d\sigma = \iint_{\hat{D}} f |\vec{w}| du dv .$$

(3.2) replaces the surface integral over S with an ordinary double integral over $\hat{D}$. To evaluate this double integral, we need only express both f and $\vec{w}$ as functions of u and v. We accomplish this as follows. Since $\vec{R}(u,v) = x(u,v)\hat{i} + y(u,v)\hat{j} + z(u,v)\hat{k}$ on $\hat{D}$, f(x,y,z) becomes f(x(u,v),y(u,v),z(u,v)). For $\vec{w}$, we can proceed in one of three possible ways: (1) we compute $\vec{w} = \frac{\partial \vec{R}}{\partial u} \times \frac{\partial \vec{R}}{\partial v}$ (this is almost always easier than using the Jacobian formula (3.4) in §21.3); or (2) in the special case where $\vec{R}(u,v) = u\hat{i} + v\hat{j} + g(u,v)\hat{k}$, S is the graph of the function g, and we can use the special formula $\vec{w} = -g_u\hat{i} - g_v\hat{j} + \hat{k}$; or (3) the parameterization $\vec{R}$ may be of a familiar and standard kind, where we can see geometrically what $\vec{w}(u,v)$ must be. The proof of (3.2) is indicated in the remark at the end of this section.

We may therefore summarize parametric evaluation of $\iint_S f d\sigma$, for a smooth surface S given by $\vec{R}$ on $\hat{D}$, as follows:

(a)   *Find parametric equations for S in the form $\vec{R}(u,v) = x(u,v)\hat{i} + y(u,v)\hat{j} + z(u,v)\hat{k}$ on $\hat{D}$.*

(b)   *Use $\vec{R}$ to express f as a function of the parameters u and v.*

(c_1)   *Find $\frac{\partial \vec{R}}{\partial u}, \frac{\partial \vec{R}}{\partial v}$, and $\vec{w} = \frac{\partial \vec{R}}{\partial u} \times \frac{\partial \vec{R}}{\partial v}$ as vector functions of u and v.*

(c_2)   *If $\vec{R} = u\hat{i} + v\hat{j} + g(u,v)\hat{k}$, take $\vec{w} = -g_u\hat{i} - g_v\hat{j} + \hat{k}$.*

(c_3)   *If $\vec{R}$ is familiar, derive $\vec{w}$ geometrically.*

(d)   *Find $|\vec{w}|$ as a function of u and v.*

(e)   *Take the product of (b) and (d).*

(f)   *Find the double integral of this product over $\hat{D}$.*

Example 4. *In E^3 with Cartesian coordinates, let S be the open-ended cylinder of radius a and height b as in example 1. ([3–3].) Let $f(x,y,z) = x^2z$. There is no obvious way to use evaluation by definition. We therefore turn to parametric evaluation.*

[3–3]

[3–4]

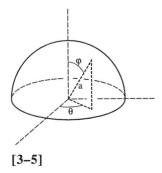

[3–5]

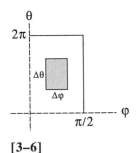

[3–6]

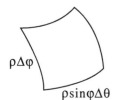

[3–7]

(a)   We use $\vec{R}(\theta,z) = a\cos\theta\hat{i} + a\sin\theta\hat{j} + z\hat{k}$ on the rectangle $\hat{D}$ given by $0 \le \theta \le 2\pi$, $0 \le z \le b$, where θ and z are the familiar cylindrical coordinates. Then

(b)   We have $f(x,y,z) = x^2z = a^2\cos^2\theta\,z$.

(c$_1$) $\dfrac{\partial\vec{R}}{\partial\theta} = -a\sin\theta\hat{i} + a\cos\theta\hat{j}$,

$\dfrac{\partial\vec{R}}{\partial z} = \hat{k}$, and

$\vec{w} = \dfrac{\partial\vec{R}}{\partial\theta} \times \dfrac{\partial\vec{R}}{\partial z} = a\cos\theta\hat{i} + a\sin\theta\hat{j}$.

(d) $|\vec{w}| = a$.

(e) $f|\vec{w}| = a^3\cos^2\theta z$.

(f) $\displaystyle\iint_S f\,d\sigma = \int_0^{2\pi}\int_0^b a^3\cos^2\theta z\,dz\,d\theta = \int_0^{2\pi}\frac{a^3b^2}{2}\cos^2\theta\,d\theta = \frac{1}{2}\pi a^3 b^2$.

In place of (c$_1$), we could have used (c$_3$) as follows.

(c$_3$)   Let ΔS be the image in S of a given small rectangle in $\hat{D}$ of sides $\Delta\theta$ and Δz. ΔS must have the form shown in [3–4]. We see that the area of ΔS is $a\Delta\theta\Delta z$ and hence that $\vec{w} = a\hat{r}$. From this it follows (d) that $|\vec{w}| = a$.

Example 5. *In* $\mathbf{E}^3$ *with Cartesian coordinates, find the average value of* z *on the upper hemisphere of a sphere of radius* a *with center at the origin. We use spherical coordinates* φ *and* θ *as parameters.* ([3–5].) *We seek* $\displaystyle\iint_S f\,d\sigma \Big/ \iint_S d\sigma$, *where* $f(x,y,z) = z$.

(a) S is given by $\vec{R}(\varphi,\theta) = a\sin\varphi\cos\theta\hat{i} + a\sin\varphi\sin\theta\hat{j} + a\cos\varphi\hat{k}$ on the rectangle in $\hat{D}$: $0 \le \varphi \le \frac{\pi}{2}, 0 \le \theta \le \pi$.

Then:

(b) $f = a\cos\varphi$.

(c$_3$)   Let ΔS be the image in S of a given small rectangle in $\hat{D}$ of sides $\Delta\varphi$ and $\Delta\theta$. ([3–6].) Then ΔS has the form shown in [3–7]; the area of $\Delta S \approx a\Delta\varphi a\sin\varphi\Delta\theta = a^2\sin\varphi\Delta\varphi\Delta\theta$, and $\vec{w} = a^2\sin\varphi\hat{\rho}$.

(d)   Hence $|\vec{w}| = a^2\sin\varphi$.

(e) $f|\vec{w}| = a^3\sin\varphi\cos\varphi$.

(f) $\iint_S f d\sigma \ = \int_0^{2\pi} \int_0^{\pi/2} a^3 \sin\varphi\cos\varphi \, d\varphi \, d\theta$

$$= 2\pi a^3 \int_0^{\pi/2} \sin\varphi\cos\varphi \, d\varphi = (2\pi a^3)\frac{\sin^2\varphi}{2}\Bigg]_0^{\pi/2} = \pi a^3 \,.$$

We also have $\iint_S d\sigma = \text{area}(S) = 2\pi a^2 \,.$

Hence the average value of z on S is $\dfrac{\pi a^3}{2\pi a^2} = \dfrac{a}{2} \,.$

Remark. As in the case of scalar line integrals in §19.3, the parametric evaluation procedure for scalar surface integrals can be justified by considering Riemann sums.

§22.4 Directed Surfaces; Vector Surface Integrals

(4.1) ***Definition.*** Let S be a two-sided, piecewise smooth, finite surface. Divide S into smooth pieces and let $\widetilde{S}$ be the set of surface-interior points of those smooth pieces. Choose a point P_0 in $\widetilde{S}$, and choose a unit normal vector $\hat{u}_0$ at P_0. (There are two choices for $\hat{u}_0$, one in each direction along the normal line at P_0.) We define a vector field $\hat{n}$ on $\widetilde{S}$ as follows: for each P in $\widetilde{S}$, $\hat{n}(P)$ is the unit normal at P obtained by a continuous motion, as in §21.6, from the unit normal $\hat{u}_0$ at P_0 to a unit normal at P. (See [4–1].) Since S is two-sided, $\hat{n}(P)$ is uniquely defined. The vector field $\hat{n}$ on $\widetilde{S}$ is called a *direction* for the surface S. There are evidently two and only two possible directions for S. If we have designated a particular direction for S, we say that S is a *directed surface*. At any surface-interior point on a directed surface S, a normal vector to S must be either *in the direction* of S or in the *opposite direction* to S. (Keep in mind that a directed surface is necessarily a two-sided surface.)

[4–1]

(4.2)   ***Definition.*** Let S be a piecewise smooth, directed, finite surface. Let $\vec{F}$ be a vector field whose domain contains S. We write the *vector surface integral* of $\vec{F}$ on S as

$$\iint_S (\vec{F} \cdot \hat{n}) d\sigma \,,$$

and we define it as follows.

$0 < i \leq n$, choose a surface-interior *sample point* P_i^* on ΔS_i, let $\hat{n}(P_i^*)$ be a unit normal vector at P_i^* in the direction of S, and let $\Delta\sigma_i$ be the area of ΔS_i. We then form the Riemann sum

$$\sum_{i=1}^{n}\left[\vec{F}(P_i^*)\cdot\hat{n}(P_i^*)\right]\Delta\sigma_i\ .$$

Finally, letting d_i be the *diameter* of ΔS_i as defined in §22.2, we see if the following limit exists and is unique:

$$\lim_{\substack{n\to\infty\\ \max d_i\to 0}}\sum_{i=1}^{n}\left[\vec{F}(P_i^*)\cdot\hat{n}(P_i^*)\right]\Delta\sigma_i\ .$$

If so, we write this limit as $\iint_S(\vec{F}\cdot\hat{n})d\sigma$; we say the integral $\iint_S(\vec{F}\cdot\hat{n})d\sigma$ *exists* for the given $\vec{F}$ and S; and we call this integral the *vector surface integral* of $\vec{F}$ on the directed surface S.

(4.3)   Note that since $\left[\vec{F}(P_i^*)\cdot\hat{n}(P_i)\right]\Delta\sigma_i = \left|\vec{F}(P_i^*)\right|\Delta\sigma_i\cos\theta_i$ where θ_i is the angle between $\vec{F}$ and $\hat{n}$ at P_i^*, the term $\left[\vec{F}(P_i^*)\cdot\hat{n}(P_i^*)\right]\Delta\sigma_i$ (in the Riemann sum) can be understood in two different ways:

(1)   we can view $\left[\vec{F}(P_i^*)\cdot\hat{n}(P_i^*)\right]\Delta\sigma_i$ as the product

$\Delta\sigma_i$ (*normal component of* $\vec{F}$ *at* P_i^*);

or   (2)   we can view $\left[\vec{F}(P_i^*)\cdot\hat{n}(P_i^*)\right]\Delta\sigma_i$ as, approximately, the product $\left|\vec{F}(P_i^*)\right|$ (*area of the projection of* ΔS_i *onto a plane perpendicular to* $\vec{F}(P_i^*)$), since, by (5.5) in Chapter 1, this projected area $\approx \Delta\sigma_i\cos\theta_i$.

Notation. We shall often write the vector surface integral of $\vec{F}$ on S as

$$\iint_S\vec{F}\cdot d\vec{\sigma}\ ,$$

where $d\vec{\sigma}$ is an abbreviation for $\hat{n}d\sigma$, and where, in a Riemann sum, we write $\Delta\vec{\sigma}_i^*$ for $\hat{n}(P_i^*)\Delta\sigma_i$.

Definition (4.2) can be expressed informally (for a smooth surface S) as follows. For the vector surface integral $\iint_S \vec{F} \cdot d\vec{\sigma}$, we have a finite smooth, directed surface S as our region of integration and a vector field $\vec{F}$ as our integrand. Because S is smooth and directed, we can assign a unique unit normal vector to each point P on S. Then we can use $\vec{F}$ to assign the scalar value $\vec{F}(P) \cdot \hat{n}(P)$ to each point P on S. Finally, defining $g(P) = \vec{F}(P) \cdot \hat{n}(P)$, we define the value of our vector surface integral $\iint_S \vec{F} \cdot d\vec{\sigma}$ to be the value of the scalar surface integral $\iint_S g d\sigma$. This informal description can be summarized even more informally: *the vector surface integral $\iint_S \vec{F} \cdot d\vec{\sigma}$ is the scalar surface integral of the normal component of $\vec{F}$ on S.*

In this text, we shall consider two main applications of vector surface integrals.

Application 1. Let S be an imaginary, directed, piecewise smooth surface, which remains stationary in the region of a continuous fluid flow. Let $\vec{F}(P)$ be the velocity vector $\vec{v}(P)$ of the moving fluid at P at a given moment. Take a decomposition of S into $\Delta S_1, ..., \Delta S_n$ with sample points $P_1^*, ..., P_n^*$. Using (2) in (4.3), we see that $[\vec{v}(P_i^*) \cdot \hat{n}(P_i^*)]\Delta\sigma_i \Delta t \approx$ the volume of fluid passing through ΔS_i, in the direction of S, during the time Δt. (See [4–2].) Hence $[\vec{v}(P_i^*) \cdot \hat{n}(P_i^*)]\Delta\sigma_i \approx$ the instantaneous rate at which fluid volume is passing through ΔS_i, and $\iint_S \vec{v} \cdot d\vec{\sigma}$

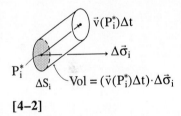

$\vec{v}(P_i^*)\Delta t$

$\Delta\vec{\sigma}_i$

P_i^*

ΔS_i $\text{Vol} = (\vec{v}(P_i^*)\Delta t) \cdot \Delta\vec{\sigma}_i$

[4–2]

gives the *instantaneous rate of net flow of volume of fluid through the directed surface S, in the direction of S,* at that moment. Similarly, if we take $\vec{F}$ to be the mass-flow vector $\vec{m}(P) = \delta(P)\vec{v}(P)$, where $\delta(P)$ is density of fluid at P at the given moment, then $\iint_S \vec{m} \cdot d\vec{\sigma}$ gives the *instantaneous rate of net flow of mass through the directed surface S, in the direction of S,* at the given moment.

Application 2. Let $\vec{F}$ be the field vector in a force field at a given moment and let S be an imaginary, directed, piecewise smooth surface in the region of the field. Then $\iint_S \vec{F} \cdot d\vec{\sigma}$ is called the *flux of $\vec{F}$ through the directed surface S* at that moment. Flux integrals are fundamental concepts in the theories of gravitational and electromagnetic force fields, as we shall see in later chapters. Vector surface integrals in general are often referred to as *flux integrals.*

The following *linearity law* (I), *additivity law* (II), and *direction-reversal law* (III) are immediate consequences of the formal definition.

(I)   *Let* $\vec{F}$ *and* $\vec{G}$ *be vector fields whose common domain contains the directed surface* S. *Let* a *and* b *be given scalar constants. Then*

$$\iint_S (a\vec{F} + b\vec{G}) \cdot d\vec{\sigma} = a\iint_S \vec{F} \cdot d\vec{\sigma} + b\iint_S \vec{G} \cdot d\vec{\sigma}.$$

(II)   *Let* $\vec{F}$ *be a vector field whose domain contains the directed surface* S. *Let* S *be divided into two directed surfaces* S_1 *and* S_2, *where the directions for* S_1 *and* S_2 *are the same as for* S. *(In this case, we sometimes write* S = $S_1 + S_2$.) *Then*

$$\iint_S \vec{F} \cdot d\vec{\sigma} = \iint_{S_1} \vec{F} \cdot d\vec{\sigma} + \iint_{S_2} \vec{F} \cdot d\vec{\sigma}.$$

(III)   *Let* $\vec{F}$ *be a vector field whose domain contains the directed surface* S, *and let* –S *be the same surface with its direction reversed. Then*

$$\iint_{-S} \vec{F} \cdot d\vec{\sigma} = -\iint_S \vec{F} \cdot d\vec{\sigma}.$$

Under what conditions for $\vec{F}$ and S can we be sure that $\iint_S \vec{F} \cdot d\vec{\sigma}$ exists? An answer is given by the following *existence theorem*:

(4.4)   *If* S *is a piecewise smooth, directed, finite surface, and* $\vec{F}$ *is a continuous vector field whose domain contains* S, *then* $\iint_S \vec{F} \cdot d\vec{\sigma}$ *exists.*

(4.4) can be proved by the methods of §14.10.

Remark. The following extended version of (4.4) can also be proved.

(4.5)   *If* $\vec{F}$ *is a vector field whose domain contains the piecewise smooth, directed, finite surface* S, *and if* $\vec{F}$ *is piecewise continuous on* S, *then* $\iint_S \vec{F} \cdot d\vec{\sigma}$

exists. Here, "piecewise continuous on S" is defined analogously to the definition (4.4) in §14.4. (4.5) is used in physical applications where, in a chosen mathematical model, there is a vector field which has a discontinuity along some curve in S.

§22.5 Evaluating Vector Surface Integrals

We describe two ways to evaluate a vector surface integral: (i) *evaluation by definition* and (ii) *parametric evaluation*.

In **evaluation by definition,** we consider specific Riemann sums and recognize directly what the value of the integral must be.

Example 1. *In* E^3 *with Cartesian coordinates, let* S *be the closed cylinder of radius* a *and height* b, *where the axis of* S *is the* z *axis and the base of* S *is in the* xy *plane. Let* S *be directed outward.* Let S_1 be the base of S, let S_2 be the lateral, curved surface, and let S_3 be the top. Then S_1 is directed down, S_2 is directed outward to the side, and S_3 is directed up. [5–1]. Let $\vec{F} = -y\hat{i} + x\hat{j}$.

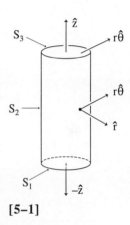

[5–1]

We use cylindrical coordinates. At every point P on S_1, we have $\vec{F}(P) = r\hat{\theta}$ and $\hat{n}(P) = -\hat{z}$; hence $\vec{F}(P) \cdot \hat{n}(P) = 0$. Similarly, at every point on S_3, we have $\vec{F}(P) = r\hat{\theta}$ and $\hat{n}(P) = \hat{z}$; hence $\vec{F}(P) \cdot \hat{n}(P) = 0$. Finally, at every point P on S_2, we have $\vec{F}(P) = r\hat{\theta}$ and $\hat{n}(P) = \hat{r}$; hence $\vec{F}(P) \cdot \hat{n}(P) = 0$. Thus every Riemann sum for $\oiint_S \vec{F} \cdot d\vec{\sigma}$ must be zero, and we have $\oiint_S \vec{F} \cdot d\vec{\sigma} = 0$. (As in §22.3, we may replace the notation $\iint$ by the notation $\oiint$ when the integral is over a closed surface.)

Example 2. Let $S = S_1 + S_2 + S_3$ *be the same directed surface as in example* 1. *Let* $\vec{F} = x\hat{i} + y\hat{j} + z\hat{k}$. By linearity, we have $\oiint_S \vec{F} \cdot d\vec{\sigma} = \oiint_S (x\hat{i} + y\hat{j}) \cdot d\vec{\sigma} + \oiint_S z\hat{k} \cdot d\vec{\sigma}$. We use cylindrical coordinates. [5–2]. Since $x\hat{i} + y\hat{j} = r\hat{r}$, we have $\iint_{S_1} (x\hat{i} + y\hat{j}) \cdot d\vec{\sigma} = \iint_{S_3} (x\hat{i} + y\hat{j}) \cdot d\vec{\sigma} = 0$. On S_2, we have $(x\hat{i} + y\hat{j}) \cdot \hat{n} = r\hat{r} \cdot \hat{r} = r = a$. Hence $\oiint_S (x\hat{i} + y\hat{j}) \cdot d\vec{\sigma} = \iint_{S_2} (x\hat{i} + y\hat{j}) \cdot d\vec{\sigma} = $

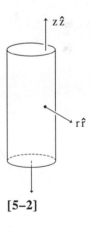

[5–2]

$a\iint_{S_2} d\sigma = a(\text{area}(S_2)) = a(2\pi ab) = 2\pi a^2 b$. Since $z\hat{k} = z\hat{z}$, we have $\iint_{S_2} z\hat{k} \cdot d\vec{\sigma} = \iint_{S_2} (z\hat{z} \cdot \hat{r})d\sigma = 0$. Since $z = 0$ on S_1, we have $\iint_{S_1} z\hat{k} \cdot d\vec{\sigma} = \iint_{S_1} \vec{0} \cdot d\vec{\sigma} = 0$. Since $z = b$ on S_3, we have $\iint_{S_3} z\hat{k} \cdot d\vec{\sigma} = \iint_{S_3} bd\sigma = b\iint_{S_3} d\sigma = b(\text{area}(S_3)) = \pi a^2 b$. We therefore have $\oiint_S \vec{F} \cdot d\vec{\sigma} = 2\pi a^2 b + \pi a^2 b = 3\pi a^2 b$.

Since $z = b$ on S_3, we have $\iint_{S_3} z\hat{k} \cdot d\vec{\sigma} = \iint_{S_3} b\,d\sigma = b\iint_{S_3} d\sigma = b(\text{area}(S_3)) =$

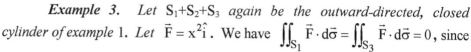

$\pi a^2 b$. We therefore have $\oiint_S \vec{F} \cdot d\vec{\sigma} = 2\pi a^2 b + \pi a^2 b = 3\pi a^2 b$.

Example 3.   *Let* $S_1 + S_2 + S_3$ *again be the outward-directed, closed cylinder of example 1. Let* $\vec{F} = x^2\hat{i}$. We have $\iint_{S_1} \vec{F} \cdot d\vec{\sigma} = \iint_{S_3} \vec{F} \cdot d\vec{\sigma} = 0$, since

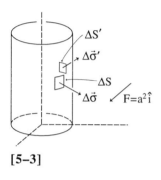

[5–3]

$\vec{F}$ is directed parallel to S_1 and S_3 and hence perpendicular to the normal vectors to S_1 and S_3. On S_2, we use symmetric cancellation as follows. For every point (a,b,c) on S_2, we have a mirror-image point $(-a,b,c)$ on S_2. Since $(-a)^2 = a^2$, $\vec{F} = a^2\hat{i}$ at both these points. If we form a Riemann sum by choosing $\Delta S'$ at the second point to be a mirror image of ΔS at the first point (see [5–3]), then the "outward flow" through ΔS must equal the "inward flow" through $\Delta S'$. Hence a Riemann sum can be formed for which the terms cancel in pairs. We have $\oiint_S \vec{F} \cdot d\vec{\sigma} = 0$.

Parametric evaluation is simpler for vector surface integrals than it is for scalar surface integrals. We assume that S has a smooth parameterization $\vec{R}(u,v)$ on $\hat{D}$ (that is to say, $\vec{R}$ is C^1 with $\vec{w} \neq \vec{0}$ on the interior of $\hat{D}$).

We use $\vec{R}(u,v)$ to evaluate $\iint_S \vec{F} \cdot d\vec{\sigma}$ as follows. In §21.3, we saw that $\vec{w}$ is normal to S at P. Thus $\hat{n} = \pm\hat{w}$ at P. From (3.1), we have $d\sigma = |\vec{w}|\,du\,dv$. Hence we have

(5.1) $d\vec{\sigma} = \hat{n}\,d\sigma = \pm\hat{w}|\vec{w}|\,du\,dv = \pm\vec{w}\,du\,dv$.

(Here, the + or − depends on whether $\vec{w}$, as calculated from $\vec{R}(u,v)$, is in the direction of S.) It follows that

(5.2) $\iint_S \vec{F} \cdot d\vec{\sigma} = \pm\iint_{\hat{D}} (\vec{F} \cdot \vec{w})\,du\,dv$.

We have replaced the vector surface integral over S by an ordinary double integral over $\hat{D}$. In order to evaluate this double integral, we must express both $\vec{F}$ and $\vec{w}$ as vector functions of u and v. For $\vec{F}(x,y,z)$, we use $\vec{R}(u,v)$ to substitute for x, y, and z in terms of u and v. For $\vec{w}$, we have the three possibilities described in §22.3. We can therefore summarize parametric evaluation for

(a)   Give a parameterization for S in the form $\vec{R}(u,v) =$ $x(u,v)\hat{i} + y(u,v)\hat{j} + z(u,v)\hat{k}$ on $\hat{D}$.

(b)   Use $\vec{R}$ to express $\vec{F}$ as a function of the parameters u and v.

(c_1)   Find $\dfrac{\partial \vec{R}}{\partial u}$, $\dfrac{\partial \vec{R}}{\partial v}$, and $\vec{w} = \dfrac{\partial \vec{R}}{\partial u} \times \dfrac{\partial \vec{R}}{\partial v}$, as vector functions of u and v.

(c_2)   If $\vec{R} = u\hat{i} + v\hat{j} + g(u,v)\hat{k}$, take $\vec{w} = -g_u\hat{i} - g_v\hat{j} + \hat{k}$.

(c_3)   If $\vec{R}(u,v)$ is familiar, derive $\vec{w}$ geometrically as a function of u and v.

(d)   Take the dot product of (b) and (c).

(e)   Find the double integral of this dot product over $\hat{D}$.

(f)   If the direction of $\vec{w}$ disagrees with the direction of S, take the negative of this result.

Example 4. *Let S be the flat, triangular surface with vertices at* $(1,0,0)$, $(0,1,0)$, *and* $(0,0,1)$, *directed down. Find the flux of* $\vec{F} = x\hat{i} + y\hat{j}$ *through* S. [5–4].

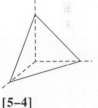

[5–4]

We use x and y as parameters. S lies in the plane $x+y+z = 1$. Hence we have S given by

(a) $\vec{R}(x,y) = x\hat{i} + y\hat{j} + (1 - x - y)\hat{k}$ on $\hat{D}$, where $\hat{D}$ is the triangle in the xy plane with vertices $(0,0)$, $(1,0)$, and $(0,1)$. We get

(b) $\vec{F} = x\hat{i} + y\hat{j}$.

(c_2) $\vec{w} = -g_x\hat{i} - g_y\hat{j} + \hat{k}$ (for $g(x,y) = 1-x-y$).

$\qquad = \hat{i} + \hat{j} + \hat{k}$.

(d) $\vec{F} \cdot \vec{w} = x + y$.

(e) $\iint_{\hat{D}} \vec{F} \cdot \vec{w}\,dxdy = \int_0^1 \int_0^{1-x} (x + y)\,dydx = \int_0^1 (xy + \frac{y^2}{2}) \Big]_0^{1-x} dx =$

$\int_0^1 \frac{1}{2}(1 - x^2)dx = \frac{1}{2}(x - \frac{x^3}{3}) \Big]_0^1 = \frac{1}{3}$.

(f)   Since the direction of $\vec{w}$ on S is up, the direction of $\vec{w}$ disagrees with the direction of S, hence $\iint_S \vec{F} \cdot d\vec{\sigma} = -\frac{1}{3}$.

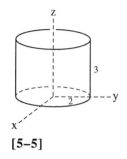

[5–5]

Example 5. *Let* S *be the open-ended cylinder of radius 2 and height 3,* *where the axis of* S *is the z axis and* S *rests on the xy plane.* S *is directed out-* *ward. Let* $\vec{F} = x\hat{i}$. *We seek* $\iint_S \vec{F} \cdot d\vec{\sigma}$. [5–5].

We have:

(a) $\vec{R}(\theta, z) = 2\cos\theta\hat{i} + 2\sin\theta\hat{j} + z\hat{k}, \;\; 0 \le \theta \le 2\pi \text{ and } 0 \le z \le 3$.

(b) $\vec{F} = 2\cos\theta\hat{i}$.

(c_1)   As in example 4 of §22.3, $\vec{w} = 2\cos\theta\hat{i} + 2\sin\theta\hat{j}$.

(d) $\vec{F} \cdot \vec{w} = 4\cos^2\theta$.

(e) $\iint_{\hat{D}} \vec{F} \cdot \vec{w} \, dz d\theta = \int_0^{2\pi} \int_0^3 4\cos^2\theta \, dz d\theta = 12 \int_0^{2\pi} \cos^2\theta \, d\theta = 12\pi$.

(f)   No change of sign, since $\vec{w} = 2\hat{r}$ is directed out. Therefore

$$\iint_S \vec{F} \cdot d\vec{\sigma} = 12\pi.$$

Remark. In Chapter 25, we state and prove the *divergence theorem.* This theorem provides a method for evaluating vector surface integrals which is often shorter and simpler than either of the methods described above. In particular, it simplifies the evaluations for examples 1 through 5 above.

§22.6 Other Integrals on Surfaces

Other kinds of integrals on surfaces can be expressed in the same vector style that we have used for vector line integrals. We give some examples below and in the problems. In each case, we shall use notation which suggests: (i) an appropriate formal definition in terms of Riemann sums and (ii) an appropriate and correct general method of parametric evaluation. In some cases, these new integrals will have vector values instead of scalar values.

Example 1. *For a given piecewise smooth, finite surface* S *(which may* *be one-sided or two-sided) and for a vector field* $\vec{F}$ *whose domain contains* S*, we* *consider the new integral*

(6.1) $\iint_S \vec{F} d\sigma$.

The integral is vector-valued and, as the notation suggests, is defined by

$$\lim_{\substack{n \to \infty \\ \max d_i \to 0}} \sum_{i=1}^{n} \vec{F}(P_i^*) \Delta \sigma_i \, ,$$

where P_i^* is a sample point in ΔS_i, where $\Delta \sigma_i$ is the surface area of ΔS_i, where $\Delta S_1, ..., \Delta S_n$ is a decomposition of S into smooth pieces, and where d_i is the *diameter* of ΔS_i, as defined in §22.2. ([6–1].) When we have a smooth parameterization $\vec{R}(u, v)$ for S on $\hat{D}$, this new integral can be parametrically evaluated by

[6–1]

$$\iint_S \vec{F} d\sigma = \iint_D \vec{F} |\vec{w}| du dv \, .$$

If we have $\vec{F} = L\hat{i} + M\hat{j} + N\hat{k}$ in Cartesian coordinates, we get $\iint_S \vec{F} d\sigma =$ $\left(\iint_S L d\sigma \right) \hat{i} + \left(\iint_S M d\sigma \right) \hat{j} + \left(\iint_S N d\sigma \right) \hat{k}$, where the scalar coefficients in the last three terms are scalar surface integrals.

If we use the position vector $\vec{R} = x\hat{i} + y\hat{j} + z\hat{k}$ (with respect to the origin) as the vector field $\vec{F}$, this new integral can be used to define and calculate the *centroid* of a given surface S. We define

$$\vec{R}_0 = \iint_S \vec{R} d\sigma \Big/ \iint_S d\sigma$$

to be the (position vector for the) centroid of S. For a thin metal surface of constant mass-density per unit surface area, this centroid will be the center of mass.

Example 2. For a given piecewise smooth, directed, finite surface S and a vector field $\vec{F}$ whose domain contains S, we consider the new integral

(6.2)
$$\iint_S \vec{F} \times d\vec{\sigma} \, ,$$

where this integral may also be written as $\iint_S (\vec{F} \times \hat{n}) d\sigma$. *The integral is vector-valued, and, as the notation suggests is defined by*

$$\lim_{\substack{n \to \infty \\ \max d_i \to 0}} \sum_{i=1}^{n} \left[\vec{F}(P_i^*) \times \hat{n}(P_i^*) \right] \Delta \sigma_i \, .$$

The integral can be parametrically evaluated by

$$\iint_S \vec{F} \times d\vec{\sigma} = \pm \iint_{\hat{D}} (\vec{F} \times \vec{w}) du dv$$

where $-$ is used if the direction of $\vec{w}$ does not agree with the direction of S. Integrals of this kind appear in the general theory of vector fields.

§22.7 Examples

We give several further illustrative examples on surface integrals.

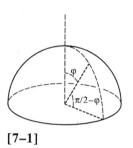

Example 1. *What is the average latitude of a point in the northern hemisphere of the earth's surface?*

[7–1]

We use spherical coordinates in E^3, and consider the upper hemisphere of radius a with center at the origin. This is our surface S. ([7–1].) We begin by finding the average value of the spherical coordinate φ on S. Let $\overline{\varphi}$ be this average value. For S, we take $\vec{R}(\varphi,\theta) = a \sin\varphi \cos\theta \hat{i} + b \sin\varphi \sin\theta \hat{j} + a \cos\varphi \hat{k}$ where $\hat{D}$ is the rectangle $0 \le \varphi \le \frac{\pi}{2}, 0 \le \theta \le 2\pi$. Then $\overline{\varphi} = \iint_S \varphi d\sigma / \iint_S d\sigma = \frac{1}{2\pi a^2} \iint_{\hat{D}} \varphi |\vec{w}| d\varphi d\theta$. $\vec{w} = \frac{\partial\vec{R}}{\partial\varphi} \times \frac{\partial\vec{R}}{\partial\theta}$ can be calculated directly from $\vec{R}(\varphi,\theta)$. The

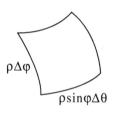

$\rho\Delta\varphi$

$\rho\sin\varphi\Delta\theta$

[7–2]

calculation is straightforward but somewhat lengthy. It is easier to note from our knowledge of spherical coordinates (see [7–2]) that $d\sigma = a^2 \sin\varphi d\varphi d\theta = |\vec{w}| d\varphi d\theta$. Hence we have

$$\overline{\varphi} = \frac{1}{2\pi a^2} \int_0^{2\pi} \int_0^{\pi/2} \varphi a^2 \sin\varphi d\varphi d\theta$$

$$= \frac{2\pi a^2}{2\pi a^2} \int_0^{\pi/2} \varphi \sin\varphi d\varphi.$$

Integrating by parts, we find that

$$\overline{\varphi} = -\varphi \cos\varphi \Big]_0^{\pi/2} + \sin\varphi \Big]_0^{\pi/2} = 1.$$

By definition, the latitude of a point (in radians) is equal to $\frac{\pi}{2} - \varphi$. It is easy to show that the average value of $(\frac{\pi}{2} - \varphi)$ on S must be $\frac{\pi}{2} - \overline{\varphi}$. (See the problems.) Hence the average latitude of S is $\frac{\pi}{2} - 1$ radians $= 32.71°$.

Example 2. *Let* S *be the closed surface* $S = S_1 + S_2$, *where* S_1 *is a disk of radius 2 lying in the plane* $z = 4$, *and* S_2 *is the surface of the paraboloid* $z = x^2 + y^2$ *between* $z = 0$ *and* $z = 4$. *Find the outward flux of* $\vec{F} = x^2\hat{i} + y^3\hat{j}$ *through this surface,* [7–3]. Since $\vec{F}$ is parallel to S_1, we have $\iint_{S_1} \vec{F} \cdot d\vec{\sigma} = 0$. For S_2,

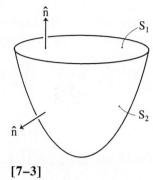

[7–3]

we have $\iint_{S_2} \vec{F} \cdot d\vec{\sigma} = \iint_{S_2} x^2\hat{i} \cdot d\vec{\sigma} + \iint_{S_2} y^3\hat{j} \cdot d\vec{\sigma}$, where S_2 is directed *down*. By symmetric cancellation, we see that $\iint_{S_2} x^2\hat{i} \cdot d\vec{\sigma} = 0$. For $\iint_{S_2} y^3\hat{j} \cdot d\vec{\sigma} = 0$, we need parametric evaluation. Since S_2 is the graph of $g(x,y) = x^2 + y^2$, we use parameters x and y and have $\vec{R}(x,y) = x\hat{i} + y\hat{j} + (x^2 + y^2)\hat{k}$ on $\hat{D}$, where $\hat{D}$ is the disk $x^2 + y^2 \le 4$ in the xy plane. We then have $\vec{w} = -g_x\hat{i} - g_y\hat{j} + \hat{k} = -2x\hat{i} - 2y\hat{j} + \hat{k}$. Hence $\iint_{S_2} y^3\hat{j} \cdot d\vec{\sigma} = -\iint_{\hat{D}} y^3\hat{j} \cdot \vec{w}\,dxdy = -\iint_{\hat{D}} -2y^4 dxdy = \iint_{\hat{D}} 2y^4 dxdy$ where the $-$ before the integrals appears because S_2 is directed down while $\vec{w}$ is directed up. To evaluate our double integral, we use polar coordinates to get $\int_0^{2\pi} \int_0^2 2r^4 \sin^4\theta\, r\,dr d\theta = \left(\int_0^{2\pi} \sin^4\theta d\theta \right)\left(\int_0^2 2r^5 dr \right) = \left(\frac{3\pi}{4}\right)\left(\frac{64}{3}\right) = 16\pi$. (For the evaluation of $\int_0^{2\pi} \sin^4\theta d\theta$, use the identities $\sin^4\theta = \sin^2\theta(1 - \cos^2\theta)$ and $\sin^2\theta\cos^2\theta = \frac{1}{4}\sin^2 2\theta$.) Hence $\oiint_S \vec{F} \cdot d\vec{\sigma} = 16\pi$.

§22.8 Problems

§22.2

1. (a) Let two points P and Q be given on a compact smooth surface ΔS. Prove that there is a smooth curve from P to Q on ΔS whose length is a minimum among the lengths of all possible smooth curves in ΔS from P to Q. (*Hint.* Let $\hat{D}$ be the parameter region for a smooth parameterization of ΔS. Use the compactness of $\hat{D}$ in E^2.)

(b) Show that for a smooth surface ΔS, the diameter of ΔS, as defined in the text, must exist. (*Hint.* Fix P, let Q vary, and apply the max-min existence theorem. Then let P vary and apply the existence theorem again.)

§22.3

1. Evaluate by definition $\iint_S xz\,d\sigma$, where S is lateral surface of cylinder $x^2 + y^2 = 1$, $0 \le z \le 1$.

2. Evaluate by definition: $\iint_S x^2 z\,d\sigma$, where S is the surface of the closed cylinder $x^2 + y^2 = 1$, $0 \le z \le 1$, including top and bottom.

3. S is the closed cylinder whose top is given by $z = 3$, whose bottom is given by $z = 0$, and whose

lateral surface is given by $x^2 + y^2 = 4$. S is directed outward. Evaluate the following surface integrals. (*Hint.* Evaluate by definition wherever possible.)

(a) $\iint_S z d\sigma$ (c) $\iint_S xyz d\sigma$

(b) $\iint_S x d\sigma$ (d) $\iint_S x^2 d\sigma$

§22.5

1. Find the value of $\iint_S \vec{F} \cdot d\vec{\sigma}$, where $\vec{F} = x\hat{i} + y\hat{j}$, and S is the surface $x^2 + y^2 = 4$, $-1 \le z \le 1$, directed out.

2. Evaluate by definition.

    (a) $\iint_S (\hat{i} + \hat{j} + \hat{k}) \cdot d\vec{\sigma}$, where S is the upper hemisphere $z = \sqrt{1 - x^2 - y^2}$, $x^2 + y^2 \le 1$, directed up.

    (b) $\iint_S (x\hat{i} + y\hat{j} + z\hat{k}) \cdot d\vec{\sigma}$, where S is as in (a).

3. S is the closed cylinder whose top is given by z = 3, whose bottom is given by z = 0, and whose lateral surface is given by $x^2 + y^2 = 4$. S is directed outward. Evaluate the following surface integrals. (*Hint.* Evaluate by definition wherever possible.)

    (a) $\iint_S (z\hat{k}) \cdot d\vec{\sigma}$ (b) $\iint_S (x\hat{i} + y\hat{j}) \cdot d\vec{\sigma}$

    (c) $\iint_S (-y\hat{i} + x\hat{j}) \cdot d\vec{\sigma}$ (d) $\iint_S (x^2\hat{i}) \cdot d\vec{\sigma}$

    (e) $\iint_S (x^3\hat{i}) \cdot d\vec{\sigma}$

4. Let S be the triangle with vertices (1,1,1), (2,2,2) and (1,2,1). S is directed upward. Let $\vec{F} = yz\hat{i} + xz\hat{j} + x^2\hat{k}$. Find $\iint_S \vec{F} \cdot d\vec{\sigma}$.

5. Find the value of $\iint_S \vec{F} \cdot d\vec{\sigma}$, where $\vec{F} = 2x\hat{i} + 2y\hat{j}$, and S is the surface $x^2 + y^2 = 9$, $-1 \le z \le 1$.

6. Let S be the curved half-cylinder surface defined by $y^2 + z^2 = 1$, $z \ge 0$, and $0 \le x \le 1$. S is directed up. Let $\vec{F} = x\hat{i} + y\hat{j} + z\hat{k}$. Find $\iint_S \vec{F} \cdot d\vec{\sigma}$.

    (Possibly useful fact: $\int \left(du / \sqrt{1 - u^2} \right) = \sin^{-1} u + C$.)

7. Evaluate by definition $\iint_S (x^2\hat{i} + y^2\hat{j} + z^2\hat{k}) \cdot d\vec{\sigma}$, where S is the closed surface of the cube $0 \le x \le 1$, $0 \le y \le 1$, $0 \le z \le 1$, directed outward.

8. Let $\vec{F} = x^2\hat{i} + xy\hat{j} + z\hat{k}$. S is the flat triangular surface with vertices (0,0,0), (1,0,0), (2,1,2). S is directed upward. Find $\iint_S \vec{F} \cdot d\vec{\sigma}$.

§22.6

1. Let S be the half-cylinder defined by $x^2 + y^2 = a^2$, $x \ge 0$, and $0 \le z \le b$. Let $\vec{F} = r\hat{\theta} = -y\hat{i} + x\hat{j}$. Evaluate $\iint_S \vec{F} d\sigma$.

2. For S and $\vec{F}$ as in problem 1, with S directed out, evaluate $\iint_S \vec{F} \times d\vec{\sigma}$.

3. Give an appropriate definition and method of evaluation for $\iint_S f d\vec{\sigma}$.

§22.7

1. For example 1, prove that the average value of $(\frac{\pi}{2} - \varphi)$ on S must be $(\frac{\pi}{2} - \bar{\varphi})$, where $\bar{\varphi}$ is the average value of φ on S.

2. Use the identities given at the end of example 2 to evaluate $\int_0^{2\pi} \sin^4 \theta d\theta$.

Chapter 23

Measures and Densities

§23.1 Elementary Regions and Regular Regions in $\mathbf{E}^2$ and $\mathbf{E}^3$

Recall the following definitions from §14.2: (i) a region R in $\mathbf{E}^2$ is said to be *elementary* if there is a simple, closed, piecewise smooth curve C such that R consists of the points on or inside C; and (ii) a region R in $\mathbf{E}^2$ is said to be *regular* if it is either elementary or can be divided into a finite number of elementary regions by a regular subdivision. In a regular subdivision, adjacent elementary regions may only touch along a single shared piece of boundary curve. (For the full definition of *regular subdivision in* $\mathbf{E}^2$, see the remark below.)

Using the Jordan curve theorem of §6.5, it is easy to see, and to prove, that there are two possible kinds of regular region R in $\mathbf{E}^2$, and that every regular region must be of one of these two kinds:

First kind. The boundary points of R form a single, simple, closed, piecewise smooth curve C. In this case, R is in fact elementary and consists of the points on or inside C. [1–1].

[1–1]

Second kind. The boundary points of R form two or more simple, closed, piecewise smooth curves. One of these is an *outer boundary curve* C. The other curves are *interior boundary curves* $C_1, C_2,...,C_k$. Each of these interior curves lies in the interior of C but is exterior to each of the other interior boundary curves. In this case, R is not elementary, and R consists of those points which are

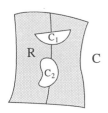

[1–2]

on or inside C and, at the same time, on or outside each of the curves $C_1, C_2, ..., C_k$. R can be thought of as "what is left" on or inside C after the interiors of $C_1, ..., C_k$ have been removed from the interior of C. More briefly put, R can be obtained by putting one or more elementary "holes" in the elementary region given by C. [1–2].

Definitions. Let D be a region in $\mathbf{E}^2$, and let R be a regular region in $\mathbf{E}^2$. If every point in R is also in D, we say that R is a regular *subregion* of D.

A regular region R can be divided into two regular subregions R_1 and R_2 in infinitely many different ways, as suggested in [1–3]. *For any such division, there must be a simple closed, piecewise smooth curve $\widetilde{C}$ such that the common boundary of R_1 and R_2 is the intersection of $\widetilde{C}$ with R.* One of the subregions will have its points on or outside $\widetilde{C}$, and the other will have its points on or inside $\widetilde{C}$. If R is divided into two subregions R_1 and R_2, we write $R = R_1 + R_2$.

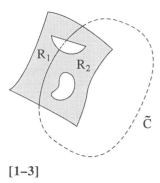

[1–3]

In order to define *elementary regions* and *regular regions* in $\mathbf{E}^3$, we begin with the following definition. A closed surface S is said to be *simple* if its interior is simply connected (thus a sphere is a simple closed surface but a torus is not). By (6.3) in §21.6, the interior of a simple closed surface must also be two-connected. A simple closed surface can be thought of as an elastically deformed version of a convex polyhedron, where this deformed version is not necessarily convex.

Definition. A region R in $\mathbf{E}^3$ is said to be *elementary* if there is a simple, closed, piecewise smooth surface S such that R consists of the points on or inside S; and a region R in $\mathbf{E}^3$ is said to be *regular* if it is either elementary or can be divided into a finite number of elementary regions by a regular subdivision. In a regular subdivision, adjacent elementary regions may only touch at a single shared piece of boundary surface. For the full definition of *regular subdivision in $\mathbf{E}^3$*, see the remark below.

Using (6.3) in §21.6, it is easy to see, and to prove, that there are three possible kinds of regular region R in $\mathbf{E}^3$, and that every regular region must be of one of these three kinds:

First kind. The boundary points of R form a single, simple, closed, piecewise smooth surface S. In this case, R is in fact elementary and consists of the points on or inside S. [1–4].

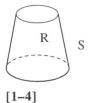

[1–4]

Second kind. The boundary points of R form a single, closed, but not simple, piecewise smooth surface S. In this case, R consists of the points on or inside S, but R is not elementary. A solid torus is an example of a regular region of this kind. [1–5].

[1–5]

Third kind. The boundary points of R form two or more closed (but not necessarily simple), piecewise smooth surfaces. One of these is an *outer boundary surface* S. The other surfaces are *interior boundary surfaces* $S_1, S_2, ..., S_k$. Each of these interior surfaces lies in the interior of S but is exterior to each of the other interior boundary surfaces. In this case, R is not elementary, and R consists of those points which are on or inside S and, at the same time, on or outside each of the surfaces $S_1, S_2, ..., S_k$. R can be thought of as "what is left" on or inside S after the interiors of $S_1, S_2, ..., S_k$ have been removed from the interior of S. More briefly put, R has been obtained by putting one or more "cavities" (of the first or second kind) in the interior of a region of the first or second kind. [1–6].

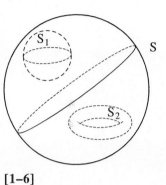

[1–6]

A regular region R in $\mathbf{E}^3$ can be divided into two regular subregions R_1 and R_2 in infinitely many different ways. *For any such division, there must be a closed, piecewise smooth surface* $\widetilde{S}$ *such that the common boundary* of R_1 *and* R_2 *is the intersection of* $\widetilde{S}$ *with* R. One of the subregions will have its points on or outside $\widetilde{S}$, and the other will have its points on or inside $\widetilde{S}$. If R is divided into two subregions R_1 and R_2, we write $R = R_1 + R_2$.

Remark. A *regular subdivision in* $\mathbf{E}^2$ of a region R into elementary subregions can be described in terms of a finite collection of *faces*, *edges*, and *vertices*, where the elementary regions are called *faces*, where the boundary of each face is made up of piecewise smooth elementary curves called *edges*, where the two endpoints of each edge are called *vertices*, and where the following five conditions analogous to the conditions in the definition of *polyhedron* in §1.6 are satisfied: (i) every edge is either shared by exactly two faces or is on the boundary of R; (ii) every vertex is either shared by three or more faces, or is on the boundary of R and shared by two or more faces; (iii) no two distinct faces can intersect except at a common edge or vertex; (iv) for every pair of non-adjacent faces R_1 and R_2, there is a chain of successively adjacent faces which joins R_1 to R_2, where we define two faces to be *adjacent* if they share a common edge; and (v) for every pair of non-adjacent faces R_1 and R_2 which share a common vertex, there is a chain of successively adjacent faces which joins R_1 to R_2 and whose members all share that vertex.

A *regular subdivision in* $\mathbf{E}^3$ of a region R into elementary subregions can be described in terms of a finite collection of *blocks, faces, edges,* and *vertices,* where the elementary regions are called *blocks,* where the boundary of each block is made up of piecewise smooth elementary surfaces called *faces,* where the boundary of each face is made up of piecewise smooth elementary curves called *edges,* where the two endpoints of each edge are called *vertices,* and where the following six conditions hold: (i) every face is either shared by exactly two blocks or is on the boundary of R; (ii) every edge is either shared by three or more blocks, or is on the boundary of R and shared by two or more blocks; (iii) every vertex is shared by four or more blocks or is on the boundary of R and shared by three or more blocks; (iv) no two distinct blocks can intersect except at a common face, edge, or vertex; (v) for every pair of non-adjacent blocks R_1 and R_2, there is a chain of successively adjacent blocks which joins R_1 to R_2, where we define two blocks to be *adjacent* if they share a common face; and (vi) for every pair of non-adjacent blocks R_1 and R_2 which share a common edge or vertex, there is a chain of successively adjacent blocks which joins R_1 to R_2 and whose members all share that edge or vertex.

§23.2 Finite Measures in $\mathbf{E}^2$ or $\mathbf{E}^3$; Integral Measures

(2.1) *Definition.* Let D be a region in $\mathbf{E}^2$ or $\mathbf{E}^3$. Let μ be a function which takes, as its individual inputs, the regular subregions of D and which gives, as its outputs, real numbers. If, for every regular subregion R of D and for every division of R into two regular subregions R_1 and R_2, we have

$$\mu(R_1) + \mu(R_2) = \mu(R) ,$$

then we say that μ is a *finite measure* on D. The region D (which need not be regular) is called the *space* of μ. For each regular subregion R of D, $\mu(R)$ is called the *measure* of R given by μ.

Example 1. Let $D = \mathbf{E}^2$. For every regular region R, define $\mu(R) = $ area(R). Then μ is a finite measure on D.

Example 2. Let $D = \mathbf{E}^3$. For every regular region R, define $\mu(R) = $ volume(R). Then μ is a finite measure on D.

Example 3. Let D in $\mathbf{E}^3$ represent a piece of solid matter, where the mass density $\delta(P)$ at each point P in D is given by a continuous scalar field δ on D. For every regular subregion R of D, we define

$$\mu(R) = \iint_R \delta dV .$$

Then μ is a measure on D, and $\mu(R)$ gives the mass in R. In this example, $\mu(R) \geq 0$ for all R in D.

Example 4. Let D in $\mathbf{E}^2$ represent a thin, flat, insulating film. Electric charge (which may be positive or negative) has been distributed over the film. Let the net charge density (per unit area) at each point P be $f(P)$, where f is a continuous scalar field on D. For every regular subregion of D, we define $\mu(R) = \iint_R fdA$. Then μ is a measure on D, and $\mu(R)$ gives the *net charge* on R. ("Net charge" means excess of positive charge over negative charge.) In this example, it is possible to have $\mu(R) < 0$.

Example 5. Let D be a region in $\mathbf{E}^2$, and let f be any continuous scalar field on D. For every regular subregion R of D, define

$$\mu_f (R) = \iint_R fdA .$$

Then, by law I for double integrals (in §14.4), μ_f is a measure on D.

The intuitive idea behind a finite measure μ is that $\mu(R)$ measures the amount contained in R of some real or hypothetical, physical or mathematical "substance" such as *area*, or *volume*, or *mass*, or *electric charge*, or *energy*.

Definition. If μ is a finite measure on a region D in $\mathbf{E}^2$ (or $\mathbf{E}^3$), and if there is a continuous scalar field f on D such that for every regular subregion R of D, $\mu(R) = \iint_R fdA$ (or $\iiint_R fdV$), then we say that μ is an *integral measure*. All of the examples above are integral measures (with $f = 1$ for examples 1 and 2). In §23.7, we shall see that not all measures are integral measures. If μ is an integral measure on D, then there must be a *unique* continuous scalar field f such that for any R in D, $\mu(R) = \iint_R fdA$ (or $\iiint_R fdV$). See (3.2) in §23.3.

Remark. In defining a finite measure, it is often convenient to include the empty region $\varnothing$, which contains no points, as an elementary region, and to specify that $\mu(\varnothing) = 0$.

§23.3 The Density of a Measure

Let D be a region in $\mathbf{E}^2$; let μ be a finite measure on D; and let P be a point in D. If we think of $\mu(R)$ as the amount contained in R of some hypothetical "substance," it is natural to try to define the *density of* μ *at* P as the "density per unit area" of that "substance."

(3.1)   *Definition.* Let ΔR be an elementary region lying in a region D, and let P be a point in ΔR. Then the maximum distance between a pair of points in ΔR is called the *diameter* of ΔR and denoted $d(\Delta R)$. Let ΔA be the area of ΔR. We use the notation $\left.\dfrac{d\mu}{dA}\right|_P$ for our desired *density of* μ *at* P. We define

$$\left.\frac{d\mu}{dA}\right|_P = \lim_{\substack{(d(\Delta R)\to 0) \\ P\ in\ \Delta R}} \frac{\mu(\Delta R)}{\Delta A}.$$

For the density $\left.\dfrac{d\mu}{dA}\right|_P$ to exist, this limit must exist. This limit exists if there are (i) a funnel function $\varepsilon(t)$ on $[0,b]$ (for some $b > 0$), and (ii) a limit value ℓ, such that *for every elementary region* ΔR *containing* P, if $d(\Delta R) \leq b$ then

$$\left| \ell - \frac{\mu(\Delta R)}{\Delta A} \right| \leq \varepsilon(d(\Delta R)).$$

If so, then $\left.\dfrac{d\mu}{dA}\right|_P = \ell$.

For a finite measure μ in $\mathbf{E}^3$, the definition of *density* is analogous (with volume ΔV in place of area ΔA), and the density of μ at P is written $\left.\dfrac{d\mu}{dV}\right|_P$.

Two fundamental facts about measures and densities are given in the following *density theorems.*

(3.2)   **Theorem.** *Let* μ *be an integral measure and* f *be continuous on a region* D *in* E^2 *such that for every regular subregion* R *of* D, $\mu(R) = \iint_R fdA$. *Then the density* $\left.\dfrac{d\mu}{dA}\right|_P$ *exists at every point* P *in* D, *and* $\left.\dfrac{d\mu}{dA}\right|_P = f(P)$.

(3.3)   **Theorem.** *Let* μ *be a finite measure on a region* D *in* E^2. *If, for all* P *in* D, *the density* $\left.\dfrac{d\mu}{dA}\right|_P$ *exists as a limit and has the value* $f(P)$, *where* f *is a continuous scalar field on* D, *and if, furthermore, this density exists on each regular region* R *in* D *as a uniform limit on* R, *then for all regular subregions* R *of* D, $\mu(R) = \iint_R fdA$. *(Here, "uniform limit on* R*" means that there is a single "master" funnel function for* R *which gives the limit for the density at each point in* R.*)*

These theorems show that the concept of density is closely related to the concept of integral measure.

   Proof of (3.2):   Take a point P in D. Let ΔR be an elementary region containing P. Let m_1 and m_2 be the minimum and maximum values of f on ΔR. (m_1 and m_2 exist by the continuity of f and the compactness of ΔR.) Let ΔA be the area of ΔR. Then by law IV in §14.4,

$$m_1 \Delta A \le \iint_{\Delta R} fdA = \mu(\Delta R) \le m_2 \Delta A.$$

Hence

$$m_1 \le \frac{\mu(\Delta R)}{\Delta A} \le m_2.$$

We also have $m_1 \le f(P) \le m_2$, by the definition of m_1 and m_2. By the continuity of f at P,

$$\lim_{d(\Delta R) \to 0} (m_1 - m_2) = 0.$$

It follows that

$$\lim_{d(\Delta R) \to 0} \frac{\mu(\Delta R)}{\Delta A} = f(P).$$

   Proof of (3.3):   Let R be a regular subregion of D. Then $\iint_R fdA$ is the limit of a proper sequence of Riemann sums, where the n-th sum has the form

$$S_n = \sum_{i=1}^{n} f(P_{in}^*)\Delta A_{in} ,$$

and ΔA_{in} is the area of an elementary region ΔR_{in}, and where $\lim\limits_{n\to\infty} \max\limits_i d(\Delta R_{in}) = 0$ and $\lim\limits_{n\to\infty} S_n = \iint_R f dA$.

For each n, consider the corresponding sum (based on the same decomposition of R into the same subregions ΔR_{in}):

$$\mu(R) = \sum_{i=1}^{n}\mu(\Delta R_{in}) = \sum_{i=1}^{n}\frac{\mu(\Delta R_{in})}{\Delta A_{in}}\Delta A_{in} .$$

Each of these corresponding sums has the exact value $\mu(R)$. Hence the sequence of these sums has the limit $\mu(R)$.

The difference between the two sums (for the same n) must be

$$S_n - \mu(R) = \sum_{i=1}^{n}\left(f(P_{in}^*) - \frac{\mu(\Delta R_{in})}{\Delta A_{in}}\right)\Delta A_{in} .$$

By the *uniform limit* assumption, there is, for each n, a value ε_n such that for each $i \le n$, $\left| f(P_{in}^*) - \dfrac{\mu(\Delta R_{in})}{\Delta A_{in}}\right| \le \varepsilon_n$ and such that $\lim\limits_{n\to\infty} \varepsilon_n = 0$. Hence $S_n - \mu(R) \le \varepsilon_n \text{area}(R)$ and $\lim\limits_{n\to\infty} S_n = \mu(R)$. Thus $\mu(R) = \iint_R f dA$. Theorems (3.2) and (3.3) together will be referred to as the *measure-density theorems* for $\mathbf{E}^2$.

For $\mathbf{E}^3$, the statements are the same as for (3.2) and (3.3), except that the measure μ is in $\mathbf{E}^3$, and volume replaces area. The proofs are the same as for (3.2) and (3.3). These are the *measure-density theorems* for $\mathbf{E}^3$.

Remark. In Chapter 24, using the measure described in §23.4 below, we shall see that *Green's theorem* is a specific instance of the measure-density theorem for $\mathbf{E}^2$. In Chapter 25, using the measure described in §23.5 below, we shall see that the *divergence theorem* is a specific instance of the measure-density theorem for $\mathbf{E}^3$. In §23.6, we shall see that the concepts of *measure* and *integral measure* for regular regions in $\mathbf{E}^2$ can be generalized, in a natural way, to concepts of *surface measure* and *vector integral surface measure* for directed, piecewise smooth, finite surfaces in $\mathbf{E}^3$. In Chapter 26, *Stokes's theorem* will show us that a particular surface measure is, in fact, a vector integral surface measure.

As the notation suggests, the density of a measure is a new kind of derivative, and the measure-density theorems are a new version (for this new derivative) of the *fundamental theorem of integral calculus*. The density of a finite measure is known as the *Radon-Nikodym derivative* of that measure.

§23.4 The Circulation Measure for a Vector Field $\vec{F}$ in E^2

We represent E^2 as a plane M in E^3, and we choose a unit vector $\hat{u}$ normal to M. Using a right-hand rule with $\hat{u}$, we can describe counterclockwise (and hence also clockwise) directions for simple closed curves in M.

Let $\vec{F}$ be a continuous vector field with domain D in E^2. Let R be a regular subregion of D. We assign directions to the boundary curves of R: the outer boundary curve C is directed counterclockwise; the interior boundary curves $C_1,\ldots,C_k$ (if R is not elementary) are directed clockwise. [4–1]. We then define the quantity $\mu_{\vec{F}}(R)$:

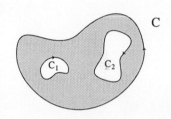

[4–1]

(4.1)   (i) if R is elementary, $\mu_{\vec{F}}(R) = \oint_C \vec{F} \cdot d\vec{R}$;

       (ii) if R is not elementary,

$$\mu_{\vec{F}}(R) = \oint_C \vec{F} \cdot d\vec{R} + \oint_{C_1} \vec{F} \cdot d\vec{R} + \cdots + \oint_{C_k} \vec{F} \cdot d\vec{R} .$$

We shall see below that μ is a finite measure on D. It is called the *circulation measure on* D *given by* $\vec{F}$.

Example 1. *In* E^2 *with Cartesian coordinates, let* $\vec{F} = -y\hat{i} + x\hat{j}$, *let* D = E^2, *and let* $\hat{u} = \hat{i} \times \hat{j}$. *Find* $\mu_{\vec{F}}(R)$ *when R is the disk of radius 2 with center at the origin.* [4–2]. Here $\mu_{\vec{F}}(R) = \oint_C \vec{F} \cdot d\vec{R}$, where C is the counterclockwise outer boundary circle. For any point P on C, $\vec{F}(P)$ is tangent to C in the direction of C, and $\left|\vec{F}(P)\right| = 2$. Evaluating by definition, we have

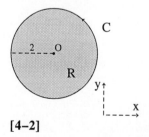

[4–2]

$$\mu_{\vec{F}}(R) = \oint_C \vec{F} \cdot d\vec{R} = 2(\text{length of C}) = 8\pi .$$

Example 2. *For* $\vec{F}$ *and D as in example 1, find* $\mu_{\vec{F}}(R)$, *where R is the result of removing the interior of the disk of radius 1/2 with center at* (1,0) *from*

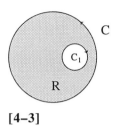

[4–3]

the disk in example 1. [4–3]. Let C be the outer counterclockwise circle, and let C_1 be the interior clockwise circle. Then, by (5.1),

$$\mu_{\vec{F}}(R) = \oint_C \vec{F} \cdot d\vec{R} + \oint_{C_1} \vec{F} \cdot d\vec{R}.$$

From example 1, $\oint_C \vec{F} \cdot d\vec{R} = 8\pi.$

For C_1, we have $\vec{R}(t) = (1 + \frac{1}{2}\cos t)\hat{i} + \frac{1}{2}\sin t\,\hat{j}$, $0 \le t \le 2\pi$, with direction of parameter opposite to direction of curve.

Then
$$\frac{d\vec{R}}{dt} = -\frac{1}{2}\sin t\,\hat{i} + \frac{1}{2}\cos t\,\hat{j}.$$

$$\vec{F} = -\frac{1}{2}\sin t\,\hat{i} + (1 + \frac{1}{2}\cos t)\hat{j}.$$

$$\vec{F} \cdot \frac{d\vec{R}}{dt} = \frac{1}{4}\sin^2 t + \frac{1}{4}\cos^2 t + \frac{1}{2}\cos t = \frac{1}{4}(1 + 2\cos t).$$

Thus
$$\oint_{C_1} \vec{F} \cdot d\vec{R} = -\frac{1}{4}\int_0^{2\pi}(1 + 2\cos t)dt = -\frac{\pi}{2}.$$

and we have $\mu_{\vec{F}}(R) = 8\pi - \frac{\pi}{2} = \frac{15}{2}\pi.$

(4.2) **Theorem.** *Let* $\vec{F}$ *be a continuous vector field on* D *in* E^2. *Let* $\mu_{\vec{F}}$ *be defined as above. Then* $\mu_{\vec{F}}$ *is a finite measure on* D.

Proof. We must show that if a regular region R is divided into two subregions R_1 and R_2, then $\mu_{\vec{F}}(R_1) + \mu_{\vec{F}}(R_2) = \mu_{\vec{F}}(R).$

Let C be the outer-directed boundary of R, and let $C_1,...,C_k$ be the interior-directed boundaries. Let $\tilde{C}$ be an undirected curve which divides R into R_1 and R_2 as described in §23.1. The curves $C, C_1,...,C_k$ reappear, segment by segment and with the same directions, in the calculations for $\mu_{\vec{F}}(R_1)$ and $\mu_{\vec{F}}(R_2)$. Certain segments of $\tilde{C}$ also appear. These segments of $\tilde{C}$ occur in oppositely directed pairs, since each such segment of $\tilde{C}$ serves as a boundary segment for R_1 and also, in the reverse direction, as a boundary segment for R_2. Thus the line integrals for the $\tilde{C}$ segments cancel when we form $\mu_{\vec{F}}(R_1) + \mu_{\vec{F}}(R_2)$. The theorem follows.

Example 3. [4–4] *illustrates this proof.* Here C' and C'' are segments of C, and C_1', and C_1'' are segments of C_1. $\tilde{C}_1'$, and $\tilde{C}_1''$ are the same segment of $\tilde{C}$, but with opposite directions. Similarly for $\tilde{C}_2'$ and $\tilde{C}_2''$.

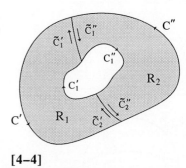

[4–4]

Thus $\quad \mu_{\vec{F}}(R) = \int_C + \int_{C_1} = \int_{C'} + \int_{C''} + \int_{C_1'} + \int_{C_1''}$.

$$\mu_{\vec{F}}(R_1) = \int_{C'} + \int_{\tilde{C}_2'} + \int_{C_1'} + \int_{\tilde{C}_1'} \ .$$

$$\mu_{\vec{F}}(R_2) = \int_{C''} + \int_{\tilde{C}_1''} + \int_{C_2'} + \int_{\tilde{C}_2''} \ .$$

Since $\int_{\tilde{C}_1'} + \int_{\tilde{C}_1''} = 0$ and $\int_{\tilde{C}_2'} + \int_{\tilde{C}_2''} = 0$, we have $\mu_{\vec{F}}(R_1) + \mu_{\vec{F}}(R_2) = \mu_{\vec{F}}(R)$.

Remark. Is the circulation measure $\mu_{\vec{F}}$ an integral measure? That is to say, given $\vec{F}$, must there be some scalar field f such that $\mu_{\vec{F}}(R) = \iint_R f\,dA$? The answer to this question depends on the smoothness of the vector field $\vec{F}$. In Chapter 24, we shall see that if $\vec{F}$ is C^1, then the answer is affirmative. The density f will be called the *rotational derivative* of $\vec{F}$. This result is known as *Green's theorem.*

§23.5 The Flux Measure for a Vector Field $\vec{F}$ in $\mathbf{E}^3$

Let $\vec{F}$ be a continuous vector field with domain D in $\mathbf{E}^3$. Let R be a regular subregion of D. Each of the boundary surfaces of R is closed and hence, by §21.6, two-sided. We assign directions to these boundary surfaces: the outer boundary surface S is directed *outward* (*away* from R), while the interior boundary surfaces (if any) $S_1,...,S_k$ are directed *inward* (also *away* from R). We then define the quantity $\mu_{\vec{F}}^f(R)$:

(5.1) (i) if R has only one boundary surface S, then $\mu_{\vec{F}}^f(R) = \oiint_S \vec{F} \cdot d\vec{\sigma}$.

(ii) if R has boundary surfaces $S,S_1,...,S_k$, then $\mu_{\vec{F}}^f(R) =$
$\oiint_S \vec{F} \cdot d\vec{\sigma} + \oiint_{S_1} \vec{F} \cdot d\vec{\sigma} + \cdots + \oiint_{S_k} \vec{F} \cdot d\vec{\sigma}$.

We shall see below that $\mu_{\vec{F}}^f$ is a finite measure on D. It is called the *flux measure on D given by* $\vec{F}$.

Example 1. *In* E^3-*with-Cartesian-coordinates, let* $\vec{F} = x\hat{i} + y\hat{j} + z\hat{k}$, *and let* $D = E^3$. *Find* $\mu_{\vec{F}}^f(R)$ *when R is a solid sphere of radius 2 with center at the origin. Here S is a sphere of radius 2, directed outward, and at every point P on S,* $\vec{F}(P)$ *is normal to S with* $\left|\vec{F}(P)\right| = 2$, [5–1]. *Hence*

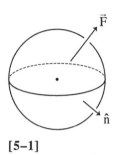

[5–1]

$$\mu_{\vec{F}}^f(R) = \oiint_S \vec{F} \cdot d\vec{\sigma} = 2(\text{Area of S}) = 2(16\pi) = 32\pi.$$

Example 2. *For* $\vec{F}$ *and D as in example 1, find* $\mu_{\vec{F}}^f(R)$ *where R is the result of removing the interior of the sphere of radius* $\frac{1}{2}$ *with center at* $(1,0)$ *from the solid sphere in example 1.* Let S be the outer (outward directed) spherical surface, and let S_1 be the interior (and inward directed) spherical surface. Then, by (5.1),

$$\mu_{\vec{F}}^f(R) = \oiint_S \vec{F} \cdot d\vec{\sigma} + \oiint_{S_1} \vec{F} \cdot d\vec{\sigma}.$$

As in example 1, $\oiint_S \vec{F} \cdot d\vec{\sigma} = 32\pi$. In the problems, we see that $\oiint_{S_1} \vec{F} \cdot d\vec{\sigma} = -\frac{\pi}{2}$. Hence

$$\mu_{\vec{F}}^f(R) = 32\pi - \frac{\pi}{2} = \frac{63}{2}\pi.$$

(5.2)   **Theorem.** *Let* $\vec{F}$ *be a continuous vector field on D in* E^3. *Let* $\mu_{\vec{F}}^f$ *be defined as above. Then* $\mu_{\vec{F}}^f$ *is a finite measure on D.*

The proof of (5.2) is exactly parallel to the proof of (4.2). Here R is divided into R_1 and R_2 by a surface $\tilde{S}$, and, for the calculations for $\mu_{\vec{F}}^f(R_1)$ and $\mu_{\vec{F}}^f(R_2)$, each shared piece of $\tilde{S}$ produces a pair of oppositely directed boundary pairs for R_1 and R_2, since each such piece of $\tilde{S}$ serves as a boundary piece for R_1 and also, with direction reversed, as a boundary piece for R_2. Thus the surface integrals for these pairs cancel when we form $\mu_{\vec{F}}^f(R_1) + \mu_{\vec{F}}^f(R_2)$. The theorem follows.

Remark. Is the flux measure $\mu_{\vec{F}}^{f}$ an integral measure? That is to say, given $\vec{F}$, must there be some scalar field f such that $\mu_{\vec{F}}^{f}(R) = \iint_{R} f\,dV$? The answer to this question depends on the smoothness of $\vec{F}$. In Chapter 25, we shall see that if $\vec{F}$ is C^1, then the answer is affirmative. The density f will be called the *divergence* of f. This result is known as the *divergence theorem.*

§23.6 Surface Measures; the Circulation Measure for $\vec{F}$ in $\mathbf{E}^3$

Elementary surfaces are defined in §21.4. Recall that a surface S is a *finite surface* if S can be divided into elementary surfaces in the same way that a regular region in $\mathbf{E}^2$ can be divided into elementary regions.

Let S be a directed (and therefore two-sided and piecewise smooth) finite surface. We say that a directed finite surface S′ is a *subsurface* of S if S′ is contained in S and if the direction of S′ agrees with the direction of S.

Surface measures. In §23.4, we considered finite measures on regular regions in $\mathbf{E}^2$. We now extend the concept of a finite measure in $\mathbf{E}^2$ to a finite measure on pieces of surface in $\mathbf{E}^3$. *Elementary, piecewise smooth, directed surfaces in $\mathbf{E}^3$* will play the role of *elementary regions*, and *finite, piecewise smooth, directed surfaces in $\mathbf{E}^3$* will play the role of *regular regions*.

Definition. Let D be a region in $\mathbf{E}^3$. Let μ^s be a function which takes finite, piecewise smooth, directed surfaces in D as inputs and which gives real numbers as its outputs. We say that μ^s is a *finite surface measure* for D if, for every finite, piecewise smooth, directed surface S in D and for every subdivision of S into two finite, piecewise smooth, subsurfaces S_1 and S_2, we have

$$\mu^s(S) = \mu^s(S_1) + \mu^s(S_2) \ .$$

We note that a finite, piecewise smooth, directed surface S can be divided into two subsurfaces S_1 and S_2 in infinitely many different ways. For any such division there must be piecewise-smooth, closed surface $\tilde{S}$ such that one of the subsurfaces has all its points on or outside $\tilde{S}$, the other has all its points on or inside $\tilde{S}$, and the common boundary of S_1 and S_2 is the intersection of $\tilde{S}$ with S.

Example 1. *For any given* D, *let* $\mu^s(S)$ = *area of* S. Then μ^s is a finite surface measure for D.

Example 2. Let $\vec{G}$ *be a given continuous vector field on* D. *Let* $\mu_{\vec{G}}^s(S) = \iint_S \vec{G} \cdot d\vec{\sigma}$. Then $\mu_{\vec{G}}^s$ is a finite surface measure on D.

Definition. If a surface measure can be defined as in example 2, we say that it is a *vector integral surface measure.*

Let $\vec{F}$ be a continuous vector field on a domain D in $\mathbf{E}^3$. Let S be a finite, piecewise smooth, directed surface in D. As noted in §21.5, S is either closed and has no boundary curves, or else the boundary of S consists of a finite number of piecewise smooth, simple, closed curves $C_1, \ldots, C_k$. Let $C_1, \ldots, C_k$ have directions *coherent* with the direction of S, where *coherence* is defined as follows.

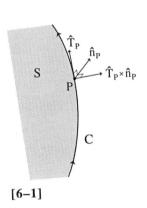

[6–1]

(6.1) *Definition.* Let C be a directed boundary curve for S. Let P be a point in a smooth segment of C. Let $\hat{n}_P$ be the unit normal vector to S at P in the direction of S, and let $\hat{T}_P$ be the unit tangent vector to C at P in the direction of C. We say that the direction of C is *coherent* with the direction of S if $\hat{T}_P \times \hat{n}_P$ is directed away from S at P. [6–1].

(Informally: the directions of S and C are coherent if, when one walks along C in the direction of C, with one's body in the direction of S, one finds that S lies immediately to one's left.)

(6.2) *Definition.* For continuous vector field $\vec{F}$ on D and a directed finite surface S in D, the quantity $\mu_{\vec{F}}^c(S)$ is defined by:

    (i)   if S is closed, then $\mu_{\vec{F}}^c(S) = 0$;

    (ii) if S is not closed, let $C_1, \ldots, C_k$ be the piecewise smooth, simple, closed boundary curves of S, and let $C_1, \ldots, C_k$ have directions coherent with the direction of S. Then

$$\mu_{\vec{F}}^c(S) = \oint_{C_1} \vec{F} \cdot d\vec{R} + \cdots + \oint_{C_k} \vec{F} \cdot d\vec{R}.$$

We shall see below that $\mu_{\vec{F}}^c$ is a finite surface measure for D. We call it the *circulation measure on* D *given by* $\vec{F}$.

Example 3. *Let* $\vec{F} = -zy\hat{i} + zx\hat{j}$. *Let* S *be the closed cylinder of radius* 2 *between* $z = -2$ *and* $z = 2$ *and with the* z *axis as its axis. Let* S *be directed outward. Since* S *is closed,* $\mu_{\vec{F}}^c(S) = 0$.

Example 4. *Let* $\vec{F}$ *be as in example* 3. *Let* S′ *be the open, outward-directed cylinder obtained from* S *in example* 3 *by removing the disks from the top and bottom of* S. *Let* C_1 *and* C_2 *be the top and bottom directed boundary curves of* S′, *with directions coherent with the direction of* S. (*See* [6–2].) *Then, by* (6.2)

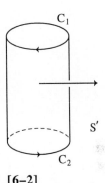

[6–2]

$$\mu_{\vec{F}}^c(S') = \oint_{C_1} \vec{F} \cdot d\vec{R} + \oint_{C_2} \vec{F} \cdot d\vec{R}.$$

The reader may verify that $|\vec{F}| = 4$ on C_1 and C_2, and that the direction of $\vec{F}$ on C_1 and C_2 is opposite to the directions of C_1 and C_2. Hence $\oint_{C_1} \vec{F} \cdot d\vec{R} = -4$ (length of C_1) $= -16\pi$, and $\oint_{C_2} \vec{F} \cdot d\vec{R} = -4$ (length of C_2) $= -16\pi$. Thus

$$\mu_{\vec{F}}^c(S) = -32\pi.$$

(6.3)   ***Theorem.*** *Let* $\vec{F}$ *be a continuous vector field on* D *in* $\mathbf{E}^3$. *Let* $\mu_{\vec{F}}^c$ *be defined as in* (6.2). *Then* $\mu_{\vec{F}}^c$ *is a finite surface measure on* S.

Proof. We must show that if a directed finite surface S is divided into two subsurfaces S_1 and S_2, then $\mu_{\vec{F}}^c(S_1) + \mu_{\vec{F}}^c(S_2) = \mu_{\vec{F}}^c(S)$. The proof is exactly parallel to the proofs of (4.2) and (5.2). The new segments of boundary curve which appear for S_1 and S_2 must occur as oppositely directed pairs which cancel when we form $\mu_{\vec{F}}^c(S_1) + \mu_{\vec{F}}^c(S_2)$.

Remarks. In the light of example 2 above, it is natural to ask the question: given a circulation surface measure $\mu_{\vec{F}}^c$ on D in $\mathbf{E}^3$ for a continuous vector field $\vec{F}$, can we find a vector field $\vec{G}$ on D such that for every finite, piecewise smooth, directed surface S in D, $\mu_{\vec{F}}^c(S) = \iint_S \vec{G} \cdot d\vec{\sigma}$? That is to say, must $\mu_{\vec{F}}^c$ be a vector integral surface measure? In Chapter 26, we shall see that the answer is affirmative provided that (1) $\vec{F}$ is C^1, and (2) we restrict our attention to surfaces which are *piecewise* C^2 *smooth*, where "piecewise C^2 smooth" means that S can be divided into a finite number of smooth elementary pieces, each of which has a C^2 parameterization. (See (3.1) in §21.3.) Moreover, we shall see that the vector field $\vec{G}$ is uniquely determined by $\vec{F}$ and is called the *curl* of $\vec{F}$. This result is known as *Stokes's theorem*.

§23.7 Singular Measures

We show that not all finite measures are integral measures.

Example. Let S be a fixed collection (finite or infinite) of isolated points in E^2, where each point P has a certain fixed mass $m(P)$. We define a measure μ on E^2 as follows. Let R be a regular region in E^2. Let $S_R = \{P_1, P_2, ..., P_k\}$ be the finite subset of S contained in R. Then $\mu(R) = h(P_1) + \cdots + h(P_k)$ where:

(i) $h(P_i) = m(P_i)$ if P_i lies in the interior of R;

(ii) $h(P_i) = \frac{1}{2}m(P_i)$ if P_i lies on a smooth piece of boundary curve R but not at a vertex.

(iii) $h(P_i) = \frac{\alpha}{2\pi}m(P_i)$ if P_i lies at a vertex which joins two smooth pieces of boundary (of R), and if α is the interior angular size of this vertex, as determined by unit tangent vectors at P_i to the smooth pieces which join at P_i. See [7–1].

It is easy to see that μ is a finite measure on E^2 and that $\mu(R)$ gives a reasonable answer to the question "what is the total mass which belongs to R?" Note that $\mu(R)$ counts only its 'fair share' of any mass on its boundary. A measure of the kind described in this example is called a *singular measure* because the hypothetical "substance" being measured is concentrated at single points.

We make several formal definitions.

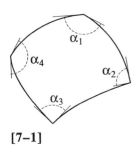

[7–1]

Definitions. Let μ be a finite measure on a region D in E^2 (or E^3). Let P be a given point in D and let ΔR be an elementary region containing P. We say that

(7.1)
$$\lim_{\substack{d(\Delta R)\to 0 \\ P\ in\ \Delta R}} \mu(\Delta R) = \ell.$$

if there is a funnel function $\varepsilon(t)$ on [0,b] such that for every regular ΔR containing P,

$$d(\Delta R) \le b \Rightarrow |\ell - \mu(\Delta R)| \le \epsilon(d(\Delta R)).$$

(7.2)   We say that μ is a *singular measure* if, for some P in D, the limit in (7.1) exists with $\ell \ne 0$.

(7.3)   We say that μ is an *absolutely continuous measure* if for every P in D, the limit in (7.1) exists with $\ell = 0$.

It is immediate from these definitions that

(7.4)   *A singular measure cannot be absolutely continuous.*

It is also easy to show that

(7.5)   *Every integral measure is absolutely continuous.*

From (7.4) and (7.5) it follows that

(7.6)   *A singular measure cannot be an integral measure.*

In §23.8, we give (1) an example of an absolutely continuous measure which is not an integral measure and (2) an example of a finite measure on a slightly restricted class of regular sets in $\mathbf{E}^2$, for which, at every point P, the limit in (7.1) fails to exist. It follows that this restricted-set measure is neither singular nor absolutely continuous.

§23.8 Counterexamples

First counterexample. Consider two statements about a finite measure μ on $\mathbf{E}^2$:
   (a) μ *is an integral measure.*
   (b) μ *is an absolutely continuous measure.*
In §23.7, we saw that (a) implies (b). We now give an example of a measure μ on $\mathbf{E}^2$ where (b) is true but (a) is not.
   Let O be the origin for a polar coordinate system in $\mathbf{E}^2$. Let R be any regular region in $\mathbf{E}^2$. We define $\mu(R)$:
   (i)   If R does not contain O, either in its interior or on its boundary, we define

$$\mu(R) = \iint_R \frac{1}{r} dA.$$

This integral exists, since $\frac{1}{r}$ is continuous on R. $\mu(R) \geq 0$, since $\frac{1}{r} > 0$ on R.

(ii)   If R contains O, then, for any $a > 0$, let R_a be the set of all points P in R such that $|\overrightarrow{OP}| \geq a$. When a is sufficiently small, R_a must be non-empty and must be a regular set in E^2 which does not contain O. (See [8–1].) Hence $\mu(R_a)$ is defined by (i). We now define

$$\mu(R) = \lim_{a \to 0} \mu(R_a).$$

It remains to prove

(1)   that $\lim_{a \to 0} \mu(R_a)$ in (ii) exists as a finite limit;

(2)   that μ, as defined by (i) and (ii), is an absolutely continuous finite measure; and

(3)   that μ is not an integral measure.

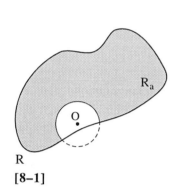

R

[8–1]

***Proof of* (1).** For each $a > 0$, let $f(a) = 0$ if R_a is empty, and let $f(a) = \mu(R_a)$ if R_a is not empty. Since $a_2 \leq a_1 \Rightarrow R_{a_2}$ contains R_{a_1}, we have

(8.1) $a_2 \leq a_1 \Rightarrow f(a_2) \geq f(a_1)$.

(8.2)   ***Lemma.*** *There exists* $k > 0$ *such that for all* $a > 0$, $f(a) \leq k$.

It will follow from (8.1) and (8.2), by the *least upper-bound principle* (see the problems) that $\lim_{a \to 0} f(a)$ exists as a finite limit. It remains to prove (8.2).

Let $b > 0$ be a fixed value such that R_b is a non-empty regular set. For any a such that $0 < a < b$, let $U_{a,b}$ be the annular ring [8–2] of all points (r,θ) with $a \leq r \leq b$. Then, by (i), $\mu(U_{a,b}) = \iint_{U_{a,b}} \frac{1}{r} dA = \int_0^{2\pi} \int_a^b \frac{1}{r} r\, dr\, d\theta =$ $2\pi(b - a)$. Since R_a is contained in $R_b + U_{a,b}$, we have

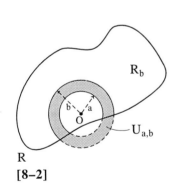

R

[8–2]

$$\mu(R_a) \leq \mu(R_b) + \mu(U_{a,b}) \leq \mu(R_b) + 2\pi(b - a).$$

Hence we have

$$f(a) = \mu(R_a) \leq \mu(R_b) + 2\pi b, \text{ for all } a > 0.$$

This proves (8.2) and completes the proof of (1).

Proof of **(2)**. We must show that for every point P, $\lim\limits_{\left(\substack{d(\Delta R)\to 0 \\ P\ \text{in}\ \Delta R}\right)}\mu(\Delta R)=0$. If $P\neq O$, this follows from the proof that an integral

measure is absolutely continuous. If $P=O$, then ΔR is a subset of $U_a(P)$ where $a=d(\Delta R)$ and $U_a(P)$ is the neighborhood of P of radius a, [8–3]. Thus $0\leq\mu(\Delta R)\leq\mu(U_a(P))$. Since $\mu(U_a(P))=\lim\limits_{c\to 0}\mu(U_{c,a})=\lim\limits_{c\to 0}2\pi(a-c)=$ $2\pi a$, and $\lim\limits_{a\to 0}\mu(U_a(P))=\lim\limits_{a\to 0}2\pi a=0$, we have $\lim\limits_{a\to 0}\mu(\Delta R)=0$. This proves (2).

Proof of **(3)**. If μ were an integral measure, we would have a continuous function g on $\mathbf{E}^3$ such that for all $a>0$, $\mu(U_a(O))=2\pi a=\iint_{U_a(O)}g\,dA$.

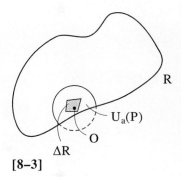

[8–3]

By the mean value law ((V) in §14.4), this would imply that for every $a>0$, there is a P_a in $U_a(O)$ such that $2\pi a=g(P_a)\pi a^2$. This gives $g(P_a)=\dfrac{2}{a}$. Since g is continuous, this would give $g(O)=\lim\limits_{a\to 0}g(P_a)=\lim\limits_{a\to 0}\dfrac{2}{a}$. This last

limit does not exist as a finite limit, since $\lim\limits_{a\to 0}\dfrac{2}{a}=\infty$, and we have a

contradiction. This proves (3).

Second counterexample. For a restricted family of regular regions in $\mathbf{E}^2$, we define a finite measure μ which is neither absolutely continuous nor singular. In particular, at every point P, $\lim\limits_{d(\Delta R)\to 0}\mu(\Delta R)$ will fail to exist

since

(8.3)　　$\lim\limits_{d(\Delta R)\to 0}\mu(\Delta R)=0$ for elementary polygonal regions ΔR, while

$\lim\limits_{d(\Delta R)\to 0}\mu(\Delta R)=2\pi$ for disk-shaped regions ΔR.

It then follows that

(8.4) $\displaystyle\lim_{d(\Delta R)\to 0}\frac{\mu(\Delta R)}{\Delta A}=0$ for   elementary   polygonal ΔR,   and

$\displaystyle\lim_{d(\Delta R)\to 0}\frac{\mu(\Delta R)}{\Delta A}=\infty$ for disk-shaped ΔR.

Definitions. We say that a finite curve is C^2-*smooth* if it has a path $R(t)$ on $[a,b]$ for which $\dfrac{d\vec R}{dt}\neq\vec 0$, and $\dfrac{d^2\vec R}{dt^2}$ is continuous on $[a,b]$. We say that an elementary region in E^2 is C^2-*elementary* if its boundary curve is piecewise C^2-smooth, and that a region in E^2 is C^2-*regular* if it can be divided into finitely many C^2-*elementary* regions by a regular subdivision.

We define a measure μ for the C^2-regular sets in E^2. We call it the *total curvature measure* in E^2. Let R be a given C^2 regular region. Let Q be a non-endpoint on a smooth piece C of a boundary curve of R. Let $\hat n_Q$ be the unit normal vector at Q directed *into* R. Let $\kappa(Q)$ be the curvature of C at Q. If $\kappa(Q)\neq 0$, let $\hat N_Q$ be the unit normal at Q as defined in §6.6. We define

$$\kappa^*(Q)=\begin{cases}0, & \text{if }\kappa(Q)=0,\\ (\hat n_Q\cdot\hat N_Q)\kappa(Q), & \text{if }\kappa(Q)\neq 0.\end{cases}$$

Thus $\kappa^*(Q)$ is positive if R is convex outward at Q and negative if R is convex inward at Q. From the compactness of each smooth piece of C, it follows that κ^* is a piecewise continuous function on C and hence that $\oint_C \kappa^*\,ds$ exists. Let $C_1,\dots,C_k$ be the simple closed boundary curves of R. We define

$$\mu(R)=\sum_{i=1}^{k}\oint_{C_i}\kappa^*\,ds.$$

By a cancellation argument similar to those used in §23.5 and §23.7, we have that μ is a finite measure.

Let R be a disk of radius a. Let C be its circular boundary. Then $\kappa^*(Q)=\dfrac{1}{a}$ at every point Q on C. Hence $\mu(R)=\oint_C\dfrac{1}{a}ds=\dfrac{1}{a}2\pi a=2\pi$. Thus every disk, regardless of size, has measure 2π.

Let R be a polygonal region. Evidently, $\mu(R)=0$ and $\kappa^*=0$ on each smooth piece of the boundary of R.

We thus have (8.3), and (8.4) follows.

Remark. An analogous "total curvature measure" can be defined in E^3, using the *mean curvature* H defined in §21.8. This measure has properties analogous to (8.3) and (8.4). (See the problems.) These total curvature measures do not, by themselves, prove that there is a non-singular and non-absolutely-continuous measure on all regular sets. They suggest, however, that any result to the contrary may be hard to prove.

§23.9 Problems

§23.1

1. Which of the following sets in E^2 are (i) elementary, (ii) regular, (iii) not regular?
   (a) E^2
   (b) Disk of radius 3 with center at origin
   (c) Disk in (b) with the interior of the concentric disk of radius 2 removed
   (d) Disk of radius 4, center at origin, from which the interiors of four disks of radius 1 with centers at (2,0), (0,2), (–2,0), and (0,–2) have been removed

2. Which of the following sets in E^3 are (i) elementary, (ii) regular, (iii) not regular?
   (a) The half-space $z \geq 0$
   (b) A solid cube
   (c) A solid torus
   (d) A solid sphere of radius 5 from which the interior of a concentric solid sphere of radius 3 has been removed

§23.2

1. Which are measures in E^2 and which are not? Give reasons. (If A and B are two sets of points, we define A $\cap$ B to be the set of all points which A and B have in common.)
   (a) $\mu(R)$ = perimeter of R (= total length of all boundary curves of R)
   (b) $\mu(R) = 5$ area(R)
   (c) $\mu(R) = [\text{area}(R)]^2$
   (d) Let H_1 and H_2 be the upper half-plane ($y \geq 0$) and the lower half-plane ($y \leq 0$).

   $\mu(R) = \text{area}(R \cap H_1) - \text{area}(R \cap H_2)$.

2. Which are measures in E^3 and which are not? Give reasons.
   (a) $\mu(R) = -2$ volume(R).
   (b) Let H_1 and H_2 be the upper half-space ($z \geq 0$) and the lower half space ($z \leq 0$). $\mu(R) = \text{volume}(R \cap H_1) - \text{volume}(R \cap H_2)$.
   (c) Let T be a given solid torus. Let T_0 be the set of points which lie on the surface of, or outside, the torus. $\mu(R) = |\text{Volume}(R \cap T)| - |\text{Volume}(R \cap T_0)|$.

§23.4

1. In each case, find $\mu_{\vec{F}}(R)$, where $\mu_{\vec{F}}$ is the circulation measure in E^2 for the given vector field $\vec{F}$.
   (a) $\vec{F} = x\hat{i} + y\hat{j}$. R is the disk of radius a with center at origin.
   (b) $\vec{F}$ as in (a). R is disk of radius 1 with center at (2,0).
   (c) $\vec{F} = xy\hat{i} + y^2\hat{j}$. R as in (a).
   (d) $\vec{F}$ as in (c). R as in (b).
   (e) $\vec{F}$ as in (c). R is a square with vertices (0,0), (3,0), (3,3), (0,3) from which we have removed the interior of the square with vertices at (1,1), (2,1), (2,2), (1,2).

§23.5

1. In each case, find $\mu_{\vec{F}}^f(R)$, where $\mu_{\vec{F}}^f$ is the flux measure in E^3 for the vector field $\vec{F} = -2x\hat{i} + y\hat{j} + 4z\hat{k}$.

(a) R is the solid cube in the first octant with diagonally opposite corners at (0,0,0) and (1,1,1).

(b) R is the solid sphere of radius a with center at the origin.

(c) R is a solid closed cylinder with z axis as axis, base in xy plane, radius 3, and height 2.

2. In example 2, show that $\oiint_{S_1} \vec{F} \cdot d\vec{\sigma} = -\frac{\pi}{2}$.

§23.6

1. Let C represent the earth's orbit around the sun. Let S be the elliptical plane region bounded by C. Let the curve C have the direction of the earth's motion. How should S be directed so that the direction of S is coherent with the direction of C? (Use north and south poles of the earth to specify sides of S.)

2. In each case, find $\mu_{\vec{F}}^c(S)$, where $\mu_{\vec{F}}^c$ is the circulation measure in E^3 for the given vector field $\vec{F}$.

(a) $\vec{F} = -y\hat{i} + x\hat{j}$ and S is the sphere of radius a, center at origin, directed out.

(b) $\vec{F}$ as in (a). S is upper hemisphere of sphere in (a), directed up.

(c) $\vec{F}$ as in (a). Let $0 < b < a$. S is the portion of the hemisphere in (b) which lies between z = 0 and $z = \sqrt{a^2 - b^2}$. S is directed out.

(d) $\vec{F}$ as in (a). Let $0 < b < a$. S is the portion of the hemisphere in (b) which lies above $z = \sqrt{a^2 - b^2}$. S is directed up.

(e) $\vec{F} = xy\hat{i} - y^2\hat{j}$. S is the open cylinder with z axis as axis, base in xy plane, radius a and height h. S is directed out.

§23.7

1. Consider cases 1(b), 1(d), 2(a), and 2(c) in the problems for §23.2.

(a) Which of these are integral measures? Give reasons.

(b) Which of these are absolutely continuous measures? Give reasons.

2. In E^2, let S_R be the set of points (x,y) such that $-4 \le x \le 4$, $-4 \le y \le 4$, and both x and y are integers. For P = (x,y) in S_R, define m(P) = x+y. Let μ be defined from m as in the example. What is $\mu(R)$

(a) when R is the rectangular region with vertices (−1,0), (−1,3), (3,3), and (3,0)?

(b) when R is the triangular region with vertices (0,−2), (0,2), and (2,0)?

3. Prove that if μ is an integral measure on D (in E^2), and if f and g are continuous scalar fields on D such that for every regular region R in D, $\mu(R) = \iint_R f dA = \iint_R g dA$, then f(P) = g(P) for all P in D.

4. Prove that every integral measure is absolutely continuous.

§23.8

1. As suggested in the remark at the end of §23.8, define an appropriate *total curvature* measure in E^3, and show that this measure has properties analogous to (8.3) and (8.4).

Chapter 24

Green's Theorem

§24.1 The Rotational Derivative in $\mathbf{E}^2$

Let $\vec{F}$ be a vector field on domain D in $\mathbf{E}^2$. Let P be a point in D. We define a new kind of derivative for $\vec{F}$ at P, called the *rotational derivative of* $\vec{F}$ *at* P. We write this derivative as $\operatorname{rot}\vec{F}\big|_P$. $\operatorname{rot}\vec{F}\big|_P$ will have a scalar value, and if $\operatorname{rot}\vec{F}\big|_P$ exists at every point P in D, these values will form a scalar field on D. We refer to this scalar field as the *rotational derivative of* $\vec{F}$ *on D* and write it as $\operatorname{rot}\vec{F}$.

In order to describe and define the rotational derivative, we treat $\mathbf{E}^2$ as a directed surface in $\mathbf{E}^3$ and assume that we are looking down on it from its positive side. We then measure counterclockwise rotations as positive and clockwise rotations as negative, [1–1].

[1–1]

$\operatorname{rot}\vec{F}\big|_P$ can be described in terms of an underlying *physical idea* and in terms of an underlying *mathematical idea*. Both ideas lead to the same mathematical definition. We begin with the underlying *physical idea*. We view the vector field $\vec{F}$ on D as the velocity, at a given time t_0, of a thin film of fluid flowing across D. Let P be a chosen fixed point in D. Let C(a,P) be the counterclockwise directed circle of radius a with center at P, and let Q be a given point on C(a,P). Let p_Q be a moving fluid particle located at Q at time t_0. (p_Q is not necessarily moving along the circle C(a,P).) Let $\hat{u}$ be a unit vector in the direction from P to the particle p_Q, and consider the angular rate of change, at time t_0, of the direction of this arrow. We call this scalar quantity the *instantaneous*

509

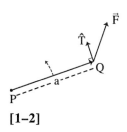

[1–2]

angular speed around **P**, *at the time* t_0, *of the fluid particle at* **Q**. By the definition of scalar product (see [1–2]), the particle's angular speed must be

(1.1) $$\frac{1}{a}\vec{F}(Q) \cdot \hat{T}(Q) \qquad \text{(in radians per second)}$$

where $\hat{T}(Q)$ is the unit tangent vector at Q to C(a,P). From (4) in §19.2, it follows that the *average* angular speed around P of all the fluid particles on C(a,P), at time t_0, must be

(1.2) $$\frac{1}{2\pi a}\oint_{C(a,P)}\frac{1}{a}\vec{F}\cdot\hat{T}\,ds = \frac{1}{2}\left(\frac{1}{\pi a^2}\oint_{C(a,P)}\vec{F}\cdot d\vec{R}\right).$$

Consider the limit

(1.3) $$\lim_{a\to 0}\frac{1}{2}\left(\frac{1}{\pi a^2}\oint_{C(a,P)}\vec{F}\cdot d\vec{R}\right).$$

It is natural to call this limit, if it exists, the *instantaneous angular* (or "rotational") *speed at* **P** *of the flow* $\vec{F}$ *around* **P** *at time* t_0. We define rot $\vec{F}\big|_P$ to be twice this angular speed. (The factor 2 will simplify the statements of later theorems.) We have:

Definition. Let $\vec{F}$ be a C^1 vector field on D in E^2. Then for any given interior point P of D, we define ([1–3])

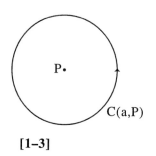

[1–3]

(1.4) $$\text{rot }\vec{F}\big|_P = \lim_{a\to 0}\frac{1}{\pi a^2}\oint_{C(a,P)}\vec{F}\cdot d\vec{R}.$$

The rotational derivative can also be described in terms of an underlying *mathematical idea*. Let P be a fixed point in D. Take $a > 0$. Let C(a,P) be as above. Let R(a,P) be the elementary region in E^2 whose boundary is C(a,P). Recall from §23.4 that

$$\mu_{\vec{F}}(R(a,P)) = \oint_{C(a,P)}\vec{F}\cdot d\vec{R}$$

gives the *circulation measure for* $\vec{F}$ (in E^2) of R(a,P). It follows, from the definition of *measure density*, that the value of the density must be

$$\left.\frac{d\mu_{\vec{F}}}{dA}\right|_P = \lim_{\substack{d(\Delta R)\to o \\ P\ in\ \Delta R}} \frac{\mu_{\vec{F}}(\Delta R)}{\Delta A} = \lim_{a\to 0} \frac{\mu_{\vec{F}}(R(a,P))}{\pi a^2}$$

$$= \lim_{a\to 0}\frac{1}{\pi a^2}\oint_{C(a,P)} \vec{F}\cdot d\vec{R} = \left.rot\ \vec{F}\right|_P\ ,$$

provided that $\mu_{\vec{F}}$ has a density at P. The existence of $\left.rot\ \vec{F}\right|_P$ does not guarantee, by itself, that the circulation measure $\mu_{\vec{F}}$ has a measure density at P, since the definition of measure density requires that $\displaystyle\lim_{\substack{d(\Delta R)\to o \\ P\ in\ \Delta R}} \frac{\mu_{\vec{F}}(\Delta R)}{\Delta A}$ exist as a unique limit for *all* possible shapes of the elementary region ΔR, while the definition of $\left.rot\ \vec{F}\right|_P$ requires only that the limit exist for certain disk-shaped regions. We shall prove in §24.6 that if $\vec{F}$ is C^1, then $\mu_{\vec{F}}$ must indeed have a density.

Let us return to the physical idea of a "thin film" fluid flow on a region of plane surface. It is often helpful, in working with a given vector field $\vec{F}$ in $\mathbf{E}^2$, to imagine that $\vec{F}$ gives the velocity of such a flow and to imagine also that we have a tiny paddlewheel of radius a (see [1–4]) whose spokes are attached rigidly to a hub which is free to rotate on an axle. The hub is held at a fixed point P in the flow, with its axle held normal to the plane of the flow. The flow interacts with the wheel by pressing on tiny circular vanes at the ends of the spokes. (We assume that the flow does not interact with the spokes themselves.) Each vane is parallel to its spoke and to the axle. Finally, we assume that if we insert the wheel in the fluid at a point P, then the wheel will rotate at the angular speed

[1–4]

(1.5) $$\omega_a = \frac{1}{2}\left(\frac{1}{\pi a^2}\oint_{C(a,P)} \vec{F}\cdot d\vec{R}\right).$$

This assumption can be justified by an elementary physical argument, provided that the density of the fluid (per unit area) is a continuous scalar field. See the problems. Then $2\omega_a$ gives an approximate value for $\left.rot\ \vec{F}\right|_P$, and this approximation will become increasingly accurate as the radius a is decreased.

Example 1. Let $\vec{F} = -y\hat{i} + x\hat{j}$ on $\mathbf{E}^2$. Let P be the point (c,d). We find $\left.rot\ \vec{F}\right|_P$ by calculating the integral in (1.4) and then taking the limit. In the integral, we have

$$\vec{R}(t) = (c+a\cos t)\hat{i} + (d+a\sin t)\hat{j}\ ,\qquad 0\le t\le 2\pi\ ;$$

$$\vec{R}(t) = (c + a\cos t)\hat{i} + (d + a\sin t)\hat{j}, \quad 0 \le t \le 2\pi;$$

$$\frac{d\vec{R}}{dt} = -a\sin t\,\hat{i} + a\cos t\,\hat{j};$$

$$\vec{F} = -(d + a\sin t)\hat{i} + (c + a\cos t)\hat{j}.$$

Hence

$$\vec{F} \cdot \frac{d\vec{R}}{dt} = da\sin t + a^2\sin^2 t + ca\cos t + a^2\cos^2 t$$

$$= da\sin t + ca\cos t + a^2, \text{ and}$$

$$\oint_{C(a,P)} \vec{F} \cdot d\vec{R} = \int_0^{2\pi} (da\sin t + ca\cos t + a^2)dt = 2\pi a^2.$$

We have $\dfrac{1}{\pi a^2} \oint_{C(a,P)} \vec{F} \cdot d\vec{R} = 2$, and $\lim\limits_{a \to 0}(2) = 2$. Thus, at every point P,

$\text{rot } \vec{F}\big|_P = 2$, and we see that $\text{rot } \vec{F}$ is the scalar field with constant value 2 over $\mathbf{E}^2$.

Example 2. Let $\vec{F} = x^2\hat{j}$ on $\mathbf{E}^2$, and let P be the point (c,d). A similar calculation gives us

$$\vec{F} \cdot \frac{d\vec{R}}{dt} = c^2 a\cos t + 2ca^2\cos^2 t + a^3\cos^3 t,$$

$$\int_0^{2\pi} \vec{F} \cdot \frac{d\vec{R}}{dt}dt = 2\pi ca^2,$$

and

$$\text{rot } \vec{F}\big|_P = \lim_{a \to 0} \frac{2\pi ca^2}{\pi a^2} = 2c.$$

Thus $\text{rot}\vec{F}$ is the scalar field $2x$ on $\mathbf{E}^2$-with-Cartesian-coordinates.

Laws. It is immediate from the linearity law for vector line integrals that the rotational derivative must satisfy the *linearity law*:

(1.6) $$\text{rot}(a\vec{F} + b\vec{G}) = a\,\text{rot }\vec{F} + b\,\text{rot } \vec{G}.$$

§24.2 The Rotational Derivative in Cartesian Coordinates

In $\mathbf{E}^2$-with-Cartesian coordinates, let $\vec{F}(x,y) = L(x,y)\hat{i} + M(x,y)\hat{j}$ be a C^1 vector field on domain D. Thus L and M are C^1 scalar functions on D. Let P be

an interior point of D. We seek a general formula for rot $\vec{F}\big|_P$ in terms of P and
the functions L and M. We begin by considering *parallel flows*.

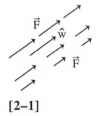

Definition. A vector field $\vec{F}$ on a domain D in $\mathbf{E}^2$ is said to be a *parallel flow* if
there are a scalar field f on D and a constant unit vector $\hat{w}$ such that $\vec{F}(P) =$
$f(P)\,\hat{w}$ for all P in D ([2–1]). Example 2 in §24.1 is a parallel flow, with $\hat{w} = \hat{j}$
and $f = f(x,y) = x^2$. Recall that we are treating $\mathbf{E}^2$ as a directed surface in $\mathbf{E}^3$.   [2–1]

(2.1)   *The parallel flow theorem for rotation. Let $\vec{F} = f\hat{w}$ on D, where f is*
C^1*, be a parallel flow in* $\mathbf{E}^2$*. Let $\hat{u}$ be a unit normal in the direction of the*
plane $\mathbf{E}^2$*. Let $\hat{v} = \hat{w} \times \hat{u}$. (Thus $\hat{v}$ is the result of rotating $\hat{w}$ in* $\mathbf{E}^2$ *by* $-\dfrac{\pi}{2}$
radians. See [2–2].) Let P be an interior point of D. *Then*

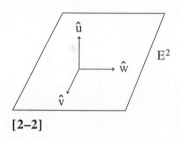

$$\text{rot }\vec{F}\big|_P = \frac{df}{ds}\bigg|_{\hat{v},P}$$

[2–2]

  Proof. We introduce Cartesian coordinates in our given $\mathbf{E}^2$ with $\hat{i} = \hat{v}$
and $\hat{j} = \hat{w}$. Then $\vec{F} = f(x,y)\hat{j}$, where f is C^1. Let C(a,P) be a counterclockwise
directed circle with radius a and center at P. Let Q be a point on C(a,P). By the
linear approximation theorem (§9.2),

(2.2) $f(Q) = f(P) + \vec{\nabla}f\big|_P \cdot \vec{PQ} + \vec{\varepsilon}(Q) \cdot \vec{PQ}$,

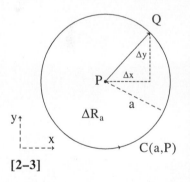

where $\displaystyle\lim_{Q \to P} \vec{\varepsilon}(Q) = \vec{0}$. See [2–3].

  Let $\vec{PQ} = \Delta x\hat{i} + \Delta y\hat{j} = a\,\widehat{PQ}$.

Then $f(Q) = f(P) + f_x(P)\Delta x + f_y(P)\Delta y + \vec{\varepsilon}(Q) \cdot \vec{PQ}$,

[2–3]

and $\vec{F}(Q) = [f(P) + f_x(P)\Delta x + f_y(P)\Delta y + \vec{\varepsilon}(Q) \cdot \vec{PQ}]\hat{j}$.

Hence $\displaystyle\oint_C \vec{F} \cdot d\vec{R} = f(P)\oint_C \hat{j} \cdot d\vec{R} + f_x(P)\oint_C \Delta x\hat{j} \cdot d\vec{R}$

       $+ f_y(P)\oint_C \Delta y\hat{j} \cdot d\vec{R} + a\oint_C (\vec{\varepsilon}(Q) \cdot \widehat{PQ})\hat{j} \cdot d\vec{R}$, where C is C(a,P). We consider the four integrals on the right-hand side of this last equation. We have: $\oint_C \hat{j} \cdot d\vec{R} = 0$ and $\oint_C \Delta y \hat{j} \cdot d\vec{R} = 0$ by symmetric cancellation. By the usual parameterization, we get $\oint_C \Delta x \hat{j} \cdot d\vec{R} = \int_0^{2\pi} a^2 \cos^2 t \, dt = \pi a^2$. Hence

$$\text{rot } \vec{F}\big|_P = \lim_{a \to 0} \frac{1}{\pi a^2} \left(\pi a^2 f_x(P) + a \oint_C (\bar{\varepsilon}(Q) \cdot \hat{PQ}) \hat{j} \cdot d\vec{R} \right)$$

$$= f_x(P) + \lim_{a \to 0} \frac{1}{\pi a} \oint_C (\bar{\varepsilon}(Q) \cdot \hat{PQ}) \hat{j} \cdot d\vec{R}$$

Let $\varepsilon^*(t)$ be a funnel function for $\lim_{Q \to P} \bar{\varepsilon}(Q) = \vec{0}$. Then

$$\left| \frac{1}{\pi a} \oint_C (\bar{\varepsilon}(Q) \cdot \hat{PQ}) \hat{j} \cdot d\vec{R} \right| \leq \frac{1}{\pi a} \oint_C \varepsilon^*(a) ds$$

$$= \frac{1}{\pi a} 2\pi a \varepsilon^*(a) = 2\varepsilon^*(a),$$

since $\left| \hat{j} \cdot d\vec{R} \right| \leq ds$. Hence $\lim_{a \to 0} \frac{1}{\pi a} \oint_C (\bar{\varepsilon}(Q) \cdot \hat{PQ}) \hat{j} \cdot d\vec{R} = 0$, and we have $\text{rot } \vec{F}\big|_P = f_x(P) = \text{rot } \vec{F}\big|_P = f_x(P) = \frac{df}{ds}\big|_{\hat{i},P} = \frac{df}{ds}\big|_{\hat{v},P}$, as desired.

Using (2.1), we now get:

(2.3)   *Formula for* rot $\vec{F}$ *in Cartesian coordinates (for* E^2). *Let* $\vec{F} = L\hat{i} + M\hat{j}$ *on D be a* C^1 *vector field in* E^2*-with-Cartesian coordinates. Then, for any interior point* P *of D,*

$$\text{rot } \vec{F}\big|_P = M_x(P) - L_y(P) = \frac{\partial M}{\partial x}\bigg|_P - \frac{\partial L}{\partial y}\bigg|_P.$$

Proof. Let $\vec{F} = \vec{F}_1 + \vec{F}_2$, where $\vec{F}_1 = L\hat{i}$ and $\vec{F}_2 = M\hat{j}$. Then each of $\vec{F}_1$ and $\vec{F}_2$ is a parallel flow. Applying the parallel flow theorem to each of $\vec{F}_1$ and $\vec{F}_2$, we have

$$\text{rot } \vec{F}_1 = -L_y \qquad (\text{using } \hat{w} = \hat{i} \text{ and } \hat{v} = -\hat{j}),$$

$$\text{rot } \vec{F}_2 = M_x \qquad \text{(with } \hat{w} = \hat{j} \text{ and } \hat{v} = \hat{i} \text{).}$$

By linearity (1.6), we have

$$\text{rot } F\big|_P = M_x(P) - L_y(P)$$

Definition. So far, we have only defined $\text{rot } \vec{F}\big|_P$ for an interior point P of D. Let $\vec{F}$ on D in E^2 be a vector field such that $\text{rot } F\big|_P$ exists at all interior points of D. Let Q be a boundary point of D. If $\lim\limits_{\substack{P \to Q \\ P \text{ int erior}}} \text{rot} F\big|_P = q$ exists, then we say that $\text{rot } F$ *exists at* Q *and has the value* q.

If $\vec{F}$ on D is C^1, then $M_x - L_y$ must be a continuous scalar field on D. Hence we have:

(2.4)   *Corollary. If* F *on D in* E^2 *is* C^1, *then* $\text{rot } \vec{F}\big|_P$ *exists at every point* P *of* D, *and* $\text{rot} \vec{F}$ *is a continuous scalar field on* D.

Example 1. As in example 1 *of §24.1, let* $\vec{F} = -y\hat{i} + x\hat{j}$. (2.3) immediately gives

$$\text{rot } \vec{F} = M_x - L_y = 1 - (-1) = 2 \, .$$

Example 2. As in example 2 *of §24.1, let* $\vec{F} = x^2\hat{j}$. (2.3) immediately gives

$$\text{rot } \vec{F} = M_x - L_y = 2x \, .$$

Example 3. Let $\vec{F} = \dfrac{-y\hat{i} + x\hat{j}}{x^2 + y^2}$ *for* (x,y) $\neq$ (0,0). (In polar coordinates, $\vec{F} = \frac{1}{r}\hat{\theta}$ for r > 0.) Using the quotient rule for derivatives, we have

$$M_x = \frac{(x^2 + y^2) - x(2x)}{(x^2 + y^2)^2} = \frac{y^2 - x^2}{(x^2 + y^2)^2} \, ,$$

and $\qquad L_y \;=\; \dfrac{-(x^2+y^2)+2y^2}{(x^2+y^2)^2} = \dfrac{y^2-x^2}{(x^2+y^2)^2}$

Therefore, $\text{rot } \vec{F} = M_x - L_y = 0.$

§24.3 Green's Theorem

In Chapter 23, we introduced the circulation measure $\mu_{\vec{F}}$ for a continuous vector field $\vec{F}$ on a domain D in $\mathbf{E}^2$. This measure assigned a value $\mu_{\vec{F}}(R)$ to every regular region R in D. In §24.1 above, we noted that if $\mu_{\vec{F}}$ is an integral measure, then its measure density must be $\dfrac{d\mu_{\vec{F}}}{dA} = \text{rot } \vec{F}$. *Green's theorem* and its corollaries now show us that if $\vec{F}$ is C^1, then $\mu_{\vec{F}}$ must, in fact, be an integral measure with $\text{rot } \vec{F}$ as its measure density. It follows that if $\vec{F}$ is C^1, then the circulation measure of any regular region in $\mathbf{E}^2$ can be calculated by integrating $\text{rot } \vec{F}$ over that region.

(3.1) ***Green's theorem.***

(a) *The circulation measure $\mu_{\vec{F}}$ on D is an integral measure.*

(b) *The density of this measure is the rotational derivative $\text{rot } \vec{F}$. That is to say, for any regular region R in D,*

$$\mu_{\vec{F}}(R) = \iint_R \text{rot } \vec{F} dA .$$

(3.1) is proved in §24.6 below. The following four corollaries are immediate from (3.1) and the definition of $\mu_{\vec{F}}$. Each corollary assumes that $\vec{F}$ is C^1.

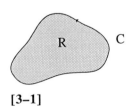

[3–1]

(3.2) ***Corollary.*** *For any elementary region R in D, let C be the counterclockwise-directed boundary of R, [3–1]. Then*

$$\iint_R \text{rot } \vec{F} dA = \oint_C \vec{F} \cdot d\vec{R} .$$

(3.3) ***Corollary.*** *In Cartesian coordinates, with $\vec{F} = L\hat{i} + M\hat{j}$, the conclusion of (3.2) becomes*

$$\iint_R (M_x - L_y)dA = \oint_C (L\hat{i} + M\hat{j}) \cdot d\vec{R} \,.$$

(3.4)   *Corollary. For any regular region* R *in* D, *with a counterclockwise directed outer-boundary curve* C *and clockwise-directed interior boundary curves* $C_1, \ldots, C_k$, [3–2], *we have*

$$\iint_R \mathrm{rot}\vec{F}dA = \oint_C \vec{F} \cdot d\vec{R} + \oint_{C_1} \vec{F} \cdot d\vec{R} + \cdots + \oint_{C_k} \vec{F} \cdot d\vec{R} \,.$$

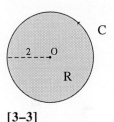

[3–2]

(3.5)   *Corollary. In Cartesian coordinates, with* $\vec{F} = L\hat{i} + M\hat{j}$, *the conclusion of* (3.4) *becomes*

$$\iint_R (M_x - L_y)dA =$$
$$\oint_C (L\hat{i} + M\hat{j}) \cdot d\vec{R} + \oint_{C_1} (L\hat{i} + M\hat{j}) \cdot d\vec{R} + \cdots + \oint_{C_k} (L\hat{i} + M\hat{j}) \cdot d\vec{R} \,.$$

Corollaries analogous to (3.3) and (3.5) can be obtained for a curvilinear coordinate system in $\mathbf{E}^2$, provided that we express $\mathrm{rot}\,\vec{F}$ in that system. (For polar coordinates, see §24.7 below.) In traditional texts, (3.3) is called *Green's theorem*, and (3.5) is called the *extended Green's theorem*.

Example 1. Recall example 1 of §23.4, where we calculated $\mu_{\vec{F}}(R)$ for $\vec{F} = -y\hat{i} + x\hat{j}$ and a disk R of radius 2 with center at the origin, [3–3]. Using Green's theorem, we can get the same result without evaluating the line integral. We have $M_x - L_y = 1 + 1 = 2$.   Hence, by Green's theorem, $\mu_{\vec{F}}(R) = 2 \cdot \mathrm{area}(R) = 2 \cdot 4\pi = 8\pi$.

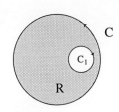

[3–3]

[3–4]

Example 2. Similarly, recall example 2 of §23.4. [3–4]. Here again we can avoid evaluating the line integrals.   We have $\mu_{\vec{F}}(R) = 2 \cdot \mathrm{area}(R) = 2(4\pi - \frac{1}{4}\pi) = 8\pi - \frac{\pi}{2} = \frac{15}{2}\pi$.

Example 3. Let $\vec{F} = -xy\hat{i} + xy\hat{j}$ *on domain* $D = \mathbf{E}^2$, *and let* C *be the counterclockwise-directed perimeter of the rectangle with vertices at* (0,0), (2,0), (2,1), *and* (0,1), [3–5]. *We verify Green's theorem for this* $\vec{F}$ *and this* C. First, we evaluate the double integral in (3.3). We have $M_x - L_y = y + x$, and

[3–5]

$$\iint_R (M_x - L_y)dA = \int_0^1 \int_0^2 (x+y)dxdy = \int_0^1 \frac{1}{2}x^2 + xy \Big]_{x=0}^{x=2} dy =$$

$$\int_0^1 (2+2y)dy = 2y + y^2 \Big]_0^1 = 3.$$

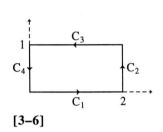

[3–6]

Then we evaluate the line integral in (3.1). We take C_1, C_2, C_3, and C_4 as in [3–6]. $\int_{C_1} \vec{F} \cdot d\vec{R} = \int_{C_4} \vec{F} \cdot d\vec{R} = 0$, since $\vec{F} = \vec{0}$ on C_1 and C_4. For C_2 and C_3, we use parametric evaluation.

For C_2: $\vec{R}(t) = 2\hat{i} + t\hat{j}$, $0 \le t \le 1$,

$$\left.\begin{array}{c} \dfrac{d\vec{R}}{dt} = \hat{j} \\[2mm] \vec{F} = -2t\hat{i} + 2t\hat{j} \end{array}\right\} \Rightarrow \vec{F} \cdot \dfrac{d\vec{R}}{dt} = 2t .$$

Thus $\int_{C_2} \vec{F} \cdot d\vec{R} = \int_0^1 2t\,dt = t^2 \Big]_0^1 = 1.$

For C_3: $\vec{R}(t) = t\hat{i} + \hat{j}$, $0 \le t \le 2$,   (directions disagree).

$$\left.\begin{array}{c} \dfrac{d\vec{R}}{dt} = \hat{i} \\[2mm] \vec{F} = -t\hat{i} + t\hat{j} \end{array}\right\} \Rightarrow \vec{F} \cdot \dfrac{d\vec{R}}{dt} = -t .$$

Thus $\int_{C_3} \vec{F} \cdot d\vec{R} = -\int_0^2 -t\,dt = \dfrac{t^2}{2} \Big]_0^2 = 2.$

Therefore $\int_C \vec{F} \cdot d\vec{R} = 0 + 1 + 2 + 0 = 3.$

Thus $\int_C \vec{F} \cdot d\vec{R} = \iint_R (M_x - L_y)dA$, and we have verified Green's theorem for our given field $\vec{F}$ and curve C.

Example 4. Let $\vec{F} = -y\hat{i} + x\hat{j}$ *on domain* $D = E^2$. *Let C be the counter-clockwise-directed perimeter of the triangle with vertices at* (0,0), (2,1), *and* (1,1), [3–7]. *We wish to find* $\int_C \vec{F} \cdot d\vec{R}$. *By using Green's theorem, we can avoid*

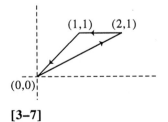

[3–7]

direct evaluation of this line integral. We have $\int_C \vec{F} \cdot d\vec{R} = \iint_R (M_x - L_y)dA =$
$\iint_R (1+1)dA = 2(\text{area of R}) = 2(\frac{1}{2}) = 1$.

Example 5. *Let* C_1 *be the upper semicircle of* $x^2+y^2 = 9$, *and let* C_2 *be the line segment between* $(-3,0)$ *and* $(3,0)$. *Let* $C = C_1+C_2$ *be directed counterclockwise,* [3–8]. *Let* $\vec{F} = (e^{x^2} + 3y)\hat{i} + (5x - e^{-y^2})\hat{j}$ *on* $D = E^2$. *We wish to find* $\int_C \vec{F} \cdot d\vec{R}$. By Green's theorem, we have

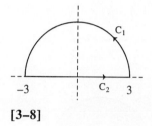

[3–8]

$$\int_C \vec{F} \cdot d\vec{R} = \iint_R (M_x - L_y)dA = \iint_R (5-3)dA = 2(\text{area of R}) = 2(\frac{9\pi}{2}) = 9\pi.$$

In examples 4 and 5, we have used the double integral in (3.3) to help evaluate the line integral in (3.3). In example 6, we shall use the line integral to help evaluate the double integral. We first observe, by Green's theorem, that for any elementary region R whose boundary is a simple, closed, counterclockwise-directed, piecewise smooth curve C,

(3.6) $\frac{1}{2}\int_C (-y\hat{i} + x\hat{j}) \cdot d\vec{R} = \frac{1}{2}\iint_R (M_x - L_y)dA$ (by Green's theorem)

$$= \frac{1}{2}\iint_R (1+1)dA = \frac{1}{2}(2)\,(area\,(R)) = area\,(R)\,.$$

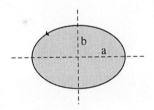

Example 6. *We wish to find the area of the region* R *enclosed by the ellipse* $(\frac{x}{a})^2 + (\frac{y}{b})^2 = 1$. Let C be this ellipse, directed counterclockwise [3–9].   [3–9]
By (3.6) , we know that

$$area\,(R) = \frac{1}{2}\int_C (-y\hat{i} + x\hat{j}) \cdot d\vec{R}\,.$$

We proceed to evaluate this line integral. C is given by the path

$$\vec{R}(t) = a\cos t\hat{i} + b\sin t\hat{j}, \qquad 0 \le t \le 2\pi.$$

We have $\frac{d\vec{R}}{dt} = -a\sin t\hat{i} + b\cos t\hat{j}$,

$$\vec{F} = -y\hat{i} + x\hat{j} = -b\sin t\hat{i} + a\cos t\hat{j},$$

and
$$\vec{F} \cdot \frac{d\vec{R}}{dt} = ab\sin^2 t\,\hat{i} + ab\cos^2 t\,\hat{j} = ab.$$

Hence
$$\frac{1}{2}\int_C \vec{F} \cdot d\vec{R} = \frac{1}{2}\int_0^{2\pi} ab\,dt = \pi ab.$$

Thus
$$\text{area of R} = \pi ab.$$

We have used Green's theorem to derive the standard formula for the area of an ellipse.

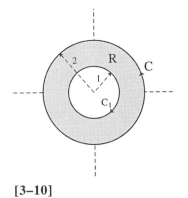

[3–10]

Example 7. Let $\vec{F} = \frac{1}{r}\hat{\theta} = \dfrac{-y\hat{i}+x\hat{j}}{x^2+y^2}$ *have the domain* $D = E^2$-*with-origin-removed. Let R have C and* C_1 *as its boundary curves, where C is the circle of radius 2 with center at the origin, and* C_1 *is the circle of radius 1 with center at the origin.* (See [3–10].) *Then R is a regular subregion of* D. *We wish to verify* (3.5).

Evaluating by definition, we have $\oint_C \vec{F} \cdot d\vec{R} = 2\pi(2)(\frac{1}{2}) = 2\pi$, and $\oint_{C_1} \vec{F} \cdot d\vec{R} = -2\pi(1)(1) = -2\pi$. Thus $\oint_C \vec{F} \cdot d\vec{R} + \oint_{C_1} \vec{F} \cdot d\vec{R} = 0$. For the double integral, recall from example 3 in §24.2 that $\text{rot}\,\vec{F} = 0$. Then $\iint_R \text{rot}\,\vec{F}\,dA = \iint_R (0)dA = 0$. We have verified (3.5) for our given field $\vec{F}$ and curves C and C_1.

§24.4 A Converse Derivative Test for E^2

In §18.5, we saw that if $\vec{F} = L\hat{i} + M\hat{j}$ on domain D in E^2-with-Cartesian-coordinates is C^1 and conservative, then we must have $M_x - L_y$ (and hence we must have $\text{rot}\,\vec{F} = 0$.) We referred to this fact as the *derivative test for a gradient field in* E^2. By (2.3), this fact can be restated: if $\vec{F}$ is C^1 and conservative on D in E^2, then $\text{rot}\,\vec{F} = 0$ on D. We shall now use Green's theorem to obtain a partial converse to this test. Recall that a set S in E^2 is *simply connected* if
    (i) S is connected, and
    (ii) for every simple closed curve C in S, all points in the interior of C are in S.

This definition implies (by the Jordan curve theorem in §6.5):

(4.1)   *Let* C *be any simple closed curve in* $\mathbf{E}^2$. *Then the set consisting of* C *together with its interior is a simply connected set,* [4–1].

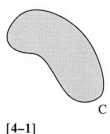

[4–1]

Using the concept of simple connectedness together with Green's theorem, we can prove the following converse for the derivative test in $\mathbf{E}^2$:

(4.2)   ***Converse derivative test for*** $\mathbf{E}^2$. *Let* $\vec{F}$ *be a* C^1 *vector field on a simply connected domain* D *in* $\mathbf{E}^2$. *If* $\mathrm{rot}\,\vec{F} = 0$, *then* $\vec{F}$ *must be conservative.*

Proof of (4.2). If $\vec{F}$ is not conservative, then, by the definition of conservative field, there must be a closed, directed, piecewise smooth curve C in D such that $\oint_C \vec{F}\cdot d\vec{R} \neq 0$. From this it follows that there must be a *simple*, closed, directed, piecewise smooth curve C′ in D such that $\oint_{C'} \vec{F}\cdot d\vec{R} \neq 0$. (See the problems.) Let R′ consist of C′ together with its interior. By the fact that D is simply connected, R′ is a subregion of D. By Green's theorem,

$$\oint_{C'} \vec{F}\cdot d\vec{R} = \pm\iint_{R'} \mathrm{rot}\,\vec{F}\,dA = 0 \ .$$

This contradicts our conclusion above that $\oint_{C'} \vec{F}\cdot d\vec{R} \neq 0$. Hence $\vec{F}$ must be conservative.

Example. Let $\vec{F} = \frac{1}{r}\hat{\theta} = L\hat{i} + M\hat{j} = \frac{-y\hat{i}+x\hat{j}}{x^2+y^2}$. In example 3 of §24.2 we have seen that $\mathrm{rot}\,\vec{F} = 0$, since $M_x - L_y = 0$. In example 5 of §20.1, we have

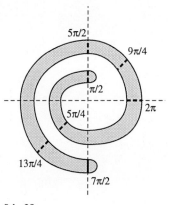

[4–2]

also seen that on the domain $D = \mathbf{E}^2$-with-origin-removed, $\vec{F}$ is not conservative. This is consistent with (4.2), since D is not simply connected. On the other hand, consider the same $\vec{F}$ on the spiral-shaped domain shown in [4–2]. By (4.1), this domain is simply connected. Hence, by (4.2), $\vec{F}$ is conservative on this domain. A scalar potential f for $\vec{F}$ is given by $f = \theta$, where the changing values of f are indicated by equipotential curves shown as radial lines in the figure.

Let $\vec{F}$ have a simply connected domain D. Assume that $\vec{F}$ is not conservative. Then, by (4.2), the condition $\mathrm{rot}\,\vec{F} = 0$ must fail on D. Green's theorem gives us a way to measure the "non-conservativeness" of $\vec{F}$ around any given piecewise-smooth simple closed curve C in D. Let C be such a curve, and let R consist of C together with the interior of C. Then, by (3.2),

$$\oint_C \vec{F} \cdot d\vec{R} = \iint_R \text{rot }\vec{F} dA = \frac{\iint_R \text{rot }\vec{F} dA}{\text{area of } R}(\text{area of } R)$$

$$= (\text{average value of } \text{rot }\vec{F} \text{ on } R)(\text{area of } R) \, .$$

Thus $\oint_C \vec{F} \cdot d\vec{R}$ is proportional to the area of R and to the average value of rot $\vec{F}$ on R.

§24.5 Irrotational Fields in E^2

Definition. If a field $\vec{F}$ on D in E^2 has rot $\vec{F} = 0$, we say that $\vec{F}$ is an *irrotational field on* D. We have noted in §24.4 that if D is not simply connected, then it is possible to have $\vec{F}$ be irrotational without being conservative.

The most important (and dramatic) applications of (3.4) are to vector fields in E^2 which are irrotational but not conservative (and hence have domains which are not simply connected).

(5.1) *Let $\vec{F}$ in E^2 be irrotational on domain D, let C_1 and C_2 be two simple closed, piecewise-smooth curves in D such that C_2 lies inside C_1 and such that the region between C_1 and C_2 is contained in D. It follows by (3.4) that if C_1 and C_2 have opposite directions, then $\oint_{C_1} \vec{F} \cdot d\vec{R} + \oint_{C_2} \vec{F} \cdot d\vec{R} = \pm \iint_R \text{rot }\vec{F} dA = 0$. Hence, if C_1 and C_2 have the same direction (both counterclockwise or both clockwise), then $\oint_{C_1} \vec{F} \cdot d\vec{R} - \oint_{C_2} \vec{F} \cdot d\vec{R} = 0$, and $\oint_{C_1} \vec{F} \cdot d\vec{R} = \oint_{C_2} \vec{F} \cdot d\vec{R}$.*

Thus, for a field $\vec{F} = L\hat{i} + M\hat{j}$ with $M_x = L_y$, (5.1) allows us to integrate $\vec{F}$ over a difficult and irregular curve C_1 by using a much simpler curve C_2 with

$$\oint_{C_2} \vec{F} \cdot d\vec{R} = \oint_{C_1} \vec{F} \cdot d\vec{R} \, .$$

Example 1. Let $\vec{F} = \frac{a}{r}\hat{\theta} = \frac{-ay\hat{i} + ax\hat{j}}{x^2 + y^2}$ on E^2-*with-origin-O-removed, where* a *is some given positive scalar constant. Let C_1 be any simple closed, piecewise smooth, counterclockwise directed curve which has O in its interior. Let C_2 be a counterclockwise-directed circle of radius* b *with center at O, where* b *is sufficiently small that* C_2 *lies entirely inside* C_1. (See [5–1].) Evaluating by definition, $\oint_{C_2} \vec{F} \cdot d\vec{R} = 2\pi a$. It follows by (5.1) that $\oint_{C_1} \vec{F} \cdot d\vec{R} = 2\pi a$. Furthermore, we see that for any simple closed curve C in D, there are exactly three possible values for $\oint_{C} \vec{F} \cdot d\vec{R}$: $2\pi a$, $-2\pi a$, and 0, depending on whether C is counter clockwise and encloses O, is clockwise and encloses O, or does not enclose O.

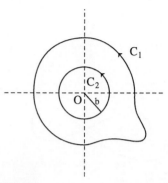

[5–1]

Using (3.4), we can also extend this analysis to piecewise-smooth curves which are closed but not simple.

Example 2. Let $\vec{F}$ *on D be as in example 1. Let* C *be a closed, piecewise smooth, directed curve in* D. *What are the possible values for* $\oint_{C} \vec{F} \cdot d\vec{R}$?

Let n be the net number of counterclockwise full turns about O made by C. This number is called the *winding number of* C *about* O. (See the problems for a more careful definition of the *winding number of a closed curve* C *about a point* P.) Examples are given in [5–2]. Consider the case of n = 3 in [5–3] (for our given $\vec{F} = \frac{a}{r}\hat{\theta}$). Here $\int_{C} \vec{F} \cdot d\vec{R} = \int_{C_1} \vec{F} \cdot d\vec{R} + \int_{C_2} \vec{F} \cdot d\vec{R} + \int_{C_3} \vec{F} \cdot d\vec{R}$, where C_1 is the inner loop of C, C_2 is the loop indicated by the dotted line, and C_3 is the loop indicated by the dashed line. Then $\oint_{C} \vec{F} \cdot d\vec{R} = 2\pi a + 2\pi a + 2\pi a = 6\pi a$. This result holds in general for our given $\vec{F}$. It is always possible to view $\oint_{C} \vec{F} \cdot d\vec{R}$ as a sum of integrals over loops leading to the value $\oint_{C} \vec{F} \cdot d\vec{R} = 2\pi an$, where n is the winding number of C about O. We conclude, for our given $\vec{F}$, that the possible values of $\oint_{C} \vec{F} \cdot d\vec{R}$ for a closed curve C in D are $2\pi an$, $n = 0, \pm 1, \pm 2, \ldots$.

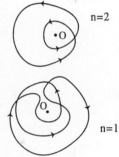

[5–2]

A similar analysis can be carried out when $\vec{F}$ is irrotational on its domain D and D = $\mathbf{E}^2$-with-several-points-removed.

Example 3. Let $\vec{F} = \vec{F}_1 + \vec{F}_2$, *where*

$$\vec{F}_1 = \frac{-ay\hat{i} + ax\hat{j}}{x^2 + y^2} \quad and \quad \vec{F}_2 = \frac{-by\hat{i} + b(x-1)\hat{j}}{(x-1)^2 + y^2}.$$

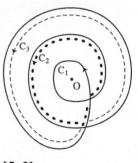

[5–3]

Here $\vec{F}_2$ has the same radial form as $\vec{F}_1$, except that $\vec{F}_2$ is centered at the point $P = (1,0)$ rather than the origin O. The domain D of $\vec{F}$ is $\mathbf{E}^2$ with O and P removed. Applying (3.4), we see that the possible values for $\oint_C \vec{F} \cdot d\vec{R}$ when C is simple and closed are $0, \pm 2\pi a, \pm 2\pi b, \pm 2\pi(a+b)$, depending on the direction of C and on whether C contains none, one, or both of the points O and P. More generally, the value of $\oint_C \vec{F} \cdot d\vec{R}$ when C is closed (but not necessarily simple) is $2\pi(na+mb)$, where n is the winding number of C about O and m is the winding number of C about P. See [5–4].

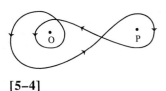

[5–4]

§24.6 Proof of Green's Theorem

In (3.1) of §23.3, we define the density of a measure μ on D (in $\mathbf{E}^2$) at a point P to be

(6.1)
$$\left.\frac{d\mu}{dA}\right|_P = \lim_{\substack{\{d(\Delta R)\to 0\} \\ \{P \text{ in } \Delta R\}}} \frac{\mu(\Delta R)}{\Delta A},$$

where ΔR is an elementary region which contains P and has area ΔA and diameter $d(\Delta R)$. Then, in (3.3) of §23.3, (the measure-density theorem), we saw that if the limit (6.1) exists uniformly for all P in D, we can conclude that μ is an integral measure.

Therefore, to complete the proof of Green's theorem, we must show (a) that the limit in (6.1) exists when μ is the circulation measure $\mu_{\vec{F}}$ and we allow, as ΔR, an elementary region of any shape (and not just a disk as in §24.2); and we must also show (b) that the limit is uniform for any given regular region in D.

That the limit must be uniform follows from the compactness of regular regions and from the uniform continuity of the continuous scalar field rot $\vec{F}$ on a compact region (see §14.9). (rot $\vec{F}$ is continuous, since $\vec{F}$ is a C^1 field.) This proves (b); see the problems.

To prove (a), we must prove that $\iint_R \text{rot}\,\vec{F} dA = \oint_C \vec{F} \cdot d\vec{R}$, for any C^1 field $\vec{F}$ and any elementary region R with counterclockwise boundary C (this is (3.2)). Thus we must find a proof for corollary (3.2) which is independent of (3.1). To do this, we need an extension of the subdivision principle (3.1) in §15.3. This extension asserts:

(6.2)   *For a regular region R in $\mathbf{E}^2$, it is always possible to subdivide R into a finite number of elementary subregions such that for each subregion there is some Cartesian coordinate system in which that region is both y-simple and x-simple.* (We do not give a proof of (6.2) in this text.)

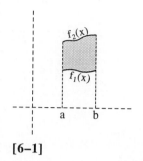

Since $\mu_{\vec{F}}$ is a finite measure, it is then enough to prove (3.3) for the case, in $\mathbf{E}^2$-with-Cartesian-coordinates, where R is both y-simple and x-simple. We accomplish this by proving (3.3) for $\vec{F}_1 = L\hat{i}$ when R is y-simple and for $\vec{F}_2 = M\hat{j}$ [6–1] when R is x-simple. We then take $\vec{F} = \vec{F}_1 + \vec{F}_2$ and use linearity.

For $\vec{F}_1 = L\hat{i}$, let R be y-simple as in [6–1]. In this case, C must consist of two piecewise smooth pieces, C_1 and C_3, and may or may not include the line segments C_2 and C_4. (See [6–2].) We proceed to evaluate both the line integral and the double integral in (3.3) and to show that they are equal.

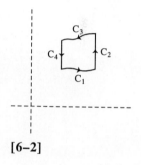

$$(6.3)\quad (a)\quad \iint_R (M_x - L_y)\,dA = \iint_R -L_y\,dA = \int_a^b \int_{f_1(x)}^{f_2(x)} -L_y\,dy\,dx$$

[6–2]

$$= \int_a^b [-L(x, f_2(x)) + L(x, f_1(x))]\,dx$$

$$(b)\quad \oint_C L\hat{i}\cdot d\vec{R} = \int_{C_1} L\hat{i}\cdot d\vec{R} + \int_{C_2} L\hat{i}\cdot d\vec{R} + \int_{C_3} L\hat{i}\cdot d\vec{R} + \int_{C_4} L\hat{i}\cdot d\vec{R}\ .$$

Since $L\hat{i}$ is perpendicular to C_2 and C_4, the integrals over C_2 and C_4 are 0. For C_1 and C_3, we use parametric evaluation.

For C_1: $\vec{R}(t) = t\hat{i} + f_1(t)\hat{j},\ a \le t \le b.$

$\dfrac{d\vec{R}}{dt} = \hat{i} + f_1'(t)\hat{j},$

$\vec{F}_1 = L(t, f_1(t))\hat{i},$

$\vec{F}_1 \cdot \dfrac{d\vec{R}}{dt} = L(t, f_1(t)),$

$\int_{C_1} \vec{F}_1 \cdot d\vec{R} = \int_a^b L(t, f_1(t))\,dt = \int_a^b L(x, f_1(x))\,dx.$

For C_3: $\vec{R}(t) = t\hat{i} + f_2(t)\hat{j},\ a \le t \le b$ (directions disagree).

$\dfrac{d\vec{R}}{dt} = \hat{i} + f_2'(t)\hat{j},$

$\vec{F}_1 = L(t, f_2(t))\hat{i},$

$$\vec{F}_1 \cdot \frac{d\vec{R}}{dt} = L(t, f_2(t)),$$

$$\int_{C_2} \vec{F}_1 \cdot d\vec{R} = -\int_a^b L(t, f_2(t))dt = \int_a^b -L(x, f_2(x))dx.$$

(6.4)   Thus $\oint_C \vec{F}_1 \cdot d\vec{R} = \int_a^b [L(x, f_1(x)) - L(x, f_2(x)]dx.$

From (6.3) and (6.4), we see that we have proved (3.3) for $\vec{F}_1$ when the case where R is y-simple. The proof of (3.3) for $\vec{F}_2 = M\hat{j}$ when R is x-simple is exact similar. Taking $\vec{F} = \vec{F}_1 + \vec{F}_2 = L\hat{i} + M\hat{j}$, we have

$$\iint_R (M_x - L_y)dA = \iint_R \text{rot}\,\vec{F}_1\,dA + \iint_R \text{rot}\,\vec{F}_2\,dA$$

$$= \oint_C \vec{F}_1 \cdot d\vec{R} + \oint_C \vec{F}_2 \cdot d\vec{R} = \oint_C (\vec{F}_1 + \vec{F}_2) \cdot d\vec{R} = \oint_C (L\hat{i} + M\hat{j}) \cdot d\vec{R}.$$

This proves (3.3) and completes the proof of Green's theorem.

§24.7 The Rotational Derivative in Polar Coordinates

Assume that we have a system of polar coordinates in $\mathbf{E}^2$, that $\vec{F}$ is a C^1 vector field on a domain D which does not include the origin, and that $\vec{F}$ is expressed as

(7.1) $$\vec{F}(r, \theta) = G(r, \theta)\hat{r} + H(r, \theta)\hat{\theta},$$

where G and H are C^1 functions of the polar coordinates r and θ. We have:

(7.2)   *Formula for the rotational derivative in polar coordinates. For* $\vec{F}$, G, *and* H *as above,*

$$\text{rot}\,\vec{F} = \frac{1}{r}\frac{\partial}{\partial r}(rH) - \frac{1}{r}\frac{\partial G}{\partial \theta}.$$

Examples. The reader may use (7.2) to verify each of the following:
(1) $\vec{F} = r\hat{r} \Rightarrow \text{rot}\,\vec{F} = 0.$

(2)   *For any given* f, $\vec{F} = f(r)\hat{r} \Rightarrow \text{rot}\,\vec{F} = 0.$

    (3) $\vec{F} = r\hat{\theta} \Rightarrow \text{rot } \vec{F} = 2$. (In the associated Cartesian system, this is example 1 of §24.3.)

    (4) $\vec{F} = \frac{1}{r}\hat{\theta} \Rightarrow \text{rot } \vec{F} = 0$. (This is example 3 in §24.3.)

    (5) $\vec{F} = r\theta\hat{r} + r\hat{\theta} \Rightarrow \text{rot } \vec{F} = 1$.

    (6) $\vec{F} = 3r^2\theta\hat{r} + r^2\hat{\theta} \Rightarrow \text{rot } \vec{F} = 0$.

    (7) $\vec{F} = 3r^2\theta\hat{r} - r^2\hat{\theta} \Rightarrow \text{rot } \vec{F} = -6r$.

    ***Proof of* (7.2).**   Take associated systems of Cartesian and polar coordinates in $\mathbf{E}^2$. The defining equations

(7.3)
$$\begin{pmatrix} x = r\cos\theta \\ y = r\sin\theta \end{pmatrix}$$

define r and θ implicitly as functions of x and y. By implicit differentiation, using the elimination method of Chapter 11 (see the problems), we find that

(7.4)
$$\frac{\partial r}{\partial x} = \cos\theta, \qquad\qquad \frac{\partial r}{\partial y} = \sin\theta,$$
$$\frac{\partial \theta}{\partial x} = -\frac{\sin\theta}{r}, \qquad\qquad \frac{\partial \theta}{\partial x} = \frac{\cos\theta}{r}.$$

From §7.4, we have

(7.5)
$$\hat{r} = \cos\theta\,\hat{i} + \sin\theta\,\hat{j},$$
$$\hat{\theta} = -\sin\theta\,\hat{i} + \cos\theta\,\hat{j}.$$

From (7.1) and (7.5), we have

$$\vec{F} = G\hat{r} + H\hat{\theta} = (G\cos\theta - H\sin\theta)\hat{i} + (G\sin\theta + H\cos\theta)\hat{j}.$$

Thus $L(x,y) = G(r,\theta)\cos\theta - H(r,\theta)\sin\theta$,

and $M(x,y) = G(r,\theta)\sin\theta + H(r,\theta)\cos\theta$.

Using the chain rule, together with (7.4), we get $M_x - L_y$ in terms of r and θ. (This is a substantial calculation; see the problems.) This gives

$$\text{rot}\ \vec{F} = M_x - L_y \quad = \frac{H}{r} + \frac{\partial H}{\partial r} - \frac{1}{r}\frac{\partial G}{\partial \theta}$$

$$= \frac{1}{r}\frac{\partial}{\partial r}(rH) - \frac{1}{r}\frac{\partial G}{\partial \theta},$$

and the proof is complete.

Remark. The proof of (7.2) above, using calculus, the chain rule, and two different coordinate systems, is an *analytical* proof. A shorter and more intuitive *geometrical* proof, using only polar coordinates, can be obtained by methods to be described in §27.2. The geometrical proof uses corollary (3.2) to Green's theorem.

§24.8 Problems

§24.1

1. Derive expression (1.1) for the angular speed with respect to P of a particle at Q, where $\vec{F}(Q)$ is the particle's velocity and $\hat{T}(Q)$ is the unit tangent vector at Q to a circle of radius $a = \left|\overrightarrow{PQ}\right|$ with center at P.

2. Assume that the paddlewheel of figure [1–4] is immersed at point P in a thin, steady-state, fluid flow across $\mathbf{E}^2$, where, for each Q, $\vec{F}(Q)$ is the velocity of the fluid at P. Assume that the force of the fluid striking a paddle vane at Q is proportional to the area of the vane and to the difference between the component of $\vec{F}(Q)$ normal to the vane and the velocity of the vane. Deduce (1.5).

3. Give details of the calculation for example 2.

§24.2

1. Let $\vec{F}$ be a given vector field in $\mathbf{E}^2$. In a given Cartesian coordinate system, we have $\vec{F} = L(x,y)\hat{i} + M(x,y)\hat{j}$ for certain functions L and M. In $\mathbf{E}^2$, we now choose a new and different Cartesian system, whose coordinates we call $\widetilde{x}$ and $\widetilde{y}$ and whose unit coordinate vectors we call $\hat{i}'$ and $\hat{j}'$.

Then, in this new system, we have $\vec{F} = \widetilde{L}(\widetilde{x},\widetilde{y})\hat{i}' + \widetilde{M}(\widetilde{x},\widetilde{y})\hat{j}'$, where $\widetilde{L}$ and $\widetilde{M}$ are appropriate functions of $\widetilde{x}$ and $\widetilde{y}$ which are, in general, different in form from L and M. Show, as a corollary to (2.3) that for every point P in the domain of $\vec{F}$,

$$\widetilde{M}_{\widetilde{x}}(\widetilde{x},\widetilde{y}) - \widetilde{L}_{\widetilde{y}}(\widetilde{x},\widetilde{y}) = M_x(x,y) - L_y(x,y).$$

§24.3

1. Verify Green's theorem for the case of example 4 by directly calculating the line integral.

2. Verify Green's theorem for the case of example 5 by calculating the line integral. (*Hint.* Use evaluation by definition for the exponential terms.)

3. Verify Green's theorem for example 6 by using direct integration to find the area.

4. Use (3.1) to find the area between one arch of the cycloid $\vec{R}(t) = a(t - \sin t)\hat{i} + a(1 - \cos t)\hat{j}$ and the x axis.

5. Use (3.6) to find the area between one arch of the sine curve $y = \sin x$ and the x axis.

6. Use (3.6) to find the area enclosed between the curve $y = x^2$ and the line $y = x$.

7. Use Green's theorem to find $\oint_C \vec{F} \cdot d\vec{R}$ when $\vec{F} = (x+y)\hat{i} + xy\hat{j}$ and C is the clockwise perimeter of the square with vertices at (0,0), (0,1), (1,1), and (1,0).

§24.4

1. For each of the following, decide whether or not it is a conservative field on $\mathbf{E}^2$. Give reasons.

   (a) $\vec{F} = 2xy^3\hat{i} + 2x^3y\hat{j}$

   (b) $\vec{F} = 2xy^3\hat{i} + 3x^2y^2\hat{j}$

   (c) $\vec{F} = \sin x\hat{i} - \sin y\hat{j}$

   (d) $\vec{F} = \sin y\hat{i} + \sin x\hat{j}$

2. In the proof of (4.2), deduce the existence of the simple closed curve C' from the existence of the closed curve C.

§24.5

1. $\vec{F} = L\hat{i} + M\hat{j}$ is defined at all points in $\mathbf{E}^2$ (with Cartesian coordinates) except the origin O. At every point P in this domain $M_x(P) = L_y(P)$. Furthermore, we are given that $\oint_C \vec{F} \cdot d\vec{R} = 4$ when C is a counterclockwise circle of radius 3 with center at O. What can we conclude about the possible values of $\oint_C \vec{F} \cdot d\vec{R}$ when C is

   (a) a clockwise circle of radius 1 with center at 0.

   (b) a counterclockwise circle of radius 2 with center at (3,0).

   (c) a clockwise circle of radius 4 with center at (3,0).

   (d) an arbitrary simple, closed curve in the domain of $\vec{F}$.

   (e) an arbitrary closed curve in the domain of $\vec{F}$.

2. $\vec{F} = L\hat{i} + M\hat{j}$ has, as its domain, all of $\mathbf{E}^2$ except for the points A = (−2,0) and B = (2,0). $M_x = L_y$ on the domain of F. On a counterclockwise circle of radius 1 with center at A, $\oint_C \vec{F} \cdot d\vec{R} = 5$. On a circle of radius 4 with center at O, $\oint_C \vec{F} \cdot d\vec{R} = 0$.

   (a) What is the value of $\oint_C \vec{F} \cdot d\vec{R}$ on a circle of radius 1 with center at B?

   (b) What are the possible values of $\oint_C \vec{F} \cdot d\vec{R}$ on an arbitrary simple closed curve in the domain of $\vec{F}$?

3. $\vec{F} = L\hat{i} + M\hat{j}$ has, as its domain, all of $\mathbf{E}^2$ except for the points $P_1 = (0,0)$, $P_2 = (0,2)$, $P_3 = (2,2)$ and $P_4 = (2,0)$. $\vec{F}$ has $M_x = L_y$ everywhere on its domain. C_1, C_2, C_3, and C_4 are counterclockwise closed curves in the domain of $\vec{F}$ with the following properties: (i) C_1 encloses P_1 and P_2 but not P_3 and P_4, and $\oint_{C_1} \vec{F} \cdot d\vec{R} = 3$; (ii) C_2 encloses P_1 and P_3 but not P_2 and P_4, and $\oint_{C_2} \vec{F} \cdot d\vec{R} = 4$; (iii) C_3 encloses P_2 and P_3 but not P_1 and P_4, and $\oint_{C_3} \vec{F} \cdot d\vec{R} = 5$; (iv) C_4 encloses P_1 and P_4 but not P_2 and P_3, and $\oint_{C_4} \vec{F} \cdot d\vec{R} = -6$. What are all the possible values of $\oint_C \vec{F} \cdot d\vec{R}$ for a simple, closed curve C in the domain of $\vec{F}$?

4. For $\mathbf{E}^2$, give a careful mathematical definition of the *winding number* of C about P, where C is a closed, piecewise smooth directed curve and P is a point not on C. (*Hint.* Use polar coordinates with origin at P; consider a parameterization r = r(t), θ = θ(t), a ≤ t ≤ b, for C.)

5. In example 3, derive the formula 2π(na+mb) for $\oint_C \vec{F} \cdot d\vec{R}$.

6. An irrotational vector field $\vec{F}$ is defined for all points in $\mathbf{E}_2$-with-Cartesian-coordinates, except for P = (0,0) and Q = (0,4). For the circle C = C(1,P)

and the circle $C' = C(5, P)$, we have $\oint_C \vec{F} \cdot d\vec{R} = 6$ and $\oint_{C'} \vec{F} \cdot d\vec{R} = 1$. Deduce the value C of $\oint_{C''} \vec{F} \cdot d\vec{R}$, where $C'' = C(1, Q)$.

7. For $\vec{F}$ is in problem 6, what are the possible values of $\oint_C \vec{F} \cdot d\vec{R}$ when C is an arbitrary simple closed curve lying in the domain of $\vec{F}$?

8. For $\vec{F}$ as in problem 6, what are the possible values of $\oint_C \vec{F} \cdot d\vec{R}$ when C is an arbitrary closed curve lying in the domain of $\vec{F}$?

9. (a) In E^2, let C be a simple, closed, piecewise smooth, directed curve which lies in the domain of a continuous vector field $\vec{F}$. Assume that $\vec{F}(P) \neq \vec{0}$ for all P on C. Give an appropriate definition for the *winding number of $\vec{F}$ on* C. (*Hint.* Consider $\vec{F}$ as a position arrow.)

(b) In (a), assume that the interior of C lies in the domain of $\vec{F}$. Show that if the winding number of $\vec{F}$ on C is not zero, then there must be some point Q inside C such that $\vec{F}(Q) \neq \vec{0}$.

§24.6

1. In the proof of Green's theorem, give further details for the proof of (b).

§24.7

1. In the proof of (7.2), give details of the calculation for $M_x - L_y$ in terms of r and θ.

2. Derive the expressions for $\frac{\partial r}{\partial x}$, $\frac{\partial r}{\partial y}$, $\frac{\partial \theta}{\partial x}$, and $\frac{\partial \theta}{\partial y}$ in (7.4).

3. Verify examples 1 through 7.

Chapter 25

The Divergence Theorem

§25.1 The Divergence in $\mathbf{E}^3$

Let $\vec{F}$ be a vector field on domain D in $\mathbf{E}^3$. Let P be a point in D. We define a new kind of derivative for $\vec{F}$ at P, called the *divergence of* $\vec{F}$ *at* P. We write this derivative as $\operatorname{div} \vec{F}\big|_P$. $\operatorname{div} \vec{F}\big|_P$ will have a scalar value, and if $\operatorname{div} \vec{F}\big|_P$ exists at every point P in D, these values will form a scalar field on D. We refer to this scalar field as the *divergence of* $\vec{F}$ and write it as $\operatorname{div} \vec{F}$.

$\operatorname{div} \vec{F}\big|_P$ can be described in terms of an underlying *physical idea* and in terms of an underlying *mathematical idea*. Both ideas lead to the same definition. We give the definition first and then consider the physical and mathematical ideas.

Definition. Let $\vec{F}$ be a continuous vector field on domain D, and let P be an interior point in D. For each $a > 0$, let S(P,a) be the outward-directed spherical surface with radius a and center at P. By the definition of *interior point*, there must be some $a > 0$ such that S(P,a) and its interior are contained in D, [1–1]. Then $\displaystyle\oiint_{S(P,a)} \vec{F} \cdot d\vec{\sigma}$ gives the net outward flux of $\vec{F}$ through S(P,a). (Note that the value of this integral can be positive, zero, or negative.) Intuitively, we can think of this net outward flux as flux "originating inside S(P,a)." The quantity

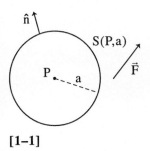

[1–1]

531

$$\frac{1}{\text{Vol}(S(P,a))} \oiint_{S(P,a)} \vec{F} \cdot d\vec{\sigma} = \frac{3}{4\pi a^3} \oiint_{S(P,a)} \vec{F} \cdot d\vec{\sigma}$$

can then be viewed as the average flux per unit volume "originating inside S(P,a)". We consider the following limit:

$$(1.1) \qquad\qquad\qquad \lim_{a \to 0} \frac{3}{4\pi a^3} \oiint_{S(P,a)} \vec{F} \cdot d\vec{\sigma}.$$

If this limit exists, then its value is called the *divergence of* $\vec{F}$ *at* **P,** and we write it $\text{div}\vec{F}\big|_P$. This limit (which can be positive, zero, or negative) can be viewed as a kind of density: the density per unit volume at which flux originates in the neighborhood of **P.**

When can we be sure that this limit exists? In §25.2, we prove the following.

(1.2)   *If* $\vec{F}$ *is* C^1 *on* **D,** *then* $\text{div}\vec{F}\big|_P$ *exists at every interior point of* **D,** *and there is a continuous scalar field on* **D** *whose value at each interior point* **P** *is* $\text{div}\vec{F}\big|_P$. As we have noted above, this scalar field is called the *divergence* of $\vec{F}$ and is written div $\vec{F}$.

The *linearity law* for the divergence is immediate from the linearity law for vector surface integrals:

(1.3)   *For vector fields* $\vec{F}$ *and* $\vec{G}$ *on a common domain, and for scalar constants* a *and* b,

$$\text{div}(a\vec{F} + b\vec{G}) = a \text{ div } \vec{F} + b \text{ div } \vec{G}.$$

How can we calculate the divergence of a given vector field at a given point? One approach is simply to evaluate the surface integral and find the limit in (1.1). A better approach, for most computational purposes, is described in §25.2, where we get a general formula for the divergence of a vector field $\vec{F}$ when $\vec{F}$ is expressed in Cartesian coordinates.

For a useful *physical idea* of the divergence imagine that our given vector field $\vec{F}$ is a *steady-state* mass-flow $\vec{m} = \delta\vec{v}$, where the density δ is a steady-state scalar field and the fluid velocity $\vec{v}$ is a steady-state vector field. As we saw in application 1 of §22.4, $\oiint_{S(P,a)} \vec{m} \cdot d\vec{\sigma}$ must be the net rate at which mass is

leaving S(P,a). (See also §18.1.) In order to sustain this steady-state outflow, there must be "spontaneous creation" of mass (at this same rate) inside S(P,a), [1–2]. (If $\oiint_{S(P,a)} \vec{m} \cdot d\vec{\sigma}$ happens to be negative, then this "spontaneous crea-

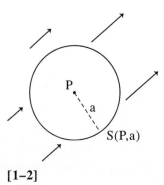

tion" must, in fact, be "spontaneous annihilation" of mass inside S(P,a).) Thus div $\vec{m}|_P$ will give the *density* of "spontaneous creation" or "annihilation" of mass near P and will have the dimensions (in c.g.s. units) gm/cm³sec.

    For an actual, physically occurring, steady-state mass-flow $\vec{m}$, we nor-

[1–2]

mally assume *conservation of mass*; that is to say, we assume that no mass is being created or annihilated, and hence that div $\vec{m}|_P = 0$ for all P. If the mass-flow $\vec{m}$ is time-dependent (and not steady-state), then we sometimes assume, instead, that at every moment t and at every point P, the *mass-density* δ must be changing in a way that compensates for the net flow of mass away from P (or towards P). This leads us to the equation

(1.4)
$$-\frac{\partial \delta}{\partial t} = \operatorname{div} \vec{m},$$

which must be satisfied at every point P and every moment t. (1.4) is known as the *equation of continuity* for a fluid flow. In §25.5, we consider further implications of (1.4) and in Chapter 28, we give a formal derivation of the physical law (1.4) from the physical law of conservation of mass.

    The divergence can also be associated with the following *mathematical idea*. Let P be a fixed point in D. Take a > 0. Let R(P,a) be the elementary region in $\mathbf{E}^3$ whose boundary is S(P,a), [1–3]. Recall from §23.6 that

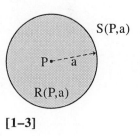

$$\mu_{\vec{F}}^f(\vec{R}(P,a)) = \oiint_{S(P,a)} \vec{F} \cdot d\vec{\sigma}$$

[1–3]

gives the *flux measure for* $\vec{F}$ (in $\mathbf{E}^3$) of R(P,a). It follows from the definition of *measure density* that if the flux measure $\mu_{\vec{F}}^f$ *has* a measure density at P, then the value of this density must be

$$\frac{d\mu_{\vec{F}}^f}{dV}\bigg|_P = \lim_{\substack{\{d(\Delta R)\to 0\} \\ P \text{ in } \Delta R}} \frac{\mu_{\vec{F}}^f(\Delta R)}{\Delta V} = \lim_{a\to 0} \frac{3}{4\pi a^3} \mu_{\vec{F}}^f(R(P,a))$$

$$= \lim_{a\to 0} \frac{3}{4\pi a^3} \oiint_{S(\vec{F},a)} \vec{F} \cdot d\vec{\sigma} = \operatorname{div} \vec{F}|_P.$$

On the other hand, the existence of $\left.\operatorname{div}\vec{F}\right|_P$ does not guarantee, by itself, that the flux measure $\mu_{\vec{F}}^f$ has a measure density at P, since the definition of measure density requires that $\lim\limits_{\left\{\substack{d(\Delta R)\to 0 \\ P\ \text{in}\ \Delta R}\right\}}\dfrac{\mu_{\vec{F}}^f(\Delta R)}{\Delta V}$ exist as a unique limit for *all* possible shapes of the elementary region ΔR, while the definition of $\left.\operatorname{div}\vec{F}\right|_P$ requires only that the limit exist for spherical regions with center at P.

Example. Let $\vec{F}=z\hat{k}$ on $\mathbf{E}^3$-with-Cartesian-coordinates. Let P be a given point (x_0,y_0,z_0) in $\mathbf{E}^3$. To find $\left.\operatorname{div}\vec{F}\right|_P$, we first calculate the integral in (1.1) and then find the limit of its value as $a\to 0$.

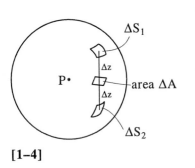

[1–4]

As suggested in [1–4], we can form a Riemann sum by using mirror-image pieces of surface in the upper and lower hemispheres. The upper piece contributes

$$(z_0+\Delta z)\hat{k}\cdot\Delta\vec{\sigma}_1,$$

while the lower piece contributes

$$(z_0-\Delta z)\hat{k}\cdot\Delta\vec{\sigma}_2$$

By geometry, $\hat{k}\cdot\Delta\vec{\sigma}_1=\Delta A$ and $\hat{k}\cdot\Delta\vec{\sigma}_2=-\Delta A$, where ΔA is the area of the projection of ΔS_1 and ΔS_2 onto the horizontal plane. We therefore have, as part of this Riemann sum,

$$(z_0+\Delta z)\Delta A+(z_0-\Delta z)(-\Delta A)=2\Delta z\Delta A.$$

Thus the entire Riemann sum has the value

$$2\sum_i\Delta z_i\Delta A_i,$$

where $\Delta A_1,\Delta A_2,\Delta A_3,\ldots,\Delta A_n$ are the distinct projections in the horizontal plane. We note that this last sum is also a Riemann sum for the volume of a sphere of radius a. Hence the value of the surface integral $\displaystyle\oiint_{S(P,a)}\vec{F}\cdot d\vec{\sigma}$ in (1.1)

Hence $\dfrac{3}{4\pi a^3}\displaystyle\iint_{S(P,a)} \vec{F}\cdot d\vec{\sigma}$ has the value 1, and the value of the limit in (1.1) must

be $\lim_{a\to 0}(1)=1$. Therefore $\operatorname{div}\vec{F}\big|_P = 1$. We conclude that the divergence of the

vector field $\vec{F}=z\hat{k}$ exists and has the constant value 1 on the domain D of $\vec{F}$.

§25.2 The Divergence in Cartesian Coordinates

In $\mathbf{E}^3$-with-Cartesian-coordinates, let $\vec{F}(x,y,z)=L(x,y,z)\hat{i}+M(x,y,z)\hat{j}$ $+N(x,y,z)\hat{k}$ be a C^1 vector field on domain D. Thus L, M, and N are C^1 scalar fields on D. Let P be an interior point of D. We seek a general formula for $\operatorname{div}\vec{F}\big|_P$ in terms of L, M, N, and P. As with the rotational derivative in Chapter 23, we consider *parallel flows*.

Definition. A vector field $\vec{F}$ on a domain D in $\mathbf{E}^3$ is said to be a *parallel flow* if there are a scalar field f on D and a constant unit vector $\hat{w}$ such that $\vec{F}(P)=$ $f(P)\hat{w}$ for all P in D, [2–1]. The example in §25.1 is a parallel flow, with $\hat{w}=\hat{k}$ and $f(x,y,z)=z$.

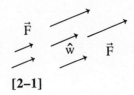

[2–1]

(2.1) *The parallel flow theorem for divergence.* Let $\vec{F}$ be a parallel flow on D *in* $\mathbf{E}^3$, where $\vec{F}=f\hat{w}$ and f *is* C^1. *Let* P *be an interior point of* D. *Then*

$$\operatorname{div}\vec{F}\big|_P = \frac{df}{ds}\bigg|_{\hat{w},P}.$$

Proof. We introduce Cartesian coordinates with $\hat{k}=\hat{w}$. Then $\vec{F}=$ $f(x,y,z)\hat{k}$, where f is C^1. Let S(P,a) be an outward-directed sphere with center at P and radius a. Let Q be a point on S(P,a). By the linear approximation theorem (§9.2),

(2.2) $$f(Q)=f(P)+\vec{\nabla}f\big|_P\cdot\overrightarrow{PQ}+\vec{\varepsilon}(Q)\cdot\overrightarrow{PQ},$$

where $\lim_{Q\to P}\vec{\varepsilon}(Q)=\vec{0}$. (See [2–2].) Let

$$\overrightarrow{PQ}=\Delta x\hat{i}+\Delta y\hat{j}+\Delta z\hat{k}=a\widehat{PQ}.$$

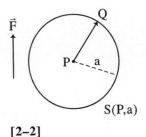

[2–2]

Then $f(Q) = f(P) + f_x(P)\Delta x + f_y(P)\Delta y + f_z(P)\Delta z + \vec{\varepsilon}(Q) \cdot \overrightarrow{PQ}$,

and $\vec{F}(Q) = [f(P) + f_x(P)\Delta x + f_y(P)\Delta y + f_z(P)\Delta z + \vec{\varepsilon}(Q) \cdot \overrightarrow{PQ}]\hat{k}$.

Hence $\displaystyle\oiint_S \vec{F} \cdot d\vec{\sigma} = f(P)\oiint_S \hat{k} \cdot d\vec{\sigma} + f_x(P)\oiint_S \Delta x \hat{k} \cdot d\vec{\sigma}$

$$+ f_y(P)\oiint_S \Delta y \hat{k} \cdot d\vec{\sigma} + + f_z(P)\oiint_S \Delta z \hat{k} \cdot d\vec{\sigma} + a\oiint_S (\vec{\varepsilon}(Q) \cdot \overset{\wedge}{\overrightarrow{PQ}})\hat{k} \cdot d\vec{\sigma},$$

where S is S(P,a).

We consider the five integrals on the right-hand side of this last equation. By symmetric cancellation, each of the first three is 0. By the example in §25.1, the fourth integral is $\frac{4}{3}\pi a^3$. Hence

$$\left. \operatorname{div} \vec{F} \right|_P = \lim_{a \to 0} \frac{3}{4\pi a^3}\left(\frac{4\pi a^3}{3} f_z(P) + a\oiint_S (\vec{\varepsilon}(Q) \cdot \overset{\wedge}{\overrightarrow{PQ}})\hat{k} \cdot d\vec{\sigma} \right)$$

$$= f_z(P) + \lim_{a \to 0} \frac{3}{4\pi a^2} \oiint_S (\vec{\varepsilon}(Q) \cdot \overset{\wedge}{\overrightarrow{PQ}})\hat{k} \cdot d\vec{\sigma} .$$

Let $\varepsilon^*(t)$ be a funnel function for $\lim\limits_{Q \to P} \vec{\varepsilon}(Q) = \vec{0}$.

Then $\displaystyle\left| \frac{3}{4\pi a^2} \oiint_S (\vec{\varepsilon}(Q) \cdot \overset{\wedge}{\overrightarrow{PQ}})\hat{k} \cdot d\vec{\sigma} \right| \le \frac{3}{4\pi a^2} \oiint_S \varepsilon^*(a)d\sigma$

$$= \frac{3}{4\pi a^2} 4\pi a^2 \varepsilon^*(a) = 3\varepsilon^*(a) , \text{ since } \left| \hat{k} \cdot d\vec{\sigma} \right| \le d\sigma .$$

Hence $\displaystyle\lim_{a \to 0} \frac{3}{4\pi a^2} \oiint_S (\vec{\varepsilon}(Q) \cdot \overset{\wedge}{\overrightarrow{PQ}})\hat{k} \cdot d\vec{\sigma} = 0$,   and   we   have $\left. \operatorname{div} \vec{F} \right|_P = f_z(P) =$

$\left. \dfrac{df}{ds} \right|_{\hat{k},P} = \left. \dfrac{df}{ds} \right|_{\hat{w},P}$, as desired.

Using (2.1), we now have:

(2.3)   *Formula for* $\operatorname{div} \vec{F}$ *in Cartesian coordinates. Let* $\vec{F} = L\hat{i} + M\hat{j} + N\hat{k}$ *on* D *be a* C^1 *vector field in* E^3-*with-Cartesian coordinates. Then for any interior point* P *of* D,

$$\left. \mathrm{div}\, \vec{F}\right|_P = L_x(P) + M_y(P) + N_z(P)$$

$$= \left.\frac{\partial L}{\partial x}\right|_P + \left.\frac{\partial M}{\partial y}\right|_P + \left.\frac{\partial N}{\partial z}\right|_P$$

Proof. Let $\vec{F} = \vec{F}_1 + \vec{F}_2 + \vec{F}_3$, where $\vec{F}_1 = L\hat{i}$, $\vec{F}_2 = M\hat{j}$, and $\vec{F}_3 = N\hat{k}$. Then each of $\vec{F}_1$, $\vec{F}_2$, and $\vec{F}_3$ is a parallel flow. Applying the parallel flow theorem (2.1) to each, we have $\mathrm{div}\,\vec{F}_1 = L_x$, $\mathrm{div}\,\vec{F}_2 = M_y$, and $\mathrm{div}\,\vec{F}_3 = N_z$. By linearity (1.3), we have

$$\mathrm{div}\,\vec{F} = L_x + M_y + N_z\ .$$

It follows that if $\vec{F}$ on D is C^1, then $L_x + M_y + N_z$ must be a continuous scalar field on D. Hence we have:

(2.4)   **Corollary.** *If $\vec{F}$ on D in E^3 is C^1, then $\left.\mathrm{div}\,\vec{F}\right|_P$ exists at every interior point P of D, and there is a continuous scalar field on D whose value at each interior point P is $\left.\mathrm{div}\,\vec{F}\right|_P$. We call this scalar field $\mathrm{div}\,\vec{F}$. (This was (1.2).)*

Examples. Using (2.3), we can verify the following ((2) and (5) exclude z axis; (6) and (7) exclude origin):

(1) $\vec{F} = r\hat{r} = x\hat{i} + y\hat{j} \Rightarrow \mathrm{div}\,\vec{F} = 2\ .$

(2) $\vec{F} = \frac{1}{r}\hat{r} = \frac{x\hat{i}+y\hat{j}}{x^2+y^2} \Rightarrow \mathrm{div}\,\vec{F} = 0\ .$

(3) $\vec{F} = r^2\hat{r} = x\sqrt{x^2+y^2}\,\hat{i} + y\sqrt{x^2+y^2}\,)\hat{j} \Rightarrow \mathrm{div}\,\vec{F} = 3\sqrt{x^2+y^2}\ .$

(4) $\vec{F} = r\hat{\theta} = -y\hat{i} + x\hat{j} \Rightarrow \mathrm{div}\,\vec{F} = 0\ .$

(5) $\vec{F} = \frac{1}{r}\hat{\theta} = \frac{-y\hat{i}+x\hat{j}}{x^2+y^2} \Rightarrow \mathrm{div}\,\vec{F} = 0\ .$

(6) $\vec{F} = \rho\hat{\rho} = x\hat{i} + y\hat{j} + z\hat{k} \Rightarrow \mathrm{div}\,\vec{F} = 3\ .$

(7) $\vec{F} = \frac{1}{\rho^2}\hat{\rho} = \frac{x\hat{i}+y\hat{j}+z\hat{k}}{(x^2+y^2+z^2)^{3/2}} \Rightarrow \mathrm{div}\,\vec{F} = 0\ .$

It is instructive to try to picture each of the examples above as a steady-state mass flow without spontaneous creation or annihilation of matter. We see that (4) and

(5) can be realized as physical mass flows but that (1), (3), and (6) cannot. (2) and (7) can be realized as mass flows only if we allow additional mass to be continuously introduced along the z axis (for (2)) or at the origin (for (7)).

In Chapter 27, we develop formulas for the divergence when $\vec{F}$ is given in cylindrical or spherical coordinates. These non-Cartesian formulas will have as special cases:

(2.5)   *for a uniform radial field in cylindrical coordinates,*

$$\vec{F} = f(r)\hat{r} \Rightarrow \operatorname{div} \vec{F} = \frac{1}{r}\frac{d}{dr}(rf)\,, \text{ and}$$

(2.6)   *for a uniform central field in spherical coordinates,*

$$\vec{F} = g(\rho)\hat{\rho} \Rightarrow \operatorname{div} \vec{F} = \frac{1}{\rho^2}\frac{d}{d\rho}(\rho^2 g)\,.$$

(2.5) and (2.6) can be used to get simpler and more direct calculations for examples 1 through 7 above. Formulas (2.5) and (2.6) can also be derived as corollaries to (2.3). See the problems.

§25.3 The Divergence Theorem

In Chapter 23, we introduced the flux measure $\mu_{\vec{F}}^{f}$ for a continuous vector field $\vec{F}$ on a domain D in $\mathbf{E}^3$. This measure assigned a value $\mu_{\vec{F}}^{f}(R)$ to every regular region R in D. In §25.1 above, we noted that if $\mu_{\vec{F}}^{f}$ has a measure density, then its measure density must be $\dfrac{d\mu_{\vec{F}}^{f}}{dV} = \operatorname{div} \vec{F}$. The *divergence theorem* and its corollaries will show us that if $\vec{F}$ is $\mathbf{C}^1$, then $\mu_{\vec{F}}^{f}$ must, in fact, be an integral measure with $\operatorname{div} \vec{F}$ as its measure density. It follows that if $\vec{F}$ is $\mathbf{C}^1$, then the flux measure of any regular region in $\mathbf{E}^3$ can be calculated by integrating $\operatorname{div} \vec{F}$ over that region.

(3.1)   ***The divergence theorem.*** *Let $\vec{F}$ be a $\mathbf{C}^1$ vector field with domain D in $\mathbf{E}^2$. Then:*

(a)   *The flux measure $\mu_{\vec{F}}^{f}$ on D is an integral measure.*

(b)   *The density of this measure is the divergence* div $\vec{F}$. *That is to say, for any regular region* R *in* D,

$$\mu_{\vec{F}}^{f}(R) = \iiint_R \text{div}\,\vec{F}\,dV\,.$$

(3.1) is proved in §25.7 below. The following four corollaries are immediate from (3.1) and the definition of $\mu_{\vec{F}}^{f}$.

(3.2)   *Corollary.*   *For any elementary region* R *in* D, *let* S *be the outward directed boundary of* R, [3–1]. *Then* $\iiint_R \text{div}\,\vec{F}\,dV = \oiint_S \vec{F}\cdot d\vec{\sigma}$.

(3.3)   *Corollary.*   *In Cartesian coordinates, with* $L\hat{i} + M\hat{j} + N\hat{k}$, *the conclusion of* (3.2) *becomes*

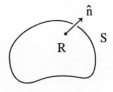

$$\iiint_R (L_x + M_y + N_z)dV = \oiint_S (L\hat{i} + M\hat{j} + N\hat{k})\cdot d\vec{\sigma}\,. \qquad [3\text{–}1]$$

(3.4)   *Corollary.*   *For any regular region* R *in* D, *with outward-directed outer boundary surface* S *and inward directed inner boundary surfaces* $S_1,\dots,S_k$, [3–2], *we have*

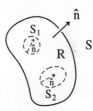

$$\iiint_R \text{div}\,\vec{F}\,dV = \oiint_S \vec{F}\cdot d\vec{\sigma} + \oiint_{S_1} \vec{F}\cdot d\vec{\sigma} + \cdots + \oiint_{S_k} \vec{F}\cdot d\vec{\sigma}\,.$$

(3.5)   *Corollary.*   *In Cartesian coordinates with* $\vec{F} = L\hat{i} + M\hat{j} + N\hat{k}$, *the conclusion of* (3.4) *becomes*   [3–2]

$$\iiint_R (L_x + M_y + N_z)dV = \oiint_S (L\hat{i} + M\hat{j} + N\hat{k})\cdot d\vec{\sigma}$$
$$+ \oiint_{S_1} (L\hat{i} + M\hat{j} + N\hat{k})\cdot d\vec{\sigma} + \cdots + \oiint_{S_k} (L\hat{i} + M\hat{j} + N\hat{k})\cdot d\vec{\sigma}\,.$$

Corollaries analogous to (3.3) and (3.5) can be obtained for a curvilinear coordinate system in $\mathbf{E}^3$, provided that we express div $\vec{F}$ in that system. In traditional texts, (3.3) is called the "divergence theorem," and (3.5) is called the "extended divergence theorem." For examples occurring in the remainder of this

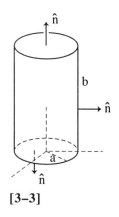

[3–3]

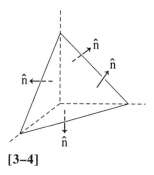

[3–4]

chapter, we shall use the phrase "divergence theorem" to mean whichever of (3.1)–(3.5) is appropriate for that example.

Example 1. *We verify the divergence theorem* (3.3) *for the case where* $\vec{F} = x\hat{i} + y\hat{j} + z\hat{k}$ *and* S *is the closed cylinder of radius* a *and height* b *with base in the* xy *plane and with* z *axis as axis,* [3–3]. In example 2 of §22.5, we evaluated the flux integral for the right-hand side of (3.3). We separately considered the three smooth pieces of S and found $\oiint_S \vec{F} \cdot d\vec{\sigma} = 3\pi a^2 b$. For the left-hand side of (3.3), we now find div $\vec{F}$ and calculate $\iiint_R \operatorname{div} \vec{F} dv$. We get div $\vec{F} = 1+1+1 = 3$, and $\iiint_R \operatorname{div} \vec{F} dv = 3(\operatorname{vol}(R)) = 3\pi a^2 b$. We have thus verified (3.3) for this example.

Example 2. *We verify the divergence theorem* (3.3) *for* $\vec{F} = x\hat{i} + y\hat{j}$, *where* S *is the surface of the tetrahedron with vertices at* (0,0,0), (1,0,0), (0,1,0), *and* (0,0,1). S *is formed of four triangular faces,* [3–4]. The outward flux on each of the three faces in the three coordinate planes is zero, since $\vec{F}$ on each of those faces is parallel to that face (and thus $\vec{F} \cdot \hat{n} = 0$). In example 4 of §22.5, we showed that the downward (and hence inward) flux of $\vec{F}$ on the triangle with vertices (1,0,0), (0,1,0), and (0,0,1) must be $-\frac{1}{3}$. Thus the outward flux through S is $\frac{1}{3}$. On the left-hand side of the theorem, we have $\iiint_R \operatorname{div} \vec{F} dv = 2(\operatorname{vol}(R)) = 2(\frac{1}{6}) = \frac{1}{3}$. We have again verified (3.3).

Example 3. *We verify* (3.3) *for* $\vec{F} = z^2\hat{k}$, *where* S *is the closed cylinder in example* 1 *above.* Evaluating by definition, we immediately find that the flux integral is $\pi a^2 b^2$. On the left-hand side, div $\vec{F} = 2z$, and $\iiint_R 2z dV = $ (average value of 2z on R)(vol(R)) = $b(\pi a^2 b) = \pi a^2 b^2$. The theorem is again verified.

Remark. Note that the divergence theorem is a *mathematical fact,* and is proved as such in §25.7. Its proof does not use, or depend on, any physical laws. In applications, the divergence theorem is often used in conjunction with given physical laws to deduce other physical laws.

§25.4 Examples and Applications

Example 1. Let us consider the uniform central field $\vec{F} = \frac{1}{\rho^2}\hat{\rho} =$

$\frac{x\hat{i}+y\hat{j}+z\hat{k}}{(x^2+y^2+z^2)^{3/2}}$, whose domain D consists of all points in E^3 except the origin.
We can apply the divergence theorem to any region R which does not include the
origin. By example 7 in §25.2, we know that $\operatorname{div}\vec{F} = 0$ at all points in the
domain $\vec{F}$. Let S be *any* outward-directed closed surface in D which does *not*
enclose the origin. Applying the divergence theorem to S and its interior, we
conclude that $\oiint_S \vec{F}\cdot d\vec{\sigma} = 0$. What about a surface S which *does* enclose the

origin? Let $\widetilde{S}$ be a sphere of radius a with center at the origin, directed inward,
where a is sufficiently small that $\widetilde{S}$ is entirely enclosed by S. Let R be the region
between $\widetilde{S}$ and S, [4–1]. Applying (3.4), we get $0 = \iiint_R \operatorname{div}\vec{F}\,dV =$

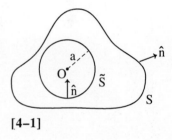

[4–1]

$\oiint_S \vec{F}\cdot d\vec{\sigma} + \iint_{\widetilde{S}} \vec{F}\cdot d\vec{\sigma} \Rightarrow \oiint_S \vec{F}\cdot d\vec{\sigma} = -\oiint_{\widetilde{S}} \vec{F}\cdot d\vec{\sigma}$. Furthermore, on $\widetilde{S}$, $\vec{F} = \frac{1}{a^2}\hat{\rho}$,

and $\oiint_{\widetilde{S}} \vec{F}\cdot d\vec{\sigma} = \oiint_S \left(\frac{1}{a^2}\right)\hat{\rho}\cdot(-\hat{\rho})d\sigma = -\frac{1}{a^2}4\pi a^2 = -4\pi$. Hence $\oiint_S \vec{F}\cdot d\vec{\sigma} =$

$-\oiint_{\widetilde{S}} \vec{F}\cdot d\vec{\sigma} = -(-4\pi) = 4\pi$.

This example shows, for our given field $\vec{F}$, that any two closed outward-
directed surfaces, each of which encloses the origin, must give the same value for
the surface integral of $\vec{F}$ and that this value must be 4π. The general fact (that the
values are the same) holds for any inverse-square central force. This result is
sometimes referred to as *Gauss's theorem*.

The divergence theorem can also be applied directly to simplify the
calculation of flux out of a closed surface S, provided that the interior of S is
contained in the domain of the given vector field.

Example 2. In Cartesian coordinates, let S *be the outward-directed
surface of the rectangular solid* $0 \le x \le 2$, $0 \le y \le 2, 0 \le z \le 3$, [4–2]. *Let* $\vec{F} =$
$\vec{F} = 5yz\hat{i} - 4y\hat{j} + 2z\hat{k}$. *Find* $\oiint_S \vec{F}\cdot d\vec{\sigma}$.

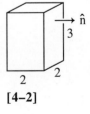

[4–2]

We have $\operatorname{div}\vec{F} = -4 + 2 = -2$. Hence $\oiint_S \vec{F}\cdot d\vec{\sigma} = \iiint_R (-2)dV =$
$-2\operatorname{Vol}(R) = -2\cdot 12 = -24$.

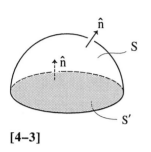

[4–3]

The divergence theorem can also simplify the calculation of a flux integral on a given elementary surface S by allowing us to use a simpler surface S', provided that S' has the same boundary curve as S. In the following three examples, S is the upper hemisphere (in Cartesian coordinates) of a sphere of radius a with center at the origin, and S' is the disk in the xy plane with radius a and center at the origin, [4–3]. Both S and S' are directed upwards. The divergence theorem tells us the following. If the domain of a vector field $\vec{F}$ contains S, S', and the region R enclosed by S and S', then

$$\iiint_R \mathrm{div}\,\vec{F}dV = \iint_S \vec{F}\cdot d\vec{\sigma} - \iint_{S'} \vec{F}\cdot d\vec{\sigma}\ .$$

Hence

(4.1) $$\iint_S \vec{F}\cdot d\vec{\sigma} = \iint_{S'} \vec{F}\cdot d\vec{\sigma} + \iiint_R \mathrm{div}\,\vec{F}dV\ .$$

Example 3. *Let* S *be as above. Let* $\vec{F} = x^2\hat{k}$. *We wish to find* $\iint_S x^2\hat{k}\cdot d\vec{\sigma}$. We use S' and R as above. We have $\mathrm{div}\,\vec{F} = 0$. Hence $\iiint_R \mathrm{div}\,\vec{F}dV = 0$. We also have

$$\iint_{S'} \vec{F}\cdot d\vec{\sigma} = \iint_{S'} x^2\hat{k}\cdot\hat{k}d\sigma = \int_0^{2\pi}\int_0^a r^2\cos^2\theta rdrd\theta = \frac{a^4}{4}\int_0^{2\pi}\cos^2\theta d\theta = \frac{\pi a^4}{4}.$$

Hence, by (4.1), we have

$$\iint_S \vec{F}\cdot d\vec{\sigma} = \frac{\pi a^4}{4} + 0 = \frac{\pi a^4}{4}.$$

Example 4. *For the same* S, *let* $\vec{F} = (x^2 + z)\hat{k}$. We again use S' and R. Here, $\mathrm{div}\,\vec{F} = 1$, and $\iiint_R \mathrm{div}\,\vec{F}dV = \iiint_R dV = \mathrm{vol}(R) = \frac{2}{3}\pi a^3$. We also have $\iint_{S'} \vec{F}\cdot d\vec{\sigma} = \iint_{S'} (x^2 + z)\hat{k}\cdot\hat{k}d\sigma = \iint_{S'} x^2\hat{k}\cdot\hat{k}d\sigma = \frac{\pi a^4}{4}$ (as in example 3), since z = 0 on S'. Thus, by (5.1), $\iint_{S'} \vec{F}\cdot d\vec{\sigma} = \frac{\pi a^4}{4} + \frac{2\pi a^3}{3}.$

Example 5. *For the same* S, *let* $\vec{F} = (x^2 + z^2)\hat{k}$. We again use S' and R. Here div $\vec{F} = 2z$. Hence

$$\iiint_R \text{div } \vec{F} dV = \iiint_R 2z dV = \int_0^{2\pi} \int_0^{\pi/2} \int_0^a 2\rho \cos\varphi \rho^2 \sin\varphi d\rho d\varphi d\theta$$

$$= 2 \left(\int_0^{2\pi} d\theta \right) \left(\int_0^{\pi/2} \sin\varphi \cos\varphi d\varphi \right) \left(\int_0^a \rho^3 d\rho \right)$$

$$= 4\pi \left(\frac{\sin^2 \varphi}{2} \right]_0^{\pi/2} \right) \frac{a^4}{4} = \frac{\pi a^4}{2}$$

As in example 4, we have $\iint_{S'} \vec{F} \cdot d\vec{\sigma} = \iint_{S'} (x^2 + z^2)\hat{k} \cdot \hat{k} d\sigma = \iint_S x^2 \hat{k} \cdot \hat{k} d\sigma = $

$\frac{\pi a^4}{4}$. Hence, by (4.1), $\iint_S \vec{F} \cdot d\vec{\sigma} = \frac{\pi a^4}{4} + \frac{\pi a^4}{2} = \frac{3\pi a^4}{4}$.

§25.5 Divergenceless Fields

Definition. Let $\vec{F}$ be a vector field on domain D. We say that $\vec{F}$ is *divergence-less* if, for every point P in D, $\text{div}\vec{F}\big|_P = 0$.

One of the first physical force-fields to be recognized and analyzed as a divergenceless field was the magnetic field inside a solenoid. For this reason, divergenceless fields are sometimes called *solenoidal fields*.

In classical (non-quantum) physics, the following assertions are accepted as laws of nature: (a) a gravitational field is divergenceless in empty space; (b) an electric field is divergenceless in empty space; and (c) a magnetic field is divergenceless everywhere. (These assertions are considered further in Chapter 28.)

Important mathematical examples of divergenceless fields include: (i) uniform central fields of the form $\frac{a}{\rho^2}\hat{\rho}$ with E^3-minus-the-origin as domain, and

(ii) uniform radial fields of the form $\frac{b}{r}\hat{r}$ with E^3-minus-the-z-axis as domain. (In

§25.2 above, we saw, in example 7, that $\text{div}\left(\frac{1}{\rho^2}\hat{\rho}\right) = 0$ and in example 5, that

$\text{div}\left(\frac{1}{r}\hat{r}\right) = 0$.) There also exist a wide variety of divergenceless fields which have

all of E^3 as their domain; for example, $\vec{F} = x^2 y\hat{i} + y^2 z\hat{j} - (z^2 y + 2xyz)\hat{k}$.

Generalized Gauss's theorem. It is immediate, as a corollary to the argument in example 1 of §25.4, that *if* S_1 *and* S_2 *are closed surfaces such that* S_1 *lies inside*

S_2, *and if* $\vec{F}$ *is divergenceless on the region between* S_1 *and* S_2, *then the flux of* $\vec{F}$ *through* S_2 *must equal the flux of* $\vec{F}$ *through* S_1.

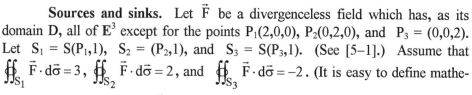

Sources and sinks. Let $\vec{F}$ be a divergenceless field which has, as its domain D, all of E^3 except for the points $P_1(2,0,0)$, $P_2(0,2,0)$, and $P_3 = (0,0,2)$. Let $S_1 = S(P_1,1)$, $S_2 = (P_2,1)$, and $S_3 = S(P_3,1)$. (See [5-1].) Assume that

$$\oiint_{S_1} \vec{F}\cdot d\vec{\sigma} = 3, \quad \oiint_{S_2} \vec{F}\cdot d\vec{\sigma} = 2, \text{ and } \oiint_{S_3} \vec{F}\cdot d\vec{\sigma} = -2.$$ (It is easy to define mathe-

matically a particular field $\vec{F}$ with exactly these properties; see the problems.)

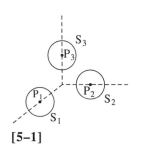

[5-1]

Let S be a closed, outward-directed surface in D such that P_1, P_2, and P_3 are all exterior to S. Then the interior of S is in D, and, by the divergence theorem, $\oiint_S \vec{F}\cdot d\vec{\sigma} = 0$.

Next, let S be a closed, outward-directed surface in D such that P_1 is interior to S, but P_2 and P_3 are exterior to S. Then there must be a sphere $\tilde{S} = S(P_1,a)$ for some sufficiently small a > 0, such that $\tilde{S}$ is interior to both S and S_1. (See [5-2].) By the divergence theorem, as illustrated in example 1 of §25.4, we get $\oiint_S \vec{F}\cdot d\vec{\sigma} = \oiint_{\tilde{S}} \vec{F}\cdot d\vec{\sigma} = \oiint_{S_1} \vec{F}\cdot d\vec{\sigma} = 3$.

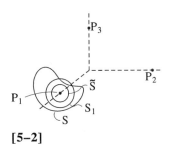

[5-2]

In the same way, for any closed, outward-directed surface S in D such that P_1 and P_2 are interior to S but P_3 is exterior to S, we can use corollary (3.4) to show that $\oiint_S \vec{F}\cdot d\vec{\sigma} = 5$.

More generally, we can show that the value of $\oiint_S \vec{F}\cdot d\vec{\sigma}$, for any given closed, outward directed S in D, must be either -2, 0, 1, 2, 3, or 5, depending on which of the points P_1, P_2, P_3 are interior to S. See the problems.

The points P_1 and P_2 in the example above are called *sources* for $\vec{F}$, because if $\vec{F}$ were a steady-state mass-flow in c.g.s. units, fluid mass would have to be steadily introduced at P_1 and P_2 at the rates of 3 gm/sec and 2 gm/sec respectively, in order to sustain the steady-state mass-flow $\vec{F}$. The point P_3 is called a *sink* for $\vec{F}$, because fluid mass would have to be steadily removed at P_3 at the rate of 2 gm/sec in order to permit the steady-state mass-flow $\vec{F}$.

Surfaces with the same boundary curve. The divergence theorem gives us the following result, as we have already seen in example 3 of §25.4.

(5.1)   *Let* S_1 *and* S_2 *be two directed finite surfaces such that each has the same directed curves as its boundary and such that the directions of* S_1 *and* S_2 *are coherent with the directions of these curves. Let* $\vec{F}$ *be a divergenceless vector field whose domain contains* S_1, S_2 *and all points lying between* S_1 *and* S_2. *Then*

$$\iint_{S_1} \vec{F} \cdot d\vec{\sigma} = \iint_{S_2} \vec{F} \cdot d\vec{\sigma}.$$ (*Coherent* *directions are defined in §23.6.*)

Integral curves in a divergenceless field. If we know the pattern of the integral curves in a *divergenceless field* $\vec{F}$, we can deduce from this pattern how the relative magnitude of $\vec{F}$ must change as we move from point to point. In particular,

(5.2) *the magnitude of* $\vec{F}$ *increases as the integral curves crowd more closely together, and the magnitude of* $\vec{F}$ *decreases as the integral curves become farther apart.*

This result is made quantitatively precise by the divergence theorem. We take, as a region of space R, a tube of finite length with flat ends, where the lateral surface is formed of integral curves of $\vec{F}$. Let the flat ends be S_1 and S_3, and let S_2 be the lateral surface (all with outward direction). (See [5-3].) By the divergence theorem:

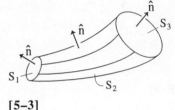

[5-3]

$$\oiint_{S_1 + S_2 + S_3} \vec{F} \cdot d\vec{\sigma} = \iiint_R (\text{div}\,\vec{F})dV = 0, \text{ since } \text{div } \vec{F} = 0.$$

But $\iint_{S_2} \vec{F} \cdot d\vec{\sigma} = 0$, since $\vec{F}$ is parallel to the surface S_2. Hence $-\iint_{S_1} \vec{F} \cdot d\vec{\sigma} = \iint_{S_3} \vec{F} \cdot d\vec{\sigma}$. From this we see that (*average value of* $\vec{F} \cdot (-\hat{n})$ *on* S_1)$\cdot$ (*area*(S_1)) must equal (*average value of* $\vec{F} \cdot \hat{n}$ *on* S_3)$\cdot$ (*area*(S_3)). But this gives our desired result.

Thus, for example, if we know the pattern of the lines of force for the electric field vector $\vec{E}$ in a region where div $\vec{E} = 0$ we can see from this pattern, how the magnitude of $\vec{E}$ changes from one part of the field to another.

Similarly, if we know the pattern of the streamlines in a steady-state mass-flow $\vec{m}$ (where by (1.4) we necessarily have div $\vec{m} = 0$), then we can see from this pattern how the magnitude of $\vec{m}$ changes from one part of the flow to another.

§25.6 The $\vec{\nabla}$ Notation for Divergence

We have now defined two differentiation operations for scalar and vector fields in $\mathbf{E}^3$:

(1)   The *gradient operation*, which takes a C^1 scalar field f to a corresponding continuous vector field $\overset{\rightarrow}{\mathrm{grad}} f$. (We temporarily write the gradient of scalar field f as $\overset{\rightarrow}{\mathrm{grad}} f$ rather than $\vec{\nabla} f$.)

(2)   The *divergence operation*, which takes a C^1 vector field $\vec{F}$ to a corresponding continuous scalar field $\mathrm{div}\,\vec{F}$.

Each of these operations can be viewed as non-numerical input-output correspondence (or "function"; see §8.1) which takes certain fields as inputs and gives certain corresponding fields as outputs. A correspondence of this kind is often called an *operator*. Evidently, the gradient operator can be applied to any C^{n+1} scalar field as input and will produce a corresponding C^n vector field as output. Similarly, the divergence operator can be applied to any C^{n+1} vector field and will yield a corresponding C^n scalar field.

We have already seen that these operators satisfy linearity laws with respect to scalar and vector addition:

(6.1)   *For any constants* a *and* b,

$$\overset{\rightarrow}{\mathrm{grad}}(af + bg) = a\,\overset{\rightarrow}{\mathrm{grad}} f + b\,\overset{\rightarrow}{\mathrm{grad}} g,$$

and

$$\mathrm{div}(a\vec{F} + b\vec{G}) = a\,\mathrm{div}\,\vec{F} + b\,\mathrm{div}\,\vec{G}.$$

These operators also satisfy a variety of other vector-algebra laws. Two of these are

(6.2) $$\overset{\rightarrow}{\mathrm{grad}}(fg) = f\,\overset{\rightarrow}{\mathrm{grad}} g + g\,\overset{\rightarrow}{\mathrm{grad}} f\,;$$

and $$\mathrm{div}(g\vec{F}) = (g\,\mathrm{div}\,\vec{F}) + \vec{F}\cdot\overset{\rightarrow}{\mathrm{grad}} g.$$

Other laws will be given in Chapters 26 and 27. The laws in (6.1) and (6.2) can be proved by using the Cartesian-coordinate formulas for gradient (§9.4) and divergence (§25.2). See the problems.

The gradient and divergence operators can be combined to form: div grad f, which carries a C^{n+2} scalar field to a C^n scalar field; and grad div $\vec{F}$, which carries a C^{n+2} vector field to a C^n vector field. Higher-order combinations, such as grad div grad f and div grad div $\vec{F}$ can also be formed.

The following abbreviations are useful for remembering and applying the laws which these operators satisfy. We introduce the symbol $\vec{\nabla}$, read as "del." In vector algebra, $\vec{\nabla}$ is treated as if it were a vector. We now abbreviate grad f as $\vec{\nabla}f$ (as if we were combining the vector $\vec{\nabla}$ with the scalar f under the vector-algebra operation of vector times scalar), and we abbreviate div $\vec{F}$ as $\vec{\nabla} \cdot \vec{F}$ (as if we were combining the vector $\vec{\nabla}$ with the vector $\vec{F}$ under the vector-algebra operation of dot product.) (Of course, we have already, since Chapter 9, been using the notation $\vec{\nabla}f$ for grad f.) For algebraic purposes, the notations $\vec{\nabla}f$ and $\vec{\nabla} \cdot \vec{F}$ are not commutative. $f\vec{\nabla}$ will not have the same meaning as $\vec{\nabla}f$, and $\vec{F} \cdot \vec{\nabla}$ will not have the same meaning as $\vec{\nabla} \cdot \vec{F}$, as we shall later see.

(6.1) and (6.2) can now be written

$$(6.3) \qquad \vec{\nabla}(af + bg) = a\vec{\nabla}f + b\vec{\nabla}g,$$

and

$$\vec{\nabla} \cdot (a\vec{F} + b\vec{G}) = a\vec{\nabla} \cdot \vec{F} + b\vec{\nabla} \cdot \vec{G}.$$

$$(6.4) \qquad \vec{\nabla}(fg) = f\vec{\nabla}g + g\vec{\nabla}f,$$

and

$$\vec{\nabla} \cdot (g\vec{F}) = g\vec{\nabla} \cdot \vec{F} + \vec{F} \cdot \vec{\nabla}g.$$

Recall the equation of continuity for a time-dependent fluid flow (1.4):

$$-\frac{\partial \delta}{\partial t} = \vec{\nabla} \cdot \vec{m} ;$$

where the vector field $\vec{m} = \delta\vec{v}$ is the mass-flow, the scalar field δ is the mass-density, and the vector field $\vec{v}$ is the fluid velocity. By (6.4) we have

$$\vec{\nabla} \cdot \vec{m} = \vec{\nabla} \cdot (\delta\vec{v}) = \delta(\vec{\nabla} \cdot \vec{v}) + \vec{v} \cdot \vec{\nabla}\delta,$$

and hence we get the following useful and informative version of the equation of continuity:

(6.5)
$$-\frac{\partial \delta}{\partial t} = \delta(\vec{\nabla} \cdot \vec{v}) + \vec{v} \cdot \vec{\nabla}\delta .$$

When we are using a system of Cartesian coordinates, $\vec{\nabla}$ may be expressed, for certain computational purposes, as

(6.6)
$$\vec{\nabla} = \hat{i}\frac{\partial f}{\partial x} + \hat{j}\frac{\partial f}{\partial y} + \hat{k}\frac{\partial f}{\partial z} .$$

For $f = f(x,y,z)$, (6.6) gives

$$\vec{\nabla}f = \left(\hat{i}\frac{\partial}{\partial x} + \hat{j}\frac{\partial}{\partial y} + \hat{k}\frac{\partial}{\partial z}\right)f = \hat{i}\frac{\partial f}{\partial x} + \hat{j}\frac{\partial f}{\partial y} + \hat{k}\frac{\partial f}{\partial z} ;$$

and for $\vec{F} = (L(x,y,z)\hat{i} + M(x,y,z)\hat{j} + N(x,y,z)\hat{k}$, (6.6) gives

$$\vec{\nabla} \cdot \vec{F} = \left(\hat{i}\frac{\partial}{\partial x} + \hat{j}\frac{\partial}{\partial y} + \hat{k}\frac{\partial}{\partial z}\right) \cdot (L\hat{i} + M\hat{j} + N\hat{k}) = \frac{\partial L}{\partial x} + \frac{\partial M}{\partial y} + \frac{\partial N}{\partial z} .$$

Using $\vec{\nabla}$, we can express

$$\text{div } \overrightarrow{\text{grad}} f = \vec{\nabla} \cdot \vec{\nabla}f ,$$

and
$$\overrightarrow{\text{grad}} \text{ div } \vec{F} = \vec{\nabla}\vec{\nabla} \cdot \vec{F} .$$

Then, for example, in Cartesian coordinates, we have

$$\vec{\nabla} \cdot \vec{\nabla}f = \hat{i}\frac{\partial^2 f}{\partial x^2} + \hat{j}\frac{\partial^2 f}{\partial y^2} + \hat{k}\frac{\partial^2 f}{\partial z^2} ,$$

and

$$\vec{\nabla}\vec{\nabla} \cdot \vec{F} = \frac{\partial}{\partial x}\left(\frac{\partial L}{\partial x} + \frac{\partial M}{\partial y} + \frac{\partial N}{\partial z}\right)\hat{i} + \frac{\partial}{\partial y}\left(\frac{\partial L}{\partial x} + \frac{\partial M}{\partial y} + \frac{\partial N}{\partial z}\right)\hat{j} + \frac{\partial}{\partial z}\left(\frac{\partial L}{\partial x} + \frac{\partial M}{\partial y} + \frac{\partial N}{\partial z}\right)\hat{k} .$$

Further uses, meanings, and laws for the symbol $\vec{\nabla}$ are given in Chapters 26 and 27. Cylindrical/spherical expressions appear in Chapter 27.

§25.7 Proof of the Divergence Theorem

Our proof of the divergence theorem is analogous to the proof of Green's theorem in §24.6. Let $\vec{F}$ be a C^1 vector field on D in E^3. Let $\mu_{\vec{F}}^f$ be the flux measure on D as defined in §23.5. The divergence theorem asserts that for any regular region R in D,

$$(7.1) \qquad \iiint_R \text{div } \vec{F} dV = \mu_{\vec{F}}^f(R).$$

That is to say, by (3.2) in §23.3, the divergence theorem asserts that $\mu_{\vec{F}}^f$ is an integral measure with $\text{div } \vec{F}$ as its measure density. By (3.3) in §23.3, it will be enough to prove that at every point P in D,

$$\left.\frac{d\mu_{\vec{F}}^f}{dV}\right|_P = \text{div } \vec{F}\Big|_P \quad , \text{ where}$$

$$(7.2) \qquad \frac{d\mu_{\vec{F}}^f}{dV} = \lim_{\substack{(d(\Delta R)\to 0 \\ P \text{ in } \Delta R)}} \frac{\mu_{\vec{F}}^f(\Delta R)}{\Delta V},$$

where ΔR is a regular region of any shape (and not just a sphere as in the definition of $\text{div } \vec{F}\big|_P$) with diameter $d(\Delta R)$ and volume ΔV, and where this limit exists uniformly on any given regular region in D.

That the limit must be uniform over any given regular R follows from the compactness of regular regions and from the uniform continuity (by §14.9) of $\text{div } \vec{F}$. (By §25.2, $\text{div } \vec{F}$ must be continuous since $\vec{F}$ is C^1.)

To prove that the limit in (7.2) exists, it will be enough to prove that

$$(7.3) \qquad \iiint_R \text{div } \vec{F} dV = \oiint_S \vec{F} \cdot d\vec{\sigma} \textit{ for any } C^1 \textit{ field } \vec{F} \textit{ and any elementary region}$$

R *with outward-directed boundary* S.

(7.3) previously appeared as corollary (3.2) to theorem (3.1). We must now find a proof for (7.3) which is independent of (3.1). To do so, we need an extension of the *subdivision principle for* E^3 ((5.1) in §15.5). We begin with a definition.

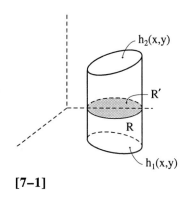

[7–1]

Definition. We say that a region R in $\mathbf{E}^3$-with-Cartesian-coordinates is *z-simple* if there exist (i) an elementary region R′ in the xy plane and (ii) two continuous functions $h_1(x,y)$ and $h_2(x,y)$ on R′, such that $h_1(x,y) \le h_2(x,y)$ for all (x,y) in R′ and such that R is the set of all points (x,y,z) for which (x,y) is in R′ and $h_1(x,y) \le z \le h_2(x,y)$. (See [7–1].) zy-simple regions and zx-simple regions are defined in §16.4. z-simple regions (as defined here) include, as special cases, both zy-simple regions and zx-simple regions. x-*simple* regions and y-*simple* regions in $\mathbf{E}^3$ are defined similarly.

Our extension of the subdivision principle for $\mathbf{E}^3$ asserts:

(7.4)   *For a regular region R in $\mathbf{E}^3$, it is always possible to subdivide R into a finite number of elementary subregions such that for each subregion there is some Cartesian coordinate system in which that subregion is simultaneously x-simple, y-simple, and z-simple.* (We do not give a proof of (7.4) in this text.)

Since $\mu_{\vec{F}}^{f}$ is a finite measure on **R,** and since $\mu(R) = \iiint_R \operatorname{div} \vec{F} dV$ also defines a finite measure on R (by law I of §14.4), it will be enough, by (7.4), to prove (7.3) for the case, in $\mathbf{E}^3$-with-Cartesian-coordinates, where R is simultaneously x-simple, y-simple, and z-simple. We accomplish this by letting $\vec{F} = L\hat{i} + M\hat{j} + N\hat{k}$ and proving (7.3) for $\vec{F}_1 = L\hat{i}$ where R is x-simple, for $\vec{F}_2 = M\hat{j}$ where R is y-simple, and for $\vec{F}_3 = N\hat{k}$ where R is z-simple. We then consider $\vec{F} = \vec{F}_1 + \vec{F}_2 + \vec{F}_3$.

For $\vec{F}_3 = N\hat{k}$: Since R is z-simple, we can use an iterated integration as follows.

$$(7.5) \quad \iiint_R \operatorname{div} \vec{F}_3 dV = \iiint_R \frac{\partial N}{\partial z} dV = \iint_{R'} dA \int_{h_1(x,y)}^{h_2(x,y)} \frac{\partial N}{\partial z} dz$$
$$= \iint_{R'} [N(x,y,h_2(x,y)) - N(x,y,h_1(x,y))] dA.$$

The fact that R is z-simple also implies that S has a bottom surface S_1, directed down, given by $R_1(x,y) = x\hat{i} + y\hat{j} + h_1(x,y)\hat{k}$, and a top surface S_2, directed up, given by $R_2(x,y) = x\hat{i} + y\hat{j} + h_2(x,y)\hat{k}$. In addition to S_1 and S_2, S may have vertical side-surfaces on which $N\hat{k} \cdot \hat{n} = 0$. Hence

$$\oiint_S N\hat{k} \cdot d\vec{\sigma} = \iint_{S_1} N\hat{k} \cdot d\vec{\sigma} + \iint_{S_2} N\hat{k} \cdot d\vec{\sigma}.$$

Evaluating parametrically, we have

$$\iint_{S_1} N\hat{k}\cdot d\vec{\sigma} = -\iint_{R'} N(x,y,h_1(x,y))\hat{k}\cdot(-h_{1x}\hat{i}-h_{1y}\hat{j}+\hat{k})dA$$

$$= -\iint_{R'} N(x,y,h_1(x,y))dA\,,$$

and similarly, $\quad \iint_{S_2} N\hat{k}\cdot d\vec{\sigma} = \iint_{R'} N(x,y,h_2(x,y))dA$. Thus

(7.6) $\qquad \oiint_S N\hat{k}\cdot d\vec{\sigma} = \iint_{R'}[N(x,y,h_2(x,y)) - N(x,y,h_1(x,y))]dA$.

Comparing (7.5) and (7.6), we find that we have proved (7.3) for $\vec{F}_3$ on R.

Similarly, we can prove (7.3) for $\vec{F}_1$ on R and for $\vec{F}_2$ on R. Finally, adding the results for $\vec{F}_1$, $\vec{F}_2$, and $\vec{F}_3$, and using the linearity of the triple integrals and surface integrals, we have (7.3) for $\vec{F}$ on R. This completes the proof of (7.3) and hence completes the proof of the divergence theorem.

§25.8 Flux and Divergence in E^2

Flux integrals for a vector field $\vec{F}$ in E^2 can be defined by using directed curves in E^2 in place of directed surfaces in E^3, and unit normal vectors to curves in place of unit normal vectors to surfaces.

We begin by assuming that E^2 itself is a directed surface and that we have a unit normal vector $\hat{k}$ in the direction of E^2. Let C be a directed, smooth finite curve in E^2. Let P be a point on C. Let $\hat{T}_P$ be the unit tangent vector to C at P. We then define $\hat{n}_P = \hat{T}_P \times \hat{k}$ to be a *unit normal vector to C at P*. ($\hat{n}_P$ is not necessarily the same as the unit normal vector $\hat{N}_P$ defined in §6.6. The existence and direction of $\hat{N}_P$ depends on $\hat{T}_P$ and the curvature of C. The direction of $\hat{n}_P$ depends only on $\hat{k}$ and $\hat{T}_P$.) See [8–1].

[8–1]

Definition. Let $\vec{F}$ be a continuous vector field whose domain contains C. We define the *flux of $\vec{F}$ through C* to be

(8.1) $\qquad \int_C (\vec{F}\cdot\hat{n})ds$. $\qquad$ See [8–2].

[8–2]

Since $\hat{n} = \hat{T} \times \hat{k}$, we get

(8.2) $\int_C (\vec{F} \cdot \hat{n}) ds = \int_C (\vec{F} \cdot \hat{T} \times \hat{k}) ds = \int_C ((\hat{k} \times \vec{F}) \cdot \hat{T}) ds = \int_C (\hat{k} \times \vec{F}) \cdot d\vec{R}$.

(See §3.6.) (8.2) expresses the flux of $\vec{F}$ through C as an ordinary vector line integral on C of the vector field $\vec{G} = \hat{k} \times \vec{F}$. In Cartesian coordinates, with $\vec{F} = L\hat{i} + M\hat{j}$, we have $\vec{G} = -M\hat{i} + L\hat{j}$, and we get

(8.3) $\int_C (\vec{F} \cdot \hat{n}) ds = \int_C \vec{G} \cdot d\vec{R} = \int_C (-M\hat{i} + L\hat{j}) \cdot d\vec{R}$

for the flux of $\vec{F}$ through C. For a fluid-flow in two dimensions, the physical interpretation of flux is similar to the interpretation of flux for a fluid flow in three dimensions.

Divergence in $\mathbf{E}^2$ can be defined in the same way as for $\mathbf{E}^3$. We have

$$\mathrm{div}_2 \vec{F}\Big|_P = \lim_{a \to 0} \frac{1}{\pi a^2} \oint_{C(a,P)} (\vec{F} \cdot \hat{n}) ds \ ,$$

where $C(a,P)$ is the counterclockwise-directed circle of radius a with center at P. The subscript 2 indicates that this is divergence for a vector field in $\mathbf{E}^2$. A Cartesian-coordinate formula for $\mathrm{div}_2 \vec{F}$ (when $\vec{F} = L\hat{i} + M\hat{j}$) can be obtained by a similar argument to §25.2. This gives

(8.4) $\mathrm{div}_2 \vec{F} = \dfrac{\partial L}{\partial x} + \dfrac{\partial M}{\partial y}$.

For a given vector field in $\vec{F}$ in $\mathbf{E}^2$, a flux measure can be defined in exact analogy to §23.5, and a corresponding two-dimensional version of the divergence theorem can be formulated and proved. As in §25.7, the proof of this theorem amounts to showing, for an arbitrary elementary region R, that

(8.5) $\iint_R \mathrm{div}_2 \vec{F} dA = \oint_C (\vec{F} \cdot \hat{n}) ds$,

where C is the counterclockwise-directed boundary of R. In Cartesian coordinates, by (8.3) and (8.4), (8.5) becomes

$$\iint_R \left(\frac{\partial L}{\partial x} + \frac{\partial M}{\partial y} \right) dA = \oint_C (-M\hat{i} + L\hat{j}) \cdot d\vec{R}.$$

But this is simply Green's theorem on R for the vector function $\vec{G} = -M\hat{i} + L\hat{j}$, which we have already proved in Chapter 24. Thus the divergence theorem in $\mathbf{E}^2$ turns out to be a restatement of Green's theorem.

§25.9 Problems

§25.1

1. S is a closed cylinder of radius 2 and height 3, with the z axis as axis. A fluid flows through the surface S. At each point on S, the mass-flow vector of the flow is $2r\hat{r}$ in cylindrical coordinates. At what total rate must mass be continuously created inside S to sustain this outward flow as a steady-state flow?

§25.2

1. Find $\text{div}\vec{F}$ for each of the following:

   (a) $\vec{F} = 2x\hat{i} - y\hat{j} - z\hat{k}$

   (b) $\vec{F} = x^2\hat{i} + yz\hat{j} - yz\hat{k}$

   (c) $\vec{F} = yz\hat{i} + xz\hat{j} + xy\hat{k}$

2. For each of the following radial fields, find the divergence by first expressing $\vec{F}$ in the associated Cartesian system, then taking the divergence, then expressing this divergence in cylindrical coordinates.

   (a) $\vec{F} = r^3\hat{r}$ (b) $\vec{F} = \frac{1}{r^3}\hat{r}$

3. Use the same approach as in problem 2 (but with spherical coordinates) to find the divergence of each of the following spherical fields.

   (a) $\vec{F} = \frac{1}{\rho}\hat{\rho}$ (b) $\vec{F} = \rho^3\hat{\rho}$

4. (a) In example 1, use the Cartesian formula for divergence to find $\text{div} \vec{F}$.

   (b) Do the same for example 2.

   (c) Do the same for example 3.

   (d) Do the same for example 4.

   (e) Do the same for example 5.

   (f) Do the same for example 6.

   (g) Do the same for example 7.

5. (a) In example 1, use (2.5) to find $\text{div} \vec{F}$.

   (b) Do the same for example 2.

   (c) Do the same for example 3.

   (d) Do the same for example 4.

   (e) Do the same for example 5.

6. (a) In example 6, use (2.6) to find $\text{div} \vec{F}$.

   (b) Do the same for example 7.

7. Derive (2.5) from (2.3). (*Hint.* Express $f(r)\hat{r}$ as $f((x^2 + y^2)^{1/2})(x\hat{i} + y\hat{j})/(x^2 + y^2)^{1/2}$.)

8. Derive (2.6) from (2.3).

§25.3

1. Use the divergence theorem to evaluate $\oiint_S \vec{F} \cdot d\vec{\sigma}$, where $\vec{F} = x\hat{i} - z^2\hat{j} + z\hat{k}$ and S is the surface of the solid defined by $x^2 + y^2 \leq 1, 0 \leq z \leq 1$, directed out.

2. Use the divergence theorem to evaluate

   (a) $\iint_S (x\hat{i} + y\hat{j} + z\hat{k}) \cdot d\vec{\sigma}$, where S is upper hemisphere $z = \sqrt{1 - x^2 - y^2}$, $x^2 + y^2 \leq 1$, directed up.

   (b) $\iint_S (\hat{i} + \hat{j} + \hat{k}) \cdot d\vec{\sigma}$, where S is as in (a).

   (c) $\iint_S (x^2\hat{i} + y^2\hat{j} + z^2\hat{k}) \cdot d\vec{\sigma}$, where S is the cube $0 \leq x \leq 1, 0 \leq y \leq 1, 0 \leq z \leq 1$, directed out.

3. S is the closed cylinder whose top and bottom are given by $z = 3$ and $z = 0$, and whose lateral surface is given by $x^2 + y^2 = 4$. S is directed out. Evaluate $\iint_S \vec{F} \cdot d\vec{\sigma}$ for

(a) $\vec{F} = z\hat{k}$ (c) $\vec{F} = x^3\hat{i}$

(b) $\vec{F} = -y\hat{i} + x\hat{j}$

4. S is the upper hemisphere of $x^2 + y^2 + z^2 = 1$, S is directed up. Find $\iint_S \vec{F} \cdot d\vec{\sigma}$ for

(a) $\vec{F} = \hat{i} + \hat{j} + \hat{k}$ (b) $\vec{F} = x\hat{i} + y\hat{j} + z\hat{k}$

5. S is the plane triangular surface with vertices $(0,0,0)$, $(1,0,0)$, and $(2,1,2)$. S is directed up. Find $\iint_S y\hat{k} \cdot d\vec{\sigma}$.

§25.4

1. S is the curved half-cylinder surface defined by $y^2 + z^2 = 1$, $z \geq 0$, and $0 \leq x \leq 1$. S is directed up. Find $\iint_S (x\hat{i} + y\hat{j} + z\hat{k}) \cdot d\vec{\sigma}$.

2. S is that portion of the cone $z^2 = x^2 + y^2$ which lies between $z = 0$ and $z = 2$. S is directed away from the z axis. Find $\iint_S y^3\hat{j} \cdot d\vec{\sigma}$.

3. In each case, use the divergence theorem to find a shorter and simpler evaluation.
(a) Example 1 in §22.5.
(b) Example 2 in §22.5.
(c) Example 3 in §22.5.
(d) Example 4 in §22.5.
(e) Example 5 in §22.5.

§25.5

1. You are given the following information:
  (i) $\vec{F}$ is a C^1 vector field whose domain is all points in $\mathbf{E}^3$ except the origin.
  (ii) $\text{div}\vec{F} = 0$ at all points in the domain of $\vec{F}$.

(iii) $\iint_S \vec{F} \cdot d\vec{\sigma} = 4$, where S is the surface of a sphere of radius 1 with center at origin, directed out.

(a) What is the value of $\iint_{S'} \vec{F} \cdot d\vec{\sigma}$, where S' is the surface of a sphere of radius 2, center at origin, and directed out?

(b) What are the possible values that can occur for $\iint_{S'} \vec{F} \cdot d\vec{\sigma}$, where S is a closed surface, directed out, lying in the domain of $\vec{F}$?

2. You are given the following information:
  (i) $\vec{F}$ is a C^1 vector field whose domain is all points in $\mathbf{E}^3$ except the origin, $(2,0,0)$, and $(4,0,0)$.
  (ii) $\text{div}\vec{F} = 0$ at all points in the domain of $\vec{F}$.
  (iii) Let S_a be the sphere of radius a with center at the origin, directed out. Then $\iint_{S_1} \vec{F} \cdot d\vec{\sigma} = 5$, $\iint_{S_3} \vec{F} \cdot d\vec{\sigma} = -1$, and $\iint_{S_5} \vec{F} \cdot d\vec{\sigma} = 7$.

What are the possible values that can occur for $\iint_S \vec{F} \cdot d\vec{\sigma}$, when S is a sphere directed out, lying in the domain of $\vec{F}$?

3. $\vec{F}$ is a divergenceless vector field whose domain is all of space except for the two points $A = (0,0,0)$ and $B = (0,1,0)$. For any point P, let $S(P,a)$ be the outward-directed sphere with radius a and center at P. You are given that the flux of through $S(A, \frac{1}{2})$ is 1 and through $S(A, \frac{3}{2})$ is –3.

(a) What is the flux through $S(B, \frac{1}{2})$?

(b) What are the possible values that can occur for the flux through $S(P,a)$ as P and a vary?

4. (a) Give a Cartesian formula for a field $\vec{F}$ on D whose properties are as given for the field $\vec{F}$ defined for figure [5–1].

(b)   Show that there can be no other fields with these properties.

5.   For the field $\vec{F}$ defined for [5–1], show that $\{-2,0,1,2,3,5\}$ are possible values, and the only possible values, for the flux of $\vec{F}$ through a closed, outward directed surface in D.

§25.6

1. $\vec{F}$ and $\vec{G}$ are vector fields, and g and h are scalar fields. Simplify

(a) $\vec{\nabla} \cdot (gh\vec{F})$ (b) $\vec{\nabla} \cdot (hg(\vec{F} - h\vec{G}))$

2.   (a)   Prove the first law in (6.2).
     (b)   Prove the second law in (6.2).

3.   Give a Cartesian formula for:
     (a)   grad div grad f $(= \vec{\nabla}\vec{\nabla} \cdot \vec{\nabla}f)$
     (b)   div grad div $\vec{F}$ $(= \vec{\nabla} \cdot \vec{\nabla}\vec{\nabla} \cdot \vec{F})$

Chapter 26

Curl and Stokes's Theorem

§26.1 Rotational Derivatives in $\mathbf{E}^3$

In §24.1, we define the concept of *rotational derivative* for a vector field in $\mathbf{E}^2$. We now extend the concept of rotational derivative to vector fields in $\mathbf{E}^3$.

Definition. Let $\vec{F}$ be a continuous vector field on a domain D in $\mathbf{E}^3$, and let P be a given interior point in D. Let the unit vector $\hat{u}$ in $\mathbf{E}^3$ define a given direction in $\mathbf{E}^3$, and let M be the directed plane in $\mathbf{E}^3$ which contains P and has its direction (as a directed surface) given by $\hat{u}$. (See [1–1].) For a > 0, let C_a be the directed circle in M which has radius a and center at P, and whose direction is right-handed with respect to $\hat{u}$. (Thus, looked at from the side of M given by $\hat{u}$, the direction of C appears counterclockwise, [1–2]. We now consider

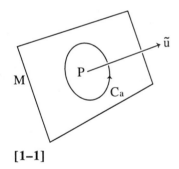

[1–1]

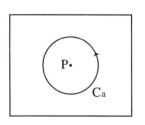

[1–2]

$$(1.1) \qquad \lim_{a \to 0} \frac{1}{\pi a^2} \oint_{C_a} \vec{F} \cdot d\vec{R}.$$

If this limit exists, we say that its value is the *rotational derivative of $\vec{F}$ in the direction $\hat{u}$ at P.* We write this derivative as $\operatorname{rot} \vec{F}\big|_{\hat{u},P}$.

An underlying physical idea for the rotational derivative in $\mathbf{E}^3$ is as follows. Let $\vec{F}$ be any continuous vector field on domain D in $\mathbf{E}^3$. We visualize $\vec{F}$ as the velocity vector of a fluid flow at a given instant. Let P be a given interior point in D, let $\hat{u}$ be a given direction in $\mathbf{E}^3$, and let L be a directed line through P in the direction $\hat{u}$. Let C_a be as defined above. Then, at any point Q on C_a, $\frac{1}{a}(\vec{F}(Q) \cdot \hat{T}_Q)$ must be the angular speed around P of a particle at Q, and

(1.2)
$$\omega_a = \frac{1}{2\pi a}\left(\frac{1}{a}\oint_{C_a}(\vec{F}\cdot\hat{T})ds\right) = \frac{1}{2}\left(\frac{1}{\pi a^2}\oint_{C_a}\vec{F}\cdot d\vec{R}\right)$$

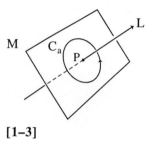

[1–3]

must be the average angular speed around L (in radians per second) of the fluid particles on C_a (where we measure non-zero angular speed to be positive if it is right-handed with respect to L, and negative if left-handed). See [1–3]. Thus the value of $\mathrm{rot}\,\vec{F}\big|_{\hat{u},P}$ is

(1.3)
$$\mathrm{rot}\,\vec{F}\big|_{\hat{u},P} = \lim_{a\to 0}(2\omega_a),$$

and $\mathrm{rot}\,\vec{F}\big|_{\hat{u},P}$ can be viewed as twice the average angular speed around L of fluid particles in the immediate neighborhood of P.

A more specific formulation of this physical idea is to imagine that we have a small paddlewheel of radius a as in [1–4]. The spokes of the wheel are attached rigidly to the hub, and the hub is free to rotate on the axle. The flowing fluid interacts with the wheel by pressing on the tiny circular vanes at the ends of the spokes. Each vane is parallel to its spoke and to the axle. We make the assumption that if we insert the wheel in the fluid with the position of its hub held at a point P and its axle held in the direction of $\hat{u}$, then the wheel will rotate at the angular speed

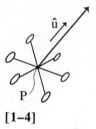

[1–4]

$$\omega_a = \frac{1}{2}\left(\frac{1}{\pi a^2}\oint_{C_a}\vec{F}\cdot d\vec{R}\right).$$

This assumption can be shown to be physically valid, provided that the fluid has constant density. If the density is not constant, but is a continuous scalar field, we see that the assumption becomes more nearly valid as $a \to 0$. Thus, for ω_a as measured by the wheel, $2\omega_a$ will be an approximate value for $\mathrm{rot}\,\vec{F}\big|_{u,P}$, and this approximation will become increasingly accurate as the radius a is decreased.

It is immediate from our definition, and from the linearity of vector line integrals, that the rotational derivative in $\mathbf{E}^3$ satisfies the *linearity law*:

(1.4)   *If $\vec{F}$ and $\vec{G}$ are continuous vector fields whose domains both contain* P *as an interior point, then, for any constants* a *and* b, *and any direction* $\hat{u}$,

$$\text{rot}(a\vec{F} + b\vec{G})\big|_{\hat{u},P} = a \text{ rot } \vec{F}\big|_{\hat{u},P} + b \text{ rot } \vec{G}\big|_{\hat{u},P} \,,$$

provided that both $\text{rot } \vec{F}\big|_{\hat{u},P}$ *and* $\text{rot } \vec{G}\big|_{\hat{u},P}$ *exist.*

In §26.2, we will see that if $\vec{F}$ is C^1 on domain D, then $\text{rot } \vec{F}\big|_{\hat{u},P}$ exists in every direction at every interior point P in D. We will also see later in this chapter and in Chapters 27 and 28 that the rotational derivative, as defined above, is a useful concept for the study of force fields as well as for the study of fluid flows.

§26.2 Rotational Derivatives in E^3 in Cartesian Coordinates

In E^3-with-Cartesian coordinates, let $\vec{F} = L(x,y,z)\hat{i} + M(x,y,z)\hat{j} + N(x,y,z)\hat{k}$ be a C^1 vector field, and let $\hat{u} = u_1\hat{i} + u_2\hat{j} + u_3\hat{k}$. We seek a formula for $\text{rot } \vec{F}\big|_{u,P}$ in terms of L, M, N, u_1, u_2, u_3 and the coordinates of P.

Let $\vec{F}_1 = L\hat{i}$, $\vec{F}_2 = M\hat{j}$, and $\vec{F}_3 = N\hat{k}$. Then $\vec{F} = \vec{F}_1 + \vec{F}_2 + \vec{F}_3$. By (1.4), it will be enough to find Cartesian formulas for $\text{rot } \vec{F}_1\big|_{\hat{u},P}$, $\text{rot } \vec{F}_2\big|_{\hat{u},P}$, and $\text{rot } \vec{F}_3\big|_{\hat{u},P}$ separately, and then take the sum of these formulas.

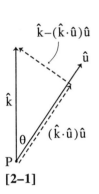

$\hat{k} - (\hat{k}\cdot\hat{u})\hat{u}$

$\hat{u}$

$\hat{k}$

$(\hat{k}\cdot\hat{u})\hat{u}$

θ

P

[2–1]

We begin with $\vec{F}_3 = N(x,y,z)\hat{k}$. Let θ be the angle between $\hat{u}$ and $\hat{k}$. We assume, for the moment, that $\pi > \theta > 0$. Let $\hat{w}$ be the unit vector in the direction of $\hat{k} - (\hat{k}\cdot\hat{u})\hat{u}$ (see [2–1]). Then $\hat{w}$ is orthogonal to $\hat{u}$ and coplanar with $\hat{u}$ and $\hat{k}$. From [2–1], we see that the field $\vec{F}_3$ can be expressed as the sum $\vec{F}_3 = \vec{G}_1 + \vec{G}_2$ of the fields $\vec{G}_1$ and $\vec{G}_2$, where

$$\vec{G}_1 = (N\cos\theta)\hat{u} \quad \text{and} \quad \vec{G}_2 = (N\sin\theta)\hat{w} \,.$$

Thus $\text{rot } \vec{F}_3\big|_{\hat{u},P} = \text{rot } \vec{G}_1\big|_{\hat{u},P} + \text{rot } \vec{G}_2\big|_{\hat{u},P}$ by   linearity.   Furthermore, $\text{rot } \vec{G}_1\big|_{\hat{u},P} = 0$ since, for each $a > 0$ in the definition of $\text{rot } \vec{G}_1\big|_{\hat{u},P}$, $\vec{G}_1(Q)$ is perpendicular to C_a at Q, for every Q on C_a. Hence $\text{rot } \vec{F}_3\big|_{\hat{u},P} = \text{rot } \vec{G}_2\big|_{\hat{u},P} \,.$

Now $\vec{G}_2$ is, by definition, a parallel flow (see §24.2) on the plane M, where M is the plane through P normal to $\hat{u}$. Moreover, by the definition of rotational derivatives in $\mathbf{E}^2$ and $\mathbf{E}^3$, rot $\vec{G}_2|_P$ (as a derivative in $\mathbf{E}^2$) must be the same as rot $\vec{G}_2|_{\hat{u},P}$ (as a derivative in $\mathbf{E}^3$). Applying the parallel flow theorem ((2.1) in §24.2), we get

$$\text{rot } \vec{G}_2|_P = \frac{d}{ds}(N\sin\theta)|_{\hat{v},P},$$

where $\hat{v} = \hat{w} \times \hat{u} = \frac{\hat{k}\times\hat{u}}{|\hat{k}\times\hat{u}|} = \frac{\hat{k}\times\hat{u}}{\sin\theta}$ since $(\hat{k} - (\hat{k}\cdot\hat{u})\hat{u}) \times \hat{u} = \hat{k}\times\hat{u}$.

By the gradient theorem,

$$\frac{d}{ds}(N\sin\theta)|_{\hat{v},P} = (\sin\theta)\vec{\nabla}N|_P \cdot \hat{v} = (\sin\theta)\vec{\nabla}N|_P \cdot \frac{\hat{k}\times\hat{u}}{\sin\theta} = \vec{\nabla}N|_P \cdot \hat{k}\times\hat{u}.$$

Thus
$$\begin{aligned}
\text{rot } \vec{F}_3|_{\hat{u},P} &= \text{rot } \vec{G}_2|_{\hat{u},P} \text{ (in } \mathbf{E}^3) = \text{rot } \vec{G}_2|_P \text{ (in } \mathbf{E}^2) \\
&= (N_x(P)\hat{i} + N_y(P)\hat{j} + N_z(P)\hat{k}) \cdot \hat{k} \times (u_1\hat{i} + u_2\hat{j} + u_3\hat{k}) \\
&= (N_x(P)\hat{i} + N_y(P)\hat{j} + N_z(P)\hat{k}) \cdot (u_1\hat{j} - u_2\hat{i}) \\
&= -u_2 N_x(P) + u_1 N_y(P).
\end{aligned}$$

We note that this last formula remains true for the special cases $\theta = 0$ and $\theta = \pi$, since, in either of these cases, (i) $u_1 = u_2 = 0$ and (ii) rot $\vec{F}_3|_{\hat{u},P} = 0$ because $\vec{F}_3$ must be perpendicular to the plane of the circle C_a in the definition of rot $\vec{F}_3|_{\hat{u},P}$.

Similar derivatives for $\vec{F}_1$ and $\vec{F}_2$ yield

$$\text{rot } \vec{F}_1|_{\hat{u},P} = u_2 L_z(P) - u_3 L_y(P)$$
and
$$\text{rot } \vec{F}_2|_{\hat{u},P} = u_3 M_x(P) - u_1 M_z(P).$$

Summing these results for $\vec{F}_1$, $\vec{F}_2$, and $\vec{F}_3$, we have our desired formula.

(2.1) **Theorem. Formula for** rot $\vec{F}|_{\hat{u},P}$ **in Cartesian coordinates.** *Let* $\vec{F} = L\hat{i} + M\hat{j} + N\hat{k}$ *be a* C^1 *vector field. Then for every interior point* $\overline{P}$ *of* D *and for every direction* $\hat{u} = u_1\hat{i} + u_2\hat{j} + u_3\hat{k}$, *we have that* $\text{rot }\vec{F}\big|_{\hat{u},P}$ *exists and that*

$$\text{rot}\vec{F}\big|_{\hat{u},P} = \left(\frac{\partial N}{\partial y}\bigg|_P - \frac{\partial M}{\partial z}\bigg|_P\right)u_1 + \left(\frac{\partial L}{\partial z}\bigg|_P - \frac{\partial N}{\partial x}\bigg|_P\right)u_2 + \left(\frac{\partial M}{\partial x}\bigg|_P - \frac{\partial L}{\partial y}\bigg|_P\right)u_3.$$

Remark. The proof of the parallel flow theorem in §24.2 requires only that $\vec{F}$ be differentiable. It follows that our theorem (2.1) above remains true if the assumption that $\vec{F}$ is C^1 is weakened to the assumption that $\vec{F}$ is differentiable. We shall see below, however, that many applications of (2.1) require that $\vec{F}$ be C^1.

§26.3 The Curl of a Vector Field

The Cartesian formula in (2.1) leads us directly to the *curl theorem* and to the definition of *curl*.

(3.1) ***The curl theorem (including the definition of* curl*).*** *Let* $\vec{F}$ *be a* C^1 *vector field on domain* D *in* $\mathbf{E}^3$.

(a) *For each interior point* P *in* D *there is a vector* $\vec{C}_P$, *called the* curl *of* $\vec{F}$ *at P and written* $\overrightarrow{\text{curl}}\,\vec{F}\big|_P$, *such that for every direction* $\hat{u}$,

$$\text{rot}\vec{F}\big|_{\hat{u},P} = \vec{C}_P \cdot \hat{u} = \overrightarrow{\text{curl}}\vec{F}\big|_P \cdot \hat{u}.$$

(*Thus* $\overrightarrow{\text{curl}}\,\vec{F}\big|_P$ *is a vector whose magnitude is the maximum value of* $\text{rot }\vec{F}\big|_{\hat{u},P}$ *as* $\hat{u}$ *varies at P and whose direction is the direction that gives this maximum value.*)

(b) *Furthermore, there is a continuous vector field on* D *whose value at every interior point* P *in* D *is* $\overrightarrow{\text{curl}}\,\vec{F}\big|_P$. *This vector field on* D *is called the* curl *of* $\vec{F}$ *and is written* $\overrightarrow{\text{curl}}\,\vec{F}$. *In Cartesian coordinates, for* $\vec{F} = L\hat{i} + M\hat{j} + N\hat{k}$, *we have:*

(3.2) $\overrightarrow{\text{curl}}\vec{F}\big|_P = \left(\dfrac{\partial N}{\partial y}\bigg|_P - \dfrac{\partial M}{\partial z}\bigg|_P\right)\hat{i} + \left(\dfrac{\partial L}{\partial z}\bigg|_P - \dfrac{\partial N}{\partial x}\bigg|_P\right)\hat{j} + \left(\dfrac{\partial M}{\partial x}\bigg|_P - \dfrac{\partial L}{\partial y}\bigg|_P\right)\hat{k}.$

Proof. Part (a) of (3.1) is immediate from (2.1) above. Part (b) is immediate from our assumption that $\vec{F}$ is C^1 and hence that the derivatives of L, M, and N are continuous scalar functions.

Part (a) of the curl theorem says that the vector $\left.\overset{\rightarrow}{\text{curl}}\,\vec{F}\right|_P$ plays the same role for rotational derivatives of $\vec{F}$ at P that the gradient vector $\left.\vec{\nabla}f\right|_P$ plays for directional derivatives of f at P. In particular, the formula $\left.\text{rot}\vec{F}\right|_{\hat{u},P} = \left.\overset{\rightarrow}{\text{curl}\vec{F}}\right|_P \cdot \hat{u}$ says that, for a given P, the value of $\left.\text{rot}\,\vec{F}\right|_{\hat{u},P}$ changes in a smooth and regular way as a result of changes in the direction $\hat{u}$.

In terms of a small paddlewheel with hub at P in a fluid-flow with velocity $\vec{F}$, we see that (in the limit as the size of the wheel gets smaller) the direction of $\left.\overset{\rightarrow}{\text{curl}}\,\vec{F}\right|_P$ is the direction of the axle that gives the maximum right-handed angular speed of the wheel, and that the magnitude of $\left.\overset{\rightarrow}{\text{curl}}\,\vec{F}\right|_P$ is twice that maximum angular speed.

The Cartesian formula for $\overset{\rightarrow}{\text{curl}}\,\vec{F}$,

$$(3.3) \qquad \overset{\rightarrow}{\text{curl}}\,\vec{F} = \left(\frac{\partial N}{\partial y} - \frac{\partial M}{\partial z}\right)\hat{i} + \left(\frac{\partial L}{\partial z} - \frac{\partial N}{\partial x}\right)\hat{j} + \left(\frac{\partial M}{\partial x} - \frac{\partial L}{\partial y}\right)\hat{k},$$

is awkward to remember and work with. Determinant notation provides a simple memory rule and a convenient format for hand-computation in Cartesian coordinates.

$$(3.4) \qquad \overset{\rightarrow}{\text{curl}}\,\vec{F} = \begin{vmatrix} \hat{i} & \hat{j} & \hat{k} \\ \frac{\partial}{\partial x} & \frac{\partial}{\partial y} & \frac{\partial}{\partial z} \\ L & M & N \end{vmatrix}.$$

Since the derivatives of the scalar functions L, M, and N in (3.3) are linear, we see that a *linearity law* holds for the curl operator (3.4).

For vector fields $\vec{F}$ and $\vec{G}$ on a common domain in $\mathbf{E}^3$, and for scalar constants a and b,

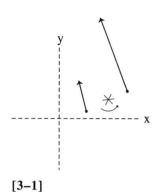

[3–1]

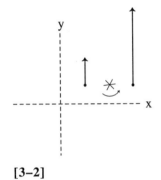

[3–2]

$$(3.5) \qquad \overrightarrow{\text{curl}}(a\vec{F} + b\vec{G}) = a\,\overrightarrow{\text{curl}}\,\vec{F} + b\,\overrightarrow{\text{curl}}\,\vec{G}.$$

Example 1. Let $\vec{F} = r\hat{\theta} = -y\hat{i} + x\hat{j}$ on E^3. (See [3–1].) Then

$$\overrightarrow{\text{curl}}\,\vec{F} = \begin{vmatrix} \hat{i} & \hat{j} & \hat{k} \\ \dfrac{\partial}{\partial x} & \dfrac{\partial}{\partial y} & \dfrac{\partial}{\partial z} \\ -y & x & 0 \end{vmatrix} = 2\hat{k}.$$

Recall that $\vec{F}$, as a fluid-flow, is a "turntable" rotation of the fluid around the z axis. We now see that for *every* point P, a small paddlewheel held with hub at P and with axle direction $\hat{k}$ will rotate with angular speed = 1 rad/sec.

Example 2. Let $\vec{F} = x\hat{j}$ *on* E^3. (See [3-2].) Then

$$\overrightarrow{\text{curl}}\,\vec{F} = \begin{vmatrix} \hat{i} & \hat{j} & \hat{k} \\ \dfrac{\partial}{\partial x} & \dfrac{\partial}{\partial y} & \dfrac{\partial}{\partial z} \\ 0 & x & 0 \end{vmatrix} = \hat{k}.$$

As a fluid-flow, $\vec{F}$ is a "shear flow." The streamlines are straight lines normal to the xz plane, and the magnitude of the velocity is proportional to x. At every point P in this flow, a wheel with hub at P and axle direction $\hat{k}$ will rotate with angular speed $= \frac{1}{2}$ rad/ sec .

Example 3. *For the flow in example 2, what will be the angular speed of a paddle-wheel held with hub at* P *and axle in the direction of* $\vec{A} = \hat{i} - \hat{k}$?

We have $\hat{A} = \dfrac{\sqrt{2}}{2}\hat{i} - \dfrac{\sqrt{2}}{2}\hat{k}$. Hence angular speed $= \frac{1}{2}\overrightarrow{\text{curl}}\,\vec{F}\Big|_{P} \cdot \hat{A} =$

$\frac{1}{2}\hat{k} \cdot \left(\dfrac{\sqrt{2}}{2}\hat{i} - \dfrac{\sqrt{2}}{2}\hat{k} \right) = -\dfrac{\sqrt{2}}{4}$.

Example 4. Let $\vec{F} = \frac{1}{r}\hat{\theta} = \dfrac{-y\hat{i}+x\hat{j}}{x^2+y^2}$ *on* E^3-*minus-the-z-axis.* Then

$$\text{curl}\,\vec{F} = \begin{vmatrix} \hat{i} & \hat{j} & \hat{k} \\ \dfrac{\partial}{\partial x} & \dfrac{\partial}{\partial y} & \dfrac{\partial}{\partial z} \\ \dfrac{-y}{x^2+y^2} & \dfrac{x}{x^2+y^2} & 0 \end{vmatrix} = \left(\dfrac{y^2-x^2}{(x^2+y^2)^2} + \dfrac{x^2-y^2}{(x^2+y^2)^2} \right)\hat{k} = \vec{0}.$$

Recall that $\vec{F}$ (as a fluid-flow) is a "vortex" flow around the z axis. The definition of rotational derivative and the fact that $\text{curl}\,\vec{F} = \vec{0}$ tells us that no matter
where a small paddlewheel is placed in D and no matter what direction is given to
its axle, the wheel will not rotate, provided that the circle of the wheel's vanes does
not enclose the z axis. (On the other hand, if the circle of vanes does enclose the
z axis, then the wheel will rotate with angular speed $\dfrac{1}{a^2}\hat{k}\cdot\hat{u}$, where a is the
radius of the wheel. (See the problems.) Note that $\left|\dfrac{1}{a^2}\right|$ becomes arbitrarily
large as a approaches 0.

   Example 5. In a uniform radial field, we have $\vec{F} = f(r)\hat{r} =$
$f\left(\sqrt{x^2+y^2}\right)\dfrac{x\hat{i}+y\hat{j}}{\sqrt{x^2+y^2}} = g\left(\sqrt{x^2+y^2}\right)\left(x\hat{i}+y\hat{j}\right)$ *on* E^2-*minus-the-z-axis. In this*
case,

$$L_y = xg'\left(\sqrt{x^2+y^2}\right)\left(x^2+y^2\right)^{-1/2}y,$$

and
$$M_x = yg'\left(\sqrt{x^2+y^2}\right)\left(x^2+y^2\right)^{-1/2}x.$$

Thus
$$M_z = L_y, \text{ and } M_z = L_z = N_x = N_y = 0.$$

Hence
$$\text{curl}\,\vec{F} = \vec{0}.$$

   A similar calculation shows that for a uniform central field $\vec{F} = h(\rho)\hat{\rho}$, we
also have $\text{curl}\,\vec{F} = \vec{0}$. See example 5 in §26.6.

   In each of the preceding examples $\text{curl}\,\vec{F}$ turned out to be a constant
vector. Most commonly, however, $\text{curl}\,\vec{F}$ will be a vector field which varies from
point to point.

Example 6. Let $\vec{F} = yz\hat{i} - xz\hat{j} + xy\hat{k}$ on E^3. Then

$$\text{curl}\,\vec{F} = \begin{vmatrix} \hat{i} & \hat{j} & \hat{k} \\ \dfrac{\partial}{\partial x} & \dfrac{\partial}{\partial y} & \dfrac{\partial}{\partial z} \\ yz & -xz & xy \end{vmatrix} = 2x\hat{i} - 2z\hat{k}.$$

The following mathematical idea also underlies the concept of *curl*. Let $\vec{F}$ be a C^1 vector field on D in E^3. Let $\mu_{\vec{F}}^c$ be the circulation measure for $\vec{F}$. In §23.6, we ask the question: can we find, for our given $\vec{F}$, a corresponding continuous vector field $\vec{G}$ such that for every finite, piecewise smooth, directed surface S in D, $\mu_{\vec{F}}^c(S) = \iint_S \vec{G} \cdot d\vec{\sigma}$? In §26.4 below, we give a general answer to this question. (The proof is given in §26.8.) The following preliminary result has an easy but instructive proof.

(3.6)   Let $\vec{F}$ be a C^1 *vector field on* D. *If there exists a continuous vector field* $\vec{G}$ *such that for every piecewise smooth, finite directed surface* S *in* D, $\mu_{\vec{F}}^c(S) = \iint_S \vec{G} \cdot d\vec{\sigma}$, *then, in fact,* $\vec{G}$ *must be* $\text{curl}\,\vec{F}$.

Proof. By elementary vector-algebra, it will be enough to show that for every interior point P in D and every unit vector $\hat{u}$,

(3.7) $$\text{curl}\,\vec{F}\big|_P \cdot \hat{u} = \vec{G} \cdot \hat{u}.$$

Let P and $\hat{u}$ be given. Let S_a be a directed disk with radius a, center at P, and direction $\hat{u}$. Then, by assumption, $\mu_{\vec{F}}^c(S_a) = \iint_{S_a} \vec{G} \cdot d\vec{\sigma}$, where by definition $\mu_{\vec{F}}^c(S_a) = \oint_{C_a} \vec{F} \cdot d\vec{R}$, and C_a is the coherently directed, circular boundary of S_a. Furthermore, we have

$$\text{curl}\,\vec{F} \cdot \hat{u} = \text{rot}\,\vec{F}\big|_{\hat{u},P} = \lim_{a \to 0} \frac{1}{\pi a^2} \oint_{C_a} \vec{F} \cdot d\vec{R}.$$

Hence $$\text{curl}\,\vec{F} \cdot \hat{u} = \lim_{a \to 0} \frac{1}{\pi a^2} \mu_{\vec{F}}^c(S_a) = \lim_{a \to 0} \frac{1}{\pi a^2} \iint_{S_a} \vec{G} \cdot d\vec{\sigma}.$$

Since $\vec{G}$ is continuous, it is readily shown from the definition of *vector surface integral* and by the use of a funnel function for the continuity of $\vec{G}$ at P, that this last limit $= \vec{G}(P) \cdot \hat{u}$. This proves (3.7) and hence proves (3.6).

Remark. In our discussion of a paddlewheel in a fluid-flow $\vec{F}$, we have assumed that the hub is maintained at a particular position in space and that the axle is maintained in a particular direction. What if we leave the paddlewheel free to drift with the flow, while at the same time maintaining a chosen direction for the axle? Let $\vec{v}\,'$ be the velocity of the wheel's drifting hub at the moment when it passes a certain point P. Then, in comparison to the case where the hub is held at P, the velocity of the fluid with respect to all parts of the wheel is altered from $\vec{F}$ to $\vec{F} - \vec{G}$, where $\vec{G} = \vec{v}'$ is a constant (in space) vector field. Since $\operatorname{curl} \vec{G} = \vec{0}$, it follows from the linearity law (3.3) that $\operatorname{curl}(\vec{F} - \vec{G}) = \operatorname{curl}\vec{F}$. Hence the angular speed of a drifting wheel when it passes P must be the same as the angular speed would be, at that moment, for a wheel held fixed at P.

Thus, for example, it is characteristic of vortex flow (a circular flow $\vec{F}$ around an axis but with $\operatorname{curl} \vec{F} = \vec{0}$ at points not on the axis) that an object which drifts around the vortex (without intersecting the axis of the vortex) will not rotate as it revolves around the axis. The reader may observe this phenomenon by floating a short (2 mm) match stick on the surface of a revolving flow of water out of a draining tub. (In this case, the flatter part of the surface is approximately a vortex flow.) [3–3].

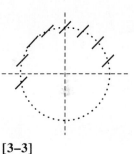

[3–3]

§26.4 Stokes's Theorem

Stokes's theorem (sometimes written Stokes' theorem) is (along with the conservative field theorem, Green's theorem, and the divergence theorem) one of the four principal theorems of multivariable integral calculus. In this section, we state Stokes's theorem, derive several corollaries, and give several examples. A proof of Stokes's theorem appears in §26.8.

Definition. A smooth elementary surface is said to be C^2-*smooth* if it has a smooth parameterization $\vec{R}$ (see (3.6) in §21.3) such that $\vec{R}$ is in C^2. A surface S is said to be *piecewise* C^2-*smooth* if it has a decomposition (see §21.5) into finitely many C^2-smooth elementary pieces.

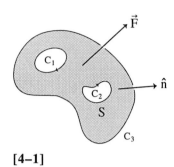

[4–1]

(4.1) **Stokes's theorem.** Let S be a piecewise C^2-smooth, directed finite surface whose boundary consists of one or more simple, closed, piecewise smooth, directed curves $C_1,...,C_k$. [4–1]. Let $\vec{F}$ be a C^1 vector field whose domain contains S (and therefore contains $C_1,..,C_k$). If the direction of each boundary curve of S is coherent with the direction of S, then

$$\iint_S \overrightarrow{\text{curl}}\,\vec{F}\cdot d\vec{\sigma} = \oint_{C_1} \vec{F}\cdot d\vec{R} + \cdots + \oint_{C_k} \vec{F}\cdot d\vec{R}.$$

(4.2) **Corollary.** Let $\vec{F}$ and S be as in (4.1), and let $\mu^c_{\vec{F}}$ be the circulation measure for $\vec{F}$ in E^2 as defined in §23.6. Then $\mu^c_{\vec{F}}(S) = \iint_S \overrightarrow{\text{curl}}\vec{F}\cdot d\vec{\sigma}$. Thus $\mu^c_{\vec{F}}$ is a vector integral measure as defined in §23.6.

We derive several further corollaries to (4.1).

(4.3) **Corollary.** Let S be a closed, piecewise C^2-smooth directed surface. Let $\vec{F}$ be a C^1 vector field whose domain contains S. Then $\oiint_S \overrightarrow{\text{curl}}\,\vec{F}\cdot d\vec{\sigma} = 0$.

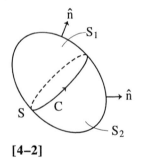

[4–2]

Proof of (4.3). Let C be a simple, closed, smooth curve lying in a smooth piece of S. Then C divides S into two piecewise smooth pieces, S_1 and S_2. (See [4–2].) Let the direction of C be coherent with S_1. Then the direction of C will not be coherent with S_2. Applying Stokes's theorem to S_1 and S_2 separately, we get

$$\iint_{S_1} \overrightarrow{\text{curl}}\vec{F}\cdot d\vec{\sigma} = \oint_C \vec{F}\cdot d\vec{R},$$

and

$$\iint_{S_2} \overrightarrow{\text{curl}}\vec{F}\cdot d\vec{\sigma} = -\oint_C \vec{F}\cdot d\vec{R}.$$

Adding these two equations, we have

$$\oiint_S \overrightarrow{\text{curl}}\,\vec{F}\cdot d\vec{\sigma} = \iint_{S_1+S_2} \overrightarrow{\text{curl}}\,\vec{F}\cdot d\vec{\sigma} = 0.$$

(4.4) **Corollary.** Let $\vec{F}$ be a C^2 vector field on D. Then $\text{div}\,\overrightarrow{\text{curl}}\,\vec{F} = 0$.

Proof of (4.4). Because $\vec{F}$ is C^2, $\overrightarrow{\text{curl}}\,\vec{F}$ must be C^1. Hence, by (4.3),

$$\oiint_{S(P,a)} \text{curl}\,\vec{F} \cdot d\vec{\sigma} = 0 \quad \text{for every sphere} \quad S(P,a) \quad \text{in D. Hence, by the definition of}$$

divergence, div $\overrightarrow{\text{curl}}\,\vec{F} = 0$.

(4.5)   ***Corollary.*** *Let* f *be a* C^2 *scalar field on* D. *Then* $\text{curl}(\vec{\nabla}f) = \vec{0}$.

Proof of (4.5). Assume $\text{curl}\,\vec{\nabla}f\big|_P \neq \vec{0}$ for some P in the domain of f.

Let M be the plane through P perpendicular to $\text{curl}\,\vec{\nabla}f\big|_P$. Let $\hat{n}_P$ be the unit vec-

tor of $\text{curl}\,\vec{\nabla}f\big|_P$. Then, by the continuity of $\text{curl}\,\vec{\nabla}f$ (which follows from the

assumption that f is in C^2), we can find a sufficiently small disk S in M with cen-

ter at P such that for every point Q on this disk, $\text{curl}\,\vec{\nabla}f\big|_Q \cdot \hat{n}_P > 0$. Let S have

the direction given by $\hat{n}_P$. Then $\iint_S \text{curl}\,\vec{\nabla}f \cdot d\vec{\sigma} = \iint_S \text{curl}\,\vec{\nabla}f \cdot \hat{n}_P d\sigma > 0$. [4–3].

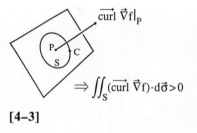

$$\Rightarrow \iint_S (\overrightarrow{\text{curl}}\,\vec{\nabla}f) \cdot d\vec{\sigma} > 0$$

[4–3]

Let C be the boundary circle of the disk. Let the direction of C be coherent with

S. By Stokes's theorem, must we have $\oint_C \vec{\nabla}f \cdot d\vec{R} > 0$. But this contradicts the

fact that a gradient field must be conservative. Hence our assumption must be

false that for some P, $\text{curl}\,\vec{\nabla}f\big|_P \neq \vec{0}$.

Shorter proof of (4.5), using Cartesian coordinates. Note that for any

vector field $\vec{F}$ in E^3 in Cartesian coordinates, the statement that $\text{curl}\,\vec{F} = \vec{0}$ is the

same as the statement that $\vec{F}$ passes the derivative test of §18.5 (for E^3). Since

$\vec{\nabla}f$ is conservative (by the conservative-field theorem), $\vec{\nabla}f$ must pass the deriva-

tive test, and hence $\text{curl}\,\vec{\nabla}f\big|_P = \vec{0}$.

Example 1. *Verify Stokes's theorem for the case where* $\vec{F} = -y\hat{i} + x\hat{j}$

and S *is the downward-directed graph of* $g(x,y) = x^2+y^2$ *on* $\hat{D}$ *where* $\hat{D}$ *is the*

disk $x^2 + y^2 \leq 4$ *in the* xy *plane.*

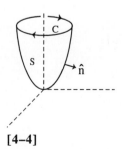

[4–4]

Then for Stokes's theorem, C must be the circle $x^2 + y^2 = 4$ in the plane

z = 4, where C is directed toward decreasing θ. See [4–4].

(a)   For the surface integral, we have $\text{curl}\,\vec{F} = 2\hat{k}$ and $\vec{w} =$ $-2x\hat{i} - 2y\hat{j} + \hat{k}$. Since $\vec{w}$ is directed upward, we get

$$\iint_S \text{curl}\,\vec{F} \cdot d\vec{\sigma} = -\iint_{\hat{D}} (\text{curl}\,\vec{F} \cdot \vec{w})dA = -\iint_{\hat{D}} 2dA = -2(\text{area}(\hat{D})) = -8\pi$$

(b)   For the line integral $\oint_C \vec{F} \cdot d\vec{R}$, we have C given by $\vec{R}(t) =$ $2\cos t\hat{i} + 2\sin t\hat{j} + 4\hat{k}$ (with parameter direction opposite to C). Then

$$\frac{d\vec{R}}{dt} = -2\sin t\hat{i} + 2\cos t\hat{j},$$
$$\vec{F} = -2\sin t\hat{i} + 2\cos t\hat{j},$$

and
$$\vec{F} \cdot \frac{d\vec{R}}{dt} = 4\sin^2 t + 4\cos^2 t = 4.$$

Hence we have $\oint_C \vec{F} \cdot d\vec{R} = -\int_0^{2\pi} 4dt = -8\pi$.

(a) and (b) give us our desired verification.

Example 2. *Verify Stokes's theorem for the case where* $\vec{F} = 2yz\hat{i} + xy\hat{k}$ *and S is the open-topped cylinder consisting of* S_1 *and* S_2 , *where* S_1 *is the disk* $x^2 + y^2 \le 4$ *is the* xy *plane and* S_2 *is the cylinder* $x^2 + y^2 = 4$ *between* $z = 0$ *and* $z = 3$. S *is directed out. See* [4–5].

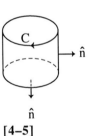

[4–5]

Then C must be the circle $x^2 + y^2 = 4$ in the plane $z = 3$, directed toward decreasing θ.

(a)   For the surface integral, we have $\text{curl}\,\vec{F} = x\hat{i} + y\hat{j} - 2z\hat{k}$.   Then $\iint_{S_1} (x\hat{i} + y\hat{j}) \cdot d\vec{\sigma} = 0$, since $(x\hat{i} + y\hat{j}) \cdot \hat{n} = 0$ on S_1; and $\iint_{S_1} -2z\hat{k} \cdot d\vec{\sigma} = 0$, since $z = 0$ on S_1. Hence $\iint_{S_1} \text{curl}\,\vec{F} \cdot d\vec{\sigma} = 0$.

On S_2, we have $x\hat{i} + y\hat{j} = 2\hat{r}$. Hence

$$\iint_{S_2} (x\hat{i} + y\hat{j}) \cdot d\vec{\sigma} = \iint_{S_2} 2\hat{r} \cdot d\vec{\sigma} = \iint_{S_2} 2\hat{r} \cdot \hat{r}d\sigma = \iint_{S_2} 2d\sigma = 2\text{area}(S_2) = 24\pi.$$

On S_2 we also have $\iint_{S_2} -2z\hat{k} \cdot d\vec{\sigma} = \iint_{S_2} -2z\hat{k} \cdot \hat{r}d\sigma = 0$, since $\hat{k} \cdot \hat{r} = 0$. Thus

$\iint_{S_2} \overrightarrow{\text{curl}}\vec{F} \cdot d\vec{\sigma} = 24\pi$, and $\iint_{S} \overrightarrow{\text{curl}}\vec{F} \cdot d\vec{\sigma} = 24\pi$.

(b) For C, we have $\vec{R}(t) = 2\cos t\hat{i} + 2\sin t\hat{j} + 3\hat{k}$, $0 \le t \le 2\pi$ (with parameter direction opposite to C).

Then
$$\frac{d\vec{R}}{dt} = -2\sin t\hat{i} + 2\cos t\hat{j},$$
$$\vec{F} = 2\sin t\hat{i} + 4\cos t\sin t\hat{k},$$

and
$$\vec{F} \cdot \frac{d\vec{R}}{dt} = -24\sin^2 t.$$

Hence we have $\oint_C \vec{F} \cdot d\vec{R} = -\int_0^{2\pi} -24\sin^2 t\,dt = 24\int_0^{2\pi} \sin^2 t\,dt = 24\pi$.

(a) and (b) give us our desired verification.

Example 3. *Verify Stokes's theorem when* $\vec{F} = y(z+1)\hat{i}$ *and S is the outward-directed open cylinder* $x^2 + y^2 = 4$ *between the planes* $z = 0$ *and* $z = 3$. Then $\overrightarrow{\text{curl}}\,\vec{F} = y\hat{j} - (z+1)\hat{k}$.

(a)   For the surface integral, we have

$$\iint_S \overrightarrow{\text{curl}}\vec{F} \cdot d\vec{\sigma} = \iint_S y\hat{j} \cdot d\vec{\sigma} - \iint_S (z+1)\hat{k} \cdot d\vec{\sigma}.$$

Here, $\iint_S (z+1)\hat{k} \cdot d\vec{\sigma} = \iint_S (z+1)\hat{k} \cdot \hat{r}d\vec{\sigma} = 0$, since $\hat{k} \cdot \hat{r} = 0$.

$$\iint_S y\hat{j} \cdot d\vec{\sigma} = \frac{1}{2}\iint_S (x\hat{i} + y\hat{j}) \cdot d\vec{\sigma} \quad (\text{since } \iint_S x\hat{i} \cdot d\vec{\sigma} = \iint_S y\hat{j} \cdot d\vec{\sigma}$$

by geometrical symmetry), and

$$\frac{1}{2}\iint_S (x\hat{i} + y\hat{j}) \cdot d\vec{\sigma} = \frac{1}{2}\iint_S 2\hat{r} \cdot \hat{r}d\vec{\sigma} = \text{area}(S) = 12\pi.$$

Thus. $\iint_S \overrightarrow{\text{curl}}\vec{F} \cdot d\vec{\sigma} = 12\pi$

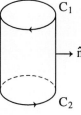

(b)   For C_1 and C_2 (which must be directed as in [4–6]), we have C_1 [4–6]
given by

$$\vec{R}_1(t) = 2\cos t\,\hat{i} + 2\sin t\,\hat{j} \qquad \text{(same direction as parameter)},$$

and C_2 given by

$$\vec{R}_2(t) = 2\cos t\,\hat{i} + 2\sin t\,\hat{j} + 3\hat{k} \qquad \text{(opposite direction to parameter)}.$$

For $\displaystyle\oint_{C_1} \vec{F}\cdot d\vec{R}$: $\qquad\qquad \dfrac{d\vec{R}}{dt} = -2\sin t\,\hat{i} + 2\cos t\,\hat{j}\,,$

$$\vec{F} = 2\sin t\,\hat{i}\,,$$

and $\qquad\qquad\qquad\qquad\qquad \vec{F}\cdot\dfrac{d\vec{R}}{dt} = -4\sin^2 t\,.$

Thus $\qquad\qquad\qquad \displaystyle\oint_{C_1}\vec{F}\cdot d\vec{R} = \int_0^{2\pi} -4\sin^2 t\,dt = -4\pi\,.$

For $\displaystyle\oint_{C_2}\vec{F}\cdot d\vec{R}$: $\qquad\qquad \dfrac{d\vec{R}}{dt} = -2\sin t\,\hat{i} + 2\cos t\,\hat{j}\,,$

$$\vec{F} = 2\sin t(4)\hat{i} = 8\sin t\,\hat{i}\,,$$

and $\qquad\qquad\qquad\qquad\qquad \vec{F}\cdot\dfrac{d\vec{R}}{dt} = -16\sin^2 t\,.$

Thus $\qquad\qquad\qquad \displaystyle\oint_{C_2}\vec{F}\cdot d\vec{R} = -\int_0^{2\pi} -16\sin^2 t\,dt = 16\pi\,.$

Hence $\qquad \displaystyle\oint_{C_1}\vec{F}\cdot d\vec{R} + \oint_{C_2}\vec{F}\cdot d\vec{R} = -4\pi + 16\pi = 12\pi\,.$

Since (a) and (b) agree, we have verified (4.1) for this example.

§26.5 Irrotational Fields in E^3

Definition. Let $\vec{F}$ be a C^1 vector field on D in E^3. $\vec{F}$ is said to be an *irrotational field* if, for every point P in D, $\left.\mathrm{curl}\,\vec{F}\,\right|_P = \vec{0}$.

Let $\vec{F}$ be expressed in a Cartesian coordinate system as $\vec{F} = L\hat{i} + M\hat{j} + N\hat{k}$. Then, from the Cartesian formula for $\mathrm{curl}\,\vec{F}$ (3.1), we see that:

(5.1) $\vec{F}$ *is irrotational* $\Leftrightarrow \left[\dfrac{\partial N}{\partial y} = \dfrac{\partial M}{\partial z} \text{ and } \dfrac{\partial L}{\partial z} = \dfrac{\partial N}{\partial x} \text{ and } \dfrac{\partial M}{\partial x} = \dfrac{\partial L}{\partial y} \right].$

In the present section, we give a few specific examples to show how Stokes's theorem can be used to help evaluate vector line integrals in an irrotational field.

Definition. We say that a piecewise smooth curve is *piecewise* C^2-*smooth* if each smooth piece has a smooth C^2 parameterization.

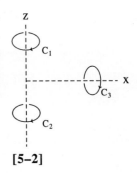

[5–1]

Example 1. Consider an irrotational field $\vec{F}$ on $D = \mathbf{E}^3$-with-the-z-axis-removed. (There are many such fields. $\vec{F} = \dfrac{1}{r}\hat{\theta}$ is one example.) Let C be an irregularly shaped, piecewise C^2-smooth, simple, closed curve which encircles the z axis once in the direction of increasing θ. Let C′ be a circle which goes around the z axis in the direction of increasing θ and is separated from C as in the figure. It is, in general, possible to find a piecewise C^2-smooth, two-sided surface S (in D) which has C and C′ as its boundary curves. (We do not prove this.) If we apply Stokes's theorem to S, we get (for S directed outward as in [5–1])

$$\iint_S \overrightarrow{\text{curl}\,\vec{F}} \cdot d\vec{\sigma} = \oint_C \vec{F} \cdot d\vec{R} - \oint_{C'} \vec{F} \cdot d\vec{R}.$$

Since $\vec{F}$ is irrotational, we have

$$\oint_C \vec{F} \cdot d\vec{R} = \oint_{C'} \vec{F} \cdot d\vec{R}.$$

We can thus obtain the value of $\oint_C \vec{F} \cdot d\vec{R}$ by using the simpler curve C′ in place of C.

Example 2. Consider an irrotational field $\vec{F}$ which has as its domain D all of $\mathbf{E}^3$ except for the z axis and the positive x axis. Such a field can be obtained by having a steady electric current flow down the positive z axis and divide at the origin and continue out the positive x and negative z axes. Let $\vec{F}$ be the magnetic field created by this current. It follows from the laws of physics that, in the absence of a time-dependent electric field, $\vec{F}$ must be irrotational (see Chapter 28). Let C_1, C_2, and C_3 be directed as in [5–2]. Assume that $\oint_{C_1} \vec{F} \cdot d\vec{R} = 5$ and $\oint_{C_3} \vec{F} \cdot d\vec{R} = 2$. What can we conclude about the value of $\oint_{C_2} \vec{F} \cdot d\vec{R}$?

[5–2]

We solve this problem by finding a two-sided piecewise smooth surface S in D whose boundary consists of C_1, C_2, and C_3. In effect, we take S' to be a balloon with holes at C_1, C_2, and C_3. If we direct the balloon outward and take account of the given directions for C_1, C_2, and C_3, then, by Stokes's theorem, we get

$$0 = \iint_S \overset{\rightarrow}{\text{curl}\vec{F}} \cdot d\vec{\sigma} = \oint_{C_1} \vec{F} \cdot d\vec{R} - \oint_{C_2} \vec{F} \cdot d\vec{R} - \oint_{C_3} \vec{F} \cdot d\vec{R}.$$

Hence, $\displaystyle\oint_{C_2} \vec{F} \cdot d\vec{R} = \oint_{C_1} \vec{F} \cdot d\vec{R} - \oint_{C_3} \vec{F} \cdot d\vec{R} = 5 - 2 = 3.$

Further examples appear in the problems.

§26.6 A Converse Derivative Test for E^3

It follows from (5.1) that

$\vec{F}$ *is irrotational* $\Leftrightarrow$ $\vec{F}$ *passes the derivative test for gradient fields in* E^3.

Recall, from §21.7, that a region R in E^2 is said to be *simply connected* if R is connected and if, for every simple piecewise smooth closed curve C in R, all points in the interior of C lie in R; and that a region R in E^3 is said to be *simply connected* if R is connected and if, for every simple closed, piecewise smooth curve C in R, there is a two-sided, piecewise smooth surface S such that (i) S has boundary C, and (ii) all points of S are in R.

Example 1. *The region* R = E^3 *is simply connected.* This is true, but a general formal proof is difficult, since it must cover cases where the curve C forms a knot in E^3. We do not give a proof. See the problems in §21.7.

Example 2. *A region* R *which consists of* E^3 *with a single point* P *removed is simply connected.* Let C be a curve in R. By example 1, there must be a two-sided surface in E^3 with boundary C. If this surface contains the removed point P (and therefore does not lie entirely in R), we simply distort it slightly near P so that it no longer contains P.

Example 3. *Let* R *consist of* E^3 *with a straight line* L *removed.* Then R is not simply connected. (Consider a curve C which encircles L.)

Example 4. *Let* R *be a solid torus.* Then R is not simply connected. (Consider the curve C shown in [6–1].)

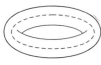

[6–1]

Stokes's theorem now gives us the following converse to the derivative test for $\mathbf{E}^3$. It is similar, in its wording, to the converse for $\mathbf{E}^2$, as stated and proved in §24.4.

(6.1) *Converse derivative test for* $\mathbf{E}^3$. *Let* $\vec{F}$ *be a vector field on a simply connected domain in* $\mathbf{E}^3$. *If* $\vec{F}$ *is irrotational, then* $\vec{F}$ *is conservative.*

Proof of (6.1). Assume that $\vec{F}$ is irrotational on a simply connected domain D. If $\vec{F}$ is not conservative, then there must be a piecewise smooth, directed, closed curve C in D such that $\oint_C \vec{F} \cdot d\vec{R} \neq 0$. From this it follows that there must be a piecewise smooth, directed, simple, closed curve C′ in D such that $\oint_{C'} \vec{F} \cdot d\vec{R} \neq 0$. (See the problems.) Since D is simply connected, we can find a two-sided surface S′ whose boundary is C′. We direct S′ so that the directions of C′ and S′ are coherent. Then by Stokes's theorem,

$$\iint_{S'} \overrightarrow{\text{curl}}\,\vec{F} \cdot d\vec{\sigma} = \oint_{C'} \vec{F} \cdot d\vec{R} \neq 0.$$

This is a contradiction, since $\vec{F}$ is irrotational and $\overrightarrow{\text{curl}}\,\vec{F} = \vec{0}$ at all points in D. This completes the proof.

Example 5. Let $\vec{F}$ *be a uniform central field* $\vec{F} = h(\rho)\hat{\rho}$ *on* $D = \mathbf{E}^3$-*with-origin-removed.* By example 2, D is simply connected. A calculation similar to that in example 5 of §26.3 shows that $\vec{F}$ is irrotational. It follows that $\vec{F}$ is conservative. (We can also see that $\vec{F}$ is irrotational by viewing $\vec{F}$ as a fluid-flow and by considering the effect of this flow on a paddle-wheel. No matter where we put the wheel, or how we orient it, the effects of $\vec{F}$ will be equal and opposite on vanes which are mirror images with respect to a plane determined by the origin and the axle of the wheel. This implies that $\overrightarrow{\text{curl}}\,\vec{F} = \vec{0}$ at all points in D.)

Example 6. Let $\vec{F}$ *be a uniform radial field on* $D = \mathbf{E}^3$-*with-z axis-removed.* In example 5 of §26.3, we saw by a calculation that a uniform radial field must be irrotational. (This also follows, as in the preceding example, by considering the effect of $\vec{F}$, as a fluid-flow, on an arbitrarily placed paddle-wheel.) In this example, D is not simply connected, so we cannot use (6.1) to prove that $\vec{F}$ is conservative. Nevertheless, a uniform radial field must be conservative, and this can be proved by using a direct geometrical argument to show that $\vec{F}$ has independence of path. See the problems.

§26.7 The $\vec{\nabla}$ Notation for Curl

The *curl* is our third differentiation operation, along with *gradient* and *divergence*, for scalar and vector fields. Analogously to operations of gradient and divergence, it can be viewed as a non-numerical *operator* (or function) which can be applied to any C^{n+1} vector field as input and will then produce a corresponding C^n vector field as output. As we have already noted, the curl operator satisfies the linearity law (3.5). Combined with other operators, it yields the identities

(7.1) $$\overrightarrow{\text{curl}} \ \overrightarrow{\text{grad}} \ f = \vec{0}, \text{ and}$$

(7.2) $$\text{div} \ \overrightarrow{\text{curl}} \ \vec{F} = \vec{0},$$

and can be used in other higher-order combinations such as

(7.3) $$\overrightarrow{\text{curl}} \ \overrightarrow{\text{curl}} \vec{F}.$$

The symbol $\vec{\nabla}$ is useful for abbreviating the curl operator. *We write* $\overrightarrow{\text{curl}} \vec{F}$ *as* $\vec{\nabla} \times \vec{F}$. (3.5), (7.1), (7.2) and (7.3) can now become

(7.4) $$\vec{\nabla} \times (a\vec{F} + b\vec{G}) = a\vec{\nabla} \times \vec{F} + b\vec{\nabla} \times \vec{G},$$

(7.5) $$\vec{\nabla} \times (\vec{\nabla}f) = \vec{0},$$

(7.6) $$\vec{\nabla} \cdot (\vec{\nabla} \times \vec{F}) = 0, \text{ and}$$

(7.7) $$\vec{\nabla} \times (\vec{\nabla} \times \vec{F}).$$

(The parentheses in (7.5), (7.6), and (7.7) can be omitted, since the notation remains unambiguous.)

Note in (7.5) how this formula, as vector-algebra, agrees with the rule for cross products that $\vec{A} \times \vec{A}c = c(\vec{A} \times \vec{A}) = \vec{0}$.

Similarly, note how (7.6) as vector-algebra, agrees with the rule for scalar triple products that $\vec{A} \cdot \vec{A} \times \vec{B} = 0$.

Finally, we note two new product rules:

$$(7.8) \qquad \vec{\nabla} \cdot (\vec{F} \times \vec{G}) = \vec{G} \cdot (\vec{\nabla} \times \vec{F}) - \vec{F} \cdot (\vec{\nabla} \times \vec{G}),$$

$$(7.9) \qquad \vec{\nabla} \times (f\vec{F}) = f(\vec{\nabla} \times \vec{F}) - \vec{F} \times \vec{\nabla}f,$$

where, in each case, the appearance of the − is consistent with rules of vector-algebra for the order of factors in scalar triple products (7.8) and in cross products (7.9).

The correctness of (7.8) and (7.9) can be verified by using the Cartesian coordinate formulas for gradient, divergence, and curl. In Chapter 27, we consider deeper theoretical and algebraic properties of the symbol $\vec{\nabla}$; we present other identities that ∇ satisfies; and we see that certain algebraic combinations, such as $\vec{\nabla}(\vec{F} \cdot \vec{G})$, $\vec{\nabla} \times (\vec{F} \times \vec{G})$, and $\vec{\nabla} \times (\vec{\nabla} \times \vec{F})$, cannot be expanded or simplified without the introduction of further special definitions and operations.

§26.8 Proof of Stokes's Theorem

In E^3 with Cartesian coordinates, let $\vec{F}$, S, and C be as assumed for the theorem. We assume, without proof, the following *subdivision principle for piecewise C^2-smooth surfaces*:

(8.1) *Let S be a finite, piecewise C^2-smooth, directed surface in E^3. Then S can be subdivided into a finite number of elementary subsurfaces such that for each subsurface, there is some Cartesian coordinate system in which that piece is the upward-directed graph of a C^2 function f(x,y) over a region in the xy plane which has a simple, closed, piecewise smooth curve as its boundary.*

Let $\widetilde{S}$ be a smooth piece of S given by (8.1) as the graph of a C^2 function f(x,y) on a region $\hat{D}$ in the xy plane of some Cartesian coordinate system. Let $\widetilde{S}$ have the simple, piecewise smooth boundary curve $\widetilde{C}$. We assume that $\widetilde{S}$ is directed up. We now prove Stokes's theorem on $\widetilde{S}$ for the two special cases $\vec{F} = N\hat{k}$ and $\vec{F} = L\hat{i}$. The proof for the special case $\vec{F} = M\hat{j}$ is similar to the proof for $\vec{F} = L\hat{i}$, and we omit it. Using the linearity laws for surface integrals and line integrals, the Stokes's theorem equations for these special cases can then be summed to give Stokes's theorem on $\tilde{S}$ for $\vec{F} = L\hat{i} + M\hat{j} + N\hat{k}$.

Case 1: $\vec{F} = N\hat{k}$. By (3.2), $\overrightarrow{\text{curl}}\,\vec{F} = N_y\hat{i} - N_x\hat{j}$.

(a) We then have

$$\iint_{\tilde{S}} \overrightarrow{\text{curl}}\,\vec{F} \cdot d\vec{\sigma}$$

$$= \iint_{\hat{D}} [N_y(x,y,f(x,y))\hat{i} - N_x(x,y,f(x,y))\hat{j}] \cdot (-f_x(x,y)\hat{i} - f_y(x,y)\hat{j} + \hat{k})dA$$

$$= \iint_{\hat{D}} (-N_y(x,y,f)f_x + N_x(x,y,f)f_y)dA .$$

(b) Let $\tilde{C} = \tilde{C}_1 + \tilde{C}_2 + \cdots + \tilde{C}_k$, where for each $\tilde{C}_i$, $i \leq k$ we have a smooth path

$$\vec{R}_i(t) = x_i(t)\hat{i} + y_i(t)\hat{j} + f(x_i(t), y_i(t))\hat{k} \quad \text{on } [a_i, b_i].$$

Then
$$\frac{d\vec{R}_i}{dt} = \dot{x}_i\hat{i} + \dot{y}_i\hat{j} + (f_x\dot{x}_i + f_y\dot{y}_i)\hat{k},$$

and
$$\vec{F} \ \ (\text{on } \tilde{C}_i) = N(x_i, y_i, f(x_i, y_i))\hat{k}.$$

We assume that parameter directions agree with curve directions. By parametric evaluation,

$$\oint_C \vec{F} \cdot d\vec{R} = \sum_i \int_{\tilde{C}_i} \vec{F} \cdot d\vec{R}$$

$$= \sum_i \int_{a_i}^{b_i} N(x_i, y_i, f)[f_x\dot{x}_i + f_y\dot{y}_i]dt$$

$$= \sum_i \int_{\hat{C}_i} (N(x,y,f)f_x\hat{i} + N(x,y,f)f_y\hat{j}) \cdot d\vec{R},$$

where $\hat{C}_i$ is the projection of $\tilde{C}_i$ onto the xy plane. (This last step is true because the $\int_{a_i}^{b_i}$ -integral happens to be the parametric formula for the $\int_{\hat{C}_i}$ -integral as well as for the $\int_{\tilde{C}_i}$ -integral.)

We then have

$$\oint_C \vec{F} \cdot d\vec{R} = \oint_{\hat{C}} (N(x,y,f)f_x\hat{i} + N(x,y,f)f_y\hat{j}) \cdot d\vec{R}$$

where $\hat{C}$ is the projection of $\tilde{C}$ onto the xy plane. We now apply Green's theorem to the line integral over $\hat{C}$ in $\mathbf{E}^2$. We get, using the chain rule,

$$\oint_{\tilde{C}} F \cdot dR = \iint_{\hat{D}} [N_x f_y + N_z f_y f_x + N f_{yx} - N_y f_x - N_z f_x f_y - N f_{xy}]dA$$

Since f is C^2, $f_{xy} = f_{yx}$, and we have

$$\oint_{\tilde{C}} F \cdot dR = \iint_{\hat{D}} (N_x(x,y,f)f_y - N_y(x,y,f)f_x)dA \, .$$

Comparing the results of (a) and (b), we have our desired result for case 1.

Case 2. $\vec{F} = L\hat{i}$. Then $\overrightarrow{\text{curl}} F = L_z\hat{j} - L_y\hat{k}$

(a) $\iint_{\tilde{S}} \overrightarrow{\text{curl}} \vec{F} \cdot d\vec{\sigma} = \iint_{\hat{D}} (L_z\hat{j} - L_y\hat{k}) \cdot (-f_x\hat{i} - f_y\hat{j} + \hat{k})dA$

$$= -\iint_{\hat{D}} (L_z f_y + L_y)dA \, .$$

(b) $\oint_C F \cdot dR = \sum \int_{\tilde{C}_i} \vec{F} \cdot d\vec{R} = \sum \int_{a_i}^{b_i} L(x_i, y_i, f)\dot{x}_i dt$

$$= \sum \int_{\hat{C}_i} L(x,y,f)\hat{i} \cdot d\vec{R} \quad \text{(for the same reason as in case 1)}$$

$$= \int_{\hat{C}} L(x,y,f)\hat{i} \cdot d\vec{R} \, .$$

We again apply Green's theorem to the integral over $\hat{C}$. We get, using the chain rule,

$$\oint_C F \cdot dR = -\iint_{\hat{D}} (L_y + L_z f_y)dA \, .$$

Comparing the results of (a) and (b), we have our desired result for case 2. By linearity, case 1 and case 2 give us

$$\mu_{\vec{F}}^c(\tilde{S}) = \iint_S \text{curl}\,\vec{F} \cdot d\vec{\sigma}.$$

By the additivity of $\mu_{\vec{F}}^c$ and of surface integrals, we get

$$\mu_{\vec{F}}^c(S) = \iint_S \text{curl}\,\vec{F} \cdot d\vec{\sigma}.$$

This is Stokes's theorem (4.1).

§26.9 Problems

§26.3

1. For each of the following vector fields $\vec{F}$ on $\mathbf{E}^3$, find $\overrightarrow{\text{curl}\,\vec{F}}$.

   (a) $\vec{F} = 2x\hat{i} - y\hat{j} - z\hat{k}$ (b) $\vec{F} = x^2\hat{i} + yz\hat{j} - yz\hat{k}$

   (c) $\vec{F} = yx\hat{i} + xz\hat{j} + xy\hat{k}$ (d) $\vec{F} = y^2\hat{i} + z^2\hat{j} + x^2\hat{k}$

2. Let $P = (a,b,c)$, and let $\hat{u} = \frac{1}{3}\hat{i} - \frac{2}{3}\hat{j} + \frac{2}{3}\hat{k}$. For each of the vector fields in problem 1, find $\overrightarrow{\text{rot}\,\vec{F}}\big|_{\hat{u},P}$.

3. $\vec{F}$ is the velocity field of a fluid-flow. A small paddlewheel is placed in the flow with its hub at P and its axle in the direction of $\hat{i} + \hat{j}$. For each vector field $\vec{F}$ in problem 1, what angular velocity will we expect for the wheel?

4. For the vortex flow in example 4, show that if the circle of vanes of a paddlewheel encloses the z axis, then the wheel will rotate with angular velocity $\frac{1}{a^2}\hat{k} \cdot \hat{u}$, where a is the radius of the wheel and $\hat{u}$ is the direction (away from the wheel) of its axle.

5. In the proof of (3.6), explain why the existence of $\vec{G}$ on D follows from the existence of $\overrightarrow{\text{curl}\,\vec{F}}$ on D.

§26.4

1. Let C be the triangle with vertices $(0,0,0)$, $(1,0,0)$, $(2,2,2)$. C is directed: $(0,0,0) \to (1,0,0) \to (2,2,2)$ $\to (0,0,0)$. Let S be the plane surface enclosed by this triangle. S is directed upward. Let $\vec{F} = x^2\hat{i} + xy\hat{j} + z\hat{k}$.

   (a) Evaluate $\oint_C \vec{F} \cdot d\vec{R}$.

   (b) Evaluate $\iint_S \vec{F} \cdot d\vec{\sigma}$.

   (c) As a check on (a) (via Stokes's theorem), evaluate $\iint_S (\overrightarrow{\text{curl}\,\vec{F}}) \cdot d\vec{\sigma}$.

2. S is union of two surfaces S_1 and S_2, where S_1 is the set of (x,y,z) with $x^2 + y^2 = 1$, $0 \le z \le 1$, and S_2 is the set of (x,y,z) with $x^2 + y^2 + (z-1)^2 = 1$, $z \ge 1$. Let $\vec{F} = (zx + z^2y + x)\hat{i} + (z^3yx + y)\hat{j} + z^4x^2\hat{k}$. Let S be directed out. Find $\iint_S (\overrightarrow{\text{curl}\,\vec{F}}) \cdot d\vec{\sigma}$. (*Hint.* Use Stokes's theorem.)

3. $\vec{F} = 3y\hat{i} - xz\hat{j} + yz^2\hat{k}$, S is the surface $2z = x^2 + y^2$ below the plane $z = 2$, directed up. Find $\iint_S (\overrightarrow{\text{curl}\,\vec{F}}) \cdot d\vec{\sigma}$ both directly and using Stokes's theorem.

4. Evaluate the following using Stokes's theorem:

   (a) $\oint_C (-3y\hat{i} + 3x\hat{j} + \hat{k}) \cdot d\vec{R}$, where C is the circle $x^2 + y^2 = 1$, $z = 2$, directed so that y increases when $x > 0$.

(b) $\oint_C (2xy^2z\hat{i} + 2x^2yz\hat{j} + (x^2y^2 - 2z)\hat{k}) \cdot d\vec{R}$,

where C is given by the parametric equations x = cos t, y = sin t, z = sin t, $0 \le t \le 2\pi$, and has the direction given by t.

5. (a) Let the surface S be the disk $x^2 + y^2 \le 4$ in the plane z = 1. S is directed upward. Let $\vec{F} = y\hat{i} + z\hat{j} + x\hat{k}$. Evaluate $\iint_S (\text{curl}\,\vec{F}) \cdot d\vec{\sigma}$.

(b) Stokes's theorem asserts that the value of this surface integral is equal to the value of a certain line integral. Set up and evaluate the line integral.

6. (a) Use Stokes's theorem to evaluate the vector line integral in example 2 of §19.5.

(b) Do as in (a) for example 5 of §19.5.

(c) Do as in (a) for example 6 of §19.5. (*Hint.* In each case, extend the given curve to a piecewise smooth, closed curve in such a way that the line integral over the added portion of curve is equal to 0.)

§26.5

1. (a) Give an example of a vector field which *is not* divergenceless, *is* irrotational, and *is not* conservative.

(b) Given an example of a vector field which *is not* divergenceless, *is not* irrotational, and *is not* conservative.

2. The surface S is a cylinder which is closed at the top but open at the bottom. The cylinder has radius 2 and altitude 3. The open end lies in the xy plane, and the positive z-axis is the axis of the cylinder. Let $\vec{F} = x\hat{j}$. Choose a direction for S, and verify that Stokes's theorem holds for S and $\vec{F}$ by separately evaluating the surface integral and the line integral that appear in the statement of the theorem. (If you wish, you may use the divergence theorem to help in evaluating the surface integral.)

3. You are given the following information:

(i) $\vec{F}$ is a vector field (continuously differentiable) whose domain is all points in $\mathbf{E}^3$ except the z axis.

(ii) $\text{curl}\,\vec{F} = \vec{0}$ at all points in the domain of $\vec{F}$.

(iii) $\oint_C \vec{F} \cdot d\vec{R} = 5$, for C: x = cos t, y = sin t, z = 0 $(0 \le t \le 2\pi)$.

(a) What is the value of $\oint_{C'} \vec{F} \cdot d\vec{R}$ for C': x = 2 cos t, y = 2 sin t, z = 0 $(0 \le t \le 2\pi)$?

(b) What are the possible values that can occur for $\oint_{C''} \vec{F} \cdot d\vec{R}$, where C'' is a simple-closed curve which lies in the xy plane and in the domain of $\vec{F}$?

4. An irrotational vector field $\vec{F}$ has as its domain all points except: points on the z axis, points on the y axis with $y \ge 1$, and points on the circle $x^2 + y^2 = 1$ in the plane z = 0. Consider the three curves C_1: $x^2 + z^2 = 4$ in the plane y = 2, C_2: $x^2 + y^2 = 4$ in the plane z = 3, and C_3: $x^2 + y^2 = 64$ in the plane z = -3. The direction of C_1 is right-handed with respect to the positive direction of the y axis, and the directions of C_2 and C_3 are right-handed with respect to the positive direction of the z axis. You are given $\oint_{C_2} \vec{F} \cdot d\vec{R} = 7$. Use Stokes's theorem to find the value of $\oint_{C_1} \vec{F} \cdot d\vec{R}$ and the value of $\oint_{C_3} \vec{F} \cdot d\vec{R}$.

5. Assume that we are given C^2 parameterizations for the curves C and C' in figure [5–1]. Using these parameterizations, find a C^2 parameterization for a surface of the kind indicated as S as in [5–1].

6. An irrotational field $\vec{F}$ has, as its domain D, all points in $\mathbf{E}^3$ except: points on the z axis; points on the y axis with $y \ge 1$; and points on the circle $x^2 + y^2 = 1$ in the plane z = 0. Five closed, directed circles C_1, C_2, C_3, C_4, and C_5 are given as shown in [9–1].

[9-1]

(a) Sketch a two-sided surface S such that S has C_1 as its only boundary curve and S lies entirely in D. Use Stokes's theorem to conclude that the circulation of $\vec{F}$ on C_1 must be 0.

(b) Sketch a two-sided surface which lies entirely in D and has C_2 and C_3 as its entire boundary. Use Stokes's theorem to conclude that the circulation of $\vec{F}$ on C_2 must equal the circulation of $\vec{F}$ on C_3.

(c) Sketch a two-sided surface which lies entirely in D and has C_2, C_4, and C_5 as its entire boundary. Use Stokes's theorem to conclude that the circulation of $\vec{F}$ on C_2 must equal the sum of the circulations of $\vec{F}$ on C_4 and C_5.

(d) It can be shown by a physical argument (see Ampere's law in §28.5) that electrical currents can be used to construct a magnetic field $\vec{F}$ such that $\vec{F}$ satisfies the conditions on $\vec{F}$ for [9–1] and such that the circulation on C_4 is 0 while the circulation on C_5 is not 0. Use Stokes's theorem to conclude that there can be no surface S which lies entirely in D and has C_1, C_4, and C_5 as its entire boundary. (*Note.* This physical argument can be translated into a purely mathematical proof.)

§26.6

1. Let D be the domain of a vector field $\vec{F}$. Assume that there is a piecewise smooth, directed, closed curve C in D such that $\oint_C \vec{F} \cdot d\vec{R} \neq 0$. Show that there must be a piecewise smooth, directed, simple closed curve C' in D such that $\oint_{C'} \vec{F} \cdot d\vec{R} \neq 0$.

2. In problem 1 for §26.3, which fields are conservative and which are not.

3. Let $\vec{F}$ be a uniform central field ($\vec{F} = h(\rho)\hat{\rho}$) on $D = \mathbf{E}^3$-with-origin-removed. Use a Cartesian computation analogous to that in example 5 of §26.3 to show that $\vec{F}$ is irrotational. (Since D is simply connected, it follows that $\vec{F}$ must be conservative. This proves the result stated in example 5.)

4. Let $\vec{F}$ be a uniform radial field ($\vec{F} = f(r)\hat{r}$) on $D = \mathbf{E}^3$-with-z-axis-removed. In example 5 of §26.3, we saw that $\vec{F}$ is irrotational. Show by a direct geometrical argument that $\vec{F}$ must have independence of path. (*Hint.* Observe that any piecewise smooth path C from a point P to a point Q in D can be approximated by a piecewise smooth path C' from P to Q whose individual smooth pieces are segments of coordinate curves for r, θ, or z. Furthermore, observe that this can always be done in such a way that the projection of the path $C + (-C')$ on the plane z = 0 has winding number 0 about the origin. It follows that C and C' lie in a simple connected subset of D. See §24.4.)

§26.7

1. $\vec{F}$ and $\vec{G}$ are vector fields, and g and h are scalar fields.
(a) Simplify $\vec{\nabla} \times gh\vec{F}$.
(b) Verify $\vec{\nabla} \cdot (\vec{F} \times \vec{G}) = \vec{G} \cdot (\vec{\nabla} \times \vec{F}) - \vec{F} \cdot (\vec{\nabla} \times \vec{G})$.

2. $\vec{F}$ is a vector field and f is a scalar field.
(a) Verify (7.1).
(b) Verify (7.2).
(c) Verify (7.9).

Chapter 27

Mathematical Applications

§27.1 Leibnitz's Rule

Let $f(x,t)$ be a given function of two variables. For any given fixed value t_0 of t, $f(x,t_0)$ defines a corresponding one-variable function. (In Chapter 8, we call this function a *partial function* of the two-variable function $f(x,t)$.) As a one-variable function, $f(x,t_0)$ can be differentiated and integrated by the familiar methods of one-variable calculus. In such calculations, we often speak of t as a *parameter* and of $f(x,t)$ as a *parameterized family of one-variable functions*. In some applications, the parameter t will represent *time*; we then speak of $f(x,t)$ as a *time-dependent one-variable function*.

When we evaluate the definite integral $\int_a^b f(x,t)dx$, we treat t as a constant. For each value of t there is a corresponding value for this integral. Hence the integral's value can be viewed as a new function $g(t)$.

Example 1. Let $f(x,t) = (x+t)^2$, and let $a = 0$ and $b = 1$. Then $\int_0^1 (x+t)^2 dx = \dfrac{(x+t)^3}{3}\Big]_{x=0}^{x=1} = t^2 + t + \frac{1}{3}$. Thus our integral defines the new function $g(t) = t^2 + t + \frac{1}{3}$.

Leibnitz's rule asserts that the derivative $g'(t)$ can be obtained either by first integrating f and then differentiating the resulting g or by first differentiating f and then integrating the resulting $\frac{\partial f}{\partial t}$.

(1.1) *Leibnitz's rule. Let* $f(x,t)$ *be a* C^1 *function of two variables defined for* $a \le x \le b$ *and* $c \le t \le d$. *Then, for* $c \le t \le d$,

$$\frac{d}{dt} \int_a^b f(x,t)dx = \int_a^b f_t(x,t)dx .$$

This rule says, in effect, that for a definite integral with a parameter t, the operation of differentiation with respect to the parameter can be "moved through" the integral sign.

Example 1 (continued).   We have $g(t) = \int_0^1 (x+t)^2 dx = t^2 + t + \frac{1}{3}.$
Hence (without Leibnitz's rule) we have

$$g'(t) = \frac{d}{dt} \int_0^1 (x+t)^2 dx = 2t + 1 .$$

If we use Leibnitz's rule, we have

$$g'(t) = \int_0^1 \frac{\partial}{\partial t}(x+t)^2 dx = \int_0^1 (2x+2t)dx = x^2 + 2xt \Big]_{x=0}^{x=1} = 1 + 2t .$$

This verifies Leibnitz's rule for our example.

The general proof of Leibnitz's rule is as follows.

Proof of **(1.1)**.   By the fundamental theorem of integral calculus, $\frac{d}{dt} \int_c^t \int_a^b f_t(x,u)dxdu = \int_a^b f_t(x,t)dx$. Hence, reversing the order in the iterated integral (by Fubini's theorem), we have

$$\int_a^b f_t(x,t)dx = \frac{d}{dt} \int_a^b \int_c^t f_t(x,u)dudx$$

$$= \frac{d}{dt} \int_a^b [f(x,t) - f(x,c)]dx \qquad (\text{since } f_t(x,u) = \frac{\partial}{\partial u}f(x,u))$$

$$= \frac{d}{dt} \int_a^b f(x,t)dx - \frac{d}{dt} \int_a^b f(x,c)dx$$

$$= \frac{d}{dt} \int_a^b f(x,t)dx \quad \left(\text{since } \int_a^b f(x,c)dx \text{ is a constant}\right).$$

This completes the proof.

It is easy to see that Leibnitz's rule must be true when the integral has other parameters besides t, since these parameters will be treated as constants by the derivatives and integrals in Leibnitz's rule. For example, we have

$$(1.2) \qquad\qquad \frac{d}{dt} \int_a^b f(x,s,t)dx = \int_a^b f_t(x,s,t)dx .$$

Analogous forms of Leibnitz's rule can be similarly proved for multiple integrals, for line integrals and surface integrals, and for other differentiation operators such as the gradient, divergence, and curl operators. We give three examples.

Example 2. If f(x,y,z,t) is a time-dependent scalar field on a fixed domain D, and if C is a fixed curve in D, then

$$(1.3) \qquad\qquad \frac{d}{dt} \int_C fds = \int_C f_t ds .$$

(1.3) can be proved by applying (1.1) to the final definite integral in the parametric evaluation of $\int_C f_t ds$. (We would, of course, use a variable other than t in the parameterization of the curve C.)

Example 3. Let f(x,y,z,t) be a time-dependent scalar field on a fixed domain D. We now take t to be the integrated variable, and x, y, and z to be parameters. We then have

$$(1.4) \qquad\qquad \vec{\nabla} \int_a^b fdt = \int_a^b \vec{\nabla}_p fdt$$

as a valid form of Leibnitz's rule. Here $\vec{\nabla}_p$ indicates that $\vec{\nabla}$ is restricted to the parameters x,y,z and does not apply to the variable t. (1.4) can be proved by applying (1.2) to each scalar component of (1.4).

Example 4. Let f(x,y,z,t) be a time-dependent mass-density function on a fixed region D in E^3. Then

(1.5)
$$\frac{d}{dt}\iiint_D f dV = \iiint_D \frac{\partial f}{\partial t} dV .$$

This tells us that the rate of change of the total mass in D can be found by integrating the rate of change of the density. (1.5) can be proved by applying (1.2) to each successive integral in an iterated integral for $\iiint f dV$.

Analogous versions of Leibnitz's rule can also be proved for vector-valued integrals. In all such cases, we can prove the rule by expressing vectors and operators in Cartesian coordinates, then using the linearity of the integrals and operators to express these integrals in terms of scalar components, then going to iterated integrals, and finally applying (1.1) or (1.2) to the successive one-variable integrals.

Leibnitz's rule is not valid when applied to the variable of integration in an indefinite integral; see (5.1) and (5.2) in §20.5. Leibnitz's rule is also not valid when the parameter t appears in the limits of integration for the given definite integral. For example, $\frac{d}{dt}\int_0^{2t} xtdx = 6t^2$, but $\int_0^{2t} \frac{\partial}{\partial t}(xt)dx = 2t^2$.

§27.2 $\vec{\nabla}f$, $\vec{\nabla}\cdot\vec{F}$, and $\vec{\nabla}\times\vec{F}$ in Curvilinear Coordinates

The formulas for *gradient*, *divergence*, and *curl* in cylindrical and spherical coordinates are as follows.

Cylindrical coordinates. A scalar field f is given as $f(r,\theta,z)$, and a vector field $\vec{F}$ is given as $\vec{F} = f_1\hat{r} + f_2\hat{\theta} + f_3\hat{z}$, where f_1, f_2, and f_3 are scalar functions of r, θ, and z. We have

(2.1)
$$\vec{\nabla}f = \frac{\partial f}{\partial r}\hat{r} + \frac{1}{r}\frac{\partial f}{\partial \theta}\hat{\theta} + \frac{\partial f}{\partial z}\hat{z},$$

(2.2)
$$\vec{\nabla}\cdot\vec{F} = \frac{1}{r}\frac{\partial}{\partial r}(rf_1) + \frac{1}{r}\frac{\partial f_2}{\partial \theta} + \frac{\partial f_3}{\partial z},$$

(2.3)
$$\vec{\nabla}\times\vec{F} = \begin{vmatrix} \frac{1}{r}\hat{r} & \hat{\theta} & \frac{1}{r}\hat{z} \\ \frac{\partial}{\partial r} & \frac{\partial}{\partial \theta} & \frac{\partial}{\partial z} \\ f_1 & rf_2 & f_3 \end{vmatrix}$$

$$= \left(\frac{1}{r}\frac{\partial f_3}{\partial \theta} - \frac{\partial f_2}{\partial z}\right)\hat{r} + \left(\frac{\partial f_1}{\partial z} - \frac{\partial f_3}{\partial r}\right)\hat{\theta} + \left(\frac{1}{r}\frac{\partial}{\partial r}(rf_2) - \frac{1}{r}\frac{\partial f_1}{\partial \theta}\right)\hat{z}.$$

Examples. (1) $\vec{\nabla}r^2 = 2r\hat{r}$.

   (2) $\vec{\nabla}\left(\dfrac{r^2\theta}{z}\right) = \dfrac{2r\theta}{z}\hat{r} + \dfrac{r}{z}\hat{\theta} - \dfrac{r^2\theta}{z^2}\hat{z}$.

   (3) $\vec{\nabla}\cdot(r\hat{r}) = 2$ (in agreement with $\vec{\nabla}\cdot(x\hat{i} + y\hat{j}) = 2$).

   (4) $\vec{\nabla}\cdot(\dfrac{1}{r}) = 0$.

   (5) $\vec{\nabla}\times r\hat{\theta} = 2\hat{z}$.

   (6) $\vec{\nabla}\times\dfrac{1}{r}\hat{\theta} = \vec{0}$.

Spherical coordinates. A scalar field g is given as $g(\rho,\varphi,\theta)$, and a vector field $\vec{G}$ is given as $\vec{G} = g_1\hat{\rho} + g_2\hat{\varphi} + g_3\hat{\theta}$, where g_1 , g_2 , and g_3 are scalar functions of ρ, φ, and θ. We have

(2.4) $\vec{\nabla}g = \dfrac{\partial g}{\partial \rho}\hat{\rho} + \dfrac{1}{\rho}\dfrac{\partial g}{\partial \varphi}\hat{\varphi} + \dfrac{1}{\rho\sin\varphi}\dfrac{\partial g}{\partial \theta}\hat{\theta}$.

(2.5) $\vec{\nabla}\cdot\vec{G} = \dfrac{1}{\rho^2}\dfrac{\partial}{\partial \rho}(\rho^2 g_1) + \dfrac{1}{\rho\sin\varphi}\dfrac{\partial}{\partial \varphi}(\sin\varphi g_2) + \dfrac{1}{\rho\sin\varphi}\dfrac{\partial g_3}{\partial \theta}$.

(2.6) $\vec{\nabla}\times\vec{G} = \begin{vmatrix} \dfrac{1}{\rho^2\sin\varphi}\hat{\rho} & \dfrac{1}{\rho\sin\varphi}\hat{\varphi} & \dfrac{1}{\rho}\hat{\theta} \\[2mm] \dfrac{\partial}{\partial\rho} & \dfrac{\partial}{\partial\varphi} & \dfrac{\partial}{\partial\theta} \\[2mm] g_1 & \rho g_2 & \rho\sin\varphi g_3 \end{vmatrix}$

$= \left(\dfrac{1}{\rho\sin\varphi}\dfrac{\partial}{\partial\varphi}(\sin\varphi g_3) - \dfrac{1}{\rho\sin\varphi}\dfrac{\partial g_2}{\partial\theta}\right)\hat{\rho} +$

$\left(\dfrac{1}{\rho\sin\varphi}\dfrac{\partial g_1}{\partial\theta} - \dfrac{1}{\rho}\dfrac{\partial}{\partial\rho}(\rho g_3)\right)\hat{\varphi} + \left(\dfrac{1}{\rho}\dfrac{\partial}{\partial\rho}(\rho g_2) - \dfrac{1}{\rho}\dfrac{\partial g_1}{\partial\varphi}\right)\hat{\theta}$.

Examples.   (7) $\vec{\nabla}(\rho^2) = 2\rho\hat{\rho}$.

   (8) $\vec{\nabla}\left(\dfrac{\rho^2\theta}{\varphi}\right) = \dfrac{2\rho\theta}{\varphi}\hat{\rho} - \dfrac{\rho\theta}{\varphi^2} + \dfrac{\rho}{\varphi\sin\varphi}$.

   (9) $\vec{\nabla}\cdot(\rho\hat{\rho}) = 3$.

   (10) $\vec{\nabla}\cdot(\dfrac{1}{\rho^2}\hat{\rho}) = 0$.

   (11) $\vec{\nabla}\times\rho\hat{\theta} = \cot\varphi\hat{\rho} - 2\hat{\varphi}$.

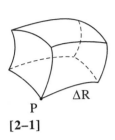

P

[2–1]

The technique for deriving these formulas is, in principle, simple. To find a formula for $\vec{\nabla}\cdot\vec{F}\Big|_P$ in a given orthogonal, curvilinear system u, v, w, we take a small region ΔR in the approximate form of a rectangular parallelepiped, where each face is a coordinate surface, where each edge is a coordinate curve, where the dimensions of ΔR are determined by chosen values of Δu, Δv, and Δw, and where one vertex of ΔR is at P. (See [2–1].) We first estimate the flux measure $\mu_{\vec{F}}^f$ of ΔR. (This calculation uses Leibnitz's rule, as we see in the example below.) We then divide our estimate of $\mu_{\vec{F}}(\Delta R)$ by the estimated volume of ΔR. This quotient gives an estimated density at P for the flux measure. In the limit, as $\Delta u, \Delta v, \Delta w$ approach 0, we find the exact density at P. By the divergence theorem, this density is $\vec{\nabla}\cdot\vec{F}\Big|_P$.

A curvilinear formula for $\vec{\nabla}\times\vec{F}\Big|_P$ can be found by a similar argument. We use Leibnitz's rule and one face of the region ΔR described above to estimate rot $\vec{F}\Big|_{\hat{u},P}$. Then, in the limit, we get the exact value of rot $\vec{F}\Big|_{\hat{u},P}$. Similarly, we find rot $\vec{F}\Big|_{\hat{v},P}$ and rot $\vec{F}\Big|_{\hat{w},P}$. Using the frame identity and the curl theorem, we get

$$\vec{\nabla}\times\vec{F}\Big|_P = \text{rot } \vec{F}\Big|_{\hat{u},P}\,\hat{u} + \text{rot } \vec{F}\Big|_{\hat{v},P}\,\hat{v} + \text{rot } \vec{F}\Big|_{\hat{w},P}\,\hat{w} .$$

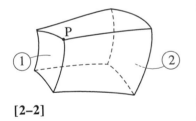

[2–2]

We illustrate these calculations by finding the first term in the spherical formula for $\nabla\cdot\vec{G}$. Let ΔR be as in [2–2]. Let $P = (\rho_0,\varphi_0,\theta_0)$. On the faces ① and ②, only f_1 contributes to the flux, since $\hat{\rho}$ is normal to these faces while $\hat{\varphi}$ and $\hat{\theta}$ are parallel to these faces. The flux through ① in the direction of $\hat{\rho}$ is $\int_{\theta_0}^{\theta_0+\Delta\theta} \int_{\varphi_0}^{\varphi_0+\Delta\varphi} f_1(\rho,\varphi,\theta)\rho^2 \sin\varphi\,d\varphi\,d\theta$, where $\rho = \rho_0$. If we let ρ increase from ρ_0 to $\rho_0 + \Delta\rho$, this flux increases by the approximate amount

$$\frac{d}{d\rho}\left(\iint f_1\rho^2 \sin\varphi\,d\varphi\,d\theta\right)\Delta\rho .$$

By Leibnitz's rule, this is

(2.7) $$\left(\iint \frac{d}{d\rho}(\rho^2 f_1) \sin\varphi\,d\varphi\,d\theta\right)\Delta\rho .$$

This increase gives the net outward flux through ① and ②. For small values of $\Delta\rho$, $\Delta\varphi$, and $\Delta\theta$, (2.7) can be approximately expressed as $\frac{d}{d\rho}(\rho^2 f_1)\sin\varphi\Delta\varphi\Delta\theta\Delta\rho$.

Dividing by the approximate volume of $\Delta V = \rho^2 \sin\varphi\Delta\rho\Delta\varphi\Delta\theta$, we get

$$\nabla \cdot (f_1\hat{\rho})\big|_P = \frac{1}{\rho^2}\frac{\partial}{\partial\rho}(\rho^2 f_1)\big|_P.$$

Remark. The informal argument above can be made logically precise by using linear approximation arguments. These arguments assume that $\vec{F}$ is C^1.

§27.3 Invariant Formulas

Let $\vec{A}$ have scalar components x,y,z in one Cartesian coordinate system and scalar components $\tilde{x}, \tilde{y}, \tilde{z}$ in another Cartesian system. Then $x^2 + y^2 + z^2 = \tilde{x}^2 + \tilde{y}^2 + \tilde{z}^2$ must hold, since $x^2 + y^2 + z^2 = \vec{A} \cdot \vec{A} = \tilde{x}^2 + \tilde{y}^2 + \tilde{z}^2$. Thus the Cartesian formula

(3.1) $$a_1^2 + a_2^2 + a_3^2$$

gives the same scalar value (for a given vector $\vec{A}$) in any chosen Cartesian system, provided we use the components of $\vec{A}$ in that system. In such a case, we say that the formula (3.1) is an *invariant scalar formula* in the components of a vector.

Similarly, for two vectors $\vec{A}$ and $\vec{B}$, we say that the Cartesian coordinate formula

(3.2) $$a_1 b_1 + a_2 b_2 + a_3 b_3$$

is an *invariant scalar formula* in the scalar components of those vectors because it gives the same value, $\vec{A} \cdot \vec{B}$, in any chosen Cartesian system, provided we use the correct components for $\vec{A}$ and $\vec{B}$ in that system.

Similarly, for two vectors $\vec{A}$ and $\vec{B}$, we say that the Cartesian formula

(3.3) $$(a_1 + b_1)\hat{i} + (a_2 + b_2)\hat{j} + (a_3 + b_3)\hat{k}$$

is an *invariant vector formula* in the scalar components of those vectors because it gives the same vector, $\vec{A} + \vec{B}$, in any chosen Cartesian system, provided we use the components for $\vec{A}$ and $\vec{B}$ in that system and take $\hat{i}, \hat{j}, \hat{k}$ to be the unit coordinate vectors in that system.

Invariance is a special and rare property. Almost all formulas fail to have it; for example, for a vector $\vec{A}$, the formula

(3.4) $a_1 + a_2 + a_3$

is not invariant, as we show in example 1 below. We know that invariance holds for the Cartesian formulas $\vec{A} \cdot \vec{A}$, $\vec{A} \cdot \vec{B}$, and $\vec{A} + \vec{B}$, because we derived those Cartesian formulas from the coordinate-free geometric definitions of those operations. In the same way, invariance holds for the Cartesian formula for $\vec{A} \times \vec{B}$.

Invariance can also be defined for Cartesian formulas of vector *calculus*. Let f on D in $\mathbf{E}^3$ be a $\mathbf{C}^1$ scalar field, and let P be a point in D. Then

(3.5) $\left.\dfrac{\partial f}{\partial x}\right|_P \hat{i} + \left.\dfrac{\partial f}{\partial y}\right|_P \hat{j} + \left.\dfrac{\partial f}{\partial z}\right|_P \hat{k}$

is an invariant vector formula in the scalar function for f and in the coordinates of P, because it gives the same vector, $\left.\vec{\nabla} f\right|_P$, in any chosen Cartesian system provided that: (i) we represent the scalar field f by the correct scalar function f(x,y,z) in that chosen system, (ii) we use the coordinates for P in that chosen system, and (iii) we take $\hat{i}, \hat{j}, \hat{k}$ to be the unit coordinate vectors in that chosen system. Invariance holds for (3.5) because we showed in §9.5 that $\left.\vec{\nabla} f\right|_P$ is the unique vector at P with certain coordinate-free, geometric properties.

Let $\vec{F}$ be a $\mathbf{C}^1$ vector field on D in $\mathbf{E}^3$, and let P be a point in D. Consider the scalar-valued Cartesian formula

(3.6) $\left.\dfrac{\partial L}{\partial x}\right|_P + \left.\dfrac{\partial M}{\partial y}\right|_P + \left.\dfrac{\partial N}{\partial z}\right|_P,$

where the scalar functions L(x,y,z), M(x,y,z), and N(x,y,z) represent the scalar components of $\vec{F}$ in some chosen Cartesian system. Then (3.6) is an *invariant scalar formula* in the scalar component functions L,M,N for $\vec{F}$ and in the coordinates of P, because it gives the same scalar value, $\left.\vec{\nabla} \cdot \vec{F}\right|_P$, in any chosen Cartesian system provided that (i) the scalar functions L(x,y,z), M(x,y,z), and N(x,y,z) give the components of $\vec{F}$ in that system, and (ii) we use the coordinates of P in that system. Invariance holds for (3.6) because the initial definition of *divergence at* P was coordinate-free; that is to say, it was purely geometric and did not depend on a particular coordinate system. (Recall that the definition of div $\vec{F}|_P$ uses surface integrals and that the definition of *surface integral* is geometric and coordinate-free.)

Let $\vec{F}$ be a C^1 vector field in D and let P be a point in D. Consider the Cartesian formula

$$(3.7) \qquad \left(\frac{\partial N}{\partial y}\bigg|_P - \frac{\partial M}{\partial z}\bigg|_P\right)\hat{i} + \left(\frac{\partial L}{\partial z}\bigg|_P - \frac{\partial N}{\partial x}\bigg|_P\right)\hat{j} + \left(\frac{\partial M}{\partial x}\bigg|_P - \frac{\partial L}{\partial y}\bigg|_P\right)\hat{k} ,$$

where, again, the scalar functions L,M,N represent the scalar components of $\vec{F}$ in a chosen Cartesian system. Then, just as in the case of (3.6), (3.7) is an *invariant vector formula* in the scalar components L,M,N for $\vec{F}$ and in the coordinates of P, because it gives the same vector value, $\vec{\nabla} \times \vec{F}|_P$, in any chosen Cartesian system, provided that L, M, and N give the components of $\vec{F}$ in that system, that we use the coordinates for P in that system, and that $\hat{i}$, $\hat{j}$, and $\hat{k}$ are the unit coordinate vectors for that system. We know that invariance holds, because our initial definition of *curl at* P was geometric and coordinate-free.

Finally, let f be a C^2 scalar field on D in E^3, and let P be a point in D. Consider the Cartesian formula

$$(3.8) \qquad \frac{\partial^2 f}{\partial x^2}\bigg|_P + \frac{\partial^2 f}{\partial y^2}\bigg|_P + \frac{\partial^2 f}{\partial z^2}\bigg|_P .$$

This must also be an *invariant scalar formula* in the scalar function for f and the coordinates of P, because it gives the same scalar value, $\vec{\nabla} \cdot \vec{\nabla} f|_P$, in every Cartesian system, provided we use the correct scalar function f(x,y,z) in that system and the correct coordinates for P in that system. We know that invariance holds for (3.8), because the formulas for *divergence* and *gradient* are invariant as noted above. In example 3 below, we verify a specific case of the invariance of (3.8). In example 4 below, we give an example of a simple symmetrical formula which is *not* invariant.

Invariance is a deep property. The invariance of (3.8) depends on functional definitions and proofs in preceding chapters. In principle, the invariance of a given formula φ involving scalar and/or vector fields can be directly checked by comparing:

(1) the result of first applying an arbitrary change of coordinates to the given fields and *then* evaluating φ for the new fields;

with   (2) the result of first evaluating φ for the given fields and then applying the change of coordinates, if appropriate, to the value obtained from φ.

If we can show that (1) and (2) always produce the same scalar or vector, then we have proved that φ is invariant.

A principal advantage of the symbol $\vec{\nabla}$ is that whenever $\vec{\nabla}$ is used in the vector-algebra of scalar and vector fields, the result must always be expressible in Cartesian coordinates by an invariant formula. Can every invariant Cartesian formula, involving derivatives of scalar and vector functions, be expressed as a formula of vector algebra with $\vec{\nabla}$? We return to this question in §27.6.

Example 1. *We show, by example, that* (3.4) *is not an invariant formula in the components of a vector* $\vec{A}$. Let Cartesian system I have unit vectors $\hat{i},\hat{j},\hat{k}$. Let system II have the same origin as I and unit vectors $\hat{i}',\hat{j}',\hat{k}'$ such that

and $\hat{i}' = \frac{\sqrt{2}}{2}\hat{i} + \frac{\sqrt{2}}{2}\hat{j}$, $\hat{j}' = -\frac{\sqrt{2}}{2}\hat{i} + \frac{\sqrt{2}}{2}\hat{j}$, and $\hat{k}' = \hat{k}$. (See §4.7.) It follows that

$$\hat{i} = (\hat{i} \cdot \hat{i}')\hat{i}' + (\hat{i} \cdot \hat{j}')\hat{j}' = \frac{\sqrt{2}}{2}\hat{i}' - \frac{\sqrt{2}}{2}\hat{j}', \text{ and } \hat{j} = (\hat{j} \cdot \hat{i}')\hat{i}' + (\hat{j} \cdot \hat{j}') = \frac{\sqrt{2}}{2}\hat{i}' + \frac{\sqrt{2}}{2}\hat{j}'.$$

Now let $\vec{A} = \hat{i}$. Evidently, the components of $\vec{A}$ in system I are $(a_1, a_2, a_3) = (1, 0, 0)$. In system II, $\vec{A} = (\vec{A} \cdot \hat{i}')\hat{i}' + (\vec{A} \cdot \hat{j}') = \frac{\sqrt{2}}{2}\hat{i}' - \frac{\sqrt{2}}{2}\hat{j}'$, and the

components of $\vec{A}$ in system II are $(a_1, a_2, a_3) = \left(\frac{\sqrt{2}}{2}, -\frac{\sqrt{2}}{2}, 0\right)$. Thus, in system I,

the value of the formula (3.4) is $a_1 + a_2 + a_3 = 1$, while in system II, the value is $a_1 + a_2 + a_3 = 0$.

Note that this same example accords with the invariance of (3.1), since $a_1^2 + a_2^2 + a_3^2 = 1 + 0 + 0 = 1$ in system I, and $a_1^2 + a_2^2 + a_3^2 = \frac{1}{2} + \frac{1}{2} + 0 = 1$ in system II.

Example 2. *Verify the invariance of* (3.5) *for the case of systems* I *and* II *in example* 1, *with* $f(x,y,z) = x^2$ *in system* I. Then in system II, f is

represented by $f(\tilde{x}, \tilde{y}, \tilde{z}) = \left(\frac{\sqrt{2}}{2}\tilde{x} - \frac{\sqrt{2}}{2}\tilde{y}\right)^2 = \frac{1}{2}(\tilde{x}^2 - 2\tilde{x}\tilde{y} + \tilde{y}^2)$. Let P be the

point with coordinates $(1,0,0)$ in system I. In system I, (3.5) gives

$$\left.\frac{\partial f}{\partial x}\right|_P \hat{i} + \left.\frac{\partial f}{\partial y}\right|_P \hat{j} + \left.\frac{\partial f}{\partial z}\right|_P \hat{k} = 2x\hat{i} = 2\hat{i} \,.$$

In system II, P has coordinates $\left(\frac{\sqrt{2}}{2}, -\frac{\sqrt{2}}{2}, 0\right)$ and we get

$$\left.\frac{\partial f}{\partial \tilde{x}}\right|_P \hat{i}' + \left.\frac{\partial f}{\partial \tilde{y}}\right|_P \hat{j}' + \left.\frac{\partial f}{\partial \tilde{z}}\right|_P \hat{k}' = (\tilde{x} - \tilde{y})\hat{i}' + (-\tilde{x} + \tilde{y})\hat{j}' = \sqrt{2}\,\hat{i}' - \sqrt{2}\,\hat{j}' \,.$$

But this last vector becomes, in system I,

$$\sqrt{2}\left(\frac{\sqrt{2}}{2}\hat{i} + \frac{\sqrt{2}}{2}\hat{j}\right) - \sqrt{2}\left(-\frac{\sqrt{2}}{2}\hat{i} + \frac{\sqrt{2}}{2}\hat{j}\right) = \hat{i} + \hat{j} + \hat{i} - \hat{j} = 2\hat{i}$$

in agreement with our calculation for system I.

Example 3. *Verify the invariance of* (3.8) *for the case described in example* 2. In system I, we have

$$\left.\frac{\partial^2 f}{\partial x^2}\right|_P + \left.\frac{\partial^2 f}{\partial y^2}\right|_P + \left.\frac{\partial^2 f}{\partial z^2}\right|_P = 2 \,.$$

In system II, we have

$$\left.\frac{\partial^2 f}{\partial \tilde{x}^2}\right|_P + \left.\frac{\partial^2 f}{\partial \tilde{y}^2}\right|_P + \left.\frac{\partial^2 f}{\partial \tilde{z}^2}\right|_P = 1 + 1 = 2 \,.$$

Example 4. *We use the case described in example* 2 *to show that*

(3.9) $$\left.\frac{\partial f}{\partial x}\right|_P + \left.\frac{\partial f}{\partial y}\right|_P + \left.\frac{\partial f}{\partial z}\right|_P$$

is not invariant.

For $f(x,y,z) = x^2$ and $P = (1,0,0)$ in system I, we have

$$\left.\frac{\partial f}{\partial x}\right|_P + \left.\frac{\partial f}{\partial y}\right|_P + \left.\frac{\partial f}{\partial z}\right|_P = 2x|_P = 2 \,.$$

For $f(\tilde{x}, \tilde{y}, \tilde{z}) = \frac{1}{2}(\tilde{x}^2 - 2\tilde{x}\tilde{y} + \tilde{y}^2)$ and $P = \left(\dfrac{\sqrt{2}}{2}, -\dfrac{\sqrt{2}}{2}, 0\right)$, we have

$$\left.\frac{\partial f}{\partial \tilde{x}}\right|_P + \left.\frac{\partial f}{\partial \tilde{y}}\right|_P + \left.\frac{\partial f}{\partial \tilde{z}}\right|_P = \tilde{x} - \tilde{y} - \tilde{x} + \tilde{y} = 0 \,,$$

and invariance fails.

§27.4 The Laplacian; Harmonic Functions; Poisson Inversion

The Laplacian. Given a C^2 scalar field f, we can form the divergence of the gradient of f. Using the operator $\vec{\nabla}$, the result can be written as $\vec{\nabla} \cdot (\vec{\nabla}f)$. More often, it is abbreviated as $\vec{\nabla} \cdot \vec{\nabla}f$ or $\vec{\nabla}^2 f$ or $\nabla^2 f$. In Cartesian coordinates, we have $\nabla^2 f = \vec{\nabla} \cdot \vec{\nabla}f = \vec{\nabla} \cdot \left(\dfrac{\partial f}{\partial x}\hat{i} + \dfrac{\partial f}{\partial y}\hat{j} + \dfrac{\partial f}{\partial z}\hat{k}\right) =$

(4.1)
$$\frac{\partial^2 f}{\partial x^2} + \frac{\partial^2 f}{\partial y^2} + \frac{\partial^2 f}{\partial z^2} \,.$$

The operator $\nabla^2 = \dfrac{\partial^2}{\partial x^2} + \dfrac{\partial^2}{\partial y^2} + \dfrac{\partial^2}{\partial z^2}$ can be used as a *scalar* operator in its own right, following the algebraic rules for multiplication by a scalar, and it can be applied to vector fields as well as scalar fields; for a vector field $\vec{F} = L\hat{i} + M\hat{j} + N\hat{k}$, we define $\nabla^2 F$ to be the vector field $(\nabla^2 L)\hat{i} + (\nabla^2 M)\hat{j} + (\nabla^2 N)\hat{k}$.

Our definition of $\nabla^2 f = \vec{\nabla} \cdot \vec{\nabla}f$ is evidently coordinate-free, since our definitions of gradient and divergence are coordinate-free. Our definition of $\nabla^2 F$, however, is based on a chosen system of Cartesian coordinates. It is easy to show that for a given vector field $\vec{F}$, we always get the same new vector field $\nabla^2 F$, regardless of the Cartesian system we choose. We merely observe that for any unit vector $\hat{u}$, $(\nabla^2\vec{F}) \cdot \hat{u} = \nabla^2(\vec{F} \cdot \hat{u})$. (We have: $(\nabla^2\vec{F}) \cdot \hat{u} = \left((\nabla^2 L)\hat{i} + (\nabla^2 M)\hat{j} + (\nabla^2 N)\hat{k}\right) \cdot \hat{u} = (\nabla^2 L)u_1 + (\nabla^2 M)u_2 + (\nabla^2 N)u_3 = \nabla^2(Lu_1 + Mu_2 + Nu_3) = \nabla^2(\vec{F} \cdot \hat{u})$, where $\hat{u} = u_1\hat{i} + u_2\hat{j} + u_3\hat{k}$.) Since $\vec{F} \cdot \hat{u}$ is a scalar field, and since the definition of ∇^2 applied to a scalar field is coordinate-free, we can conclude that the value of $(\nabla^2 \vec{F}) \cdot \hat{u}$ must be independent of our choice of Cartesian coordinates. But this implies that $\vec{\nabla}^2 F$ itself is independent of our choice of coordinates. (Recall that if $\vec{A} \cdot \hat{u} = \vec{B} \cdot \hat{u}$ for all unit vectors $\hat{u}$, then $\vec{A} = \vec{B}$.)

Harmonic fields. A scalar field (or function) is said to be *harmonic* if $\nabla^2 f = \vec{\nabla} \cdot \vec{\nabla} f = 0$. The equation $\nabla^2 f = 0$ is called *Laplace's equation*. Thus f is a harmonic field (or function) $\Leftrightarrow$ f is a solution to Laplace's equation. To say that f is harmonic is the same as saying that f is a scalar potential for a divergenceless vector field. (Not all divergenceless fields have scalar potentials, since not all divergenceless fields are conservative. For example, $\vec{F} = r\hat{\theta}$ has $\vec{\nabla} \cdot \vec{F} = 0$ but $\vec{\nabla} \times \vec{F} = 2\hat{z}$. See (2.2) and (2.3).) In Cartesian coordinates, Laplace's equation takes the form

(4.2)
$$\frac{\partial^2 f}{\partial x^2} + \frac{\partial^2 f}{\partial y^2} + \frac{\partial^2 f}{\partial z^2} = 0 .$$

Harmonic fields occur extensively in mathematics and theoretical science. One of the best known examples is the *potential* field of an electric field in a simply connected region of empty space where there is no time-varying magnetic field. (We consider this further in Chapter 28.) Another example of a harmonic field is the potential field for a gravitational field in empty space.

The following theorem describes an important special feature of harmonic fields. It asserts that the value of a harmonic field f, at the center of any sphere in the domain of f, is exactly equal to the average value of f over the surface of that sphere. This theorem is not true, in general, for closed surfaces other than spheres. For example, it is not true for non-spherical ellipsoids.

(4.3) *Theorem. The mean-value property of harmonic fields. Let f be harmonic on domain* D. *Let* S *be a sphere of radius* a > 0 *such that* S *and its interior* R *lie in* D. *Let* Q *be the center of the sphere. Then*

$$f(Q) = \frac{1}{4\pi a^2} \oiint_S f\, d\sigma .$$

Proof. We use spherical coordinates with origin at Q. Let P be a point on the sphere with coordinates a, φ, θ. Then, by the results of §20.3,

$$f(P) = f(Q) + \int_Q^P \vec{\nabla} f \cdot d\vec{R} = f(Q) + \int_C \vec{\nabla} f \cdot d\vec{R} \,,$$

where C is the line segment from Q to P. Thus

$$f(a, \varphi, \theta) = f(P) + \int_0^a \vec{\nabla} f(\rho, \varphi, \theta) \cdot \hat{\rho}(\varphi, \theta) d\rho \,,$$

where ρ serves as parameter for C and $\hat{\rho}$ depends only on φ and θ. Then the average value of f on S is

$$(4.4) \qquad \frac{1}{4\pi a^2} \oiint_S f d\sigma = \underbrace{\frac{1}{4\pi a^2} \oiint_S f(Q) d\sigma}_{= f(Q)} + \underbrace{\frac{1}{4\pi a^2} \oiint_S \left[\int_0^a \vec{\nabla} f \cdot \hat{\rho} d\rho \right] d\sigma}_{= g(a)}$$

The first integral $= f(Q)$, since $\dfrac{1}{4\pi a^2} \iint_S d\sigma = 1$. We abbreviate the second integral, whose value depends only on a, as $g(a)$. Using Leibnitz's rule, we calculate $g'(a)$, for $a > 0$. We first express $g(a)$ as a multiple integral

$$g(a) = \frac{1}{4\pi a^2} \int_0^{2\pi} \int_0^\pi \int_0^a \vec{\nabla} f \cdot \hat{\rho}(a^2 \sin\varphi) d\rho d\varphi d\theta$$

$$= \frac{1}{4\pi} \int_0^{2\pi} \int_0^\pi \int_0^a \vec{\nabla} f \cdot \hat{\rho} \sin\varphi d\rho d\varphi d\theta \,.$$

Then, by Leibnitz's rule and the FTIC,

$$g'(a) = \frac{1}{4\pi} \int_0^{2\pi} \int_0^\pi \vec{\nabla} f(a, \varphi, \theta) \cdot \hat{\rho} \sin\varphi d\varphi d\theta$$

$$= \frac{1}{4\pi a^2} \int_0^{2\pi} \int_0^\pi \vec{\nabla} f \cdot \hat{\rho} a^2 \sin\varphi d\varphi d\theta$$

$$= \frac{1}{4\pi a^2} \oiint_S \vec{\nabla} f \cdot \hat{\rho} d\sigma = \frac{1}{4\pi a^2} \oiint_S \vec{\nabla} f \cdot d\vec{\sigma} \qquad \text{(S directed out)}$$

$$= \frac{1}{4\pi a^2} \iiint_R \vec{\nabla} \cdot \vec{\nabla} f dV \qquad\qquad \text{(by the divergence theorem)}$$

$$= 0 \,, \text{ since } \nabla^2 f = 0 \,, \text{ by assumption.}$$

Since $g'(a) = 0$ for $a > 0$, $g(a)$ must be constant for $a > 0$. It remains to show that $\lim\limits_{a \to 0} g(a) = 0$. By (4.4), this is the same as showing that

$$\lim_{a \to 0} \frac{1}{4\pi a^2} \oiint_S f \, d\sigma = f(Q).$$

The latter follows from the continuity of the scalar field f. Therefore since $\lim_{a \to 0} g(a) = 0$, and g(a) is constant for a > 0, we have g(a) = 0 for a ≥ 0.

Hence $\frac{1}{4\pi a^2} \oiint_S f \, d\sigma = f(Q)$. This completes the proof.

The converse of (4.3) is also true: *If a* C^2 *scalar field* f *possesses the mean value property of (4.3), then* f *must be harmonic.* See the problems.

Harmonic fields have many useful properties. Here is one example.

(4.5)   *If* f *on domain* D *is harmonic, then all local extreme points of* f, *if any, must occur at the boundary of* D.

   Proof. If, for example, a local maximum occurred at an interior point Q, then we could surround Q by a small sphere in D such that, at all points P on the sphere, f(P) < f(Q). But this would contradict (4.3), which asserts that f(Q) must be the average of the values f(P).

   Harmonic fields in E^2 are defined in E^2 in the same way as in E^3: f is said to be *harmonic* if $\dfrac{\partial^2 f}{\partial x^2} + \dfrac{\partial^2 f}{\partial y^2} = 0$. (4.3) can be proved, in the same way as before, as a theorem about circles. It states that the value of a harmonic function f at the center of any circle is exactly equal to its average value on the circumference of that circle.

   Poisson inversion. Let g be a given continuous scalar field on a domain D. The equation $\nabla^2 f = g$ is called *Poisson's equation for* g. A C^2 scalar field f on D is called a *solution* for this equation if $\nabla^2 f = g$. If f is a solution, then, for every harmonic field h on D, f+h must also be a solution (since $\nabla^2 h = 0$ and $\nabla^2(f + h) = \nabla^2 f + \nabla^2 h = \nabla^2 f = g$). Hence, for a given g, Poisson's equation may have many different solutions. By a similar argument, any two solutions for a Poisson's equation must differ by a harmonic function. It is possible to show that for every regular region R and every continuous scalar field g on R, there must be at least one field f on R such that $\nabla^2 f = g$. In fact, the following integral formula gives such a solution when g is sufficiently smooth. (See the remark below.)

(4.6)   *Theorem. Poisson Inversion. Let* g *be a sufficiently smooth scalar field on a regular region* R. *Then*

$$f(Q) = \frac{1}{4\pi} \iiint_R \frac{g(P)}{|\overrightarrow{PQ}|} dV_P$$

defines a field f *such that* $\nabla^2 f = g$, *where the notation* dV_P *indicates that Riemann sums are to be formed with sample points for the variable point* P.

The integral in (4.6) does not satisfy the condition for the existence theorem for multiple integrals (§14.4), since this integrand is not a continuous scalar field. Indeed the integrand becomes infinite when $g(Q) \neq 0$ and P approaches the fixed point Q. Such an integral, with such an integrand, is called an *improper integral*.

The particular improper integral in (4.6) does have a unique value for each point Q, provided that we use a restricted definition of Riemann sum; and special techniques exist for calculating this value. Not all improper integrals give unique values. Moreover, improper integrals which do have unique values may not satisfy some of the standard theorems for multiple integrals such as Fubini's theorem and Leibnitz's rule.

Improper integrals are considered in §28.1. With regard to Poisson's equation, we note that if there are additional conditions which a solution must satisfy (often called *boundary conditions*), these conditions may require a solution different from the solution given by (4.6). For this reason, (4.6) is of limited practical usefulness. It has, however, theoretical importance in that it indicates how at least one solution can be found.

Remark The integral in (4.6) has a natural physical meaning, which we describe in §28.2. The condition in (4.6) that g be "sufficiently smooth" will not be described in this text. This condition is met by virtually all scalar fields used in applied mathematics as models of physical phenomena. Poisson inversion is true for every continuous field g, in the sense that there must exist *some* field f such that $\nabla^2 f = g$. The smoothness condition is only needed to ensure that such a field f is given by the particular integral formula in (4.6).

§27.5 Vector Potentials

Let $\vec{F}$ be a continuous vector field on D in E^3. If $\vec{A}$ is a C^1 vector field on D such that $\vec{F} = \vec{\nabla} \times \vec{A}$, we say that $\vec{A}$ is a *vector potential for* $\vec{F}$.

We note the following immediate facts about vector potentials.

(5.1)   *If* $\vec{A}$ *is a vector potential for* $\vec{F}$ *on* D *and* f *is a* C^1 *scalar field on* D, *then* $\vec{A}' = A + \vec{\nabla}f$ *is also a vector potential for* $\vec{F}$ (*because* $\vec{\nabla} \times \vec{A}' = \vec{\nabla} \times (\vec{A} + \vec{\nabla}f) = \vec{\nabla} \times \vec{A} + \vec{\nabla} \times \vec{\nabla}f = \vec{\nabla} \times \vec{A} + \vec{0} = \vec{F}$.)

(5.2)   *If* $\vec{A}$ *and* $\vec{A}'$ *are both vector potentials for* $\vec{F}$ *on* D, *and* D *is simply connected, then* $\vec{A} - \vec{A}'$ *must be the gradient of some scalar field* (*because* $\vec{\nabla} \times \vec{A} = \vec{\nabla} \times \vec{A}' \Rightarrow \vec{\nabla} \times (\vec{A} - \vec{A}') = \vec{0} \Rightarrow$ there is a scalar potential for $\vec{A} - \vec{A}'$ since D is simply connected.)

(5.3)   *If* $\vec{F}$ *has a vector potential, then* $\vec{F}$ *must be divergenceless* (*because* $\vec{F} = \vec{\nabla} \times \vec{A} \Rightarrow \vec{\nabla} \cdot \vec{F} = \vec{\nabla} \cdot \vec{\nabla} \times \vec{A} = 0$).

(5.1), (5.2), and (5.3) are analogous to previously proved facts for *scalar* potentials: (a) If f is a scalar potential for $\vec{F}$ on D, and c is a constant, then f + c is also a scalar potential for $\vec{F}$. (b) If f and g are both scalar potentials for $\vec{F}$ on D, then f − g must be a constant on D. (c) If $\vec{F}$ has a scalar potential, then $\vec{F}$ is irrotational.
The following theorem gives a systematic method for finding a vector potential for a given divergenceless field $\vec{F}$ on a convex domain D. The method is analogous to the *line-integral method* (in §20.4) for finding a scalar potential for a given irrotational vector field $\vec{F}$ on a convex domain.

(5.4)   *Theorem. Curl inversion. Let* $\vec{F}$ *be a divergenceless vector field on a convex domain* D. (*Thus* $\vec{\nabla} \cdot \vec{F} = 0$.) *Choose a fixed point* P_0 *in* D. *For any point* P = (a,b,c) *in* D, *define* $\vec{A}(P)$ *by the vector-valued line integral*

$$\vec{A}(a,b,c) = \frac{1}{\left| \overrightarrow{P_0 P} \right|} \int_C (\vec{F} \times \vec{R})ds \,,$$

where C *is the line segment between* P_0 *and* P *and* $\vec{R}$ *is defined as* $\overrightarrow{P_0 Q}$ *for any point* Q *on* C. *Then, the new vector field* $\vec{A}$ *satisfies:* $\vec{\nabla} \times \vec{A} = \vec{F}$.

For the proof of this theorem, see the problems. Using Cartesian coordinates and a parameterization of C, we can evaluate the integral in (5.4) in the usual way. For simplicity, let $P_0 = (0,0,0)$, and let $\vec{R}(t) = ta\hat{i} + tb\hat{j} + tc\hat{k}$, $0 \le t \le 1$. Then $\frac{ds}{dt} = \sqrt{a^2 + b^2 + c^2} = |\overrightarrow{P_0 P}|$, and we have

$$(5.5) \qquad \vec{A}(a,b,c) = \int_0^1 \vec{F}(ta, tb, tc) \times (ta\hat{i} + tb\hat{j} + tc\hat{k}) dt,$$

after canceling $\frac{ds}{dt}$. $\vec{A}(a,b,c)$ can now be calculated by evaluating the cross-product and then evaluating the resulting scalar-component integrals in t.

Example. Let $\vec{F} = x\hat{i} - y\hat{j}$ on $D = E^3$. Let $P_0 = (0,0,0)$. Then we have

$$\vec{A}(a,b,c) = \int_0^1 (ta\hat{i} - tb\hat{j}) \times (ta\hat{i} + tb\hat{j} + tc\hat{k}) dt$$
$$= \int_0^1 (-bct^2\hat{i} - act^2\hat{j} + 2abt^2\hat{k}) dt$$
$$= -\frac{1}{3} bc\hat{i} - \frac{1}{3} ac\hat{j} + \frac{2}{3} ab\hat{k}.$$

Therefore, $\vec{A}(x,y,z) = -\frac{1}{3} yz\hat{i} - \frac{1}{3} xz\hat{j} + \frac{2}{3} xy\hat{k}$. Checking, we find $\vec{\nabla} \times \vec{A} = x\hat{i} - y\hat{j}$, as desired.

In the problems, we present and illustrate an even simpler form of line-integral calculation of vector potential. We shall also present an indefinite-integral method for curl inversion, analogous to the indefinite-integral method (in §20.5) for finding a scalar potential.

In the problems, we also outline a proof of the following theorem, which shows that a vector potential for a divergenceless field must always exist, not just on a convex domain but on any simply connected domain. It is the analogue, for existence of a vector potential, to the converse derivative test (in §26.7) for existence of a scalar potential.

(5.6) **Converse derivative test for existence of a vector potential.** *If $\vec{F}$ is defined on a simply connected domain D and $\vec{\nabla} \cdot \vec{F} = 0$, then $\vec{F}$ has a vector potential on D.*

The following easily proved fact improves the conclusion of (5.6). We use it in §27.8.

(5.7) *If* $\vec{F}$ *has a vector potential, then it has a vector potential which is itself divergenceless.*

Proof. Let $\vec{A}'$ be a given vector potential for $\vec{F}$. Let $g = \vec{\nabla} \cdot \vec{A}'$. By Poisson inversion, we find an f such that $\nabla^2 f = g$. Define $\vec{A} = \vec{A}' - \vec{\nabla}f$. Then $\vec{\nabla} \times \vec{A} = \vec{\nabla} \times (\vec{A}' - \vec{\nabla}f) = \vec{\nabla} \times \vec{A}' - \vec{\nabla} \times \vec{\nabla}f = \vec{\nabla} \times \vec{A}' = \vec{F}$, and $\vec{\nabla} \cdot \vec{A} = \vec{\nabla} \cdot (\vec{A}' - \vec{\nabla}f) = \vec{\nabla} \cdot \vec{A}' - \nabla^2 f = \vec{\nabla} \cdot \vec{A}' - \vec{\nabla} \cdot \vec{A}' = 0$.

§27.6 Algebraic Rules for $\vec{\nabla}$

In §27.4, we define $\nabla^2 \vec{F}$, for $\vec{F} = L\hat{i} + M\hat{j} + N\hat{k}$, as $\nabla^2 L\hat{i} + \nabla^2 M\hat{j} + \nabla^2 N\hat{k}$, and then note that the vector field $\nabla^2 \vec{F}$ is independent of our choice of the frame $\{\hat{i}, \hat{j}, \hat{k}\}$. Similarly, for a given vector field $\vec{G}$, we can define a new scalar operator which we write as $[\vec{G} \cdot \vec{\nabla}]$ and which can be applied to any scalar field or vector field. For a scalar field f, $[\vec{G} \cdot \vec{\nabla}]f$ is simply the scalar field given by $\vec{G} \cdot \vec{\nabla}f$. For a vector field $\vec{F} = L\hat{i} + M\hat{j} + N\hat{k}$, $[\vec{G} \cdot \vec{\nabla}]\vec{F}$ is a new vector field defined by $[\vec{G} \cdot \vec{\nabla}]\vec{F} = (\vec{G} \cdot \vec{\nabla}L)\hat{i} + (\vec{G} \cdot \vec{\nabla}M)\hat{j} + (\vec{G} \cdot \vec{\nabla}N)\hat{k}$. By a similar argument to the argument in §27.4 for $\nabla^2 \vec{F}$, we can show that $[\vec{G} \cdot \vec{\nabla}]\vec{F}$ is independent of our choice of a Cartesian coordinate frame $\{\hat{i}, \hat{j}, \hat{k}\}$ by noting (i) that for any given unit vector $\hat{u}$, $([\vec{G} \cdot \vec{\nabla}]\vec{F}) \cdot \hat{u} = \vec{G} \cdot \vec{\nabla}(\vec{F} \cdot \hat{u})$, and (ii) that the definitions used to define $\vec{G} \cdot \vec{\nabla}(\vec{F} \cdot \hat{u})$ are coordinate free.

The operator $[\vec{G} \cdot \vec{\nabla}]$ now provides us with a coordinate-free formula for the *directional derivative* of a vector field. (See §18.9.) Just as we had

$$\left.\frac{df}{ds}\right|_{\hat{u},P} = \hat{u} \cdot \vec{\nabla}f\Big|_{P}$$

for a scalar field f, we now get

$$\left.\frac{d\vec{F}}{ds}\right|_{\hat{u},P} = [\hat{u} \cdot \vec{\nabla}]\vec{F}$$

for a vector field $\vec{F}$.

The following algebraic identities hold for the gradient, divergence, curl, and Laplacian operators. Some are directly suggested by vector algebra together with the idea of $\vec{\nabla}$ as the "vector" $\hat{i}\frac{\partial}{\partial x} + \hat{j}\frac{\partial}{\partial y} + \hat{k}\frac{\partial}{\partial z}$. Others are less obvious. All may be verified by detailed calculation.

(6.1) $\vec{\nabla}(\vec{F}\cdot\vec{G}) = [\vec{G}\cdot\vec{\nabla}]\vec{F} + [\vec{F}\cdot\vec{\nabla}]\vec{G} + \vec{F}\times(\vec{\nabla}\times\vec{G}) + \vec{G}\times(\vec{\nabla}\times\vec{F})$.

(6.2) $\vec{\nabla}\times(f\vec{F}) = f(\vec{\nabla}\times\vec{F}) - \vec{F}\times\vec{\nabla}f$.

(6.3) $\vec{\nabla}\times(\vec{F}\times\vec{G}) = [\vec{G}\cdot\vec{\nabla}]\vec{F} - [\vec{F}\cdot\vec{\nabla}]\vec{G} + \vec{F}(\vec{\nabla}\cdot\vec{G}) - \vec{G}(\vec{\nabla}\cdot\vec{F})$.

(6.4) $\vec{\nabla}\cdot(f\vec{F}) = f\vec{\nabla}\cdot\vec{F} + \vec{F}\cdot\vec{\nabla}f$.

(6.5) $\vec{\nabla}\cdot(\vec{F}\times\vec{G}) = \vec{G}\cdot(\vec{\nabla}\times\vec{F}) - \vec{F}\cdot(\vec{\nabla}\times\vec{G})$.

(6.6) $\vec{\nabla}\times(\vec{\nabla}\times\vec{F}) = \vec{\nabla}(\vec{\nabla}\cdot\vec{F}) - \nabla^2\vec{F}$.

(6.7) $\vec{\nabla}\times\vec{\nabla}f = \vec{0}$.

(6.8) $\vec{\nabla}\cdot(\vec{\nabla}\times\vec{F}) = 0$.

(6.9) $\vec{\nabla}(fg) = f\vec{\nabla}g + g\vec{\nabla}f$.

(6.10) $\nabla^2(fg) = f\nabla^2 g + g\nabla^2 f + 2(\vec{\nabla}f\cdot\vec{\nabla}g)$.

Remark. Vector equations with $\vec{\nabla}$ can be complex, and there are no simple rules for automatically simplifying such equations. (We saw in Chapter 3 that the scalar and vector triple products give us such simplification rules for ordinary vector algebra.) Indeed, for vector algebra with $\vec{\nabla}$, we needed to define the new operator $[\vec{G}\cdot\vec{\nabla}]$ in order to find product rules for $\vec{\nabla}(\vec{F}\cdot\vec{G})$ and $\vec{\nabla}\times(\vec{F}\times\vec{G})$. A more general, systematic, and comprehensive system of definitions and notation, called *tensor analysis*, is used in advanced mathematics and theoretical physics. Using this system, we can express vector equations with $\vec{\nabla}$ and much else as well.

§27.7 Properties of Vector Fields

We illustrate, in simple and familiar cases, some of the combinations of vector-field properties that can occur. Each example is in $\mathbf{E}^3$. In each example, we give definitions in cylindrical and Cartesian coordinates; we consider the domain D and its connectivities; we find the divergence and curl; we state whether or not the field is conservative; and, as appropriate, we give scalar and vector potentials. We use C as an arbitrary constant, and h as an arbitrary C^2 scalar field on the domain D.

Example 1. $\vec{F} = r\hat{r} = x\hat{i} + y\hat{j}$.

Domain D: $\mathbf{E}^3$; simply connected.

$\vec{\nabla} \cdot \vec{F} = 2$; $\vec{\nabla} \times \vec{F} = \vec{0}$; irrotational but not divergenceless.

$\vec{F}$ is conservative. Scalar potential: $f = \frac{1}{2}r^2 + C = \frac{1}{2}(x^2 + y^2) + C$.

No vector potential, since $\vec{\nabla} \cdot \vec{F} \neq 0$.

Example 2. $\vec{F} = \frac{1}{r}\hat{r} = \frac{x\hat{i} + y\hat{j}}{x^2 + y^2}$.

Domain D: $r > 0$ in $\mathbf{E}^3$; not simply connected.

$\vec{\nabla} \cdot \vec{F} = 0$; $\vec{\nabla} \times \vec{F} = 0$; irrotational and divergenceless.

$\vec{F}$ is conservative. Scalar potential: $f = \ln r + C = \frac{1}{2}\ln(x^2 + y^2) + C$.

Vector potential: $\vec{A} = \theta\hat{z} + \vec{\nabla}h$.

Example 3. $\vec{F} = r\hat{\theta} = -y\hat{i} + x\hat{j}$.

Domain D: $\mathbf{E}^3$; simply connected.

$\vec{\nabla} \cdot \vec{F} = 0$; $\vec{\nabla} \times \vec{F} = 2\hat{z}$; divergenceless but not irrotational.

$\vec{F}$ is not conservative; no scalar potential.

Vector potential: $\vec{A} = rz\hat{r} + \vec{\nabla}h = xz\hat{i} + yz\hat{j} + \vec{\nabla}h$.

Example 4. $\vec{F} = \frac{1}{r}\hat{r} = \frac{x\hat{i} + y\hat{j}}{x^2 + y^2}$.

Domain D: $r > 0$ in $\mathbf{E}^3$; not simply connected.

$\vec{\nabla} \cdot \vec{F} = 0$; $\vec{\nabla} \times \vec{F} = 0$; divergenceless and irrotational.

$\vec{F}$ is not conservative, can have $\oint_C \neq 0$ on some closed curve C.

Vector potential: $\vec{A} = \frac{z}{r}\hat{r} + \vec{\nabla}h = \frac{z}{x^2+y^2}(x\hat{i} + y\hat{j}) + \vec{\nabla}h$.

Example 5. $\vec{F} = r\hat{r} + r\hat{\theta} = (x - y)\hat{i} + (x + y)\hat{j}$.

Domain D: E^3; simply connected.

$\vec{\nabla}\cdot\vec{F} = 2$; $\vec{\nabla}\times\vec{F} = 2\hat{z}$; not divergenceless and not irrotational.

$\vec{F}$ is not conservative; has neither scalar nor vector potential.

$\vec{F}$ can be expressed as the sum of an irrotational field $\vec{G} = x\hat{i} + y\hat{j}$ and a divergenceless field $\vec{H} = -y\hat{i} + x\hat{j}$.

$\vec{G}$ has scalar potential $\frac{1}{2}r^2 + C$ as in example 1; $\vec{H}$ has vector potential $rz\hat{r} + \vec{\nabla}h$ as in example 3.

§27.8 Helmholtz's Theorem

Let $\vec{F}$ be a given vector field on a simply connected domain D. Assume that we have mathematical expressions for the scalar field $\vec{\nabla}\cdot\vec{F}$ and for the vector field $\vec{\nabla}\times\vec{F}$. Can we use these expressions to obtain a mathematical expression for $\vec{F}$ itself? The answer to this question is easily seen to be negative, since other vector fields than $\vec{F}$ can have the same divergence and curl as $\vec{F}$. For example, if h is any harmonic scalar field on D, then $\vec{\nabla}\times(\vec{F} + \vec{\nabla}h) = \vec{\nabla}\times\vec{F}$ and $\vec{\nabla}\cdot(\vec{F} + \vec{\nabla}h) = \vec{\nabla}\cdot\vec{F}$.

On the other hand, the following easily proved theorem shows that if two vector fields on D have the same curl and divergence, then they must be sufficiently similar that they differ only by the gradient of a harmonic function.

(8.1) **Theorem.** *Let* $\vec{F}$ *and* $\vec{G}$ *be given* C^1 *vector fields on a simply connected domain* D. *Then* $\vec{\nabla}\times\vec{F} = \vec{\nabla}\times\vec{G}$ *and* $\vec{\nabla}\cdot\vec{F} = \vec{\nabla}\cdot\vec{G}$ *if and only if there exists a harmonic scalar field* h *such that* $\vec{F} - \vec{G} = \vec{\nabla}h$.

Proof. $\vec{\nabla}\times\vec{F} = \vec{\nabla}\times\vec{G} \Leftrightarrow \vec{\nabla}\times(\vec{F} - \vec{G}) = \vec{0} \Leftrightarrow \vec{F} - \vec{G} = \vec{\nabla}h$, where h is a scalar potential for $\vec{F} - \vec{G} \Leftrightarrow \nabla^2 h = \vec{\nabla}\cdot(\vec{F} - \vec{G}) = \vec{\nabla}\cdot\vec{F} - \vec{\nabla}\cdot\vec{G} = 0 \Leftrightarrow$ h is harmonic.

We can now reformulate our initial question: Let $\vec{F}$ be a given vector field on a simply connected domain D and assume that we have *mathematical expressions* for $\vec{\nabla} \cdot \vec{F}$ and $\vec{\nabla} \times \vec{F}$. Can we use these expressions to find a mathematical expression for *some* field $\vec{G}$ such that $\vec{F}$ and $\vec{G}$ differ only by the gradient of a harmonic field? The answer to this question is affirmative. Indeed, we can prove the following.

(8.2)   *Helmholtz's theorem. Given the existence of a C^1 vector field $\vec{F}$ on a simply connected domain D, and given expressions for $\vec{\nabla} \cdot \vec{F}$ and $\vec{\nabla} \times \vec{F}$, we can find expressions for vector fields $\vec{G}$ and $\vec{H}$ such that* (i) *$\vec{G}$ is irrotational and $\vec{H}$ is divergenceless;* (ii) *$\vec{\nabla} \times (\vec{G} + \vec{H}) = \vec{\nabla} \times \vec{H} = \vec{\nabla} \times \vec{F}$, and $\vec{\nabla} \cdot (\vec{G} + \vec{H}) = \vec{\nabla} \cdot \vec{G} = \vec{\nabla} \cdot \vec{F}$; and* (iii) *$\vec{F} - (\vec{G} + \vec{H}) = \vec{\nabla} h$ for some harmonic function h.*

Proof. We are given $\vec{\nabla} \times \vec{F}$ and $\vec{\nabla} \cdot \vec{F}$. By Poisson inversion, we find a scalar field g such that $\nabla^2 g = \vec{\nabla} \cdot \vec{F}$. We take $\vec{G} = \vec{\nabla} g$. Hence $\vec{\nabla} \cdot \vec{G} = \nabla^2 g = \vec{\nabla} \cdot \vec{F}$ and $\vec{\nabla} \times \vec{G} = \vec{\nabla} \times \vec{\nabla} g = \vec{0}$.

We then find a vector potential $\vec{A}$ for the given field $\vec{\nabla} \times \vec{F}$. Hence $\vec{\nabla} \times \vec{A} = \vec{\nabla} \times \vec{F}$. Moreover, by (5.6) we can find such an $\vec{A}$ so that $\vec{A}$ is divergenceless. We take $\vec{H} = \vec{A}$. Then $\vec{\nabla} \times \vec{A} = \vec{\nabla} \times \vec{F}$ and $\vec{\nabla} \cdot \vec{H} = 0$.

It follows that $\vec{\nabla} \cdot (\vec{G} + \vec{H}) = \vec{\nabla} \cdot \vec{G} + \vec{\nabla} \cdot \vec{H} = \vec{\nabla} \cdot \vec{G} = \vec{\nabla} \cdot \vec{F}$, and $\vec{\nabla} \times (\vec{G} + \vec{H}) = \vec{\nabla} \times \vec{G} + \vec{\nabla} \times \vec{H} = \vec{\nabla} \times H = \vec{\nabla} \times \vec{F}$. The existence of h follows by (8.1) (although there is no way to find an expression for h without having more information about $\vec{F}$).

§27.9   Other Theorems and Definitions

Let an elementary region R in E^3 with outward-directed boundary surface S, the divergence theorem (also known as *Gauss's theorem*) relates the triple integral of the divergence of a vector field $\vec{F}$ on R to the surface integral of $\vec{F}$ on S by the Gaussian formula

(9.1) $$\iiint_R \vec{\nabla} \cdot \vec{F} dV = \oiint_S \vec{F} \cdot d\vec{\sigma}.$$

Analogous *Gaussian formulas,* using vector valued integrals, relate the triple integrals on R of the gradient and the curl to certain corresponding surface integrals on S.

(9.2)   ***Theorem.*** (a) *Let* f *be a* C^1 *scalar field in* E^3 *whose domain contains the elementary region* R. *Let* S *be the outward-directed boundary surface of* R. *Then*

$$\iiint_R \vec{\nabla} f dV = \oiint_S f d\vec{\sigma}.$$

(*Here, both integrals are vector-valued.*)

     (b) *Let* $\vec{F}$ *be a* C^1 *vector field in* E^3 *whose domain contains the elementary region* R. *Let* S *be the outward-directed boundary surface of* R. *Then*

$$\iiint \vec{\nabla} \times \vec{F} dV = -\oiint_S \vec{F} \times d\vec{\sigma}.$$

(*Here again, both integrals are vector-valued.*)

     The formulas in (9.2) can be deduced directly from the divergence theorem. We illustrate with (a).
     At each point P on a smooth piece of S, let $\hat{n}(P)$ be the unit outward normal vector to S at P. Hence we can write

$$\oiint_S f d\vec{\sigma} = \oiint_S f \hat{n} d\sigma$$

To show the desired equality between vector-valued integrals, it will be enough to show that for an arbitrary constant unit vector $\hat{u}$, we must have

$$\hat{u} \cdot \oiint_S f \hat{n} d\sigma = \hat{u} \cdot \iiint_R \vec{\nabla} f dV.$$

We have

$$\hat{u} \cdot \oiint_S f \hat{n} d\sigma = \oiint_S f \hat{n} \cdot \hat{u} d\sigma = \oiint_S f \hat{u} \cdot d\vec{\sigma}$$
$$= \iiint_R \vec{\nabla} \cdot (f \hat{u}) dV \quad \text{(by the divergence theorem)}$$

$$= \iiint_R \hat{u} \cdot \vec{\nabla} f \, dV \qquad \text{(since } \vec{\nabla} \cdot (f\hat{u}) = f(\vec{\nabla} \cdot \hat{u}) + \hat{u} \cdot \vec{\nabla} f = \hat{u} \cdot \vec{\nabla} f \text{ by}$$

$$(6.4) \text{ in §25.6)}$$

$$= \hat{u} \cdot \iiint_R \vec{\nabla} f \, dV \qquad \text{as desired.}$$

The proof for (b) is analogous and is given as a problem.
The following corollary to (9.2) is immediate.

(9.3)   **Corollary.** *For* f *and* $\vec{F}$ *as in* (9.2),

(a) $\quad \vec{\nabla} f\big|_P = \lim_{a \to 0} \dfrac{3}{4\pi a^3} \oiint_{S(P,a)} f \, d\vec{\sigma} \; ;$

(b) $\quad \vec{\nabla} \times \vec{F}\big|_P = \lim_{a \to 0} \dfrac{3}{4\pi a^3} \oiint_{S(P,a)} -(\vec{F} \times d\vec{\sigma}) \, .$

These *Gaussian definitions* of *gradient* and *curl* give alternative, coordinate-free ways to visualize and to physically interpret the *gradient* of a scalar field and the *curl* of a vector field.

For a directed, piecewise C^2-smooth, elementary surface S with a coherently directed boundary curve C, Stokes's theorem relates the surface integral of the curl of a vector field $\vec{F}$ to the line integral of $\vec{F}$ around C by the *Stokesian formula*

(9.4) $$\iint_S \vec{\nabla} \times \vec{F} \cdot d\vec{\sigma} = \oint_C \vec{F} \cdot d\vec{R} \, .$$

An analogous *Stokesian formula* for a scalar field f in $\mathbf{E}^3$ is given by the following theorem.

(9.5)   **Theorem.** *Let* f *be a* C^1 *scalar field in* $\mathbf{E}^3$ *whose domain includes the piecewise* C^2*-smooth directed elementary surface* S, *where the boundary of* S *consists of a simple, closed, coherently directed curve* C. *Then*

$$\iint_S \vec{\nabla} f \times d\vec{\sigma} = -\oint_C f \, d\vec{R} \, .$$

(Here, both integrals are vector-valued.)

The proof for (9.5) uses Stokes's theorem and is otherwise analogous to the proof above for (9.2a). It is given as a problem.

The definition of finite (scalar-valued) measure in §23.2 can be imme-diately extended to a definition of *finite vector-valued measure* by taking the value of a measure to be a vector. As a "measure of content," such a measure gives both a magnitude and a direction as its value. When a regular region R is subdivided into two subregions, the measure of R must be the *vector sum* of the vector values for the two subregions. In each of (9.2a) and (9.2b), the surface integral defines a vector-valued measure in $\mathbf{E}^3$, and the triple integrals show that these are integral measures each with its own appropriate vector-valued measure-density.

This concept of measure allows us to extend the formulas in (9.2a) and (9.2b) from elementary regions with single boundary surfaces to regular regions with multiple boundary surfaces.

In (9.5), a vector-valued surface measure can be defined from the line integral in an analogous way. (9.5) then shows that this surface measure is an integral measure. This concept of surface measure allows us to extend the formula in (9.5) from elementary surfaces with single boundary curves to finite surfaces with multiple boundary curves.

Each of these vector-valued measures has its own natural significance. (9.2a) gives a "net-variation" measure in $\mathbf{E}^3$. (9.2b) gives a "net vorticity" mea-sure in $\mathbf{E}^3$. (9.5) gives a "net-vorticity" measure on surfaces.

§27.10   Problems

§27.1

1. Explain why (1.2) follows from (1.1).

2. Use parametric evaluation to prove (1.3).

3. Use the Cartesian formula for gradient to prove (1.4).

4. Use the suggestion in the text to prove (1.5).

5. (a) Let $u = u(t)$ be a function of t. Prove that
$$\frac{d}{dt}\int_0^u f(x,t)\,dx = f(u,t)\frac{du}{dt} + \int_0^u f_t(x,t)\,dx.$$

   (*Hint.* Use the chain rule and the fundamental theorem of integral calculus.)

   (b) As a check, apply the result of (a) to
$$\frac{d}{dt}\int_0^{2t} xt\,dx.$$

§27.2

1. Find

   (a) $\vec{\nabla}(r + \theta)$ (c) $\vec{\nabla}(\rho + \varphi)$

   (b) $\vec{\nabla}(\rho + \theta)$

2. Find

   (a) $\vec{\nabla} \cdot (r\hat{r} + r\theta\hat{\theta})$ (b) $\vec{\nabla} \cdot (\rho\hat{\rho} + \rho\varphi\hat{\varphi} + \rho\theta\hat{\theta})$

3. Find

   (a) $\vec{\nabla} \times (r\hat{r} + r\theta\hat{\theta})$

   (b) $\vec{\nabla} \times (\rho\hat{\rho} + \rho\varphi\hat{\varphi} + \rho\theta\hat{\theta})$

4. The unit vectors $\hat{r}, \hat{\rho}, \hat{\varphi},$ and $\hat{\theta}$ may themselves be considered as vector fields. $\hat{\rho}$ and $\hat{\varphi}$ are defined at all points except the origin. $\hat{r}$ and $\hat{\theta}$ are defined at all points except points on the z axis.

   (a) Use (2.2) and (2.3) to find $\vec{\nabla} \cdot \hat{r}$ and $\vec{\nabla} \times \hat{r}$.

(b) Use (2.5) and (2.6) to find $\vec{\nabla} \cdot \hat{\rho}$, $\vec{\nabla} \times \hat{\rho}$, $\vec{\nabla} \cdot \hat{\varphi}$, and $\vec{\nabla} \times \hat{\varphi}$.

(c) Use (2.2) and (2.3) to find $\vec{\nabla} \cdot \hat{\theta}$ and $\vec{\nabla} \times \hat{\theta}$.

(d) Use (2.5) and (2.6) to find $\vec{\nabla} \cdot \hat{\theta}$ and $\vec{\nabla} \times \hat{\theta}$, and show that these are the same scalar field and vector field as were obtained in (c).

5. Prove (2.4).

6. Prove (2.5).

7. Prove (2.3).

8. Verify examples 2, 5, and 6.

9. Verify examples 8, 10, and 11.

§27.4

1. Prove that a C^2 scalar field in $\mathbf{E}^3$ which possesses the mean value property must be harmonic.

2. Let f and g be harmonic functions in C^2, and let R be a regular region which is contained in the domain of f and also in the domain of g. Assume that for each point Q on the boundary of R, f(Q) = g(Q). Prove that for each point P in R, f(P) = g(P). (*Hint.* Use (4.5) and the max-min existence theorem.)

§27.5

1. Prove that (5.5) gives the same vector field $\vec{A}$ as (5.4).

2. Prove (5.4). (*Hint.* Take the curl of (5.5) with respect to the variables a,b,c. Apply Leibnitz's rule (with a,b,c as parameters), simplify, and use the assumed fact that $\vec{\nabla} \cdot \vec{F} = 0$.)

3. Use (5.5) to find a vector potential for
    (a) $\vec{F} = -z\hat{j} + y\hat{k}$ (c) $\vec{F} = x^2\hat{i} - xy\hat{j} - xz\hat{k}$
    (b) $\vec{F} = \hat{i} + xz\hat{j} + x^2\hat{k}$

4. A vector field $\vec{F}(x,y,z)$ is said to be *homogeneous of degree* k if $\vec{F}(tx,ty,tz) = t^k\vec{F}(x,y,z)$. (In problem 3a, $\vec{F}$ is homogeneous of degree 1 and in problem

3c of degree 2.) In many models of physical fields, the field is homogeneous. Prove that in this case

(10.1) $\vec{A}(P) = \frac{1}{k+2}(\vec{F}(P) \times \vec{R}(P))$

gives the same vector potential as (5.5), but without a need for integration.

5. Check that (10.1) gives the same potentials as were found in problems 3a and 3c.

6. Prove that for a divergenceless field $\vec{F} = L\hat{i} + M\hat{j} + N\hat{k}$ the following *indefinite integral formulas* give a vector potential $\vec{A} = A_1\hat{i} + A_2\hat{j} + A_3\hat{k}$.

(10.2) $A_1 = 0$

   $A_2 = \int N\,dx$

   $A_3 = -\int M\,dx + \int \left[L + \frac{\partial}{\partial y} \int M\,dx + \frac{\partial}{\partial z} \int N\,dx \right] dy$

7. Use (10.2) to find a vector potential $\vec{A}$ for
    (a) $\vec{F} = -z\hat{j} + y\hat{k}$ (c) $\vec{F} = x\hat{i} - y\hat{j}$
    (b) $\vec{F} = \hat{i} + xz\hat{j} + x^2\hat{k}$ (d) $\vec{F} = x^2\hat{i} - xy\hat{j} - xz\hat{k}$

8. The vector potentials found in problem 7 are different from (and simpler than) those calculated in problems 3 and 5 (or, for (d), in the text). The difference between a potential found by (5.4) and a potential found by (10.2) must be, by (5.2), a conservative field. Check that this is true.

9. In connection with the choice of $A_1 = 0$ in (10.2), prove that for any function f(x,y,z) in C^1, there is a gradient field $\vec{\nabla}g$ (for some g) such that $\vec{\nabla}g \cdot \hat{i} = f$. (It follows from (5.2) that for any given divergenceless $\vec{F}$, there must be a vector potential $\vec{A}$ with $\vec{A} \cdot \hat{i} = 0$.)

10. Prove (5.6). (*Hint.* The proof is lengthy but not unduly complex. The methods of (5.4) and (10.2) require that D be convex. Given $\vec{F}$, we can use (5.4) to define a vector potential on the interior of

any given sphere in our simply connected D. Given two overlapping spheres S_1 and S_2 with vector potentials $\vec{A}_1$ and $\vec{A}_2$, their vector potentials must differ on their overlap by a gradient field $\vec{\nabla}g$. The scalar potential g can be extended in a natural way to a potential on the whole interior of S_2. We can then find a common vector potential $\vec{A}$ for both S_1 and S_2 by letting $\vec{A} = \vec{A}_1$ on S_1 and $\vec{A} = \vec{A}_2 + \vec{\nabla}g$ on the remaining portion of S_2. Iteration of this process enables us to define a single potential $\vec{A}$ over all of D.)

§27.6

1. (a) Explain in detail why the vector computed as $[\vec{G} \cdot \vec{\nabla}]\vec{F}$ is independent of the Cartesian coordinate system used.

   (b) The *invariance* (independence of choice of Cartesian coordinates) in (a) is also immediate from the identity

$$([\vec{G} \cdot \vec{\nabla}]\vec{F})\big|_P = \vec{\nabla} \cdot \vec{F}\big|_P \vec{G}(P) - \vec{\nabla} \times (\vec{G}(P) \times \vec{F})\big|_P$$

because the operations on the right-hand side all have invariant geometric definitions. Deduce this identity from (6.3).

   (c) Prove (6.3).

2. Find the directional derivative of the vector field $\vec{R} = x\hat{i} + y\hat{j} + z\hat{k}$ in the direction $\hat{u} = \hat{j}$ at the point $P = (1,1,1)$.

3. For any scalar fields f and g in C^2, show that $\vec{\nabla}f \times \vec{\nabla}g$ is divergenceless.

§27.8

1. Assume, for a given vector field $\vec{F}$ in E^3, that $\vec{\nabla} \cdot \vec{F} = y + z$ and $\vec{\nabla} \times \vec{F} = -z\hat{i} + x\hat{j} - x\hat{k}$. Find $\vec{G}$ and $\vec{H}$ as described in (8.2). (*Hint.* It is not necessary to use Poisson inversion.)

§27.9

1. Prove part (b) of (9.2).

2. Prove (9.5).

Chapter 28

Physical Applications

§28.1 Improper Integrals

Given a scalar field f (or vector field $\vec{F}$) as integrand over a given compact region R, we can form Riemann sums for f (or $\vec{F}$) on R. For these sums to converge to a unique limit, it is necessary that $|f|$ (or $|\vec{F}|$) be bounded. If, however, we make a minor modification in the definition of Riemann sum (to be described below), we can find integrands over a compact R for which f (or $|\vec{F}|$) is unbounded but for which the Riemann sums may still converge to a unique limit. An integral over a compact R with an unbounded integrand is called an *improper integral*. If its modified Riemann sums converge to a unique limit, we say that the integral is *convergent*. If not, we say that it is *divergent*. (An integral, over a compact R, whose integrand is continuous on R and hence bounded, will be called a *proper integral*.)

We shall be concerned with the case of integrals for which the integrand f (or $\vec{F}$) is *continuous except at a single point* Q_0 *in* R, where $\lim\limits_{P \to Q_0} |f(P)| = \infty$ (or $\lim\limits_{P \to Q_0} |\vec{F}(P)| = \infty$). Q_0 is then called the *singular point* of the integrand. For such an integrand, ordinary Riemann sums will not approach a unique limit, since for any decomposition $(\Delta R_1, \Delta R_2, ..., \Delta R_n)$ of R into elementary subregions a Riemann sum on that decomposition can be made arbitrarily large by taking, in the subregion ΔR_i containing Q_0, the sample point P_i^* sufficiently close to Q_0.

We now modify the definition of Riemann sum as follows. *If a sub-region* ΔR_i *happens to contain* Q_0, *we require that the sample point* P_i^* *be chosen to minimize* $|f|$ (*or* $|\vec{F}|$) *on* ΔR_i; *if a subregion* ΔR_j *does not contain* Q_0, *we choose* P_j^* *arbitrarily in* ΔR_j (*as usual*). With this new definition of Riemann sum, it is possible for an improper integral to be convergent, as we see in an example.

Example 1. Consider $\int_0^1 \frac{1}{x^{1/2}} dx$. Here $x = 0$ is the singular point. It is easy to show that the modified Riemann sums for this improper integral have a unique limit, and that this limit is the same as $\lim_{\varepsilon \to 0} \int_\varepsilon^1 \frac{1}{x^{1/2}} dx = \lim_{\varepsilon \to 0} \left(2x^{1/2} \Big]_\varepsilon^1 \right) =$ $\lim_{\varepsilon \to 0} (2 - 2\varepsilon^{1/2}) = 2$. Thus our integral is convergent with value $= 2$.

Example 2. Similarly, consider $\int_{-1}^1 \frac{1}{|x|^{1/2}} dx$. $x = 0$ is again the singular point. By an argument similar to example 1, we find that this integral is convergent and has the value 4.

Example 3. On the other hand, consider $\int_0^1 \frac{1}{x} dx$. Here we have $\lim_{\varepsilon \to 0} \int_\varepsilon^1 \frac{1}{x} dx = \lim_{\varepsilon \to 0} \left(\ln x \Big]_\varepsilon^1 \right) = \lim_{\varepsilon \to 0} (-\ln \varepsilon) = \infty$. In this case, the modified Riemann sums become arbitrarily large (as is easily shown) and the integral is divergent.

More generally, we have the following definition and theorem.

Definition. Let Q_0 be the single singular point for an integrand f (or $\vec{F}$) on a compact region R. Let ΔR equal the intersection of R with U_a, where U_a is a neighborhood of Q_0 of radius a. Let ΔV be the length, area, or volume of ΔR (depending on whether R lies on a line, in a plane, or in E^3), and let P^* be a point in ΔR chosen to minimize $|f|$ (or $|\vec{F}|$) on ΔR. We say that Q_0 is a *proper singular point* if $\lim_{a \to 0} f(P^*)\Delta V = 0$ (or $\lim_{a \to 0} |\vec{F}(P^*)| \Delta V = 0$).

In the problems, we prove the following:

(1.1)   *Theorem. An improper integral whose integrand is continuous except at a single singular point is convergent $\Leftrightarrow$ its singular point is proper.*

Finally, if f is continuous on a compact region R except at a single, proper, singular point Q_0, then $\int_{R'} f \, dV$ must have a unique value for every regular subregion R' of R, and $\mu(R') = \int_{R'} f \, dV$ must give an absolutely continuous measure on R. (It is this last fact which gives convergent improper integrals their usefulness. We can use a convergent integral to measure "content" (of some kind) on any subregion of a given region R.)

On the other hand, as we shall see in §28.3, there are other basic mathematical facts about proper integrals which may fail to hold for convergent improper integrals. In particular, Fubini's theorem and Leibnitz's rule are not always true.

Remark. The phrase "improper integral" is commonly used also to include integrals over regions which are closed but unbounded (and are therefore not compact), for example $\int_1^\infty \frac{1}{x^2} \, dx$. For such integrals, one must allow Riemann sums in the form of infinite series. We do not consider such integrals in the present text.

§28.2 Gravitational Fields in Empty Space

Many physical processes and situations can be represented mathematically by scalar fields or vector fields. Among these are gravitational forces, fluid-flows, electromagnetic forces, heat flow, and quantum mechanical phenomena. Our goal in this chapter is to give examples of how vector calculus can be used to express natural laws for these phenomena in terms of *differential equations*. Differential equation formulations of natural laws have two advantages: (1) they are *local* statements, that is to say, they concern the behavior of a given field in the immediate vicinity of any chosen point, and (2) they are usually *universal*, that is to say, they apply at all points in all specific instances of the field or process in question. We shall also be interested in a third desirable property of a differential equation formulation of physical laws. (3) We say that a formulation by differential equations is *complete* if every mathematical field satisfying the differential equations represents a physically possible instance of the physical fields being described. In what follows, Helmholtz's theorem (§27.8) will be helpful in establishing the universality and completeness of the differential equation formulations which we get for gravitational and electromagnetic fields.

In this section and the following, we consider physical laws for gravitational fields as identified and described by Newton. In the mid-17th century, Newton found a general method for describing and predicting the motions of the sun, the planets, and the planetary satellites. He showed that, given the positions, velocities, and masses of these bodies at a certain moment, their future motions can be logically deduced from four simple and universal physical laws. These are the laws which we now know as Newton's *law of gravity* and Newton's three *laws of motion*. To assist in carrying out these deductions, Newton developed the basic concepts of differential and integral calculus.

In this chapter, we consider Newton's theory of gravitational attraction. We shall assume, without further discussion, the concept of *mass*, the concept of *force*, and the concept of *force due to a particular cause*. (These concepts are less simple than they may at first appear and have been further clarified in the 19th and 20th centuries.) We shall also assume that all forces and motions are observed and measured in an inertial frame. An *inertial frame* is a Cartesian coordinate system in physical space with respect to which observed motions of objects satisfy Newton's laws of motion. (Newton's discovery of his laws can be equivalently described as the discovery that inertial frames exist.) In such a frame, for example, a particle unaffected by forces must move in a straight line. An inertial frame is sometimes (misleadingly perhaps) described as a set of axes which is not rotating with respect to the average position of the fixed stars and whose origin has constant velocity with respect to the average position of the fixed stars. (Such a description is based on ideas from general relativity.)

Newton's theory begins with *point masses*. Given a point mass of mass M and a point mass of mass m, with distance d between M and m, the theory asserts that

(i)   The gravitational force exerted by M on m is an attractive force acting in the direction of a straight line from m to M.

(ii)   The magnitude of this force is directly proportional to M, directly proportional to m, and inversely proportional to d^2.

(iii)   The magnitude and direction of this force do not depend on current or previous states of motion of M or m.

(iv)   The magnitude of this force does not depend on the direction in space of the line from m to M.

(v)   Hence, the magnitude of this force depends only on the quantities M, m, and d.

If we use spherical coordinates with origin at M and let $\bar{F}$ be the force on m due to M, we get the formula

(2.1)
$$\vec{F} = -\frac{\gamma Mm}{\rho^2}\hat{\rho} \; ,$$

where γ is a universal constant (the *gravitational constant*) and ρ is the ρ coordinate of m. This formula summarizes assumptions (i)–(v).

A more general and useful way to describe the gravitational force which M can exert on another point mass is to introduce the *gravitational field* $\vec{G}$ *due to* M. $\vec{G}$ is a vector field, defined at all points other than the point occupied by M, such that at each point P, $\vec{G}(P)$ is the gravitational force which M would exert on a *unit point mass* at P. Thus we have the formula in spherical coordinates

(2.2)
$$\vec{G}(\rho,\varphi,\theta) = -\frac{\gamma M}{\rho^2}\hat{\rho} \; .$$

Newton's theory now makes two further assumptions:

(vi)   Given several point masses $M_1, M_2, M_3,\ldots$, with gravitational fields $\vec{G}_1, \vec{G}_2, \vec{G}_3,\ldots$, then the combined (simultaneous) gravitational field due to $M_1, M_2, M_3,\ldots$ is given by the vector sum

$$\vec{G} = \vec{G}_1 + \vec{G}_2 + \vec{G}_3 + \ldots \; .$$

(This is called the *superposition* assumption.)

(vii)   Finally, we can extend the mathematical theory from point masses to *continuous distributions* of matter as follows. Let D be a region of space occupied by a continuous distribution of matter whose density (in mass per unit volume) at each point P in D is given by $\delta(P)$, where δ is a continuous scalar field. We assume that the matter occupying a small piece ΔR of D acts more and more like a point mass as $d(\Delta R) \to 0$ (where $d(\Delta R)$ is the diameter of ΔR), and that the combined effect of a multitude of such pieces will be given by superposition. (This is called the *continuity assumption.*)

The continuity assumption can be made more precise as follows. Let Q be a fixed point in empty space and let P be a point in a small piece of D with volume ΔV. Then the mass in this piece is approximately $\delta(P)\Delta V$. It follows that the field at Q due to the matter in ΔV will be given by

$$\Delta\vec{G}(Q) \approx \frac{\gamma\delta(P)\Delta V}{|\overrightarrow{QP}|^2}\widehat{QP} \; ,$$

where $\overset{\wedge}{QP}$ is the unit vector of $\overset{\rightarrow}{QP}$. Hence the combined gravitational field at Q due to all parts of the mass distribution in D can be expressed as

$$\vec{G}(Q) \approx \sum \Delta\vec{G}(Q) \approx \sum_i \frac{\gamma\delta(P_i)}{|\overset{\rightarrow}{QP_i}|^2} \overset{\wedge}{QP_i}\, \Delta V_i \,,$$

or

(2.3)
$$\vec{G} = \gamma \iiint_D \left(\frac{\delta}{\rho^2}\right) \hat{\rho} dV \,,$$

where we have used a vector triple integral of the form $\iiint \vec{F} dV$ with a spherical coordinate system whose origin is at Q. In this integral, P varies, but Q remains fixed. By the existence theorem for integrals this integral exists if δ is continuous and D is compact.

Newton's theory can hence be summarized as follows:

(a)   The gravitational field of a point mass M at the origin is given by the formula

$$\vec{G} = -\frac{\gamma M}{\rho^2} \hat{\rho} \,.$$

(b)   The superposition assumption.

(c)   The continuity assumption.

Let R_1 and R_2 be the compact regions of two mass-density distributions, and let δ_1 and δ_2 be their respective density fields. Then, by (2.3), at each point Q in R_2 the gravitational field due to R_1 is

(2.4)
$$\vec{G}_1(Q) = \iiint_{R_1} \frac{\gamma\delta_1}{\rho^2} \hat{\rho} dV_1 \,,$$

where ρ is a spherical coordinate with origin Q. Hence the total force on R_2 due to R_1 must be

(2.5)
$$\vec{F}_{R_2 R_1} = \iiint_{R_2} \delta_2 \vec{G}_1 dV_2 \,.$$

It is evident, by Fubini's theorem and in accord with Newton's laws of motion, that the force $\vec{F}_{R_2 R_1} = -\vec{F}_{R_1 R_2}$, where $\vec{F}_{R_1 R_2}$ is the force on R_1 due to R_2.

The integral in (2.5) is really a 6-tuple integral, since it uses (2.4). In attempting to calculate the force exerted on the earth by the sun, Newton faced the problem of evaluating this 6-tuple integral. He assumed that each of the two bodies was a solid sphere whose density field was *radially symmetric* (that is to say, $\delta(P) = h(\rho(P))$, where h is some one-variable function and $\rho(P)$ is the distance of P from the body's center). He then used integral calculus (which he is said to have invented for this purpose) to calculate (2.4). Under the assumption of radial symmetry, he discovered that:

(2.6) *The gravitational field of the sun, in space outside the sun, is exactly the same as it would be if the extended mass distribution of the sun were replaced, at the center point of the sun, by a point-mass of the same total mass as the sun.*

It then also followed, by a similar calculation of (2.5), that the total force on the earth caused by this field of the sun must be exactly the same as if the earth's extended mass distribution were replaced, at its center, by a point-mass of the same total mass as the earth. This fortunate discovery greatly simplified Newton's subsequent calculations of planetary and satellite orbits. In the problems, using the tools of integral calculus, we shall see that Newton's calculation is (for us) lengthy but not difficult. In §28.3, we shall use the divergence theorem to obtain a simpler proof of (2.6). This simpler proof avoids Newton's detailed calculation.

We now obtain a more general and comprehensive result about the mathematical properties of a gravitational field in empty space.

(2.7) **Theorem.** *Let $\vec{G}_e$ be the gravitational field in empty space due to a continuous density field δ on a regular region R. (Thus $\vec{G}_e$ is the gravitational field outside R caused by the matter inside R.) Then*

(i) $\vec{G}_e$ *is divergenceless;*

(ii) $\vec{G}_e$ *is irrotational;*

(iii)   *a scalar potential g_e for $\vec{G}_e$ is given by*

$$g_e(Q) = \iiint_R \frac{\gamma\delta}{\rho} dV ,$$

where the spherical coordinate ρ has Q as its origin (g_e is called the gravitational potential *or "Newtonian potential" due to the density δ in R), hence $\vec{G}_e$ is conservative;*

(iv)   *the potential g_e is a harmonic function.*

(Note that for Q outside R, the integrals for $\vec{G}_e(Q)$ and $g_e(Q)$ are proper.)

Proof. (iv) is immediate from (i) and (iii), since $\vec{G}_e = \vec{\nabla} g_e$ by (iii), and $\nabla^2 g_e = \vec{\nabla} \cdot \vec{G}_e = 0$ by (i).

Each of (i), (ii), and (iii) is obtained by an application of Leibnitz's rule. For (i) and (ii), we first express the integral (2.3) for $\vec{G}_e(Q)$ in Cartesian coordinates (with an arbitrary, fixed origin different from Q). Let (a_1, a_2, a_3) be the coordinates of Q, we have

$$\vec{G}_e(Q) = \vec{G}_e(a_1, a_2, a_3) = \iiint_R \gamma \frac{\delta(x,y,z)[(x-a_1)\hat{i}+(y-a_2)\hat{j}+(z-a_3)\hat{k}]}{((x-a_1)^2+(y-a_2)^2+(z-a_3)^2)^{3/2}} dxdydz ,$$

where a_1, a_2, and a_3 serve as parameters in the integral. For the case of (i) we now apply Leibnitz's rule to get

$$\vec{\nabla} \cdot \vec{G}_e \big|_Q = \vec{\nabla}_a \cdot \iiint_R \cdots dxdydz ,$$

where $\vec{\nabla}_a$ indicates differentiation with respect to the parameters a_1, a_2, and a_3. Moving $\vec{\nabla}_a$ through the integral signs, we have

$$(2.8) \quad \vec{\nabla} \cdot \vec{G}_e \big|_Q = \iiint_R \gamma \delta \vec{\nabla}_a \cdot \left(\frac{(x-a_1)\hat{i}+(y-a_2)\hat{j}+(z-a_3)\hat{k}}{((x-a_1)^2+(y-a_2)^2+(z-a_3)^2)^{3/2}} \right) dxdydz .$$

Here, in the action of $\vec{\nabla}_a$, x, y, and z are treated as constants. To see the result of this action, we temporarily introduce spherical coordinates with origin at (x,y,z). We get $\vec{\nabla}_a \cdot \left(\frac{(x-a_1)\hat{i}+\cdots}{((x-a_1)^2+\cdots)^{3/2}} \right) = \vec{\nabla} \cdot \left(\frac{-\rho\hat{\rho}}{\rho^3} \right) = -\vec{\nabla} \cdot \left(\frac{1}{\rho^2}\hat{\rho} \right) = 0$. Thus the integrand in (2.7) must be 0, and we get $\vec{\nabla} \cdot \vec{G}_e \big|_Q = 0$. Thus $\vec{G}_e$ is a divergenceless field.

For (ii), by an exactly similar application of Leibnitz's rule using $\vec{\nabla} \times \left(\frac{1}{\rho^2}\hat{\rho} \right) = \vec{0}$, we find that $\vec{\nabla} \times \vec{G}_e \big|_Q = \vec{0}$. Hence $\vec{G}_e$ is an irrotational field.

We prove (iii) in the same way that we proved (i) and (ii). We express the integral for g_e in *Cartesian* coordinates with (a_1, a_2, a_3) as the coordinates of Q. We take the gradient, apply Leibnitz's rule, go to temporary spherical coordinates with origin at (x,y,z), then calculate the gradient, and finally return to spherical coordinates with origin at Q. We get

$$\vec{\nabla}\cdot\vec{G}_e\big|_Q = \vec{\nabla}_a \iiint_R \frac{\gamma\delta}{((x-a_1)^2+(y-a_2)^2+(z-a_3)^2)^{1/2}}dxdydz$$

$$= \iiint_R \gamma\delta\vec{\nabla}(\tfrac{1}{\rho})dV = \iiint_R \gamma\delta\left(-\frac{1}{\rho^2}\hat{\rho}\right)dV = -\iiint_R \frac{\gamma\delta}{|\overrightarrow{PQ}|^2}\overset{\wedge}{PQ}dV$$

$$= \iiint_R \frac{\gamma\delta}{|\overrightarrow{QP}|^2}\overset{\wedge}{QP}dV = \iiint_R \frac{\gamma\delta}{\rho^2}\hat{\rho}dV \quad \text{(with origin at Q)}$$

$$= \vec{G}_e(Q), \quad \text{by (2.3).}$$

This completes the proof of (2.7).

§28.3 Gravitational Fields within Matter

Let R be a regular region of a mass-density field δ . What can we say about the gravitational field at a point Q, caused by matter in R, when Q itself is a point *within* R? We call this field, which has R as its domain, $\vec{G}_w$. By the same superposition argument as for (2.3), we can obtain the same integral formula

(3.1)
$$\vec{G}_w(Q) = \iiint_R \frac{\delta}{\rho^2}\hat{\rho}dV .$$

Now, however, since Q is in R, the integral in (3.1) is an improper integral with Q itself as a singular point. Does this improper integral express a unique value for $\vec{G}_w(Q)$? That is to say, is this improper integral convergent? We use (1.1) to answer this question.

(3.2)   ***Theorem.*** *The improper integral in (3.1) is convergent.*

Proof. Consider a small spherical region ΔR with center at Q and radius ε. See [3–1]. Let Q* be a point on the boundary of ΔR. Then ΔR has volume $\Delta V = \frac{4}{3}\pi\varepsilon^3$. Let a sample point P* be chosen to minimize $\dfrac{\gamma\delta(P*)}{|\overrightarrow{QP*}|^2}$ in ΔR and let M be the maximum value of the continuous field δ on R. Then we have

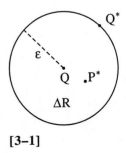

[3–1]

$$\frac{\gamma\delta(P*)}{|\overrightarrow{QP*}|^2}\Delta V \le \frac{\gamma\delta(Q*)}{\varepsilon^2}\frac{4}{3}\pi\varepsilon^3 \le \frac{4}{3}\pi\gamma M\varepsilon .$$

Hence, as $\varepsilon \to 0$, $\dfrac{\gamma\delta(P^*)}{\left|\overrightarrow{QP^*}\right|^2}\Delta V$ must approach the limit 0. Thus, by definition, Q

is a proper singular point and, by (1.1), the integral (3.1) gives a value for
$\vec{G}_w(Q)$. In this way, the integral (3.1) defines a field $\vec{G}_w$ on all of R. This field
is called the *gravitational field due to R within R.* (3.1) and (2.3) together, as
proper and improper integrals, define a gravitational field $\vec{G}_s$, due to R, on all of
space. It is easy to prove, as a corollary to the proofs above, that this vector field
is continuous.

In analogy to (2.7), we obtain the following general mathematical properties of the gravitational field $\vec{G}_w$ within R.

(3.3)   **Theorem.** *Let* δ *be a continuous density field on a regular region* R,
and let $\vec{G}_w$ *be the gravitational field due to R within R. Then*

    (i)   *for any given point* Q *in* R, $\vec{\nabla}\cdot\vec{G}_w\big|_Q = -4\pi\gamma\delta(Q)$;

    (ii)   *for any given point* Q *in* R, $\vec{\nabla}\times\vec{G}_w\big|_Q = \vec{0}$, *and* $\vec{G}_w$ *is conser-*
vative;

    (iii)   *if* g_w *is a scalar potential for* $\vec{G}_w$, *then* $\nabla^2 g_w = -4\pi\gamma\delta$;

    (iv)   *if* δ *is sufficiently smooth (in the sense of the remark in §27.4),*
then a scalar potential g_w *(for* $\vec{G}_w$ *) is given by*

$$g_w(Q) = \iiint_R \frac{\gamma\delta}{\rho}dV ,$$

where for each Q *in* R, *this integral is improper but convergent and has* Q *as
the origin for* ρ.

Before proving (3.3), we prove a special result about gravitational fields
in empty space.

(3.4)   **Gauss's theorem.** *Let* $\vec{G}_e$ *be the gravitational field in empty space due
to a continuous mass-density* δ *on a regular region* R. *Let* S *be any closed,
outward-directed surface such that* R *lies inside* S. *Then for the total flux of*
$\vec{G}_e$ *through* S, *we have*

$$\oiint_S \vec{G}_e \cdot d\vec{\sigma} = -4\pi\gamma M ,$$

where M *is the total mass in* R.

Proof of (3.4). It will be enough to obtain this result for the special case where S is a sphere. The full result (3.4) then follows by (2.7) and the divergence theorem. See §25.5.

Let a be the radius of the sphere S. Then

$$(3.5) \qquad \oint\!\!\!\oint_S \bar{G}_e \cdot d\bar{\sigma} = \iint_{\hat{D}} \bar{G}_e(a, \varphi, \theta) \cdot \hat{\rho}(\varphi, \theta) a^2 \sin\varphi\, d\varphi\, d\theta,$$

where we use spherical coordinates on S, and where we use φ and θ on $\hat{D}$ as parameters for S. The value of $\bar{G}_e$ at a point (a, φ, θ) on S can be expressed by the triple integral

$$(3.6) \qquad \bar{G}_e(a, \varphi, \theta) = \iiint_R \frac{\gamma\delta(x, y, z)}{|\overrightarrow{QP}|^2} \widehat{QP}\, dx\, dy\, dz,$$

where P is the point in R with Cartesian coordinates (x,y,z), and Q is the point on S with spherical coordinates (a, φ, θ).

If we now substitute (3.6) into (3.5) and carry out the dot product $\widehat{QP} \cdot \hat{\rho}$ in Cartesian coordinates by expressing Cartesian components for $\hat{\rho}$ in terms of the spherical coordinates (a, φ, θ), then (3.5) can be expressed as a 5-tuple iterated integral. We use Fubini's theorem to interchange the integration over S with the integration over R. In this new iterated integral, the integration over $\hat{D}$ is a surface integral giving the flux on S of a divergenceless field (outside R) due to a mass of $\delta(x,y,z)\Delta x\Delta y\Delta z$ located at the point (x,y,z). Considering this mass as a point-mass, and first finding the flux due to this mass through a large sphere centered at (x,y,z), and then using the divergence theorem, we see that the flux of this point-mass through S must be $-4\pi\gamma\delta(x,y,z)\Delta x\Delta y\Delta z$. (See the problems.) Performing the final triple integral on R, we now get

$$\oint\!\!\!\oint_S \bar{G}_e \cdot d\bar{\sigma} = -4\pi\gamma M,$$

where M is the total mass inside R. This completes the proof of Gauss's theorem.

Newton's theorem on spheres now follows as a corollary to Gauss's theorem.

(3.7) Corollary. *Let* R *be a spherical region with a radially symmetric density field. Let* M *be the total mass in* R, *then the field due to* R *at any point* Q *outside* R *must, by symmetry, be* $\vec{G}_e(Q) = \dfrac{-4\pi\gamma M}{4\pi |OQ|^2} = \dfrac{-\gamma M}{|\overrightarrow{OQ}|^2} \widehat{OQ}$, *where* O *is*

the center point of R.

Further corollaries to Gauss's theorem will be considered in the problems. We now return to the proof of (3.3).

Proof of (3.3).   (iii) is immediate from (i), since $\nabla^2 g_w = \vec{\nabla} \cdot \vec{G}_w$ by (ii) and $\vec{\nabla} \cdot \vec{G}_w = -4\pi\gamma\delta$ by (i).

We cannot obtain (i) and (ii) by Leibnitz's rule since the rule may not hold for an improper integral. Indeed, if we try to apply Leibnitz's rule to finding the divergence of the convergent improper integral (3.1), we get a new improper integral which converges to 0, as the reader may verify. On the other hand, we shall now show, by using the fundamental definition of *divergence*, that $\vec{\nabla} \cdot \vec{G}_w \big|_Q = -4\pi\gamma\delta(Q)$ as asserted in (i). (Our proof of (i) will therefore also serve as a demonstration that Leibnitz's rule when applied to an improper integral, can lead to an incorrect answer.)

We evaluate $\vec{\nabla} \cdot \vec{G}_w \big|_Q$ by finding the flux of $\vec{G}_w$ through a small sphere S_a of radius a with center at Q, then dividing this flux value by the volume of S_a, and finally finding the limit of this quotient as a goes to 0. By superposition, the field $\vec{G}_w$ is the sum of two fields, $\vec{G}_i$ and $\vec{G}_o$ where $\vec{G}_i$ is the field due to mass inside S_a, and $\vec{G}_o$ is the field due to mass inside R but outside S_a. By (2.6), $\vec{\nabla} \cdot \vec{G}_o = 0$ inside S_a, since $\vec{G}_o$, by itself, is an empty-space field inside S_a. Thus by the divergence theorem, the flux of $\vec{G}_w$ through S_a must equal the flux of $\vec{G}_i$ through S_a. Since $\vec{G}_i$, by itself, is an empty-space field outside S_a, we can apply Gauss's theorem. We find that the flux of $\vec{G}_i$ through $S_a = -4\pi\gamma M_a$, where M_a is the total mass inside S_a. Thus $-4\pi\gamma M_a$ is the total flux of $\vec{G}_w$ through S_a. Let V_a be the total volume of S_a. Then

$$\vec{\nabla} \cdot \vec{G}_w \big|_Q = \lim_{a \to 0} \frac{-4\pi\gamma M_a}{V_a} = -4\pi\gamma\delta(Q).$$

The proof of (ii) is similar. As in the proof of (i), we evaluate $\vec{\nabla} \times \vec{G}_w \big|_Q$ by using the fundamental definition of curl. We do so by choosing an arbitrary direction $\hat{u}$ and evaluating $\mathrm{rot}\,\vec{G}_w\big|_{\hat{u},Q}$. This proves to be 0, and hence $\vec{\nabla}\times\vec{G}_w\big|_Q$ must be $\vec{0}$. (To evaluate $\mathrm{rot}\,\vec{G}_w\big|_{\hat{u},Q}$, we take an appropriate circle C of radius a with center at Q and find the circulation of $\vec{G}_w$ on C. We do this by using the sphere S_a (on which C lies) and expressing $\vec{G}_w$ as $\vec{G}_i+\vec{G}_o$ as for (i). By (2.7), $\vec{G}_i$ is irrotational outside S_a and $\vec{G}_o$ is irrotational inside S_a. Hence $\vec{G}_i$ and $\vec{G}_o$ each has zero circulation on C, and so also does $\vec{G}_w$.

Then $\mathrm{rot}\,\vec{G}_w\big|_{\hat{u},Q} = \lim_{a\to 0}\left(\dfrac{\oint_C \vec{G}_w\cdot d\vec{R}}{\text{Area in C}}\right) = 0$. Hence $\vec{\nabla}\times\vec{G}_w\big|_Q = \vec{0}$.)

To show that $\vec{G}_w$ is conservative, let R* be a simply connected region which contains R, and the matter in R, but no other matter outside R. See [3–1]. Let $\vec{G}_s$ on R* be defined by

$$\vec{G}_s = \begin{cases} \vec{G}_w \text{ inside R,} \\ \vec{G}_e \text{ outside R but inside } R^*. \end{cases}$$

Then $\vec{G}_s$ is irrotational on R* and hence conservative on R*. It follows that $\vec{G}_w$ must be conservative on R.

To prove (iv), we must use the fundamental definition of gradient in the same way that we used the fundamental definitions of divergence and curl in (i) and (ii). To be successful, this requires the additional assumption of smoothness for the field δ. (See §27.4.) We do not give this proof.

Returning to Helmholtz's theorem in §27.8, we see that (i) and (ii) in (3.3) form a universal and complete set of differential equations for $\vec{G}_w$. If two fields $\vec{G}_1$ and $\vec{G}_2$ on R satisfy these equations, then $\vec{G}_1$ and $\vec{G}_2$ must differ, at most, by the gradient of a harmonic field. That is to say, they must only differ by what can be viewed as the empty-space field of other mass distributions outside R. (See the problems.)

§28.4 Fluid Flow

As we have already indicated, vector field concepts are useful in providing a continuous model for fluid-flow. We assume Cartesian coordinates. Given a fluid (say a gas) in some region, let $\delta(x,y,z,t)$ be the density of the fluid at position (x,y,z) at time t, and let $\vec{v}(v,y,z,t)$ be the velocity. We assume that both δ and $\vec{v}$ are C^1. Let R be a small region in the fluid around a fixed point P. Let S be the surface of R. Then, as we have seen, $\oiint_S \delta\vec{v} \cdot d\vec{\sigma}$ is the rate at which fluid mass is leaving the small region R. By the divergence theorem and by the continuity of δ and $\vec{v}$,

$$(4.1) \qquad \oiint_S \delta\vec{v} \cdot d\vec{\sigma} = \iiint_R \vec{\nabla} \cdot (\delta\vec{v})dV \approx \vec{\nabla} \cdot (\delta\vec{v})\big|_P \Delta V \,,$$

where ΔV is the volume of R. On the other hand, in a short time interval Δt, $\left[\oiint_S \delta\vec{v} \cdot d\vec{\sigma}\right]\Delta t \approx$ the amount of mass leaving R, and hence $\frac{1}{\Delta V}\left[\oiint_S \delta\vec{v} \cdot d\vec{\sigma}\right]\Delta t \approx$ $-\Delta\delta$, the change in density of fluid in R during the interval Δt. Thus we have

$$(4.2) \qquad \oiint_S \delta\vec{v} \cdot d\vec{\sigma} = -\frac{\partial\delta}{\partial t}\big|_P \Delta V \,.$$

Comparing (4.1) and (4.2), we have the equation

$$(4.3) \qquad -\frac{\partial\delta}{\partial t} = \vec{\nabla} \cdot (\delta\vec{v}) \,.$$

This equation, known as the *equation of continuity* for fluid-flow, must be satisfied by δ and $\vec{v}$ in any flow where a continuous model applies. (A flow for which no continuous model applies is called a *turbulent* flow.)

By a $\vec{\nabla}$ calculation,

$$(4.4) \qquad \vec{\nabla} \cdot (\delta\vec{v}) = \delta(\nabla \cdot \vec{v}) + \vec{\nabla}\delta \cdot \vec{v} \,.$$

If δ is constant in space but varies in time (as in a uniformly expanding gas) we have $\vec{\nabla}\delta = \vec{0}$ and

$$(4.5) \qquad -\frac{\partial\delta}{\partial t} = \delta(\nabla \cdot \vec{v}) \,.$$

Hence $\vec{\nabla} \cdot \vec{v} = \frac{1}{\delta}\left(-\frac{\partial\delta}{\partial t}\right) = -\frac{\partial}{\partial t}(\log\delta)$ is constant in space (as for an "expanding universe").

If δ is constant in time but varies in space (as in a slow motion of earth's atmosphere) we have

(4.6)
$$\delta(\vec{\nabla}\cdot\vec{v}) + \vec{\nabla}\delta\cdot\vec{v} = 0$$

and hence $\nabla\cdot\vec{v} = -\vec{\nabla}(\log\delta)\cdot\vec{v}$.

If δ is constant in both time and space (as is often assumed for water flows), we have

(4.7)
$$\vec{\nabla}\cdot\vec{v} = 0.$$

Note that $\vec{v}$ can be time dependent in each of formulas (4.1) through (4.7).

More generally, a flow is defined to be *incompressible* if $\vec{\nabla}\cdot\vec{v} = 0$ (whether or not δ is constant in time or space). For an incompressible flow, we must have

(4.8)
$$-\frac{\partial\delta}{\partial t} = \vec{\nabla}\delta\cdot\vec{v}$$

If a flow $\vec{v}(x,y,z,t)$ is *irrotational* ($\vec{\nabla}\times\vec{v} = \vec{0}$) in a simply connected region, then we know that at each instant t a scalar potential $f(x,y,z,t)$ for $\vec{v}$ must exist such that $\vec{v} = \vec{\nabla}f$. f is called a *scalar velocity potential* for $\vec{v}$. If in addition, $\vec{v}$ is incompressible, then we must have $\nabla^2 f = 0$. Thus the scalar velocity potential $f(x,y,z,t)$ for an incompressible, irrotational flow $\vec{v}(x,y,z,t)$ must be *harmonic* at each instant t.

If a flow $\vec{v}$ is incompressible ($\vec{\nabla}\cdot\vec{v} = 0$) in a simply connected region, then we know that a *vector velocity potential* $\vec{A}(x,y,z,t)$ must exist for $\vec{v}$ such that $\vec{v} = \vec{\nabla}\times\vec{A}$.

If the density δ is constant in time (this is the case in steady-state flow), we get $\frac{\partial\delta}{\partial t} = 0$, and by (4.3) $\vec{\nabla}\cdot(\delta\vec{v}) = 0$; hence, in a simply connected region, a vector potential for $\delta\vec{v}$ must exist. If δ is constant in space as well, then by (4.3) and (4.4) the flow is incompressible, and, in a simply connected region, a vector velocity potential must exist for $\vec{v}$.

§28.5 Electromagnetic Fields; Maxwell's Equations

We consider the elementary mathematics of electric and magnetic fields. The intensive study of these fields, and of electric charge, began in the late 18th century. It was observed that the motion of electric charge gives rise to magnetic forces, and that changes in the strength of a magnetic field give rise to electric forces. An important initial discovery was that electric charge can be viewed as of two kinds, called *positive* and *negative*, and that like charges repel each other, while unlike charges attract each other.

When there are no moving charges and no changing magnetic field (the ideal case of "electrostatic" fields), the laws for the attraction or repulsion of charges have exactly the form of Newton's laws for gravitational fields. Point charges attract or repel each other according to an inverse-square law (Coulomb's law). Furthermore the unit for measuring charge can be chosen so that the gravitational constant γ is replaced by 1 or -1 dependent on like or unlike charges. The unit of force (in CGS units) as the repulsive force between two unit positive point charges at a distance of 1 cm, and the unit of magnetic force is defined as the magnetic force (right handed) at a distance of one cm from a long, straight wire carrying a current of one unit of positive charge per cm per sec. As with gravity, we can make useful mathematical models with continuous distributions of electric charge. We then have scalar fields δ of *charge density*, where δ can have negative as well as positive values. Steady magnetic force fields are also observed to have the same general properties as gravitational fields in empty space, although no distributions of magnetic charge density have as yet been observed. It follows that the entire mathematical apparatus of §28.2 and §28.3 can be applied. This leads to differential equations for electrostatic fields, where we denote the electric field as $\vec{E}$ and the magnetic field as $\vec{B}$. From (3.3) we get

(5.1)
$$
\begin{aligned}
&\text{(i)} \quad \vec{\nabla} \cdot \vec{E} = 4\pi\delta, \\
&\text{(ii)} \quad \vec{\nabla} \cdot \vec{B} = 0, \\
&\text{(iii)} \quad \vec{\nabla} \times \vec{E} = \vec{0}, \\
&\text{(iv)} \quad \vec{\nabla} \times \vec{B} = \vec{0}.
\end{aligned}
$$

Moreover, on a simply connected region, we can find a harmonic scalar potential for $\vec{B}$ and, in empty space, a harmonic scalar potential for $\vec{E}$.

In the early 19th century, research turned from static fields to the careful study of the effect on $\vec{E}$ of changes in $\vec{B}$ and of the effect on $\vec{B}$ of moving charges. Two major facts were discovered and formulated. To describe these facts, known as *Faraday's law* and *Ampere's law*, we must consider $\vec{E}$ and $\vec{B}$ as time-dependent vector fields and δ as a time-dependent scalar field. In Cartesian coordinates, we have $\vec{E}(x,y,z,t)$, $\vec{B}(x,y,z,t)$, and $\delta(x,y,z,t)$.

(5.2)   *Faraday's law. Let* S *be a directed two-sided surface with coherently directed boundary* C. *Then*

$$\frac{\partial}{\partial t} \iint_S \vec{B} \cdot d\vec{\sigma} = -c \oint_C \vec{E} \cdot d\vec{R},$$

where c *is an empirically observed physical constant.* For this law, we can think of C as a loop of wire and of S as an imaginary surface spanning C. See [5–1].

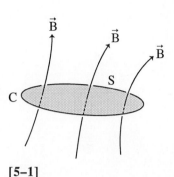

[5–1]

For Ampere's law, we introduce the time-dependent vector field $\vec{J} = \delta \vec{v}$, where $\vec{v}$ is the time-dependent vector field of charge-velocity. Thus $\vec{J}$ is a "charge-flow" vector analogous to our previously considered mass-flow vector for a fluid. $\vec{J}$ is usually called the *current density* vector.

(5.3)   *Ampere's law. Let* S *be a directed two-sided surface with coherently directed boundary* C. *Then*

$$\iint_S \vec{J} \cdot d\vec{\sigma} = \frac{c}{4\pi} \oint_C \vec{B} \cdot d\vec{R},$$

where c *is observed to be the same physical constant as in* (5.2).

For this law, we can think of C as an imaginary curve and of S as an imaginary surface through which electric charge is flowing. See [5–2].

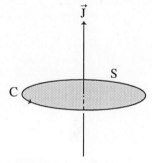

[5–2]

The technology of electrical generators and motors began to evolve in the 19th century. It was, and is, based on the two laws (5.2) and (5.3).

In the mid-19th century, the British scientist Maxwell undertook a careful reconsideration of the theory of electric and magnetic fields, using new empirical knowledge and mathematical techniques. He first concluded that the laws expressed above as (5.1(i)) and (5.1(ii)) were well supported by empirical evidence and were true for the case of moving charges and time-varying fields, as well as for the case of electrostatic fields. He then used multivariable calculus to deduce differential-equation forms of Faraday's law and Ampere's law. For Faraday's law, he argued

$$\frac{\partial}{\partial t} \iint_S \vec{B} \cdot d\vec{\sigma} = \iint_S \frac{\partial \vec{B}}{\partial t} \cdot d\vec{\sigma} \qquad \text{(Leibnitz's rule)}$$

$$= -c \oint \vec{E} \cdot d\vec{R} \qquad \text{(Faraday's law).}$$

$$= -\iint_S \vec{\nabla} \times \vec{E} \cdot d\vec{\sigma} \qquad \text{(Stokes's theorem).}$$

Since this must be true for any surface S, however small, we have

(5.4)
$$\vec{\nabla} \times \vec{E} = -\frac{1}{c}\frac{\partial \vec{B}}{\partial t}$$

as an equation which must hold at every point in space.

Similarly, for Ampere's law Maxwell got

$$\iint_S \vec{J} \cdot d\vec{\sigma} = \frac{c}{4\pi} \oint_C \vec{B} \cdot d\vec{R} \qquad \text{(Ampere's law)}$$

$$= \frac{c}{4\pi} \oint_C \vec{\nabla} \times \vec{B} \cdot d\vec{\sigma} \qquad \text{(Stokes's theorem)}.$$

Since this must be true for any surface S, however small, we have

(5.5)
$$\vec{\nabla} \times \vec{B} = \frac{4\pi}{c}\vec{J}$$

at every point in space.

Maxwell then noticed that moving charges should obey the basic laws of fluid-flow, in particular the equation of continuity (4.3):

$$\vec{\nabla} \cdot \vec{J} = -\frac{\partial \delta}{\partial t}.$$

By (5.1(i)) and using equality of cross-derivatives, we get

(5.6)
$$\vec{\nabla} \cdot \vec{J} = -\frac{1}{4\pi}\frac{\partial}{\partial t}(\vec{\nabla} \cdot \vec{E}) = -\frac{1}{4\pi}\vec{\nabla} \cdot \frac{\partial \vec{E}}{\partial t}.$$

On the other hand, taking the divergence of both sides of Ampere's law (5.5) we get

$$\vec{\nabla} \cdot \vec{J} = \frac{c}{4\pi}\vec{\nabla} \cdot \vec{\nabla} \times \vec{B} = 0.$$

Maxwell thus found that Ampere's law, as stated in (5.5), is inconsistent with the equation of continuity in (5.6). ((5.6) is sometimes called the law of *conservation of charge*.) It followed that Ampere's law is not entirely correct. Maxwell saw that he could reconcile Ampere's law (5.5) with conservation of charge (5.6) by adding the term $\frac{1}{c}\frac{\partial \vec{E}}{\partial t}$ to Ampere's law to get

(5.7)
$$\vec{\nabla} \times \vec{B} = \frac{4\pi}{c}\vec{J} + \frac{1}{c}\frac{\partial \vec{E}}{\partial t}.$$

This achieves the desired correction, since

$$\vec{\nabla}\cdot\left(\frac{4\pi}{c}\vec{J}+\frac{1}{c}\frac{\partial\vec{E}}{\partial t}\right) = \frac{4\pi}{c}\vec{\nabla}\cdot\vec{J}+\frac{1}{c}\vec{\nabla}\cdot\frac{\partial\vec{E}}{\partial t}$$

$$= -\frac{4\pi}{c}\frac{1}{4\pi}\vec{\nabla}\cdot\frac{\partial\vec{E}}{\partial t}+\frac{1}{c}\vec{\nabla}\cdot\frac{\partial\vec{E}}{\partial t} \qquad \text{(by (5.6))}$$

$$= 0 \,.$$

For any directed surface S, $\iint_S \vec{J}\cdot d\vec{\sigma}$ is called the *current* through S. Thus, in

(5.7), $\iint_S \frac{1}{4\pi}\frac{\partial\vec{E}}{\partial t}\cdot d\vec{\sigma}$ plays the role of another kind of "current" through S. It is

called by Maxwell the *displacement current* through S. The term "displacement" arose from Maxwell's study of electric and magnetic fields inside matter, where he found empirical confirmation for his corrected form of Ampere's law.

Maxwell now had four basic differential equations for $\vec{E}$ and $\vec{B}$. We can write them as

(5.8) (i) $\vec{\nabla}\cdot\vec{E} = 4\pi\delta\,,$

(ii) $\vec{\nabla}\cdot\vec{B} = 0\,,$

(iii) $\vec{\nabla}\times\vec{E} = -\frac{1}{c}\frac{\partial\vec{B}}{\partial t}\,,$

(iv) $\vec{\nabla}\times\vec{B} = \frac{4\pi}{c}\vec{J}+\frac{1}{c}\frac{\partial\vec{E}}{\partial t}\,.$

These equations are known today as *Maxwell's equations.* By Helmholtz's theorem, they form a universal and complete set of differential equations for the vector fields $\vec{E}$ and $\vec{B}$, and they are accepted as a correct and precise model for the macroscopic (non-quantum mechanical) behavior of these fields. The identification of these equations by Maxwell was an epochal event in the modern history of science and technology.

While considering the case of $\vec{E}$ and $\vec{B}$ in empty space, where $\delta = 0$ and $\vec{J} = 0$, Maxwell made two further deductions: Differentiating (5.8(iv)) with respect to time gives

$$\frac{\partial}{\partial t}(\vec{\nabla}\times\vec{B}) = \vec{\nabla}\times\frac{\partial\vec{B}}{\partial t} = \frac{1}{c}\frac{\partial^2\vec{E}}{\partial t^2}\,.$$

Using (5.8(iii)) to substitute for $\frac{\partial\vec{B}}{\partial t}$ gives

$$\vec{\nabla}\times(-c\vec{\nabla}\times\vec{E}) = \frac{1}{c}\frac{\partial^2\vec{E}}{\partial t^2}\,.$$

But $$\vec{\nabla} \times (\vec{\nabla} \times \vec{E}) = \vec{\nabla}(\vec{\nabla} \cdot \vec{E}) - \vec{\nabla}^2 \vec{E} .$$

Since $\vec{\nabla} \cdot \vec{E} = 0$ in empty space, we have

(5.9)
$$\vec{\nabla}^2 \vec{E} = \frac{1}{c^2} \frac{\partial^2 \vec{E}}{\partial t^2} .$$

By a similar argument, Maxwell also got

(5.10)
$$\vec{\nabla}^2 \vec{B} = \frac{1}{c^2} \frac{\partial^2 \vec{B}}{\partial t^2} .$$

(5.9) and (5.10) have the mathematical form of *wave equations* for a wave like pattern of values for $\vec{E}$ and $\vec{B}$ which moves through space with velocity c.

This deduction of wave equations led scientists to search for the as yet unnoticed physical phenomenon of electromagnetic waves. They were soon successful (Helmholtz was the first). Moreover, the observed constant c in (5.8) turned out to equal the observed speed of light in a vacuum.

§28.6 Heat Flow

Given a homogeneous substance in some region, let $T(x,y,z,t)$ be the temperature of the substance at position (x,y,z) at time t. Let $\vec{H}(x, y, z, t)$ be a vector giving the direction and magnitude of the flow of heat energy at position (x,y,z) at time t. The fundamental physical law known as the *law of heat conduction* asserts that

(6.1)
$$\vec{H} = -\mu \vec{\nabla} T ,$$

where μ is a constant called the *conductivity* of the substance. A second physical law known as the *law of specific heat* asserts that when the temperature of a small region of volume ΔV is increased by ΔT, a quantity of heat ΔQ must enter that region, where

(6.2)
$$\Delta Q = c \Delta T \Delta V,$$

and c is a quantity called the *specific heat* of the substance in that region (c may vary as T varies). Let P be a point in the substance. Consider a small region R of volume ∇V around P. Taking the divergence of both sides of the conduction equation (6.1), we get

$$\vec{\nabla} \cdot \vec{H} = -\mu \nabla^2 T .$$

Let S be the surface of R. By the divergence theorem

$$\oiint_S \vec{H} \cdot d\vec{\sigma} = \iiint_R \vec{\nabla} \cdot \vec{H} dV \approx \vec{\nabla} \cdot \vec{H}\big|_P \Delta V = -\mu \nabla^2 T\big|_P \Delta V .$$

On the other hand, the flux integral $\oiint_S \vec{H} \cdot d\vec{\sigma}$ is also the rate at which heat is leaving the region R. By the specific heat equation, for a small time interval Δt,

$$\oiint_S \vec{H} \cdot d\vec{\sigma} \approx -\frac{\Delta Q}{\Delta t} = -c \frac{\Delta T}{\Delta t} \Delta V \approx -c \frac{\partial T}{\partial t}\big|_P \Delta V .$$

Thus we have $-c \dfrac{\partial T}{\partial t}\big|_P \Delta V \approx -\mu \nabla^2 T\big|_P \Delta V$, and, as $\Delta V \to 0$, we get

(6.3) $$c \frac{\partial T}{\partial t}\big|_P = \mu \nabla^2 T\big|_P .$$

The temperature distribution in a homogeneous substance must satisfy this basic equation, which is called the *heat equation*. For the special case where T is *steady-state*, that is to say, where T does not depend on t, we see that T must satisfy Laplace's equation $\nabla^2 T = 0$. Hence, in this case, the temperature distribution T(x,y,z) must be a harmonic function. (Note that a steady-state temperature distribution need not be constant in space, since a steady flow of heat may be occurring, with heat being supplied and removed at various locations on the boundary of the substance.)

§28.7 Quantum Mechanics

The mathematics of vector fields has played a central role in the development of quantum mechanics. The *wave equation* for quantum mechanics (known as *Schrödinger's equation*) is commonly expressed in the general form

(7.1) $$\nabla^2 \psi + V \psi = k \frac{\partial \psi}{\partial t} ,$$

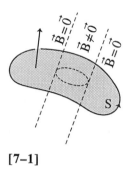

[7–1]

where V is a scalar field and ψ is a complex-valued, time-dependent scalar field in $\mathbf{E}^3$.

Vector-field mathematics is central in *quantum field theory* where, in particular, vector potentials are prominent in the theoretical treatment of interactions between electromagnetic fields and matter. We note one particular use here. Let $\vec{B}$ be a magnetic field. See [7–1]. From Maxwell's equations, we must have $\vec{\nabla} \cdot \vec{B} = 0$. Hence a vector potential $\vec{A}$ must always exist for $\vec{B}$. Moreover, by Stokes' theorem a flux integral for $\vec{B}$ over a surface S can be expressed as the line integral of $\vec{A}$ around the boundary of S. It is possible to make an experimental arrangement where the boundary of S lies in a region of space where $\vec{B} = \vec{0}$ even though there is a non-zero flux of $\vec{B}$ through a central portion of S. In this case, the value of the flux of $\vec{B}$ is given by the circulation of the vector potential $\vec{A}$ in the region where $\vec{B} = \vec{0}$. This raises the question of whether the vector potential $\vec{A}$ has a physical reality of its own or is merely a convenient mathematical fiction. Recent experimental evidence (the *Bohm-Aharanov effect*) shows that in such an experimental arrangement, changes in the central magnetic field can affect the behavior of matter (in this case the wave properties of electrons as measured by interference) in the region where $\vec{B} = \vec{0}$. This suggests that matter is being directly affected by $\vec{A}$, and that $\vec{A}$ (though not unique) has a physical reality of its own. (The observations made in the Bohm-Aharanov effect are predicted by current physical theory. The vector potential $\vec{A}$ has a central role in this theory.)

§28.8 Problems

§28.1

1. Prove (1.1).

§28.2

1. Prove (2.6) by taking the origin as the center of the sun, letting a be the radius of the sum, and calculating the gravitational field at the point (b,0,0) for some $b > a$ by evaluating the integral in (2.3). Use spherical coordinates and assume that $\delta = h(\rho)$ for some function h. (*Hint.* By symmetry and superposition, it will be enough to integrate only over φ and θ.)

§28.3

1. Prove that the field $\vec{G}_S$ (in the proof of (3.2)) is continuous.

2. In the proof of (3.4), use a large sphere together with the divergence theorem to get the flux of a small mass through S.

3. Show that Leibnitz's rule fails if we try to use it to get the divergence of the improper integral in (3.1).

4. Use (3.4) to show that the gravitational field caused by a radically symmetric solid sphere with a concentric spherical cavity must be $\vec{0}$ at every point inside the cavity.

5. Assume that a solid sphere S has radius 1, mass M, and constant density. Use (3.4) to find the magnitude of the gravitational field caused by S at a point P inside S, where P is at distance x from the center of S.

6. Prove the final sentence in §28.3.

§28.4

1. Consider a flow in $\mathbf{E}^3$ of a frictionless fluid of constant density δ. ("Frictionless" means that there is no loss of kinetic energy into heat due to internal friction ("viscosity") of the fluid.) Let $\vec{v}(t, x, y, z)$ in C^2 give the velocity of the flow. Let p be a particular particle of the fluid. Let $\vec{R}(x_p(t), y_p(t), z_p(t))$ be the path of p. Then $\vec{v}(t, x_p(t), y_p(t), z_p(t))$ gives the velocity of p as a function of t.

(a) Use the chain rule to find the acceleration $\vec{a}$ of p.   Show that $\vec{a}$ can be expressed as $\frac{\partial \vec{v}}{\partial t} + [\vec{v} \cdot \vec{\nabla}]\vec{v}$. (See §18.9 and §27.6.)

(b) Let U be a small portion of fluid of volume ΔV containing p. Let the scalar field P(t,x,y,z) in C^2 give the pressure of the fluid at point (x,y,z) at time t. Use (9.2) in §27.9 to conclude that the net force of pressure on U is $\approx -\vec{\nabla}P\Delta V$.

(c) Assume that no other force acts on the fluid. Use Newtons second law to conclude that the flow must obey *Euler's equation*

$$\delta(\frac{\partial \vec{v}}{\partial t} + [\vec{v} \cdot \vec{\nabla}]\vec{v}) + \vec{\nabla}P = \vec{0}.$$

§28.5

1. Give the argument for (5.10).

Chapter 29

Vectors and Matrices

§29.1 The Space $\mathbf{R}^n$; n-vectors

So far, we have confined our development of multivariable calculus to functions of two and three variables and to geometrical spaces of two and three dimensions. In order to permit the broadest use of multivariable calculus, we must extend our concepts and methods to spaces of more than three dimensions and to functions of more than three variables. In Chapters 29 through 34, we begin this extension by considering the geometry of higher dimensional spaces and by developing, at the same time, the mathematics needed to describe and work with these spaces.

When we study the geometry of two and three dimensions, we make direct use of spatial intuitions acquired from everyday experience in the physical world. In the study of geometry in higher dimensions, however, this intuitive help is not available, and we must rely, at first, on algebraic ideas and techniques. These algebraic ideas permit us (i) to define higher dimensional spaces, (ii) to prove theorems about them, and (iii) to develop, in due course, an intuitive understanding that will be analogous, in many respects, to our intuitive understanding of two- and three-dimensional spaces.

We begin our study by defining the n-dimensional space $\mathbf{R}^n$. The definition (§29.1) is based on, and generalizes, our use of Cartesian coordinates in $\mathbf{E}^2$ and $\mathbf{E}^3$. We then introduce, in succession, two new and distinct algebraic systems. The first of these systems (§29.2-§29.4) is n-*dimensional vector algebra*. It is a direct generalization of the concepts, notations, and operations of vector algebra for $\mathbf{E}^2$ and $\mathbf{E}^3$ (as presented in Chapters 2 and 3). n-dimensional vector algebra will enable us later to express many of the facts of n-dimensional multivariable calculus in forms that closely parallel the forms used for three-dimensional calculus.

The second algebraic system is *matrix algebra*. Matrix algebra will be less familiar, but will incorporate, in new notation, most of the concepts and facts of n-dimensional vector algebra. In addition, it will permit us to express and use a variety of other mathematical operations, concepts, and relationships.   For example, matrix algebra will enable us to express, in a simple and economical way, n-dimensional versions of such geometrical ideas as *line, plane, parallelism, perpendicularity, projection, area,* and *volume.*

Matrix algebra also has a variety of other applications in mathematics, science, and technology. Matrix algebra is part of the more general mathematical field known as *linear algebra.*

In the three-dimensional Euclidean space $\mathbf{E}^3$, we can choose Cartesian axes and use them to assign to each point P an ordered triple (x_P, y_P, z_P) of coordinates. Two different points must have different triples assigned, and for every possible triple of real numbers, there must be some point P having that triple as its coordinates. We thus have a complete and exact correspondence between *points* in $\mathbf{E}^3$-with-Cartesian-axes and *ordered triples of real numbers.*

In $\mathbf{E}^3$-with-axes, we can also associate with each point P the three-dimensional vector represented by the arrow $\overrightarrow{OP}$ from the origin. For every three dimensional vector $\vec{A}$, there is some point P such that $\vec{A} = \overrightarrow{OP}$. Hence, as we saw in Chapter 3, we can use the ordered triple of coordinates for this point P to represent the vector $\vec{A}$. We refer to the members of this triple as the *scalar components* of the vector $\vec{A}$ in the given coordinate system. We thus have a complete and exact correspondence between *vectors* in $\mathbf{E}^3$-with-axes and *ordered triples of real numbers*. Therefore, in $\mathbf{E}^3$-with-Cartesian-axes, we use ordered triples of real numbers to represent *points*, and we also use ordered triples to represent *vectors*.

Recall that an ordered triple of real numbers is different from a *set* of real numbers. For example, $\{2,3,-1\}$ and $\{-1,2,3\}$ denote the same set, but $(2,3,-1)$ and $(-1,2,3)$ are different as ordered triples (and they represent different points and different vectors). Recall also that the numbers appearing in an ordered triple need not be distinct. Thus $(1,0,1)$ is an ordered triple and so is $(3,3,3)$.

Definition. Let n be a given positive integer. An *ordered* n-*tuple* is a sequence, $(a_1,...,a_n)$, of n real numbers in order, with repetitions allowed. We say that two ordered n-tuples $(a_1,...,a_n)$ and $(b_1,...,b_n)$ are the *same* (or *equal*) if and only if $[a_1 = b_1$ and $a_2 = b_2$ and ... $a_n = b_n]$. We shall usually refer to an ordered n-tuple simply as an n-*tuple*.

We define the n-dimensional space $\mathbf{R}^n$ as follows.

(1.1)   *Definition.* For each $n > 0$, $\mathbf{R}^n$ is the set of all ordered n-tuples of real numbers. We sometimes refer to $\mathbf{R}^n$ as n-*space* and to n-tuples in $\mathbf{R}^n$ as *points* in n-space.

If we choose axes for $\mathbf{E}^2$, we can represent $\mathbf{E}^2$ as $\mathbf{R}^2$. Similarly, if we choose axes for $\mathbf{E}^3$, we can represent $\mathbf{E}^3$ as $\mathbf{R}^3$.

To define distance between points in $\mathbf{R}^n$, we use a formula analogous to the formulas already used for $\mathbf{R}^2$ and $\mathbf{R}^3$.

Definition. Given points $P = (p_1,...,p_n)$ and $Q = (q_1,...,q_n)$ in $\mathbf{R}^n$, the *distance* between P and Q is defined to be $\sqrt{(p_1 - q_1)^2 + \cdots + (p_n - q_n)^2}$.

We now introduce vector terminology and notation in $\mathbf{R}^n$.

(1.2)   *Definition.* An n-*vector* ("n-dimensional vector") $\vec{A}$ is an n-tuple $(a_1,...,a_n)$ of real numbers. For $1 \leq j \leq n$, a_j is called the j-th *component* (or j-th *coordinate*) of the n-vector $\vec{A}$.

Thus we shall sometimes treat (and think of) an n-tuple in $\mathbf{R}^n$ as a *point* in the space $\mathbf{R}^n$, and we shall sometimes treat (and think of) an n-tuple in $\mathbf{R}^n$ as a *vector in the space* $\mathbf{R}^n$. (As noted above, we already do this in $\mathbf{R}^3$: we use ordered triples to represent points, and we also use them to represent vectors.)

In §29.2 and §29.3, we develop *vector algebra* for n-vectors.

§29.2 Vector Operations in $\mathbf{R}^n$

We defined a *vector* in $\mathbf{E}^3$ to be a direction in space together with a numerical magnitude, and our *vector operations* were defined by simple geometric pictures. Later, after choosing coordinate axes, we were able to represent a vector as a triple of numbers (a_1,a_2,a_3) and to find equivalent descriptions of vector operations as computational operations on triples of numbers. In $\mathbf{R}^n$ with $n > 3$, however, we no longer have geometric pictures of a simple and familiar kind, and we must *start* by defining the basic operations of vector algebra as operations on n-tuples of numbers.

(2.1) **Definitions.** In **R**^{n}, let $\vec{0}$ be the vector $(0,\ldots,0)$. We define the following operations on n-vectors. These operations are analogous to previously defined operations on 3-vectors.

Multiplication of a vector by a scalar:

$$\left.\begin{array}{l} k \text{ a scalar} \\ \vec{A} = (a_1,\ldots,a_n) \end{array}\right\} \to k\vec{A} = (ka_1,\ldots,ka_n) \;.$$

Addition of two vectors:

$$\left.\begin{array}{l} \vec{A} = (a_1,\ldots,a_n) \\ \vec{B} = (b_1,\ldots,b_n) \end{array}\right\} \to \vec{A} + \vec{B} = (a_1 + b_1,\ldots,a_n + b_n) \;.$$

Scalar (dot) product of two vectors:

$$\left.\begin{array}{l} \vec{A} = (a_1,\ldots,a_n) \\ \vec{B} = (b_1,\ldots,b_n) \end{array}\right\} \to \vec{A} \cdot \vec{B} = a_1 b_1 + \cdots + a_n b_n \;.$$

Magnitude of a vector:

$$\vec{A} = (a_1,\ldots,a_n) \to |\vec{A}| = \sqrt{a_1^2 + \cdots + a_n^2} = \sqrt{\vec{A} \cdot \vec{A}} \;.$$

Unit vectors. When $\vec{A} \neq \vec{0}$, $\left(\dfrac{1}{|\vec{A}|}\right)\vec{A}$ has magnitude = 1. We sometimes write $\left(\dfrac{1}{|\vec{A}|}\right)\vec{A}$ as $\hat{A}$.

Direction of a vector. Given a vector $\vec{A} \neq \vec{0}$, we sometimes refer to the unit vector $\hat{A}$ as the *direction* of $\vec{A}$.

As in three dimensions, $-\vec{A}$ is defined as $(-1)\,\vec{A}$, and $\vec{A} - \vec{B}$ is defined as $\vec{A} + (-\vec{B})$. If $\vec{A} \cdot \vec{B} = 0$, we say that $\vec{A}$ and $\vec{A}$ are *orthogonal* (or "perpendicular") vectors. As before, we sometimes write $\vec{A} \cdot \vec{A}$ as $\vec{A}^2$.

Calculations with n-vectors are carried out exactly as in the three-dimensional case.

Examples.

(1) $\vec{A} = (1,2,2,0)$. *Find* $\hat{A}$.

Answer. $|\vec{A}| = \sqrt{1+4+4+0} = \sqrt{9} = 3$. $\therefore \hat{A} = \dfrac{\vec{A}}{|\vec{A}|} = \left(\dfrac{1}{3}, \dfrac{2}{3}, \dfrac{2}{3}, 0\right)$.

(2)   *Let* $\vec{A} = (1,2,2,0)$ *and* $\vec{B} = (1,1,1,1)$. *Express* $\vec{A}$ *as* $\vec{C} + \vec{D}$, *where* $\vec{C}$ *is a scalar multiple of* $\vec{B}$, *and* $\vec{D}$ *is orthogonal to* $\vec{B}$.

Answer. We seek $\vec{A} = \vec{C} + \vec{D}$, where $\vec{C} = k\vec{B}$ for some k, and $\vec{B} \cdot \vec{D} = 0$. We assume the same algebraic laws as for vectors in $\mathbf{E}^3$ and $\mathbf{E}^2$ (see §29.3 below).   If $\vec{A} = \vec{C} + \vec{D}$, then $\vec{A} \cdot \vec{B} = (\vec{C} + \vec{D}) \cdot \vec{B} = \vec{C} \cdot \vec{B} = k(\vec{B} \cdot \vec{B})$. Hence $k = \dfrac{\vec{A} \cdot \vec{B}}{\vec{B} \cdot \vec{B}} = \dfrac{5}{4}$, and $\vec{C} = \dfrac{5}{4}\vec{B} = \left(\dfrac{5}{4}, \dfrac{5}{4}, \dfrac{5}{4}, \dfrac{5}{4}\right)$.   Finally, $\vec{D} = \vec{A} - \vec{C} = \left(-\dfrac{1}{4}, \dfrac{3}{4}, \dfrac{3}{4}, -\dfrac{5}{4}\right)$.

As a check, we see that $\vec{B} \cdot \vec{D} = -\dfrac{1}{4} + \dfrac{3}{4} + \dfrac{3}{4} - \dfrac{5}{4} = 0$.

Note that this algebraic calculation has the same general form as was used earlier in three dimensions for the vector projection of $\vec{A}$ on $\vec{B}$. (See (4.8) in §2.4.)

The operations defined above provide us with a system of *vector algebra* for $\mathbf{R}^n$. We consider this algebra in §29.3 below.

We shall sometimes refer to a vector $(a_1, ..., a_n)$ of specific given real numbers as a *numerical* vector. We shall also use notations such as $(x_1, ..., x_n)$ for *variable* vectors where $x_1, ..., x_n$ are variables representing real numbers. The operations of vector algebra can be applied to variable vectors as well as to numerical vectors.

Remark. A counterpart to the operation of *vector (cross) product* in three-dimensional space can also be defined in n-dimensional space. It is an operation which combines n–1 given n-vectors to form a certain new n-vector which is orthogonal to each of the n–1 given n-vectors. We consider this operation in §31.5.

§29.3 Vector Algebra in $\mathbf{R}^n$

It is easy to show, from our coordinate definitions, that the same **algebraic laws** hold for operations on n-vectors as hold for vector operations in $\mathbf{E}^3$. We thus have

(3.1)
$$\vec{A} + \vec{B} = \vec{B} + \vec{A} \qquad\qquad \vec{A} \cdot \vec{B} = \vec{B} \cdot \vec{A}$$
$$(\vec{A} + \vec{B}) + \vec{C} = \vec{A} + (\vec{B} + \vec{C}) \qquad k(\vec{A} + \vec{B}) = k\vec{A} + k\vec{B}$$
$$\vec{A} \cdot (\vec{B} + \vec{C}) = \vec{A} \cdot \vec{B} + \vec{A} \cdot \vec{C} \qquad (k + \ell)\vec{A} = k\vec{A} + \ell\vec{A}$$
$$\vec{A} + \vec{0} = \vec{A} \qquad\qquad (k\ell)\vec{A} = k(\ell\vec{A})$$
$$\vec{A} \cdot \vec{0} = 0 \qquad\qquad \vec{A} \cdot \vec{A} \geq 0$$

The proofs follow directly from the definitions.

The $\hat{i}, \hat{j}, \hat{k}$ -representation for vectors in $\mathbf{R}^3$ has allowed us, when we carry out theoretical or numerical calculations, to take advantage of laws of vector algebra. We can use a similar representation for vectors in $\mathbf{R}^n$ and gain similar advantages.

(3.2)   We define the unit vectors

$$\hat{e}_1 = (1,0,0,\ldots,0),$$
$$\hat{e}_2 = (0,1,0,\ldots,0),$$
$$\cdots\cdots\cdots$$
$$\hat{e}_n = (0,0,0,\ldots,1).$$

Using multiplication by a scalar together with vector addition, we immediately have

(3.3) $$\vec{A} = (a_1, a_2, \ldots, a_n) = a_1\hat{e}_1 + a_2\hat{e}_2 + \cdots + a_n\hat{e}_n \ .$$

$\hat{e}_1, \ldots, \hat{e}_n$ can be used in $\mathbf{R}^n$ in the same way that $\hat{i}, \hat{j}, \hat{k}$ are used in $\mathbf{R}^3$.

We define a *frame* in $\mathbf{R}^n$ to be a set of n mutually orthogonal unit vectors. If $\{\hat{B}_1, \ldots, \hat{B}_n\}$ is a frame in $\mathbf{R}^n$, it follows (as in $\mathbf{E}^3$) that any vector A can be uniquely expressed as

(3.4) $$\vec{A} = c_1\hat{B}_1 + \cdots + c_n\hat{B}_n ,$$

where, for each $i \leq n$, $c_i = \vec{A} \cdot \hat{B}_i$. (This is the *frame identity* for $\mathbf{R}^n$. See §3.4.)

§29.4 Vector Geometry in $\mathbf{R}^n$

The following general inequalities were geometrically obvious in $\mathbf{E}^3$. We now state and prove them for $\mathbf{R}^n$.

(4.1)   **Theorem.** *For any given vectors $\vec{A}$ and $\vec{B}$ in $\mathbf{R}^n$:*
    (a) $\left| \hat{A} \cdot \hat{B} \right| \le 1$, *and*
    (b) $\left| \vec{A} + \vec{B} \right| \le \left| \vec{A} \right| + \left| \vec{B} \right|$.

((b) is obvious for $\mathbf{E}^3$ because it asserts that the length of one side of a triangle is less than the sum of the lengths of the other two sides. (a) is obvious for $\mathbf{E}^3$ because $\hat{A} \cdot \hat{B} = \cos\theta$ and $\left| \cos\theta \right| \le 1$, where θ is the angle between $\vec{A}$ and $\vec{B}$.) We shall first prove (a) and then deduce (b) from (a).

Proof of (a). By our algebraic laws, we have

$$0 \le (\hat{A} + \hat{B})^2 = \hat{A}^2 + 2\hat{A} \cdot \hat{B} + \hat{B}^2 = 2 + 2\hat{A} \cdot \hat{B} \quad (\text{since} \quad \hat{A}^2 = \hat{B}^2 = 1).$$

Thus $-(\hat{A} \cdot \hat{B}) \le 1$.
Similarly,

$$0 \le (\hat{A} - \hat{B})^2 = \hat{A}^2 - 2\hat{A} \cdot \hat{B} + \hat{B}^2 = 2 - 2\hat{A} \cdot \hat{B}.$$

Thus $(\hat{A} \cdot \hat{B}) \le 1$.
Hence $\left| \hat{A} \cdot \hat{B} \right| \le 1$. This completes the proof.

(4.2)   Inequality (a) is known as the *Cauchy-Schwartz inequality*. It is sometimes written in the equivalent form $\dfrac{\left| \vec{A} \cdot \vec{B} \right|}{\left| \vec{A} \right| \left| \vec{B} \right|} \le 1$ or in the equivalent form $(\vec{A} \cdot \vec{B})^2 \le \vec{A}^2 \vec{B}^2$.

Proof of (b). From (a), we have $\vec{A} \cdot \vec{B} \le \left| \vec{A} \right| \left| \vec{B} \right|$.
Thus $2\vec{A} \cdot \vec{B} \le 2\left| \vec{A} \right| \left| \vec{B} \right|$,

$$\vec{A}^2 + 2\vec{A} \cdot \vec{B} + \vec{B}^2 \le \vec{A}^2 + 2\left| \vec{A} \right| \left| \vec{B} \right| + \vec{B}^2,$$

and
$$(\vec{A} + \vec{B})^2 \leq (|\vec{A}| + |\vec{B}|)^2,$$

(since $\vec{A}^2 = |\vec{A}|^2$ and $\vec{B}^2 = |\vec{B}|^2$).

Hence $|\vec{A} + \vec{B}| \leq |\vec{A}| + |\vec{B}|$.

(4.3)   Inequality (b) is known as the *triangle inequality*.

Example 1. *Let* $\vec{A} = (1,2,2,0)$ *and* $\vec{B} = (1,1,1,1)$. Then (a) asserts that $(\vec{A} \cdot \vec{B})^2 \leq \vec{A}^2 \vec{B}^2$. Here $\vec{A} \cdot \vec{B} = 5$, $\vec{A}^2 = 9$, and $\vec{B}^2 = 4$. Since $(\vec{A} \cdot \vec{B})^2 = 5^2 = 25 \leq \vec{A}^2 \vec{B}^2 = 9 \cdot 4 = 36$, the Cauchy-Schwartz inequality is satisfied.

(b) asserts that $|A + B| \leq |A| + |B|$. Here $|A| = 3$, $|B| = 2$, $A+B = (2,3,3,1)$, and $|A + B| = \sqrt{23}$. Since $\sqrt{23} \leq 5 \ (= \sqrt{25})$, the triangle inequality is satisfied.

We can now begin to think of $\mathbf{R}^n$ as a geometrical space. As in $\mathbf{E}^3$, the *distance* between two points whose position vectors are $\vec{A}$ and $\vec{B}$ is $|\vec{A} - \vec{B}|$. We also know what it means for two vectors to be *perpendicular*, namely $\vec{A} \cdot \vec{B} = 0$. We can define two non-zero vectors $\vec{A}$ and $\vec{B}$ to be *parallel* if $\vec{A} = k\vec{B}$ for some scalar k. In fact, because of the Cauchy-Schwartz inequality, we can define the *angle* between two vectors $\vec{A}$ and $\vec{B}$ to be $\cos^{-1}(\hat{A} \cdot \hat{B}) = \cos^{-1}(\frac{\vec{A} \cdot \vec{B}}{|\vec{A}| \, |\vec{B}|})$.

Example 2. What is the angle between $\vec{A} = (1,2,2,0)$ and $\vec{B} = (1,1,1,1)$?

Answer. $\cos^{-1}(\frac{\vec{A} \cdot \vec{B}}{|\vec{A}| \, |\vec{B}|}) = \cos^{-1}(\frac{5}{3 \cdot 2}) = \cos^{-1}(\frac{5}{6}) = 33.6°$.

We can give *curves* in $\mathbf{R}^n$ by means of parametric equations, just as in $\mathbf{R}^2$ and $\mathbf{R}^3$. For example, we can define a *curve* in $\mathbf{R}^n$ by a *parameterization*

$$\vec{R}(t) = x_1(t)\hat{e}_1 + x_2(t)\hat{e}_2 + \cdots + x_n(t)\hat{e}_n, \quad a \leq t \leq b$$

where $x_1(t), \ldots, x_n(t)$ are given continuous scalar functions of the parameter t on $[a,b]$. In particular, a *straight line* in $\mathbf{R}^n$ can be given by

$$\vec{R}(t) = \vec{A}t + \vec{R}_0, \quad -\infty < t < \infty,$$

for appropriate constant vectors $\vec{A}$ and $\vec{R}_0$.

Much of the geometric, vector mathematics of $\mathbf{E}^3$ can be generalized to $\mathbf{R}^n, n > 3$. We consider these matters further in Chapters 33 and 34. In particular, in Chapter 33, we shall consider *vector subspaces* and *affine subspaces*. *Vector subspaces* of $\mathbf{R}^n$ are the subsets of $\mathbf{R}^n$ that serve as counterparts in $\mathbf{R}^n$ to (in $\mathbf{R}^3$) the origin, lines and planes through the origin, and all of $\mathbf{R}^3$. *Affine subspaces* of $\mathbf{R}^n$ are the subsets of $\mathbf{R}^n$ that serve as counterparts in $\mathbf{R}^n$ to (in $\mathbf{R}^3$) individual points, arbitrary lines and planes, and all of $\mathbf{R}^3$.

For a given vector $\vec{A} = (a_1,\ldots,a_n)$, the set of vectors $\vec{X} = (x_1,\ldots,x_n)$ which satisfies the equation $\vec{A}\cdot\vec{X} = 0$ (this equation can also be written $a_1 x_1 + \cdots + a_n x_n = 0$) is the set of all vectors $\vec{X}$ that are perpendicular to $\vec{A}$. Considered as points in $\mathbf{R}^n$, these vectors form an "(n-1)-dimensional plane" through the origin, with $\vec{A}$ as "normal vector." (In §34.2, we define and consider "k-dimensional planes" in $\mathbf{R}^n$ for each value of k, $1 < k < n$.)

The language of n-vectors is used widely in science and engineering and for a variety of purposes. Here is a simple example from applied economics: A company makes 57 varieties of canned goods. The per-unit wholesale prices that it sets for its products form a 57-vector $\vec{P} = (p_1, p_2, \ldots, p_{57})$ called the *price vector*. The numbers of units of each variety sold in a given day form a *sales vector* $\vec{S} = (s_1,\ldots,s_{57})$. It follows that the total of dollar sales for that day is given by $\vec{S}\cdot\vec{P}$.

Higher dimensional spaces are used widely in mathematics and physics. One example is the four-dimensional *space-time* of special relativity. Another is the 6n-dimensional *phase space* of theoretical physics, in which the 6n coordinates of a single position vector in $\mathbf{R}^{6n}$ represent the simultaneous position coordinates and momentum components of n given particles in three-dimensional physical space with chosen axes. The changing state (in time) of the collection of n particles is then represented by a single curve in this phase space.

§29.5 Matrices; Column-vectors and Row-vectors

A rectangular array (or "table") of m rows and n columns of numbers is called an m×n *matrix*. We use long parentheses or long square brackets to indicate a single matrix. (The plural of "matrix" is "matrices," pronounced *may-triss-ease*.)

Examples.
$$\begin{bmatrix} 0 & 1 \\ -1 & 0 \\ 2 & \frac{3}{4} \end{bmatrix}$$
is a 3×2 matrix.

$$\begin{bmatrix} 1 & 0 & 0 \\ 0 & 2 & 1 \\ 0 & 3 & 2 \end{bmatrix}$$
is a 3×3 matrix.

$$\begin{bmatrix} 4 \\ 2 \end{bmatrix}$$
is a 2×1 matrix.

$$\begin{bmatrix} -1, 0, \sqrt{2} \end{bmatrix}$$
is a 1×3 matrix.

In matrix algebra, an n×1 matrix is called a *column-vector* or n-*dimensional column-vector*, and a 1×n matrix is called a *row-vector* on n-*dimensional row-vector*. In the examples above, $\begin{bmatrix} 4 \\ 2 \end{bmatrix}$ is a two-dimensional column vector; $\begin{bmatrix} -1, 0, \sqrt{2} \end{bmatrix}$ is a three-dimensional row-vector. A 1×1 matrix will usually be viewed and treated simply as a real number.

If a matrix has the same number of rows as columns, we say that the matrix is *square*. We say that two matrices are *equal* if

(i)   they have the same size, and

(ii)   corresponding entries have the same numerical value. Thus two matrices are equal if they are identical as matrices of numbers.

The individual numerical entries in a matrix are called the *elements* of the matrix. We shall use capital letters A, B, C,... to stand for matrices. As a general notation, we sometimes use the following expression for a matrix:

(5.1)
$$A = \begin{bmatrix} a_{11} & a_{12} & a_{13} & \cdots & a_{1n} \\ a_{21} & a_{22} & a_{23} & \cdots & a_{2n} \\ \vdots & \vdots & \vdots & & \vdots \\ a_{m1} & a_{m2} & \cdot & \cdots & a_{mn} \end{bmatrix}$$

where a_{ij} is the element occurring in the i-th row and j-th column.

Note that the phrases "column-vector" and "row-vector" are used in matrix algebra to indicate matrices of a special kind. They should not be confused with "n-vectors" in vector algebra. The matrix and vector concepts are, of course, closely related, since the elements of an n×1 column-vector or a 1×n row-vector can also be thought of as the components of an n-vector in $\mathbf{R}^n$.

§29.6 Matrix Operations: Multiplication by a Scalar; Addition

We define the following two operations:

(6.1) *Multiplication: a scalar times a matrix.*

$$k \text{ scalar}$$

$$A = \begin{bmatrix} a_{11} & a_{12} & \cdot & \cdot \\ a_{21} & \cdot & \cdot & \cdot \\ \cdot & \cdot & \cdot & a_{mn} \end{bmatrix} \rightarrow kA = \begin{bmatrix} ka_{11} & ka_{12} & \cdot & \cdot \\ ka_{21} & \cdot & \cdot & \cdot \\ \cdot & \cdot & \cdot & ka_{mn} \end{bmatrix}.$$

Example 1.
$$3 \begin{bmatrix} 1 & 0 \\ 0 & 2 \\ -1 & 1 \end{bmatrix} = \begin{bmatrix} 3 & 0 \\ 0 & 6 \\ -3 & 3 \end{bmatrix}.$$

(6.2) *Addition of two matrices.*

$$\begin{bmatrix} a_{11} & a_{12} & \cdot & \cdot \\ \cdot & \cdot & \cdot & \cdot \\ \cdot & \cdot & \cdot & a_{mn} \\ b_{11} & b_{12} & \cdot & \cdot \\ \cdot & \cdot & \cdot & \cdot \\ \cdot & \cdot & \cdot & b_{mn} \end{bmatrix} \rightarrow A + B = \begin{bmatrix} a_{11}+b_{11} & a_{12}+b_{12} & \cdot & \cdot \\ \cdot & \cdot & \cdot & \cdot \\ \cdot & \cdot & \cdot & a_{mn}+b_{mn} \end{bmatrix}.$$

Example 2.
$$\begin{bmatrix} 1 & 0 \\ 0 & 1 \end{bmatrix} + \begin{bmatrix} 2 & 1 \\ -1 & 3 \end{bmatrix} = \begin{bmatrix} 3 & 1 \\ -1 & 4 \end{bmatrix}.$$

Note that matrices may be added only if they are exactly the same size (same number of rows and same number of columns). In this case, we say that the matrices are *conformable for addition.*

$-A$ is defined as $(-1)A$, and $A - B$ is defined as $A + (-B)$.

Example 3. $\begin{bmatrix} 1 & 0 \\ 0 & 1 \end{bmatrix}$ and $\begin{bmatrix} 1 & 0 & 2 \\ 2 & 4 & 1 \end{bmatrix}$ cannot be added.

§29.7 Matrix Operation: Multiplication of a Matrix times a Column-vector

We also have the following operation:

(7.1)   *Multiplication: a matrix times a column-vector.*

$$A = \begin{bmatrix} a_{11} & a_{12} & \cdots & \cdots \\ \cdots & \cdots & \cdots & \cdots \\ \cdots & \cdots & \cdots & a_{mn} \end{bmatrix}$$

$$C = \begin{bmatrix} c_1 \\ c_2 \\ \vdots \\ c_n \end{bmatrix}$$

$$\rightarrow AC = \begin{bmatrix} a_{11}c_1 + a_{12}c_2 + \cdots + a_{1n}c_n \\ a_{21}c_1 + a_{22}c_2 + \cdots + a_{2n}c_n \\ \cdots \quad \cdots \quad \cdots \quad \cdots \\ a_{m1}c_1 + \cdots \quad \cdot + a_{mn}c_n \end{bmatrix}.$$

Example 1. $\begin{bmatrix} 1 & 0 \\ 0 & 2 \\ -1 & 1 \end{bmatrix} \begin{bmatrix} 4 \\ 2 \end{bmatrix} = \begin{bmatrix} 4 \cdot 1 & + & 2 \cdot 0 \\ 4 \cdot 0 & + & 2 \cdot 2 \\ 4 \cdot (-1) & + & 2 \cdot 1 \end{bmatrix} = \begin{bmatrix} 4 \\ 4 \\ -2 \end{bmatrix}.$

Note that:

(i)   The product only exists if the number of rows in C is equal to the number of columns in A. In this case we say that A and C are *conformable for multiplication.*

(ii)   The product AC is a column-vector with the same number of rows as the matrix A.

(iii)   The j-th element of AC is the scalar product of the n-vector whose components form the j-th row of A with the n-vector whose components form the column-vector C.

(iv)   We always write the matrix as the first factor and the column-vector as the second factor.

Example 2. $\begin{bmatrix} 1 & 0 \\ 0 & 2 \\ -1 & 1 \end{bmatrix}$ and $\begin{bmatrix} 0 \\ 1 \\ 1 \end{bmatrix}$ cannot be multiplied.

§29.8 Matrix Operations: Transposition; Inner Product

(8.1)   *Definition.* Let A be any given m×n matrix. We can obtain a new matrix B by taking the matrix A and doing the following. For each i, $1 \le i \le n$, we take the elements of the i-th *column* of A (reading from top to bottom) and use them to form the i-th *row* of B (reading from left to right). The matrix B that we obtain from A in this way is called the *transpose of* A. We use the notation A^T to denote the transpose of A. Since A is an m×n matrix, A^T must be an n×m matrix.

Examples.
$$A = \begin{bmatrix} 1 & 3 \\ 3 & -1 \\ 4 & 0 \end{bmatrix} \Rightarrow A^T = \begin{bmatrix} 1 & 3 & 4 \\ 3 & -1 & 0 \end{bmatrix}$$

$$A = \begin{bmatrix} 2 & 1 \\ 0 & 3 \end{bmatrix} \Rightarrow A^T = \begin{bmatrix} 2 & 0 \\ 1 & 3 \end{bmatrix}$$

$$A = \begin{bmatrix} 1 \\ 2 \\ 3 \end{bmatrix} \Rightarrow A^T = \begin{bmatrix} 1 & 2 & 3 \end{bmatrix}$$

$$A = \begin{bmatrix} 2 & -2 \end{bmatrix} \Rightarrow A^T = \begin{bmatrix} 2 \\ -2 \end{bmatrix}$$

$$A = \begin{bmatrix} 3 & 1 & 2 \\ 1 & 0 & 5 \\ 2 & 5 & 4 \end{bmatrix} \Rightarrow A^T = A$$

If $A^T = A$ (as in the last example), we say that the matrix A is *symmetric.* Note that a symmetric matrix is necessarily square.

The operation of forming A^T from A is called *transposition.* We can give a brief and simple description of transposition as follows: the transpose of an m×n matrix A is an n×m matrix B such that for all i ≤ n and j ≤ m, $b_{ij} = a_{ji}$.

Thus an n×n symmetric matrix A is a matrix such that for all i,j ≤ n, $a_{ij} = a_{ji}$.

(8.2) *Definition.* If C and D are both n×1 column-vectors, then C^T is a 1×n row-vector, and the matrix product $C^T D$ gives a real number (1×1 matrix) as its result. This real number is called the *inner product* of the two column-vectors C and D.

We note that $C^T D$, the inner product of C and D, yields the same real number as the scalar product (in vector algebra) of the two n-vectors whose components are the elements of the column vectors C and D. We also note that the matrix operations of addition of column-vectors and of multiplication of a column-vector by a scalar give the same results for column-vectors that the corresponding vector-algebra operations give for n-vectors. We thus see that all the operations and equations of the vector algebra in §29.2 can be translated into operations and equations of matrix algebra, provided that we translate n-vectors into n-dimensional column-vectors (n×1 matrices). Matrix algebra therefore includes, as a special case, a translated version of vector algebra. The two systems, *vector algebra* and *matrix algebra*, are distinct algebraic systems, however, and use different notations and terminology. The vector algebra of $\mathbf{R}^n$ is closely analogous to the familiar vector algebras of $\mathbf{E}^2$ and $\mathbf{E}^3$. Matrix algebra is richer and more flexible, and permits us to express a wider variety of operations and relationships. We shall make use of both systems. It will always be clear, from notation and from context, which system is being used at a given moment. For example, the notation $\vec{A} \cdot \vec{B}$ will indicate a scalar product (of two n-vectors) in vector algebra, while the notation $A^T B$ will indicate an inner product (of two column-vectors) in matrix algebra.

For convenience, we carry over two special notations from vector algebra to matrix algebra. Given a column vector $C = \begin{bmatrix} c_1 \\ \vdots \\ c_n \end{bmatrix}$, we define

$$|C| = \sqrt{C^T C} = \sqrt{c_1^2 + \cdots + c_n^2} \; ,$$

and we define

$$\hat{C} = (\tfrac{1}{|C|}) C \, .$$

We speak of $|C|$ as the *magnitude* of the column-vector C and of $\hat{C}$ as the *unit column-vector* for C. Sometimes, when it is clear that we are using matrix algebra and that there is no danger of confusion, we shall speak of column-vectors simply as "vectors."

§29.9 Special Matrices; Laws of Matrix Algebra

If a matrix consists entirely of zeroes, we call it a *zero matrix* and use O to denote it. (Note that there is a different zero matrix for each choice of m and n.)

If an n×n square matrix consists of all 1's on the diagonal from upper left to lower right and 0's everywhere else, we call it an *identity matrix* and use I to stand for it. (Note that there is a different identity matrix for each choice of n.)

In a square matrix, the diagonal of elements from the upper left corner to the lower right corner is called the *main diagonal* of the matrix. Hence, an identity matrix is a square matrix whose elements on the main diagonal are all 1 and whose other elements are all 0.

$$\text{Thus} \quad \begin{bmatrix} 0 & 0 \\ 0 & 0 \\ 0 & 0 \end{bmatrix} \text{ is the } 3\times 2 \text{ zero matrix, and } \begin{bmatrix} 1 & 0 & 0 \\ 0 & 1 & 0 \\ 0 & 0 & 1 \end{bmatrix} \text{ is the } 3\times 3 \text{ identity}$$

matrix.

It is easy to verify that the following *laws* are satisfied by the operations of matrix algebra: (X and Y are column vectors. In each law, we assume conformability as needed.)

$$
\begin{aligned}
(9.1) \qquad & A + B = B + A & & (A+B)X = AX + BX \\
& A + (B+C) = (A+B) + C & & A(X+Y) = AX + AY \\
& A + O = A & & k(AX) = (kA)X \\
& k(A+B) = kA + kB & & A(kX) = k(AX) \\
& (k+\ell)A = kA + \ell A & & (A^T)^T = A \\
& OX = O & & (A+B)^T = A^T + B^T \\
& \quad \uparrow \quad \uparrow \\
& \text{(matrix O)} \ \ \text{(column vector O)} & & (kA)^T = kA^T \\
& \qquad IX = X & & X^T Y = Y^T X
\end{aligned}
$$

We return to the study of matrix algebra in Chapter 32. In §32.1 we define a more general *multiplication* operation for matrices, and in §32.3 we define an *inversion* operation for square matrices. The inversion operation will then lead us to a general *division* operation for matrices.

§29.10 Problems

§29.2

1. Let $\vec{A} = (1,-2,0,1,3)$ and $\vec{B} = (2,0,4,0,1)$. Find $2\vec{A} - 3\vec{B}$.

2. Find vectors $\vec{C}$ and $\vec{D}$ such that $\vec{B} = \vec{C} + \vec{D}$, $\vec{C} = k\vec{A}$, and $\vec{D} \cdot \vec{A} = 0$, for $\vec{A}$ and $\vec{B}$ in 1.

§29.3

1. For the vector operations defined in $\mathbf{R}^n$, prove
   (a) $\vec{A} \cdot \vec{B} = \vec{B} \cdot \vec{A}$
   (b) $k(\ell\vec{A}) = (k\ell)\vec{A}$
   (c) $\vec{A} \cdot (\vec{B} + \vec{C}) = \vec{A} \cdot \vec{B} + \vec{A} \cdot \vec{C}$

2. (a) Find unit vectors in $\mathbf{R}^4$ for $(3,-4,0,0)$, $(4,3,0,0)$, $(0,0,3,4)$, $(0,0,4,-3)$.
   (b) Show that these unit vectors form a frame.
   (c) Find the scalar components of $(1,1,1,0)$ in this frame.
   (d) Check your answers to (c) by showing that the frame identity holds.

3. In this problem, X and Y are unknown matrices, not necessarily column vectors.

(a) Solve for X and Y in terms of A and B.

$$\begin{pmatrix} 3X = A \\ X + Y = B \end{pmatrix}$$

(b) solve for X, Y, and k in terms of A and B.

$$\begin{pmatrix} kX = A \quad (k \neq 0) \\ X + Y = B \\ X + A = 2A + 2A \end{pmatrix}$$

(c) What can you conclude for (b) if k = 0?

§29.4

1. Find the angle between $\vec{A}$ and $\vec{B}$, for $\vec{A}$ and $\vec{B}$ in problem 1 for §29.2.

2. Let $\vec{A} = (a_1, a_2, \ldots, a_n)$ and $\vec{B} = (b_1, b_2, \ldots, b_n)$. What does the Cauchy-Schwartz inequality assert about the quantities $a_1, \ldots, a_n, b_1, \ldots, b_n$?

§29.7

1. Let M be the matrix $\begin{bmatrix} 1 & 4 & -2 & -1 & 0 \\ 3 & -1 & 2 & 0 & \frac{1}{2} \end{bmatrix}$. Let A be the 5×1 column vector for the 5-vector $\vec{A}$ given in problem 1 for §29.2. Find MA.

§29.9

1. In each case, if the given equation is an identity (law) of matrix algebra, prove it; if not give an example for which the equation fails.

(a) $IX = X$

(b) $A^T + B = A + B^T$

(c) $(AX)^T Y = (AY)^T X$

(d) $(A+B)X = AX + BX$

Chapter 30

Solving Simultaneous Equations by Row-reduction

§30.1 Systems of Simultaneous Linear Equations

In this chapter, we present a computational technique called *row-reduction* which will be fundamental for our subsequent work on vectors and matrices. We shall first use the technique for numerical calculation. Later, we shall use it as a conceptual tool for proving general facts about vectors and matrices. We shall also use it to illustrate the concepts of n-dimensional geometry and to develop further the relationship of these geometric concepts to the algebra of vectors and matrices.

Row-reduction is the basis for many standard computer programs for vectors and matrices. We shall present row-reduction as a method for solving systems of simultaneous linear equations. In this form, row-reduction can be viewed as a systematic and efficient algorithm for the solution of linear systems by the *method of substitution*. Row-reduction is sometimes referred to as *Gaussian elimination* or as *Gauss-Jordan reduction*.

We consider systems of linear equations of the general kind that students first meet in elementary-school algebra. In school algebra, students study systems of two equations in two unknowns, such as

$$\begin{pmatrix} 3x + 2y = 4 \\ 9x + 6y = 11 \end{pmatrix}.$$

We now consider systems of m equations in n unknowns such as

$$\begin{pmatrix} 2x + y = 4 \\ 3x - y = 7 \\ 7x + y = 15 \end{pmatrix}$$

$$[m = 3, n = 2],$$

or

$$\begin{pmatrix} 3x - 7y + 4z = 0 \\ 2x - y + 3z = 0 \\ x + 5y + 2z = 0 \end{pmatrix}$$

$$[m = n = 3],$$

or

$$\begin{pmatrix} x + y - z = 4 \\ x - y + 2z = 8 \end{pmatrix}$$

$$[m = 2, n = 3].$$

We give a general procedure for finding the *set of all solutions* to such a system.

As standard notation, we use $x_1, x_2, \dots, x_n$ for the n unknowns. Thus, for the final example above, we would write

$$\begin{pmatrix} x_1 + x_2 - x_3 = 4 \\ x_1 - x_2 + 2x_3 = 8 \end{pmatrix}.$$

As notation for a general system, we write the constants on the right-hand side as $d_1, d_2, \dots, d_m$, and we represent the coefficients on the left-hand side as an $m \times n$ matrix A with elements a_{ij}. Thus the system becomes:

$$(1.1) \quad \begin{pmatrix} a_{11}x_1 + a_{12}x_2 + \cdots + a_{1n}x_n = d_1 \\ a_{21}x_1 + a_{22}x_2 + \cdots + a_{2n}x_n = d_2 \\ \cdots\cdots\cdots\cdots \\ a_{m1}x_1 + a_{m2}x_2 + \cdots + a_{mn}x_n = d_m \end{pmatrix},$$

where the coefficients and constants are given real numbers.

$$\text{Finally, let } X = \begin{bmatrix} x_1 \\ x_2 \\ \vdots \\ x_n \end{bmatrix}, \ D = \begin{bmatrix} d_1 \\ d_2 \\ \vdots \\ d_m \end{bmatrix}, \text{ and } A = \begin{bmatrix} a_{11} \ a_{12} \cdots a_{1n} \\ \cdots\cdots\cdots \\ a_{m1} \ a_{m2} \cdots a_{mn} \end{bmatrix}. \qquad \text{Here,}$$

A is an $m \times n$ matrix, D is an $m \times 1$ column-vector, and X is a *variable* $n \times 1$ column-vector. The elements of X are the *variables* (or "unknowns") $x_1, x_2, \ldots, x_n$. We operate with X algebraically in exactly the same way that we operate with a numerical column-vector. We see that the product AX is the $m \times 1$ column vector

$$\begin{bmatrix} a_{11}x_1 + a_{12}x_2 + \cdots + a_{1n}x_n \\ a_{21}x_1 + a_{22}x_2 + \cdots + a_{2n}x_n \\ \cdots\cdots\cdots\cdots \\ a_{m1}x_1 + a_{m2}x_2 + \cdots + a_{mn}x_n \end{bmatrix}.$$

The matrix equation $AX = D$ asserts that two column-vectors are equal:

$$\begin{bmatrix} a_{11}x_1 + \cdots + a_{1n}x_n \\ \cdots\cdots\cdots \\ a_{m1}x_1 + \cdots + a_{mn}x_n \end{bmatrix} = \begin{bmatrix} d_1 \\ \vdots \\ d_m \end{bmatrix}.$$

The matrix equation $AX = D$ thus expresses and abbreviates the given system of simultaneous equations. As we shall see, it provides us with a simple, useful, and compact notation for the system. Henceforth, we shall often use the language and notation of matrices when we consider systems of linear equations and their solutions.

Example. *In the system* $\begin{pmatrix} x_1 + x_2 - x_3 = 4 \\ x_1 - x_2 + 2x_3 = 8 \end{pmatrix}$ *we have*

$$A = \begin{bmatrix} 1 & 1 & -1 \\ 1 & -1 & 2 \end{bmatrix}, \ X = \begin{bmatrix} x_1 \\ x_2 \\ x_3 \end{bmatrix}, \text{ and } D = \begin{bmatrix} 4 \\ 8 \end{bmatrix}.$$

Given an $m \times n$ system $AX = D$, the *solution-set* of the system is defined to be the set of numerical column-vectors which satisfy the system when substituted for the variable column-vector X. We shall sometimes think of and refer to these column-vector solutions as n-*vectors* rather than as *column-vectors*. These n-vectors, in turn, can be thought of as a set of *points* in $\mathbf{R}^n$. We observe that the following possibilities can occur for the solution-set of a linear system.

(1)   There is *no solution*. (The solution-set is empty.)

$\begin{pmatrix} 2x_1 + 3x_2 = 4 \\ 4x_1 + 6x_2 = 5 \end{pmatrix}$ is an example of a system with no solution. If

we multiply the first equation by 2 and subtract from the second, we get $0 = -3$, showing that no solution can exist.

(2)   There is a *unique solution*. (The solution-set has exactly one member.)

$\begin{pmatrix} 2x_1 + 3x_2 = 4 \\ x_1 + 3x_2 = 3 \end{pmatrix}$ is an example of a system with a unique solu-

tion. Applying elementary-school algebra, we find that $x_1 = 1$, $x_2 = \frac{2}{3}$.

(3)   There is *more than one solution*. (The solution-set has more than one member.)

$\begin{pmatrix} 2x_1 + 3x_2 = 4 \\ 4x_1 + 6x_2 = 8 \end{pmatrix}$ is an example of a system with infinitely many

solutions. In fact, any point on the straight line $\begin{bmatrix} x_1 \\ x_2 \end{bmatrix} =$

$\begin{bmatrix} 2 \\ 0 \end{bmatrix} + t \begin{bmatrix} -\frac{3}{2} \\ 1 \end{bmatrix}$ is a solution, as we can verify by substituting into

the given equations. (On the left-hand side of the first equation, for example, we get $2(2 - \frac{3}{2}t) + 3(t) = 4 - 3t + 3t = 4$.)

In the remainder of this chapter, we describe *row-reduction*, and we shall see how, for a general $m \times n$ system, row-reduction enables us (i) to decide which of the above three cases holds; and (ii) to find (in cases (2) and (3)) an expression for the solution set, where this expression will be similar in nature to the expression for the straight line in case (3) above. In Chapters 31 and 32, we give two further special-purpose methods for solving systems for which $m = n$. These methods are *Cramer's rule* (§31.4) and *matrix inversion* (§32.3).

(1.2)   **Definition.** In a system $AX = D$, if $D = O$ (*all* elements in D are zero), we say that the system is *homogeneous*. If $D \neq O$ (*some* element in D is not zero), we call the system *inhomogeneous*. We note the following obvious fact: *A homogeneous system $AX = O$ always has at least one solution, namely $X = O$.*

$X = O$ is sometimes called the *trivial solution* (of a given homogeneous system). Any other solutions, if they exist, are called *non-trivial solutions*.

§30.2 Geometry of Solution-sets for Systems with Two or Three Unknowns

The case n = 2.

For n = 2, the solution-set can be viewed as a set of points in E^2-with-Cartesian coordinates. For a single equation $ax_1 + bx_2 = d$ (where a and b are not both 0), the solution-set is a *straight line* in the plane. Hence for m such equations, the solution-set must be the set of points forming the common intersection of the m straight lines given by the m equations. Thus, for example, when m = n = 2, the possibilities are as follows:

Homogeneous case. Here both lines must go through the origin. Two possibilities occur:
- The lines have different slopes and have the origin as their common intersection. We have a *unique solution*; the solution is the *point* (0,0).
- The lines have the same slope and hence coincide. We have *more than one solution*; the solution-set is the entire common *line*.

Inhomogeneous case. At least one line does not include the origin. Three possibilities occur:
- The lines have different slopes and therefore have a single point (not the origin) as their common intersection. We have a *unique solution*; the solution is the *point* of intersection.
- The lines have the same slope and coincide. We have *more than one solution*; the solution-set is the common *line*.
- The lines have the same slope, but have no common point. Here we have two distinct parallel lines; the solution-set is *empty*.

The case n = 3.

In this case, the solution-set can be viewed as a set of points in E^3-with-Cartesian-coordinates. For a single equation $ax_1 + bx_2 + cx_3 = d$ (where a, b, and c are not all 0), the solution-set is a *plane*. Hence for m such equations, the solution-set must be the set of points forming the common intersecion of the m planes given by the equations. Thus, for example, when m = n = 3, the possibilities are as listed below.

Recall that a set of vectors in $\mathbf{E}^3$-with-coordinates is *coplanar* if the position arrows for these vectors *lie in a common plane through the origin*. We also define a set of vectors to be *collinear* if their position arrows *lie on a common line through the origin*.

Homogeneous case. Here all three planes go through the origin. Three possibilities occur:
- The normal vectors to the three planes are not coplanar; hence the planes have the origin as their common intersection. We have a *unique solution*; the solution is the *point* (0,0,0).
- The normal vectors are coplanar but not collinear, and hence the planes have a single straight line as their common intersection (the line through the origin perpendicular to the common plane of the normal position arrows). We have *more than one solution*; the solution-set is this *line*.
- The normal vectors are collinear, and hence all three planes coincide. We have *more than one solution*; the solution-set is this common *plane*.

Inhomogeneous case. At least one plane does not include the origin. Five possibilities occur:
- The normal vectors are not coplanar, and the planes therefore have a single point (not the origin) as common intersection. We have a *unique solution*; the solution is this *point* of intersection.
- The normal vectors are coplanar but not collinear, and all three planes intersect in a single common line. We have *more than one solution*; the solution-set is this common *line*.
- The normal vectors are coplanar but not collinear, and the three planes have no common intersection. The solution-set is *empty*. This case arises when two planes intersect in a line parallel to the third plane.
- The normal vectors are collinear, and all three planes coincide. We have *more than one solution*; the solution-set is this common *plane*.
- The normal vectors are collinear, but at least two of the planes are parallel and distinct, so that the three planes have no common intersection. The solution-set is *empty*.

Our approach to simultaneous linear equations for $n > 3$ will, in effect, provide results that are analogous (in n-dimensional geometry) to the geometric pictures above for $n = 2$ and $n = 3$. Note that in both the case $n = 2$ and the case $n = 3$, a system of equations with more than one solution-point necessarily has infinitely many solution-points. We shall see below that this is true for all systems of simultaneous linear equations.

§30.3 Elementary Operations and Row-reduction

Row-reduction is a procedure for putting any given system of equations into a simple form which has the same solution-set as the original system and in which the geometric nature of that solution-set (in n-dimensional geometry) is made obvious. We begin with some examples.

Example 1. $\begin{pmatrix} 2x_1 + 3x_2 = 1 \\ x_1 + 3x_2 = -1 \end{pmatrix} \underset{(\alpha)}{\to} \begin{pmatrix} x_1 + \frac{3}{2}x_2 = \frac{1}{2} \\ x_1 + 3x_2 = -1 \end{pmatrix}$

$\underset{(\beta)}{\to} \begin{pmatrix} x_1 + \frac{3}{2}x_2 = \frac{1}{2} \\ \frac{3}{2}x_2 = -\frac{3}{2} \end{pmatrix} \underset{(\alpha)}{\to} \begin{pmatrix} x_1 + \frac{3}{2}x_2 = \frac{1}{2} \\ x_2 = -1 \end{pmatrix} \underset{(\beta)}{\to} \begin{pmatrix} x_1 \quad\quad = 2 \\ x_2 = -1 \end{pmatrix}.$

Here we have performed operations of two kinds:

 (α)   *alter an equation by multiplying it by a non-zero constant;*
 (β)   *alter an equation by adding to it some multiple of another equation.*

(In the first application of β in the example above, the multiplying factor is –1; in the second application, it is $-\frac{3}{2}$.) We shall also allow a third kind of operation:

 (γ)   *alter the system by interchanging two equations.*

Operations of these three kinds are called *elementary operations.*

Example 2. $\begin{pmatrix} 2x_2 - x_3 = 0 \\ x_1 - x_2 + x_3 = 0 \\ 2x_1 \quad\quad + x_3 = 0 \end{pmatrix} \underset{(\gamma)}{\to} \begin{pmatrix} x_1 - x_2 + x_3 = 0 \\ 2x_2 - x_3 = 0 \\ 2x_1 \quad\quad + x_3 = 0 \end{pmatrix} \underset{(\beta)}{\to}$

$\begin{pmatrix} x_1 - x_2 + x_3 = 0 \\ 2x_2 - x_3 = 0 \\ 2x_2 - x_3 = 0 \end{pmatrix} \underset{(\alpha)}{\to} \begin{pmatrix} x_1 - x_2 + x_3 = 0 \\ x_2 - \frac{1}{2}x_3 = 0 \\ 2x_2 - x_3 = 0 \end{pmatrix} \underset{(\beta)}{\to}$

$\begin{pmatrix} x_1 \quad\quad + \frac{1}{2}x_3 = 0 \\ x_2 - \frac{1}{2}x_3 = 0 \\ 2x_2 - x_3 = 0 \end{pmatrix} \underset{(\beta)}{\to} \begin{pmatrix} x_1 \quad\quad + \frac{1}{2}x_3 = 0 \\ x_2 - \frac{1}{2}x_3 = 0 \\ 0 = 0 \end{pmatrix}.$

In both examples, the final system is said to be in *row-reduced form*. (We define this more precisely in §30.4 below.) We note that when elementary operations are applied to a homogeneous system, the system remains homogeneous.

The usefulness of elementary operations for solving systems of equations rests on the following fact:

(3.1)   Theorem. *If a system of equations is altered by a succession of elementary operations, the new system has the same solution-set as the original system.*

Proof. It is enough to show that a single elementary operation leaves the solution-set unchanged. It then follows that any finite number of elementary operations must leave the solution set unchanged. To see that a single elementary operation leaves the solution set unchanged, we observe:
(i) that if an n-vector is a solution of a system before an elementary operation is performed, it must remain a solution of the new system obtained by that operation; and (ii) that any elementary operation can be reversed by an elementary operation (in fact, by an elementary operation of the same kind). See the problems. These two facts logically imply that an elementary operation must leave the solution-set unchanged: (i) implies that every solution of the old system must be a solution of the new system; (i) and (ii) together imply that every solution of the new system must be a solution of the old system. Hence the old system and the new system have exactly the same solution-set. This completes the proof. Note that this theorem applies to both homogeneous and inhomogeneous systems.

§30.4 Row-reducing a Matrix

It is easier, in carrying out the row-reduction process for a *homogeneous* system $AX = 0$, just to work with the matrix A. The operations then take the form:

(α)   *alter a row by multiplying it by a non-zero constant;*
(β)   *alter a row by adding to it some multiple of another row;*
(γ)   *interchange two rows.*

Matrix operations of these three kinds are called *elementary row-operations*. We do an example in matrix form.

$$\textbf{\textit{Example 1.}} \quad \begin{pmatrix} x_1 + 2x_2 - x_3 = 0 \\ x_1 + 2x_2 + x_3 = 0 \\ x_1 + 2x_2 - 2x_3 = 0 \end{pmatrix} \rightarrow \begin{bmatrix} 1 & 2 & -1 \\ 1 & 2 & 1 \\ 1 & 2 & -2 \end{bmatrix} \xrightarrow{(\beta)(\beta)}$$

$$\begin{bmatrix} 1 & 2 & -1 \\ 0 & 0 & 2 \\ 0 & 0 & -1 \end{bmatrix} \underset{(\alpha)(\beta)(\beta)}{\rightarrow} \begin{bmatrix} 1 & 2 & 0 \\ 0 & 0 & 1 \\ 0 & 0 & 0 \end{bmatrix} \rightarrow \begin{pmatrix} x_1 + 2x_2 & = 0 \\ & x_3 = 0 \\ & 0 = 0 \end{pmatrix}.$$

Note that at the end, in this example and in the second example in §30.3, we have a matrix of the following special form, which we call *row-reduced form*:

(4.1)   (i)   *every non-zero row has a 1 as its first non-zero element; (we refer to these initial 1's as* pivots*);*
  (ii) *each pivot is the only non-zero element in its column;*
  (iii) *pivots occur farther and farther to the right as we move down from row to row;*
  (iv) *each row consisting entirely of 0's (if there are any such rows) occurs below all the rows containing one or more non-zero elements.*

If a homogeneous system has its matrix in row-reduced form, we say that *the homogeneous system is in row-reduced form.*

(4.2)   *Our **general procedure** for getting a matrix to row-reduced form is to work through the columns from the left.*

In the first non-zero column, we take a non-zero element and make it 1 by an α-operation. We then make all other elements in that column 0 by β-operations. We then move our 1 to the first row by a γ-operation.

In the resulting matrix, we move to the first column which has a non-zero element below its first row. We take such an element, make it 1 by an α-operation, make all other elements in the column 0 by β-operations, and move the 1 up to the second row by a γ-operation.

In the resulting matrix, we move to the first column having a non-zero element below its second row, take such an element and make it 1 by an α-operation, make all other elements in the column 0 by β-operations, and move the 1 up to the third row by a γ-operation.

And so forth.

We continue in this way until, in our final matrix, every non-zero row has a pivot as its first non-zero element.

***Example 2.** Put* $\begin{bmatrix} 0 & 0 & 1 & 5 & 0 \\ 2 & 0 & 0 & 0 & 3 \\ 4 & 0 & 4 & 0 & 1 \\ 1 & 0 & 2 & 0 & 1 \end{bmatrix}$ *in row-reduced form.*

$$\xrightarrow[(\beta)(\beta)]{}
\begin{bmatrix}
0 & 0 & 1 & 5 & 0 \\
0 & 0 & -4 & 0 & 1 \\
0 & 0 & -4 & 0 & -3 \\
1 & 0 & 2 & 0 & 1
\end{bmatrix}
\xrightarrow[(\gamma)]{}
\begin{bmatrix}
1 & 0 & 2 & 0 & 1 \\
0 & 0 & -4 & 0 & 1 \\
0 & 0 & -4 & 0 & -3 \\
0 & 0 & 1 & 5 & 0
\end{bmatrix}$$

$$\xrightarrow[(\beta)(\beta)(\beta)]{}
\begin{bmatrix}
1 & 0 & 0 & -10 & 1 \\
0 & 0 & 0 & 20 & 1 \\
0 & 0 & 0 & 20 & -3 \\
0 & 0 & 1 & 5 & 0
\end{bmatrix}
\xrightarrow[(\gamma)]{}
\begin{bmatrix}
1 & 0 & 0 & -10 & 1 \\
0 & 0 & 1 & 5 & 0 \\
0 & 0 & 0 & 20 & -3 \\
0 & 0 & 0 & 20 & 1
\end{bmatrix}$$

$$\xrightarrow[(\alpha)(\beta)(\beta)(\beta)]{}
\begin{bmatrix}
1 & 0 & 0 & 0 & \frac{3}{2} \\
0 & 0 & 1 & 0 & -\frac{1}{4} \\
0 & 0 & 0 & 0 & -4 \\
0 & 0 & 0 & 1 & \frac{1}{20}
\end{bmatrix}
\xrightarrow[(\gamma)(\alpha)(\beta)(\beta)(\beta)]{}
\begin{bmatrix}
1 & 0 & 0 & 0 & 0 \\
0 & 0 & 1 & 0 & 0 \\
0 & 0 & 0 & 1 & 0 \\
0 & 0 & 0 & 0 & 1
\end{bmatrix}.$$

In §30.5, we see that we can get a simple and informative expression for the *solution-set* of any homogeneous system $AX = O$ from the row-reduced form of the matrix A. In §30.6, we consider inhomogeneous systems.

Remark. The choice of elementary row-operations for row-reducing a given matrix is not unique. In the last example above, for instance, we could have begun with the second row instead of the fourth row. Even so, as we prove in §30.7, the final row-reduced form of any matrix *is* unique. *That is to say, for a given matrix, every possible choice and order of elementary row-operations which leads to a row-reduced matrix must lead to the same final row-reduced matrix.*

On the other hand, for a given *system of equations*, if the *indexing* (the subscript numbering) of the unknowns is changed, this may result in a different matrix for the system, and this matrix may have a different row-reduced form. For example, consider the system

$$(x + 2y = 0)$$

of 1 equation in 2 unknowns. If we take x_1 for x and x_2 for y, the matrix is $[1,2]$ and the row-reduced form is $[1,2]$. If, instead, we take x_1 for y and x_2 for x, the matrix is $[2,1]$ and the row-reduced form is $[1,\frac{1}{2}]$.

§30.5 Solving Homogeneous Systems

We begin with an example. Assume that after row-reduction, the matrix for a system of 3 equations in 4 unknowns has the form

$$\begin{bmatrix} 1 & 2 & 0 & -3 \\ 0 & 0 & 1 & 4 \\ 0 & 0 & 0 & 0 \end{bmatrix}.$$

This gives the row-reduced system of equations

$$\begin{pmatrix} x_1 + 2x_2 & - 3x_4 = 0 \\ & x_3 + 4x_4 = 0 \\ & 0 = 0 \end{pmatrix}.$$

In such a row-reduced system, the unknowns occurring in the pivot positions of the row-reduced matrix are called *pivot-variables* (or sometimes "basic" or "dependent" variables), and the remaining unknowns are called *parameter-variables* (or sometimes "free" or "independent" variables.) In the example above, x_1 and x_3 are pivot-variables, and x_2 and x_4 are parameter-variables.

(5.1)   *We proceed to get our solution set from the row-reduced equations by the following four steps.*
    First step. *Transpose the parameter-variables to the right-hand side.* In the example, this gives

$$\begin{pmatrix} x_1 = -2x_2 + 3x_4 \\ x_3 = \quad\ \ - 4x_4 \end{pmatrix}.$$

Second step. *Insert equations of the form* $x_j = x_j$ *for each parameter-variable* x_j. *Evidently, this does not alter the solution-set.* In the example, this gives

$$\begin{pmatrix} x_1 = -2x_2 + 3x_4 \\ x_2 = \quad x_2 \\ x_3 = \qquad\ \ - 4x_4 \\ x_4 = \qquad\quad x_4 \end{pmatrix}.$$

Third step. *Write the system as a single equation in column-vectors, with the parameter-variables used as scalar coefficients.* In the example, this gives

$$\begin{bmatrix} x_1 \\ x_2 \\ x_3 \\ x_4 \end{bmatrix} = x_2 \begin{bmatrix} -2 \\ 1 \\ 0 \\ 0 \end{bmatrix} + x_4 \begin{bmatrix} 3 \\ 0 \\ -4 \\ 1 \end{bmatrix}.$$

Observe that this expression gives a solution for each particular choice of values for the parameter-variables.

Final step. *Replace each parameter-variable by a symbol for a parameter on* $(-\infty,\infty)$. In the example, this gives

$$\begin{bmatrix} x_1 \\ x_2 \\ x_3 \\ x_4 \end{bmatrix} = t_1 \begin{bmatrix} -2 \\ 1 \\ 0 \\ 0 \end{bmatrix} + t_2 \begin{bmatrix} 3 \\ 0 \\ -4 \\ 1 \end{bmatrix}, \quad \text{where } -\infty < t_1, t_2 < \infty.$$

Evidently, we can always apply these steps to a row-reduced homogeneous system. We have the following theorem:

(5.2) **Theorem.** *For any given homogeneous system of simultaneous linear equations, the solution-set can be expressed either as:* $X = O$ *(when there are no parameter-variables), or as* $X = t_1 B_1 + t_2 B_2 + \cdots t_\ell B_\ell$, *where* ℓ *is the number of parameter-variables, where* $t_1, t_2, \ldots, t_\ell$ *are parameters on* $(-\infty,\infty)$, *and where* $B_1, \ldots, B_\ell$ *are specific numerical vectors produced by the row-reduction process as described above.*

Example 1. $\begin{pmatrix} 2x_1 + 3x_2 = 0 \\ 4x_1 + 6x_2 = 0 \end{pmatrix}.$

Row reducing, we get $\begin{bmatrix} 2 & 3 \\ 4 & 6 \end{bmatrix} \to \begin{bmatrix} 1 & \frac{3}{2} \\ 0 & 0 \end{bmatrix}$. Thus we have $(x_1 + \frac{3}{2}x_2 = 0)$, and x_2 is a parameter variable.

Step one: Transposing, we get

$$(x_1 = -\frac{3}{2}x_2).$$

Step two: Adding parameter-variable equations, we get

$$\begin{pmatrix} x_1 = -\frac{3}{2}x_2 \\ x_2 = \quad x_2 \end{pmatrix}.$$

Step three: Going to column-vector form, we get

$$\begin{bmatrix} x_1 \\ x_2 \end{bmatrix} = x_2 \begin{bmatrix} -\frac{3}{2} \\ 1 \end{bmatrix}.$$

Final step: Introducing a parameter-symbol, we get

$$\begin{bmatrix} x_1 \\ x_2 \end{bmatrix} = t \begin{bmatrix} -\frac{3}{2} \\ 1 \end{bmatrix}, \quad -\infty < t < \infty.$$

We now see that the solution set is the straight line through the origin given by all scalar multiples of $\begin{bmatrix} -\frac{3}{2} \\ 1 \end{bmatrix} = -\frac{3}{2}\hat{i} + \hat{j}.$

Example 2. Assume that we have reached the row-reduced matrix

$$\begin{bmatrix} 1 & 2 & 0 & -1 & 0 \\ 0 & 0 & 1 & 1 & 0 \\ 0 & 0 & 0 & 0 & 1 \end{bmatrix}.$$

From this we get

$$\begin{pmatrix} x_1 + 2x_2 & & -x_4 & = 0 \\ & & x_3 + x_4 & = 0 \\ & & & x_5 = 0 \end{pmatrix}.$$

The free variables are x_2 and x_4. Thus we have

$$\begin{pmatrix} x_1 = -2x_2 + x_4 \\ x_2 = \quad x_2 \\ x_3 = \qquad -x_4 \\ x_4 = \qquad x_4 \\ x_5 = 0 \end{pmatrix} \rightarrow \begin{bmatrix} x_1 \\ x_2 \\ x_3 \\ x_4 \\ x_5 \end{bmatrix} = t_1 \begin{bmatrix} -2 \\ 1 \\ 0 \\ 0 \\ 0 \end{bmatrix} + t_2 \begin{bmatrix} 1 \\ 0 \\ -1 \\ 1 \\ 0 \end{bmatrix},$$

where $-\infty < t_1, t_2 < \infty$.

Remark. $t_1 B_1 + \cdots + t_\ell B_\ell$ can be a correct expression for the solution set (of a given homogeneous system) for other choices of the numerical vectors $B_1, \ldots, B_\ell$ besides those obtained by row-reduction. In example 1 above, any scalar multiple of $\begin{bmatrix} -\frac{3}{2} \\ 1 \end{bmatrix}$ will serve, and in example 2, for instance, the vectors

$\begin{bmatrix} -2 \\ 1 \\ 0 \\ 0 \\ 0 \end{bmatrix}$ and $\begin{bmatrix} -1 \\ 1 \\ -1 \\ 1 \\ 0 \end{bmatrix}$ (where this last vector is the sum of the two vectors previously

obtained) can serve.

§30.6 Solving Inhomogeneous Systems

Given an inhomogeneous system $AX = D$, we can use the same elementary operations and the same row-reduction process as for the homogeneous case.

Example 1. $\qquad \begin{pmatrix} x_1 + 2x_2 - x_3 = 1 \\ x_1 + 3x_2 + 2x_3 = 2 \\ 2x_1 + 5x_2 + x_3 = 6 \end{pmatrix}.$

Row-reducing we get

$$\begin{pmatrix} x_1 + 2x_2 - x_3 = 1 \\ x_2 + 3x_3 = 1 \\ x_2 + 3x_3 = 4 \end{pmatrix} \rightarrow \begin{pmatrix} x_1 \qquad - 7x_3 = -1 \\ x_2 + 3x_3 = 1 \\ 0 = 3 \end{pmatrix}.$$

We see that the final system has *no* solutions, since the equation $0 = 3$ is false, and no vector X can make this equation true. Hence the original system has no solutions.

In the inhomogeneous case, our solution procedure for a system $AX = D$ must make use of both the matrix A and the column-vector D. In order to include both A and D in a matrix row-reduction, we work with the matrix $[A \vdots D]$ formed by adding D to A as an additional $n+1$st column. $[A \vdots D]$ is called the *augmented matrix* of the system. Our procedure is simply to row-reduce the augmented matrix $[A \vdots D]$. For the example above, this row-reduction gives

$$\begin{bmatrix} 1 & 2 & -1 \cdot 1 \\ 1 & 3 & 2 \cdot 2 \\ 2 & 5 & 1 \cdot 6 \end{bmatrix} \rightarrow \begin{bmatrix} 1 & 0 & -7 \cdot 0 \\ 0 & 1 & 3 \cdot 0 \\ 0 & 0 & 0 \cdot 1 \end{bmatrix}.$$

(6.1)   *After row-reducing the augmented matrix of a given inhomogeneous system, we can draw conclusions about the solution set as follows:*

    (i)   *if the final column of the row-reduced matrix contains a pivot* (as in the example above.), *then the original system has no solution,* since this pivot must correspond to an equation $0 = 1$;

    (ii)   *if the final column has no pivot, then the original system has a solution,* and the solution can be found by the same steps as in the homogeneous case. We illustrate case (ii) in the following examples. We use the same four steps as for a homogeneous system.

Example 2. $\quad \begin{pmatrix} x_1 + 2x_2 - x_3 = 1 \\ x_1 + 3x_2 + 2x_3 = 2 \\ 2x_1 + 5x_2 + x_3 = 3 \end{pmatrix} \rightarrow \begin{bmatrix} 1 & 2 & -1 & \cdot & 1 \\ 1 & 3 & 2 & \cdot & 2 \\ 2 & 5 & 1 & \cdot & 3 \end{bmatrix} \rightarrow$

$$\begin{bmatrix} 1 & 0 & -7 & \cdot & -1 \\ 0 & 1 & 3 & \cdot & 1 \\ 0 & 0 & 0 & \cdot & 0 \end{bmatrix} \rightarrow \begin{pmatrix} x_1 & -7x_3 = -1 \\ & x_2 + 3x_3 = 1 \end{pmatrix}.$$

First step: Transpose the parameter-variable

$$\rightarrow \begin{pmatrix} x_1 = -1 + 7x_3 \\ x_2 = 1 - 3x_3 \end{pmatrix}.$$

Second step: Add an equation for the parameter-variable

$$\rightarrow \begin{pmatrix} x_1 = -1 + 7x_3 \\ x_2 = 1 - 3x_3 \\ x_3 = x_3 \end{pmatrix}.$$

Third step: Write as a single equation in column-vectors

$$\rightarrow \begin{bmatrix} x_1 \\ x_2 \\ x_3 \end{bmatrix} = \begin{bmatrix} -1 \\ 1 \\ 0 \end{bmatrix} + x_3 \begin{bmatrix} 7 \\ -3 \\ 1 \end{bmatrix}.$$

Final step: Use a parameter-symbol

$$\rightarrow \begin{bmatrix} x_1 \\ x_2 \\ x_3 \end{bmatrix} = \begin{bmatrix} -1 \\ 1 \\ 0 \end{bmatrix} + t \begin{bmatrix} 7 \\ -3 \\ 1 \end{bmatrix}, \quad -\infty < t < \infty.$$

In the vector notation of §4.4, this final expression gives the solution set as the straight line $\vec{X} = \vec{X}_0 + t\vec{B}$, where $\vec{X} = (x_1, x_2, x_3)$, $\vec{X}_0 = (-1,1,0)$, and $\vec{B} = (7,-3,1)$.

Example 3. $\begin{pmatrix} x_1 - x_2 + x_3 = 1 \\ x_1 + x_2 - x_3 = 1 \\ x_1 + 2x_2 - x_3 = 2 \end{pmatrix} \rightarrow \begin{bmatrix} 1 & -1 & 1 & \cdot & 1 \\ 1 & 1 & -1 & \cdot & 1 \\ 1 & 2 & -1 & \cdot & 2 \end{bmatrix} \rightarrow$

$$\rightarrow \begin{bmatrix} 1 & -1 & 1 & \cdot & 1 \\ 0 & 2 & -2 & \cdot & 0 \\ 0 & 3 & -2 & \cdot & 1 \end{bmatrix} \rightarrow \begin{bmatrix} 1 & 0 & 0 & \cdot & 1 \\ 0 & 1 & -1 & \cdot & 0 \\ 0 & 0 & 1 & \cdot & 1 \end{bmatrix} \rightarrow \begin{bmatrix} 1 & 0 & 0 & \cdot & 1 \\ 0 & 1 & 0 & \cdot & 1 \\ 0 & 0 & 1 & \cdot & 1 \end{bmatrix} \rightarrow \begin{bmatrix} x_1 \\ x_2 \\ x_3 \end{bmatrix} = \begin{bmatrix} 1 \\ 1 \\ 1 \end{bmatrix}.$$

Here there are no parameter-variables, and the solution is unique.

In the following theorem, we summarize the possible results of row-reduction for an inhomogeneous system:

(6.2)   *Theorem. For any given inhomogeneous system of simultaneous linear equations, row-reduction gives the solution-set either as:*
   (i)   *empty; or*
   (ii) $X = X_0$, *where X_0 is a specific numerical vector produced by the row-reduction process; or*
   (iii) $X = X_0 + t_1 B_1 + t_2 B_2 + \cdots t_\ell B_\ell$, *where ℓ is the number of parameter-variables, where $t_1, \ldots, t_\ell$ are parameters on $(-\infty, \infty)$, and where $X_0, B_1, \ldots, B_\ell$ are specific numerical vectors produced by the row-reduction process.*

Remark. It is possible for the expression $X_0 + t_1 B_1, \ldots, t_\ell B_\ell$ to be a correct expression for the solution-set of a given inhomogeneous system where the numerical vectors $X_0, B_1, \ldots, B_\ell$ are different from those produced by row-reduction. In example 2 above, for instance, any scalar multiple of $\begin{bmatrix} 7 \\ -3 \\ 1 \end{bmatrix}$ can serve in place of B, and any solution at all (of the given inhomogeneous system) can serve in place of X_0.

§30.7 The Rank of a Matrix; Uniqueness of Row-reduced Form

(7.1)   *Definition.* The *rank of a matrix* A is defined to be the number of pivots appearing when A is put in row-reduced form. Equivalently, the *rank of a system of equations* is defined to be the number of pivot-variables appearing after the system is row-reduced. We shall prove that the row-reduced form of a matrix is unique. It follows that the rank of any given matrix is a uniquely defined non-negative integer. We write the rank of a matrix A as $r(A)$.

We shall also later prove (in §33.5) that *if the indexing* (subscript-numbering) *of the unknowns in a given system is changed, the rank of the system remains unchanged*: that is to say, no matter which unknowns we choose as $x_1, x_2, \ldots$, we always end up, after row reduction, with the same number of pivots.

$r(A)$, the rank of the matrix A, is a single integer which presents and summarizes important information about the matrix A. For example, if we know the rank of a matrix, we can draw the following conclusions immediately.

For a homogeneous system $AX = O$, where A is an m×n matrix:

   (i)   *the system has a unique solution if* $r(A) = n$;
   (ii)   *the system has more than one solution if* $r(A) < n$;
and (iii)   *the number of parameter-variables* $= n - r(A)$.

Similarly, for an inhomogeneous system $AX = D$, where A is an m×n matrix:

   (i)   *the system has no solution if* $r(A) < r([A \vdots D])$;
   (ii)   *the system has a unique solution if* $r(A) = r([A \vdots D]) = n$;
   (iii)   *the system has more than one solution if* $r(A) = r([A \vdots D]) < n$.
and (iv)   *if the system has a solution, the number of parameter-variables is* $n - r(A)$.

We now prove:

(7.2)   *Theorem. Given a matrix* A, *row-reduction of* A *must always lead to the same final row-reduced matrix, regardless of the particular choice and order of elementary row-operations used.*

Proof. We give a proof by contradiction. Assume that some matrix A has two distinct row-reduced forms, A_1 and A_2. Then, by (3.1), the homogeneous systems $A_1 X = O$ and $A_2 X = O$ must have the same solution-set. There are two possible cases: (1) A_1 gives the same set of parameter-variables as A_2; (2)

A_1 gives a different set of parameter-variables from A_2. In each case, we shall obtain a contradiction. This will be a proof of our desired conclusion that A cannot have two distinct row-reduced forms.

Case (1). (Same set of parameter-variables.) In this case, A_1 and A_2 must also give the same set of pivot-variables. Take each equation in $A_2 X = O$ and subtract it from the equation in $A_1 X = O$ with the same pivot variable. In the resulting equations, no pivot-variable can appear. Since A_1 and A_2 are distinct, however, some parameter-variable x_k must appear with a non-zero coefficient in at least one of these resulting equations. Dividing through by this coefficient, we can get an equation of the form $x_k = 0$ or $x_k =$ some linear expression in other parameter-variables. Since $A_1 X = O$ and $A_2 X = O$ have the same solution set, this last equation is satisfied by all solutions of the original system. Hence, once we have assigned values to the other parameter variables, we have determined a corresponding unique value for x_k. This contradicts the fact that, in the solution-set of a system, the values associated with a given parameter are independent of the values associated with other parameter variables.

Case (2). (Distinct sets of parameter-variables.) In this case, the sets of pivot-variables for $A_1 X = O$ and $A_2 X = O$ must also be distinct. Let k be the largest subscript such that x_k is a pivot variable in one system but not in the other. Assume that x_k is a pivot variable in $A_1 X = O$ but not in $A_2 X = O$. Then in $A_1 X = O$, we have an equation of the form $x_k = 0$ or $x_k = $ a linear expression in certain parameter-variables which are also parameter-variables in $A_2 X = O$. On the other hand, in $A_2 X = O$, x_k is a parameter-variable along with those same other parameter-variables. Thus we have the same contradiction as in case (1).

The contradictions in cases (1) and (2) show that the row-reduced matrices A_1 and A_2 cannot be distinct; they must be the same matrix.

§30.8 General Solutions

(8.1) *Definition.* An expression in parameters and n-dimensional numerical column-vectors, such as $X_0 + t_1 B_1 + \cdots t_\ell B_\ell$, is called a *general solution* of a given system if, as the parameters range independently over all possible parameter values, the expression gives all possible solutions of the given system. The expression $X_0 + t_1 B_1 + \cdots t_\ell B_\ell$ in (6.2) is a general solution, because (i) any numerical solution gives corresponding values for $t_1, \ldots, t_\ell$, and (ii) every choice of values for $t_1, \ldots, t_\ell$ gives a numerical solution.

If we are given an inhomogeneous system $AX = D$, the system $AX = O$ (with the same matrix A) is called the *associated homogeneous system* of the system $AX = D$. We immediately see, from our row-reduction process, that if $X_0 + t_1 B_1 + \cdots t_\ell B_\ell$ is obtained by row reduction as a general solution to the system $AX = D$, then $t_1 B_1 + \cdots t_\ell B_\ell$ must be a general solution to the associated homogeneous system $AX = O$. In the remark at the end of §30.5, we note that there can be various other general solutions as well (to the homogeneous system $AX = 0$). We now prove the following theorem:

(8.2)   **Theorem.** *Let* X^* *be any fixed specific solution to the system* $AX = D$. *Let the expression* $t_1 B_1 + \cdots t_\ell B_\ell$ *be any general solution to the associated homogeneous system* $AX = O$. *Then the expression* $X^* + t_1 B_1 + \cdots t_\ell B_\ell$ *must be a general solution to the inhomogeneous system* $AX = D$.

Proof.   By the laws of matrix algebra, $A(X^* + t_1 B_1 + \cdots t_\ell B_\ell) = AX^* + A(t_1 B_1 + \cdots t_\ell B_\ell) = D + O = D$. Hence, for all values of $t_1, \ldots, t_\ell$, the expression $X^* + t_1 B_1 + \cdots t_\ell B_\ell$ gives solutions to the system $AX = D$. Conversely, let X' be any given solution vector for $AX = D$. We must show that there are parameter values $t_1, \ldots, t_\ell'$ such that $X' = X^* + t_1' B_1 + \cdots t_\ell' B_\ell$.

Since $AX' = D$ and $AX^* = D$, we have $AX' - AX^* = 0$. Hence $A(X' - X^*) = O$ (by the laws given in §29.9). Thus $X' - X^*$ must be a solution to $AX = O$. Hence, by assumption, $X' - X^* = t_1' B_1 + \cdots t_\ell' B_\ell$ for some choice of parameter values $t_1', \ldots, t_\ell'$. But this gives $X' = X^* + t_1' B_1 + \cdots t_\ell' B_\ell$ as desired. This completes the proof.

This theorem provides us with the following *principle on general solutions:*

(8.3)   *To get a general solution to an inhomogeneous system, it is enough to find:*

      (1)   *some particular solution to the inhomogeneous system, and*
      (2)   *some general solution to the associated homogeneous system.*
A general solution to the inhomogeneous system can then be obtained as the sum of (1) and (2).

Remark. In the study of elementary linear differential equations, analogous concepts of homogeneous and inhomogeneous systems are defined for differential equations, and analogous forms of (8.2) and (8.3) are proved.

§30.9 Integer Solutions; Balancing Chemical Equations

We define a numerical column-vector B to be an *integer vector* if the elements of B are all integers. We define a numerical column-vector B to be a *rational vector* if the elements of B are all rational numbers (quotients of integers).

The following theorem is a direct consequence of the row-reduction process.

(9.1) **Theorem.** *Let* $AX = O$ *be a homogeneous system for which all the elements in the coefficient matrix* A *are integers. If the system has a non-trivial solution, then the general solution to the system can be expressed as* $X = t_1 C_1 + \cdots t_\ell C_\ell$, *where the numerical vectors* $C_1, \ldots, C_\ell$ *are all integer vectors.*

Proof. Assume that the elements in A are all integers. Successive elementary row-operations combine matrix elements under arithmetical addition, subtraction, multiplication, and division. These row-operations can only lead from A to a row-reduced matrix whose elements are rational numbers. Hence we get a general solution $X = t_1 B_1 + \cdots t_\ell B_\ell$, where each numerical vector is a rational vector.

For each rational vector B_j, let d_j be the least common denominator of the elements of B_j. Then $C_j = d_j B_j$ is an integer vector and

$$X = t_1 B_1 + \cdots + t_\ell B_\ell \qquad \text{(where } -\infty < t_1, \ldots, t_\ell < \infty)$$

$$= \frac{t_1}{d_1} d_1 B_1 + \cdots + \frac{t_\ell}{d_\ell} d_\ell B_\ell$$

$$= \frac{t_1}{d_1} C_1 + \cdots + \frac{t_\ell}{d_\ell} C_\ell$$

$$= t_1' C_1 + \cdots + t_\ell' C_\ell \qquad \text{(where } -\infty < t_1', \ldots, t_\ell' < \infty)$$

yields a general solution, in integer vectors, to the original system.

The theorem above provides us with a systematic and universal algorithm for balancing chemical equations. We illustrate with an example.

Example 1. *Consider a reaction in which boron, water, and osmium tetroxide produce osmium and boric acid:*

$$B + H_2O + OsO_4 \rightarrow Os + H_3BO_3 .$$

We seek to obtain a balanced chemical equation by finding integer coefficients x,y,z,u,v such that when we put

$$x\,B + y\,H_2O + z\,OsO_4 = u\,Os + v\,H_3BO_3 ,$$

we then obtain, for each element, the same number of atoms on each side of the equation. That is to say, in this example, x,y,z,u,v must satisfy the equations:

$$
\begin{array}{ll}
x = v & \text{(B balance)} \\
2y = 3v & \text{(H balance)} \\
y + 4z = 3v & \text{(O balance)} \\
z = u & \text{(Os balance)} .
\end{array}
$$

This is a homogeneous linear system in the unknowns x,y,z,u,v. By the preceding theorem on integer solutions, we can use row-reduction to look for an integer-vector non-trivial solution. This will give us a balanced chemical equation.

We let $X = \begin{bmatrix} x \\ y \\ z \\ u \\ v \end{bmatrix}$. We have the coefficient matrix

$$
\begin{bmatrix}
1 & 0 & 0 & 0 & -1 \\
0 & 2 & 0 & 0 & -3 \\
0 & 1 & 4 & 0 & -3 \\
0 & 0 & 1 & -1 & 0
\end{bmatrix} .
$$

Row-reducing, we have

$$
\begin{bmatrix}
1 & 0 & 0 & 0 & -1 \\
0 & 1 & 0 & 0 & -\frac{3}{2} \\
0 & 0 & 4 & 0 & -\frac{3}{2} \\
0 & 0 & 1 & -1 & 0
\end{bmatrix}
\rightarrow
\begin{bmatrix}
1 & 0 & 0 & 0 & -1 \\
0 & 1 & 0 & 0 & -\frac{3}{2} \\
0 & 0 & 1 & 0 & -\frac{3}{8} \\
0 & 0 & 0 & 1 & -\frac{3}{8}
\end{bmatrix} .
$$

Hence

$$X = \begin{bmatrix} x \\ y \\ z \\ u \\ v \end{bmatrix} = t \begin{bmatrix} 1 \\ \frac{3}{2} \\ \frac{3}{8} \\ \frac{3}{8} \\ 1 \end{bmatrix}.$$

To get an integer vector, we multiply the numerical vector by the least common denominator of its elements. This least common denominator is 8. We have $X =$

$$\frac{t}{8} \begin{bmatrix} 8 \\ 12 \\ 3 \\ 3 \\ 8 \end{bmatrix} = t' \begin{bmatrix} 8 \\ 12 \\ 3 \\ 3 \\ 8 \end{bmatrix}, \text{ and we find that the chemical equation is balanced by } \begin{bmatrix} 8 \\ 12 \\ 3 \\ 3 \\ 8 \end{bmatrix}.$$

Thus our final balanced equation is

$$8(B) + 12(H_2O) + 3(OsO_4) = 3(Os) + 8(H_3BO_3) .$$

We define a numerical column-vector to be *positive* if all its elements are positive. We define a numerical column vector to be *semipositive* if at least one of its elements is positive and none of its elements is negative.

In the previous example, we found a positive vector which was a non-trivial solution of the given homogeneous system. This positive vector enabled us to balance the chemical equation. If a homogeneous system obtained from a stated chemical reaction has no non-trivial solution, or if its non-trivial solutions do not include any positive or semipositive vector, then we must conclude that the chemical equation cannot be balanced as stated.

Sometimes a general solution to a homogeneous system will be expressed in terms of more than one given numerical vector. (This occurs when $r(A) < n-1$, where n is the number of unknowns.) In such a case, there may be a positive-vector solution even when none of the numerical vectors obtained is even semipositive. In a problem for §33.6, we shall see that the following theorem is true:

(9.2)   *If a homogeneous system has at least one positive-vector solution, then we can find a general-solution expression* $X = t'_1 B_1 + \cdots t'_\ell B_\ell$ *in which all of the numerical vectors* $B'_1, \ldots, B'_\ell$ *are positive.*

Such a solution-expression will show us that the given chemical equation can be thought of (at least from an algebraic point of view) as representing a combination of ℓ different reactions. We illustrate this with another example.

Example 2. Consider a reaction in which a mixture of hydrogen and methane is burned:

$$H_2 + CH_4 + 2O_2 \rightarrow CO_2 + H_2O$$

Applying our procedure for balancing, we have

$$x\, H_2 + y\, CH_4 + z\, O_2 = u\, CO_2 + v\, H_2O$$

Hence

$$
\begin{array}{ll}
2x + 4y = 2v & \text{(H balance)} \\
y = u & \text{(C balance)} \\
2z = 2u + v & \text{(O balance)}
\end{array}
$$

This gives the matrix

$$
\begin{bmatrix}
2 & 4 & 0 & 0 & -2 \\
0 & 1 & 0 & -1 & 0 \\
0 & 0 & 2 & -2 & -1
\end{bmatrix}
\rightarrow
\begin{bmatrix}
1 & 2 & 0 & 0 & -1 \\
0 & 1 & 0 & -1 & 0 \\
0 & 0 & 1 & -1 & -\frac{1}{2}
\end{bmatrix}
\rightarrow
\begin{bmatrix}
1 & 0 & 0 & 2 & -1 \\
0 & 1 & 0 & -1 & 0 \\
0 & 0 & 1 & -1 & -\frac{1}{2}
\end{bmatrix}.
$$

Hence

$$
\begin{bmatrix} x \\ y \\ z \\ u \\ v \end{bmatrix}
= t_1 \begin{bmatrix} -2 \\ 1 \\ 1 \\ 1 \\ 0 \end{bmatrix}
+ t_2 \begin{bmatrix} 1 \\ 0 \\ -\frac{1}{2} \\ 0 \\ 1 \end{bmatrix}
= t_1' \begin{bmatrix} -2 \\ 1 \\ 1 \\ 1 \\ 0 \end{bmatrix}
+ t_2' \begin{bmatrix} 2 \\ 0 \\ 1 \\ 0 \\ 2 \end{bmatrix}.
$$

Since we have at least one positive solution (take $t_1' = 1$ and $t_1' = 2$), we know, from the theorem stated above about positive vectors, that there is a general-solution expression with positive numerical vectors. In chemical examples, it is enough to find a general-solution expression with semipositive vectors. This is usually easy to do. In our example, we add the two given vectors to get

$$
\begin{bmatrix} -2 \\ 1 \\ 1 \\ 1 \\ 0 \end{bmatrix}
+ \begin{bmatrix} 2 \\ 0 \\ 1 \\ 0 \\ 2 \end{bmatrix}
= \begin{bmatrix} 0 \\ 1 \\ 2 \\ 1 \\ 2 \end{bmatrix}.
$$

We can then show (see the concept of vector subspace in §33.2) that

$$\begin{bmatrix} x \\ y \\ z \\ u \\ v \end{bmatrix} = t''_1 \begin{bmatrix} 0 \\ 1 \\ 2 \\ 1 \\ 2 \end{bmatrix} + t''_2 \begin{bmatrix} 2 \\ 0 \\ 1 \\ 0 \\ 2 \end{bmatrix}$$

is also a general solution.

In this solution, $\begin{bmatrix} 0 \\ 1 \\ 2 \\ 1 \\ 2 \end{bmatrix}$ gives $CH_4 + 2O_2 = CO_2 + 2H_2O$ and corresponds to

the burning of methane, while $\begin{bmatrix} 2 \\ 0 \\ 1 \\ 0 \\ 2 \end{bmatrix}$ gives $2H_2 + O_2 = 2H_2O$ and corresponds to

the burning of hydrogen.

Remark. The theorem stated above for positive vectors does not hold in general when "positive" is replaced by "semipositive." For example, the system

$AX = O$ with $A = [1 \ 1 \ 0]$ has the semipositive solution $\begin{bmatrix} 0 \\ 0 \\ 1 \end{bmatrix}$ but does not have a

general solution $X = t_1 B_1 + t_2 B_2$ in which both B_1 and B_2 are semipositive.

§30.10 Problems

§30.1

1. Given the system

$$\left(\begin{array}{rcl} x_1 + 3x_2 + x_3 - 2x_4 &=& 2 \\ 4x_1 + 7x_2 + 8x_3 - x_4 &=& -1 \\ 3x_1 + 2x_2 - x_3 + 4x_4 &=& 0 \\ x_2 + 2x_3 &=& 3 \end{array} \right),$$

what are the matrices A, X, and D such that the matrix equation $AX = D$ expresses the system?

§30.3

1. Show that every elementary operation on a system of equations can be reversed by an elementary operation of the same kind.

2. Find the solution to the system in problem 1 of §30.1 by row-reduction.

§30.4

1. Take the matrix A (in problem 1 of §30.1) and put it in row-reduced form.

§30.5

1. Give the general solution to the system $AX = O$, where A is in problem 1 of §30.1.

§30.6

1. An inhomogeneous system of three linear equations in four unknowns has the following augmented matrix in row-reduced form,

$$\begin{bmatrix} 1 & 2 & 0 & 1 & \cdot & 7 \\ 0 & 0 & 1 & -4 & \cdot & -3 \\ 0 & 0 & 0 & 0 & \cdot & 0 \end{bmatrix}.$$

Given the general solution of the system.

2. (a) The coefficient matrix for a given homogeneous system is $\begin{bmatrix} 0 & 1 & 2 & 3 \\ 1 & 2 & 4 & 1 \\ 1 & 3 & 6 & 4 \end{bmatrix}$. Row-reduce this matrix and find the general solution.

   (b) For a given inhomogeneous system, the augmented matrix row-reduces to

$$\begin{bmatrix} 1 & -2 & 0 & 3 & \cdot & 9 \\ 0 & 0 & 1 & 8 & \cdot & 2 \\ 0 & 0 & 0 & 0 & \cdot & 0 \end{bmatrix}.$$

   Give the general solution.

3. An inhomogeneous system of three linear equations in four unknowns has the following augmented matrix in row reduced form,

$$\begin{bmatrix} 1 & 2 & -1 & 0 & \cdot & 6 \\ 0 & 0 & 0 & 1 & \cdot & -3 \\ 0 & 0 & 0 & 0 & \cdot & 0 \end{bmatrix}.$$

Give the general solution of the system.

4. An inhomogeneous system of three linear equations in four unknowns has the following augmented matrix in row-reduced form,

$$\begin{bmatrix} 1 & 2 & 0 & 1 & \cdot & 7 \\ 0 & 0 & 1 & -4 & \cdot & -3 \\ 0 & 0 & 0 & 0 & \cdot & 0 \end{bmatrix}.$$

Give the general solution of the system.

5. Given the system of simultaneous equations

$$\begin{aligned} w - x &= -4 \\ x + y &= 2 \\ y - z &= 1 \\ z - w &= 3, \end{aligned}$$

solve the system by row-reduction.

6. Given the system of simultaneous equations

$$\begin{aligned} w - x &= -4 \\ x + y &= 2 \\ y - z &= 1 \\ z + w &= -3, \end{aligned}$$

   (a) solve the system by row-reduction;
   (b) check that the solution obtained satisfies the given equations.

7. Given the system of simultaneous equations

$$\begin{aligned} w - x &= -4 \\ x + y &= 2 \\ y - z &= 1 \\ z + w &= 0, \end{aligned}$$

solve the system by row-reduction.

8. Consider the system

$$\begin{pmatrix} x_1 + x_2 + ax_3 = b \\ x_1 - x_2 + x_3 = 3 \\ x_1 + 2x_2 + x_3 = 1 \end{pmatrix}.$$

Use row-reduction to answer the following questions:

   (a) For what values of a and b is there exactly one solution?
   (b) For what values of a and b is there no solution?
   (c) For what values of a and b is there more than one solution?

§30.7

1. What is the rank of the matrix A in problem 1 of §30.1?

2. In row-reduced form, a homogeneous system of equations $AX = O$ has the coefficient matrix

$$\begin{bmatrix} 1 & 1 & 0 & 2 \\ 0 & 0 & 1 & -1 \\ 0 & 0 & 0 & 0 \end{bmatrix}.$$

(a)   What is the rank of A?

(b)   Give the general solution of the system.

3.   An inhomogeneous system of three linear equations in five unknowns has the following augmented matrix in row-reduced form,

$$\begin{bmatrix} 1 & 0 & 1 & 0 & 2 & \cdot & 2 \\ 0 & 1 & 3 & 0 & 1 & \cdot & -1 \\ 0 & 0 & 0 & 1 & 2 & \cdot & 6 \end{bmatrix}.$$

(a)   What is the rank of this matrix?

(b)   Give a general solution of the system.

4.   Each of the following statements refers to a system of m linear equations in n unknowns with coefficient matrix A and column-vector of constants D. Mark each statement true or false.

(a)   When m < n, a homogeneous system necessarily has more than one solution.

(b)   When m < n, an inhomogeneous system necessarily has more than one solution.

(c)   If $r(A) = r([A \colon D])$, the system must have at least one solution.

(d) $r([A \colon D]) - r(A)$ necessarily equals either 0 or 1.

(e)   When m > n, an inhomogeneous system cannot have any solution.

§30.9

1.   When steam is passed over heated iron filings to produce hydrogen, the reaction is

$$Fe + H_2O \rightarrow Fe_3O_4 + H_2.$$

Show how the algorithm in §30.9 can be used to balance this equation.

2.   To recover the iron in problem 1, carbon monoxide can be passed over the heated iron oxide. The reaction is

$$Fe_3O_4 + CO \rightarrow Fe + CO_2.$$

Show how the algorithm in §30.9 can be used to balance this equation.

Chapter 31

Determinants

§31.1 Laplace Expansions and n×n Determinants

In our previous work on vector algebra and multivariable calculus in $\mathbf{E}^2$ and $\mathbf{E}^3$, we have seen a variety of uses for 2×2 and 3×3 determinants. Determinants were used to describe the vector product operation, to define the Jacobian function, to calculate areas and volumes, and to solve systems of linear equations in two and three unknowns.

Given a 2×2 matrix, $A = \begin{bmatrix} a & b \\ c & d \end{bmatrix}$, we defined $|A|$, the *determinant of* A, to be a certain function of the elements of A; namely,

$$|A| = \begin{vmatrix} a & b \\ c & d \end{vmatrix} = ad - bc.$$

For a 3×3 matrix A, we saw that we could define $|A|$, the *determinant of* A, to be the result obtained by carrying out a Laplace expansion (see §3.7). No matter which row or column we used for the expansion, we always reached the same final value for $|A|$. For example, for

$$A = \begin{bmatrix} a & b & c \\ d & e & f \\ g & h & i \end{bmatrix},$$

$$|A| = \begin{vmatrix} a & b & c \\ d & e & f \\ g & h & i \end{vmatrix}$$ could be calculated as

$$|A| = a\begin{vmatrix} e & f \\ h & i \end{vmatrix} - b\begin{vmatrix} d & f \\ g & i \end{vmatrix} + c\begin{vmatrix} d & e \\ g & h \end{vmatrix},$$

or as
$$|A| = -b\begin{vmatrix} d & f \\ g & i \end{vmatrix} + e\begin{vmatrix} a & c \\ g & i \end{vmatrix} - h\begin{vmatrix} a & c \\ d & f \end{vmatrix}.$$

We shall now define n×n determinants, for any integer n > 3, by extending these ideas to n×n matrices. As we might expect, n×n determinants (also known as *n-th order* determinants) will be useful in the algebra of n-vectors, in the geometry of $\mathbf{R}^n$ (especially in connection with measures of area and volume in higher dimensions), in the solution of systems of equations in n unknowns, in the further development of matrix algebra, and in the development of multivariable calculus in n dimensions. We shall encounter some of these applications in the later sections of this chapter and in Chapters 32, 33, and 34.

Determinants have a variety of other applications in mathematics, science, and engineering, which are not covered in this book. These further applications range from the solution of differential equations in mathematics to formulation of theories in such diverse fields as mathematical economics and quantum mechanics.

An *n-th order* (or n×n) *determinant* (for n ≥ 2) will be expressed as an n×n square matrix written between vertical bars:

$$\begin{vmatrix} a_{11} & a_{12} & \cdots & a_{1n} \\ a_{21} & \cdots & \cdots & \cdots \\ \cdots & \cdots & \cdots & \cdots \\ a_{n1} & a_{n2} & \cdots & a_{nn} \end{vmatrix}.$$

The bars indicate that a certain numerical value (defined below) is to be associated with the given matrix. In §3.7, we saw how this value is defined for 2×2 and 3×3 determinants, and we noted some facts concerning 2×2 and 3×3 determinants. In particular, we defined the *Laplace expansions* of a third-order determinant. We now define (inductively) the values of higher-order determinants $(4 \times 4, 5 \times 5, \ldots)$ by using Laplace expansions in terms of already defined, smaller-size determinants.

Example. $\begin{vmatrix} 1 & 3 & 1 & -2 \\ 4 & 7 & 8 & -1 \\ 3 & 2 & -1 & 4 \\ 0 & 1 & 2 & 0 \end{vmatrix}$. The Laplace expansion in the first row is

$$1\begin{vmatrix} 7 & 8 & -1 \\ 2 & -1 & 4 \\ 1 & 2 & 0 \end{vmatrix} - 3\begin{vmatrix} 4 & 8 & -1 \\ 3 & -1 & 4 \\ 0 & 2 & 0 \end{vmatrix} + 1\begin{vmatrix} 4 & 7 & -1 \\ 3 & 2 & 4 \\ 0 & 1 & 0 \end{vmatrix} - (-2)\begin{vmatrix} 4 & 7 & 8 \\ 3 & 2 & -1 \\ 0 & 1 & 2 \end{vmatrix}$$

$$= -29 + 114 - 19 + 4 = 70 .$$

The expansion may be on any row or column. It has alternating signs as indicated in the example, except that for an even-numbered row or column, we begin the alternating signs with a minus. Thus the expansion of the determinant above in its second column is

$$-3\begin{vmatrix} 4 & 8 & -1 \\ 3 & -1 & 4 \\ 0 & 2 & 0 \end{vmatrix} + 7\begin{vmatrix} 1 & 1 & -2 \\ 3 & -1 & 4 \\ 0 & 2 & 0 \end{vmatrix} - 2\begin{vmatrix} 1 & 1 & -2 \\ 4 & 8 & -1 \\ 0 & 2 & 0 \end{vmatrix} + 1\begin{vmatrix} 1 & 1 & -2 \\ 4 & 8 & -1 \\ 3 & -1 & 4 \end{vmatrix}$$

$$= 114 - 140 + 68 + 28 = 70 .$$

In the same way, a Laplace expansion for a given 5×5 determinant can be done in terms of 4×4 determinants, and then those 4×4 determinants can be expanded in terms of 3×3 determinants. Similarly for 6×6, 7×7, ... determinants. The following theorem justifies our use of Laplace expansions.

(1.1)   *Fundamental theorem on Laplace expansions.* *Given an* n×n *determinant, no matter what rows and/or columns we use for successive Laplace expansions (in successively smaller determinants), we eventually obtain the same final value for the determinant.*

The proof of this theorem appears in §31.6.

     We shall use the following terminology in connection with Laplace expansions.

Definitions. Given an element in an n×n matrix A, the *minor* of that element is defined to be the determinant of the (n–1)×(n–1) matrix that remains after we delete, from A, the row and the column of the given element. The *cofactor* of a given element is defined to be the value of the minor of the given element, multiplied by +1 or −1, depending on the position of the given element in the n×n checkerboard pattern

$$\begin{vmatrix} + & - & + & - & \cdots \\ - & + & - & + & \cdots \\ + & - & + & - & \cdots \\ \cdot & \cdot & \cdot & \cdot & \cdots \end{vmatrix}.$$

In other words, if m is the value of the *minor* of the element a_{ij} of A, then

$(-1)^{i+j}m$ is the *cofactor* of a_{ij} . Thus in the 4×4 determinant above, the cofactor

of the first element in the first row is $\begin{vmatrix} 7 & 8 & -1 \\ 2 & -1 & 4 \\ 1 & 2 & 0 \end{vmatrix} = -29$, while the cofactor of the

second element in the first row is $-\begin{vmatrix} 4 & 8 & -1 \\ 3 & -1 & 4 \\ 0 & 2 & 0 \end{vmatrix} = 38$.   Using the concept of

cofactor, we can describe the Laplace expansion on any given row by the following simple statement: *multiply each element in the row by its cofactor, then take the sum of those products.* Similarly for the Laplace expansion on any given column.

    Remarks. (1) Note that only a *square* matrix can have a determinant. (2) We have defined the *value* $|A|$ of the determinant of an n×n matrix A, for $n \geq 3$. In future work, we shall also refer to the determinant $|A|$ of a 1×1 matrix A = [a]. (Here, a is a single number.) In such cases, the *determinant of* [a] is defined simply to be the number a. We thus have for A = [a], $|A| = |[a]| = a$.

    For example, if A = [−3], we have $|A| = |[-3]| = -3$.

    We do not use the notation $|a|$ for the 1×1 determinant $|[a]|$, since this notation could be confused with (and might not agree with) the familiar absolute-value notation. For example, $|[-3]| = -3$ but, as absolute value, $|-3| = 3$.

    We can now take the Laplace expansion of a 2×2 determinant in terms of 1×1 determinants, and the result will be correct. For example, expanding on the first row, we have

$$\begin{vmatrix} a & b \\ c & d \end{vmatrix} = a|[d]| - b|[c]| = ad - bc \ .$$

§31.2 Basic Properties of Determinants

(2.1)   *The following properties, previously stated in §3.7 for 2×2 and 3×3 determinants, also hold for* n×n *determinants.*

      (1)   *Altering a determinant, by interchanging two rows, multiplies the value of the determinant by* -1.

      (2)   *If a determinant has two identical rows, its value is* 0.

      (3)   *Altering a determinant, by multiplying each element in some chosen row by the same constant, multiplies the value of the determinant by that constant.*

      (4)   *Altering a determinant, by adding to a given row any multiple of another row, leaves the determinant's value unchanged.*

      (5)   *A determinant's value* $= 0 \Leftrightarrow$ *there is some row that can be expressed as a linear combination of other rows.*

               (Let $A, B_1, ..., B_k$ be (row-vectors for) rows in a determinant. A is a *linear combination* of $B_1, ..., B_k$ if $A = \alpha_1 B_1 + \cdots + \alpha_k B_k$ for some scalars $\alpha_1, ..., \alpha_k$.)

      (6)   *Statements* (1)–(5) *all hold with "column" substituted for "row."*

Proofs of these properties are given in §31.7.

§31.3 Evaluating Determinants

If we evaluate a determinant of order n by doing a succession of Laplace expansions which go to smaller and smaller determinants, we eventually obtain a sum of n! terms. This means, for a 50×50 determinant (not uncommon in mathematical economics, for example), $50! \geq 10^{64}$ terms. (This number exceeds the number of elementary particles in the known universe.) Such a calculation would exceed the capacity of any computer which calculated the different terms at different times or stored them at different locations within itself. Calculating a determinant's value by successive Laplace expansions is sometimes called the *brute-force method*. In practice, for determinants of order higher than 4, the brute-force method is seldom used. Instead, a method known as the *row-reduction method* (sometimes called the *condensation* method or the method of *pivotal condensation*) is used. In the row-reduction method, *the matrix of the determinant is simplified by elementary row-operations of the following two kinds:*

      (β)   *alter any row by adding to it a multiple of some other row,* (by basic property (4) this leaves the value of the determinant unchanged);

(γ) *interchange two rows* (by basic property (1) this changes the sign of the determinant).

The simplification procedure is the same as the previous row-reduction procedure in Chapter 30 except that:

(i) *pivot-elements will not necessarily have the value 1, since we do not use* α-*operations;*

and (ii) *in the column of a non-zero pivot, we only need to get 0's below the pivot.*

The result of this modified row-reduction process is called a *row-echelon matrix*.

Examples.

(1) $\begin{vmatrix} 2 & 1 \\ 3 & 2 \end{vmatrix} \xrightarrow[(\beta)]{} \begin{vmatrix} 2 & 1 \\ 0 & \frac{1}{2} \end{vmatrix}$

(2) $\begin{vmatrix} 1 & 2 & 4 \\ 0 & 2 & -1 \\ 3 & 2 & -2 \end{vmatrix} \xrightarrow[(\beta)]{} \begin{vmatrix} 1 & 2 & 4 \\ 0 & 2 & -1 \\ 0 & -4 & -14 \end{vmatrix} \xrightarrow[(\beta)]{} \begin{vmatrix} 1 & 2 & 4 \\ 0 & 2 & -1 \\ 0 & 0 & -16 \end{vmatrix}$

(3) $\begin{vmatrix} 1 & 2 & 1 & -2 \\ 2 & 1 & 0 & 1 \\ 0 & 4 & 2 & -1 \\ 3 & 7 & 3 & -2 \end{vmatrix} \xrightarrow[(\beta,\beta)]{} \begin{vmatrix} 1 & 2 & 1 & -2 \\ 0 & -3 & -2 & 5 \\ 0 & 4 & 2 & -1 \\ 0 & 1 & 0 & 4 \end{vmatrix}$

$\xrightarrow[(\beta,\beta)]{} \begin{vmatrix} 1 & 2 & 1 & -2 \\ 0 & -3 & -2 & 5 \\ 0 & 0 & -\frac{2}{3} & \frac{17}{3} \\ 0 & 0 & -\frac{2}{3} & \frac{17}{3} \end{vmatrix} \xrightarrow[(\beta)]{} \begin{vmatrix} 1 & 2 & 1 & -2 \\ 0 & -3 & -2 & 5 \\ 0 & 0 & -\frac{2}{3} & \frac{17}{3} \\ 0 & 0 & 0 & 0 \end{vmatrix}$

Let A′ be the final row-echelon matrix obtained from an original matrix A. Evidently two cases can occur.

(i) The bottom row of A′ is all 0's. In this case, by the Laplace expansion on that row, $|A'| = 0$. Hence the original determinant must be $|A| = 0$.

(ii) The bottom row of A′ is not all 0's. In this case the matrix A′ must have *no* 0's on the main diagonal and *all* 0's below the main diagonal (as the reader may verify). Doing successive Laplace expansions by columns, we see that $|A'|$ must be the product of the elements on the main diagonal of A' and that $|A'|$ must be $\neq 0$. So the original determinant $|A|$ must be $= (-1)^k |A'|$, where k is the number of γ-operations used in the reduction procedure.

We thus find in the examples above that the first determinant $= 1$, the second $= -32$, and the third $= 0$.

Note that the number of non-zero rows in A' must equal the number of pivots in the fully row-reduced form of the matrix A, since each non-zero pivot in A' becomes a 1 when an α-operation is used. Hence we have the following fact, which relates the rank of a square matrix to the value of its determinant:

(3.1)   *Theorem.* *Given an* n×n *square matrix* A,

$$|A| \neq 0 \Leftrightarrow r(A) = n,$$

and $$|A| = 0 \Leftrightarrow r(A) < n.$$

(3.2)   *Definition.* Let A be a square matrix. If $|A| = 0$, we say that A is a *singular* matrix. If $|A| \neq 0$, we say that A is a *non-singular* matrix.

(3.1) leads us to an important and useful fact about systems of linear equations:

(3.3)   *Corollary.* *We say that a system* AX = D *is square if the matrix* A *is square. For a square system* AX = D,

A is nonsingular $\Leftrightarrow |A| \neq 0 \Leftrightarrow$ *the solution of the system is unique.*

Hence: (i)   *for an inhomogeneous system,*

A is singular $\Leftrightarrow |A| = 0 \Leftrightarrow$ *either there is no solution*
or there are infinitely many solutions;

and   (ii)   *for a homogeneous system,*

A is singular $\Leftrightarrow |A| = 0 \Leftrightarrow$ *there are infinitely many solutions*
$\Leftrightarrow$ *there is a non-trivial (non-zero) solution.*

The last line of this corollary gives us the widely used *determinant test* for the existence of non-trivial solutions for square homogeneous systems: *non-trivial solutions exist if and only if the determinant of the coefficient matrix is zero.*

How does the brute-force calculation compare with condensation? One way of measuring the length of such a calculation is to count the maximum number of individual arithmetical operations (additions and multiplications) that may be required. It is not difficult to show that, in general, for an $n \times n$ determinant, brute force requires $n(n!)-1$ operations, but condensation requires $\frac{1}{3}(2n^3 + n - 3)$. These results are compared in the following table.

n	Brute Force	Condensation
2	3	5
3	17	18
4	95	43
5	599	84
10	36,287,999	669
20	$\sim 10^{29}$	5,339
40	$\sim 10^{47}$	42,679
50	$\sim 10^{64}$	83,349

Remark. The general formulas above imply that the condensation method is a polynomial-time algorithm and that the brute-force method is a superexponential-time algorithm (since $n! > \sqrt{n}\left(\frac{n}{e}\right)^n$). The concepts *polynomial-time algorithm* and *superexponential-time algorithm* are defined in theoretical computer science and have been proposed as precise meanings for the phrases: "fast algorithm" and "very slow algorithm."

§31.4 Cramer's Rule

Consider the determinant $\begin{vmatrix} 1 & 2 & 1 \\ 4 & 0 & 2 \\ 3 & 1 & 1 \end{vmatrix}$. If we make a Laplace expansion on the first row, we get

$$1\begin{vmatrix} 0 & 2 \\ 1 & 1 \end{vmatrix} - 2\begin{vmatrix} 4 & 2 \\ 3 & 1 \end{vmatrix} + 1\begin{vmatrix} 4 & 0 \\ 3 & 1 \end{vmatrix} = -2 + 4 + 4 = 6.$$

What if, in error, we had used the cofactors of the second row in place of the cofactors of the first row? We would have obtained:

$$-1\begin{vmatrix}2 & 1\\1 & 1\end{vmatrix}+2\begin{vmatrix}1 & 1\\3 & 1\end{vmatrix}-1\begin{vmatrix}1 & 2\\3 & 1\end{vmatrix}.$$

An incorrect expansion of this kind, which uses the cofactors of a wrong row, is called a *false Laplace expansion*. Note that the false expansion just given is, in fact, a correct Laplace expansion (in the second row) for the determinant $\begin{vmatrix}1 & 2 & 1\\1 & 2 & 1\\3 & 1 & 1\end{vmatrix}$. In this last determinant, the second row of the original determinant has been replaced by the first row of the original determinant. Basic property (2) now shows us that our false expansion must give the value 0. Indeed, checking by direct calculation, we get $-1 - 4 + 5 = 0$. The argument above holds, in general. *A false Laplace expansion must always give the value 0.*

We are now in a position to state and prove *Cramer's rule* for a system of linear equations. As we shall see, Cramer's rule can only be applied to a system $AX = D$ for which:

        (i)   A is square,

and   (ii)   A is non-singular ($|A| \neq 0$).

By (3.3), a system with these properties must have a unique solution. Cramer's rule uses determinants to express this unique solution.

(4.1)   *Cramer's rule for solving square systems with determinant $\neq$ **0**. Let $AX = D$ be a system of* n *equations in* n *unknowns with $|A| \neq 0$. Let $A_1, A_2, ..., A_n$ be the columns of* A. *Then the unique solution of this system is given by the formulas*

$$x_1 = \frac{\left|[D \vdots A_2 \vdots A_3 \vdots \cdots \vdots A_n]\right|}{|A|},$$

$$x_2 = \frac{\left|[A_1 \vdots D \vdots A_3 \vdots \cdots \vdots A_n]\right|}{|A|},$$

$$x_3 = \frac{\left|[A_1 \vdots \cdots \vdots A_{n-1} \vdots D]\right|}{|A|},$$

where, for example, $\|[D \vdots A_2 \vdots A_3 \vdots \cdots \vdots A_n]\|$ *is the* $n \times n$ *matrix whose columns are* D, $A_2, \ldots, A_n$.

Since $|A|$ appears in the denominators, the formulas of Cramer's Rule cannot be applied when $|A| = 0$.

Examples.

(1) $\begin{bmatrix} 2x_1 - x_2 = 3 \\ x_1 + 3x_2 = 5 \end{bmatrix}$. Then

$$x_1 = \frac{\begin{vmatrix} 3 & -1 \\ 5 & 3 \end{vmatrix}}{\begin{vmatrix} 2 & -1 \\ 1 & 3 \end{vmatrix}} = \frac{14}{7} = 2; \quad x_2 = \frac{\begin{vmatrix} 2 & 3 \\ 1 & 5 \end{vmatrix}}{7} = \frac{7}{7} = 1.$$

(2) $\begin{bmatrix} x_1 + x_2 - x_3 = 2 \\ x_1 - x_2 + x_3 = 0 \\ x_1 + x_2 + x_3 = 4 \end{bmatrix}$. Then $x_1 = \frac{\begin{vmatrix} 2 & 1 & -1 \\ 0 & -1 & 1 \\ 4 & 1 & 1 \end{vmatrix}}{\begin{vmatrix} 1 & 1 & -1 \\ 1 & -1 & 1 \\ 1 & 1 & 1 \end{vmatrix}} = \frac{-4}{-4} = 1,$

$$x_2 = \frac{\begin{vmatrix} 1 & 2 & -1 \\ 1 & 0 & 1 \\ 1 & 4 & 1 \end{vmatrix}}{-4} = 2, \quad x_3 = \frac{\begin{vmatrix} 1 & 1 & 2 \\ 1 & -1 & 0 \\ 1 & 1 & 4 \end{vmatrix}}{-4} = 1.$$

Proof of (4.1). Let $\tilde{x}_1, \ldots, \tilde{x}_n$ be the values given by Cramer's rule. We need only show that each equation of the system is satisfied by these values. For each element a_{ij} in A, let α_{ij} be the cofactor of a_{ij}. Then, by Laplace expansion,

$$\tilde{x}_1 = \frac{1}{|A|}(d_1 \alpha_{11} + d_2 \alpha_{21} + \cdots + d_n \alpha_{n1}),$$
$$\tilde{x}_2 = \frac{1}{|A|}(d_1 \alpha_{12} + d_2 \alpha_{22} + \cdots + d_n \alpha_{n2}),$$

and so forth.

Substituting these values in the left-hand side of the first equation of the system, we get

$$a_{11}\tilde{x}_1 + a_{12}\tilde{x}_2 + \cdots a_{1n}\tilde{x}_n =$$

$$\frac{a_{11}}{|A|}(d_1\alpha_{11} + d_2\alpha_{21} + \cdots) + \frac{a_{12}}{|A|}(d_1\alpha_{12} + d_2\alpha_{22} + \cdots) + \cdots .$$

Regrouping terms, we have

$$\frac{d_1}{|A|}(a_{11}\alpha_{11} + a_{12}\alpha_{12} + \cdots) + \frac{d_2}{|A|}(a_{11}\alpha_{21} + a_{12}\alpha_{22} + \cdots) + \cdots .$$

But in this last expression the coefficient of $\dfrac{d_1}{|A|}$ is $(a_{11}\alpha_{11} + \cdots) = |A|$, while the coefficients of $\dfrac{d_2}{|A|}, \dfrac{d_3}{|A|}, \cdots$ are false Laplace expansions of A and must hence be 0. Thus we have $a_{11}\tilde{x}_1 + \cdots + a_{1n}\tilde{x}_n = d_1$, and we see that the first equation is satisfied by $\tilde{x}_1, \ldots, \tilde{x}_n$. Similarly for the remaining equations. Hence the values of $x_1, \ldots, x_n$ given by Cramer's Rule are the correct unique solution of the given system.

§31.5 Vector Product for n-vectors.

In our discussion (in Chapter 29) of vector operations and vector algebra in $\mathbf{R}^n$, we do not define an n-dimensional version of the vector product. We now use determinants to do this. We use the notations of §29.3 for vectors in $\mathbf{R}^n$.

Recall that, in three dimensions, we get a coordinate formulation for the vector product of $\vec{A} = a_1\hat{i} + a_2\hat{j} + a_3\hat{k}$ and $\vec{B} = b_1\hat{i} + b_2\hat{j} + b_3\hat{k}$ by taking

$$\vec{A} \times \vec{B} = \begin{vmatrix} \hat{i} & \hat{j} & \hat{k} \\ a_1 & a_2 & a_3 \\ b_1 & b_2 & b_3 \end{vmatrix} .$$

We see that the vector product of two 3-vectors has the following properties:

(5.1)   (i) $\vec{A} \times \vec{A} = \vec{0}$;

        (ii) $\vec{A} \times \vec{0} = \vec{0}$;

(iii) $(a\vec{A}) \times (b\vec{B}) = (ab)(\vec{A} \times \vec{B})$;

(iv) $\vec{B} \times \vec{A} = -(\vec{A} \times \vec{B})$;

(v) $\vec{A} \times (\vec{B} + \vec{C}) = (\vec{A} \times \vec{B}) + (\vec{A} \times \vec{C})$;

(vi) $\vec{A} \times \vec{B}$ *is orthogonal to* $\vec{A}$ *and to* $\vec{B}$;

(vii) $\vec{A} \times \vec{B} = \vec{0} \Leftrightarrow [\vec{A}$ *is a scalar multiple of* $\vec{B}$ *or* $\vec{B}$ *is a scalar multiple of* $\vec{A}$].

In a similar way, we can define a vector product in $\mathbf{R}^4$ by using 4×4 determinant-notation and taking three 4-vectors as the given factors for the product. If these factors are $\vec{A} = (a_1, a_2, a_3, a_4)$, $\vec{B} = (b_1, b_2, b_3, b_4)$, and $\vec{C} = (c_1, c_2, c_3, c_4)$, we write the product as

$$\vec{A} \times \vec{B} \times \vec{C} ,$$

and we define it to be

(5.2)
$$\vec{A} \times \vec{B} \times \vec{C} = \begin{vmatrix} \hat{e}_1 & \hat{e}_2 & \hat{e}_3 & \hat{e}_4 \\ a_1 & a_2 & a_3 & a_4 \\ b_1 & b_2 & b_3 & b_4 \\ c_1 & c_2 & c_3 & c_4 \end{vmatrix} .$$

For example, if $\vec{A} = (4,7,8,-1)$, $\vec{B} = (3,2,-1,4)$, and $\vec{C} = (0,1,2,0)$, then

$$\vec{A} \times \vec{B} \times \vec{C} = \begin{vmatrix} \hat{e}_1 & \hat{e}_2 & \hat{e}_3 & \hat{e}_4 \\ 4 & 7 & 8 & -1 \\ 3 & 2 & -1 & 4 \\ 0 & 1 & 2 & 0 \end{vmatrix} = -29\hat{e}_1 + 38\hat{e}_2 - 19\hat{e}_3 - 2\hat{e}_4$$
$$= (-29, 38, -19, -2) .$$

The vector product in $\mathbf{R}^n$ is defined in the same way, where we use $n-1$ factors and $n\times n$ determinant-notation.

From the basic properties of determinants, we immediately have the following facts and laws for the vector product in $\mathbf{R}^n$. We state these facts and laws for the case $n = 4$.

(5.3)   (i) $\vec{A} \times \vec{B} \times \vec{C} = \vec{0}$ *if any two factors are equal.*

(ii) $\vec{A} \times \vec{B} \times \vec{C} = \vec{0}$ *if any factor is* $\vec{0}$.

(iii) $(a\vec{A}) \times (b\vec{B}) \times (c\vec{C}) = (abc)(\vec{A} \times \vec{B} \times \vec{C})$.

(iv)   *Interchanging any two factors in the product* $\vec{A} \times \vec{B} \times \vec{C}$ *is the same as multiplying the product by* -1.

(v) $\vec{A} \times \vec{B} \times (\vec{C}_1 + \vec{C}_2) = \vec{A} \times \vec{B} \times \vec{C}_1 + \vec{A} \times \vec{B} \times \vec{C}_2$.

(vi) $\vec{A} \times \vec{B} \times \vec{C}$ *is orthogonal to each of its factors.*

It also follows from (5) in (2.1) that:

(vii) $\vec{A} \times \vec{B} \times \vec{C} = \vec{0}$ $\Leftrightarrow$ *there is at least one of the factors which can be expressed as a linear combination of the others.*

(5.4)   Given a vector $\vec{D}$ (in $\mathbf{R}^4$), we can also form the dot product $(\vec{A} \times \vec{B} \times \vec{C}) \cdot \vec{D}$. This product of four vectors in $\mathbf{R}^4$ is analogous to the scalar triple product of three vectors in $\mathbf{R}^3$. We write this *scalar quadruple product* of four 4-vectors as $[\vec{A}, \vec{B}, \vec{C}, \vec{D}]$. Its value is given by the 4×4 determinant whose rows are the components of $\vec{A}, \vec{B}, \vec{C},$ and $\vec{D}$. An application of this product for n-vectors appears in §31.8.

§31.6 Proof of the Fundamental Theorem on Laplace Expansions

In §31.1, the value of an n×n determinant is defined to be the value given by a Laplace expansion in terms of (n–1)×(n–1) determinants. Each of these determinants, in turn, ha its value given by a Laplace expansion in terms of (n–2)×(n–2) determinants; and so forth. The *fundamental theorem on Laplace expansions* asserts that *any two ways of expanding a given determinant must lead to the same final value.* We now prove this theorem inductively on the order of the given determinant. That is to say, we assume that the assertion is true for all determinants of order n or less, and we then prove that it must hold for all determinants of order n+1. We give the proof in outline.

Basis for induction. We observe that our assertion is true for second-order determinants. In a 2×2 determinant $\begin{vmatrix} a & b \\ c & d \end{vmatrix}$, there are four possible Laplace expansions in terms of 1×1 determinants, and each leads to the value ad–bc.

The induction. We assume that our assertion is true for all determinants of order n or less. We show that the assertion must then be true for all determinants of order n+1. Consider a given determinant of order n+1. We first show that Laplace expansions on two different rows must lead to the same value. Let the upper row be $a_1, a_2, \ldots, a_{n+1}$ and the lower row be $b_1, b_2, \ldots, b_{n+1}$. If we do a Laplace expansion on the a's and then, for each minor in that expansion, do a Laplace expansion on the b's, we get a sum of terms of the form $\pm a_i b_j C_{ij}$, where $i \neq j$ and C_{ij} is the n-first-order determinant obtained when the rows and columns of both a_i and b_j are deleted from the original determinant. If, instead, we do a Laplace expansion on the b's and then, for each minor in that expansion, do a Laplace expansion on the a's, we get the sum (in a different order) of the same terms with the same signs. (To show that the signs are the same, the cases $i < j$ and $i > j$ must be treated separately. In the first case $(i < j)$, this sign (for either order of expansions) is the product of the checkerboard signs (see §31.1) for a_i and b_j. In the second case, it is the negative of this product.) By our inductive assumption, the values of the C_{ij} are uniquely defined. Hence we have shown, for our $(n+1) \times (n+1)$ determinant, that we get the same value no matter what row we use for our first Laplace expansion. The proof for two different columns is exactly similar. Finally, we show that a Laplace expansion on the first row leads to the same value as a Laplace expansion on the first column. Let $a_1, a_2, \ldots, a_{n+1}$ be the first row, and let $b_1, b_2, \ldots, b_{n+1}$ be the first column (thus $a_1 = b_1$). If we do a Laplace expansion on the a's and then, for each minor after the first, do a Laplace expansion on the b's, we get the term $a_1 C_1$ plus terms of the form $\pm a_i b_j C_{ij}$, where C_1 is the $n \times n$ minor of a_1, and the C_{ij} are as before. If instead, we do a Laplace expansion on the b's and then, after the first minor, do a Laplace expansion on the a's, we get the same terms with the same signs. Again, by our inductive assumption we have that the values of C_1 and the C_{ij} are uniquely defined. Hence we get the same value for our determinant whether we start with the first row or with the first column.

We have thus shown that we get the same value for our determinant of order $n+1$ no matter what row or column we use in our first Laplace expansion. This completes the induction and hence completes the proof.

§31.7 Proofs of the Basic Properties of Determinants

The proofs of the six basic properties of determinants are straightforward.

Property **(1).** *Interchanging two rows changes the sign of the determinant's value.*

Proof. We prove this by induction on the determinant's order.

Basis. For n = 2, the truth of the assertion is immediate from the formula for the value of a 2×2 determinant.

Induction. Assume the assertion is true for all determinants of order n. Given a determinant of order n+1, let $a_1, a_2, \ldots, a_{n+1}$ be some row other than the two rows being interchanged. Do a Laplace expansion on the a's. Then interchanging the two rows in our original determinant interchanges corresponding rows in each minor in the expansion. By inductive assumption, this changes the sign of each minor. Hence it changes the sign of the expansion and therefore changes the sign of the value of the given determinant of order n+1. This completes the proof.

Property (2).   *If two rows are identical, the determinant's value is 0.*

Proof. Interchanging two identical rows must leave the determinant's value unchanged. But, by property (1), it must also change the sign of the value. It follows that the value must be 0. (For let v be the value. Then $v = -v \Rightarrow 2v = 0 \Rightarrow v = 0$.)

Property (3).   *Altering a determinant, by multiplying each element in some chosen row by the same constant, multiplies the value of the determinant by that constant.*

Proof. Immediate by considering the Laplace expansion on the given row.

Property (4).   *Altering a determinant, by adding to a given row any multiple of another row, leaves the determinant's value unchanged.*

Proof. By induction on the order of the determinant.

Basis. For n = 2, we have $\begin{vmatrix} a & b \\ c & d \end{vmatrix}$. Adding a multiple of the second row to the first gives $\begin{vmatrix} a + \lambda c & b + \lambda d \\ c & d \end{vmatrix} = (a + \lambda c)d - (b + \lambda d)c = ad - bc$.

Similarly for adding a multiple of the first row to the second.

Induction. Assume true for determinants of order n. Do Laplace expansions of the given determinant (of order n+1) and of the new determinant on some row other than the rows being used for the addition. Then each minor in the expansion of the new determinant must have the same value as the corresponding minor of the original determinant, since it is obtained from the original minor by adding to some row a multiple of some other row, and, by inductive assumption, this change leaves the value of the minor unchanged. It follows that the two Laplace expansions must give the same value.

Property (5). *A determinant's value $= 0 \Leftrightarrow$ there is a row that can be expressed as a linear combination of other rows.*

Proof. Let A be the matrix of the given determinant. Let $B = A^T$. By Laplace expansions, $|A| = |B|$. If $|A| = 0$, then $|B| = 0$. By (3.3) (which did not use basic property (5)), if $|B| = 0$, then the homogeneous system $BX = O$ must have a

non-trivial solution $X = \begin{bmatrix} x_1 \\ \vdots \\ x_n \end{bmatrix}$. Suppose that, in fact, $x_1 \neq 0$. Let $A_1, A_2, \ldots, A_n$

be the column-vectors of B. Hence $A_1, \ldots, A_n$ are column-vectors giving the *rows* of A. Then $BX = O$ asserts that $x_1 A_1 + \cdots + x_n A_n = 0$. Hence we have that

$$A_1 = \left(-\frac{x_2}{x_1} A_2 \right) + \cdots + \left(-\frac{x_n}{x_1} A_n \right),$$

and we see that A_1 is a linear combination of $A_2, \ldots, A_n$.

Conversely, if A_1 is a linear combination of $A_2, \ldots, A_n$, then we have $A_1 = \alpha_2 A_2 + \cdots \alpha_n A_n$ for scalars $\alpha_2, \ldots, \alpha_n$. Hence $x_1 = -1$, $x_2 = \alpha_2, \ldots, x_n = \alpha_n$ is a non-trivial solution to $BX = O$. Thus, again by (3.2), we must have $|B| = 0$; and since $|A| = |B|$, we have $|A| = 0$. This completes the proof.

Property (6). *Properties (1)–(5) hold with "column" substituted for "row."*

Proof. The proofs of (1)–(5) carry over directly. (The proof of (5) is shorter, since we can use the matrix A directly.)

§31.8 Hypervolume of Regions in $\mathbf{R}^n$

We have seen that, in $\mathbf{E}^2$-with-axes, the absolute value of $\begin{vmatrix} a_1 & a_2 \\ b_1 & b_2 \end{vmatrix}$ gives the area of the parallelogram with the two position-vectors $a_1\hat{i} + a_2\hat{j}$ and $b_1\hat{i} + b_2\hat{j}$ as adjacent edges. Similarly, in $\mathbf{E}^3$-with-axes, we have seen that the absolute value of

$$\begin{vmatrix} a_1 & a_2 & a_3 \\ b_1 & b_2 & b_3 \\ c_1 & c_2 & c_3 \end{vmatrix}$$

gives the volume of the parallelepiped with the three vectors $a_1\hat{i}+a_2\hat{j}+a_3\hat{k}$, $b_1\hat{i}+b_2\hat{j}+b_3\hat{k}$, and $c_1\hat{i}+c_2\hat{j}+c_3\hat{k}$ as adjacent edges. These results generalize directly to $\mathbf{R}^n$. We need three definitions, which we give in outline, omitting details. For these definitions, we use the notations of §29.3 for vectors in $\mathbf{R}^n$.

Definitions.

(1)   If $\vec{B}$ is a given vector in $\mathbf{R}^n$, and $\vec{A}_1,...,\vec{A}_n$ are n given vectors in $\mathbf{R}^n$, then the set of all points (vectors) in $\mathbf{R}^n$ of the form $\vec{B}+t_1\vec{A}_1+\cdots+t_n\vec{A}_n$, $0 \le t_1 \le 1, 0 \le t_2 \le 1,..., 0 \le t_n \le 1$, is called the *parallelepiped, with vertex at* $\vec{B}$, *determined by the edges* $\vec{A}_1,...,\vec{A}_n$.

(2)   If $\vec{A}_1,...,\vec{A}_n$ are mutually orthogonal vectors of the special form $\vec{A}_1 = a_1\hat{e}_1, \vec{A}_2 = a_2\hat{e}_2,..., \vec{A}_n = a_n\hat{e}_n$, where $a_1, a_2,..., a_n \ge 0$, we say that the parallelepiped determined by the edges $\vec{A}_1,...,\vec{A}_n$ with vertex at $\vec{B}$ is a *basic rectangular parallelepiped*. We associate with this basic rectangular parallelepiped, as its *hypervolume*, the value $a_1 a_2 ... a_n$.

(3)   Given an arbitrary region D in $\mathbf{R}^n$, we can approximate the region D by a collection of adjacent and non-overlapping basic rectangular parallelepipeds, where this collection either

   (a)   contains all of D;
   (b)   is fully contained in D.

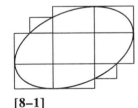

[8–1]

[8–2]

[8–1] and [8-2] suggest an example in $\mathbf{R}^2$. Considering the sum of the hypervolumes of these basic parallelepipeds, we can look for a limit of such sums as smaller and smaller basic-rectangular parallelepipeds are used in the approximation. (By "smaller and smaller" we mean a succession of approximations in which the maximum edge length occurring decreases towards 0.) If a unique limit exists for sums of type (a), and a unique limit also exists for sums of type (b), and if these two limits are equal, we define the *hypervolume* of the original region D to be the value of this common limit.

It can be shown that this definition has the following desirable *additivity property*: if a region D is subdivided into a finite number of subregions which have hypervolumes $V_1, V_2, \ldots, V_n$, then D itself has the hypervolume $V_1 + V_2 + \cdots + V_n$.

For an arbitrary (and not necessarily basic rectangular) parallelepiped D, it is not difficult to show that the limit considered in (3) must always exist and that such a parallelepiped must hence possess a hypervolume. The generalized result for determinants and hypervolume now states:

(8.1) *Given an arbitrary parallelepiped in $\mathbf{R}^n$ determined by the edges $\vec{A}_1, \ldots, \vec{A}_n$ with vertex at $\vec{B}$, then the hypervolume of this region is given by the absolute value of the determinant $|A|$, where A is the matrix whose rows are given by the vectors $\vec{A}_1, \ldots, \vec{A}_n$.* In the algebra of n-vectors, this hypervolume can be expressed as the absolute value of $[\vec{A}_1, \ldots, \vec{A}_n]$, where $[\vec{A}_1, \ldots, \vec{A}_n]$ is the scalar n-tuple product of $\vec{A}_1, \ldots, \vec{A}_n$ (see (5.4)). This result is analogous to the result in $\mathbf{E}^3$ which gives the volume of a parallelepiped as the absolute value of a scalar triple product (see §3.6).

Remark. For $\mathbf{R}^2$ and $\mathbf{R}^3$, *hypervolume* makes precise the usual concepts of *area* in $\mathbf{E}^2$ and *volume* in $\mathbf{E}^3$.

§31.9 Problems

§31.3

1. Evaluate each of the following determinants by Laplace expansions and then by row reduction.

(a) $\begin{vmatrix} 1 & 0 & 1 & 1 \\ 1 & 1 & 0 & 1 \\ 1 & 0 & 1 & 0 \\ 0 & 1 & 0 & 0 \end{vmatrix}$
(b) $\begin{vmatrix} 1 & 0 & 1 & 1 \\ 1 & 1 & 0 & 1 \\ 1 & 0 & 0 & 1 \\ 0 & 1 & 0 & 0 \end{vmatrix}$

2. Find the value of the determinant

$$\begin{vmatrix} 1 & 2 & 1 & 2 \\ 0 & 1 & 2 & 0 \\ 1 & 1 & 1 & 1 \\ 0 & 1 & 1 & 0 \end{vmatrix}$$

3. Consider the system of equations

$$\begin{aligned} x_1 + x_2 + ax_3 &= 1 \\ x_1 - x_2 + x_3 &= b \\ x_1 + 2x_2 + x_3 &= 1 \end{aligned}$$

(a) For what values of a and b is there exactly one solution?
(b) For what values is there no solution?
(c) For what values is there more than one solution?

4. Consider the system of equations

$$\begin{aligned} x_1 + x_2 + ax_3 &= b \\ x_1 - x_2 + x_3 &= 3 \\ x_1 + 2x_2 + x_3 &= 1 \end{aligned}$$

Use determinants to answer the following questions:

(a) For what value of a and b is there exactly one solution?

(b) For what values is there no solution?

(c) For what values is there more than one solution?

§31.4

1. Solve the following systems by Cramer's Rule.

(a) $\begin{pmatrix} 2x + 3y = 3 \\ x - 4y = 5 \end{pmatrix}$
(b) $\begin{pmatrix} x + y - z = 1 \\ x - y + z = 2 \\ x - y - z = 0 \end{pmatrix}$

2. Solve by Cramer's Rule:

$$\begin{pmatrix} x + w = 4 & z + y = -1 \\ x - y = 2 & w + z = 3 \end{pmatrix}$$

3. Apply Cramer's Rule to the system:

$$\begin{pmatrix} x + w = 4 & z + y = -1 \\ x - y = 2 & w - z = 3 \end{pmatrix}$$

Explain the significance of the result you get.

§31.5

1. Deduce the results of (5.3) from (2.1).

Chapter 32

Matrix Algebra

§32.1 Matrix Multiplication

In Chapter 29, we define *vectors* in $\mathbf{R}^n$ and extend the notations and operations of *vector-algebra* to vectors in $\mathbf{R}^n$. We then define *matrices* and describe several operations of *matrix algebra*. Matrix algebra is an algebraic system, distinct and separate from vector-algebra, with its own notations and operations. If we represent n-vectors as n×1 matrices (which we also call *column-vectors*), and if we represent the dot product $\bar{C} \cdot \bar{D}$ of two n-vectors as the inner product $C^T D$ of the corresponding column-vectors, we then find that much of vector-algebra in $\mathbf{R}^n$ can be expressed within matrix algebra. In this chapter, we complete our description of matrix algebra by defining two further operations: *matrix multiplication* (in its general form), and *matrix inversion*.

Matrix algebra is a broad and useful system with a variety of techniques and applications. We describe two of these in later sections of this chapter: the technique for finding eigenvalues and eigenvectors (§32.8), and the use of matrices to change from one set of Cartesian axes to another (§32.9).

In §29.7, we define *multiplication (for a matrix* A *times a column vector* C), when A is m×n and C is n×1. We saw that the product AC is an m×1 column-vector, and that each element in AC is the dot product of C (considered as an n-vector) with the corresponding row in A (considered as an n-vector).

Definition. We now define an operation of *multiplication (for a matrix* A *times a matrix* B), where A is m×n and B is n×p. Let $B_1, \ldots, B_p$ be the columns of B.

Each of these is an $n \times 1$ column vector. We write the product of A and B as AB, and we define AB to be *the $m \times p$ matrix whose columns are the products* $AB_1, \ldots, AB_p$. Thus:

(1.1) *if* $B = [B_1 \vdots \cdots \vdots B_p]$, *then* $AB = [AB_1 \vdots \cdots \vdots AB_p]$.

For the product AB to be defined, the number of rows in B must equal the number of columns in A. In this case, we say that A and B (in that order) are *conformable for multiplication.*

If $AB = C$, where A is $m \times n$, B is $n \times p$, and C is $m \times p$, we can summarize our definition of multiplication by the formula

(1.2) $$c_{ij} = \sum_{k=1}^{n} a_{ik} b_{kj} .$$

This states that the element c_{ij} of C is the dot product of the i-th row of A with the j-th column of B (where this row and this column are considered as n-vectors).

Example.

$$\begin{bmatrix} 1 & 2 \\ 2 & 1 \\ 3 & -1 \end{bmatrix} \begin{bmatrix} 1 & 5 \\ 0 & 4 \end{bmatrix} = \begin{bmatrix} \begin{bmatrix} 1 & 2 \\ 2 & 1 \\ 3 & -1 \end{bmatrix} \cdot \begin{bmatrix} 1 \\ 0 \end{bmatrix} & \begin{bmatrix} 1 & 2 \\ 2 & 1 \\ 3 & -1 \end{bmatrix} \begin{bmatrix} 5 \\ 4 \end{bmatrix} \end{bmatrix} = \begin{bmatrix} 1 & 13 \\ 2 & 14 \\ 3 & 11 \end{bmatrix}.$$

It is sometimes helpful to picture the multiplication of two matrices in a *multiplication diagram* of the following kind:

$$B \rightarrow \begin{bmatrix} 1 & 5 \\ 0 & 4 \end{bmatrix}$$

$$A \rightarrow \begin{bmatrix} 1 & 2 \\ 2 & 1 \\ 3 & -1 \end{bmatrix} \cdot \begin{bmatrix} 1 & 13 \\ 2 & 14 \\ 3 & 11 \end{bmatrix} \leftarrow AB.$$

Here, each element of AB appears in line with both the row of A and the column of B that we use to calculate it. (For example, 14 is the dot product of (2,1) and

§32.2 Properties of Matrix Multiplication

The following **general laws** hold for matrix multiplication. (For each law, we assume that we have conformability as needed.) The laws follow directly from corresponding laws for the multiplication of a matrix times a column-vector. Note that the properties of the matrices O and I in matrix multiplication are analogous to the properties of 0 and 1 in multiplication of real numbers.

(2.1)
$$OA = O \qquad A(B + C) = AB + AC$$
$$AO = O \qquad (A+B)C = AC + BC$$
$$IA = A \qquad (kA)B = k(AB)$$
$$AI = A \qquad (A(kB) = k(AB).$$

In addition to the laws above, the law of *associativity* (that $(AB)C = A(BC)$) also holds for matrix multiplication. We state and prove this law as a theorem.

(2.2) **Theorem.** *Let* A, B, *and* C *be matrices such that* A *is* $m{\times}n$, B *is* $n{\times}p$, *and* C *is* $p{\times}r$. *Then*

$$(AB)C = A(BC).$$

Proof. (i) We first prove associativity for the case, where C is a column-vector $(r = 1)$. Let $C = \begin{bmatrix} c_1 \\ \vdots \\ c_p \end{bmatrix}$, and let $B = [B_1 \vdots B_2 \vdots \cdots \vdots B_p]$.

Then $(AB)C = [AB_1 \vdots AB_2 \vdots \cdots \vdots AB_p]C$ (by the definition of matrix multiplication);

$= c_1 AB_1 + \cdots + c_p AB_p$ (by the definition of multiplying a matrix times a column vector);

$= A(c_1 B_1) + \cdots + A(c_p B_p)$ (by the law that $k(AB) = A(kB)$);

$= A(c_1 B_1 + \cdots + c_p B_p)$ (by the distributivity law that $A(B+C) = AB+AC$);

$= A(BC)$ (by the definition of the product BC).

(ii) We now complete the proof by proving associativity for the general case, where C is a matrix with r columns. Let $C = [C_1 \vdots \cdots \vdots C_r]$.

Then $(AB)C = [(AB)C_1 \vdots \cdots \vdots (AB)C_r]$ (by the definition of matrix multiplication);

$= [A(BC_1) \vdots \cdots \vdots A(BC_r)]$ (by (i) above);

$= A[BC_1 \vdots \cdots \vdots BC_r]$ (by the definition of matrix multiplication);

$= A(BC)$ (by the definition of matrix multiplication).

This completes the proof.

The laws stated above for matrix-algebra are analogous to laws which hold for the algebra of real numbers. There are two further laws for real numbers, however, which do not hold, in their analogous versions, as laws of matrix-algebra. These laws are the *commutative law for multiplication* and the *zero-divisor law*.

The *commutative law* for real numbers asserts that for every pair of real numbers a and b, ab = ba. This law does not hold as a general law of matrix-multiplication, since it is possible to have the products AB and BA both defined (A and B must be conformable in either order), but to have AB ≠ BA. For example,

$$\begin{bmatrix} 1 & 1 \\ 0 & 1 \end{bmatrix}\begin{bmatrix} 1 & 0 \\ 1 & 1 \end{bmatrix} = \begin{bmatrix} 2 & 1 \\ 1 & 1 \end{bmatrix},$$

but

$$\begin{bmatrix} 1 & 0 \\ 1 & 1 \end{bmatrix}\begin{bmatrix} 1 & 1 \\ 0 & 1 \end{bmatrix} = \begin{bmatrix} 1 & 1 \\ 1 & 2 \end{bmatrix}.$$

It is also possible, when A and B are not square, to have AB and BA both defined but of different sizes. For example,

$$\begin{bmatrix} 1 \\ 1 \end{bmatrix}[1 \ \ 1] = \begin{bmatrix} 1 & 1 \\ 1 & 1 \end{bmatrix}, \text{ and } [1 \ \ 1]\begin{bmatrix} 1 \\ 1 \end{bmatrix} = [2] .$$

The *zero-divisor law* for real numbers asserts that the product of two non-zero real numbers must always be non-zero. This implies that if a product of real-number factors is zero, then at least one of the factors must be zero. (We use the zero-divisor law, in the elementary algebra of real numbers, when we find the roots of a factored polynomial. For example, if we know that $(x-7)(x+3) = 0$, then we can conclude that either $x-7 = 0$ or $x+3 = 0$.)

The zero-divisor law does not hold as a general law of matrix multiplication. It is possible to have two matrices A and B where $A \neq O$, and $B \neq O$, but $AB = O$. For example,

$$\begin{bmatrix} 1 & -1 \\ 2 & -2 \end{bmatrix}\begin{bmatrix} 1 & 2 \\ 1 & 2 \end{bmatrix} = \begin{bmatrix} 0 & 0 \\ 0 & 0 \end{bmatrix}.$$

A pair of matrices A and B with $A \neq O$, $B \neq O$, and $AB = O$ is called a *pair of zero-divisors*.

A weak form of the zero-divisor law holds for square matrices and will be proved in §32.3:

(2.3) *If* A *and* B *are non-singular* n×n *matrices, then* AB *must be non-singular.*

Since commutativity of multiplication can fail, we must pay careful attention to the order of factors in a product of matrices. In the product AB, we say that A *premultiplies* B or that B is *premultiplied* by A. Similarly, we say that B *postmultiplies* A or that A is *postmultiplied* by B.

The following *transposition law for multiplication* relates the transposition operation to the multiplication operation:

(2.4) *If* A *and* B *are conformable for multiplication, then* B^T *and* A^T *are conformable for multiplication, and* $(AB)^T = B^T A^T$.

We prove this result by "transposing" the entire *multiplication diagram* for the product AB. Thus, for example, the diagram at the end of §32.1 transposes to:

$$A^T \rightarrow \begin{bmatrix} 1 & 2 & 3 \\ 2 & 1 & -1 \end{bmatrix}$$

$$B^T \rightarrow \begin{bmatrix} 1 & 0 \\ 5 & 4 \end{bmatrix} \cdot \begin{bmatrix} 1 & 2 & 3 \\ 13 & 14 & 11 \end{bmatrix} \leftarrow B^T A^T = (AB)^T$$

The transposition law implies that the transpose of the product of any number of factors is equal to the product, in reverse order, of the transposed factors. For example, using the transposition law twice, we have

$$(ABC)^T = ((AB)C)^T = C^T (AB)^T = C^T B^T A^T.$$

§32.3 Inverses of Square Matrices

Definition. Let A be an n×n *square* matrix. We say that an n×n matrix B is an *inverse* of A if BA = I.

$$\textit{Example 1.} \quad \begin{bmatrix} 1 & 0 \\ 1 & 1 \end{bmatrix} \textit{ is an inverse of } \begin{bmatrix} 1 & 0 \\ -1 & 1 \end{bmatrix}, \textit{ since } \begin{bmatrix} 1 & 0 \\ 1 & 1 \end{bmatrix}\begin{bmatrix} 1 & 0 \\ -1 & 1 \end{bmatrix} = \begin{bmatrix} 1 & 0 \\ 0 & 1 \end{bmatrix}.$$

The following theorem gives us some basic information about inverses.

(3.1)   ***Theorem.*** *For a square matrix* A:
     (1) $BA = I \Leftrightarrow AB = I$,
     (2) $[BA = I$ *and* $CA = I] \Rightarrow B = C$,
     (3)   A *has an inverse* $\Leftrightarrow |A| \neq 0$.

(1) says that if a matrix B gives the product I from one side, it must also give the product I from the other. (2) says that if A has an inverse, then that inverse is unique. (Thus (2) permits us to speak of *the* inverse of a matrix A.) (3) tells us exactly which matrices have inverses: the matrices with non-zero determinants. The theorem is proved at the end of the present section.

(3.2)   ***Corollary.*** *If* B *is the inverse of* A, *then* A *is the inverse of* B.

(Immediate from (1), since $BA = I \Rightarrow AB = I$.)

Definitions. If A is an n×n matrix, we sometimes say that A is a *square matrix of order* n. If A has an inverse, we say that A is *invertible*. If A is invertible, we write A^{-1} for the inverse of A. (It follows from the corollary that if A^{-1} exists, then $(A^{-1})^{-1}$ exists and $(A^{-1})^{-1} = A$.) (3) of (3.1) can be restated: *for a square matrix* A, A *is invertible* $\Leftrightarrow$ A *is non-singular*.

    The operation of finding A^{-1} from A (when A is square and invertible) is called the *inversion* operation or the operation of *inverting* A. In §32.4 and §32.5, we give systematic procedures for carrying out this operation.

    If A is square and invertible, the inverse matrix A^{-1} can be used to solve the system AX = D.

     (i)   Assume that there is a solution X_0 such that $AX_0 = D$. Multiply the equation $AX_0 = D$ by A^{-1} from the left. Then

$$A^{-1}(AX_0) = A^{-1}D \Rightarrow (A^{-1}A)X_0 = A^{-1}D \Rightarrow IX_0 = A^{-1}D \Rightarrow X_0 = A^{-1}D.$$

Thus, if a solution exists for the given system, it must be the column-vector $A^{-1}D$.

(ii)   Conversely, let $X_1 = A^{-1}D$. Then

$$AX_1 = A(A^{-1}D) = (AA^{-1})D = ID = D,$$

and we see that X_1 must, in fact, be a solution for the given system.

Therefore, if A has an inverse A^{-1}, the solution to the system $AX = D$ is unique, and this solution is the product $A^{-1}D$. This method of solution (where we obtain $A^{-1}D$ by matrix multiplication) is especially useful when we have a single given A and want to solve the system for a variety of different D's. We simply form $A^{-1}D$ for each D. (Of course, in order to do this, we must first have A^{-1}.)

Example 2. $\left(\begin{array}{l} x_1 - x_2 = 2 \\ x_1 + x_2 = 5 \end{array} \right)$. Here $A = \begin{bmatrix} 1 & -1 \\ 1 & 1 \end{bmatrix}$.

Assume that we know that $A^{-1} = \begin{bmatrix} \frac{1}{2} & \frac{1}{2} \\ -\frac{1}{2} & \frac{1}{2} \end{bmatrix}$.

(We will learn below how to find A^{-1}. The reader may check that $\begin{bmatrix} \frac{1}{2} & \frac{1}{2} \\ -\frac{1}{2} & \frac{1}{2} \end{bmatrix} \begin{bmatrix} 1 & -1 \\ 1 & 1 \end{bmatrix} = \begin{bmatrix} 1 & 0 \\ 0 & 1 \end{bmatrix}$.) Then the solution is

$$A^{-1}D = \begin{bmatrix} \frac{1}{2} & \frac{1}{2} \\ -\frac{1}{2} & \frac{1}{2} \end{bmatrix} \begin{bmatrix} 2 \\ 5 \end{bmatrix} = \begin{bmatrix} \frac{7}{2} \\ \frac{3}{2} \end{bmatrix}.$$

Thus $x_1 = \frac{7}{2}$ and $x_2 = \frac{3}{2}$.

Similarly, for $\left(\begin{array}{l} x_1 - x_2 = -1 \\ x_1 + x_2 = -3 \end{array} \right)$, we get

$$A^{-1}D = \begin{bmatrix} \frac{1}{2} & \frac{1}{2} \\ -\frac{1}{2} & \frac{1}{2} \end{bmatrix} \begin{bmatrix} -1 \\ -3 \end{bmatrix} = \begin{bmatrix} -2 \\ -1 \end{bmatrix}.$$

Thus $x_1 = -2$ and $x_2 = -1$.

The following *inversion law for multiplication* relates the operation of inverting a matrix to the operation of multiplication.

(3.3)   *If A and B are invertible square matrices of the same order, then AB is invertible, and* $(AB)^{-1} = B^{-1}A^{-1}$.

To show this, we need only consider the product $(B^{-1}A^{-1})(AB)$. Using associativity, we have

$$(B^{-1}A^{-1})(AB) = B^{-1}(A^{-1}A)B = B^{-1}(I)B = B^{-1}B = I.$$

This shows that AB has $B^{-1}A^{-1}$ as its inverse and this, in turn, shows that AB is invertible. Thus (2.3) is an immediate corollary to (3.3).

The inversion law implies that the inverse of the product of any number of invertible factors is equal to the product, in reverse order, of the inverses of those factors. For example,

$$(ABC)^{-1} = ((AB)C)^{-1} = C^{-1}(AB)^{-1} = C^{-1}B^{-1}A^{-1}.$$

We conclude with a proof of (3.1), the basic theorem on inverses stated at the beginning of this section. We begin by proving, as a lemma, two facts that will be helpful in our final proof. (Recall that a *lemma* gives facts which are used to prove other facts.)

(3.4)   *Lemma.*   (a) $BA = I \Rightarrow |A| \neq 0.$
                            (If A has an inverse, then A is non-singular.);
                    (b) $|A| \neq 0 \Rightarrow AB' = I,$ *for some* B'.
                            (If A is non-singular, then A is an inverse of some matrix B'.)

Proof of (3.4a).   Assume BA = I. Let X_0 be a solution to the homogeneous system AX = O. Then

$$AX_0 = O \Rightarrow BAX_0 = O \Rightarrow IX_0 = O \Rightarrow X_0 = O .$$

This shows that the system $AX = O$ has the unique solution $X = O$. Hence $|A| \neq 0$ (by (3.3) in §31.3, which asserts that $AX = O$ has a unique solution only when $|A| \neq 0$.)

Proof of (3.4b). If $|A| \neq 0$, then (again by (3.3) in §31.3) we can find a unique solution to $AX = D$ for any D. In particular, we can find $X_1, X_2, \ldots, X_n$ such that

$$AX_1 = \begin{bmatrix} 1 \\ 0 \\ 0 \\ \cdot \\ \cdot \\ \cdot \\ 0 \end{bmatrix}, \quad AX_2 = \begin{bmatrix} 0 \\ 1 \\ 0 \\ \cdot \\ \cdot \\ \cdot \\ 0 \end{bmatrix}, \quad \cdots, \quad AX_n = \begin{bmatrix} 0 \\ \cdot \\ \cdot \\ \cdot \\ \cdot \\ 0 \\ 1 \end{bmatrix} .$$

But then $A[X_1 \vdots X_2 \vdots \cdots \vdots X_n] = [AX_1 \vdots AX_2 \vdots \cdots \vdots AX_n] = I$. Hence $[X_1 \vdots \cdots \vdots X_n]$ is the desired B'. This completes the proof of the lemma.

Proof of (3.1).

    Proof of (1) *in* (3.1). ((1) asserts that $BA = I \Leftrightarrow AB = I$.) If $BA = I$, then $|A| \neq 0$, by (a) in the lemma. Hence $AB' = I$ for some B', by (b) in the lemma. Using associativity: $B = BI = B(AB') = (BA)B' = IB' = B'$. Thus $AB = I$. Hence $BA = I \Rightarrow AB = I$. But the same proof, with A and B interchanged, shows that $AB = I \Rightarrow BA = I$. Hence $BA = I \Leftrightarrow AB = I$.

    Proof of (2) *in* (3.1). ((2) asserts that $[BA = I$ and $CA = I] \Rightarrow B = C$.) By (1), $CA = I \Rightarrow AC = I$. Hence $BA = I \Rightarrow B = BI = B(AC) = (BA)C = IC = C$.

    Proof of (3) *in* (3.1). ((3) asserts that A has an inverse $\Leftrightarrow$ $|A| \neq 0$.) If A has the inverse B, then $|A| \neq 0$, by (a) in the lemma. On the other hand, if $|A| \neq 0$, then, by (b) in the lemma, $AB' = I$. Hence by (1), $B'A = I$, and A has the inverse B'. Thus we have shown that A has an inverse $\Leftrightarrow |A| \neq 0$. This completes the proof of (3.1).

§32.4 Inverting by Formula

Let A be given with $|A| \neq 0$. Let α_{ij} be the cofactor of the element a_{ij} in A, for $1 \leq i,j \leq n$. We form the matrix $\begin{bmatrix} \alpha_{11} & \alpha_{12} & \cdots & \alpha_{1n} \\ \alpha_{21} & & & \\ \cdots & \cdots & \cdots & \cdots \end{bmatrix}$ of correspond-

ing cofactors. The transpose of this matrix is called the *adjugate matrix* of A and is written $Adj(A)$. Thus we have:

$$Adj(A) = \begin{bmatrix} \alpha_{11} & \alpha_{21} & \cdots & \alpha_{n1} \\ \alpha_{12} & & & \\ \cdots & \cdots & \cdots & \cdots \end{bmatrix}.$$

The following theorem gives us our formula for the inverse of A:

(4.1)   **Theorem.** $A^{-1} = \dfrac{1}{|A|}(Adj(A)) = \dfrac{1}{|A|} \begin{bmatrix} \alpha_{11} & \alpha_{21} & \cdots & \alpha_{n1} \\ \alpha_{12} & \cdots & \cdots & \cdots \\ \cdots & \cdots & \cdots & \cdots \end{bmatrix}.$

Proof. Consider the product

$$A(Adj(A)) = \begin{bmatrix} a_{11} & a_{12} & \cdots & a_{1n} \\ a_{21} & \cdots & \cdots & \cdots \\ \cdots & \cdots & \cdots & \cdots \\ a_{n1} & \cdots & \cdots & a_{nn} \end{bmatrix} \begin{bmatrix} \alpha_{11} & \alpha_{21} & \cdots & \alpha_{n1} \\ \alpha_{12} & \cdots & \cdots & \cdots \\ \cdots & \cdots & \cdots & \cdots \\ \alpha_{1n} & \cdots & \cdots & \alpha_{n1} \end{bmatrix}.$$

This product must be $\begin{bmatrix} |A| & 0 & \cdots & \cdots & 0 \\ 0 & |A| & 0 & \cdots & 0 \\ 0 & 0 & |A| & \cdots & 0 \\ \cdots & \cdots & \cdots & \cdots \\ 0 & \cdots & 0 & \cdots & |A| \end{bmatrix}$, since each element on the

main diagonal of the product is given by a correct Laplace expansion for A and each element off the diagonal is given by a false Laplace expansion for A. (See §31.4.) Hence

$$A\left(\frac{1}{|A|}\right)(Adj(A)) = \frac{1}{|A|}A(Adj(A)) = I,$$

and the proof is complete.

Examples. Invert each of the following by formula.

(1) $A = \begin{bmatrix} 1 & 4 \\ 1 & 3 \end{bmatrix}$. Here $|A| = -1$, and the matrix of cofactors is

$\begin{bmatrix} 3 & -1 \\ -4 & 1 \end{bmatrix}$. Therefore $\text{Adj}(A) = \begin{bmatrix} 3 & -4 \\ -1 & 1 \end{bmatrix}$.

We get $A^{-1} = -\text{Adj}(A) = \begin{bmatrix} -3 & 4 \\ 1 & -1 \end{bmatrix}$.

(2) $A = \begin{bmatrix} 1 & 2 & 1 \\ 1 & 0 & 1 \\ 2 & 1 & 2 \end{bmatrix}$. Here $|A| = 4 + 1 - 1 - 4 = 0$. Thus A is singu-

lar, and no inverse exists.

(3) $A = \begin{bmatrix} 1 & 2 & 1 \\ 1 & 0 & 1 \\ 2 & 2 & 1 \end{bmatrix}$. Here $|A| = 2$, the matrix of cofactors is

$\begin{bmatrix} -2 & 1 & 2 \\ 0 & -1 & 2 \\ 2 & 0 & -2 \end{bmatrix}$, $\text{Adj}(A) = \begin{bmatrix} -2 & 0 & 2 \\ 1 & -1 & 0 \\ 2 & 2 & -2 \end{bmatrix}$, and $A^{-1} = \begin{bmatrix} -1 & 0 & 1 \\ \frac{1}{2} & -\frac{1}{2} & 0 \\ 1 & 1 & -1 \end{bmatrix}$.

The reader may check, by calculating $A^{-1}A$, that we have indeed found the correct inverses in examples 1 and 3.

§32.5 Inverting by Row-reduction

This procedure is described in the following theorem.

(5.1) **Theorem.** *Given an* $n \times n$ *matrix* A, *form the* $n \times 2n$ *matrix* $[A \vdots I]$. *Row-reduce this matrix. If* A *is invertible, the final row-reduced matrix will be of the form* $[I \vdots B]$ *for some B, and the matrix B will be* A^{-1}, *the inverse of* A. *If* A *is not invertible, the final row-reduced matrix will not be of the form* $[I \vdots B]$.

We begin with some examples and then go on to prove the theorem.

Examples.

(1) $A = \begin{bmatrix} 1 & 4 \\ 1 & 3 \end{bmatrix}$. We form $[A \vdots I]$ and row-reduce:

$$\begin{bmatrix} 1 & 4 & 1 & 0 \\ 1 & 3 & 0 & 1 \end{bmatrix} \rightarrow \begin{bmatrix} 1 & 4 & 1 & 0 \\ 0 & -1 & -1 & 1 \end{bmatrix} \rightarrow \begin{bmatrix} 1 & 0 & -3 & 4 \\ 0 & 1 & 1 & -1 \end{bmatrix}.$$

Thus $A^{-1} = \begin{bmatrix} -3 & 4 \\ 1 & -1 \end{bmatrix}$. This agrees with the result previously obtained by formula in §32.4.

(2) $A = \begin{bmatrix} 1 & 2 & 1 \\ 1 & 0 & 1 \\ 2 & 1 & 2 \end{bmatrix}$. We form $[A \vdots I]$ and row-reduce:

$$\begin{bmatrix} 1 & 2 & 1 & \cdot & 1 & 0 & 0 \\ 1 & 0 & 1 & \cdot & 0 & 1 & 0 \\ 2 & 1 & 2 & \cdot & 0 & 0 & 1 \end{bmatrix} \rightarrow \begin{bmatrix} 1 & 2 & 1 & \cdot & 1 & 0 & 0 \\ 0 & -2 & 0 & \cdot & -1 & 1 & 0 \\ 0 & -3 & 0 & \cdot & -2 & 0 & 1 \end{bmatrix} \rightarrow$$

$$\begin{bmatrix} 1 & 0 & 1 & \cdot & 0 & 1 & 0 \\ 0 & 1 & 0 & \cdot & \frac{1}{2} & -\frac{1}{2} & 0 \\ 0 & 0 & 0 & \cdot & \frac{1}{2} & -\frac{3}{2} & 1 \end{bmatrix}.$$

Here A is not invertible, since the row-reduction of A, on the left, shows that $|A| = 0$.

(3) $A = \begin{bmatrix} 1 & 2 & 1 \\ 1 & 0 & 1 \\ 2 & 2 & 1 \end{bmatrix}$. We form $[A \vdots I]$ and row-reduce:

$$\begin{bmatrix} 1 & 2 & 1 & \cdot & 1 & 0 & 0 \\ 1 & 0 & 1 & \cdot & 0 & 1 & 0 \\ 2 & 2 & 1 & \cdot & 0 & 0 & 1 \end{bmatrix} \rightarrow \begin{bmatrix} 1 & 2 & 1 & \cdot & 1 & 0 & 0 \\ 0 & -2 & 0 & \cdot & -1 & 1 & 0 \\ 0 & -2 & -1 & \cdot & -2 & 0 & 1 \end{bmatrix} \rightarrow$$

$$\begin{bmatrix} 1 & 0 & 1 & \cdot & 0 & 1 & 0 \\ 0 & 1 & 0 & \cdot & \frac{1}{2} & -\frac{1}{2} & 0 \\ 0 & 0 & -1 & \cdot & -1 & -1 & 1 \end{bmatrix} \rightarrow \begin{bmatrix} 1 & 0 & 0 & \cdot & -1 & 0 & 1 \\ 0 & 1 & 0 & \cdot & \frac{1}{2} & -\frac{1}{2} & 0 \\ 0 & 0 & 1 & \cdot & 1 & 1 & -1 \end{bmatrix}.$$

Thus $A^{-1} = \begin{bmatrix} -1 & 0 & 1 \\ \frac{1}{2} & -\frac{1}{2} & 0 \\ 1 & 1 & -1 \end{bmatrix}$, agreeing with the result previously obtained by formula in §32.4.

Proof of the theorem (5.1). We saw in §32.3, in the proof of (3.4b), that the inverse of a given invertible matrix A will be

$$A^{-1} = [X_1 \vdots X_2 \vdots \cdots \vdots X_n],$$

where $X_1, X_2, \ldots, X_n$ are the unique solutions to the systems

$$AX_1 = \begin{bmatrix} 1 \\ 0 \\ 0 \\ \cdot \\ \cdot \\ \cdot \\ 0 \end{bmatrix}, \ AX_2 = \begin{bmatrix} 0 \\ 1 \\ 0 \\ \cdot \\ \cdot \\ \cdot \\ 0 \end{bmatrix}, \ \cdots, \ AX_n = \begin{bmatrix} 0 \\ 0 \\ \cdot \\ \cdot \\ \cdot \\ 0 \\ 1 \end{bmatrix}.$$

Assume that A is invertible and let C_i be the i-th column of I in $[A \vdots I]$. Our row-reduction of $[A \vdots I]$ will provide a solution by row-reduction of the system $AX = C_i$. It will do so by transforming the column C_i of I into the desired solution X_i. Thus we obtain the desired solutions $X_1, X_2, \ldots, X_n$ as the columns of the final matrix B, and we have $B = A^{-1}$.

On the other hand, if A is not invertible, then $r(A) < n$, and the final row-reduced $n \times 2n$ matrix must have fewer than n pivots in its left half. Hence the final matrix cannot have the form $[I \vdots B]$. This completes the proof.

§32.6 Calculating with Determinants and Matrices

Matrix-inversion is a commonly occurring, and often lengthy, computational problem. Usually, for matrices of size 4×4 or larger, row-reduction is the best method. In the table following, we compare three methods:

(I)   *Inversion by formula, where cofactors are evaluated by Laplace expansion;*

(II)   *Inversion by formula, where cofactors are evaluated by row-reduction;*

(III)   *Inversion by row-reduction.*

In each case, we measure length of computation by the maximum number of additions and multiplications required to produce the inverse matrix.

n	(I)	(II)	(III)
2	7	9	12
3	53	72	45
4	383	347	112
5	2,999	1,184	225
20	9.7×10^{20}	770,272	15,600
40	1.3×10^{51}	63,337,079	126,400
50	7.6×10^{67}	196,205,849	247,500
General formulas:	$n^2(n!) - 1$	$\frac{1}{3}(2n^5 - 6n^4 + 9n^3 - 3n^2 + n - 3)$	$2n^3 - n^2$

To invert a 40×40 matrix by sequential computation, at a rate of one arithmetical operation per microsecond, would take 4×10^{28} billions of years by method (I), 1 minute by method (II), and 1/8 second by method (III). Standard computer programs for general matrix inversion are based on (III).

Matrices in which most of the elements are zero are referred to as *sparse* matrices. If a square matrix A of order > 3 is sparse, it is sometimes easier to find the value of $|A|$ by Laplace expansions, or by a clever combination of Laplace expansions and row-reduction, rather than by row-reduction alone. Similarly, for a sparse matrix A, it may be easier to find A^{-1} by a clever use of (I) or (II) rather than by (III). In hand calculation, these differences can be significant.

Example. Let $A = \begin{bmatrix} 1 & 0 & 0 & 0 \\ 0 & 1 & 0 & 0 \\ 0 & 2 & 0 & 2 \\ 0 & 0 & 1 & 0 \end{bmatrix}$. We immediately have $|A| = \begin{vmatrix} 0 & 2 \\ 1 & 0 \end{vmatrix} =$

−2, by successive top-row Laplace expansions. Similarly, again by Laplace expansions, the matrix of cofactors is quickly seen to be

$$\begin{bmatrix} -2 & 0 & 0 & 0 \\ 0 & -2 & 0 & 2 \\ 0 & 0 & 0 & -1 \\ 0 & 0 & -2 & 0 \end{bmatrix}.$$

Hence
$$\text{Adj}(A) = \begin{bmatrix} -2 & 0 & 0 & 0 \\ 0 & -2 & 0 & 0 \\ 0 & 0 & 0 & -2 \\ 0 & 2 & -1 & 0 \end{bmatrix},$$

and
$$A^{-1} = \begin{bmatrix} 1 & 0 & 0 & 0 \\ 0 & 1 & 0 & 0 \\ 0 & 0 & 0 & 1 \\ 0 & -1 & \frac{1}{2} & 0 \end{bmatrix}.$$

For this example, the hand-calculation of A^{-1} by row-reduction would be longer and less convenient.

§32.7 Determinants and Matrix Multiplication

The following theorem for square matrices gives a useful relationship between matrix multiplication and determinants.

(7.1)   *Theorem. If* A *and* B *are* n×n *matrices, then* $|AB| = |A|\,|B|$.

Thus the value of the determinant of the matrix product of A and B must equal the product of the values of the determinants of A and B.

Outline of Proof. The proof is simple and occurs in four steps. We omit details.

(a)   We show that if we take any n×n matrix C and form the larger 2n×2n matrix $\begin{bmatrix} A & \cdot & O \\ \cdot & \cdot & \cdot \\ C & \cdot & B \end{bmatrix}$, then $\begin{vmatrix} A & \cdot & O \\ \cdot & \cdot & \cdot \\ C & \cdot & B \end{vmatrix} = |A|\,|B|$. The proof is by induction on n, using Laplace expansions.

(b)   Similarly, we show that $\begin{vmatrix} O & \cdot & A \\ \cdot & \cdot & \cdot \\ B & \cdot & C \end{vmatrix} = |-B|\,|A|$. We go from $\begin{bmatrix} O & \cdot & A \\ \cdot & \cdot & \cdot \\ B & \cdot & C \end{bmatrix}$ to $\begin{bmatrix} A & \cdot & O \\ \cdot & \cdot & \cdot \\ C & \cdot & B \end{bmatrix}$ by interchanging columns, and then applying (a).   (See properties (1) and (6) of determinants in §31.2.) The cases n even and n odd are treated separately.

(c)   We show that appropriate elementary row operations of type β will carry
$$\begin{bmatrix} A & \cdot & O \\ \cdot & \cdot & \cdot \\ -I & \cdot & B \end{bmatrix} \text{ to } \begin{bmatrix} O & \cdot & D \\ \cdot & \cdot & \cdot \\ -I & \cdot & B \end{bmatrix}, \text{ where } D = AB.$$

(d)   Since β-operations leave the value of the determinant $\begin{vmatrix} A & \cdot & O \\ \cdot & & \cdot \\ -I & \cdot & B \end{vmatrix}$

unchanged, we get, from (a), (b), and (c):

$$|A|\,|B| \underset{(a)}{=} \begin{vmatrix} A & \cdot & O \\ \cdot & \cdot & \cdot \\ -I & \cdot & B \end{vmatrix} \underset{(c)}{=} \begin{vmatrix} O & \cdot & AB \\ \cdot & \cdot & \cdot \\ -I & \cdot & B \end{vmatrix} \underset{(b)}{=} |I|\,|AB| = |AB| \ .$$

This completes the proof.

§32.8 Eigenvectors and Eigenvalues for Square Matrices

Definitions. Let A be an n-th-order square matrix. Let X be any $n \times 1$ column-vector such that $X \neq O$. Then the product AX is a new $n \times 1$ column-vector. If it turns out that the new vector AX is itself a scalar multiple of X – that is to say, if it turns out that $AX = \lambda X$ for some scalar λ – then we say that the non-zero vector X is an *eigenvector* of the matrix A. Since $X \neq O$, the value of the scalar λ must be unique. (Otherwise, if λ_1 and λ_2 were two different values for λ, we would have $(\lambda_2 - \lambda_1)X = O$, which would imply X = O.) λ is said to be an *eigenvalue of* the matrix A and is called the *associated eigenvalue of* the eigenvector X.

Let X be an eigenvector of A with the associated eigenvalue λ. Let kX be a scalar multiple of X for any $k \neq 0$. Then kX must also be an eigenvector of A with the same associated eigenvalue λ, since

$$A(kX) = k(AX) = k(\lambda X) = \lambda(kX).$$

Example 1. Let $\begin{bmatrix} 2 & -1 \\ 4 & -3 \end{bmatrix}$. Then $X = \begin{bmatrix} 1 \\ 1 \end{bmatrix}$ is an eigenvector of A with

eigenvalue 1, since $\begin{bmatrix} 2 & -1 \\ 4 & -3 \end{bmatrix} \cdot \begin{bmatrix} 1 \\ 1 \end{bmatrix} = \begin{bmatrix} 1 \\ 1 \end{bmatrix} = (1)\begin{bmatrix} 1 \\ 1 \end{bmatrix}$. It follows that for any $k \neq 0$,

$\begin{bmatrix} k \\ k \end{bmatrix}$ must also be an eigenvector of A with eigenvalue 1. Similarly, $X = \begin{bmatrix} 1 \\ 4 \end{bmatrix}$ is

an eigenvector of A with eigenvalue –2, since $\begin{bmatrix} 2 & -1 \\ 4 & -3 \end{bmatrix} \cdot \begin{bmatrix} 1 \\ 4 \end{bmatrix} = \begin{bmatrix} -2 \\ -8 \end{bmatrix} = (-2)\begin{bmatrix} 1 \\ 4 \end{bmatrix}$.

It follows that for any $k \neq 0$, $\begin{bmatrix} k \\ 4k \end{bmatrix}$ must also be an eigenvector of A with

eigenvalue –2.

We shall see below that the eigenvectors above are the only eigenvectors of A and that $\lambda = 1$, and $\lambda = -2$ are the only eigenvalues of A.

Let A be a given n-th order square matrix. Then for any given n×1 column-vector X and any scalar λ, we have the following equivalent statements, by the laws of matrix algebra: $AX = \lambda X \Leftrightarrow AX-\lambda X = O \Leftrightarrow AX-\lambda(IX) = O \Leftrightarrow$ $(A-\lambda I)X = O = \Leftrightarrow X$ *is a solution to the square homogeneous system whose coefficient matrix is* $A-\lambda I$.

It follows from this that:

(8.1)   X *is an eigenvector of* A *with eigenvalue* $\lambda \Leftrightarrow X$ *is a non-trivial solution of the homogeneous system* $(A-\lambda I)X = 0$.

Hence, by the corollary in §31.3:

(8.2) λ *is an eigenvalue of* A $\Leftrightarrow$ *the determinant* $|A - \lambda I| = 0$.

If, for the moment, we treat the symbol λ as a variable, and if we evaluate the determinant $|A - \lambda I|$, we obtain a polynomial of degree n in the variable λ. This polynomial is called the *characteristic polynomial* of the matrix A. The equation that we get by setting this polynomial equal to 0 is called the *characteristic equation* of A. Hence *the set of eigenvalues of* A *must be the set of real solutions for the characteristic equation of* A. Since a polynomial of degree n can have at most n distinct real solutions, we see that an n-th order square matrix A can have at most n distinct eigenvalues. We shall usually indicate the

characteristic polynomial of A by writing $\left|A - \lambda I\right|$, and the characteristic equation
by writing $\left|A - \lambda I\right| = 0$.

Example 2. Let $A = \begin{bmatrix} 2 & -1 \\ 4 & -3 \end{bmatrix}$, as in the previous example.   Then

$A - \lambda I = \begin{bmatrix} 2-\lambda & -1 \\ 4 & -3-\lambda \end{bmatrix}$. The characteristic polynomial of A is $\left|A - \lambda I\right| =$
$(2-\lambda)(-3-\lambda)+4 = \lambda^2+\lambda-2$. Hence, $\lambda^2+\lambda-2 = 0$ is the characteristic equation of A.
Solving this equation, we have

$$\lambda^2+\lambda-2 = (\lambda+2)(\lambda-1) = 0 .$$

Hence the eigenvalues of A are $\lambda = -2$ and $\lambda = 1$.
We thus have the following two-stage procedure for finding the eigenvectors and eigenvalues of a given matrix A:

Stage 1. *Find the characteristic polynomial of* A *by evaluating the determinant* $\left|A - \lambda I\right|$ *to get a polynomial in the variable* λ. *Then set this polynomial equal to* 0 *to get the characteristic equation of* A. *The real solutions of this equation will be the eigenvalues of* A. (The solutions of the characteristic equation are sometimes called the *roots* of the characteristic polynomial.) In certain simple cases, as in the examples below, solutions can be found by factoring the polynomial.   Often, however, computer-approximation methods will be required.

Stage 2. *Consider, in turn, each of the eigenvalues found in stage 1. For each eigenvalue* λ, *find the non-trivial solutions of the homogeneous system* $(A-\lambda I)X = 0$. *These non-trivial solutions will be the eigenvectors of* A *for the given* λ. We can find these non-trivial solutions by the row-reduction method of §30.5.

We illustrate this two-stage procedure with the following examples.

Example 3. *Let* $A = \begin{bmatrix} 2 & -1 \\ 4 & -3 \end{bmatrix}$, *as in the previous example above.*

Stage 1. We carried out stage 1 in the previous example. We found the eigenvalues $\lambda = -2$ and $\lambda = 1$.
Stage 2. We consider each eigenvalue in turn:

$\underline{\lambda = -2}$: Then $A-\lambda I = A+2I$. We seek the non-trivial solutions to the homogeneous system $(A+2I)X = 0$. We have the coefficient matrix $A + 2I = \begin{bmatrix} 2+2 & -1 \\ 4 & -3+2 \end{bmatrix} = \begin{bmatrix} 4 & -1 \\ 4 & -1 \end{bmatrix}$. Row-reducing the coefficient matrix, we get

$$\begin{bmatrix} 1 & -\frac{1}{4} \\ 0 & 0 \end{bmatrix}.$$

Hence the solutions must be

$$X = t \begin{bmatrix} \frac{1}{4} \\ 1 \end{bmatrix},$$

and the eigenvectors must be

$$X = t \begin{bmatrix} \frac{1}{4} \\ 1 \end{bmatrix} \quad \text{for } t \neq 0.$$

$\underline{\lambda = 1}$: We have $A - \lambda I = A - I = \begin{bmatrix} 2-1 & -1 \\ 4 & -3-1 \end{bmatrix} = \begin{bmatrix} 1 & -1 \\ 4 & -4 \end{bmatrix}$.

Row-reduction gives $\begin{bmatrix} 1 & -1 \\ 0 & 0 \end{bmatrix}$.

Hence the eigenvectors must be

$$X = t \begin{bmatrix} 1 \\ 1 \end{bmatrix} \quad \text{for } t \neq 0.$$

We have thus found all eigenvalues and all eigenvectors for the given matrix A.

Example 4. *Let* $A = \begin{bmatrix} 1 & -1 \\ 1 & 1 \end{bmatrix}$.

Stage 1. The characteristic polynomial is

$$|A - \lambda I| = \begin{bmatrix} 1-\lambda & -1 \\ 1 & 1-\lambda \end{bmatrix} = (1-\lambda)^2 + 1 = \lambda^2 - 2\lambda + 2.$$

The characteristic equation, $\lambda^2 - 2\lambda + 2 = 0$, has no real roots. Hence A has no eigenvalues or eigenvectors, and we cannot go on to stage 2. (In this case, the characteristic equation does have the complex-number solutions $y = 1 + i, 1 - i$. Such non-real, complex-number solutions of the characteristic equation of a matrix are sometimes referred to as *complex eigenvalues* of the matrix. Complex eigenvalues have an important role in the solution of differential equations.

Example 5. Let $A = \begin{bmatrix} 2 & 1 & 0 \\ 1 & 1 & 1 \\ 0 & 1 & 2 \end{bmatrix}$.

Stage 1. The characteristic polynomial is

$$|A - \lambda I| = \begin{bmatrix} 2 - \lambda & 1 & 0 \\ 1 & 1 - \lambda & 1 \\ 0 & 1 & 2 - \lambda \end{bmatrix} = (2 - \lambda)^2 (1 - \lambda) - 2(2 - \lambda)$$

$$= (2 - \lambda)((2 - \lambda)(1 - \lambda) - 2) = (2 - \lambda)(-3\lambda + \lambda^2) = -\lambda(\lambda - 2)(\lambda - 3).$$

Hence the eigenvalues of A are $\lambda = 0, 2, 3$.

Stage 2. We consider each eigenvalue in turn:

$\underline{\lambda = 0}$: A $-\lambda I = A$. Row-reducing, we obtain $\begin{bmatrix} 1 & 0 & -1 \\ 0 & 1 & 2 \\ 0 & 0 & 0 \end{bmatrix}$.

Hence the eigenvectors for $\lambda = 0$ are $X = t \begin{bmatrix} 1 \\ -2 \\ 1 \end{bmatrix}$ for $t \neq 0$.

$\underline{\lambda = 2}$: A $-\lambda I$ = A$-2I$ = $\begin{bmatrix} 0 & 1 & 0 \\ 1 & -1 & 1 \\ 0 & 1 & 0 \end{bmatrix}$. Row-reducing, we obtain

$\begin{bmatrix} 1 & 0 & 1 \\ 0 & 1 & 0 \\ 0 & 0 & 0 \end{bmatrix}$. Hence the eigenvectors for $\lambda = 2$ are $X = t \begin{bmatrix} -1 \\ 0 \\ 1 \end{bmatrix}$

for $t \neq 0$.

$$\underline{\lambda = 3}: \quad A\text{-}\lambda I = A - 3I = \begin{bmatrix} -1 & 1 & 0 \\ 1 & -2 & 1 \\ 0 & 1 & -1 \end{bmatrix}. \quad \text{Row-reducing, we obtain}$$

$$\begin{bmatrix} 1 & 0 & -1 \\ 0 & 1 & -1 \\ 0 & 0 & 0 \end{bmatrix}. \quad \text{Hence the eigenvectors for } \lambda = 3 \quad \text{are} \quad X = t\begin{bmatrix} 1 \\ 1 \\ 1 \end{bmatrix}$$

for $t \neq 0$.

The concepts of *eigenvector* and *eigenvalue* and the techniques for finding eigenvectors and eigenvalues are widely used. They are central to the solution of linear differential equations and appear, as well, in many applications of matrix algebra to physical science, to engineering, and to other parts of mathematics. We consider some of these applications in Chapter 34.

The eigenvalues of a matrix A are sometimes referred to as the *characteristic values* or *latent values* of A. The eigenvectors are sometimes referred to as the *characteristic vectors* of A.

Remark. In the last example above, the given matrix A was symmetric. The reader may verify that if we take any two distinct eigenvalues for this matrix, and then choose an eigenvector for each of these eigenvalues, we get a pair of orthogonal vectors. We shall see in Chapter 34 that this is always true for a symmetric matrix. If X_1 and X_2 are eigenvectors associated with distinct eigenvalues of a symmetric matrix, then X_1 and X_2 must be orthogonal vectors. In fact, we shall show that for any $n \times n$ symmetric matrix, we can always find a set of n mutually orthogonal eigenvectors.

§32.9 Changing Cartesian Axes; Orthogonal Matrices

Let $\hat{i}, \hat{j}, \hat{k}$ be the frame of unit coordinate vectors for a set of Cartesian axes with origin O in $\mathbf{E}^3$. We call these axes the x, y, and z *axes*. Let another set of Cartesian axes be given with the *same origin* O. We call the new axes the x', y', and z' *axes*. Let $\hat{u}, \hat{v}, \hat{w}$ be the frame of unit coordinate vectors for the new axes. Then every point P has a triple of *old* coordinates (x,y,z) and a triple of *new* coordinates (x',y',z'), and every vector $\vec{A}$ has a triple of *old* scalar components (giving $\vec{A} = a_1\hat{i} + a_2\hat{j} + a_3\hat{k}$) and a triple of new scalar components (giving $\vec{A} = a_1'\hat{u} + a_2'v + a_3'\hat{w}$). We wish to study the mathematical relationship between the old and new coordinate systems.

We begin by forming the 3×3 matrix Q, where

(9.1)
$$Q = \begin{bmatrix} \hat{u}\cdot\hat{i} & \hat{v}\cdot\hat{i} & \hat{w}\cdot\hat{i} \\ \hat{u}\cdot\hat{j} & \hat{v}\cdot\hat{j} & \hat{w}\cdot\hat{j} \\ \hat{u}\cdot\hat{k} & \hat{v}\cdot\hat{k} & \hat{w}\cdot\hat{k} \end{bmatrix}.$$

From our knowledge of frames (§3.4), we immediately see that Q is, in fact, the matrix

$$Q = [U \vdots V \vdots W],$$

where the column-vectors U,V,W give the components, in the *old* coordinate system, of the *new* unit coordinate vectors $\hat{u}, \hat{v}, \hat{w}$.

In the same way, we see that Q is also, in fact, the matrix

$$Q = \begin{vmatrix} I^T \\ \cdots \\ J^T \\ \cdots \\ K^T \end{vmatrix},$$

where the column-vectors I,J,K give the components, in the *new* coordinate system, of the *old* coordinate vectors $\hat{i}, \hat{j}, \hat{k}$.

Let some vector $\vec{A}$ be given. Let the column-vector A give the components of $\vec{A}$ in the *old* system, and let the column vector A′ give the components of $\vec{A}$ in the *new* system. Consider the matrix product

$$QA' = \begin{bmatrix} I^T \\ \cdots \\ J^T \\ \cdots \\ K^T \end{bmatrix} \cdot [A'] = \begin{bmatrix} I^T A' \\ J^T A' \\ K^T A' \end{bmatrix}.$$

The inner products in the final column-vector are calculated in the new coordinate system. We see, however, that this column-vector must be

$$\begin{bmatrix} \vec{A} \cdot \hat{i} \\ \vec{A} \cdot \hat{j} \\ \vec{A} \cdot \hat{k} \end{bmatrix} ,$$

which gives the components of $\vec{A}$ in the old system. Hence this last column-vector is A, and we have obtained the matrix equation

$$(9.2) \qquad\qquad A = QA' ,$$

which relates new components A' to old components A, for any given vector $\vec{A}$. Thus, *premultiplication by Q carries us from new components to old components.*

Let X and X' be column-vectors giving the old and new coordinates for some point P. By the same argument, using the position vector for P, we have:

$$(9.3) \qquad\qquad X = QX' .$$

Thus *premultiplication by Q carries us directly from the new coordinates of any given point to the old coordinates of that same point.*

We now form the matrix

$$Q^T = \begin{bmatrix} U^T \\ \cdots \\ V^T \\ \cdots \\ W^T \end{bmatrix} .$$

Q^T is the transpose of Q. In the same way as before, we see that

$$(9.4) \qquad\qquad Q^T A = \begin{bmatrix} U^T A \\ \cdots \\ V^T A \\ \cdots \\ W^T A \end{bmatrix} = \begin{bmatrix} \vec{A} \cdot \hat{u} \\ \vec{A} \cdot \hat{v} \\ \vec{A} \cdot \hat{w} \end{bmatrix} = A' ,$$

and that

$$(9.5) \qquad\qquad Q^T X = X' .$$

Thus *premultiplication by* Q^T *carries us from old components to new components and from old coordinates to new coordinates.*

Recall that an *orthonormal set* is a set of mutually orthogonal unit-vectors. Because the vectors $\hat{u}, \hat{v}, \hat{w}$ form a frame, they are an orthonormal set. It follows that the product Q^TQ must have the multiplication diagram:

$$[U \vdots V \vdots W]$$

$$Q^TQ = \begin{bmatrix} U^T \\ \cdots \\ V^T \\ \cdots \\ W^T \end{bmatrix} \begin{bmatrix} 1 & 0 & 0 \\ 0 & 1 & 0 \\ 0 & 0 & 1 \end{bmatrix}.$$

Hence we have $Q^TQ = I$. *Thus* Q *has an inverse, and* $Q^{-1} = Q^T$. *It follows that we must also have* $QQ^T = I$.

Our discussion of changing axes in three dimensions leads us to the following general definition in n dimensions:

Definition. We say that an $n \times n$ matrix Q is an *orthogonal matrix* if $Q^TQ = I$.

Evidently a matrix is orthogonal if and only if and only if its columns, as n-vectors, are an orthonormal set. Since $Q^TQ = I \Rightarrow QQ^T = I$ (by theorem (3.1)), we see that the rows of Q, as n-vectors, must also be an orthonormal set. Each of these two orthonormal sets gives a frame in $\mathbf{R}^n$, and, as before, the matrix equations

$$A = QA' \quad \text{and} \quad A' = Q^TA$$

give us a means for transforming from components in one frame to components in the other. We see that *every orthogonal matrix describes a change of frames, and that every change of frames can be described by some orthogonal matrix.* Moreover for $n = 2$ or $n = 3$, as we have already noted, *every orthogonal matrix describes a change of Cartesian axes with fixed origin, and every change of Cartesian axes with fixed origin can be described by some orthogonal matrix.*

The following theorem provides some useful general facts about orthogonal matrices.

(9.6)   *Theorem.* Let Q be an $n \times n$ *orthogonal matrix.*

(i)   *If* X *and* Y *are two column-vectors such that* X = QY, *then*
        $X^T X = Y^T Y$. (Thus premultiplication by Q leaves the magnitude
        of a column-vector unchanged, where the *magnitude* of a column-
        vector is defined to be its magnitude as an n-vector.)

(ii)   *Let* Q_1 *be an* n×n *orthogonal matrix. Then the product* QQ_1 *is
        also an orthogonal matrix.*

(iii)   *The determinant of* Q *must have the value* +1 *or* −1.

(iv)   *The only possible real-number eigenvalues for* Q *are* +1 *and* −1.

(v)   *If* n *is odd,* Q *must have at least one real-number eigenvalue,
        and hence must have an eigenvector.* (If n is even, it is possible
        for Q to have no real-number eigenvalues.   Consider   Q =
        $\begin{bmatrix} 0 & -1 \\ 1 & 0 \end{bmatrix}$. In this case, Q gives a change of axes in E^2, where the

        old axes are rotated by π/2 to give the new axes.)

Proof. For (i), we note that $X = QY \Rightarrow X^T = Y^T Q^T \Rightarrow X^T X = Y^T Q^T Q Y =$
$Y^T(I)Y = Y^T Y$. For (ii) we note that $(QQ_1)^T QQ_1 = Q_1^T Q^T QQ_1 = Q_1^T (I) Q_1 =$
$Q_1^T Q_1 = I$. Thus QQ_1 satisfies the definition of an orthogonal matrix.

For (iii), we note that $|Q^T| = |Q|$, since repeated Laplace expansions by
rows for Q^T coincide with repeated Laplace expansions by columns for Q. Hence,
by the theorem (7.1), we have $|Q^T||Q| = |Q|^2$. Since $Q^T Q = I$, we have
$|Q^T Q| = 1$. Therefore $|Q|^2 = 1$ (by (7.1)), and $|Q| = +1$ or $|Q| = -1$.

For (iv), we note, by (i), that for any eigenvector X of Q, the magnitude
of QX must equal the magnitude of X. Hence for any eigenvalue λ, $|\lambda| = 1$, and
we must have λ = +1 or λ = −1.

For (v), we note that if n is odd, the characteristic equation of Q must
have odd degree. Since, from elementary algebra, a polynomial equation of odd
degree must have at least one real solution, the result follows.

This completes the proof.

Remarks. (1) For n = 3, part (v) of the theorem gives unexpected geometric information. It tells us that whenever we go from one frame to another, there
must either be at least one non-zero vector whose components remain unchanged
or else at least one non-zero vector whose new components are, respectively, the
negatives of its old components.

(2) For each n, (iii) tells us that the n×n orthogonal matrices fall into
two classes: those with determinant = +1 and those with determinant = −1. The

reader may verify that in the case n = 3, $|Q| = 1$ means that Q either takes us from a right-handed frame to a right-handed frame or from a left-handed frame to a left-handed frame, while $|Q| = -1$ means that Q either takes us from a right-handed frame to a left-handed frame or vice versa.

(3)   In Chapter 34, we shall see that it is sometimes possible to use an orthogonal matrix to simplify the equation $F(x,y,z) = 0$ of a given surface by changing to a new Cartesian system. (See also §4.7.)

(4)   Let n = 3. If we go from given Cartesian coordinates (x,y,z) to new Cartesian coordinates (x',y',z') with the same origin, then the defining equations for the new coordinate system are given by the matrix equation

$$X = QX'.$$

(These defining equations are

$$x = q_{11}x' + q_{12}y' + q_{13}z', \quad y = q_{21}x' + q_{22}y' + q_{23}z',$$
$$z = q_{31}x' + q_{32}y' + q_{33}z',$$

where the q_{ij} are the elements of Q.) As defining equations, they express the old coordinates in terms of the new. To express the new coordinates in terms of the old, we have

$$X' = Q^{T}X.$$

If, instead, we wish to change to new Cartesian coordinates with a *new origin*, then the defining equations for the new coordinates are given by

$$X = QX' + R,$$

where Q describes the change in frames (if any), and R is the column-vector of the components, in the old coordinate system, of the position vector $\vec{R}$ of the new origin with respect to the old origin [9–1].

(5)   We conclude with a discussion of orthogonal matrices for the case n = 2:

(a) $Q = \begin{bmatrix} \cos\theta & -\sin\theta \\ \sin\theta & \cos\theta \end{bmatrix}$

gives a new coordinate system in which both axes have been rotated through an angle θ. In this case, $|Q| = 1$.

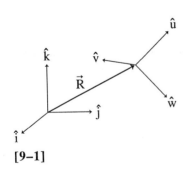

[9–1]

(b) $$Q = \begin{bmatrix} \cos\theta & \sin\theta \\ \sin\theta & -\cos\theta \end{bmatrix}$$

gives a new coordinate system in which the x'-axis has been rotated through an angle θ and the direction of the y'-axis has then been reversed. In this case, $|Q| = -1$.

The reader may verify that every 2×2 orthogonal matrix must be of one or the other of these two forms, for some θ.

§32.10 Problems

§32.1

1. Let $A = \begin{bmatrix} 1 & 1 \\ 1 & 2 \\ 2 & 1 \end{bmatrix}$, $B = \begin{bmatrix} -1 & -1 & 1 \\ 1 & 0 & 1 \end{bmatrix}$. Find the products AB and BA.

2. In the following statements, A and B are 3×3 matrices and O is the 3×3 *zero matrix*. Which statements are true and which are false, where "true" means true for all matrices A and B, and "false" means false for some choice of A and B.

   (a) If $A^2 = 0$, then $A = O$. (Here $A^2 = AA$.)

   (b) If $A^2 = 0$, then $r(A) < 3$.

   (c) If $r(A) < 3$, then $A^2 = 0$.

   (d) $(A + B)^2 = A^2 + 2AB + B^2$.

§32.4

1. (a) Which of the following matrices is invertible?

   (i) $\begin{bmatrix} 1 & 0 & 0 \\ 0 & 0 & 4 \\ 0 & 2 & 0 \end{bmatrix}$ (ii) $\begin{bmatrix} 1 & 0 & 0 \\ 1 & 4 & 2 \\ 0 & 2 & 1 \end{bmatrix}$

   (b) In each invertible case, find the inverse.

2. Let $A = \begin{bmatrix} 1 & 0 & 0 \\ 1 & 1 & 0 \\ 0 & 1 & 1 \end{bmatrix}$.

   (a) Find A^{-1}.

(b) Use A^{-1} to solve the system $AX = D$, where

   $$D = \begin{bmatrix} 1 \\ 2 \\ 1 \end{bmatrix}.$$

3. Check that the inverses in examples 1 and 3 are correct.

§32.5

1. Let $A = \begin{bmatrix} 1 & 1 & -1 \\ 1 & -1 & 1 \\ 1 & -1 & -1 \end{bmatrix}$.

   (a) Find A^{-1} by formula.

   (b) Find A^{-1} by row reduction.

   (c) Use A^{-1} to solve the system in problem 1b for §31.4.

2. (a) Calculate A^{-1} when $A = \begin{bmatrix} 1 & 0 & -2 \\ 0 & 1 & 1 \\ 1 & 0 & -1 \end{bmatrix}$.

   (b) Check your result by matrix multiplication.

   (c) Use your result to solve $AX = D$, when $D = \begin{bmatrix} 2 \\ -1 \\ 1 \end{bmatrix}$.

3. Consider the system $AX = D$, where

   $$A = \begin{bmatrix} -1 & 0 & 0 & 1 \\ 1 & 1 & 0 & 0 \\ 0 & 1 & -1 & 0 \\ 0 & 0 & 1 & -1 \end{bmatrix} \text{ and } D = \begin{bmatrix} -4 \\ 2 \\ 1 \\ 3 \end{bmatrix}.$$

(This is the system of problem 5 in §30.6.)

(a) Find A^{-1} by row reduction, and use it to solve the system.

(b) Find solutions (for the same A) when

$$D = \begin{bmatrix} 1 \\ 0 \\ 1 \\ 0 \end{bmatrix} \text{ and when } D = \begin{bmatrix} 1 \\ 1 \\ 1 \\ 0 \end{bmatrix}.$$

4. For A as in 3 above, find the second row of A^{-1} by the inversion formula.

5. Let $A = \begin{bmatrix} 1 & 0 & 1 \\ 0 & 1 & 0 \\ 0 & 1 & 2 \end{bmatrix}$.

(a) Find $|A|$.

(b) Find A^{-1}.

(c) Check your result for A^{-1} by showing that $A^{-1}A = 1$.

(d) Solve the system $A^{-1}X = \begin{bmatrix} 1 \\ 2 \\ 3 \end{bmatrix}$.

§32.8

1. Find eigenvalues and eigenvectors for the matrix

$$\begin{bmatrix} 3 & 1 & 1 \\ 0 & 1 & 0 \\ 0 & 1 & 2 \end{bmatrix}.$$

§32.9

1. Verify that any given 2×2 orthogonal matrix can either be expressed in the form (a), or else in the form (b), as given in remark 5.

Chapter 33

Subspaces and Dimension

§33.1 Vector Subspaces

We now turn to the concepts of *subspace, linear dependence,* and *dimension*. These concepts are more abstract than the computational concepts (such as *row-reduction*) that have been stressed in Chapters 29–32. These new concepts will provide deeper mathematical insights—and new and better computational methods.

In this chapter, we primarily use matrix algebra rather than vector algebra, with column-vectors to represent points in $\mathbf{R}^n$. Recall that $\mathbf{R}^3$ may be viewed as $\mathbf{E}^3$-with-coordinate-axes. When, in §30.2, we took a geometric view of the solution-set of a system of linear equations in three unknowns, we saw that the solution-set must always be either: the empty set, a single point, the points on a line, the points on a plane, or all of $\mathbf{E}^3$, and that for a homogeneous system, the solution-set must always be either: the origin, points on a line through the origin, points on a plane through the origin, or all of $\mathbf{E}^3$. In §33.1, we define sets of points in $\mathbf{R}^n$ called *vector subspaces*, and in §33.2 we define sets of points in $\mathbf{R}^n$ called *affine subspaces*. Vector subspaces in $\mathbf{R}^n$ will be analogous to, in $\mathbf{E}^3$: the origin, lines through the origin, planes through the origin, and all of $\mathbf{E}^3$. Affine subspaces in $\mathbf{R}^n$ will be analogous to, in $\mathbf{E}^3$: arbitrary points, lines, planes, and all of $\mathbf{E}^3$. Thus a vector subspace will be a special kind of affine subspace.

Definition. A vector A is said to be a *linear combination* of vectors $B_1,\ldots,B_k$, if there is some choice of scalars $t_1,\ldots,t_k$ such that $A = t_1 B_1 + \cdots + t_k B_k$.

Examples. (1) *In $\mathbf{R}^3$, can we express $\hat{i}$ as a linear combination of the three vectors $\hat{i}+\hat{j}, \hat{i}+\hat{k}, \hat{j}+\hat{k}$?*

Answer: We seek $\hat{i} = t_1(\hat{i}+\hat{j}) + t_2(\hat{i}+\hat{k}) + t_3(\hat{j}+\hat{k})$.

Going to coordinates, we get

$$\begin{bmatrix} 1 \\ 0 \\ 0 \end{bmatrix} = t_1 \begin{bmatrix} 1 \\ 1 \\ 0 \end{bmatrix} + t_2 \begin{bmatrix} 1 \\ 0 \\ 1 \end{bmatrix} + t_3 \begin{bmatrix} 0 \\ 1 \\ 1 \end{bmatrix}.$$

This is the simultaneous system

$$\begin{pmatrix} 1 = t_1 + t_2 \\ 0 = t_1 \quad\;\; + t_3 \\ 0 = \quad\;\; t_2 + t_3 \end{pmatrix}.$$

Solving by row-reduction, we obtain the unique solution

$$t_1 = \tfrac{1}{2}, \;\; t_2 = \tfrac{1}{2}, \;\; t_3 = -\tfrac{1}{2}\,.$$

Hence $\hat{i} = \tfrac{1}{2}(\hat{i}+\hat{j}) + \tfrac{1}{2}(\hat{i}+\hat{k}) - \tfrac{1}{2}(\hat{j}+\hat{k})$, and we have the desired expression.

(2) *In $\mathbf{R}^3$, can we express $\hat{i}$ as a linear combination of the three vectors $\hat{i}+\hat{j}, \hat{i}+\hat{k}, \hat{j}-\hat{k}$?*

Answer: Here we must solve the system

$$\begin{pmatrix} 1 = t_1 + t_2 \\ 0 = t_1 \quad\;\; + t_3 \\ 0 = \quad\;\; t_2 - t_3 \end{pmatrix}.$$

Applying row-reduction, we find that the system has no solution. Hence $\hat{i}$ cannot be expressed as a linear combination of $\hat{i}+\hat{j}, \hat{i}+\hat{k}, \hat{j}-\hat{k}$.

(3) *In $\mathbf{R}^3$, can we express $4\hat{i} + 2\hat{j} + 2\hat{k}$ as a linear combination of the three vectors $\hat{i}+\hat{j}, \hat{i}+\hat{k}, \hat{j}-\hat{k}$?*

Answer: Here, we must solve the system

$$\begin{pmatrix} 4 = t_1 + t_2 \\ 2 = t_1 \quad\;\; + t_3 \\ 2 = \quad\;\; t_2 - t_3 \end{pmatrix}.$$

Row-reducing, we find that there are infinitely many solutions. Thus, for instance, $4\hat{i} + 2\hat{j} + 2\hat{k}$ can be expressed as $2(\hat{i} + \hat{j}) + 2(\hat{i} + \hat{k})$, or as $3(\hat{i} + \hat{j}) + (1)(\hat{i} + \hat{k}) + (-1)(\hat{j} - \hat{k})$, or as $4(\hat{i} + \hat{j}) + (0)(\hat{i} + \hat{k}) + (-2)(\hat{j} - \hat{k})$.

Definition. Given a non-empty, finite set of vectors $\{B_1,...,B_k\}$ in $\mathbf{R}^3$, the set of all vectors that can be expressed as linear combinations of $B_1,...,B_k$ is called the *span* of $\{B_1,...,B_k\}$ and is written as $\text{span}(B_1,...,B_k)$. We also define the *span of the empty set* of vectors to be $\{O\}$, the set consisting of the zero vector. (This is consistent with the usual mathematical convention of defining an algebraic sum of no addends to be zero.) Note that the span of any non-empty set of vectors $\{B_1,...,B_k\}$ always includes the vector O, since $O = (0)B_1 + (0)B_2 + \cdots + (0)B_k$.

Example 4. *In $\mathbf{R}^3$ with x,y,z as coordinates, the span of $\{\hat{i}, \hat{j}\}$ is the xy plane; the span of $\{\hat{i}\}$ is the x axis; the span of $\{\hat{i}, \hat{i} + \hat{j}, \hat{i} - \hat{j}\}$ is also the xy plane; and the span of $\{\hat{i}, \hat{j}, \hat{k}\}$ is all of $\mathbf{R}^3$.*

Definition. Let $\mathcal{V}$ be a set of vectors in $\mathbf{R}^n$. $\mathcal{V}$ is called a *vector subspace*, if $\mathcal{V}$ is the span of some finite set of non-zero vectors. We include the empty set as a finite set; hence, by our definition of *span*, $\mathcal{V} = \{O\}$ is a vector subspace.

If $\mathcal{V}$ is the span of some finite set, we refer to that finite set as a *generating set* for $\mathcal{V}$. We note that different generating sets can have, as their spans, the same set of vectors in $\mathbf{R}^n$. For example, each of $\{\hat{i}, \hat{j}\}$, $\{\hat{i}, \hat{i} + \hat{j}, \hat{i} - \hat{j}\}$, and $\{\hat{i} + \hat{j}, \hat{i} - \hat{j}\}$ is a generating set for the xy plane in $\mathbf{R}^3$. We shall speak of two vector subspaces $\mathcal{V}_1$ and $\mathcal{V}_2$ as the *same subspace* if every vector in $\mathcal{V}_1$ is in $\mathcal{V}_2$ and every vector in $\mathcal{V}_2$ is in $\mathcal{V}_1$. If $\mathcal{V}_1$ and $\mathcal{V}_2$ are the same subspace, we write $\mathcal{V}_1 = \mathcal{V}_2$. Thus, for example, we say that $\{\hat{i}, \hat{j}\}$ and $\{\hat{i}, \hat{i} + \hat{j}, \hat{i} - \hat{j}\}$ are generating sets for the same subspace.

It is evident from geometry that in $\mathbf{R}^2$, a vector subspace is either $\{O\}$, a line through the origin, or all of $\mathbf{R}^2$. Similarly in $\mathbf{R}^3$, a vector subspace must be either $\{O\}$, a line through the origin, a plane through the origin, or all of $\mathbf{R}^3$. These observations, taken together with our earlier discussion on the geometry of solutions of a homogeneous system (in §30.2), lead us to the following theorem:

(1.1)   **Theorem.** *The solution-set of a homogeneous system of linear equations in* n *unknowns is a vector subspace in* $\mathbf{R}^n$.

Proof. If a given system of equations has non-trivial solutions, then the general solution $X = t_1 B_1 + \cdots t_\ell B_\ell$ (obtained by row-reduction) expresses the solution-set directly as $\mathrm{span}(B_1,\ldots,B_\ell)$ and hence as a vector subspace. If there is no non-trivial solution, then the solution-set is $\{O\}$, which is also a vector subspace. This completes the proof.

The converse of the theorem (1.1) is also true:

(1.2)   *Every vector subspace is the solution-set of some homogeneous system.*

In §33.6, we prove this converse by showing how to obtain, for any given set of vectors $\{B_1,\ldots,B_k\}$, a matrix A such that $AX = 0$ has $\mathrm{span}(B_1,\ldots,B_\ell)$ as its solution set.

The following theorem gives two fundamental properties of vector subspaces.

(1.3)   **Theorem.** *Let* $\mathcal{V}$ *be a vector subspace in* $\mathbf{R}^n$. *Then:*
    (i)   *The vector* O *must be in* $\mathcal{V}$.
    (ii)   *Any linear combination of vectors from* $\mathcal{V}$ *must be in* $\mathcal{V}$.

Proof. (i) We noted in our definition of *span* that the span of a finite set always includes the vector O.
    (ii) Let D be given as a linear combination of vectors $C_1,\ldots,C_k$, where $C_1,\ldots,C_k$ are vectors in $\mathrm{span}(B_1,\ldots,B_\ell)$, and let the vectors $C_1,\ldots,C_k$ be given as linear combinations of the vectors $B_1,\ldots,B_\ell$. If these latter linear combinations are substituted for $C_1,\ldots,C_k$ in the expression for D, we can get an expression for D as a linear combination of $B_1,\ldots,B_\ell$. This completes the proof.

§33.2 Affine Subspaces

Definition. Let $\mathcal{V}$ be a given vector subspace of $\mathbf{R}^n$, and let X_0 be a given vector in $\mathbf{R}^n$. We then use the notation $\mathcal{A} = X_0 + \mathcal{V}$ to represent the following set of vectors: a vector X is in $\mathcal{A} \Leftrightarrow$ there is some vector in V in $\mathcal{V}$ such that $X = X_0 + V$. If $\mathcal{A} = X_0 + \mathcal{V}$ for some fixed vector X_0 and some vector subspace $\mathcal{V}$, we say that $\mathcal{A}$ is an *affine subspace* of $\mathbf{R}^n$.

> **Examples.** *In* $\mathbf{R}^3$, *affine subspaces are single points (when* $V = \{O\}$*), lines (not necessarily through the origin), planes (not necessarily through the origin), and the entire space.*

Affine subspaces are sometimes called "linear subspaces," although mathematicians sometimes use the phrase "linear subspace" to mean a vector subspace.

As with vector subspaces, we say that two affine subspace $\mathcal{A}_1$ and $\mathcal{A}_2$ are the *same* subspace (we write: $\mathcal{A}_1 = \mathcal{A}_2$) if every vector in $\mathcal{A}_1$ is in $\mathcal{A}_2$ and every vector in $\mathcal{A}_2$ is in $\mathcal{A}_1$. Note that different choices of the fixed vector X_0 can be used to define the same affine subspace. For example, $\mathcal{A}_1 = \hat{i} + \text{span}(\hat{i} - \hat{j})$ and $\mathcal{A}_2 = \hat{j} + \text{span}(\hat{i} - \hat{j})$ give us the same affine subspace in $\mathbf{R}^2$ (the line with the equation $x+y = 1$). In fact we have the following theorem:

(2.1)　**Theorem.** *Let* $\mathcal{A} = X_0 + \mathcal{V}$ *be an affine subspace of* $\mathbf{R}^n$, *and let* X_1 *be any vector in* $\mathcal{A}$. *Then we also have* $\mathcal{A} = X_1 + \mathcal{V}$.

Proof. Let $\mathcal{A}_1 = X_1 + \mathcal{V}$. Since X_1 is in $\mathcal{A}$, we have $X_1 = X_0 + V_1$, for some vector V_1 in $\mathcal{V}$. Hence $X_0 = X_1 - V_1$. Let X be any vector in $\mathcal{A}$. Then $X = X_0 + V$ for some V in $\mathcal{V}$. Hence $X = X_1 - V_1 + V = X_1 + (V - V_1)$. Since $V - V_1$ must be in $\mathcal{V}$ by (1.2), X must be in $\mathcal{A}_1$.

Conversely, let X be any vector in $\mathcal{A}_1$. Then $X = X_1 + V$ for some V in $\mathcal{V}$. Since X_1 is in $\mathcal{A}$, let $X_1 = X_0 + V_1$ for some V_1 in $\mathcal{V}$. Then $X = X_0 + (V_1 + V)$. Since $V_1 + V$ must be in $\mathcal{V}$, X must be in $\mathcal{A}$. Thus $\mathcal{A}$ and $\mathcal{A}_1$ are the same set of vectors. This completes the proof.

On the other hand, we have the following:

(2.2)   **Theorem.** *Assume that $\mathcal{A} = X_0 + \mathcal{V}_0$ and also that $\mathcal{A} = X_1 + \mathcal{V}_1$. Then $\mathcal{V}_0$ and $\mathcal{V}_1$ must be the same vector subspace.*

Proof. By the previous theorem, we have $\mathcal{A} = X_0 + \mathcal{V}_1$. Let V_0 be any vector in $\mathcal{V}_0$. Then $X = X_0 + V_0$ is in $\mathcal{A}$. Since $\mathcal{A} = X_0 + \mathcal{V}_1$, we also have $X = X_0 + V_1$ for some V_1 in $\mathcal{V}_1$. But $X_0 + V_0 = X_0 + V_1 \Rightarrow V_0 = V_1$. Hence V_0 must also be in $\mathcal{V}_1$. By a similar argument, every vector V_1 in $\mathcal{V}_1$ must also be in $\mathcal{V}_0$. Hence $\mathcal{V}_0 = \mathcal{V}_1$. This completes the proof.

From the theorem above, we see that every affine subspace $\mathcal{A}$ has a unique *associated vector subspace* $\mathcal{V}$ such that $\mathcal{A}$ is expressible in the form $\mathcal{A} = X_0 + \mathcal{V}$. In the case of a line or plane in $\mathbf{R}^3$, this associated vector subspace is the line or plane through the origin parallel to (or identical with) the given line or plane.

The following theorem is proved in the same way as theorem (1.1) above on solution-sets of homogeneous systems.

(2.3)   **Theorem.** *If the solution-set of a system of linear equations (not necessarily homogeneous) in n unknowns is non-empty, then it is an affine subspace in $\mathbf{R}^n$.*

In §33.6, we shall see that the converse of (2.3) is also true: *given any affine subspace $\mathcal{A}$ in $\mathbf{R}^n$, there must be a system $AX = D$ in n unknowns such that $AX = D$ has $\mathcal{A}$ as its solution-set.*

In §34.2, we shall use the concepts of *affine subspace* and *associated vector subspace* to develop n-dimensional Euclidean geometry.

§33.3 Linear Independence

Definitions. We say that a finite set of vectors in $\mathbf{R}^n$ is *linearly independent* if there is no vector in the set that is O or that is a linear combination of other vectors in the set. We note from this definition that the members of a linearly independent set must all be non-zero, and that the empty set is a linearly independent set.

We say that a finite set of vectors in $\mathbf{R}^n$ is *linearly dependent* if there is at least one vector in the set that is either O or expressible as a linear combination of other vectors in the set. We note that a linearly dependent set must be non-empty.

Examples in $\mathbf{R}^n$.

(1) *The set $\{\hat{i}, \hat{i}+\hat{j}\}$ is linearly independent.* (In fact, it follows from our definition of linear dependence that: a set of *two* vectors is linearly dependent ⇔ one of the vectors is a scalar multiple of the other.)

(2) *The set $\{\hat{i}, \hat{i}+\hat{j}, \hat{j}-\hat{k}\}$ is linearly independent.* (In fact, it follows from our definition of linear independence and from our geometric definition of vector addition in $\mathbf{E}^3$, that: a set of three vectors in $\mathbf{R}^3$ is linearly independent ⇔ the position arrows for the three vectors do not lie in a common plane through the origin.)

(3) *The set $\{\hat{i}, \hat{i}+\hat{j}, \hat{j}-\hat{k}, \hat{i}-2\hat{k}\}$ is linearly dependent, since, for example:*

$$\hat{j}-\hat{k} = -\tfrac{2}{3}\hat{i} + 1(\hat{i}+\hat{j}) + \tfrac{1}{2}(\hat{i}-2\hat{k}) \ .$$

(We shall later see that every set of four vectors in $\mathbf{R}^3$ must be linearly dependent.)

It is customary to drop the phrase "the set" in a statement like (1), and to say "the vectors $\hat{i}$ and $\hat{i}+\hat{j}$ are linearly independent" or to speak of "the linearly independent vectors $\hat{i}$ and $\hat{i}+\hat{j}$." The reader should keep in mind, however, that a statement of linear independence or linear dependence is always, strictly speaking, a statement about a set of vectors.

The concepts of *linear independence* and *linear dependence* are central in matrix algebra and appear in a variety of applications. In the following theorem, we give several further (and equivalent) formulations of linear dependence. These formulations will be used in later work. (Formulation (2) will be especially useful.)

(3.1) **Theorem.** *The following statements are equivalent:*
  (1) *$\{A_1,...,A_k\}$ is a linearly dependent set.*
  (2) *There exist scalars $t_1,...,t_k$, not all 0, such that $t_1 A_1 + \cdots + t_k A_k = O.$*
  (3) *The homogeneous system $AX = O$ has a non-trivial solution, where $A = [A_1 \vdots \cdots \vdots A_k]$ is the matrix whose columns are the vectors $A_1,...,A_k$.*

Proof. We prove (1) ⇔ (2) for the case $k \geq 2$. The case $k = 1$ is easier and is left to the reader. (Note, from our definitions, that the set $\{A_1\}$ is linearly independent if and only if $A_1 \neq O$.)

(1) $\Rightarrow$ (2) because if $A_1 = s_2 A_2 + \cdots + s_k A_k$, for example, then $-A_1 + s_2 A_2 + \cdots + s_k A_k = O$, and we have $-1, s_2, \ldots, s_k$ as our desired $t_1, \ldots, t_k$, not all 0.

Conversely, (2) $\Rightarrow$ (1) because $t_1 A_1 + \cdots + t_k A_k = O$, and $t_1 \neq 0$, for example, then $A_1 = \left(-\frac{t_2}{t_1}\right) A_2 + \cdots + \left(-\frac{t_k}{t_1}\right) A_k$ gives the desired linear combination.

(2) $\Rightarrow$ (3), because if $t_1 A_1 + \cdots + t_k A_k = O$ and $t_1, \ldots, t_k$ are not all 0, then, for the vector $X_0 = \begin{bmatrix} t_1 \\ \vdots \\ t_k \end{bmatrix}$, we have $AX_0 = t_1 A_1 + \cdots t_k A_k = O$, and X_0 is a non-trivial solution to the system $AX = O$.

Conversely, (3) $\Rightarrow$ (2), since the components of any non-trivial solution for $AX = O$ give the desired $t_1, \ldots, t_k$. This completes the proof.

Definition. An equation of the form $t_1 A_1 + \cdots t_k A_k = O$ with $t_1, \ldots, t_k$ not all 0 (as in (2) above) is called a *linear dependence among the vectors* $A_1, \ldots, A_k$. Note that for $A_1, \ldots, A_k$ in $\mathbf{R}^n$, the equivalence of (1) and (3) accords with basic properties (5) and (6) of determinants in §31.2 and with (3.3) in §31.3.

§33.4 Bases for Vector Subspaces

Definition. Let $\mathcal{V}$ be a vector subspace $\mathbf{R}^n$. A set of vectors $\mathcal{B}$ in $\mathbf{R}^n$ is called a *basis* for $\mathcal{V}$, if

    (i)   the set $\mathcal{B}$ is linearly independent, and

    (ii) $\mathcal{V}$ is the span of $\mathcal{B}$.

Thus a basis for $\mathcal{V}$ is simply a linearly independent generating set for $\mathcal{V}$.

Examples (1) *In* $\mathbf{R}^3$, $\{\hat{i}, \hat{j}, \hat{k}\}$ *is a basis for* $\mathcal{V} = \mathbf{R}^3$, *since none of the vectors in* $\{\hat{i}, \hat{j}, \hat{k}\}$ *can be expressed as a linear combination of the others.*

    (2)   *In* $\mathbf{R}^3$, $\{\hat{i}, \hat{j}\}$ *is a basis for* $\mathcal{V} =$ *the* xy *plane.*

    (3)   *In* $\mathbf{R}^3$, $\{\hat{i}, \hat{i} + \hat{j}, \hat{i} - \hat{j}\}$ *is not a basis since, for example,* $\hat{i} = \frac{1}{2}(\hat{i} + \hat{j}) + \frac{1}{2}(\hat{i} - \hat{j})$, *and the set is linearly dependent.* If we delete $\hat{i}$, the remaining vectors $\{\hat{i} + \hat{j}, \hat{i} - \hat{j}\}$ *are a basis and are, in fact, a basis for* $\mathcal{V} =$ the xy *plane.*

(4)   *Given any non-empty, finite set of vectors* $\{A_1,...,A_k\}$, *we can always (as in (3) above) find a basis for* $\mathcal{V} = \text{span}(A_1,...,A_k)$ *by successively deleting from* $\{A_1,...,A_k\}$ *each vector that can be expressed as a linear combination of other vectors not yet deleted.* Each deletion leaves the span unchanged, otherwise the deleted vector would violate theorem (1.2), which asserts that a linear combination of vectors from a vector subspace $\mathcal{V}$ must also be in $\mathcal{V}$. See the problems.

We note that every vector in a basis must be non-zero and that the empty set of vectors is a basis for the vector subspace $\{O\}$.

Examples (2) and (3) show that it is possible for a given vector subspace $\mathcal{V} \neq \{O\}$ to have more than one basis. The following theorem states a fundamental property possessed by all bases. (The accepted plural for "basis" is "bases," pronounced *base-ease*.)

(4.1)   **Theorem.** *Let* $\{B_1,...,B_k\}$ *be a given basis for a vector subspace* $\mathcal{V} \neq \{O\}$. *Then for each vector X in* $\mathcal{V}$, *there is one and only one set of coefficients for expressing X as a linear combination of the basis vectors* $B_1,...,B_k$.

Proof. Let $\mathcal{B} = \{B_1,...,B_k\}$. Since $\mathcal{B}$ is a basis for $\mathcal{V}$, $\mathcal{B}$ is a generating set for $\mathcal{V}$. Hence, for any X in $\mathcal{V}$, X must be expressible as a linear combination of vectors in $\mathcal{B}$ in at least one way.

Assume that there are two different ways to express the same vector X as a linear combination of vectors in $\mathcal{B}$. Then we have:

$$X = c_1 B_1 + \cdots + c_k B_k$$

and
$$X = d_1 B_1 + \cdots + d_k B_k.$$

Subtracting these two equations, we have

$O = t_1 B_1 + \cdots + t_k B_k$, where, for each i, $t_i = c_i - d_i$. Since the two equations for X are different, there must be at least one i such that $c_i - d_i \neq 0$. We now have a linear dependence among the basis vectors $B_1,...,B_k$, contrary to our assumption (as part of our definition of *basis*) that the basis vectors are linearly independent. This completes the proof.

Example 5. *Let* $B_1,...,B_\ell$ *be the constant vectors that we obtained by row-reduction for the general solution of a homogeneous system* $AX = O$ *(see*

§30.5 and §30.8). These vectors evidently form a generating set for the solution set. For each i, B_i has a component 1 in a position (given by the parameter-variable equation $x_i = x_i$) where all the other B_j, $j \neq i$, have component 0. It follows that there can be no linear dependence among the vectors $B_1, \ldots, B_k$. Hence the vectors $B_1, \ldots, B_k$ form a basis for the vector subspace of all solutions of $AX = 0$.

The various examples given in this section suggest that any two bases for the same given vector subspace $\mathcal{V}$ must contain the same number of vectors. This statement is, in fact, true, and we prove it in §33.5. The number of vectors appearing in any basis of $\mathcal{V}$ is called the *dimension* of $\mathcal{V}$.

§33.5 Linear Dependence in Row-reduction; Dimension

We begin with a useful lemma about row-reduction.

(5.1)   *Row-reduction lemma. Let A be an m×n matrix. Let A′ be a matrix that we get by applying an elementary row-operation to A.*

(a)   *If we consider the rows of A and of A′ as n-vectors, then the rows of A′ span the same vector subspace in $\mathbf{R}^n$ as the rows of A.*

(b)   *If we consider the columns of A and of A′ as m-vectors, then a linear dependence that holds for the columns of A must hold for the columns of A′ and vice versa.*

Proof of the lemma. (a) The result is immediate for an elementary row operation of type α or γ. For an operation of type β, we observe that the new row in A′ is a linear combination of rows in A, and that the to-be-altered row in A is a linear combination of rows in A′. Hence every row in A′ is a linear combination of rows in A, and every row in A is a linear combination of rows in A′. Hence every linear combination of rows in A′ is a linear combination of rows in A and vice versa. This proves (a).

(b) As noted in proving theorem (3.1), every linear dependence among columns of A gives a non-trivial solution of the homogeneous system $AX = 0$, and every non-trivial solution of this homogeneous system gives a linear dependence among the columns of A. Thus (b) is a restatement of the previously proved fact that an elementary row-operation leaves the solution set of a system unchanged.

Definition. We shall show that any two bases for a given vector subspace $\mathcal{V}$ must have the same number of vectors. This number is called the *dimension of* $\mathcal{V}$. We abbreviate the dimension of a vector subspace $\mathcal{V}$ as dim($\mathcal{V}$).

The following examples show how this concept agrees with the previous and familiar use of the word "dimension" in the geometry of $\mathbf{R}^2$ and $\mathbf{R}^3$.

Examples in $\mathbf{R}^3$.

(1) $\mathcal{V} = \mathrm{span}(\hat{i} + \hat{j})$ *has dimension 1 and is a straight line through the origin.*

(2) $\mathcal{V} = \mathrm{span}(\hat{i} + \hat{j}, \hat{i} + \hat{k}, \hat{j} - \hat{k})$ *has dimension 2, since* $\hat{j} - \hat{k} = 1(\hat{i} + \hat{j}) + (-1)(\hat{i} + \hat{k})$, *and* $\{\hat{i} + \hat{j}, \hat{i} + \hat{k}\}$ *is linearly independent.* $\mathcal{V}$ *is a plane through the origin.*

(3) $\mathcal{V} = \mathrm{span}(\hat{i} + \hat{j}, \hat{i} + \hat{k}, \hat{j} + \hat{k})$ *has   dimension   3,   since* $\{\hat{i} + \hat{j}, \hat{i} + \hat{k}, \hat{j} + \hat{k}\}$ *is linearly independent.* $\mathcal{V}$ *is all of* $\mathbf{R}^3$.

We now prove our fundamental theorem on dimension.

(5.2)   **Dimension theorem.**   *For a vector subspace $\mathcal{V}$ in $\mathbf{R}^n$, any two bases must have the same number of vectors.*

Proof.   The theorem is immediate for $\mathcal{V} = \{O\}$, since the empty set of vectors is the only possible basis for $\{O\}$.

Consider the following statement:

(5.3)   *Let $A_1, \ldots, A_k$ be given vectors. Then any set $\{B_1, \ldots, B_{k+1}\}$ of $k+1$ vectors in $\mathrm{span}(A_1, \ldots, A_k)$ must be linearly dependent.*

Two bases of different sizes for the same vector subspace $\mathcal{V} \neq \{O\}$ would necessarily violate (5.3). Hence to prove (5.2) for $\mathcal{V} \neq \{O\}$, it will be enough to prove (5.3). We form the matrix $[A_1 \vdots \cdots \vdots A_k \vdots B_1 \vdots \cdots \vdots B_{k+1}]$ and row-reduce it. Since each B_i is in $\mathrm{span}(A_1, \ldots, A_k)$, there can be (by (b) of (5.1)) no pivot in a B_i column. (Otherwise, that pivot would have to be a linear combination of the zeroes which precede it in its row—which is impossible.) Thus all pivots must occur in the first k columns. Hence our row-reduced matrix must have the form

$$\begin{bmatrix} A' & \cdot & B' \\ \cdot & \cdot & \cdot \\ O & \cdot & O \end{bmatrix},$$

where A' has k columns, B' has k+1 columns, and both A' and B' have at most k rows. Consider the simultaneous system B'Y = O, where X is a (k+1)-vector of variables: $y, \ldots, y_{k+1}$.   Since B' has more columns than rows, the system must have a non-trivial solution (because there must be at least one parameter variable). This solution gives a linear dependence among the columns of B'. By (b) of (5.1), since B' was obtained from B by elementary row-operations, the same linear dependence must hold for the columns $B_1, B_2, \ldots, B_{k+1}$.   Hence the set $\{B_1, \ldots, B_{k+1}\}$ is linearly dependent.   This completes the proof. Note that the subspace $\mathcal{V} = \{O\}$ has dimension 0.

The following corollary is immediate from our proof.

(5.4)   *Corollary. If a vector subspace $\mathcal{V}$ has dimension* k, *then no set of more than* k *vectors in $\mathcal{V}$ can be linearly independent.*

Recall from (2.2) that every affine subspace has a unique associated vector subspace. We can therefore define the *dimension* of an affine subspace to be the dimension of its associated vector subspace. This again accords with previous and familiar use of the word "dimension" in the geometry of $\mathbf{R}^2$ and $\mathbf{R}^3$. We abbreviate the dimension of an affine subspace $\mathcal{A}$ as dim($\mathcal{A}$).

The dimension theorem has a further useful corollary.

Definition. If $\mathcal{A}$ and $\mathcal{B}$ are sets, we define "$\mathcal{A} \subseteq \mathcal{B}$" to mean that every member of $\mathcal{A}$ is also a member of $\mathcal{B}$. (Thus if $\mathcal{A} \subseteq \mathcal{B}$ and $\mathcal{B} \subseteq \mathcal{A}$, then $\mathcal{A} = \mathcal{B}$.)

Definition. We say that two vector subspaces $\mathcal{V}_1$ and $\mathcal{V}_2$ are *comparable,* if either $\mathcal{V}_1 \subseteq \mathcal{V}_2$ or else $\mathcal{V}_2 \subseteq \mathcal{V}_1$. Similarly, we say that two affine subspaces $\mathcal{A}_1$ and $\mathcal{A}_2$ are *comparable,* if either $\mathcal{A}_1 \subseteq \mathcal{A}_2$ or else $\mathcal{A}_2 \subseteq \mathcal{A}_1$.

(5.5)   *Corollary on comparability.*
   (a)   *If two comparable vector subspaces have the same dimension, then they must be the same vector subspace.*
   (b)   *If two comparable affine subspaces have the same dimension, then they must be the same affine subspace.*

Proof. (a): Assume $\mathcal{V}_1 \subseteq \mathcal{V}_2$. It will be enough to show that if $\mathcal{V}_2$ is not the same as $\mathcal{V}_1$, then $\mathcal{V}_2$ must have greater dimension than $\mathcal{V}_1$. Let k be the dimension of $\mathcal{V}_1$ and let $\{B_1, \ldots, B_k\}$ be a basis for $\mathcal{V}_1$. If $\mathcal{V}_2$ is not the same as $\mathcal{V}_1$, then $\mathcal{V}_2$ must contain a vector C which is not in $\mathcal{V}_1$. But this implies that the set of vectors $\{B_1, \ldots, B_k, C\}$ in $\mathcal{V}_2$ is linearly independent. Then, by (5.4), the dimension of $\mathcal{V}_2$ must be greater than k. This completes the proof of (a). The proof of (b) is similar and is left to the reader.

§33.6 Subspaces and Solution-sets

Let $\{B_1, \ldots, B_k\}$ be the vectors obtained by row-reduction for the general solution of an $m \times n$ homogeneous system AX = O. In example 4 of §33.4, we noted that the set $\{B_1, \ldots, B_k\}$ must be a basis for the solution-set of the system AX = O. We can now conclude that the dimension of this solution-set, as a vector subspace, must be ℓ. Since $\ell = n - r(A)$, we know the value of this dimension ℓ as soon as we know $r(A)$.

Given a homogeneous system of equations, such as

$$\begin{pmatrix} x + y - 2z = 0 \\ x - y + z = 0 \\ 2x + 3y - z = 0 \end{pmatrix},$$

the result of the row-reduction process that we carry out will depend on the order in which we take the unknowns x, y, and z as x_1, x_2, and x_3. If we take $x_1 = x$, $x_2 = y$, $x_3 = z$, in the system above, we get a different row-reduced matrix than if we take $x_1 = z$, $x_2 = x$, $x_3 = y$. We see, however, that every choice of variables for x_1, x_2, x_3 must lead to the same value of ℓ, since ℓ is the dimension of the solution set in $\mathbf{R}^3$, and this dimension, by its definition, does not depend on the order of variables chosen for x_1, x_2, x_3. *Since the rank of the coefficient matrix is $n - \ell$, it follows that this rank must also be independent of how we choose x_1, x_2, x_3.* Different choices may lead to different row-reduced matrices, but these matrices must all have the same rank.

Definition. Let $\{A_1, \ldots, A_k\}$ be a basis for a given vector subspace $\mathcal{V} \neq \{O\}$ of dimension k in $\mathbf{R}^n$. Let A be the $k \times n$ matrix whose rows are $A_1^T, \ldots, A_k^T$. Consider the solution-set of the homogeneous system AX = O. This is the set of all vectors B such that B is orthogonal to each of the vectors $A_1,...,A_k$. (Hence B is orthogonal to each vector in $\mathcal{V}$, as the reader can readily show.) This solution-set must be a vector subspace of $\mathbf{R}^n$ and is called the *orthogonal complement* of $\mathcal{V}$. We write it $\mathcal{V}^\perp$. ($\mathcal{V}^\perp$ can be equivalently defined by: B is in $\mathcal{V}^\perp \Leftrightarrow$ B is orthogonal to every vector in $\mathcal{V}$.) Let ℓ be the dimension of $\mathcal{V}^\perp$. By the dimension theorem (5.2) and (a) of the row-reduction lemma (5.1), $r(A) = k$. Hence, from our discussion of rank at the beginning of this section, $k + \ell = n$. Note that $\mathcal{V} = \mathbf{R}^n \Rightarrow \mathcal{V}^\perp = \{O\}$ and that $\mathcal{V} = \{O\} \Rightarrow \mathcal{V}^\perp = \mathbf{R}^n$.

We can go a step further and form $\mathcal{V}^{\perp\perp}$, the orthogonal complement of $\mathcal{V}^\perp$. By the same argument as for $\mathcal{V}^\perp$, the vector subspace $\mathcal{V}^{\perp\perp}$ must have dimension k' with $\ell + k' = n$. Hence $k = k'$. Moreover, each vector X in $\mathcal{V}$ must be orthogonal to all the vectors in $\mathcal{V}^\perp$, and $\mathcal{V}^{\perp\perp}$ is the set of *all* vectors that are orthogonal to all vectors in $\mathcal{V}^\perp$. It follows that every vector X in $\mathcal{V}$ must also be in $\mathcal{V}^{\perp\perp}$. Thus $\mathcal{V}$ and $\mathcal{V}^{\perp\perp}$ are comparable. Applying the corollary (5.5) above on comparable vector subspaces, we conclude that $\mathcal{V}$ and $\mathcal{V}^{\perp\perp}$ must be the same vector subspace. We state this as a theorem:

(6.1)   **Theorem.** *Let $\mathcal{V}$ be a vector subspace of $\mathbf{R}^n$. Then $\mathcal{V}^{\perp\perp} = \mathcal{V}$.*

Since $\mathcal{V}^{\perp\perp} = \mathcal{V}$ is obtained from $\mathcal{V}^\perp$ in the same way that $\mathcal{V}^\perp$ is obtained from $\mathcal{V}$, namely as the solution space of a certain homogeneous system, we can now prove the following theorem, which is the converse to our earlier theorem (1.1) that every solution set of a homogeneous system is a vector subspace.

(6.2)   **Theorem.** *Every vector subspace in $\mathbf{R}^n$ is the solution set for some homogeneous system.*

Proof. Let $\mathcal{V}$ be a vector subspace in $\mathbf{R}^n$. If $\mathcal{V}$ has dimension 0, then $\mathcal{V} = \{O\}$, and $\mathcal{V}$ is the solution set of $AX = O$, where $A = I$, the $n \times n$ identity matrix. If $\mathcal{V}$ has dimension n, then $\mathcal{V} = \mathbf{R}^n$ and $\mathcal{V}$ is the solution set of $AX = O$, where $A = O$, the $n \times n$ zero matrix.

If $\mathcal{V}$ has dimension k, $0 < k < n$, let $\{A_1,...,A_k\}$ be a basis for $\mathcal{V}$. Let the matrix A have, as its rows, $A_1^T,...,A_k^T$. Then the solution set of $AX = 0$ is $\mathcal{V}^\perp$. Let $\{B_1,\cdots,B_\ell\}$ be a basis for $\mathcal{V}^\perp$. Let the matrix B have, as its rows,

$B_1^T, \ldots, B_\ell^T$. Then the solution set of $BX = 0$ is $\mathcal{V}^{\perp\perp}$. Since $\mathcal{V}^{\perp\perp} = \mathcal{V}$ by (6.1), this completes the proof of (6.2).

Examples. In $\mathbf{R}^2$, two perpendicular lines through the origin will be orthogonal complements to each other. Also $\mathcal{V}_1 = \{0\}$ and $\mathcal{V}_2 = \mathbf{R}^2$ are orthogonal complements.

In $\mathbf{R}^3$, a plane through the origin and its perpendicular line through the origin will be orthogonal complements. Also $\mathcal{V}_1 = \{0\}$ and $\mathcal{V}_2 = \mathbf{R}^3$ are orthogonal complements.

Definition. For any two sets of vectors $\mathcal{A}$ and $\mathcal{B}$ in $\mathbf{R}^n$, we define $\mathcal{A} \cap \mathcal{B}$ (called the *intersection* of $\mathcal{A}$ and $\mathcal{B}$) to be the set of all vectors (in $\mathbf{R}^n$) that are in both $\mathcal{A}$ and $\mathcal{B}$. We have the following theorem.

(6.3)   *Theorem. Let $\mathcal{V}_1$ and $\mathcal{V}_2$ be vector subspaces of $\mathbf{R}^n$. Then $\mathcal{V}_1 \cap \mathcal{V}_2$ is also a vector subspace of $\mathbf{R}^n$.*

Proof. By (6.2), let $\mathcal{V}_1$ be the solution-set for $B_1 X = O$, and $\mathcal{V}_2$ be the solution-set for $B_2 X = O$. It immediately follows that $\mathcal{V}_1 \cap \mathcal{V}_2$ is the solution-set for $BX = O$, where B is the matrix $\begin{bmatrix} B_1 \\ \cdots \\ B_2 \end{bmatrix}$. Since $\mathcal{V}_1 \cap \mathcal{V}_2$ is the solution-set of a homogeneous system, $\mathcal{V}_1 \cap \mathcal{V}_2$ must be a vector subspace. This completes the proof.

We leave the proofs of the following three corollaries to the reader. These corollaries are used in §34.2.

(6.4)   *Corollary. Let $\mathcal{V}_1$ and $\mathcal{V}_2$ be vector subspaces of $\mathbf{R}^n$, with $d_1 = \dim(\mathcal{V}_1)$ and $d_2 = \dim(\mathcal{V}_2)$. Then*

$$\max\{d_1 + d_2 - n, 0\} \leq \dim(\mathcal{V}_1 \cap \mathcal{V}_2) \leq \min\{d_1, d_2\}.$$

(6.5)   *Corollary. Let $\mathcal{A}_1$ and $\mathcal{A}_2$ be intersecting affine subspaces of $\mathbf{R}^n$. Then $\mathcal{A}_1 \cap \mathcal{A}_2$ is also an affine subspace of $\mathbf{R}^n$.*

(6.6)   *Corollary.* Let $\mathcal{A}_1$ and $\mathcal{A}_2$ be intersecting affine subspaces of $\mathbf{R}^n$, and let $d_1 = \dim(\mathcal{A}_1)$, and $d_2 = \dim(\mathcal{A}_2)$. Then

$$\max\{d_1 + d_2 - n, 0\} \leq \dim(\mathcal{A}_1 \cap \mathcal{A}_2) \leq \min\{d_1, d_2\}.$$

The reader may also show, by making appropriate choices of $\mathcal{V}_1$, $\mathcal{V}_2$, $\mathcal{A}_1$, and $\mathcal{A}_2$, that all the permitted values in (6.4) and (6.6) can be achieved.

§33.7 Column-space and Row-space; Theorems on Rank

Definitions. Let A be an m×n matrix. We define the *column-space* of A to be the vector subspace of $\mathbf{R}^m$ generated by the columns of A. We define the *row-space* of A to be the vector subspace of $\mathbf{R}^n$ generated by the transposed rows of A. We have the following results about the rank of A.

(7.1)   *Theorem.* Given an m×n matrix A,
      (i)   r(A) is the dimension of the column-space of A.
      (ii)   r(A) is also the dimension of the row-space of A.

    Proof. We use the row-reduction lemma (5.1).
    (i) Let A' be the row-reduced form of A. Then in A', the columns with pivots (there are r(A) of them) evidently form a basis for the column-space of A'. By (b) of the lemma, the corresponding columns in A form a basis for the column-space of A.
    (ii) Let A' be the row-reduced form of A. By (a) of the lemma, the rows of A' span the same subspace as the rows of A. But the number of non-zero rows in A' is r(A), and these rows are linearly independent (by the positions of the pivots). Hence r(A) is the dimension of the row-space. This completes the proof.
    We note that in the case of a square matrix, the row space may be quite different from the column-space, even though both must have the same dimension. For example, for the matrix $\begin{bmatrix} 1 & 0 \\ 2 & 0 \end{bmatrix}$, the row space (in $\mathbf{R}^2$) is the line $y = 0$, while the column space is the line $x = 2y$.
    We also have the following result, which is useful for determining the rank of a matrix.

Definition. A *submatrix* of a given matrix A is a matrix that results from A by deleting zero or more rows and zero or more columns.

(7.2)   *Theorem.* r(A) *is the order (the number of rows or of columns) of the largest square submatrix of* A *that is non-singular.*

   Proof. Let A, and a row-reduction of A, be given. During the row-reduction, it is possible to keep track of a given element as it is altered by α- or β-operations or is shifted from row to row by a γ-operation. If an element of A eventually becomes a pivot, we say that is a *pivot-element* of A (for the given row-reduction). Let M be the submatrix of A formed by deleting the rows and columns of A that do not contain pivot-elements. Since there are r(A) pivot-elements of A, and these elements fall in distinct rows and distinct columns, M must be a square submatrix of order r(A). Moreover, if we consider the row-reduction of A, we see that M itself must have the identity matrix (of order r(A)) as its own row-reduced form. It follows from §31.3 that $|M| \neq 0$. On the other hand, any square submatrix of order greater than r(A) must have linearly dependent columns, since the column space A has dimension r(A). Hence, by property (5) of determinants of §31.2, such a larger submatrix must have determinant 0. This completes the proof.

§33.8 Linear Dependence Calculations

   The results in §33.5 and §33.7 suggest computational procedures for testing linear independence, for determining dimension, and for finding bases. We describe several of these procedures.

   (I)   Given vectors $C_1,\ldots,C_q$, we *can find the dimension* of $\mathcal{V} = $ span$(C_1,\ldots,C_q)$ by forming a q×n matrix A with $C_q^T,\ldots,C_q^T$ as its *rows*. Row-reducing to A′ gives r(A) as the desired dimension. Moreover, the non-zero rows of A′ give a basis for $\mathcal{V}$.

   Thus, in particular, we can *test a set of vectors* $\{C_1,\ldots,C_q\}$ *for linear independence* by seeing if the dimension d of $\mathcal{V} = $ span$(C_1,\ldots,C_q)$ is q. If d = q, we have linear independence. If d < q, we have linear dependence. (See (a) of the row-reduction lemma (5.1).)

   (II)   Given $C_1,\ldots,C_q$, we can find a *subset* of $C_1,\ldots,C_q$ which is a basis for $\mathcal{V} = $ span$(C_1,\ldots,C_q)$ as follows. We form an n×q matrix B whose *columns* are $C_1,\ldots,C_q$. Row-reducing this matrix to B′, we find, as the desired basis, the columns in B corresponding to pivot columns in B′. Moreover, if B′ is fully row-reduced, each non-pivot column in B′ gives, as its components, coefficients for expressing itself as a linear combination of the pivot columns of B′. The same dependencies must hold for the corresponding columns in B (by (b) of the row-reduction lemma).

Example. Take

$$C_1 = \begin{bmatrix} 1 \\ 1 \\ 2 \\ 2 \end{bmatrix}, \quad C_2 = \begin{bmatrix} 1 \\ 2 \\ 4 \\ 3 \end{bmatrix}, \quad C_3 = \begin{bmatrix} 2 \\ 3 \\ 6 \\ 5 \end{bmatrix},$$

For (I):

$$\begin{bmatrix} 1 & 1 & 2 & 2 \\ 1 & 2 & 4 & 3 \\ 2 & 3 & 6 & 5 \end{bmatrix} \rightarrow \begin{bmatrix} 1 & 1 & 2 & 2 \\ 0 & 1 & 2 & 1 \\ 0 & 1 & 2 & 1 \end{bmatrix} \rightarrow \begin{bmatrix} 1 & 1 & 2 & 2 \\ 0 & 1 & 2 & 1 \\ 0 & 0 & 0 & 0 \end{bmatrix}.$$

Thus we see that the dimension of $\text{span}(C_1, C_2, C_3) = 2$ and that

$$\begin{bmatrix} 1 \\ 1 \\ 2 \\ 2 \end{bmatrix} \quad \text{and} \quad \begin{bmatrix} 0 \\ 1 \\ 2 \\ 1 \end{bmatrix} \quad \text{form a basis.}$$

In particular, we see that $\{C_1, C_2, C_3\}$ is a linearly dependent set.

For (II):

$$\begin{bmatrix} 1 & 1 & 2 \\ 1 & 2 & 3 \\ 2 & 4 & 6 \\ 2 & 3 & 5 \end{bmatrix} \rightarrow \begin{bmatrix} 1 & 1 & 2 \\ 0 & 1 & 1 \\ 0 & 2 & 2 \\ 0 & 1 & 1 \end{bmatrix} \rightarrow \begin{bmatrix} 1 & 1 & 2 \\ 0 & 1 & 1 \\ 0 & 0 & 0 \\ 0 & 0 & 0 \end{bmatrix}.$$

Thus we see that $\{C_1, C_2\}$ is also a basis for $\text{span}(C_1, C_2, C_3)$.

Completing the row-reduction, we get

$$\begin{bmatrix} 1 & 0 & 1 \\ 0 & 1 & 1 \\ 0 & 0 & 0 \\ 0 & 0 & 0 \end{bmatrix}.$$

Here, the third column shows us that $C_3 = (1)C_1 + (1)C_2$.

Remark. To find the dimension in (I) or to identify a basis in (II), it is enough to row-reduce only as far as is needed to identify the pivot columns. Note that (II) gives more information than (I). On the other hand, when $q < n$, (I) may be a shorter and easier calculation than (II), since (I) will usually require fewer β operations than (II).

§33.9 Orthonormal Bases

Definitions. A basis $\{B_1,...,B_k\}$ for a vector subspace $\mathcal{V}$ is said to be an *orthogonal basis* for $\mathcal{V}$ if the vectors in the basis are mutually orthogonal (that is, if $i \neq j \Rightarrow B_i^T B_j = 0$.)

A basis $\{B_1,...,B_k\}$ for a vector subspace $\mathcal{V}$ is said to be an *orthonormal basis* for $\mathcal{V}$ if it is an orthogonal basis and if, in addition, all vectors in the basis are unit vectors (that is, if $B_i^T B_i = 1$ for each i).

An orthonormal basis $\{B_1,...,B_k\}$ for $\mathcal{V}$ in $\mathbf{R}^n$ plays the same role in $\mathcal{V}$ that a frame plays in $\mathbf{E}^2$ or $\mathbf{E}^3$. In particular, any given vector A in $\mathcal{V}$ can be uniquely expressed as a linear combination. $A = a_1 B_1 + \cdots + a_k B_k$, of the basis vectors, where each coefficient a_i can be obtained as the inner product $a_i = A^T B_i$.

The following theorem states a useful fact, and its proof gives a computational procedure.

(9.1) **Theorem.** *Every vector subspace $\mathcal{V} \neq \{O\}$ of $\mathbf{R}^n$ has an orthonormal basis.*

Proof. Let $\mathcal{V} \neq \{O\}$ be a vector subspace. Let $\{C_1,...,C_\ell\}$ be a generating set of $\mathcal{V}$. We show how to find, from $\{C_1,...,C_\ell\}$, an orthogonal basis $\{B_1,...,B_k\}$ for $\mathcal{V}$. We can then take the unit-vectors $\{\hat{B}_1,...,\hat{B}_k\}$ as our desired orthonormal basis for $\mathcal{V}$. We obtain the orthogonal vectors $B_1,...,B_k$ inductively as follows:

Step 1. Let i_1 be the smallest i such that $C_i \neq O$. Define $B_1 = C_{i_1}$. Then B_1 is in $\mathcal{V}$.

Step 2. Let i_2 be the smallest i such that $i_1 < i \leq \ell$ and $C_i - \left(\dfrac{B_1^T C_i}{B_1^T B_1} \right) B_1 \neq$

0.   Define $B_2 = (B_1^T B_1) \left(C_{i_2} - \left(\dfrac{B_1^T C_{i_2}}{B_1^T B_1} \right) B_1 \right) = (B_1^T B_1) C_{i_2} - (B_1^T C_{i_2}) B_1.$

Then B_2 is in $\mathcal{V}$, and $B_1^T B_2 = 0$ (as the reader may verify immediately).

Step 3.   Let i_3 be the smallest i such that $i_2 < i \leq \ell$ and

$C_i - \left(\dfrac{B_1^T C_i}{B_1^T B_1} \right) B_1 - \left(\dfrac{B_2^T C_i}{B_2^T B_2} \right) B_2 \neq 0.$ Define

$$B_3 = (B_1^T B_1)(B_2^T B_2) \left(C_{i_3} - \left(\dfrac{B_1^T C_{i_3}}{B_1^T B_1} \right) B_1 - \left(\dfrac{B_2^T C_{i_3}}{B_2^T B_2} \right) B_2 \right)$$

$$- (B_1^T B_1)(B_2^T B_2) C_{i_3} - (B_2^T B_2)(B_1^T C_{i_3}) B_1 - (B_1^T B_1)(B_2^T C_{i_3}) B_2.$$

Then B_3 is in $\mathcal{V}$, $B_1^T B_3 = 0$, and $B_2^T B_3 = 0$ (as the reader may verify immediately).

Step 4. . . . .

We continue in this way until we fail to find an i with the desired properties. Let $B_1, \ldots, B_k$ be the new vectors obtained. (The initial coefficients $(B_1^T B_1)$, $(B_1^T B_1)(B_2^T B_2)$, ... in the definitions for $B_2, B_3, \ldots$ may be omitted. We have used them to get expressions for $B_2, B_3, \ldots$, which are free of fractions.)

The following facts are evident from the construction:

(i)   The vectors $B_1, B_2, \ldots, B_k$ are in $\mathcal{V}$; each vector B_i is non-zero; and the vectors $B_1, B_2, \ldots$ are mutually orthogonal.

(ii)   Each of the vectors in $\{C_1, \ldots, C_\ell\}$ is a linear combination of vectors in $\{B_1, \ldots, B_k\}$, since each C_i is either discarded because it is already a linear combination, or else is used to make a new linear combination. It follows that $\mathcal{V} = \text{span}(B_1, \ldots, B_k)$.

(iii)   The vectors $B_1, \ldots, B_k$ form a linearly independent set. If $t_1 B_1 + \cdots + t_k B_k = 0$ were a linear dependence with $t_i \neq 0$ for some i, then multiplying both sides by B_i^T, we would get $t_i B_i^T B_i = 0$, contrary to the fact that $t_i \neq 0$ and $B_i \neq 0$.

From (i), (ii), and (iii), we see that the set $\{B_1, \ldots, B_k\}$ is an orthogonal basis for $\mathcal{V}$ and that k must be the dimension of $\mathcal{V}$. Finally, we obtain an orthonormal basis for $\mathcal{V}$ by replacing each vector B_i by its own unit vector,

$$\hat{B}_i = \left(\frac{1}{\sqrt{B_i^T B_i}} \right) B_i .$$ This completes the proof.

The proof of the theorem supplies us with a computational procedure for converting any generating set into an orthonormal basis. This procedure is known as the *Gram-Schmidt orthonormalization procedure*. We illustrate it by the following example:

Let C_1, C_2, C_3 be the vectors $\begin{bmatrix} 1 \\ 1 \\ 2 \\ 2 \end{bmatrix}, \begin{bmatrix} 1 \\ 2 \\ 4 \\ 3 \end{bmatrix}, \begin{bmatrix} 2 \\ 3 \\ 6 \\ 5 \end{bmatrix}$. (These vectors were used in the previous example.) Then we get

$$B_1 = \begin{bmatrix} 1 \\ 1 \\ 2 \\ 2 \end{bmatrix} .$$

$$B_2 = (B_1^T B_1)C_2 - (B_1^T C_2)B_1 = 10 \begin{bmatrix} 1 \\ 2 \\ 4 \\ 3 \end{bmatrix} - 17 \begin{bmatrix} 1 \\ 1 \\ 2 \\ 2 \end{bmatrix} = \begin{bmatrix} -7 \\ 3 \\ 6 \\ -4 \end{bmatrix} .$$

The procedure stops here, since an attempt to get B_3 gives

$$\begin{bmatrix} 2 \\ 3 \\ 6 \\ 5 \end{bmatrix} - \frac{27}{10} \begin{bmatrix} 1 \\ 1 \\ 2 \\ 2 \end{bmatrix} - \frac{11}{110} \begin{bmatrix} -7 \\ 3 \\ 6 \\ -4 \end{bmatrix} = \begin{bmatrix} 0 \\ 0 \\ 0 \\ 0 \end{bmatrix} .$$

Finally, forming unit vectors, we have $\hat{B}_1 = \frac{1}{\sqrt{10}} B_1$ and $\hat{B}_2 = \frac{1}{\sqrt{110}} B_2$ as our desired orthonormal basis.

The following corollary to our theorem is known as the *basis-extension theorem.*

(9.2) ***Corollary. (Basis-extension theorem.)*** *Let $\mathcal{V}_1$ and $\mathcal{V}_2$ be vector subspaces in $\mathbf{R}^n$, with dimensions k and ℓ, such that $\mathcal{V}_1 \subset \mathcal{V}_2$ and $k < \ell$. Let $\{B_1,\ldots,B_k\}$ be some orthonormal basis for $\mathcal{V}_1$. Then we can find vectors $B_{k+1},\ldots,B_\ell$ such that $\{B_1,\ldots,B_k,B_{k+1},\ldots,B_\ell\}$ is an orthonormal basis for $\mathcal{V}_2$.*

Proof. Let $\{C_1,\ldots,C_\ell\}$ be a basis for $\mathcal{V}_2$. Apply the Gram-Schmidt procedure to the set of vectors $\{B_1,\ldots,B_k,C_1,\ldots,C_\ell\}$, in the order given. From the definition of the procedure, we find that the first k vectors produced must be the vectors $B_1,\ldots,B_k$ themselves, and that the set of vectors yielded by the entire procedure must be an orthonormal basis for $\mathcal{V}_2$. This completes the proof.

Remark. Let $\mathcal{V}$ be a vector subspace in $\mathbf{R}^n$ of dimension k. By the above theorem, $\mathcal{V}$ has an orthonormal basis $\{B_1,\ldots,B_k\}$. Every vector A in $\mathcal{V}$ can be expressed, in a unique way, as a linear combination of the vectors in this basis. This allows us to associate (or "identify") each vector $A = t_1 B_1 + \cdots + t_k B_k$ with

the corresponding vector $\begin{bmatrix} t_1 \\ \vdots \\ t_k \end{bmatrix}$ in $\mathbf{R}^k$. Under this identification, vector-algebra

operations in $\mathcal{V}$ and in $\mathbf{R}^k$ produce the same results, as the reader may readily verify from the orthonormality of $B_1,\ldots,B_k$. Thus the orthonormal basis $B_1,\ldots,B_k$ allows us to view $\mathcal{V}$ as a *copy* of $\mathbf{R}^k$. For example, if $\mathcal{V}$ happens to be a three-dimensional vector subspace of $\mathbf{R}^{10}$, we can choose some orthonormal basis $\{B_1,B_2,B_3\}$ and then treat $\mathcal{V}$ as a copy of $\mathbf{R}^3$, where each vector $A = t_1 B_1 + t_2 B_2 + t_3 B_3$ in $\mathcal{V}$ is identified with (t_1,t_2,t_3) in $\mathbf{R}^3$. Since $\mathbf{R}^3$ can itself be viewed as $\mathbf{E}^3$ with a chosen origin and chosen Cartesian axes, we can view $\mathcal{V}$, together with its orthonormal basis, as a copy of $\mathbf{E}^3$-with-coordinate-axes. It follows that within $\mathcal{V}$, we can use the concepts and techniques of geometry and calculus that we have previously used in $\mathbf{E}^3$. Similarly, given a two-dimensional vector subspace $\mathcal{V}$ of $\mathbf{R}^n$, we can choose an orthonormal basis for $\mathcal{V}$ and then treat $\mathcal{V}$ as a copy of $\mathbf{E}^2$-with-axes.

§33.10 Problems

§33.3

1. Verify examples 1 and 2.

§33.4

1. Give further details of the proof for example 4.

2. Prove (9.2) in §30.9.

§33.5

1. Given the four 4-vectors $(1,0,2,1)$, $(3,-1,4,2)$, $(0,1,0,0)$, and $(5,0,8,4)$.
   (a) What is the dimension of the subspace spanned by the vectors?
   (b) Are these vectors linearly independent?
   (c) If they are linearly dependent, find a linear dependence among them.

2. Given the four 4-vectors $(1,0,0,1)$, $(1,1,0,0)$, $(1,0,1,1)$, and $(1,1,1,0)$.
   (a) Are they linearly independent?
   (b) Find a linearly independent subset with the same span.

§33.6

1. In row reduced form, a homogeneous system of equations $AX = O$ has the coefficient matrix

$$\begin{bmatrix} 1 & 1 & 0 & 2 \\ 0 & 0 & 1 & -1 \\ 0 & 0 & 0 & 0 \end{bmatrix} .$$

   (a) What is the rank of A?
   (b) Give a basis for the row space of A.
   (c) What is the dimension of the solution-set?
   (d) Give a basis for the solution-set.

2. In each case, give an example of an inhomogeneous system that fits the given set of facts. (The system is given as $AX = D$.) In each case, state what you can conclude about the solution-set from the given facts. If the solution-set is not empty, be sure to give its dimension.
   (a) $n = 3$, $r(A) = 3$.
   (b) $n = 3$, $r([A \vdots D]) = 3$, $r(A) = 2$.
   (c) $n = 4$, $r([A \vdots D]) = 2$, $r(A) = 2$.

3. Prove (6.4).

4. Prove (6.5).

5. Prove (6.6).

6. For $n = 3$ and $d_1 = d_2 = 2$, give examples of vector subspaces for all permitted values in (6.4).

7. For $n = 3$ and $d_1 = d_2 = 1$, give examples of affine subspaces for all permitted values in (6.6).

§33.8

1. Given the following four 5-vectors: $(0,0,0,1,1)$, $(0,1,0,1,1)$, $(1,1,1,0,1)$, $(1,0,1,0,1)$,
   (a) test for linear independence;
   (b) give dimension of their span;
   (c) find a subset forming a basis for the span.

2. Given the following four 5-vectors: $(1,0,0,1,0)$, $(1,1,1,1,1)$, $(0,1,1,0,1)$, $(0,0,0,1,1)$,
   (a) test for linear dependence;
   (b) give dimension of span;
   (c) find subset forming basis for span.

§33.9

1. Verify step 2 in the proof of (9.1).

2. Verify step 3 in the proof of (9.1).

Chapter 34

Topics in Linear Algebra

§34.1 Introduction

In §34.2, we use the concepts of Chapter 33 to define $\mathbf{E}^n$, a coordinate-free version of $\mathbf{R}^n$. We refer to $\mathbf{E}^n$ as n-*dimensional Euclidean space*. For $\mathbf{E}^n$, we obtain n-dimensional versions of *lines*, *planes*, *parallelism*, and *perpendicularity*. In §34.3, we consider an alternative definition for the concept of *vector subspace*. This definition is more abstract, but also more concise and economical, than the definition given in §33.1. In §34.4, we consider an alternative approach to the concept of matrix and to the operations of matrix algebra. This approach is based on the idea of *linear transformation*. Linear transformations serve as coordinate-free, geometric versions of matrices. Finally, in §34.5 on *quadratic forms and principal axes*, we present several deeper results of matrix algebra which have a variety of applications in science and engineering.

§34.2 The Space $\mathbf{E}^n$; n-dimensional Euclidean Geometry

In this section, we show how concepts and results of Euclidean geometry can be adapted to $\mathbf{R}^n$ (for any n). At the end of §33.9, we saw that each k-dimensional vector subspace of $\mathbf{R}^n$ can be viewed as a copy of $\mathbf{R}^k$. In much the same way, a k-dimensional affine subspace $\mathcal{A}$ of $\mathbf{R}^n$ can be viewed as a copy of $\mathbf{R}^k$. For let $\mathcal{A}$ be given. Let $\mathcal{V}$ be the associated vector subspace of $\mathcal{A}$ (see §33.2). We choose a fixed vector X_0 in $\mathcal{A}$, and call it the *origin* of $\mathcal{A}$. We then associate (or "identify") each vector X in $\mathcal{A}$ with the corresponding vector $X-X_0$ in $\mathcal{V}$. In particular, we associate the vector X_0 in $\mathcal{A}$ with the vector O in $\mathcal{V}$. Under this association, $\mathcal{A}$ can be viewed as a copy of $\mathcal{V}$. If we then choose an orthonormal basis for $\mathcal{V}$, $\mathcal{V}$ can, in its turn, be viewed as a copy of $\mathbf{R}^k$ (as we have noted in §33.9). Hence we can view $\mathcal{A}$ as a copy of $\mathbf{R}^k$. In the special cases k = 2 and k = 3, we can view the vectors in $\mathcal{A}$ as *points* in $\mathbf{E}^2$ or $\mathbf{E}^3$. We can apply the geometry of $\mathbf{E}^2$ and $\mathbf{E}^3$ to sets of points in $\mathcal{A}$, and we can apply the calculus of $\mathbf{E}^2$ and $\mathbf{E}^3$ to scalar and vector fields in $\mathcal{A}$. Thus, for example, $X{-}X_0$ and $Y{-}X_0$ would give

position-vectors (in $\mathbf{E}^2$ or $\mathbf{E}^3$) for the points X and Y, and $\cos^{-1} \dfrac{(X{-}X_0)^T (Y{-}X_0)}{|X{-}X_0|\,|Y{-}X_0|}$

would be the *angle* between those position-vectors.

Let $\mathbf{R}^n$ be given. Each 0-dimensional affine subspace of $\mathbf{R}^n$ is simply a single vector in $\mathbf{R}^n$. We refer to these 0-dimensional subspaces as *points*. The *distance* between two points X_1 and X_2 is defined to be $|X_2 - X_1|$, the magnitude of the vector $X_2 - X_1$.

For $0 < k < n$, a k-dimensional affine subspace of $\mathbf{R}^n$ is called a k-*plane*. We shall also sometimes refer to 1-planes in $\mathbf{R}^n$ as *lines* and to (n–1)-planes in $\mathbf{R}^n$ as *hyperplanes*.

Finally, the only n-dimensional affine subspace of $\mathbf{R}^n$ is $\mathbf{R}^n$ itself.

As in the cases of the Euclidean spaces $\mathbf{E}^2$ and $\mathbf{E}^3$, a variety of geometric concepts and facts may be stated and proved for affine subspaces of $\mathbf{R}^n$, without reference to coordinates. (We might say that these concepts and facts belong to the *synthetic geometry of* $\mathbf{R}^n$.) In particular, we consider *incidence relationships, parallelism, perpendicularity,* and *projection* for the affine subspaces of $\mathbf{R}^n$. We show how coordinate-free geometric ideas can be used to define various *Cartesian coordinate systems* in these subspaces, just as, in Chapter 3, we used geometric constructions based on perpendicularity, and projections to define Cartesian coordinates for $\mathbf{E}^2$ and $\mathbf{E}^3$. When we consider the synthetic geometry of $\mathbf{R}^n$, we sometimes refer to $\mathbf{R}^n$ as n-*dimensional Euclidean space* and use the notation $\mathbf{E}^n$ for $\mathbf{R}^n$.

Incidence relationships describe the geometrical ways in which planes can intersect with planes of equal or different dimension, and the ways in which planes and points can define planes of higher dimension. (In these statements, lines are included as "planes," since lines are also called "1-planes.") For example, for $\mathbf{E}^3$, we know that two distinct intersecting 1-planes must intersect at a single point, that two distinct intersecting 2-planes must intersect in a 1-plane, and that a 1-plane intersecting a 2-plane must either lie in that plane or intersect it at a single point. We also know, for $\mathbf{E}^3$, that a point P and a 1-plane not containing P determine a unique 2-plane, and that it is possible for two 1-planes not to lie in any common 2-plane. Corresponding results for $\mathbf{E}^n$ are easy to formulate and prove. We give several examples in the following theorems.

(1.1)   *Theorem*. *If two planes (of any dimensions) intersect, their intersection must be a plane or a point.*

   Proof. By (6.5) in §33.6.

(1.2)   *Theorem.* *If $\mathcal{A}$ is a line, and two distinct points on $\mathcal{A}$ lie in a k-plane $\mathcal{A}'$, (k $\geq$ 1) then the entire line $\mathcal{A}$ is contained in $\mathcal{A}'$.*

   Proof. Assume that the distinct points X_1 and X_2 are in $\mathcal{A}$ and also in $\mathcal{A}'$. Then $\mathcal{A} = X_1 + \mathrm{span}(X_2 - X_1)$. Let $\mathcal{A}' = X_1 + \mathcal{V}'$. Since $X_2 = X_1 + (X_2 - X_1)$, we see that $(X_2 - X_1)$ must be in $\mathcal{V}'$. Hence $\mathrm{span}(X_2 - X_1) \subseteq \mathcal{V}'$, and $\mathcal{A} \subseteq \mathcal{A}'$. This completes the proof.

(1.3) *Theorem.* *If two distinct hyperplanes ((n–1)-planes) intersect, their intersection must be an (n–2)-plane.*

   Proof. Let $\mathcal{A}_1 = X_0 + \mathcal{V}_1$ and $\mathcal{A}_2 = X_0 + \mathcal{V}_2$ be the given planes. By (6.6) in §33.6,

$$(2n-2) - n = n - 2 \leq \dim(\mathcal{V}_1 \cap \mathcal{V}_2) = \dim(\mathcal{A}_1 \cap \mathcal{A}_2) \leq n - 1 .$$

If $\dim(\mathcal{V}_1 \cap \mathcal{V}_2) = n - 1$, then $\mathcal{V}_1 \cap \mathcal{V}_2 = \mathcal{V}_1 = \mathcal{V}_2$, by (5.5) in §33.5, and the given planes are not distinct. Hence $\mathcal{A}_1 \cap \mathcal{A}_2$ must have dimension $n - 2$. This completes the proof.

(1.4)   *Theorem.* *For $n \geq 4$, the intersection of two distinct and intersecting 2-planes in $\mathbf{E}^n$ must be either a line or a point.*

   Proof. For $n \geq 4$, this follows directly from (6.6) in §33.6, since

$$2 + 2 - n \leq 0 \leq \dim(\mathcal{A}_1 \cap \mathcal{A}_2) \leq 2 .$$

$\dim(\mathcal{A}_1 \cap \mathcal{A}_2) = 2$ can only occur if $\mathcal{A}_1$ and $\mathcal{A}_2$ are the same 2-plane. This completes the proof.

(As an example of two 2-planes intersecting in a single point, consider $\mathrm{span}(\hat{e}_1, \hat{e}_2)$ and $\mathrm{span}(\hat{e}_3, \hat{e}_4)$ in $\mathbf{R}^4 = \mathbf{E}^4$.)

(1.5)   Theorem. *If* X_1 *is a point and* $\mathcal{A}$ *is a* k*-plane in* $\mathbf{E}^n$*, and* X_1 *does not lie in* $\mathcal{A}$*, then there is a unique* (k+1)*-plane* $\mathcal{A}'$ *containing both* X_1 *and* $\mathcal{A}$*.*

Proof. Let $\mathcal{A} = X_0 + \mathcal{V}$, and let $\{B_1,...,B_k\}$ be a basis for $\mathcal{V}$. Since X_1 is not in $\mathcal{A}$, $X_1 - X_0$ is not in $\mathcal{V}$, and $\{B_1,...,B_k,X_1 - X_0\}$ must be a lin-early independent set. Then $\mathcal{A}' = X_0 + \text{span}\{B_1,...,B_k,X_1 - X_0\}$ is a (k+1)-plane containing both $\mathcal{A}$ and X_1. Any other plane containing X_1 and $\mathcal{A}$ must contain $\mathcal{A}'$ and must therefore, by the comparability corollary ((5.5) in §33.5), be of higher dimension than $\mathcal{A}'$. This competes the proof.

A variety of other incidence relationships can be obtained by similar arguments.

We now turn to definitions of *parallelism* and *perpendicularity* in $\mathbf{E}^n$.

Definition. We say that two planes in $\mathbf{E}^n$ (of possibly different dimensions) are *parallel* if (i) they do not intersect and (ii) their associated vector sub-spaces are comparable.   In particular, two distinct planes of the same dimension are parallel if and only if they have the same associated vector subspace. The following theorem for $\mathbf{E}^n$ is a counterpart and generalization of the *parallel axiom* for $\mathbf{E}^2$ and $\mathbf{E}^3$.

(1.6) Theorem. *If* $\mathcal{A}$ *is a* k*-plane of* $\mathbf{E}^n$ *and* X_1 *is a point not in* $\mathcal{A}$*, then there is a unique* k*-plane* $\mathcal{A}'$ *such that* X_1 *is in* $\mathcal{A}'$ *and* $\mathcal{A}'$ *is parallel to* $\mathcal{A}$*.*

Proof. The theorem is proved immediately by taking $\mathcal{A}' = X_1 + \mathcal{V}$, where $\mathcal{V}$ is the associated vector subspace of $\mathcal{A}$. To see that $\mathcal{A}'$ is unique, note that any other k-plane $\mathcal{A}''$ that satisfied the theorem would have to have its associated vector subspace $\mathcal{V}''$ comparable with $\mathcal{V}$. By the comparability corollary in §33.5, this would give $\mathcal{V}'' = \mathcal{V}$, and we would have $\mathcal{A}'' = \mathcal{A}$. This completes the proof.

Our definition for *perpendicularity* in $\mathbf{E}^n$ is less obvious.

Definition. We say that two intersecting planes $\mathcal{A}_1 = X_0 + \mathcal{V}_1$ and $\mathcal{A}_2 = X_0 + \mathcal{V}_2$ (of possibly different dimensions) are *perpendicular* if $\mathcal{V}_1 \subseteq \mathcal{V}_2^{\perp}$. (It is immediate from the definition of $\mathcal{V}^{\perp}$ that $\mathcal{V}_1 \subseteq \mathcal{V}_2^{\perp} \Leftrightarrow \mathcal{V}_2 \subseteq \mathcal{V}_1^{\perp}$.) The definition above of *perpendicularity* agrees with the earlier definitions for planes and lines in $\mathbf{E}^3$. The reader may verify this by considering the various possible combinations of dimension that can occur. We have the following uniqueness theorem for perpendicularity.

(1.7)   **Theorem.** *If $\mathcal{A}$ is a k-plane of $\mathbf{E}^n$, and if X_1 is any point in $\mathbf{E}^n$, then there is a unique (n–k)-plane $\mathcal{A}'$ such that X_1 is in $\mathcal{A}'$, and $\mathcal{A}'$ is perpendicular to $\mathcal{A}$.*

  Proof. Let $\mathcal{V}$ be the associated vector subspace of $\mathcal{A}$. Take $\mathcal{A}'$ to be the affine subspace $\mathcal{A}' = X_1 + \mathcal{V}^\perp$. To see that $\mathcal{A}'$ is unique, note that any other (n–k)-plane $\mathcal{A}''$ perpendicular to $\mathcal{A}$ must have its associated vector subspace $\mathcal{V}'' \subseteq \mathcal{V}^\perp$. Since $\mathcal{V}^\perp$ also has dimension $n-k$, this would give $\mathcal{V}'' = \mathcal{V}^\perp$. If X_1 is in $\mathcal{A}''$, we must have $\mathcal{A}'' = \mathcal{A}'$. This completes the proof.

  If $\mathcal{A}$ is a k-plane in $\mathbf{E}^n$, and X is a point in $\mathbf{E}^n$, we define the *projection* of X on $\mathcal{A}$ to be the intersection of $\mathcal{A}$ with the unique $(n-k)$-plane $\mathcal{A}'$ which contains X and is perpendicular to $\mathcal{A}$. This intersection must be a single point, as the reader may easily verify.

  Let X_0 be any chosen point in $\mathbf{E}^n$, and let $\{B_1,...,B_k\}$ be any chosen orthonormal basis for $\mathbf{E}^n = \mathbf{R}^n$. We can use X_0 and $\{B_1,...,B_k\}$ to define *Cartesian coordinates* in $\mathbf{E}^n$ as follows. X_0 will be our new *origin*. The lines $X_0 + \mathrm{span}(B_1),...,X_0 + \mathrm{span}(B_2)$ will be our new *coordinate axes*. For any point X, we obtain the *coordinates* of X (in our new coordinate system) by finding the projections of X on the coordinate axes. Let these projections be the points $X_1,...,X_n$. We then take the coordinates of X to be the scalars $x_1' = (X_i - X_0)^T B_i$ for $1 \le i \le n$. Each x_1' can be thought of as the "directed distance" of X_i from the origin X_0 along the i-th axis, where the *positive direction* of the i-th axis is given by B_i.

  Since $\mathbf{E}^n = \mathbf{R}^n$, the simplest way to introduce Cartesian coordinates into $\mathbf{E}^n$ is to use O as origin and $\{\hat{e}_1,...,\hat{e}_n\}$ as basis. The axes are then $\mathrm{span}(\hat{e}_1),...,\mathrm{span}(\hat{e}_n)$, and the coordinates of any given X are simply the components of X as an n-vector. This particular (and natural) coordinate system in $\mathbf{E}^n$ is called the *canonical coordinate system*.

§34.3 Non-constructive Definitions

Some of our basic definitions can be given in a more general abstract form. We illustrate with the definition of *vector subspace*.

Alternative definition. Let $\mathcal{V}$ be a subset of $\mathbf{R}^n$. $\mathcal{V}$ is a *vector subspace*, if

    (i)   O is in $\mathcal{V}$; and

    (ii)   for every choice of scalars t_1 and t_2 and for every pair of vectors A_1 and A_2 in $\mathcal{V}$, the vector $t_1A_1 + t_2A_2$ is also in $\mathcal{V}$.

(i) and (ii) are called *closure rules*. (ii) implies that $\mathcal{V}$ is closed under the formation of linear combinations. This definition of vector subspace differs from our earlier definition in that it makes no reference to a generating set.

The proof that the alternative definition is equivalent to the earlier definition is simple.

    (1)   Let $\mathcal{V}$ be a vector subspace in $\mathbf{R}^n$ according to the earlier definition. By (1.3) in §33.1, $\mathcal{V}$ satisfies (i) and (ii) in the alternative definition. Hence $\mathcal{V}$ satisfies the alternative definition.

    (2)   Let $\mathcal{V}$ be a vector subspace in $\mathbf{R}^n$ according to the alternative definition. Choose some vector $B_1 \neq 0$ in $\mathcal{V}$. (If there is none, then $\mathcal{V} = \{0\}$, which is a vector subspace according to the earlier definition.) See if $\mathcal{V} = \text{span}(B_1)$. If not, choose a vector B_2 that is in $\mathcal{V}$ but not in $\text{span}(B_1)$. See if $\mathcal{V} = \text{span}(B_1, B_2)$. If not, choose a vector B_3 that is in $\mathcal{V}$ but not in $\text{span}(B_1, B_2), \dots$, and so forth. By the first corollary to the dimension theorem ((5.4) in §33.5), we must eventually reach B_k such that $\mathcal{V} = \text{span}(B_1, \dots, B_k)$, and this must occur before $k = n+1$. Hence $\mathcal{V}$ is a vector subspace according to the earlier definition. This completes the proof.

Since no specific procedure for finding $B_1, B_2, \dots$ is given in (2), we say that the proof in (2) is *non-constructive*, and that the alternative definition is itself a *non-constructive definition*. We also say that the alternative definition is *invariant,* since it does not use, or depend on, a particular generating set. Invariant definitions can provide economy and elegance in the way we think about problems and in the way we solve them. Here, as examples, are new proofs for two earlier theorems.

    Examples.

(1)   *Theorem. The solution set to a homogeneous system* $AX = O$ *is a vector subspace.* (This is (1.1) in §33.1.)

New proof. Our previous proof used the general solution obtained by row reduction. For our new proof, we simply observe:

(i)   that $AO = O$; and

(ii)   that if X_1 and X_2 are solutions to the system (that is, if $AX_1 = O$ and $AX_2 = O$), and t_1, t_2 are any scalars, then $t_1 X_1 + t_2 X_2$ is a solution, since $A(t_1 X_1 + t_2 X_2) = t_1 A X_1 + t_2 A X_2 = O$ by the laws of matrix multiplication.

(i) and (ii) show that the alternative definition of *vector subspace* is satisfied. This completes the proof.

(2)   *Theorem.* *Every vector subspace is the solution set for some homogeneous system.* (This is (6.2) in §33.6.)

New proof. In our previous proof, we began with a basis for $\mathcal{V}$ and used row-reduction to find a basis for $\mathcal{V}^\perp$. For our new proof, we find the existence of such a basis as follows.

(i) O is orthogonal to every vector in $\mathcal{V}$.

(ii)   If A_1 and A_2 are each orthogonal to every vector in $\mathcal{V}$, and t_1, t_2 are any scalars, then $t_1 A_1 + t_2 A_2$ must be orthogonal to every vector in $\mathcal{V}$, since

$$[A_1^T B = O \text{ and } A_2^T B = O] \Rightarrow (t_1 A_1 + t_2 A_2)^T B = t_1 A_1^T B + t_2 A_2^T B = O, \quad \text{by}$$

algebraic laws for transposition and matrix multiplication.

By (i) and (ii), we have immediately that $\mathcal{V}^\perp$ is a vector subspace, and hence that $\mathcal{V}^\perp$ has a basis.

The reader may give, as a third example, a new proof for the theorem: *If $\mathcal{V}_1$ and $\mathcal{V}_2$ are vector subspaces in $\mathbf{R}^n$, then $\mathcal{V}_1 \cap \mathcal{V}_2$ is a vector subspace.* In all three examples, however, the reader should note that the earlier constructive proofs give us information (about dimension) that the non-constructive proofs do not.

§34.4 Linear Transformations

The concepts of *matrix* and *matrix multiplication* can themselves be put in invariant form. We indicate this briefly and in outline.

Definition. A function f defined on $\mathbf{R}^n$ and mapping into $\mathbf{R}^m$ is called a *linear transformation from $\mathbf{R}^n$ to $\mathbf{R}^m$* if, for every choice of scalars t_1 and t_2, and of vectors X_1 and X_2 in $\mathbf{R}^n$, we have

$$f(t_1 X_1 + t_2 X_2) = t_1 f(X_1) + t_2 f(X_2) .$$

(This condition implies that $f(O) = O$.)

(4.1)   **Theorem.** *Let* f *be a linear transformation from* $\mathbf{R}^n$ *to* $\mathbf{R}^m$, *and let* g *be a linear transformation from* $\mathbf{R}^m$ *to* $\mathbf{R}^p$, *then the function* gf *from* $\mathbf{R}^n$ *to* $\mathbf{R}^p$ *defined by*

$$gf(X) = g(f(X)) \ \textit{for all} \ X \ \textit{in} \ \mathbf{R}^n$$

must be a linear transformation.

   Proof. We observe that

$$gf(t_1 X_1 + t_2 X_2) = g\big(t_1 f(X_1) + t_2 f(X_2)\big) = t_1 gf(X_1) + t_2 gf(X_2),$$

and the proof is complete.

   gf is called the *composition* of the linear transformations g and f. The theorem above shows that the composition of two linear transformations must be a linear transformation.

   The following theorem shows that every matrix defines a linear transformation:

(4.2)   **Theorem.** *Let* A *be an* m×n *matrix. Then the function* g *defined by*

$$g(X) = AX \ \textit{for} \ X \ \textit{in} \ \mathbf{R}^n$$

is a linear transformation from $\mathbf{R}^n$ *to* $\mathbf{R}^m$.

   Proof. By the laws of matrix multiplication,

$$g(t_1 X_1 + t_2 X_2) = A(t_1 X_1 + t_2 X_2) = t_1 AX_1 + t_2 AX_2 = t_1 g(X_1) + t_2 g(X_2),$$

and the proof is complete.
   We can also prove the converse:

(4.3)   **Theorem.** *Let* g *be a linear transformation from* $\mathbf{R}^n$ *to* $\mathbf{R}^m$. *Then there must be an* m×n *matrix* A *such that*

$$g(X) = AX \ \textit{for all} \ X \ \textit{in} \ \mathbf{R}^n.$$

Proof. For any X in $\mathbf{R}^n$, we can write $X = x_1\hat{e}_1 + \cdots + x_n\hat{e}_n$. Since g is a linear transformation, we have

$$g(X) = x_1 g(\hat{e}_1) + \cdots + x_n g(\hat{e}_n).$$

Let A be the m×n matrix $[g(\hat{e}_1) \vdots \cdots \vdots g(\hat{e}_n)]$. Evidently, by the definition of matrix multiplication,

$$g(X) = AX.$$

This completes the proof.

Finally, we note the following:

(4.4)   *Theorem.* Different m×n *matrices give different linear transformations from* $\mathbf{R}^n$ *to* $\mathbf{R}^m$.

Proof. If two matrices differ, there must be some j, $1 \le j \le n$, such that their j-th columns differ. Hence the linear transformations defined by the two matrices must map $\hat{e}_j$ to different vectors in $\mathbf{R}^m$. This completes the proof.

The theorems above show that there is a natural one-to-one correspondence between linear transformations from $\mathbf{R}^n$ to $\mathbf{R}^m$ and m×n matrices. The next theorem shows that the matrix for the composition of two linear transformations can be obtained by multiplying the matrices for the two given transformations.

(4.5)   *Theorem.* Let f *be a linear transformation from* $\mathbf{R}^n$ *to* $\mathbf{R}^m$, *and let* g *be a linear transformation from* $\mathbf{R}^m$ *to* $\mathbf{R}^p$. *Let* A *and* B *be matrices such that*

$$f(X) = AX \qquad \textit{for X in } \mathbf{R}^n,$$

and $g(Y) = BY \qquad \textit{for Y in } \mathbf{R}^m.$

Then $gf(X) = (BA)X \qquad \textit{for X in } \mathbf{R}^n.$

Proof. The theorem is immediate from the associativity of matrix multiplication.

The concept of *linear transformation* provides a useful and invariant way to approach matrices and matrix algebra. We illustrate this invariant approach in the following results. The proofs are easy, and we omit them.

Definitions. Let f be a linear transformation from $\mathbf{R}^n$ to $\mathbf{R}^m$. The set of all Y in $\mathbf{R}^m$ such that $Y = f(X)$ for some X in $\mathbf{R}^n$ is called the *range of* f. The set of all X in $\mathbf{R}^n$ such that $f(X) = 0$ is called the *kernel* of f.

(4.6)   *Theorem. Let* f *be a linear transformation from* $\mathbf{R}^n$ *to* $\mathbf{R}^m$, *and let* A *be its matrix. Then the range of* f *is a vector subspace of* $\mathbf{R}^m$ *with dimension* r(A), *and the kernel of* f *is a vector subspace of* $\mathbf{R}^n$ *with dimension* n–r(A). *Moreover, the range of* f *is the column-space of* A, *and the kernel of* f *is the orthogonal complement of the row-space of* A.

(4.7)   *Theorem. Let* $\mathcal{V}$ *be a given vector subspace of* $\mathbf{R}^n$, *and let* $\mathcal{W}$ *be a vector subspace of* $\mathbf{R}^m$ *such that the sum of the dimension of* $\mathcal{V}$ *and the dimension of* $\mathcal{W}$ *is* n. *Then there exists a linear transformation from* $\mathbf{R}^n$ *to* $\mathbf{R}^m$ *having* $\mathcal{W}$ *as its range and* $\mathcal{V}$ *as its kernel.*

(The reader may prove this last theorem (i) by noting that $\mathcal{W}$ and $\mathcal{V}^\perp$ have the same dimension, (ii) by using a basis for $\mathbf{R}^n$ that includes a basis for $\mathcal{V}$ and a basis for $\mathcal{V}^\perp$, (iii) by choosing a basis for $\mathcal{W}$, and (iv) by then defining a linear transformation from $\mathbf{R}^n$ to $\mathbf{R}^m$ which maps the basis vectors of $\mathcal{V}$ to the vector 0 and the basis vectors of $\mathcal{V}^\perp$ to the basis vectors of $\mathcal{W}$.)

We give a further example of the advantages that this coordinate-free way of thinking can provide. Let f be a linear transformation from $\mathbf{R}^n$ to $\mathbf{R}^m$, and let g be a linear transformation from $\mathbf{R}^m$ to $\mathbf{R}^p$. Let matrix A correspond to f and matrix B correspond to g. Evidently, the range of the composition gf must be contained in the range of g. Hence we have the following result: *given any two matrices* B *and* A *which are conformable for multiplication, it must be the case that* r(BA) ≤ r(B). In terms of linear transformations, this is a natural and easy result to think of and to prove. (By considering transposes, this result is readily extended to r(BA) ≤ min{r(B),r(A)}.)

§34.5 Quadratic Forms and Principal Axes

Definitions. If an algebraic expression has the form of a constant multiplied by positive-integer powers of one or more of the variables $x_1,...,x_n$, then the expression is called a *monomial in* $x_1,...,x_n$. The *degree of a monomial* is the sum of the exponents occurring in it. If an algebraic expression is the finite sum of one or more monomials in $x_1,...,x_n$, where each monomial is of degree 2, this expression is called a *quadratic form in* $x_1,...,x_n$.

Examples. $-2x^2y$ and $3xyz^3$ are monomials in x,y,z. ($-2x^2y$ is also a monomial in x and y.) $-2x^2y$ has degree 3, and $3xyz^3$ has degree 5. The polynomial $x^2-2y^2+z^2-3xz+xy$ is a quadratic form. The polynomial $x^2-2yz+4z$ is not a quadratic form, since the monomial $4z$ has degree 1.

(5.1) **Theorem.** (a) *Let* A *be a* 3×3 *matrix, and let* $X = \begin{bmatrix} x \\ y \\ z \end{bmatrix}$ *be a variable column vector. Then the product* X^TAX *is a quadratic form in* x,y,z.

(b) *Let* q *be a quadratic form in* x,y,z. *Then there is a* 3×3 *matrix* A *such that* (i) A *is symmetric, and* (ii) X^TAX *is the given form* q.

Proof. (a) Let $A = \begin{bmatrix} a & d & e \\ g & b & f \\ h & i & c \end{bmatrix}$. Then X^TAX must be $ax^2 + by^2 + cz^2 +$

$(d+g)xy + (e+h)xz + (f+i)yx$. This is a quadratic form in x,y,z.

(b) Let $q = ax^2 + by^2 + cz^2 + dxy + exy + fyz$ be a given quadratic form.

Take A to be the symmetric matrix $\begin{bmatrix} a & \frac{d}{2} & \frac{e}{2} \\ \frac{d}{2} & b & \frac{f}{2} \\ \frac{e}{2} & \frac{f}{2} & c \end{bmatrix}$. Then X^TAX is the given form q.

This completes the proof.

In the present section, we develop interrelated theorems concerning (i) 3×3 symmetric matrices; (ii) quadratic forms in x,y,z; and (iii) scalar fields defined on E^3-with-axes by quadratic forms in x,y,z. The reader will see that our theorems and proofs (including the preceding theorem and proof) can be generalized directly to n×n symmetric matrices; to quadratic forms in $x_1,...,x_n$; and to scalar fields on E^n-with-axes defined by quadratic forms in $x_1,...,x_n$.

Consider E^3 with a given system of Cartesian coordinates x,y,z. Let a new system of coordinates x',y',z' be given by new Cartesian axes at the same origin. Then, as we saw in §32.9, there is a certain 3×3 orthogonal matrix Q with the property that for every point P in E^3, if $X = \begin{bmatrix} x \\ y \\ z \end{bmatrix}$ and $X' = \begin{bmatrix} x' \\ y' \\ z' \end{bmatrix}$ give, respectively, the old and new coordinates for P, then $X = QX'$. Let $X = \begin{bmatrix} x \\ y \\ z \end{bmatrix}$ be a

variable vector, and let $q = q(x,y,z) = X^T A X$ be a quadratic form, where A is a symmetric matrix. We can, if we wish, treat this quadratic form q as a function of the three variables x,y,z, and we can use q, in the old coordinate system, to define a scalar field f on E^3. If we substitute QX' for X (and hence $X'^T Q^T$ for X^T), we see that the same scalar field is given in the new coordinate system by the new quadratic form:

$$q' = q'(x',y',z') = (X'^T Q^T)A(QX') = X'^T(Q^T A Q)X' \, .$$

Here $Q^T A Q$ is also a symmetric matrix, since $A = A^T \Rightarrow (Q^T A Q)^T = Q^T A^T Q^{TT} = Q^T A Q$. Thus,

(5.2)   *if a scalar field f is defined by a quadratic form in some given Cartesian system, then, for any other Cartesian system at the same origin,* f *must also be defined by some quadratic form.*

      We now ask the following question: If a scalar field on E^3 is defined for given Cartesian axes by a certain quadratic form, to what extent and in what way can we find new Cartesian axes (at the same origin) for which the same scalar field is given by a *simpler* quadratic form? We answer this question with the *diagonalization theorem* stated and proved below.

Definitions. The *diagonal elements* of a square matrix are the elements that occur on the main diagonal. A square matrix is a *diagonal matrix* if all elements not on the main diagonal are 0. (The diagonal elements themselves may or may not be 0.)

(5.3)   ***Diagonalization theorem.*** *Given a 3×3 symmetric matrix* A, *we can find a 3×3 orthogonal matrix* Q *such that* $D = Q^T A Q$ *is a diagonal matrix. Moreover, the diagonal elements $\lambda_1, \lambda_2, \lambda_3$ of* D *will be the eigenvalues of* A *(with possible repetitions), and the columns of* Q *will be eigenvectors of A with $\lambda_1, \lambda_2, \lambda_3$ as their respective eigenvalues.* (Since Q is an orthogonal matrix, these columns must be an orthonormal basis for $\mathbf{R}^3$.)

(5.4)   ***Corollary.*** *If* $q = X^T A X$ *is a quadratic form in x,y,z, where x,y,z are Cartesian coordinates for* E^3, *and* A *is symmetric, then we can find, from q, a new Cartesian coordinate system x',y',z' and a new quadratic form q' in x',y',z',* *such that* q' *gives the same scalar field as* q, *and such that* q' *has the special form*

$$q' = \lambda_1(x')^2 + \lambda_2(y')^2 + \lambda_3(z')^2,$$

where $\lambda_1, \lambda_2, \lambda_3$ *are eigenvalues of* A. (The axes for the new coordinate system are known as *principal axes* for the given quadratic form q.)

We first prove the corollary from the theorem. We then prove the theorem.

Proof of corollary. Take Q as in the theorem. Let x', y', z' be a new coordinate system given by $X = QX'$. Then $q' = X'^T(Q^TAQ)X'$ gives the same scalar field as q, and the quadratic form q' must be $\lambda_1(x')^2 + \lambda_2(y')^2 + \lambda_3(z')^2$.

Proof of the theorem. Eigenvalues and eigenvectors are defined in §32.8. In the problems, we shall show that all the roots of the characteristic polynomial of a symmetric matrix must be real numbers (see §32.8). Thus if c_A is the characteristic polynomial of a $3{\times}3$ symmetric matrix, then c_A can be expressed in the form $c_A(\lambda) = -(\lambda - \lambda_1)(\lambda - \lambda_2)(\lambda - \lambda_3)$ where $\lambda_1, \lambda_2, \lambda_3$ are *real numbers* and are also the *eigenvalues* of A. (If A is not symmetric, then the roots of c_A may include non-real complex numbers.) The roots $\lambda_1, \lambda_2, \lambda_3$ need not be distinct. The number of times that a particular eigenvalue λ occurs in the list $\lambda_1, \lambda_2, \lambda_3$ is called the *multiplicity* of λ. The corresponding fact for $2{\times}2$ symmetric matrices is also true: if B is a $2{\times}2$ symmetric matrix, then the characteristic polynomial c_B can be expressed in the form $c_B(\lambda) = (\lambda - \lambda_1)(\lambda - \lambda_2)$, where λ_1 and λ_2 are *real numbers*.

We proceed to prove the theorem. Let A be the given symmetric matrix. Let $c_A(\lambda) = -(\lambda - \lambda_1)(\lambda - \lambda_2)(\lambda - \lambda_3)$. Find a unit eigenvector $\hat{U}_1$ for the eigenvalue λ_1. Let $\hat{U}_2$ and $\hat{U}_3$ be vectors which extend $\{\hat{U}_1\}$ to an orthonormal basis $\{\hat{U}_1, \hat{U}_2, \hat{U}_3\}$ of $\mathbf{R}^3$ (by the extension-of-basis corollary in §33.9). $\hat{U}_2$ and $\hat{U}_3$ need not be eigenvectors. Define

$$Q_1 = \left[\hat{U}_1 : \hat{U}_2 : \hat{U}_3\right].$$

Q_1 is an orthogonal matrix, since its columns are an orthonormal basis for $\mathbf{R}^3$. Let $A' = Q_1^T A Q_1$. Using multiplication diagrams for this matrix product, we find that A' is a 3×3 matrix and that the elements of A' are given by $a'_{ij} = \hat{U}_i^T A \hat{U}_j$.

Since $\hat{U}_1$ is an eigenvector with eigenvalue λ_1, we have

$$a'_{11} = \hat{U}_1^T A \hat{U}_1 = \hat{U}_1^T \lambda_1 \hat{U}_1 = \lambda_1 \hat{U}_1^T \hat{U}_1 = \lambda_1,$$

$$a'_{21} = \hat{U}_2^T A \hat{U}_1 = \lambda_2 \hat{U}_2^T \hat{U}_1 = 0,$$

and $\qquad a'_{31} = \hat{U}_3^T A \hat{U}_1 = \lambda_1 \hat{U}_3^T \hat{U}_1 = 0.$

We noted earlier that [A symmetric and Q orthogonal] $\Rightarrow Q^T A Q$ symmetric. Hence A' is symmetric and we have

$$A' = \begin{bmatrix} \lambda_1 & 0 & 0 \\ 0 & \hat{U}_2^T A \hat{U}_2 & \hat{U}_2^T A \hat{U}_3 \\ 0 & \hat{U}_3^T A \hat{U}_2 & \hat{U}_3^T A \hat{U}_3 \end{bmatrix},$$

where the 2×2 matrix

$$B = \begin{bmatrix} \hat{U}_2^T A \hat{U}_2 & \hat{U}_2^T A \hat{U}_3 \\ \hat{U}_3^T A \hat{U}_2 & \hat{U}_3^T A \hat{U}_3 \end{bmatrix}$$

must be symmetric.

By a Laplace expansion, $c_{A'}(\lambda) = |A' - \lambda I| = (\lambda_1 - \lambda)|B - \lambda I| = -(\lambda - \lambda_1)(\lambda - \lambda'_1)(\lambda - \lambda'_2)$, where λ'_1 and λ'_2 are the roots of c_B. We also observe that for every value of λ, $c_{A'}(\lambda) = |A' - \lambda I| = |Q_1^T A Q_1 - \lambda Q_1^T Q_1| = |Q_1^T (A - \lambda I) Q_1| = |Q_1^T| |A - \lambda I| |Q_1| = |Q_1^T Q_1| |A - \lambda I| = |I| |A - \lambda I| = |A - \lambda I| = c_A(\lambda)$. Hence the roots λ'_1 and λ'_2 of c_B must be the same as the roots λ_2 and λ_3 of c_A. Thus B has λ_2 and λ_3 as eigenvalues.

Let $\hat{W}_2$ (in $\mathbf{R}^2$) be a unit eigenvector of B for λ_2. Let $\hat{W}_3$ be a vector which extends $\{\hat{W}_2\}$ to an orthonormal basis $\{\hat{W}_2, \hat{W}_3\}$ of $\mathbf{R}^2$. We define

$$Q_2 = \begin{bmatrix} 1 & 0 & 0 \\ 0 & & \\ 0 & [\hat{W}_2 \vdots \hat{W}_3] \end{bmatrix}.$$

The columns of Q_2 are unit vectors in R^3. We denote these vectors as $\hat{V}_1, \hat{V}_2, \hat{V}_3$. These vectors evidently form an orthonormal set. Hence $Q_2 = [\hat{V}_1 : \hat{V}_2 : \hat{V}_3]$ is an orthogonal matrix. Let

$$D = Q_2^T A' Q_2.$$

Then D must be symmetric, and the elements of D are given by $d_{ij} = \hat{V}_i^T A' \hat{V}_j$. Using multiplication diagrams, the symmetry of D, and the fact that $\hat{W}_2$ is an eigenvector of B with eigenvalue λ_2, we find that

$$D = \begin{bmatrix} \lambda_1 & 0 & 0 \\ 0 & \lambda_2 & 0 \\ 0 & 0 & \mu \end{bmatrix}, \text{ where } \mu = \hat{V}_3^T A' \hat{V}_3.$$

Thus D is a diagonal matrix. By the same argument as for A', we find that $c_D(\lambda) = c_A(\lambda)$ for all λ. Since $c_D(\lambda) = -(\lambda - \lambda_1)(\lambda - \lambda_2)(\lambda - \mu)$, we have $\mu = \lambda_3$.

Let $Q = Q_1 Q_2$. Because Q is the product of orthogonal matrices, Q must itself be orthogonal (see §32.9). Hence $D = Q_2^T A' Q_2 = Q_2^T Q_1^T A Q_1 Q_2 = Q^T A Q$. We have therefore found an orthogonal matrix Q such that $Q^T A Q$ is diagonal and has the eigenvalues of A as its diagonal elements. We note that an eigenvalue of A of multiplicity k must occur k times on the diagonal of D.

Finally, let C_1, C_2, C_3 be the columns of Q. Then we can express

$$Q^T A Q = \begin{bmatrix} C_1^T A C_1 & C_1^T A C_2 & C_1^T A C_3 \\ C_2^T A C_1 & \cdots & \cdots \\ C_3^T A C_1 & \cdots & \cdots \end{bmatrix}$$

Here the first column gives the components of AC_1 in the orthonormal basis $\{C_1, C_2, C_3\}$. Since this first column is, in fact, $\begin{bmatrix} \lambda_1 \\ 0 \\ 0 \end{bmatrix}$, we see that $AC_1 = \lambda_1 C_1$.

Similarly, for C_2 and C_3, we get $AC_2 = \lambda_2 C_2$ and $AC_3 = \lambda_3 C_3$. Thus the columns of Q are eigenvectors of A with $\lambda_1, \lambda_2, \lambda_3$ as eigenvalues.

This completes the proof of the diagonalization theorem. We complete our study of eigenvalues and eigenvectors for 3×3 symmetric matrices with the following further corollary.

(5.5)   **Corollary.**   *Let* A *be a* 3×3 *symmetric matrix.*   *Let* $\lambda_1, \lambda_2, \lambda_3$ *be the roots of* c_A. *Then:*

(i)   *There is an orthonormal basis (for* $\mathbf{R}^3$*) whose members are eigenvectors for* $\lambda_1, \lambda_2, \lambda_3$.

(ii)   *If* X_1 *and* X_2 *are any eigenvectors (of* A*) whose eigenvalues are unequal, then* X_1 *and* X_2 *are orthogonal (that is to say,* $X_2^T X_1 = 0$*).*

(iii)   *If* X_1 *and* X_2 *are eigenvectors (of* A*) with the same eigenvalue, then any non-zero linear combination of* X_1 *and* X_2 *is also an eigenvector with that eigenvalue.*

(iv)   *If an eigenvalue* λ *(of* A*) has multiplicity* m, *then there is a vector subspace of* $\mathbf{R}^3$ *of dimension* m *whose non-zero vectors constitute the set of all eigenvectors (of* A*) with eigenvalue* λ.

   Proof.   *For* (i).   The columns C_1, C_2, C_3 of Q serve as the desired basis.

   For (ii).   Let the eigenvalues of X_1 and X_2 be λ_1 and λ_2 with $\lambda_1 \neq \lambda_2$.
Then

$$AX_1 = \lambda_1 X_1 \; ,$$

and

$$AX_2 = \lambda_2 X_2 \; .$$

Hence

$$X_2^T AX_1 = X_2^T \lambda_1 X_1 = \lambda_1 X_2^T X_1,$$

and

$$X_1^T AX_2 = X_1^T \lambda_2 X_2 = \lambda_2 X_1^T X_2 \; .$$

Taking transposes in the last equation, we have

$$(X_1^T AX_2)^T = X_2^T AX_1 = \lambda_2 X_2^T X_1.$$

Subtracting this last equation from the third equation, we have

$$0 = (\lambda_1 - \lambda_2)X_2^T X_1 \; .$$

Since $\lambda_1 - \lambda_2 \neq 0$, we must have $X_2^T X_1 = 0$.

   For (iii).   Let λ be the common eigenvalue.   Then

$$A(t_1 X_1 + t_2 X_2) = t_1(AX_1) + t_2(AX_2) = t_1\lambda X_1 + t_2\lambda X_2 = \lambda(t_1 X_1 + t_2 X_2) \; .$$

   For (iv).   By (i), we have an orthonormal basis of eigenvectors (the columns of Q).   By the diagonalization theorem,   m   of these vectors must have eigenvalue λ. Let $\mathcal{M}$ be the set of basis vectors with eigenvalue λ. Then the span of $\mathcal{M}$ is an m-dimensional vector space $\mathcal{V}$. By (iii), every member of $\mathcal{V}$ is an eigenvector for λ. Conversely, let X be any eigenvector for λ. By (ii), the components of X along the basis vectors not in $\mathcal{M}$ must all be 0. Hence X must be in the span of $\mathcal{M}$, which is $\mathcal{V}$. This proves (iv). This completes the proof of the corollary.

The foregoing theorems and corollaries remain true if we replace $\mathbf{E}^3$ and $\mathbf{R}^3$ by $\mathbf{E}^2$ and $\mathbf{R}^2$, 3×3 matrices by 2×2 matrices, and quadratic forms in x,y,z by quadratic forms in x,y. The proofs are similar (and shorter).

We now define two further properties that a quadratic form may or may not possess. These properties arise in applications of matrix algebra to mathematics, engineering, and other sciences.

Definitions. We say that a quadratic form in x,y,z is *positive-definite* if it defines a scalar field (on $\mathbf{R}^3$) that is positive at all points other than the origin. (At the origin, a quadratic form necessarily has the value 0.) We say that a 3×3 symmetric matrix A is *positive-definite* if X^TAX is a positive definite quadratic form in x,y,z.

We say that a quadratic form in x,y,z is *negative definite* if it defines a scalar field (on $\mathbf{R}^3$) that is negative at all points other than the origin. We say that a 3×3 symmetric matrix A is *negative definite* if X^TAX is a negative definite quadratic form in x,y,z.

Similar definitions can be given for quadratic forms in x and y, scalar fields in $\mathbf{R}^2$, and 2×2 symmetric matrices.

Examples.
(1) $q(x,y,z) = x^2 + 2y^2 + 3z^2$ is evidently positive-definite.
(2) $q(x,y,z) = -x^2 - 2y^2 - 3z^2$ is evidently negative-definite.
(3) $q(x,y,z) = x^2 + y^2 - z^2$ is neither positive-definite nor negative-definite, since $q(0,0,1) = -1$ and $q(1,0,0) = 1$.
(4) $q(x,y,z) = x^2 + y^2 + z^2 + 2xy + xz$ is neither positive-definite nor negative-definite, since $q(1,-1,-\frac{1}{2}) = -\frac{1}{4}$ and $q(1,1,1) = 6$.
(5) $q(x,y,z) = x^2 + y^2 + z^2 + xy + xz$ is, however, positive-definite, as we shall see below.

The first three examples show us that we can test a quadratic form for positive-definiteness by expressing it as a sum of squares (by diagonalization) and seeing if the coefficients in this sum are all positive. We thus have the theorem:

(5.6) **Theorem.** *Let* A *be a* 2×2 *or* 3×3 *symmetric matrix. Then:* A *is positive-definite* ⇔ *all the eigenvalues of* A *are positive; and,* A *is negative-definite* ⇔ *all the eigenvalues of* A *are negative.*

In order to make direct use of this theorem, we have to obtain the eigenvalues of A by finding the roots of the characteristic polynomial c_A. (In the fourth example above, we get the eigenvalues 1, $1 + \frac{\sqrt{5}}{2}$, and $1 - \frac{\sqrt{5}}{2}$. These values show that the form is neither positive nor negative definite. In the fifth example, we get the eigenvalues 1, $1 + \frac{\sqrt{2}}{2}$, and $1 - \frac{\sqrt{2}}{2}$. These values show that the form is positive definite.)

We now describe a method which enables us to test the positive-definiteness of A without requiring that we solve for the roots of c_A.

Definition. For any square matrix A, the submatrix D_i, obtained by deleting from A all but the first i rows and first i columns, is called the i-th *order discriminant matrix* of A. The determinant $|D_i|$ is then called the i-th *order discriminant* of A. Note that for a 3×3 matrix A, the first-order discriminant is simply the 1×1 determinant $|D_1| = a_{11}$, and the third-order discriminant is simply the 3×3 determinant $|D_3| = |A|$. We prove the following theorem:

(5.7) **Discriminant theorem.** *Let* A *be a* 3×3 *symmetric matrix. Then* A *is positive-definite* ⇔ *each of* $|D_1|$, $|D_2|$, *and* $|D_3|$ *is positive.*

In order to prove this theorem, we first prove a lemma.

(5.8) **Lemma.** *Let* A *be a* 2×2 *or* 3×3 *matrix. Then the product of the roots of* c_A *is* $|A|$.

Proof of the lemma. We prove the lemma for a 3×3 matrix A. The proof for a 2×2 matrix is similar. We have the identity $c_A(\lambda) = |A - \lambda I| = -(\lambda - \lambda_1)(\lambda - \lambda_2)(\lambda - \lambda_3)$ where $\lambda_1, \lambda_2, \lambda_3$ are the roots of c_A. Setting $\lambda = 0$ in this identity, we get $c_A(0) = |A| = -(-\lambda_1)(-\lambda_2)(-\lambda_3) = \lambda_1 \lambda_2 \lambda_3$. This proves the lemma.

Proof of the discriminant theorem.

$\Rightarrow$: Assume that A is positive-definite. Then $X \neq 0 \Rightarrow X^T A X > 0$. Let

$$X = \begin{bmatrix} 1 \\ 0 \\ 0 \end{bmatrix}.$$ Then $X^T A X = a_{11} = |D_1|$, and we see that $|D_1| > 0$.

Let $\begin{bmatrix} x \\ y \end{bmatrix}$ be any non-zero vector in $\mathbf{R}^2$. Let $X = \begin{bmatrix} x \\ y \\ 0 \end{bmatrix}$. Since A is positive-definite, $X^T A X > 0$. But $X^T A X = \begin{bmatrix} x \\ y \end{bmatrix}^T D_2 \begin{bmatrix} x \\ y \end{bmatrix}$, as the reader may verify from a multiplication diagram. Hence D_2 must be a positive-definite matrix. By the previous theorem, D_2 has positive eigenvalues. Hence, by the lemma above, $|D_2| > 0$.

Finally, again by the previous theorem, the eigenvalues of A must be positive, and hence, by the lemma above, $|A| = |D_3| > 0$. We thus have $|D_1|, |D_2|$, and $|D_3|$ positive. This proves $\Rightarrow$.

$\Leftarrow$: We prove this by contradiction. Assume that $|D_1|, |D_2|$, and $|D_3|$ are all positive, but that A is not positive-definite. By the lemma, since $|D_3| = |A| > 0$, A cannot have zero as an eigenvalue. Since A is not positive definite, c_A must have two negative roots, λ_1 and λ_2, and one positive root, λ_3. By (ii) and (iv) in the second corollary to the diagonalization theorem, there must be two mutually orthogonal unit eigenvectors $\hat{U}_1$ and $\hat{U}_2$ for λ_1 and λ_2 (whether or not $\lambda_1 = \lambda_2$). Let u' and u'' be the third components, respectively, of $\hat{U}_1$ and $\hat{U}_2$. Define constants c_1 and c_2 as follows: If u' and u'' are not both zero, let $c_1 = u''$ and $c_2 = -u'$; if $u' = u'' = 0$, let $c_1 = 1$ and $c_2 = 0$. Let X be the vector $c_1 \hat{U}_1 + c_2 \hat{U}_2$. Then X must have the form $X = \begin{bmatrix} x \\ y \\ 0 \end{bmatrix}$. Moreover, $X \neq 0$, because $\hat{U}_1$ and $\hat{U}_2$, as non-zero orthogonal vectors, must be linearly independent. Therefore

$$X^T A X = (c_1 \hat{U}_1^T + c_2 \hat{U}_2^T) A (c_1 \hat{U}_1 + c_2 \hat{U}_2)$$
$$= (c_1 \hat{U}_1^T + c_2 \hat{U}_2^T)(c_1 \lambda_1 \hat{U}_1 + c_2 \lambda_2 \hat{U}_2) = c_1^2 \lambda_1 + c_2^2 \lambda_2.$$

Hence we have $X^T A X < 0$, since λ_1 and λ_2 are negative. It follows (as in the proof of $\Rightarrow$) that $X^T A X = \begin{bmatrix} x \\ y \end{bmatrix}^T D_2 \begin{bmatrix} x \\ y \end{bmatrix} < 0$. Then, by the previous theorem, c_{D_2} has at least one negative root. Since $|D_2| > 0$, c_{D_2} must have both roots negative (by the lemma). Hence D_2 is negative-definite. Hence, for any $x \neq 0$, we must have $\begin{bmatrix} x \\ 0 \end{bmatrix}^T D_2 \begin{bmatrix} x \\ 0 \end{bmatrix} < 0$. But $\begin{bmatrix} x \\ 0 \end{bmatrix}^T D_2 \begin{bmatrix} x \\ 0 \end{bmatrix} = x D_1 x = a_{11} x^2$. Thus $a_{11} = |D_1| < 0$, contrary to our assumption that $|D_1| > 0$. This proves $\Leftarrow$ and completes the proof of the theorem.

Examples. This discriminant theorem gives us correct results for the fourth and fifth examples above, without requiring us to solve for the eigenvalues. In the fourth example, we get $|D_1| = 1$, $|D_2| = 0$, $|D_3| = -\frac{1}{4}$, and we conclude that q is not positive-definite. In the fifth example, we get $|D_1| = 1$, $|D_2| = \frac{3}{4}$, $|D_3| = \frac{1}{2}$, and we conclude that q must be positive-definite.

Definition. Let A be a 3×3 symmetric matrix. D is a *principal submatrix* of A, if D can be obtained from A by deleting both the i-th row and the i-th column of A, for zero or more values of i. (Thus a principal submatrix of A is a square submatrix whose main diagonal is contained in the main diagonal of A.) If D is a principal submatrix of A, then $|D|$ is called a *principal minor* of A.

The following corollary is immediate.

(5.9)   *Corollary. If A is positive-definite, then all principal minors of A are positive.*

Proof. Let D be a principal submatrix of A. Let X be any vector in $\mathbf{R}^3$. Simultaneously reorder the rows and columns of A to form a matrix $\tilde{A}$ for which D is now a discriminant matrix. Let $\tilde{X}$ be obtained by the same reordering of the components of X. Then $X^T A X = \tilde{X}^T \tilde{A} \tilde{X}$. Hence $\tilde{A}$ must be positive-definite. Hence $|D| > 0$. This proves the corollary. (The converse of the corollary is also true and is an immediate consequence of the discriminant theorem.)

This corollary gives us a powerful negative test for positive-definiteness. If we can find a single principal minor which is not positive, then we can conclude that A is not positive-definite.

From the discriminant theorem, we get analogous discriminant and principal-minor tests for negative-definiteness. We state these as a corollary.

(5.10) *Corollary. Let A be a 3×3 symmetric matrix. Then:*

(i)   A *is negative-definite* ⇔ [*each discriminant of odd order* ($|D_1|$ *and* $|D_3|$) *is negative, and each discriminant of even order* ($|D_2|$) *is positive.*)

(ii)   *If* A *is negative-definite, then every principal minor of odd order is negative, and every principal minor of even order is positive.*

Proof. Observe that A is negative-definite ⇔ −A is positive-definite. The desired results are immediate when we apply our previous results on positive-definiteness to the matrix −A, observing that for any n×n matrix D: n even ⇒ $|-D| = |D|$; and n odd ⇒ $|-D| = -|D|$. This proves the corollary.

In the problems, we use the diagonalization and discriminant theorems to prove the derivative test for unconstrained relative extreme points for a quadratic function of three variables (see §12.5) and to obtain derivative tests for constrained maximum-minimum problems for quadratic functions with quadratic constraints. (By the use of infinite series, these results can be extended to C^2 functions and C^2 constraints.) In the spirit of (5.8), we also prove

(5.11) *Theorem.   For any 2×2 or 3×3 matrix A, the sum of the main diagonal elements of A* (*called the trace of A*) *must equal the sum of the eigenvalues of A.*

We conclude the present section with a geometric application of diagonalization.

Definitions. A curve in $\mathbf{R}^2$ is called a *central conic* if it is given by an equation of the form $ax^2 + by^2 + cxy = 1$. A surface in $\mathbf{R}^3$ is called a *central quadric* if it is given by an equation of the form $ax^2 + by^2 + cz^2 + dxy + exz + fyz = 1$. The diagonalization and discriminant theorems enable us to get geometric information from an equation for a central conic or a central quadric, as we now show.

In the case of a central conic curve, let A be the 2×2 symmetric matrix of the given quadratic form. We know from the diagonalization theorem that we can find principal axes (in new coordinates x',y') that give, for the same curve, an equation of the form

$$\lambda_1(x')^2 + \lambda_2(y')^2 = 1,$$

where λ_1 and λ_2 are the roots of c_A. If both roots are positive, we have an *ellipse*. If one is positive and one is negative, we have a *hyperbola*. If one is positive and the other is zero, we have a pair of *parallel lines*. These are the only possible curves. Moreover, we can use the discriminants of A to decide whether one of the first two cases holds, and if so, which one. The reader may verify that $[|D_1| > 0$ and $|D_2| > 0] \Leftrightarrow$ ellipse, and $|D_2| < 0 \Leftrightarrow$ hyperbola.

Similarly, in the case of a central quadric surface, let A be the 3×3 symmetric matrix of the given quadratic form. Principal axes must give, for the same surface, an equation of the form

$$\lambda_1(x')^2 + \lambda_2(y')^2 + \lambda_3(z')^2 = 1,$$

where $\lambda_1, \lambda_2, \lambda_3$ are the roots of c_A. If all three roots are positive, we have an *ellipsoid*. If two are positive and one is negative, we have a *hyperboloid of one sheet*. If one is positive and two are negative, we have a *hyperboloid of two sheets*. If one or more coefficients are zero, we may have a *cylindrical surface* whose cross section takes the form of a central conic. These are the only possible surfaces. Moreover, we can use the discriminants of A to decide whether one of the first three cases holds, and if so, which one. The reader may verify that $[|D_1| > 0$ and $|D_2| > 0$ and $|D_3| > 0] \Leftrightarrow$ ellipsoid, $[|D_3| < 0$, and either $|D_2| < 0$ or both $|D_1|$ and $|D_2| > 0] \Leftrightarrow$ hyperboloid of one sheet; and $[|D_3| > 0$, and either $|D_2| < 0$ or else $|D_2| > 0$ and $|D_1| < 0] \Leftrightarrow$ hyperboloid of two sheets. These results can be used to complete the proof of (6.1) in §4.6.

Remark. The theorems and corollaries of the present section hold in analogous form for n×n symmetric matrices and for quadratic forms in $x_1, \ldots, x_n$. Our proofs for n = 3 can be generalized to all values of n by straightforward inductive constructions. The statements above and proofs for n = 3 have been arranged to suggest and permit such generalization.

§34.6 Problems

§34.1

1. Verify that the general definitions for *parallelism* and *perpendicularity* agree with earlier definitions as given in §1.3 and §1.4 for planes and lines in $\mathbf{E}^3$.

2. In the proof of (1.7) verify that the intersection of $\mathscr{A}$ with $\mathscr{A}'$ must be a single point.

§34.2

1. (a) In $\mathbf{R}^4 = \mathbf{E}^4$, give a minimal necessary and sufficient condition for given points P_1, P_2, P_3, and P_4 to be the vertices of a parallelogram in some 2-plane.
    (b) Show that this 2-plane is uniquely determined by these points.

2. (a) Do the same as in the problem 1a for given points P_1, P_2, P_3, and P_4 to have the property that P_1 is a vertex of a parallelepiped in some 3-plane, and P_2, P_3, P_4 are the vertices of that parallelepiped adjacent to P_1.
    (b) Show that this 3-plane is uniquely determined by these points.

3. For $\mathbf{E}^4$, state and prove a generalized version of (7.2) in §2.7.

4. In $\mathbf{E}^4$, what possible intersections can occur between a 2-plane and a 3-plane? (*Hint.* See (6.6) in §33.6.)

5. In $\mathbf{E}^4$, prove that two 2-planes are parallel if and only if there is another 2-plane which is perpendicular to them both.

§34.3

1. Use the alternative definition of *vector subspace* to prove the theorem in the last paragraph of §34.3.

§34.4

1. Prove (4.6).

2. Prove (4.7). (*Hint.* See the comment following the statement of (4.7).)

3. For a matrix product BA, prove that

$$r(BA) \le \min\{r(B), r(A)\}.$$

§34.5

1. Prove that all roots of the characteristic polynomial of a real, symmetric matrix A must be real. (*Hint.* If λ is a root (possibly complex) of the characteristic polynomial of A, then, by the argument in §32.8, we can find a non-zero eigenvector X (with possibly complex entries) such that $AX = \lambda X$. Replace λ by its complex conjugate, each element in A by its complex conjugate (since A is real, this leaves A unchanged), and each element in X by its complex conjugate. This carries λ to its conjugate $\overline{\lambda}$ and X to the vector $\overline{X}$ of its conjugates. Since arithmetical identities always remain true when we take complex conjugates, we have $A\overline{X} = \overline{\lambda}\,\overline{X}$. We now take transposes of this equation. By the symmetry of A, we get $\overline{X}^T A = \overline{\lambda}\,\overline{X}^T$. Postmultiply both sides by X, and, recalling that the product of a complex number with its conjugate must be real, conclude that $\overline{\lambda} = \lambda$ and hence that λ is real.)

2. Prove (5.11).

3. Using (5.9) and (5.10),
    (a) Prove the unconstrained three-variable derivative test ((4.7) in §12.4) for the case of a quadratic (at most second-degree polynomial) function of three variables.
    (b) Prove the constrained two-variable derivative test ((4.5) in §13.4) for the case of a quadratic function with a quadratic constraint.
    (c) Formulate and prove appropriate three-variable derivative tests for the case of a quadratic function with one quadratic constraint and for the case of a quadratic function with two quadratic constraints. (See §13.4.)

Answers to Selected Odd-Numbered Problems

§1.6   **11.** d, $d\cos\theta$, $\dfrac{d^2}{4}\pi\cos\theta$; πab

§2.4   **3a.** $\vec{G} = \vec{A} - \vec{B} - \vec{E}$

       5. $\hat{A}_1 + \hat{A}_2$

      11.   sphere of diameter $\left|\vec{A} - \vec{B}\right|$ through points $\vec{R} = \vec{A}$ and $\vec{R} = \vec{B}$

      13.   always true

§3.3   **1.** 7; 5; $4\hat{i} - \hat{j} + 3\hat{k}$; $5\hat{i} + 2\hat{j} - \hat{k}$

       3. $\sqrt{3}$; $\dfrac{\sqrt{3}}{3}(\hat{i} + \hat{j} + \hat{k})$; 2; $\hat{i} - \hat{j}$; $35.3°$

       5. 5; 25

       7. $6\vec{B}$; 2; $-\dfrac{2}{3}$

§3.4   **1b.** $-\dfrac{1}{2}\sqrt{6}$; $-\dfrac{1}{2}\sqrt{2}$; $2\sqrt{3}$

      3b. $a = -1$, $b = -1$, $c = -5$

       5. $\left\{\dfrac{\sqrt{2}}{2}(\hat{i} + \hat{j}), \dfrac{\sqrt{6}}{6}(-\hat{i} + \hat{j} + 2\hat{k}), \dfrac{\sqrt{3}}{3}(\hat{i} - \hat{j} + \hat{k})\right\}$

§3.5   **1a.** $\dfrac{3}{2}\sqrt{2}$

       3. $31.0°$, $64.6°$, $73.4°$

       5. $73.2°$

       7. $109.5°$

§3.6   **1.** $-\hat{k}$; -3

       3. 1, $-\hat{k}$, $-\hat{i} + \hat{j}$

       5. 5

       7. $[\vec{A}, \vec{C}, \vec{B}](\vec{A}^2 + \vec{A} \cdot \vec{B})$; $\vec{A} \cdot \vec{B} - \vec{A}^2 - 2[\vec{A}, \vec{B}, \vec{C}]$

      11. $\vec{X} = \dfrac{1}{\vec{A} \cdot \vec{B}}(k\vec{B} + (\vec{A} \times \vec{C}))$

      13. $\vec{X} = \dfrac{1}{D}[((\vec{A} \cdot \vec{B})^2 + (\vec{A} \cdot \vec{B})\vec{B}^2)\vec{A} - (\vec{A}^2\vec{B}^2 + (\vec{A} \cdot \vec{B})\vec{A}^2)\vec{B}]$,

           $\vec{Y} = \vec{A} + \vec{B} - \vec{X}$, where $D = \vec{A}^2\vec{B}^2 - (\vec{A} \cdot \vec{B})^2$

§3.7   **3.** $x = 2$, $y = 1$

§4.2   **1.** $2x - y - z = 1$

§4.3   **1.** $3x - 5y + z = 0$
              3. $x + 2y + 3z = 3$
              5. $x + 3y + z = 7$; $\frac{7}{11}\sqrt{11}$; $\left(-\frac{1}{11}\sqrt{11}, -\frac{3}{11}\sqrt{11}, -\frac{1}{11}\sqrt{11}\right)$

              7. $\dfrac{|Aa + Bb + Cc - D|}{(A^2 + B^2 + C^2)^{1/2}}$

              9. $2x + y + 4z = 2$

§4.4   **1.** $x = 1 + (3\sqrt{2} - 2\sqrt{3})t$, $y = (3\sqrt{2} - 2\sqrt{2})t$, $z = -2\sqrt{3}\,t$
              3. $c = -2$, $d = -1$

§4.5   **1.** $(2, 0, -4)$
              3. $(1, 4, 17)$, $(-3, 0, 9)$
              5. $(1, 2, 4)$

              9. $\vec{A} \cdot \vec{N} \neq 0$; $\vec{R} = \dfrac{d - \vec{B} \cdot \vec{N}}{\vec{A} \cdot \vec{N}} \vec{A} + \vec{B}$

              11. $\vec{R} = ((\vec{A} \times \vec{N}) \times \vec{N})t + \left(\dfrac{d - \vec{B} \cdot \vec{N}}{\vec{N} \cdot \vec{N}} \vec{N} + \vec{B}\right)$, single point when

                       $(\vec{A} \times \vec{N}) \times \vec{N} = \vec{0}$

              13.   condition: $\vec{A}_1 \times \vec{A}_2 \neq \vec{0}$ and $\vec{A}_1 \times \vec{A}_2 \cdot (\vec{B}_2 - \vec{B}_1) = 0$

                       point: $\vec{R} = \vec{A}_1 t_1 + \vec{B}_1$, where $t_1 = \dfrac{\vec{A}_2 \times \vec{A}_1 \cdot (\vec{A}_2 \times (\vec{B}_2 - \vec{B}_1))}{(\vec{A}_2 \times \vec{A}_1)^2}$

§4.6   **1.**   For $b + \vec{A}^2 > 0$, sphere with center at $\vec{R} = \vec{A}$ and radius $\sqrt{b + \vec{A}^2}$

§4.7   **5bd.** parallel lines; hyperbola

§5.3   **1bdf.** $\hat{i} + 4\hat{j} - 4\hat{k}$; $2\hat{j} - 2\hat{k}$; $\sqrt{33}$

§5.4   **3.** $(\vec{R} \times \ddot{\vec{R}}) \times \dot{\vec{R}} + (\vec{R} \times \dot{\vec{R}}) \times \ddot{\vec{R}}$

              5. $\ddot{f}\vec{B} + 2\dot{f}\dot{\vec{B}} + f\ddot{\vec{B}}$

§5.7   **1a.** $\vec{v} = -\hat{i} + (2 + \frac{3}{2}t^2)\hat{j} - (2 + 2t)\hat{k}$

                       $\vec{R} = -t\hat{i} + (2t + \frac{1}{2}t^3)\hat{j} - (2t + 2t^2)\hat{k}$

§5.8   **1b.** $-2\hat{i} + 2\hat{j} + \frac{3}{2}\pi\hat{k}$

§6.2   **1.** $\cos t\,\hat{i} + 3\sin t\,\hat{j}$

3. $(a\cos t - \frac{bt}{\sqrt{2}}(\sin t + \cos t))\hat{i} + (a\sin t - \frac{bt}{\sqrt{2}}(\cos t - \sin t))\hat{j}$

5. $((a-b)\cos\varphi + b\cos(\frac{a}{b}\varphi))\hat{i} + ((a-b)\sin\varphi - b\sin(\frac{a}{b}\varphi))\hat{j}$

§6.3   **1.** $2\cos t\hat{i} + 3\sin t\hat{j}$

§6.4   **1.** $\sqrt{2}$

       3. $\vec{v} = (1+e^{2x})^{-1/2}\hat{i} + (1+e^{2x})^{-1/2}e^{x}\hat{j}$,

           $\vec{a} = -(1+e^{2x})^{-2}e^{2x}\hat{i} + (1+e^{2x})^{-2}e^{x}\hat{j}$, $\left(\frac{dx}{dt} = (1+e^{2x})^{1/2}\right)$

       5. $\frac{8}{27}(10^{3/2} - 1) = 9.37$

       7. $8a$

§6.7   **1.** $2\hat{i} - 2\hat{j} - \hat{k}$; 3; $\frac{2}{3}\hat{i} - \frac{2}{3}\hat{j} - \frac{1}{3}\hat{k}$; $-2\hat{j} + 2\hat{k}$; $\frac{2}{27}\sqrt{17}$;

           $\frac{1}{51}\sqrt{17}(-2\hat{i} - 7\hat{j} + 10\hat{k})$

       3a. $-4t^{3}\hat{i} + (\frac{1}{2} + 3t^{2})\hat{j}$

       5.   see example 1

      11. $\sec x$

      13. $((\ddot{y}\dot{z} - \ddot{z}\dot{y})^{2} + \ddot{z}^{2} + \ddot{y}^{2})^{1/2}(1 + \dot{y}^{2} + \dot{z}^{2})^{-3/2}$

      15. $\hat{i} + f'(x)\hat{j}$; $(1 + f'(x)^{2})^{1/2}$; $(1 + f'(x)^{2})^{-1/2}(\hat{i} + f'(x)\hat{j})$; $f''(x)\hat{j}$;

           $|f''(x)|(1 + f'(x)^{2})^{-3/2}$; $\pm(1 + f'(x)^{2})^{-1/2}(-f'(x)\hat{i} + \hat{j})$

           [+ if $f'' > 0$, – if $f'' < 0$]; $(1 + f'(x)^{2})^{3/2}|f''(x)|^{-1}$

§6.8   **1.** $\frac{2}{9}$; $4,2$; $\frac{2}{9}$

§7.1   **1.** $(-\frac{3}{2}\sqrt{2}, \frac{3}{2}\sqrt{2}, -1)$

§7.4   **1.** $(-2, 2\sqrt{3}, 1)$; $-\frac{1}{2}\hat{i} + \frac{\sqrt{3}}{2}\hat{j}, -\frac{\sqrt{3}}{2}\hat{i} - \frac{1}{2}\hat{j}, \hat{k}$; $\frac{3}{2}\sqrt{3}, -\frac{3}{2}, 0$

       3. $(-\frac{6}{5}\sqrt{5}, \frac{3}{5}\sqrt{5}, -4)$

       5. $(2, 0, 1)$

§7.5   **1.** $a\hat{r} + a\hat{\theta}$, $-a\hat{r} + a\hat{\theta} - \hat{z}$

       3. $(2t\hat{r} + 2t^{2}\hat{\theta} + 3t^{2}\hat{z})(4t^{2} + 13t^{4})^{-1/2}$

       5. $2(4 + t^{2}e^{2t})^{-1/2}$, $te^{t}(4 + t^{2}e^{2t})^{-1/2}, 0$

       7. $\hat{r} + 2\hat{\theta} + \hat{z}$; $-4\hat{r} + 6\hat{\theta}$; $\frac{1}{9}\sqrt{93}$

9. $(a\hat{r} + \hat{\theta})(a^2 + 1)^{-1/2}$, $(-\hat{r} + a\hat{\theta})(a^2 + 1)^{-1/2}$; $e^{-a\theta}(a^2 + 1)^{-1/2}$

11. $(\hat{r} + \theta\hat{\theta})(1 + \theta^2)^{-1/2}$, $(-\theta\hat{r} + \hat{\theta})(1 + \theta^2)^{-1/2}$, $(2 + \theta^2)(1 + \theta^2)^{-3/2}$

§8.1

1. xy plane for M, $f = |z|$

3. set $Q_1 = (c,0)$, $Q_2 = (-c,0)$, then

$$f = ((x - c)^2 + y^2)^{1/2} - ((x + c)^2 + y^2)^{1/2}$$

5. Q_1, Q_2 as in 3. $f = (((x - c)^2 + y^2)^{1/2} - ((x + c)^2 + y^2)^{1/2})^2$

§8.2

1. $\dfrac{x^2}{a^2} + \dfrac{y^2}{b^2} = 1$, where $a^2 > c^2$ and $b = (a^2 - c^2)^{1/2}$

3. $\dfrac{x^2}{a^2} - \dfrac{y^2}{b^2} = 1$, where $a^2 < c^2$ and $b = (c^2 - a^2)^{1/2}$

§8.3

1. $1; 0 ; -1; \frac{1}{2}\sqrt{2}$

§8.5

3ac. point (0,0,0), sphere with center (0,0,0) and radius 1; two-sheeted cone on x axis, hyperboloid of two sheets on x axis

5d. all (x,y) such that $|x| = |y| > 0$

§8.6

1a. $f_x = \dfrac{-2xy}{(x^2 + y^2)^2}$, $f_y = \dfrac{x^2 - y^2}{(x^2 + y^2)^2}$

3. h

5b. $-\dfrac{1}{z}\sin\left(\dfrac{xy}{z}\right) - \dfrac{xy}{z^2}\cos\left(\dfrac{xy}{z}\right)$

7ac. x; y

9. $\dfrac{1}{u+v}\hat{i} + 2v\sin(2uv)\hat{j} - \dfrac{1}{u^2 v}\hat{k}$, $\dfrac{1}{u+v}\hat{i} + 2u\sin(2uv)\hat{j} - \dfrac{1}{uv^2}\hat{k}$,

$\dfrac{-1}{(u+v)^2}\hat{i} + 2v^2\cos(2uv)\hat{j} + \dfrac{2}{u^3 v}\hat{k}$,

$\dfrac{-1}{(u+v)^2}\hat{i} + (2\sin(2uv) + 4uv\cos(2uv))\hat{j} + \dfrac{1}{u^2 v^2}\hat{k}$

§9.2

1. 0.4

3. $6x + 8y - z = 25$

5. 6%

7. $6x - 3y + e^6 z = 1$

9. 1.3×10^{-4} mm

§9.4

1. $-\dfrac{10}{3}$

3. 0

5. $-\dfrac{4\sqrt{3}}{3}$

7. $\pm\dfrac{\sqrt{2}}{2}(\hat{i}+\hat{j})$

9. $\dfrac{1}{2}\sqrt{2}$

11. $\pm\dfrac{\sqrt{5}}{5}(2\hat{i}-\hat{j})$

13bcd. $\sqrt{5}$, 0, 0

15ac. $z,\ \dfrac{2}{3}\hat{i}-\dfrac{2}{3}\hat{j}+\hat{k}$

§9.6 **1b.** $2x-z=1$

       3. $x-5y+z=3$

       5. $12x-6y-25z=-20$

§10.2 **1.** $\dfrac{df}{du}\dfrac{\partial g}{\partial x}$

       3. $2yu+xv,\ 2yv+xu;\ 5,\ 2,\ 10;\ 14,\ 13$

       5. 2

       9. cannot find, 27, –31

       11b. $u^2+u=\dfrac{x^2+xy}{y^2},\ -\dfrac{v^2}{y^2}=-\dfrac{(x+y)^2}{y^2}$

§10.5 **1.** $T_x\dot{x}+T_y\dot{y}+T_z\dot{z}+T_t$

       3. $f_xg_vh_t+f_t,\ f_xg_u+f_xg_vh_u$ [using chain rule on $w=f(g(u,h(u,t)),t)$]

       5. 8, 2, 4; 48; 12

       7. $f_ug_y+f_y;\ f_ug_x$

       9. $uf_u^2+\dfrac{u^2}{v}f_v^2$

§10.6 **1b.** $\dfrac{\partial^2 w}{\partial x\partial y}=f_{uu}g_xg_y+f_{uv}h_xg_y+f_{vu}g_xh_y+f_{vv}h_xh_y+f_ug_{yx}+f_vh_{yx}$

       3. $f_{uu}h_x^2+(f_{uv}+f_{vu})(g_uh_x+g_x)h_x+f_{vv}(g_uh_x+g_x)^2+f_uh_{xx}+$

            $f_v(g_{uu}h_x^2+(g_{ux}+g_{xu})h_x+g_uh_{xx}+g_{xx})$

       5. $f_{uu}g_v^2+(f_{uv}+f_{vu})g_v+f_{vv}+f_ug_{vv}$

§11.1 **1ac.** $x\,dy+y\,dx+z\,dy+y\,dz=x\,dz+z\,dx\ ;\ du+v\,dx+x\,dv=$

           $e^{xy}du+xue^{xy}dy+yue^{xy}dx$

§11.2 **1a.** $f_uh_x+f_vg_uh_x+f_vg_x$

§11.4   **1.** $\dfrac{1}{2}$

       3. $\dfrac{a}{bc \sin A}$, $\dfrac{c \cos A - b}{bc \sin A}$

       5. $\dfrac{-g_u}{g_y}$; $\dfrac{-f_y g_u}{g_y}$

       7. $f_x + \dfrac{f_y}{g_y}$

       9. $-\dfrac{1}{4}, \dfrac{9}{4}$

     11. $\cos\theta$, $\dfrac{\cos\theta}{r}$

     13. $\dfrac{-f_v h_x g_y}{g_x} + f_v h_y$

     15. $\dfrac{-x}{2x+u}$, $\dfrac{1}{2x+u}$; $\dfrac{-2}{(2x+u)^3}$

     17. $\pm 1, 0, \pm\dfrac{1}{2}$

     19. $\dfrac{-x}{2x+u}$, $\dfrac{2x}{2x+u}$; $-\dfrac{1}{5}$ or $-\dfrac{4}{5}$, $\dfrac{2}{5}$ or $\dfrac{8}{5}$

     21. $-1, \dfrac{1}{4}$

     23. $\dfrac{y-x}{y^2}$, $\dfrac{x-y}{y}$

    25a. $\dfrac{2u^2 x}{3x^2 - 2u^2 v}$

     27. $\dfrac{2u^2 x^2}{3x^2 - 4u^2 vx - yu^2}$; -1, $\dfrac{1}{3}$, $\dfrac{8}{3}$

     29. $\cos\theta \dfrac{\partial w}{\partial r} - \dfrac{1}{r}\sin\theta \dfrac{\partial w}{\partial\theta}$, $\sin\theta \dfrac{\partial w}{\partial r} + \dfrac{1}{r}\cos\theta \dfrac{\partial w}{\partial r}$, $\dfrac{\partial w}{\partial z}$

     31. 4

     33. $\dfrac{1}{F_x^3}(F_{zy}F_x^2 + F_{zx}F_z F_y - F_{zy}F_y F_x - F_{xy}F_z F_x)$

§11.5   **1.** $(g_y(a,b) - f_y(a,b))\hat{i} + \ (f_x(a,b) - g_x(a,b))\hat{j} +$

              $(f_x(a,b)g_y(a,b) - f_y(a,b)g_x(a,b))\hat{k}$

      3. $\dfrac{bc \sin A}{c \cos A - b}$, $\dfrac{a}{b - c \cos A}$, $\dfrac{b \cos A - c}{b - c \cos A}$ (b may not be unique)

§11.6   **1.** $11\hat{i} - 13\hat{j} + 8\hat{k}$

    3cd. $2x - y + 2z = 9$; 7.82

      5. $(3\hat{i} + 2\hat{j}) + t(-\hat{i} + \hat{j})$, $(3\hat{i} + 2\hat{j}) + t(\hat{i} + \hat{j})$

§12.1 **1.** global: $\frac{\pi}{2}$ max, $\frac{3\pi}{2}$ min, $\frac{5\pi}{2}$ max; values: 1, –1, 1; local: 0 min,

$\frac{\pi}{2}$ max, $\frac{3\pi}{2}$ min, $\frac{5\pi}{2}$ max, 3π min; values: 0, 1, –1, 1, 0

3. global: a max, b min; values: f(a), f(b); local: a max, b min; values: f(a), f(b)

§12.2 **5.** set of points (x,y,z) where x, y, and z are rational numbers and $x^2 + y^2 + z^2 < 1$

7. i and ii closed, iii open, iv neither, i compact

§12.3 **1.** F, T, T

§12.4 **1a.** local min (–1, –2), value –18; global/local max (1,2), value 18
1c. global/local max (–8,–23), value 59
1e. saddle (–2,1)
1g. global/local max (0,0,0), value 1
1i. global/local min (0,0,0), value 0
1k. saddles: $(0,0,0)$, $(-\frac{2}{3},\frac{2}{3},0)$

1m. saddles: (1,–1), (–1,1)
1o. global/local min (0,0,0), val 0
1q. saddle (0,0,0)

§13.2 & §13.3 Max and min points are global unless qualified.

1. max $(1,0,\frac{1}{2})$, min $(-1,0,-\frac{1}{2})$

3. $\pm\frac{1}{9}\sqrt{6}$

5. max (2,–2,0), min (–2,2,0)

7a. $\frac{1}{8}$

9. max (2,1), min (–2,–1), by max-min theorem

11. $\frac{65}{16}$,1; max $\left(\frac{1}{8} \pm \frac{3\sqrt{7}}{4}\right)$, min (–1,0); local min (1,0)

13. max $(\sqrt{3},1)$, $(-\sqrt{3},-1)$, min $(\sqrt{3},-1)$, $(-\sqrt{3},1)$, non-extreme $(0,\pm 2)$

15. max $(\frac{1}{2}\sqrt{2},-\frac{1}{2}\sqrt{2},0)$, min $(-\frac{1}{2}\sqrt{2},\frac{1}{2}\sqrt{2},0)$; max $(\pm 1,0,0)$, min $(0,\pm 1,0)$, non-extreme $(0,0 \pm 1)$

17. max (–1,–1,2), (–1,2,–1), (2,–1,–1), min (1,1,–2), (1,–2,1), (–2,1,1)

19. closest $(0,\sqrt{2},\sqrt{2})$, farthest $(0,-\sqrt{2},-\sqrt{2})$

21. radius = height = $\frac{10}{\pi^{1/3}}$ cm

23. (1,1,–1); $x + zg_x = 0$, $y + zg_y = 0$, $z - g = 0$

§13.6 **1.** max $\left(\frac{\sqrt{2}}{2},\frac{1}{2}\right)$, $\left(-\frac{\sqrt{2}}{2},-\frac{1}{2}\right)$, min $\left(\frac{\sqrt{2}}{2},-\frac{1}{2}\right)$, $\left(-\frac{\sqrt{2}}{2},\frac{1}{2}\right)$

3. max $(\sqrt{3},\pm\sqrt{6})$, min $(-\sqrt{3},\pm\sqrt{6})$, local max $(-3,0)$, local min $(3,0)$

5a. min value 3 at $(1,1)$

7. local max $\left(\frac{2}{9},\frac{4}{9}\right)$, local min $(2,0)$

9. $\frac{2}{9}\sqrt{3}$

11a. max at $(1,2)$

§13.7 **1.** max $(1,0)$, min $(-1,0)$, local max $(-\frac{1}{3},-\frac{1}{3})$, local min $(0.248,-0.969)$

§14.3 **1ac.** $\iiint_R (x^2+y^2+z^2)dV$; $\iiint_R (x^2+y^2+z^2)(y^2+z^2)dV$

3. let $P = (x,y,z)$, then $d = \iiint_R \dfrac{|\overrightarrow{P_0P}|\,dV}{\iiint_R dV}$

§14.4 **1.** no; no; yes

§14.5 **1a.** 0, by cancellation

3. $\frac{3}{4}a$

§14.6 **1.** $4\pi\hat{j}$

§14.7 **1a.** $\left(\dfrac{1}{M}\iiint_R x(x^2+y^2+z^2)dV,\ \dfrac{1}{M}\iiint_R y(x^2+y^2+z^2)dV,\right.$

$\left.\dfrac{1}{M}\iiint_R z(x^2+y^2+z^2)dV\right)$, where $M = \iiint_R (x^2+y^2+z^2)dV$

3. $\vec{C} = \frac{1}{2}\hat{k}$

§15.1 **1a.** $8\ln 8 - 16 + e$

3a. $\frac{15}{72}$

§15.2 **1a.** $\int_0^1 \int_0^{x^2} dy\,dx$

3. $\frac{4}{3}$

§15.4 **1.** $\frac{1}{6}abc$

§15.5 **1.** $\int_{-1}^2 \int_{-\sqrt{2+y-y^2}}^{\sqrt{2+y-y^2}} \int_{x^2+y^2}^{y+2} dz\,dx\,dy$

§15.6   **1.** $\dfrac{63}{20}$

3. $(\frac{3}{7}, 0, \frac{2}{7})$

5a. $\dfrac{1}{10} Ma^2$

§16.1   **1ac.** $r^2 = \sec 2\theta$; $r^2 = (\cos^2\theta + 2\sin^2\theta)^{-1}$

§16.2   **1.** $M = \displaystyle\int_{-\pi/2}^{\pi/2} \int_0^{2\cos\theta} r^3\, dr d\theta$, $\bar{x}_m = \dfrac{1}{M} \iint r^4 \cos\theta\, dr d\theta$,

$\bar{y}_m = \dfrac{1}{M} \iint r^4 \sin\theta\, dr d\theta$

3. $(\frac{4}{3}, 0)$

§16.4   **1.** $\displaystyle\int_0^1 \int_0^{\sqrt{1-x^2}} \int_0^{1-\sqrt{x^2+y^2}} xz\, dz dy dx$

3. $16\displaystyle\int_0^{\pi/4} \int_0^1 \int_0^{\sqrt{4-r^2\cos^2\theta}} r\, dz dr d\theta$

5. $\left(0, 0, \dfrac{56\sqrt{2}+49}{158} a\right) = (0, 0, 0.81a)$

7a. $\displaystyle\int_0^{2\pi} \int_0^{\sqrt{2}/2} \int_0^{\sqrt{1-r^2}} r\, dz dr d\theta$

§16.6   **1.** $\displaystyle\int_0^{2\pi} \int_0^{\pi} \int_1^2 (\frac{1}{4})(\rho\sin\varphi)^2 \rho^2 \sin\varphi\, d\rho d\varphi d\theta$; $\dfrac{62\pi}{15}$

5. $\dfrac{1}{4}\pi^2 a^3$

§16.8   **1.** $\dfrac{7}{5} Ma^2$

§17.2   **1ad.** $\dfrac{\sqrt{2}}{2}\hat{i} + \dfrac{\sqrt{2}}{2}\hat{j}$, $\dfrac{\sqrt{2}}{2}\hat{i} - \dfrac{\sqrt{2}}{2}\hat{j}$, yes; $\dfrac{2u^3\hat{i} - v^2\hat{j}}{(4u^6+v^4)^{1/2}}$, $\dfrac{-u^3\hat{i} + v^2\hat{j}}{(u^6+v^4)^{1/2}}$, no

§17.3   **1.** (x,y) such that $xy^2 = \pm\frac{1}{2}$

§17.5   **1.** $\dfrac{1}{3}\displaystyle\int_1^2 \int_{-\frac{1}{2}(u-3)}^u \sin(\frac{u}{v})\, dv du$

3. $2 \log 2$

5. $\dfrac{1}{2}$

7. $\dfrac{17}{9}$

9. $\dfrac{1}{2}\displaystyle\int_1^2\int_{-u}^{u-2}\dfrac{v}{u}\,dvdu$; $\log 2 - 1$

11. $2\log 3$

§18.2   1. $\dfrac{1}{8}$ gm/sec

3. xyz

§18.5   1. $f = z^3 x + x^2 y + y^2$

3. $f = \dfrac{1}{2}x^2 z - \dfrac{1}{2}y^2$

5. $f = x^2 yz + y^2 z$

§18.6   1a. $(\theta^2 + z^2)\hat{r} + 2\theta\hat{\theta} + 2rz\hat{z}$

2a. $\varphi\theta\hat{\rho} + \theta\hat{\varphi} + \dfrac{\varphi}{\sin\varphi}\hat{\theta}$

§18.7   1. yes, $f = \theta$

3. no, integral curves: $r = Ce^{\theta}$

§18.8   1. $(zu_2 + yu_3)\hat{i} + (zu_1 + xu_3)\hat{j} + (yu_1 + xu_2)\hat{k}$

§19.3   1a. 0

3. $\dfrac{a^2}{2}$

§19.5   1. 2

3. $2\pi + 2\pi^2$

§19.6   1. 0

§19.7   1. $\dot{\rho}\hat{\rho} + \rho\dot{\varphi}\hat{\varphi} + \rho\sin\varphi\dot{\theta}\hat{\theta}$

§20.3   1ac. 2; 0

3. $\ln\rho - \dfrac{1}{\rho}$

§20.4   1. i; $x^2 yz^2 + 3x + y^2 z$

3. $f = xyz + e^y z + \dfrac{1}{2}z$

§20.5   1. $e + 1$

1ac. $\frac{1}{2}x^2 + xy = C$; $\frac{x}{y} + \ln y = C$

§21.1   **3.** $\frac{1}{2}z\cos\theta\hat{i} + \frac{1}{2}z\sin\theta\hat{j} + z\hat{k}$, $0 \le z \le 6$, $0 \le \theta \le 2\pi$

      5. $(z^2+1)^{1/2}\cos\theta\hat{i} + (z^2+1)^{1/2}\sin\theta\hat{j} + z\hat{k}$, $-2 \le z \le 2$, $0 \le \theta \le 2\pi$

      7. $(25-z^2)^{1/2}\cos\theta\hat{i} + (25-z^2)^{1/2}\sin\theta\hat{j} + z\hat{k}$, $-5 \le z \le 5$, $0 \le \theta \le 2\pi$

§21.3   **1bd.** $\frac{1}{2}z\cos\theta\hat{i} + \frac{1}{2}z\sin\theta\hat{j} - \frac{1}{4}z\hat{k}$;

           $(z^2+1)^{1/2}\cos\theta\hat{i} + (z^2+1)^{1/2}\sin\theta\hat{j} - z\hat{k}$

§21.4   **1bd.** $\int_0^{2\pi}\int_0^6 \frac{\sqrt{5}}{4}z\,dz\,d\theta = 9\pi\sqrt{5}$; $\int_0^{2\pi}\int_{-2}^2 \sqrt{2z^2+1}\,dz = \pi(12 + \sqrt{2}\ln(3+2\sqrt{2}))$

§22.3   **1.**   0

      3ac. 30π; 0

§22.5   **1.** 16π

      3ace. 12π; 0; 36π

      5. 72π

      7.   3

§22.6   **1.** $2a^2 b\hat{j}$

§23.1   **1ac.** iii; ii

§23.2   **1ac.** no; no

§23.4   **1ace.** 0; 0; -12

§23.5   **1ac.** 3; 54π

§23.6   **1.**   north

§23.7   **1a.**   yes, no, yes, no

§24.3   **5.**   2

      7. $\frac{1}{2}$

§24.4   **1ac.** no; yes

§24.5   **1ace.** -4; -4; $\pm 4n$, $n = 0,1,2\ldots$

      3.   all integers n, $-7 \le n \le 7$

      7.   0, ± 1, ± 5, ± 6

§25.1   **1.** 48π

§25.2   **1ac.** 0; 0

3a. $\dfrac{1}{\rho^2}$

§25.3   **1.** 2π

   3ac. 12π; 12π

   5. $\dfrac{1}{6}$

§25.4   **1.** π

§25.5   **1.** 4; 0, 4

   3. -4; 0, 1, -4, -3

§25.6   **1a.** $gh\vec{\nabla}\cdot\vec{F} + h\vec{F}\cdot\vec{\nabla}g + g\vec{F}\cdot\vec{\nabla}h$

   3a. $\dfrac{\partial}{\partial x}\left(\dfrac{\partial^2 f}{\partial x^2} + \dfrac{\partial^2 f}{\partial y^2} + \dfrac{\partial^2 f}{\partial z^2}\right)\hat{i} + \dfrac{\partial}{\partial y}\left(\dfrac{\partial^2 f}{\partial x^2} + \dfrac{\partial^2 f}{\partial y^2} + \dfrac{\partial^2 f}{\partial z^2}\right)\hat{j} + \dfrac{\partial}{\partial z}\left(\dfrac{\partial^2 f}{\partial x^2} + \dfrac{\partial^2 f}{\partial y^2} + \dfrac{\partial^2 f}{\partial z^2}\right)\hat{k}$

§26.3   **1bd.** $-(y+z)\hat{i}$; $-2z\hat{i} - 2x\hat{j} - 2y\hat{k}$

   3bd. $-\dfrac{\sqrt{2}}{4}(y+z)$; $-\dfrac{\sqrt{2}}{2}(x+z)$

§26.4   **1.** $\dfrac{2}{3}$

   3. -20π

   5. -4π

§26.5   **1a.** $\vec{F} = r\hat{r} + \dfrac{1}{r}\hat{\theta}$ on $\mathbf{E}^3$-except-the-z-axis

   3. 5; 0, 5, -5

§26.7   **1a.** $gh\vec{\nabla}\times\vec{F} - g\vec{F}\times\vec{\nabla}h - h\vec{F}\times\vec{\nabla}g$

§27.2   **1ac.** $\hat{r} + \dfrac{1}{r}\hat{\theta}$; $\hat{\rho} + \dfrac{1}{\rho}\hat{\phi}$

   3a. $2\theta\hat{z}$

§27.5   **3a.** $-\dfrac{1}{3}(y^2 + z^2)\hat{i} + \dfrac{1}{3}xy\,\hat{j} + \dfrac{1}{3}xz\,\hat{k}$

   7a. $xy\,\hat{j} + xz\,\hat{k}$

§27.8   **1.** $\vec{G} = \dfrac{1}{2}y^2\hat{j} + \dfrac{1}{2}z^2\hat{k}$, $\vec{H} = -\dfrac{1}{2}(x^2 - y^2)\hat{j} - \dfrac{1}{2}(x^2 + 2yz)\hat{k}$ (use

   $g = \dfrac{1}{6}(y^3 + z^3)$ from $\nabla^2 g = y + z$)

§28.3   **5.** γMx

§29.2   **1.** $(-4, -4, -12, 2, 3)$

§29.3   **3.** $X = \frac{1}{3}A$, $Y = B - \frac{1}{3}A$; *either* $A \neq O$, $X = 3A$, $k = \frac{1}{3}$, $Y = B - 3A$ *or*

                 $A = O$, $X = O$, k arbitrary, $Y = B$; $A = O$, $X = O$, $Y = B$

§29.4   **1.** $\cos^{-1}\left(\dfrac{\sqrt{35}}{21}\right) = 73.6°$

§29.7   **1.** $\begin{bmatrix} -8 \\ \frac{13}{2} \end{bmatrix}$

§30.1   **1.** $A = \begin{bmatrix} 1 & 3 & 1 & -2 \\ 4 & 7 & 8 & -1 \\ 3 & 2 & -1 & 4 \\ 0 & 1 & 2 & 0 \end{bmatrix}$, $X = \begin{bmatrix} x_1 \\ x_2 \\ x_3 \\ x_4 \end{bmatrix}$, $D = \begin{bmatrix} 2 \\ -1 \\ 0 \\ 3 \end{bmatrix}$

§30.4   **1.** $\begin{bmatrix} 1 & 0 & 0 & 0 \\ 0 & 1 & 0 & 0 \\ 0 & 0 & 1 & 0 \\ 0 & 0 & 0 & 1 \end{bmatrix}$

§30.5   **1.** $\begin{bmatrix} 0 \\ 0 \\ 0 \\ 0 \end{bmatrix}$

§30.6   **1.** $\begin{bmatrix} 7 & 0 & -3 & 0 \end{bmatrix}^T + t_1 \begin{bmatrix} -2 & 1 & 0 & 0 \end{bmatrix}^T + t_2 \begin{bmatrix} -1 & 0 & 4 & 1 \end{bmatrix}^T$

         3. $\begin{bmatrix} 6 & 0 & 0 & -3 \end{bmatrix}^T + t_1 \begin{bmatrix} -2 & 1 & 0 & 0 \end{bmatrix}^T + t_2 \begin{bmatrix} 1 & 0 & 1 & 0 \end{bmatrix}^T$

         5. $\begin{bmatrix} x & y & z & w \end{bmatrix}^T = \begin{bmatrix} 1 & 1 & 0 & -3 \end{bmatrix}^T$

         7.   no solution

§30.7   **1.**   4

         3. $3, \begin{bmatrix} 2 & -1 & 0 & 6 & 0 \end{bmatrix}^T + t_1 \begin{bmatrix} -1 & -3 & 1 & 0 & 0 \end{bmatrix}^T + t_2 \begin{bmatrix} -2 & -1 & 0 & -2 & 1 \end{bmatrix}^T$

§30.9   **1.**   get: $3Fe + H_2O = Fe_3O_4 + 4H_2$

§31.3   **1a.**   1

         3. $a \neq 1$, all b; $a = 1$, $b \neq 1$; $a = 1$, $b = 1$

§31.4   **1a.** $(\frac{27}{11}, -\frac{7}{11})$

         3.   no unique solution (in fact; more than one solution)

§32.1 1. $\begin{bmatrix} 0 & -1 & 2 \\ 1 & -1 & 3 \\ -1 & -2 & 3 \end{bmatrix}$, $\begin{bmatrix} 0 & -2 \\ 3 & 2 \end{bmatrix}$

§32.4 1. i, $\begin{bmatrix} 1 & 0 & 0 \\ 0 & 0 & \frac{1}{2} \\ 0 & \frac{1}{4} & 0 \end{bmatrix}$

§32.5 1. $\begin{bmatrix} \frac{1}{2} & \frac{1}{2} & 0 \\ \frac{1}{2} & 0 & -\frac{1}{2} \\ 0 & \frac{1}{2} & -\frac{1}{2} \end{bmatrix}$; $\begin{bmatrix} \frac{3}{2} \\ \frac{1}{2} \\ 1 \end{bmatrix}$

3. $\begin{bmatrix} -\frac{1}{2} & \frac{1}{2} & -\frac{1}{2} & -\frac{1}{2} \\ \frac{1}{2} & \frac{1}{2} & \frac{1}{2} & \frac{1}{2} \\ \frac{1}{2} & \frac{1}{2} & -\frac{1}{2} & \frac{1}{2} \\ \frac{1}{2} & \frac{1}{2} & -\frac{1}{2} & -\frac{1}{2} \end{bmatrix}$ $\begin{bmatrix} 1 \\ 1 \\ 0 \\ -3 \end{bmatrix}$; $\begin{bmatrix} -1 \\ 1 \\ 0 \\ 0 \end{bmatrix}$, $\begin{bmatrix} -\frac{1}{2} \\ \frac{3}{2} \\ \frac{1}{2} \\ \frac{1}{2} \end{bmatrix}$

§32.8 1. eigenvalues: 3, 2, 1; eigenvectors: $t\begin{bmatrix} 1 \\ 0 \\ 0 \end{bmatrix}$, $t\begin{bmatrix} 1 \\ 0 \\ -1 \end{bmatrix}$, $t\begin{bmatrix} 0 \\ 1 \\ -1 \end{bmatrix}$ for $t \neq 0$

§33.5 1. 3; no; $2V_1 + V_2 + V_3 = V_4$

§33.6 1. 2; {(1 1 0 2), (0 0 1 −1)}; 2;
{(−1 1 0 0),(−2 0 −1 1)}

§33.8 1. dependent; 3; first three vectors

§34.2 1a. condition: that the four points P_1, P_2, P_3, P_4 can be relabeled, in some order, as A,B,C,D so that span(B−A) is parallel to span(C−D), and span(D−A) is parallel to span(C−B)

Index